AF292280

Prostaglandins, Leukotrienes, and Lipoxins

Biochemistry, Mechanism of Action, and Clinical Applications

**GWUMC Department of Biochemistry
Annual Spring Symposia**
Series Editors:
Allan L. Goldstein, Ajit Kumar, George V. Vahouny, and J. Martyn Bailey
The George Washington University Medical Center

DIETARY FIBER IN HEALTH AND DISEASE
Edited by George V. Vahouny and David Kritchevsky

EUKARYOTIC GENE EXPRESSION
Edited by Ajit Kumar

THYMIC HORMONES AND LYMPHOKINES
Basic Chemistry and Clinical Applications
Edited by Allan L. Goldstein

PROSTAGLANDINS, LEUKOTRIENES, AND LIPOXINS
Biochemistry, Mechanism of Action, and Clinical Applications
Edited by J. Martyn Bailey

Prostaglandins, Leukotrienes, and Lipoxins

Biochemistry, Mechanism of Action, and Clinical Applications

Edited by

J. Martyn Bailey

The George Washington University Medical Center
Washington, D.C.

Springer Science+Business Media, LLC

Library of Congress Cataloging in Publication Data

Main entry under title:

Prostaglandins, leukotrienes, and lipoxins.

(GWUMC Department of Biochemistry annual spring symposia)
Based on the 4th International Spring Symposium on Health Sciences held in
Washington, D.C. in May 1984.
Includes bibliographies and index.
1. Prostaglandins—Congresses. 2. Leukotrienes—Congresses. I. Bailey, J. Martyn.
II. International Spring Symposium on Health Sciences (4th: 1984: Washington, D.C.)
III. Title: Lipoxins. IV. Series. [DNLM: 1. Leukotrienes B—congresses. 2. Lipoxy-
genases—congresses. 3. Prostaglandins—congresses. 4. SRS-A—congresses. QU
90 P9685 1984]
QP801.P68P727 1985   599′.01927   85-16941
ISBN 978-1-4684-4948-8

ISBN 978-1-4684-4948-8   ISBN 978-1-4684-4946-4 (eBook)
DOI 10.1007/978-1-4684-4946-4

Preface

The family of known essential fatty acid metabolites continues to grow. Synthesis of the prostaglandins from essential fatty acids was first described by Bergstrom and Samuelsson in 1964. The thromboxanes were discovered in 1975, the prostacyclins, by Moncada and Vane, in 1976, and the leukotrienes by Samuelsson in 1979. The discovery of a new class of biologically active arachidonic acid metabolites named lipoxins was announced by Bengt Samuelsson at the IVth International Spring Symposium on Health Sciences held in Washington D.C., May 1984. This volume, *Prostaglandins, Leukotrienes and Lipoxins,* contains most of the papers presented in the plenary sessions of the Washington Symposium. The book is divided into six parts, each covering a different aspect of this rapidly expanding field, and contains a total of 63 chapters by an internationally recognized group of authors in each area.

Part I contains 11 chapters and covers the basic biochemistry and enzymology of prostaglandins, leukotrienes, and lipoxins. Chapter 1 by Professor Samuelsson details the discovery of lipoxins. The enzymatic synthesis and biological activities of the first two members of this series, lipoxins A and B, are described. They appear to have selective secretagogue activity for human neutrophils. The lipoxins contain a conjugated trihydroxytetraene structure and thus differ significantly from the previously described leukotrienes and THETEs and, in contrast to these substances, are derived by the integrated activity of two different lipoxygenase pathways.

Chapter 2 in this section, also by Samuelsson, describes a highly significant advance in the biology of the leukotrienes, documenting the synthesis and cytochemical distribution of these substances in the brain. The first evidence for a neuroendocrine role for leukotriene C_4 is presented. Four chapters by Kulmacz, Smith, Hickok, and McNamara follow and outline properties of prostaglandin-biosynthesizing enzymes including the purification of prostaglandin synthases, the subcellular distribution of the enzymes, and the first isolation of the human cyclooxygenase. The last five chapters in this section by Hammarstrom, Lagarde, Fiskum, Shak, and Pace-Asciak describe properties of leukotriene and lipoxygenase enzyme systems including new information on their roles in platelet function and reticulocyte maturation. Pace-Asciak contributes a new mechanism of formation for the γ-HEPAs via hematin-catalyzed intramolecular oxygen transfer.

Part II covers regulatory mechanisms in eicosanoid release and metabolism and comprises 13 chapters including contributions by Hirata, Braquet, Metz, Gerritsen, Dennis, Sha'afi, and Galli on the regulation of prostaglandin release, and chapters by Beckner and Owen on the cell biology of prostaglandins. The quantitative assessment of the relative roles of the phospholipase A_2 and phospholipase C pathways in platelets is detailed by Smith. Several chapters describing other new discoveries, including the role of EGF in cyclooxygenase synthesis by Bailey, the postphospholipase activation of lipoxygenases by Vanderhoek, and the provocative finding by Wasner that prostaglandylinositol cyclic phosphate may serve as an antagonist to cyclic AMP, complete part II of the book.

Part III is concerned with cardiovascular–pulmonary interactions of the prostaglandins, prostacyclins, and leukotrienes and begins with chapters by Weksler and by Gerritsen on the role of prostacyclin in platelet–blood vessel interactions. Chapters by Ramwell, Kaley, Brace, Seiss, and Rao comprehensively outline prostaglandin interactions with the cardiovascular system, and others by Levi, Kadowitz, and Roth give authoritative surveys of the responses of the cardiovascular system to leukotrienes. Feuerstein describes the role of platelet-activating factor in modulating coronary function. Austen reports the discovery of three distinct sulfidopeptide leukotriene receptors, a finding that will help rationalize the wide variety of tissue responses to these compounds. Two important nutritional chapters, one by Schoene on the effects of selenium deficiency on arachidonic acid metabolism, which is mediated by the selenium-containing enzyme glutathione peroxidase, and the other by Vahouny on the comparative absorption of (ω3) fatty acids similar to those found in fish oil diets, complete this section of the book.

Part IV contains seven chapters describing recent clinical applications of the eicosanoids. Harker authoritatively summarizes the status of recent clinical trials that evaluated the role of eicosanoids and inhibitory drugs in heart attacks and thrombosis, and Goldstein describes relationships between coronary leukotriene release and myocardial ischemia. This work expands the developing evidence implicating leukotriene-induced vascular contraction in "sudden death" situations. Bygdeman, Norman, and Lindblom contribute chapters relating to prostaglandin levels in relation to human fertility, to delayed labor, and to differential contractile effects of prostaglandins on the human uterus. Foegh describes recent developments in the novel use of eicosanoids to monitor allograft rejection. New findings on the role of arachidonic acid metabolites in psoriasis and related indications for therapy of this disorder are described in a chapter by Camp.

Part V of the book details the role of eicosanoids in immune reactions, inflammatory processes, and cancer. This section contains chapters by Hadden, Coffey, Goodwin, Rhodes, and Rola-Pleszczynski relative to the role played by lipoxygenase metabolites of arachidonic acid in lymphocyte activation and blastogenesis. These important contributions detail the role of eicosanoids in the generation of immune responses, macrophage activation, and in helper and suppressor T-cell activation and function. They comprise an excellent summary of current knowledge in this rapidly expanding field. Additional chapters by Ishizaka, Salari, Jones, and

Uotila describe the biochemical cascade involved in mast-cell activation and the role of leukotrienes and other arachidonate metabolites in anaphylaxis and asthma.

The concluding section of the book—Part VI—contains nine chapters that summarize recent advances that have been made in the development of antiinflammatory drugs and inhibitors of cyclooxygenase, lipoxygenase, and leukotriene systems. This includes a chapter by Fitzgerald detailing novel aspects of aspirin action and a thoughtful overview by Flower on the historical development of cyclooxygenase inhibitors and the present status of these important drugs. Chapters by Egan and Coutts outline the current knowledge and state of development of lipoxygenase and leukotriene inhibitors. Fischer, Kort, McMillan, and Rainsford describe some biological effects of cyclooxygenase and lipoxygenase inhibitors, and in the final chapter Magolda describes the synthesis and properties of new and potent site-specific phospholipase A_2 inhibitors as possible future generalized inhibitors of eicosanoid release.

J. Martyn Bailey

Contents

**PART IV—RECENT CLINICAL APPLICATIONS
OF EICOSANOIDS**

PART V—IMMUNE REACTIONS, NEOPLASIA, AND INFLAMMATORY PROCESSES

General Biochemistry and Enzymology of Prostaglandins, Leukotrienes, and Lipoxins

Lipoxins
A Novel Series of Biologically Active Compounds

CHARLES N. SERHAN, MATS HAMBERG, and
BENGT SAMUELSSON

1. INTRODUCTION

A variety of cells release and oxygenate arachidonic acid during their interaction with specific stimuli. In its nonesterified form, arachidonic acid may be subjected to oxygenation by either the cyclooxygenase or lipoxygenase pathways. The lipoxygenase pathways of mammalian tissues (i.e., 5-, 12-, or 15-) (Hamberg, 1984) transform arachidonate into a number of biologically potent compounds, which include several mono- and dihydroxyeicosatetraenoic acids. For example, via the 5-lipoxygenase pathway, arachidonic acid is converted to 5(S)-hydroperoxy-6,8,11,14-eicosatetraenoic acid (5-HPETE), which may be further transformed into leukotrienes (Borgeat et $al.$, 1976; Borgeat and Samuelsson, 1979). Leukotrienes, particularly LTB_4 and LTC_4, are believed to serve as mediators in both immediate hypersensitivity reactions and inflammation (Samuelsson, 1983). In fact, leukotriene B_4 serves as a complete secretagogue in human neutrophils, stimulating aggregation, mobilization of Ca^{2+}, degranulation, and the generation of active oxygen species within seconds of its addition (Serhan et $al.$, 1982).

In view of the potential importance of lipoxygenase products in both normal and pathophysiological states, we have studied interactions among the major lipoxygenase pathways and examined the products formed. In this chapter we describe

CHARLES N. SERHAN, MATS HAMBERG, and BENGT SAMUELSSON ● Department of Physiological Chemistry, Karolinska Institutet, S-104 01 Stockholm, Sweden. *Present address of C.N.S.:* Visiting Scientist, Fellow of the Arthritis Foundation, Department of Medicine, Division of Rheumatology, New York University, New York, New York 10016.

a new series of oxygenated derivatives of arachidonic acid that contain a conjugated tetraene structure as a characteristic feature of the group and report the structures and some of the biological activities of two of the major compounds of this series. Since these products appear to arise via the interaction(s) of distinct lipoxygenase pathways, we propose the trivial name of "lipoxins" (lipoxygenase interaction products) (Serhan *et al.*, 1984b).

2. MATERIALS AND METHODS

Cytochalasin B, cytochrome *c*, superoxide dismutase, and N-*t*-BOC-L-alanine-*p*-nitrophenyl ester (BOC-Ala-ONp) were purchased from Sigma (St. Louis, MO). Arachidonic acid was purchased from Nu-Chek Prep. (Elysian, MN), and soybean lipoxygenase (E.C. 1.13.11.12) type I from Sigma; 15-HPETE was obtained by incubation of arachidonic acid with soybean lipoxygenase. Reference compounds for oxidative ozonolysis, i.e., the (−)menthoxycarbonyl derivatives of methyl-2-DL-hydroxy- and 2-L-hydroxyheptanoates and of dimethyl-2-DL-hydroxy and 2-L-hydroxyadipates, were prepared as previously described (Borgeat *et al.*, 1976; 1979; Hamberg, 1971). The HPLC equipment was from Waters Associated (Milford, MA) (pump 6000A, injector U6K) and LDC (Riviera Beach, FL) (UV detector, LDC-III).

2.1. Cell Preparation and Incubation Conditions

Human leukocytes from peripheral blood were prepared as described by Lundberg *et al.* (1981). These preparations represent a mixed population of leukocytes (neutrophils, basophils, eosinophils, etc.) in which the neutrophil contribution represents >90% as determined by Giemsa stain and light microscopy. Cells were washed and suspended in a buffered salt solution (pH 7.45) at 100×10^6 cells/ml.

Leukocytes (100–500 ml of 100×10^6 cells/ml) were warmed to 37°C in a water bath with slow continuous stirring for 5 min. Then, either 15-HPETE (100 μM) or 15-HPETE (100 μM) plus A_{23187} (5 μM) were added in ethanol (<1% final vol/vol), and the incubations continued for an additional 30 min. Incubations were stopped by addition of 2 volumes of methanol (Serhan *et al.*, 1984a,b).

2.2. Extractions and Purification

Procedures for ether extraction and silicic acid chromatography were as described (Borgeat *et al.*, 1976; Borgeat and Samuelsson, 1979). The ethyl acetate fraction from silicic acid chromatography was evaporated, dissolved in methanol, treated with diazomethane, and then subjected to thin-layer chromatography. This

TLC step was essential since nonenzymatic products of 15-HPETE interfered with both the structural analysis and bioassay of the compounds of interest (see Section 3). Thus, methyl-$[1-^{14}C]$-11,12,15-trihydroxy-5,8,13-eicosatetraenoate and methyl-$[1-^{14}C]$-11,14,15-trihydroxy-5,8,12-eicosatrienoate were prepared (Bryant and Bailey, 1981) and added to the material eluting in the ethyl acetate fractions. Thin-layer chromatography was carried out using plates coated with silica gel G and ethyl acetate : 2,2,4-trimethylpentane (5 : 1, v/v) as solvent. A Berthold Dünn-schichtsscanner II was used for localization of labeled material on TLC plates. The zones containing methyl esters exhibiting tetraene UV spectra (i.e., λ_{max} 301, see Section 3) but not methyl-$[1-^{14}C]$-11,14,15-trihydroxy-5,8,12-eicosatrienoate or methyl-$[1-^{14}C]$-11,12,15-trihydroxy-5,8,13-eicosatrienoate were scraped off, and the material was recovered from the silica gel by elution with methanol. These samples were extracted with ether, dried under N_2, and injected into a reverse-phase HPLC column. The column (length 50 $\times$ 10mm; C_{18}) was eluted with methanol/water (70 : 30, v/v) at 3.0 ml/min. A UV detector set at 301 nm recorded the absorption of the eluate. Fractions showing a tetraene UV spectrum were collected separately and rechromatographed in the same HLPC system. The UV spectra of separated components were recorded in methanol using a Hewlett-Packard 8450A spectrophotometer.

2.3. Analytical Procedures

Analytical procedures were performed as described elsewhere (Borgeat and Samuelsson, 1979; Hamberg, 1971; Serhan *et al.*, 1984a,b). Human neutrophils were prepared for aggregation, superoxide anion generation, and lysosomal enzyme release studies as described previously (Serhan *et al.*, 1982, 1983, 1984b).

3. RESULTS

To examine interactions among the major lipoxygenase pathways of mammalian tissues (i.e., 5-, 12- 15-lipoxygenases), 15-HPETE was added to human leukocytes (mixed leukocyte suspensions), and the products formed were characterized. Ethyl acetate fractions obtained after silicic acid chromatography contained strongly absorbing material at 243 nm (5,15-DHETE; cf. Maas *et al.*, 1982) and 301 nm (trihydroxytetraenes; Fig. 1). Analysis by GC/MS revealed that this fraction also contained large amounts of non-UV-absorbing materials derived from 15-HPETE (i.e., 11,12,15-THETE and 11,14,15-THETE). These compounds, which can be formed *inter alia* in the presence of heme proteins (Bryant and Bailey, 1981), interfered with both structural and biological studies. Therefore, it was necessary to include a TLC separation in the purification scheme (Fig. 2).

Both $[1-^{14}C]$-11,12,15-THETE and $[1-^{14}C]$-11,14,15-THETE were mixed with

                                    CHARLES N. SERHAN *et al.*

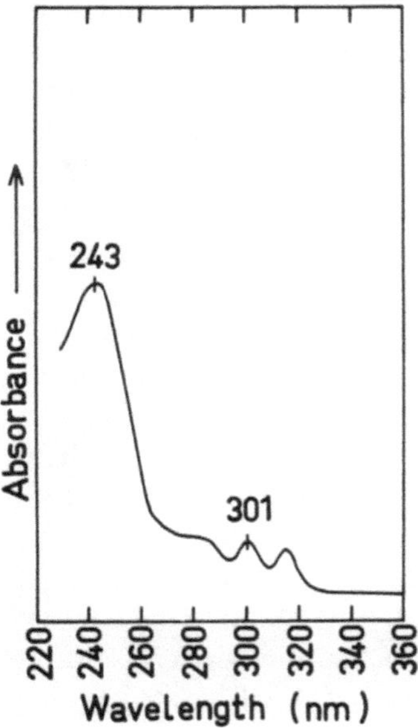

FIGURE 1. Ultraviolet spectrum of material eluted with ethyl acetate during silicic acid chromatography. Solvent: methanol.

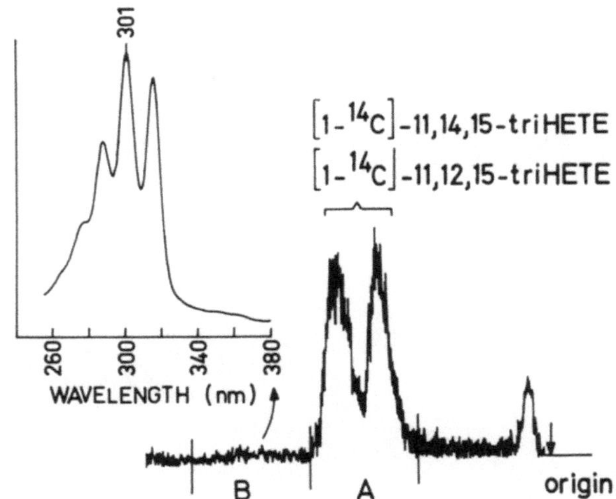

FIGURE 2. Thin-layer chromatographic separation of methyl esters of [1-¹⁴C]THETE and compounds I and II. Material eluted with ethyl acetate was mixed with approximately 25 nCi of [1-¹⁴C];-11,12,15-trihydroxy-5,8,13-eicosatrienoic acid and [1-¹⁴C]-11,14,15-trihydroxy-5,8,12-eicosatrienoic acid. The mixtures were esterified by treatment with diazomethane, spotted, and subjected to TLC (solvent system, ethyl acetate–2,2,4-trimethylpentane (5 : 1, v/v). Methyl esters of compounds I and II appeared in the zone labeled B just above the zone containing 11,14,14-THETE and 11,12,15-THETE.

an ethyl acetate fraction obtained from incubations of 15-HPETE with human leukocytes, and the mixtures were treated with diazomethane and subjected to TLC (Fig. 2; Serhan *et al.*, 1984a). The tetraene compounds were in the region labeled B (Fig. 2). Simultaneous addition of the divalent cation ionophore A_{23187} (5 μM) and 15-HPETE led to a 100-fold increase in tetraene-containing compounds. Thus, following TLC, 1.02 ± 0.23 μg of tetraene-containing material/1×10^8 human leukocytes could be obtained ($n = 13$) (Serhan *et al.*, 1984b).

3.1. Isolation of Compounds I and II

Following TLC, the material migrating in region B (Fig. 2) was eluted from the silica gel with methanol, concentrated, and injected onto a RP-HPLC (MeOH : H_2O 7 : 3, v/v) with the UV detector monitoring at 301 nm. The HPLC profile of this material is given in Fig. 3. Two major components showing strong absorption at 301 nm were found. Materials eluting in the HPLC profile labeled I and II were collected, rechromatographed, and examined by UV spectroscopy. Their UV spectra are shown in the insets of Fig. 3.

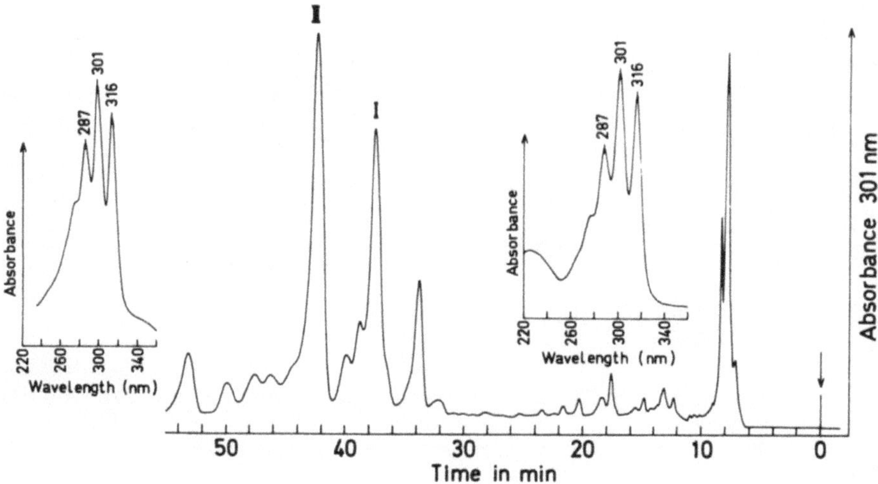

FIGURE 3. Reversed-phase HPLC chromatogram of products obtained from incubation of human leukocytes with 15-HPETE plus ionophore A23187. The incubation (30 min) was terminated by addition of two volumes of methanol. After removal of precipitated proteins, acidic ether extraction, and purification by silicic acid column chromatography, samples were treated with ethereal diazomethane and further purified by TLC. Products were eluted from silica gel, extracted, and injected. The UV detector was set at 301 nm, and the column was eluted with methanol/water (70/30, v/v) at 3.0 ml/min. Inset A: Ultraviolet spectrum of compound I in methanol. Inset B: Ultraviolet spectrum of compound II in methanol.

3.2. Structures of Compounds I and II

The ultraviolet spectrum of compound I showed a triplet of absorption bands at 287, 301, and 316 nm (in MeOH). Compound II also showed λ_{max} at 287, 301, and 316 nm (see inset). These findings suggested that each of the compounds contained a conjugated tetraene structure.

Samples of compound I were converted into the Me_3Si derivative and analyzed by GC/MS. The material eluted as a relatively broad peak on a 1% OV-1 column with an equivalent chain length corresponding to $C_{24.0}$ to $C_{24.1}$ (Serhan *et al.*, 1984b). Its mass spectrum showed prominent ions at *m/e* 173 (base peak; Me_3SiO^+ $=CH-(CH_2)_4-CH_3$) and *m/e* 203 ($Me_3SiO^+=CH-(CH_2)_3-COOCH_3$) with weaker ions at *m/e* 582 (*M*), 492 (*M* − 90; loss of Me_3SiOH), 482 (*M* − 100; rearrangement followed by loss of O $=HC-(CH_2)_4-CH_3$), 409 (*M* − 173), 379 (*M* − 203), 319 [*M* − (173 + 90)], 301 [*M* − (101 + 2 × 90); loss of $\cdot CH_2-(CH_2)_2-COOCH_3$ plus 2 Me_3SiOH], 275 [$Me_3SiO^+=CH-CH(OSiMe_3)-(CH_2)_4-CH_3$], 229 [*M* − (173 + 2 × 90)], and 171 (203 − 32) (Fig. 4). The mass spectrum of the Me_3Si derivative of hydrogenated compound I ($C_{25.3}$) showed ions of high intensity at *m/e* 575 (*M* − 15), 490 [*M* − 100; rearrangement followed by loss of O $=HC-(CH_2)_4-CH_3$)], 417 (*M* − 173), 399 [*M* − (101 + 90)], 297 [*M* − (203 + 90)], 203 (base peak; $Me_3SiO^+=CH-(CH_2)_3-COOCH_3$), and 173 [$Me_3SiO^+=CH-(CH_2)_4-CH_3$] (Fig. 5).

To prove the presence of an allylic hydroxyl group at C-5 of compound I and to determine its absolute configuration, the (−)menthoxycarbonyl derivative of compound I was subjected to oxidative ozonolysis. Gas–liquid chromatographic analysis (column, 5% QF-1) of the ozonolysis product showed the presence of the (−)menthoxycarbonyl derivative of methyl hydrogen 2-L-hydroxyadipate as well

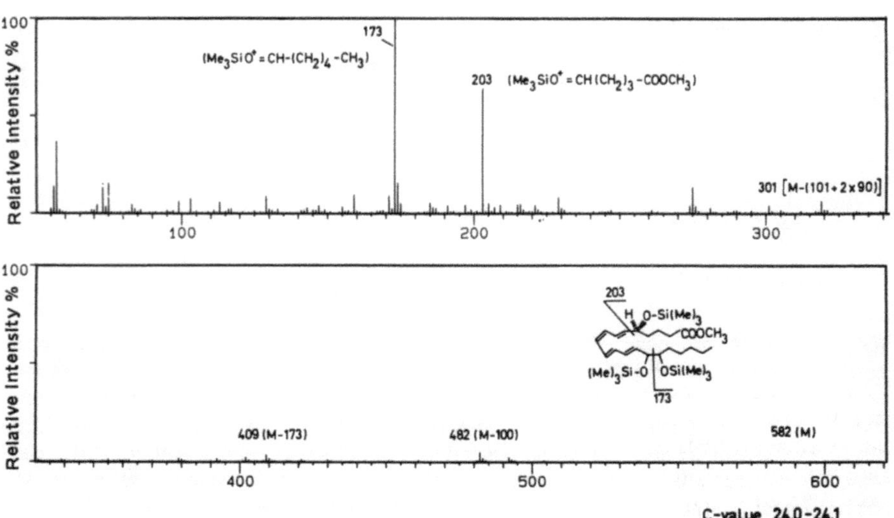

FIGURE 4. Mass spectrum of the Me_3Si derivative of the methyl ester of compound I.

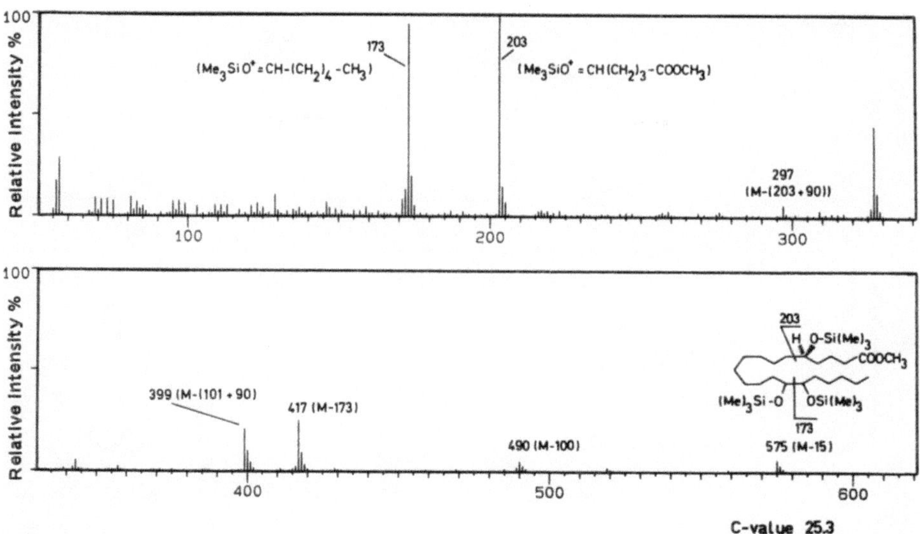

FIGURE 5. Mass spectrum of the Me₃Si derivative of the methyl ester of hydrogenated compound I.

as variable amounts (less than 20%) of 2-L-hydroxyheptanoic acid. These results showed that compound I was mainly a methyl eicosatetraenoate containing a hydroxy group at C-5 (D-configuration) and one of its four double bonds at Δ^6. On the basis of these findings, the parent acid of the major component present at peak I (Fig. 3) was assigned the structure 5-D-14,15-trihydroxy-6,8,10,12-eicosatetraenoic acid.

We next examined the structure of the compound eluting in peak II of the HPLC profile (Fig. 3). An aliquot of compound II was converted into the Me₃Si derivative and subjected to gas–liquid chromatographic–mass spectrometric analysis. Its C value was 24.1, and the mass spectrum showed prominent ions at m/e 379 [M − 203; loss of ·CH(OSiMe₃)-(CH₂)₃-COOCH₃], 289 (379 − 90; elimination of Me₃SiOH), 203 [Me₃SiO⁺=CH-(CH₂)₃-COOCH₃], 173 [Me₃SiO⁺=CH-(CH₂)₄-CH₃], and 171 (203 − 32; elimination of CH₃OH). Ions of low intensity were observed *inter alia* at m/e 582 (M), 492 (M − 90; elimination of Me₃SiOH), and 482 [M − 100; rearrangement followed by elimination of O=HC-(CH₂)₄-CH₃] (Fig. 6).

The product formed on catalytic hydrogenation of compound II was also examined (Fig. 6). The major peak of the gas chromatogram (C₂₅.₃) was the saturated derivative of compound II (Me₃Si derivative). Ions of high intensity were observed at m/e 575 (M − 15; loss of ·CH₃), 519 [M − 71; loss of ·(CH₂)₄-CH₃], =490 [M−100; rearrangement followed by loss of OHC-(CH₂)₄-CH₃], 297 [M − (203 + 90); loss of ·CH(OSiMe₃)-(CH₂)₃-COOCH₃ plus Me₃SiOH], 276 [M − 314; rearrangement followed by loss of O=HC-(CH₂)₈-CH(OSiMe₃)-(CH₂)₄-CH₃], 203 (base peak; Me₃SiO⁺=CH-(CH₂)₃-COOCH₃), and 173 [Me₃SiO⁺=CH-(CH₂)₄-CH₃]. A minor peak (C₂₃.₄) also appeared in the gas chromatogram of hydrogenated compound II. The mass spectrum showed prominent

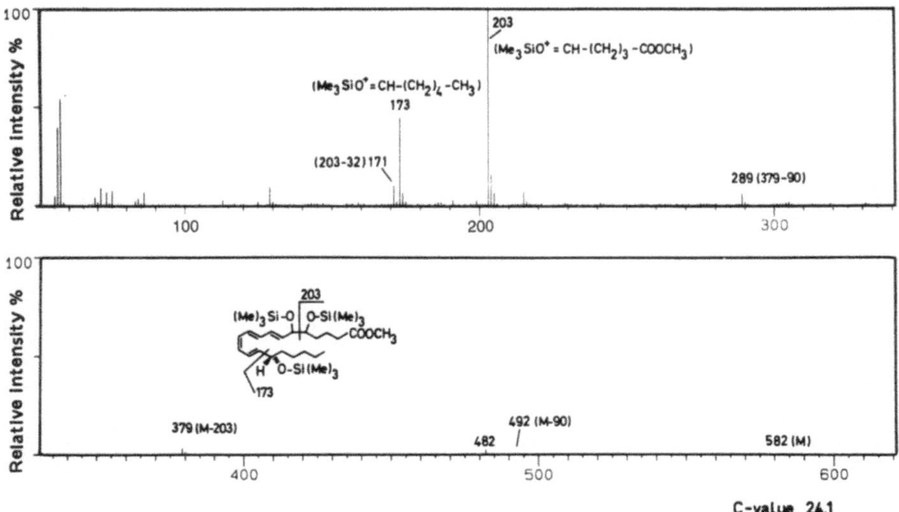

FIGURE 6. Mass spectrum of the Me₃Si derivative of compound II.

ions at *m/e* 487 ($M - 15$; loss of ·CH₃), 471 ($M - 31$; loss of ·OCH₃), 299 [$M - 203$; loss of ·CH-(OSiMe₃)-(CH₂)₃-COOCH₃], 276 [$M - 226$; rearrangement followed by loss of O=HC-(CH₂)₁₃-CH₃], and 203 [base peak; Me₃SiO⁺=CH-(CH₂)₃-COOCH₃]. Material present in the minor peak was assigned the structure methyl-5,6-dehydroxyeicosanoate (Me₃Si derivative) based on the C value and the mass spectrum.

Together, these findings suggested that compound II was a methyleicosatetraenoate carrying hydroxyl groups at C-5, C-6, and C-15. The location of the hydroxyl groups was supported by the high intensity of the ion at *m/e* 203 (base peak in the spectra) and by the formation of methyl-5,6-dihydroxyeicosanoate as a hydrogenolysis product during catalytic hydrogenation of compound II. In order to exclude alternative structures and to prove the presence of a vicinal diol structure in compound II, an aliquot was treated with *n*-butylboronic acid in acetone followed by treatment with hexamethyldisilazane and trimethylchlorosilane in pyridine. This resulted in the formation of an *n*-butylboronate-Me₃Si derivative of compound II (C₂₅.₃). Mass spectrum showed a prominent ion at *m/e* 173 [Me₃SiO⁺=CH-(CH₂)₄-CH₃], demonstrating that the nonvicinal hydroxyl group was located at C-15 (Fig. 7). Ions were also observed at *m/e* 504 (*M*), 489 ($M - 15$), 414 ($M - 90$), 404 ($M - 100$), and 199 {possibly [CH=CH-CH(OSiMe₃)-(CH₂)₄-CH₃]⁺}.

In order to determine whether the stereochemistry at C-15 was retained during the biosynthesis of compound II, the (−)menthoxycarbonyl detrivative of compound II (cf. Serhan *et al.*, 1984a,b) was subjected to oxidative ozonolysis. Gas–liquid chromatographic analysis of the ozonolysis product showed the presence of the (−)menthoxycarbonyl derivative of 2-L-hydroxyheptanoic acid, thus demonstrating that the configuration of the hydroxyl group at C-15 of compound II was "L" and that one of the four double bonds of compound II was located at Δ^{13}. The latter

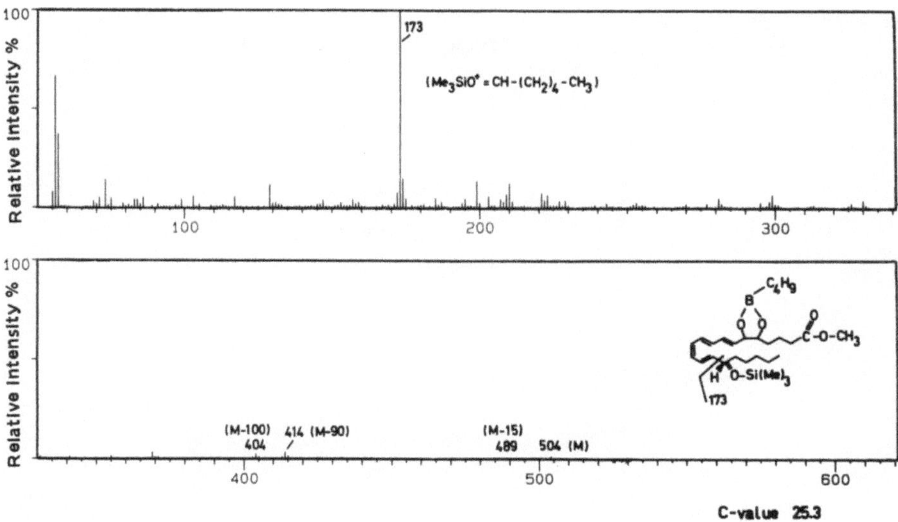

FIGURE 7. Mass spectrum o the *n*-butylboronate-Me₃Si derivative of Compound II.

finding together with the untraviolet spectrometry data, which showed that the four double bonds of compound II were conjugated (see above), demonstrated that the locations of the four double bonds were at Δ^7, Δ^9, Δ^{11}, and Δ^{13}. On the basis of these results, the parent acid of compound II was assigned the structure 5,6,15-L-trihydroxy-7,9,11,13-eicosatetraenoic acid. These studies indicate that compounds I and II are positional isomers. This is in agreement with their similar behavior on HPLC [compound I being only slightly more polar than compound II (Fig. 3)].

Following saponification and purification (RP-HPLC) of the free acid of compound II, samples were treated with diazomethane and subjected to RP-HPLC. The reesterified compound comigrated with the methyl ester of compound II both on RP-HPLC and on a straight-phase (chiral) column.

3.3. Human Neutrophil Responses to Lipoxin A

5,6,15-L-Trihydroxy-7,9,11,13-eicosatetraenoic acid (lipoxin A) was added to suspensions of human neutrophils, and we examined the generation of oxygen radicals, the release of lysosomal enzymes, and aggregation of neutrophils exposed to the free acid of this compound as markers of neutrophil activation. Continuous recording techniques were utilized to examine the kinetics of neutrophil responses upon addition of lipoxin A. Results of these experiments with cytochalasin-B-treated neutrophils are shown in Fig. 8.

As depicted in Fig. 8A, lipoxin A (5×10^{-7} M) induced a rapid burst in the generation of superoxide anion and stimulated the release of lysosomal elastase, although at the same concentration it exerted little to no effect in provoking aggregation. Figure 8A shows representative tracings obtained for aggregation, $O_2^{-\cdot}$

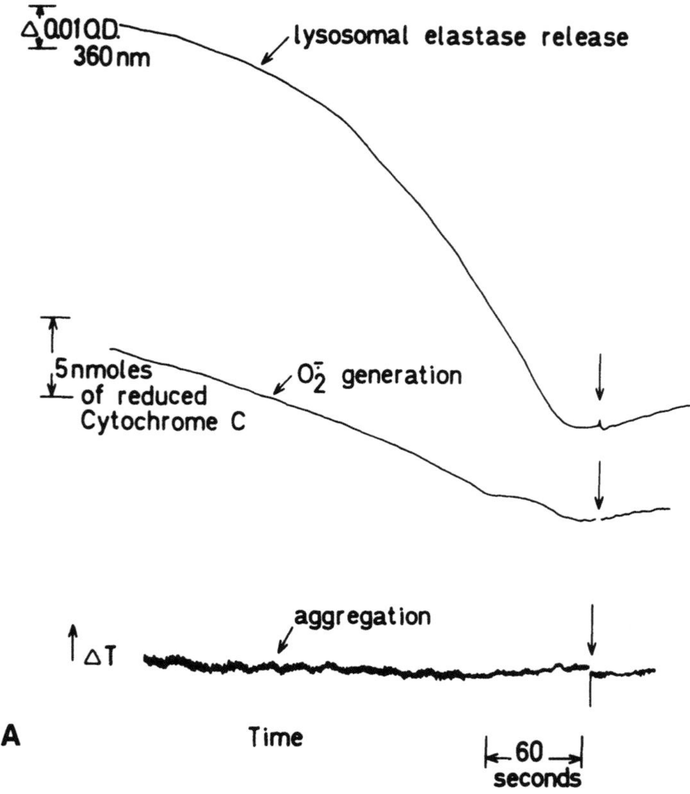

FIGURE 8. Human neutrophil aggregation, superoxide anion generation, and elastase release. A, Upper tracing: Representative tracing of lysosomal elastase release. Neutrophils (3 × 10⁶ cells/ml) were incubated with cytochalasin B (5 μg/ml) 3 min before addition of BOC-Ala-ONp (10 μM) followed by addition of compound II (5 × 10⁻⁷ M). Middle tracing: Representative continuous recording of superoxide anion generation. Neutrophils (3 × 10⁶ cells/ml) were incubated with cytochalasin B (5 μg/ml) 3 min at 37°C. Lower tracing: Representative aggregation

ester provoked aggregation (10^{-6}–10^{-10} M). However, lipoxin A proved to be a potent stimulator of $O_2^-\cdot$ generation (Fig. 8). At concentrations $>10^{-7}$ M, lipoxin A provoked $O_2^-\cdot$ generation and in this respect proved to be as potent as leukotriene B_4 (Table I). Under these conditions, the synthetic chemotactic peptide f-Met-Leu-Phe (10^{-7} M) proved to be the most potent of the three agents studied.

Both f-Met-Leu-Phe (fMLP) and LTB_4 are potent stimulators of elastase release in human neutrophils (cf. Serhan *et al.*, 1983). When lipoxin A was compared to the effects of these agents, lipoxin A proved to be approximately 2 log orders of magnitude less potent than either fMLP or LTB_4. A group of representative tracings obtained from the same donor are shown in Fig. 4B. Here, the response to fMLP

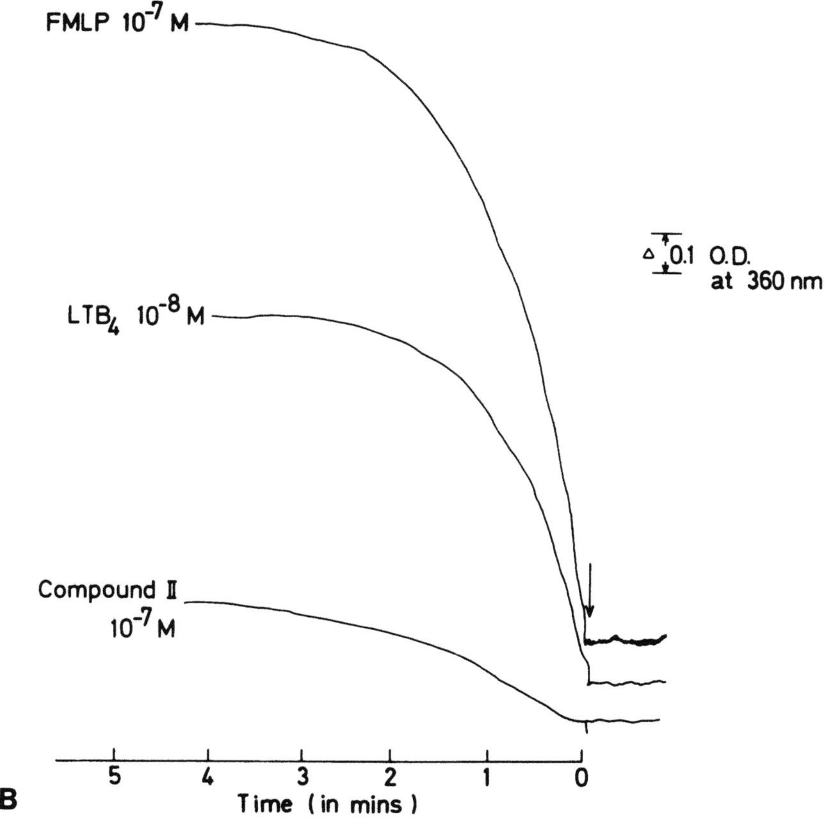

tracing. Neutrophils (3 × 10^6 cells/0.1 ml) were incubated with cytochalasin B (5 μg/ml) 3 min at 37°C before addition of compound II (5 × 10^{-7} M). Compound II (5 × 10^{-7} M) was added to the sample cuvette at the time indicated by the arrow. B: Continuous recording of lysosomal elastase release from human neutrophils. Tracings are representative of those obtained from individual donors.

TABLE I. Superoxide Anion Generation by Human Neutrophils[a]

Agent	Cytochrome c reduced (nmol/5 min)
fMLP (10^{-7} M)	14.5 ± 5.5
5,6,15-Trihydroxy-7,9,11,13-eicosatetraenoic acid (5 × 10^{-7} M)	8.1 ± 3.7
LTB$_4$ (5 × 10^{-7} M)	6.5 ± 3.6

[a] Neutrophils (3 × 10^6 cells/ml) were incubated with cytochalasin B (5 μg/ml) 3 min at 37°C before addition of the agents. Data represent 12 separate experiments ± S.D.

generation, and elastase release by human neutrophil exposed to the free acid of lipoxin A (5×10^{-7} M) (Serhan *et al.*, 1984b).

Dose–response studies revealed that neither lipoxin A (free acid) nor its methyl and LTB$_4$ are shown for purposes of comparison. In each experiment, lipoxin A displayed a longer lag phase ($\sim$30 sec) than either fMLP or LTB$_4$ in inducing lysosomal elastase release.

4. DISCUSSION

In this chapter, we describe the structures of two new compounds isolated from human leukocytes: 5-D,14,15,-triydroxy-6,8,10,12-eicosatetraenoic acid (lipoxin B; LX-B, the parent acid of compound I) and 5,6,15-L-trihydroxy-6,8,10,12-eicosatetraenoic acid (lipoxin A; LX-A, the parent acid of compound II).

The compounds have four conjugated double bonds, three hydroxyl groups, and are positional isomers with respect to positions of double bonds and hydroxyl groups. These compounds appear to be the major products of a new series of compounds that contain four conjugated double bonds as a distinguishing feature of the group (Fig. 9). Since the hydroxyl group at C5 had the D-configuration, it seems likely that oxygenation by a 5-lipoxygenase is involved in their formation. A number of biosynthetic routes may lead to the formation of lipoxin A and lipoxin B. However, further studies are required to establish whether the compounds are formed via multiple lipoxygenations or via other intermediates.

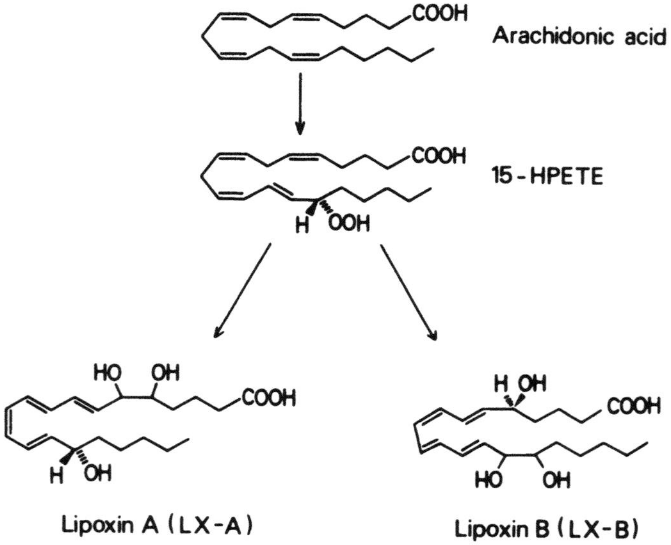

FIGURE 9. Scheme of formation of compounds I and II. The geometric configurations of the double bonds have not been determined and thus are tentative assignments.

Addition of lipoxin A to purified human neutrophils provoked a rapid and sustained generation of superoxide anion and induced degranulation without provoking substantial aggregation. These findings suggest that lipoxin A stimulates oxygen metabolism and the generation of active oxygen species in human neutrophils. With respect to oxygen radical production, lipoxin A proved to be as potent as LTB_4. However, lipoxin A proved to be two orders of magnitude less potent than LTB_4 in inducing degranulation (Fig. 9B). Thus, lipoxin A, unlike LTB_4, is a selective secretagogue in human neutrophil. Whether lipoxin A will prove to be an intracellular rather than extracellular signal remains to be determined. Nevertheless, results of the present study suggest a role for interactions among the major lipoxygenases of mammalian tissues in regulating specific cellular responses. Moreover, these findings may provide additional or alternative means by which the oxygenation of arachidonic acid by various cells can exert an effect on allergic reactions, inflammation, thrombosis, and host defense. The biological activities of these and related compounds are currently being investigated in other tissues.

5. SUMMARY

The interactions have been studied between the 5-lipoxygenase and 15-lipoxygenase pathways in human leukocytes. Addition of 15(S)-hydroperoxy-5,8,11,13-eicosatetraenoic acid (15-HPETE) to human leukocytes led to the formation of a novel series of compounds containing four conjugated double bonds. Following purification by silicic acid chromatography, a fraction containing several unidentified tetraenes from leukocytes was obtained. The material was esterified, separated by thin-layer chromatography, and analyzed by reversed-phase high-pressure liquid chromatography. The structures of the two major compounds were elucidated by ultraviolet spectrometry, gas chromatography–mass spectrometry, and oxidative ozonolysis. One compound was identified as 5,6,15-L-trihydroxy-7,9,11,13-eicosatetraenoic acid, and the other as 5-D,14,15-trihydroxy-6,8,10,12-eicosatetraenoic acid. When added to human neutrophils, kinetic and dose–response studies showed that 5,6,15-L-trihydroxy-7,9,11,13-eicosatetraenoic acid stimulated these cells with respect to superoxide anion generation and the release of lysosomal enzymes but had little ability to induce aggregation. Together, these results demonstrate that interaction(s) between the 5- and 15-lipoxygenase pathways of human leukocytes lead to the formation of a new series of arachidonic acid-derived compounds. Moreover, they suggest that such products may be involved in regulating specific cellular responses.

ACKNOWLEDGMENTS. We wish to thank Mrs. G. Hamberg, Mrs. A. Öhrström, Ms. Lena Hörlin, and Ms. Lena Delér for excellent technical assistance. The research was supported by the Swedish Medical Research Council (03X-00217 Samuelsson, 03X-05170 Hamberg).

REFERENCES

Borgeat, P., and Samuelsson, B., 1979, Metabolism of arachidonic acid in polymorphonuclear leukocytes, *J. Biol. Chem.* **254:**7865–7869.

Borgeat, P., Hamberg, M., and Samuelsson, B., 1976, Transformation of arachidonic acid and homo-γ-linolenic acid by rabbit polymorphonuclear leukocytes, *J. Biol. Chem* **254:**7816–7820.

Bryant, R., and Bailey, J. M., 1981, Rearrangement of 15-hydroperoxyeicosatetraenoic acid (15-HPETE) during incubations with hemoglobin. A model for platelet lipoxygenase metabolism, *Prog. Lipid Res.* **20:**279–281.

Hamberg, M., 1971, Steric analysis of hydroperoxides formed by lipoxygenase oxygenation of linoleic acid, *Anal. Biochem.* **43:**515–526.

Hamberg, M., 1984, Studies on the formation and degradation of unsaturated fatty acid hydroperoxides, *Prostaglandins, Leukotrienes Med.* **13:**27–34.

Lundberg, U., Rådmark, O., Malmsten, C., and Samuelsson, B., 1981, Transformation of 15-hydroperoxy-5,9,11,13-eicosatetraenoic acid into novel leukotrienes, *FEBS Lett.* **126:**127–132.

Maas, R. L., Turk, J., Oates, J. A., and Brash, A. R., 1982, Formation of a novel dihydroxy acid from arachidonic acid by lipoxygenase-catalyzed double oxygenation in rat mononuclear cells and human leukocytes, *J. Biol. Chem.* **257:**7056–7067.

Samuelsson, B., 1983, Leukotrienes: Mediators of immediate hypersensitivity reactions and inflammation, *Science* **220:**568–575.

Serhan, C. N., Radin, A., Smolen, J. E., Korchak, H. M., Samuelsson, B., and Weissmann, G., 1982, Leukotriene B$_4$ is a complete secretagogue in human neutrophils. A kinetic analysis, *Biochem. Biophys. Res. Commun.* **107:**1006–1012.

Serhan, C. N., Smolen, J. E., Korchak, H. M., and Weissmann, G., 1983, Leukotriene B$_4$ is a complete secretagogue in human neutrophils: Ca^{2+}-translocation in liposomes and kinetics of neutrophil activation, in: *Advances in Prostaglandin, Thromboxane and Leukotriene Research*, Vol. 11 (B. Samuelsson, R. Paoletti, and P. Ramwell, eds), Raven Press, New York, pp. 53–62.

Serhan, C. N., Hamberg, M., and Samuelsson, B., 1984a, Trihydroxytetraenes: A novel series of compounds formed from arachidonic acid in human leukocytes, *Biochem. Biophys. Res. Commun.* **118:**943–949.

Serhan, C. N., Hamberg, M., and Samuelsson, B., 1984b, Lipoxins: A novel series of biologically active compounds formed from arachidonic acid in human leukocytes, *Proc. Natl. Acad. Sci. U.S.A.* **81:**5335–5339.

2

Occurrence of Leukotrienes in Rat Brain

Evidence for a Neuroendocrine Role of Leukotriene C_4

JAN ÅKE LINDGREN, ANNA-LENA HULTING, TOMAS HÖKFELT, SVEN-ERIK DAHLÉN, PETER ENEROTH, SIGBRITT WERNER, CARLO PATRONO, and BENGT SAMUELSSON

1. INTRODUCTION

Leukotrienes (LT) are bioactive compounds with proposed roles as mediators of inflammation and allergy (Samuelsson, 1983). Production of LT is initiated by 5-lipoxygenation of arachidonic acid to $5(S)$-hydroperoxyeicosatetraenoic acid (5-HPETE), which is further transformed to an unstable epoxide (LTA_4) (Borgeat *et al.*, 1976; Borgeat and Samuelsson, 1979a). This intermediate can be enzymatically hydrolyzed to LTB_4 (Borgeat and Samuelsson, 1979b; Corey *et al.*, 1981). Alter-

JAN ÅKE LINDGREN and BENGT SAMUELSSON ● Department of Physiological Chemistry, Karolinska Institutet, S-104 01 Stockholm, Sweden. ANNA-LENA HULTING and SIGBRITT WERNER ● Department of Endocrinology, Karolinska Hospital, S-104 01 Stockholm, Sweden. TOMAS HÖKFELT ● Department of Histology, Karolinska Institutet, S-104 01 Stockholm, Sweden. SVEN-ERIK DAHLÉN ● Department of Physiology, Karolinska Institutet, S-104 01 Stockholm, Sweden. PETER ENEROTH ● Department of Obstetrics and Gynecology, Karolinska Hospital, S-104 01 Stockholm, Sweden. CARLO PATRONO ● Department of Pharmacology, Catholic University, I-00168 Rome, Italy.

natively, LTA_4 is converted, by addition of glutathione at C-6, into LTC_4 (Murphy *et al.*, 1979; Morris *et al.*, 1980). Stepwise enzymatic elimination of glutamic acid and glycine in the peptide side chain leads to formation of LTD_4 and LTE_4, respectively (Örning and Hammarström, 1980; Bernström and Hammarström, 1980). The proposed mediator of allergic reactions, slow-reacting substance of anaphylaxis (SRS-A) (Orange and Austen, 1969), has been identified as an entity composed of LTC_4, LTD_4, and LTE_4 (Samuelsson, 1983).

Leukotriene formation has almost exclusively been observed in various leukocytes and lung tissue (Samuelsson, 1983), whereas production of arachidonic acid metabolites such as prostaglandins and thromboxanes is widely distributed in various organ systems including the central nervous system (Wolfe, 1982).

This report describes the isolation of LTC_4, LTD_4, and LTE_4 from incubations of rat brain *in vitro* (Lindgren *et al.*, 1984). Identification of the compounds was carried out using high-performance liquid chromatography, UV spectroscopy, radioimmunoassay, and bioassay. Most regions of the brain produced leukotriene C_4 as judged by radioimmunoassay. The highest levels were obtained in the hypothalamus and the median eminence. By the indirect immunofluorescence technique, nerve endings in the median eminence and cell bodies in the preoptic area reacting with LTC_4 antiplasma were observed. Furthermore, LTC_4 at concentrations below the picomolar range released luteinizing hormone from dispersed rat anterior pituitary cells *in vitro* (Hulting *et al.*, 1984).

2. METHODS

Male rats (body weight 150–200 g, pathogen-free strain, Anticimex, Stockholm) were anesthetized with pentobarbital (40 mg/kg, i.p.) and perfused with 100 ml oxygenated Tyrode's solution at room temperature. The brains were rapidly removed and placed in ice-cold phosphate-buffered saline (PBS).

Frontal sections (thickness about 0.5 mm) from the entire brain were sliced with two razor blades on a cooled stage. Alternatively, thin slices (about 0.3 mm) from various brain regions were prepared by hand on a cooled stage with the help of a razor blade and a frosted object slide, and small round slices (diameter 3 mm) were punched out from the larger original slices.

The incubation procedure, purification of the products, analytical methods, bioassay (using lung parenchymal strips and longitudinal ileum muscle from guinea pig), and immunohistochemical methods were described recently (Lindgren *et al.*, 1984). Radioimmunoassay of LTC_4 (Lindgren *et al.*, 1983) and LH (Andersson *et al.*, 1983) was carried out as described. Dispersed rat anterior pituitary cells were prepared and cultured as described (Hulting *et al.*, 1984). Synthetic LTC_4 was a kind gift from Dr. J. Pike, the Upjohn Company; LTB_4 was biosynthetically prepared as described previously (Lindgren *et al.*, 1981).

3. RESULTS

3.1. Isolation of Leukotrienes from Rat Brain

Incubation of sliced rat brains with arachidonic acid (75 μM) and ionophore A23187 (5 μM) led to leukotriene formation. Figure 1 shows the ultraviolet absorbance pattern (at 280 nm) of eluted products. Peaks I, II, and III had elution times corresponding to standards of LTC_4, LTD_4, and LTE_4, respectively. Peak I also coeluted with $[^3H]LTC_4$ added to the sample after the incubation. Ultraviolet spectroscopy of compound I showed maximal absorbance at 280 nm with shoulders at 270 and 290 nm (Fig. 1). The material that cochromatographed with LTD_4 (peak II) or LTE_4 (peak III) was dissolved in 100 μl of ethanol/water (1 : 1, v/v) after evaporation of the HPLC solvent. Samples of both compounds elicited dose-related contractions of the lung strip (Fig. 2). These contractions were unaffected by indomethacin or by receptor antagonists for biogenic amines but were susceptible to antagonism by FPL 55712. Quantitation on the lung strip indicated that peaks II and III contained approximately 60 and 5 pmol of LTC_4-like activity, respectively (Fig. 2). The ileum assay indicated that peak II contained 60–70 pmol of LTD_4. Finally, on the ileum the material in peak III was considerably less active than

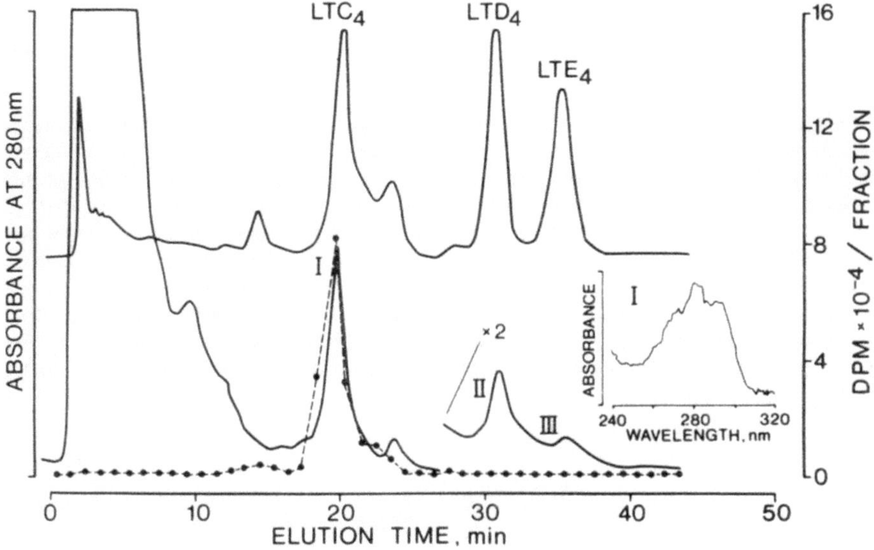

FIGURE 1. The upper curve shows a RP-HPLC chromatogram of an injected mixture of leukotrienes C_4, D_4, and E_4. The lower curve shows a RP-HPLC chromatogram obtained after injection of the methanol fraction of a silicic acid column chromatograph of the chloroform extract. The dotted line represents content of radioactivity in each HPLC fraction. Inset: Ultraviolet spectrum of material eluted in peak I.

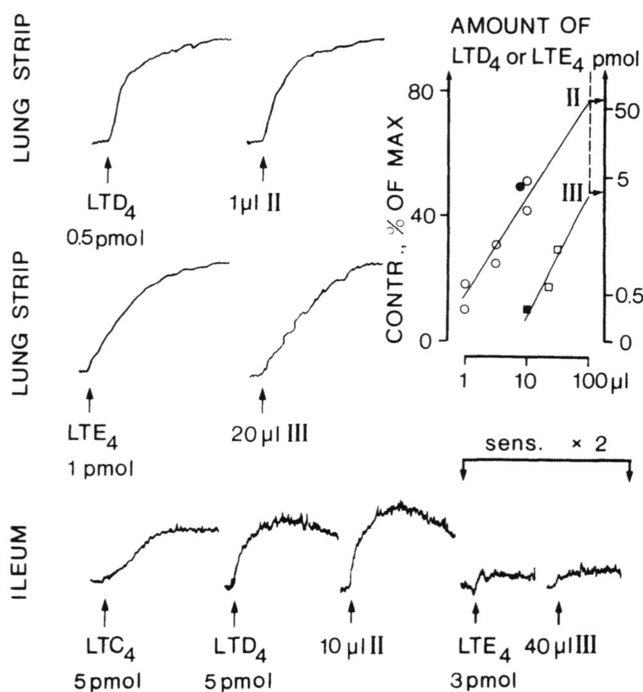

FIGURE 2. Bioassay of material eluted in peaks II and III. Upper two tracings show typical contractions evoked by the samples and standards of LTD_4 or LTE_4 on the guinea pig lung strip. The quantitation on the lung strip is presented in the upper right hand corner diagram. Abscissa: Volumes of samples tested (both peaks were dissolved in a total volume of 100 µl). Left ordinate: Contractions (expressed as percentage of the maximal responses in each preparation) evoked by samples of peak II (○) and III (□), respectively. Filled symbols show responses in indomethacin (10 µM)-pretreated preparations. Right ordinate: Contraction responses evoked by standards of LTD_4 or LTE_4. The amount of LTD_4- or LTE_4-like activity in peak II or III was extrapolated from the linear volume response curves of each sample. Lower tracing shows contractions elicited by the samples and standards of leukotrienes on the longitudinal muscle of the guinea pig ileum.

equal amounts of material eluted as peak II, giving further evidence that peak III contained LTE_4, because this leukotriene is less potent than LTD_4 in the ileum but not in the lung strip (Fig. 2) (Dahlén, 1983; Björck *et al.*, 1984). Therefore, when considered together, it is reasonable to conclude that peaks II and III contained 60 and 5 pmol of LTD_4 and LTE_4, respectively.

After rechromatography of peak I on RP-HPLC, the fractions were analyzed for LTC_4 by radioimmunoassay. As shown in Fig. 3A, the UV-absorbing (280 nm)

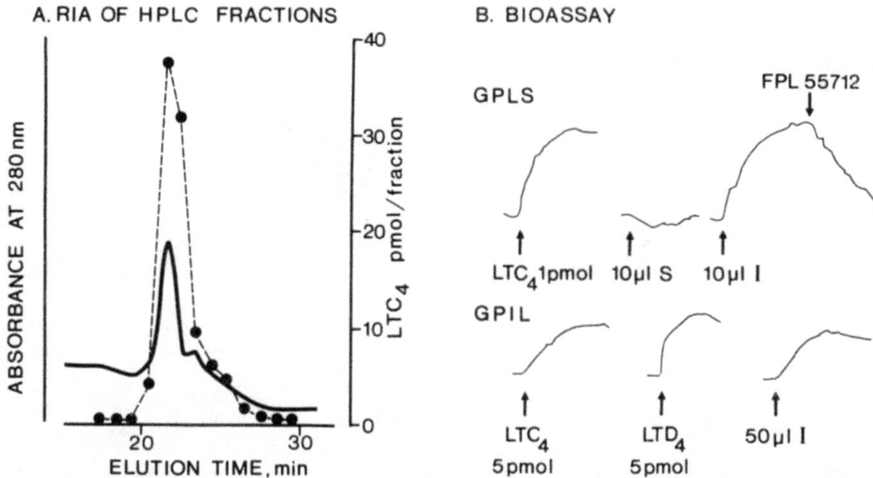

FIGURE 3. A: Reversed-phase HPLC chromatogram (partially shown) obtained after rechromatography of peak I (Fig. 1). The dotted line represents the LTC_4 content in each fraction as judged by radioimmunoassay. The amounts of $[^3H]LTC_4$ added in aliquots of the diluted HPLC fractions never exceeded 5% of $[^3H]LTC_4$ used in the assay procedure. B: Contractions evoked by samples of peak I on the guinea pig lung strip (GPLS) and ileum (GPIL), respectively. S, solvent control $[EtOH-H_2O, 1 : 1 (v/v)]$. Concentration of FPL 55712, 1 μM.

fractions also contained immunologically active material binding to the antibody. The total amount of LTC_4 remaining after rechromatography was around 90 pmol as judged by radioimmunoassay. The fractions containing LTC_4 were pooled, and the HPLC-solvent was evaporated. Thereafter, the sample was dissolved in 1 ml ethanol–water (1 : 1, v/v) and subjected to bioassay (Fig. 3B).

The material had a pattern of contractile activity on the lung strip and ileum that was indistinguishable from that of LTC_4. The total amount of LTC_4-like activity in peak I as estimated by final quantitation on the lung strip (not shown) was very close (around 100 pmol) to the amount determined by radioimmunoassay.

The recovery of LTC_4 during the purification procedure was estimated by the use of radiolabeled compound (Lindgren *et al.*, 1984). Assuming equal recovery of leukotrienes C_4, D_4, and E_4, the average production under the present conditions was accordingly calculated to be 25, 8, and 0.7 pmol/g brain tissue (ww) of LTC_4, LTD_4, and LTE_4, respectively.

3.2. Regional Distribution of LTC_4 Production

Slices from various regions of the rat brain were incubated with ionophore A23187 (5 μM) and arachidonic acid (75 μM), and the LTC_4 formation was measured by radioimmunoassay (Fig. 4). Synthesis of LTC_4 was observed in all regions tested. Highest levels of LTC_4 were produced by hypothalamus (2.9 ± 1.32

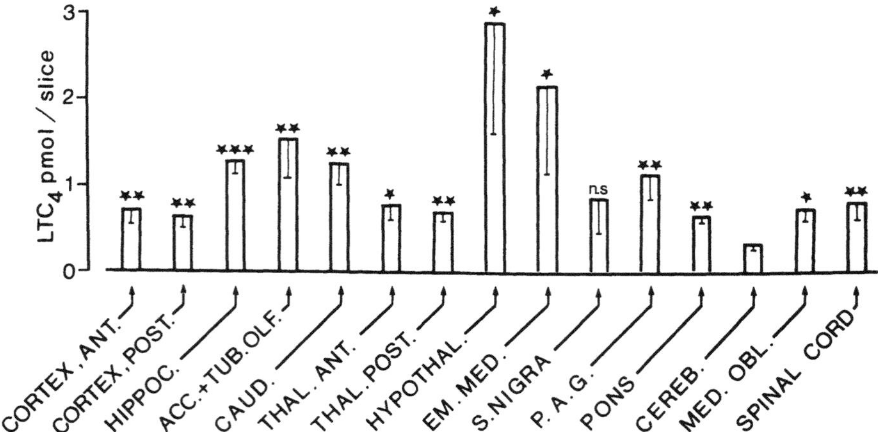

FIGURE 4. Regional distribution of LTC$_4$ biosynthesis in various parts of the rat brain. One (Em. Med.) or two slices from each of the indicated regions were preincubated in 0.5 ml PBS (pH 7.4) for 5 min prior to a 10-min incubation with ionophore A23187 (5 μM) and arachidonic acid (75 μM). Each LTC$_4$ value is the mean ± S.D. of duplicate determinations on samples from three different animals ($n = 3$). Statistical analyses were performed using Student's t-test: n.s., not significant; *$P < 0.05$; **$P < 0.01$; ***$P < 0.001$ (differences in LTC$_4$ values compared to the value obtained in samples from the cerebellum). Cortex, ant., anterior cortex; cortex, post. posterior cortex; Hippoc., hippocampus; Acc. + tub. olf., nucleus accumbens plus tuberculum olfactorium; Caud., nucleus caudatus; Thal. ant., anterior thalamus; Thal. post., posterior thalamus; Hypothal., hypothalamus; Em. Med., median eminence; S. Nigra, substantia nigra; P.A.G., periaqueductal gray matter; Cereb., cerebellum; Med. Obl., medulla oblongata.

pmol/slice), median eminence (2.1 ± 0.92 pmol/slice), and nucleus accumbens/ tuberculum olfactorium (1.6 ± 0.44 pmol/slice), whereas the production in slices from cerebellum was small (0.37 ± 0.06 pmol/slice) (mean ± S.D.). Background levels obtained with unstimulated tissue incubated in the absence of ionophore and exogenous arachidonic acid never exceeded 0.15 ± 0.03 (S.D.) pmol/slice.

3.3. Immunohistochemical Determination of LTC$_4$ in Rat Brain

Two patterns of immunoreactivity were observed after incubation of sections with LTC$_4$ antiplasma. In the lateral part of the external layer of the median eminence, a medium-dense plexus of LTC$_4$ immunoreactive fibers was seen with antiplasma dilutions up to 1 : 3200 (Fig. 5A). The fibers were strongly varicose and terminated close to the portal vessels. Single fibers with the same appearance were also observed in other brain regions including many areas in the hypothalamus, septum, and subfornical organ. In addition, weakly fluorescent cell bodies were seen in the medial preoptic area (Fig. 5C). Both patterns of fluorescence were completely abolished by preabsorption of the LTC$_4$ antiplasma with LTC$_4$–BSA conjugate (Fig. 5B) or glutathione disulfide (8 × 10^{-4} M) but not with LTC$_4$ alone.

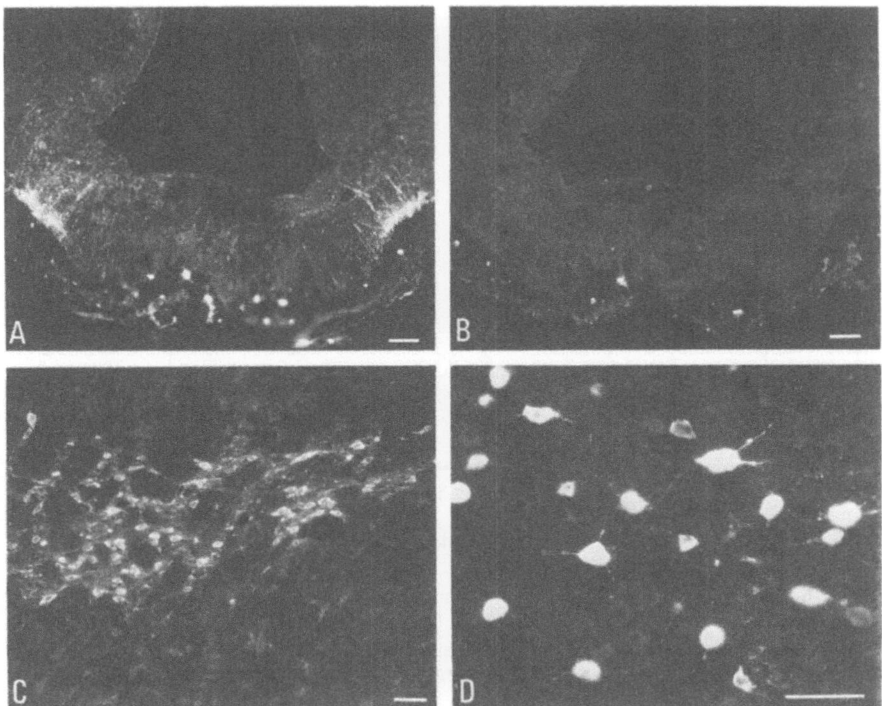

FIGURE 5. Immunofluorescence micrographs of the median eminence (A,B: consecutive sections), reticular thalamic nucleus (C), and parietal cortex (D) after incubation with LTC$_4$ antiplasma (A,C,D) or LTC$_4$ antiplasma preabsorbed with LTC$_4$–BSA conjugate (B).

With higher concentrations of antibodies (up to 1 : 800 dilution), numerous fluorescent cell bodies were seen in many brain regions (Fig. 5C,D). This fluorescence was not affected by absorption with LTC$_4$ or glutathione disulfide (up to 8×10^{-4} M), but a detectable decrease was seen when antiplasma preabsorbed with LTC$_4$–BSA conjugate was used. Absorption with BSA had no effect on the LTC$_4$-like immunoreactivity.

3.4. Effect of Leukotrienes on LH Release from Dispersed Rat Anterior Pituitary Cells

Leukotriene C$_4$ at concentrations of 10^{-14} to 10^{-9} M significantly stimulated basal luteinizing hormone (LH) release from dispersed rat anterior pituitary cells after 0.5 hr in culture (Fig. 6). At 10^{-14} M, LTC$_4$ caused a 90% increase of LH levels. Maximal effects were seen at concentrations of 10^{-13} to 10^{-12} M (123–129% stimulation). After 3 hr, the stimulatory effect of LTC$_4$ on LH release could no

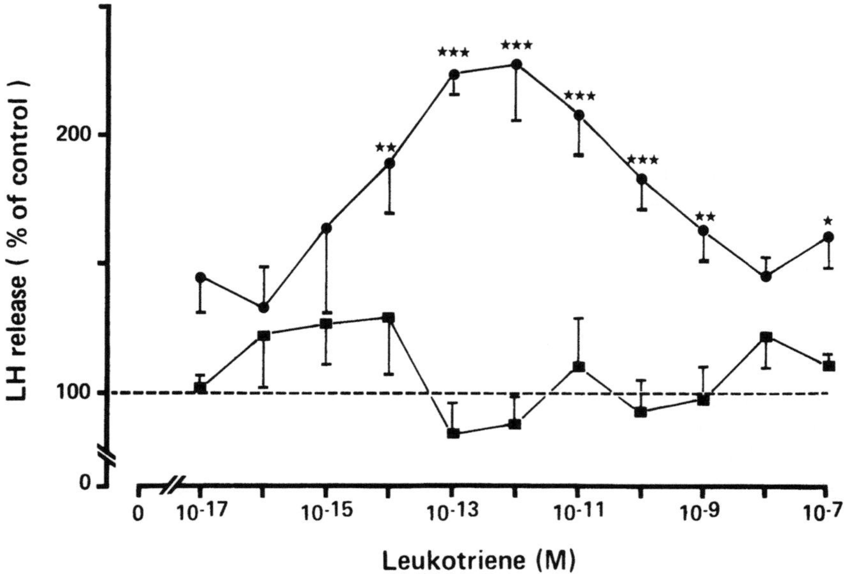

FIGURE 6. Effects of leukotrienes B_4 (■——■) and C_4 (●——●) on LH secretion from dispersed rat anterior pituitary cells after 0.5 hr in culture. Luteinizing hormone values are expressed as percentage of basal release. The curves show the mean ± S.E.M. from four different experiments with ten identical cultures in each experiment. $^*P < 0.02$; $^{**}P < 0.01$; $^{***}P < 0.001$ (Student's t-test).

longer be seen. At higher concentrations, the effect decreased: LTB_4 at concentrations of 10^{-17} to 10^{-7} M had no significant effect on basal LH release (Fig. 1). Luteinizing hormone-releasing hormone (10^{-9} or 10^{-6} M) had no stimulatory effect on LH release after 0.5 hr. However, after 3 hr, a dose-dependent stimulation of LH release was observed (Table I).

TABLE I. Release of LH (ng/ml ± S.E.) from
Cultures of Rat Anterior Pituitary in the Presence
of LTC_4 or LHRH[a]

	Incubation time	
	0.5 hr	3 hr
Control	97 ± 11	238 ± 12
LTC_4, 10^{-14}M	183 ± 21**	259 ± 18
LHRH, 10^{-9} M	97 ± 16	407 ± 38**
LHRH 10^{-6} M	129 ± 40	913 ± 117***

[a] Figures represent the mean of two experiments; $n = 10$ in each experiment. $^{**}P < 0.01$; $^{***}P < 0.001$ (Student's t-test).

4. DISCUSSION

The present report demonstrates the isolation of LTC_4, LTD_4, and LTE_4 from rat brain incubations. The capacity to produce LTC_4 was widely distributed in the central nervous system as indicated by radioimmunologic determinations. However, the detected amounts of LTC_4 varied markedly among different brain regions, with the highest levels found in the hypothalamus and the median eminence. The results may suggest distinct differences in biosynthesis between the investigated brain regions; however, differences in catabolism of LTC_4 may also contribute.

Evidence for the presence of LTC_4 was also obtained as an immunohisto-chemical analysis demonstrating immunoreactive fibers in the lateral part of the external layer of the median eminence and cell bodies in the preoptic area. However, further investigations are clearly required to fully elucidate the identity of the immunoreactivity. Thus, the immunofluorescence could only be abolished by absorption with LTC_4–BSA conjugate but not by LTC_4 alone. Control experiments revealed that this lack of effect by LTC_4 alone was not attributable to breakdown of the compound during the incubation (unpublished data). The fluorescence on widely distributed cell bodies could not be abolished by absorption by LTC_4 and only to a certain degree by the LTC_4–BSA conjugate. On the other hand, the immunoreaction in the median eminence could also be abolished by high concentrations of glutathione disulfide (ten times higher than LTC_4–BSA conjugate).

The highest amounts of LTC_4 were detected in the hypothalamus and median eminence. Therefore, it is of great interest that low concentrations of LTC_4 (10^{-14} M) specifically evoked release of LH from dispersed anterior pituitary cells. Leukotriene C_4 stimulated LH secretion through a mechanism with rapid onset and short duration. This fast action of LTC_4 was in contrast to the apparently slow action of LHRH as seen in our model. Thus, an effect of LHRH on LH release was not obvious after incubation for 0.5 hr in our system but was seen after 3 hr. At this time interval, a marked effect was observed with high concentrations of LHRH (10^{-6} M), but at 10^{-9} M, the increase of LH levels was only about 70%, i.e., less than the effect observed at 0.5 hr with LTC_4 at a concentration of 10^{-14} M. These findings suggest that LTC_4 is a potent stimulator of LH release. The results indicate that LTC_4 and LHRH stimulate LH release at least partly through different mechanisms. An interesting hypothesis is that LTC_4 and LHRH cooperate in the control of LH release in that LTC_4 may be preferentially responsible for initial release and LHRH for a more long-lasting effect. In this context, it is of interest that the LTC_4-immunoreactive nerve endings in the lateral part of the external layer of the median eminence were identical to those containing LHRH (Hulting *et al.*, 1985).

The observation that LTC_4 was formed in the cerebellum may have functional implications because LTC_4 and LTD_4 cause a conspicuously prolonged excitation of cerebellar Purkinje cells (Palmer *et al.*, 1980, 1981). Taken together, the present findings suggest that the leukotrienes may have a role as messengers or modulators of neuroendocrine events and central nervous activity.

ACKNOWLEDGMENTS. We wish to thank Ms. Margareta Maxe, Anne Peters, Waldtraut Hiort, Sandra Kleinau, Agneta Hilding, Gaby Åström, Kerstin Englund, and Lilian Franzén for skillful technical assistance. This work was supported by grants from the Swedish Medical Research Council (projects 03X-06805, 03P-6347, 04X-02887, 03P-06949, 19X-06865, and 03X-00217), Karolinska Institutet Research Funds, Alice and Knut Wallenberg Foundation, and The Söderberg Foundation.

REFERENCES

Andersson, K., Nielsen, O. G., Toftgård, R., Eneroth, P., Gustavsson, J.-Å., Battistini, N., and Agnati, L. S., 1983, Increased amine turnover in several hypothalamic noradrenaline nerve terminal systems and changes in prolactic secretion in the male rat by exposure to various concentrations of toluene, *Neurotoxicology* **4(4):** 43–56.

Bernström, K., and Hammarström, S., 1981, Metabolism of leukotriene D by porcine kidney, *J. Biol. Chem.* **256:**9579–9582.

Björck, T., Dahlén, S.-E., Gustavsson, L.-E., and Hedqvist, P., 1985, Characterization of the contractile responses to leukotrienes C_4, D_4, and E_4 in the guinea pig ileum, *Acta Physiol. Scand.* (in press).

Borgeat, P., Hamberg, M., and Samuelsson, B., 1976, Transformation of arachidonic acid and homo-γ-linolenic acid by rabbit polymorphonuclear leukocytes, *J. Biol. Chem.* **251:**7816–7820. Correction, 1977, **252:**8872.

Borgeat, P., and Samuelsson, B., 1979a, Arachidonic acid metabolism in polymorphonuclear leukocytes: Unstable intermediate in formation of dihydroxy acid, *Proc. Natl. Acad. Sci. U.S.A.* **76:**3213–3217.

Borgeat, P., and Samuelsson, B., 1979b, Arachidonic acid metabolism in polymorphonuclear leukocytes. Effects of ionophore A23187, *Proc. Natl. Acad. Sci. U.S.A.* **76:**2148–2152.

Corey, E. J., Marfat, A., Goto, G., and Brion, F., 1980, Leukotriene B: Total synthesis and assignment of stereochemistry, *J. Am. Chem. Soc.* **102:**7984–7985.

Dahlén, S.-E., 1983, Pulmonary effects of leukotrienes, *Acta Physiol. Scand.* **118 (Suppl. 512):**1–51.

Hulting, A. L., Lindgren, J. Å., Hökfelt, T., Heidvall, K., Eneroth, P., Werner, S., Patrono, C., and Samuelsson, B., 1984, Leukotriene C_4 stimulates LH secretion from rat pituitary cells in vitro, *Eur. J. Pharmacol.* **106:**459–460.

Hulting, A. L., Lindgren, J. Å., Hökfelt, T., Eneroth, P., Werner, S., and Samuelsson, B., 1985, Leukotriene C_4 as a mediator of LH release from rat anterior pituitary cells, *Nature Proc. Natl. Acad. Sci. U.S.A.* (in press).

Lindgren, J. Å., Hammarström, S., and Goetzl, E. J., 1983, A sensitive and specific radioimmunoassay for leukotriene C_4, *FEBS Lett.* **152:**83–88.

Lindgren, J. Å., Hansson, G., and Samuelsson, B., 1981, Formation of novel hydroxylated eicosatetraneoic acids in preparations of human polymorphonuclear leukocytes, *FEBS Lett.* **128:**329–335.

Lindgren, J. Å., Hökfelt, T., Dahlén, S.-E., Patrono, C., and Samuelsson, B., 1984, Leukotrienes in the rat central nervous system, *Proc. Natl. Acad. Sci. U.S.A.* **81:**6212–6216.

Morris, H. R., Taylor, G. W., Piper, P. J., and Tippins, J. R., 1980, Structure of slow reacting substance of anaphylaxis from guinea-pig lung, *Nature* **285:**104–106.

Murphy, R. C., Hammarström, S., and Samuelsson, B., 1979, Leukotriene C: A slow-reacting substance from murine mastocytoma cells, *Proc. Natl. Acad. Sci. U.S.A.* **76:**4275–4279.

Orange, R. P., and Austen, F., 1969, Slow reacting substance of anaphylaxis, *Adv. Immunol.* **10:**105–144.

Örning, L., and Hammarström, S., 1980, Inhibition of leukotriene C and leukotriene D biosynthesis. *J. Biol. Chem.* **255:**8023–8026.

Palmer, M. R., Mathews, W. R., Murphy, R. C., and Hoffer, B. J., 1980, Leukotriene C elicits a prolonged excitation of cerebellar Purkinje neurons, *Neurosci. Lett.* **18:**173–180.

Palmer, M. R., Mathews, W. R., Hoffer, B. J., and Murphy, R. C., 1981, Electrophysiological response of cerebellar Purkinje neurons to leukotriene D_4 and B_4. *J. Pharmacol. Exp. Ther.* **219:**91–96.

Samuelsson, B., 1983, Leukotrienes: Mediators of immediate hypersensitivity reactions and inflammation, *Science* **220:**568–575.

Wolfe, L. S., 1982, Eicosanoids: Prostaglandins, thromboxanes, leukotrienes and other derivatives of carbon-20 unsaturated fatty acids. *J. Neurochem.* **38:**1–14.

3

Properties of Prostaglandin-Biosynthesizing Enzymes

RICHARD J. KULMACZ

1. INTRODUCTION

The biosynthetic conversion of the esterified arachidonate in membrane lipids to the various eicosanoids can be divided into three stages. First, arachidonate must be released from tissue esters as the free acid. It can then react with one of the fatty acid oxygenases (the various lipoxygenases and cyclooxygenase are in this category) to form a hydroperoxy acid. In the third stage, the intermediate hydroperoxy acid is chemically modified to furnish the leukotrienes, classical prostaglandins, thromboxanes, and prostacyclin.

The fatty acid oxygenases occupy a crucial position in the arachidonate cascade, as they catalyze the first committed steps in the pathways to the eicosanoids. Information about the regulation of these enzymes would clearly further our understanding of the physiological control of eicosanoid biosynthesis. One important feature common to all fatty acid oxygenases examined so far is a requirement for lipid hydroperoxide activator. Because the catalytic products are themselves hydroperoxides, the fatty acid oxygenases show autoaccelerating reaction kinetics. In addition, there is the possibility of crossover activation between different fatty acid oxygenases in a tissue.

The biosynthetic enzymes in the PGH synthase branch of the cascade are discussed in this chapter in reverse order, first touching on some interesting aspects of the PGD, PGE, TXA, and PGI synthases, all of which have PGH as their substrate. Prostaglandin H synthase, the enzyme that produces PGH_2 from arachidonic acid, is then considered in more detail.

RICHARD J. KULMACZ ● Department of Biological Chemistry, University of Illinois at Chicago, Chicago, Illinois 60612.

2. PROSTAGLANDIN D SYNTHASE

Prostaglandin D synthase has been purified to homogeneity from rat brain and spleen. It is interesting to note that whereas both enzymes appear in the cytosolic fraction, they differ in physical characteristics and in their response to added glutathione. The brain enzyme is a monomer with a molecular weight of about 82,000, and its activity is not stimulated by glutathione (Shimizu *et al.*, 1979). On the other hand, the native PGD synthase from spleen has an apparent molecular weight of 34,000. When examined by SDS-polyacrylamide gel electrophoresis, its migration indicated a molecular weight of 26,000 (Christ-Hazelhof and Nugteren, 1979). This discrepancy is rather large, and it is possible that a subunit of about 8,000 is present but migrated out of the gel during electrophoresis. The spleen enzyme is stimulated approximately twofold by glutathione.

3. PROSTAGLANDIN E SYNTHASE

Prostaglandin E synthase has not yet been purified to homogeneity. Seminal vesicles are a rich source of this microsomal protein, which, although it requires GSH to catalyze the conversion of PGH to PGE, does not oxidize a detectable amount of GSH in the process (Ogino *et al.*, 1977). It is not known whether the GSH serves as a direct catalytic intermediary or simply maintains a sulfhydryl group on the synthase in a reduced state.

4. THROMBOXANE A SYNTHASE

Thromboxane A synthase is an interesting and complex enzyme. Although not yet purified to homogeneity, the microsomal enzyme appears to catalyze two distinct transformations of PGH_2: into thromboxane A_2 and, in a reaction that splits off a molecule of malondialdehyde, to hydroxyheptadecatrienoic acid (HHT; Yoshimoto *et al.*, 1977; Hammarström and Falardeau, 1977). Unsaturation at C-5 appears to influence the relative rates of the two reactions. Whereas PGH_2 is converted to HHT and TXA_2 at roughly the same rate, PGH_1 and Δ^4-PGH_1 are converted almost exclusively to hydroxyheptadecadienoic acid (HHD) and Δ^4-HHD (Hammarström, 1982).

5. PROSTAGLANDIN I SYNTHASE

Two alternative catalytic fates of PGH_2 are also a characteristic of PGI synthase. Like TXA synthase, PGI synthase will catalyze the synthesis of MDA and HHT as well as its "primary" product, PGI_2 (Watanabe *et al.*, 1979). Prostaglandin H_1, lacking unsaturation at C-5, is converted exclusively to HHD. Prostaglandin synthase has recently been purified to homogeneity from bovine aorta microsomes by DeWitt and Smith (1983). The enzyme has a subunit of molecular weight 52,000. Self-inactivation is observed during catalysis by PGI synthase, and activity is also lost during incubation with lipid hydroperoxides (DeWitt and Smith, 1983). The appearance of the absorbance spectrum of PGI synthase suggests the presence of heme. The exact stoichiometry of heme in the protein is not well established, but heme has been implicated in catalysis by the apparent coincidence between loss of activity (either during catalysis or on incubation with hydroperoxide) and bleaching of the chromophore (DeWitt and Smith, 1983). There has recently been some speculation (Ullrich *et al.*, 1981; Ullrich and Haurand, 1983), based on spectroscopic and inhibitor studies on crude preparations, that both TXA synthase and PGI synthase are cytochromes P-450, but the matter does not appear to have been resolved as yet.

6. PROSTAGLANDIN H SYNTHASE

All of the four enzymes just discussed utilize PGH_2 as their substrate. I should like to turn attention now to the enzyme responsible for the synthesis of PGH_2 from arachidonic acid and oxygen.

An examination of the properties of PGH synthase is of importance not only to the understanding of how the biosynthesis of the various prostaglandins, thromboxanes, and prostacyclin are regulated but also because PGH synthase represented an accessible model for the study of fatty acid oxygenases. Prostaglandin synthase has been purified to homogeneity from ovine and bovine seminal vesicles (Hemler *et al.*, 1976; Miyamoto *et al.*, 1976; van der Ouderaa *et al.*, 1977) and can be prepared in milligram quantities. The detergent-solubilized enzyme is a dimer of two similar 70,000-Da subunits (van der Ouderaa *et al.*, 1979; Roth *et al.*, 1980) and catalyzes both the *bis*-dioxygenation of arachidonic acid to PGG_2 and the peroxidative reduction of PGG_2 to PGH_2.

Both enzyme activities of PGH synthase have an absolute requirement for heme (Miyamoto *et al.*, 1976; van der Ouderaa *et al.*, 1977). We have recently found (Kulmacz and Lands, 1984) that full cyclooxygenase activity is obtained with only one heme bound per synthase dimer. This result implies that the two subunits are functionally different.

The two enzymatic activities of PGH synthase also have a requirement for hydroperoxide: the cyclooxygenase needs hydroperoxide continuously as activator (Hemler and Lands, 1980), and the peroxidase uses hydroperoxide as substrate. The requirement of peroxide for activation of the cyclooxygenase and the presence of endogenous peroxidase activity in the synthase posed a paradox: How can cyclooxygenase activity be expressed in the presence of peroxidase, which can potentially remove a necessary cyclooxygenase activator? We suspected that the answer lay in a difference between the peroxide concentration needed to effectively activate the cyclooxygenase (K_p) and that needed for efficient peroxidase activity

(K_m). When the relative sensitivities of the cyclooxygenase and peroxidase activities to inhibition by an added peroxide scavenger, glutathione peroxidase, were examined, it was found that the peroxidase activity of the synthase was more readily suppressed than the cyclooxygenase activity (Kulmacz and Lands, 1983). Further analysis of this result allowed calculation of a K_p value of 20 nM for the cyclooxygenase and a K_m value of 3 µM for the peroxidase. This large difference in peroxide requirements means that the cyclooxygenase can be activated effectively at peroxide levels too low to be removed effectively by the peroxidase.

We have recently investigated the ability of cyanide to inhibit the peroxidase activity of PGH synthase (Kulmacz and Lands, 1985). The inhibition, apparently noncompetitive with respect to peroxide, was characterized by a K_I value of 0.3 mM NaCN and probably resulted from liganding of synthase heme by cyanide, because the K_d value of the synthase–cyanide complex was determined to be 0.2 mM (Kulmacz and Lands, 1985).

Cyanide also has distinct effects on the cyclooxygenase activity of PGH synthase and has provided a tool to investigate the process of activation by peroxide. The time course of oxygen consumption in a typical control cyclooxygenase reaction (Hemler and Lands, 1980) shows a rapid accelerative phase, with the optimal velocity reached in about 7 sec, followed by steadily falling velocity as self-inactivation sets in. Addition of 20 mM cyanide to such a system resulted (Kulmacz and Lands, 1985) in a 90% inhibition of the cyclooxygenase optimal velocity. The initial acceleration was retarded, and the optimal velocity was reached only after more than 3 min of lag time. The eventual extent of oxygen consumption before complete self-inactivation was not appreciably reduced, however, so the presence of cyanide seems to have altered the rate-determining step in the reaction mechanism.

The slow cyclooxygenase reaction kinetics induced by cyanide appear to be related to an impairment of the activation of the cyclooxygenase by hydroperoxide. This can be demonstrated by the ability of added hydroperoxide to reverse the inhibitory effects of cyanide (Marshall *et al.*, 1984). This graded response of the cyanide-inhibited system has furnished an assay for low levels of hydroperoxide (Marshall *et al.*, 1984). It has also encouraged us to interpret experimental alterations in cyclooxygenase kinetics in terms of changes in the rates of peroxide generation and removal and in the activator effectiveness of peroxide.

The stimulatory effect of phenolic antioxidants on the cyclooxygenase activity of PGH synthase can also be related to the activation of the cyclooxygenase by peroxide. When the levels of activator peroxide are lowered by glutathione peroxidase (Hanel and Lands, 1982), or the activator effectiveness of the peroxide is decreased by cyanide (Kulmacz and Lands, 1985), the ability of phenol to stimulate the cyclooxygenase is lost. Thus, the stimulatory effect of phenolic agents may occur by some facilitation of the activation process.

The reaction kinetics of the cyclooxygenase can also be changed by treating the synthase with indomethacin (Fig. 1). In comparison with the control reaction,

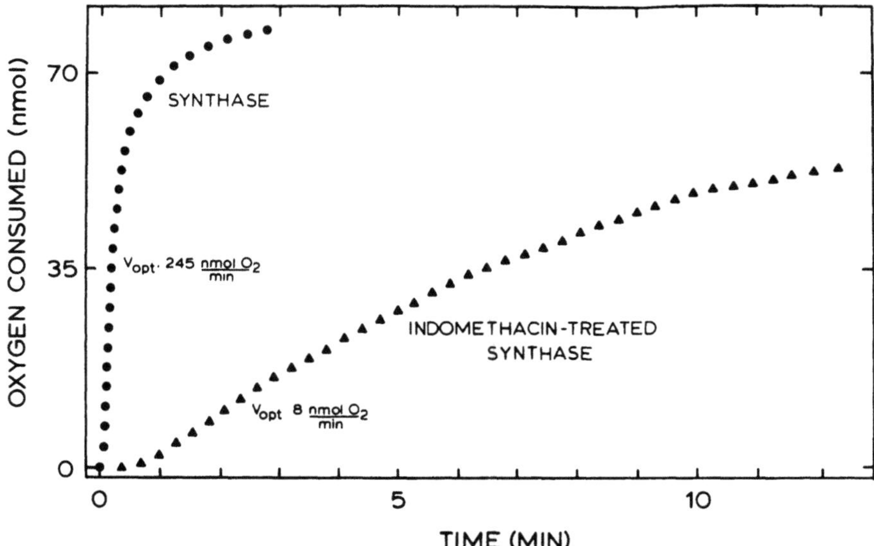

FIGURE 1. Effect of indomethacin on cyclooxygenase reaction kinetics. The untreated PGH synthase (●) or synthase previously incubated with excess indomethacin (▲) was injected into a reaction mixture containing 0.1 M potassium phosphate (pH 7.2), 1mM phenol, and 0.1 mM arachidonic acid. The concentration of synthase subunit in the reaction was 11 nM. Oxygen consumption was monitored with a polarographic oxygen electrode (Rome and Lands, 1975).

the oxygen consumption time course of the indomethacin-treated enzyme is sluggish and prolonged, and the optimal velocity is decreased 97%. The slow kinetics of the indomethacin-treated enzyme indicate an impairment of the activation of cyclo-oxygenase and are consistent with a decreased availability of the peroxidase acti-vator. This decreased availability is the expected consequence of the selective inhibition of activator generation by indomethacin. Thus, the changes in cyclo-oxygenase reaction kinetics observed in the perturbed systems mentioned above can be interpreted in terms of changes in either the availability of activator hydro-peroxide or the relative effectiveness of hydroperoxide as activator.

The ability to manipulate the cyclooxygenase activity of PGH synthase *in vitro* by changes in activator peroxide availability or effectiveness suggested that similar controls might modulate PG biosynthesis *in vivo*. The results of a preliminary examination of the balance between lipid hydroperoxide biosynthetic capacity and glutathione peroxidase capacity in several rat tissues are shown in Table I. Pros-taglandin synthetic capacities were derived from the results of Pace-Asciak and Rangaraj (1977). The ratio of the hydroperoxide synthetic capacity to glutathione peroxidase capacity can be used to give some indication of the availability of hydroperoxide in a tissue. A ranking of the tissues with respect to this ratio parallels a ranking of their recognized tendencies to form prostaglandins under physiological

TABLE I. Rat Tissue Capacities for Synthesis and
Removal of Lipid Hydroperoxides

	Synthesis (mmol PGG_2/min per g)	Glutathione peroxidase[a] (mmol ROOH/min per g)	Synthesis/removal
Resident peritoneal cells	0.05	0.7	0.07
Lung	0.09[b]	4.7	0.02
Spleen	0.07[b]	7.9	0.009
Kidney	0.02[b]	9.2	0.003
Heart	0.02[b]	8.1	0.002
Liver	0.03[b]	33.0	0.001

[a] Glutathione peroxidase was assayed as described by Lawrence *et al.* (1974).
[b] Estimated from Pace-Asciak and Rangaraj (1977).

conditions. This correlation supports a role for the tissue peroxide level in the modulation of prostaglandin biosynthesis.

Results from tissue homogenates do not, of course, accurately reflect the situation in intact cells, where the subcellular localization of the prostaglandin biosynthetic enzymes (which W. L. Smith and A. Garcia-Perez address in Chapter 4) and of the peroxidases must be considered. It is therefore important for an assessment of the role of peroxides in control of eicosanoid biosynthesis to determine the actual peroxide levels *in vivo*. In this regard, we have used the cyanide-inhibited cyclooxygenase system to assay the concentration of hydroperoxides in human plasma. The average for five separate samples was 0.53 ± 0.21 µM. The fact that this concentration is considerably higher than that needed to activate the purified cyclooxygenase *in vitro* serves to emphasize once again the importance of understanding how the localization of the various fatty acid oxygenases influences the ability of their circulating peroxides to reach them and stimulate their activity.

ACKNOWLEDGMENT. This work has been supported in part by Public Health Service grant GM 30509.

REFERENCES

Christ-Hazelhof, E., and Nugteren, D. H., 1979, Purification and characterization of prostaglandin endoperoxide D-isomerase, a cytoplasmic, glutathione-requiring enzyme, *Biochim. Biophys. Acta* **572:**43–51.

DeWitt, D. L., and Smith, W. L., 1983, Purification of prostacyclin synthase from bovine aorta by immunoaffinity chromatography, *J. Biol. Chem.* **258:**3285–3293.

Hammarström, S., and Falardeau, P., 1977, Resolution of prostaglandin endoperoxide synthase and thromboxane synthase of human platelets, *Proc. Natl. Acad. Sci. U.S.A.* **74:**3691–3695.

Hammarström, S., 1982, Biosynthesis and biological actions of prostaglandins and thromboxanes, *Arch. Biochem. Biophys.* **214:**431–445.

Hanel, A. M., and Lands, W. E. M., 1982, Modification of anti-inflammatory drug effectiveness by ambient lipid peroxides, *Biochem. Pharmacol.* **31:**3307–3311.

Hemler, M., Lands, W. E. M., and Smith, W. L., 1976, Purification of the cyclooxygenase that forms prostaglandins, *J. Biol. Chem.* **251:**5575–5579.

Hemler, M. E., and Lands, W. E. M., 1980, Evidence for a peroxide-initiated free radical mechanism of prostaglandin biosynthesis, *J. Biol. Chem.* **255:**6253–6261.

Kulmacz, R. J., and Lands, W. E. M., 1983, Requirements for hydroperoxide by the cyclooxygenase and peroxidase activities of prostaglandin H synthase, *Prostaglandins* **25:**531–540.

Kulmacz, R. J., and Lands, W. E. M., 1984, Prostaglandin H synthase. Stoichiometry of heme cofactor, *J. Biol. Chem.* **259:**6358–6363.

Kulmacz, R. J., and Lands, W. E. M., 1985, Quantitative similarities in the several actions of cyanide on prostaglandin H synthase, *Prostaglandins* **29:** 175–190.

Lawrence, R. A., Sunde, R. A., Schwartz, G. L., and Hoekstra, W. G., 1974, Glutathione peroxidase activity in rat lens and other tissues in relation to dietary selenium intake, *Exp. Eye Res.* **18:**563–569.

Marshall, P. J., Kulmacz, R. J., and Lands. W. E. M., 1984, Hydroperoxides, free radicals and prostaglandin biosynthesis, in: *Oxygen Radicals in Chemistry and Biology* (W. Bors, M. Saran, and D. Tait, eds.), Walter de Gruyter, Berlin, pp. 299–304.

Miyamoto, T., Ogino, N., Yamamoto, S., and Hayaishi, O., 1976, Purification of prostaglandin endoperoxide synthetase from bovine vesicular gland microsomes, *J. Biol. Chem.* **251:**2629–2636.

Ogino, N., Miyamoto, T., Yamamoto, S., and Hayaishi, O., 1977, Prostaglandin endoperoxide E isomerase from bovine vesicular gland microsomes, a glutathione-requiring enzyme, *J. Biol. Chem.* **252:**890–895.

Pace-Asciak, C. R., and Rangaraj, G., 1977, Distribution of prostaglandin biosynthetic pathways in several rat tissues. Formation of 6-ketoprostaglandin $F_{1\alpha}$, *Biochim. Biophys. Acta* **486:**579–582.

Rome, L. H., and Lands, W. E. M., 1975, Properties of a partially-purified preparation of the prostaglandin-forming oxygenase from sheep vesicular gland, *Prostaglandins* **10:**813–824.

Roth, G. J., Siok, C. J., and Ozols, J., 1980, Structural characteristics of prostaglandin synthetase from sheep vesicular gland, *J. Biol. Chem.* **255:**1301–1304.

Shimizu, T., Yamamoto, S., and Hayaishi, O., 1979, Purification and properties of prostaglandin D synthetase from rat brain, *J. Biol. Chem.* **254:**5222–5228.

Ullrich, V., Castle, L., and Weber, P., 1981, Spectral evidence for the cytochrome P450 nature of prostacyclin synthetase, *Biochem. Pharmacol.* **30:**2033–2036.

Ullrich, V., and Haurand, M., 1983, Thromboxane synthetase as a cytochrome P450 enzyme, in: *Advances in Prostaglandin, Thromboxane, and Leukotriene Research,* Vol. 11 (B. Samuelsson, R. Paoletti, and P. Ramwell, eds.), Raven Press, New York, pp. 105–110.

van der Ouderaa, F. J., Buytenhek, M., Nugteren, D. H., and van Dorp, D. A., 1977, Purification and characterization of prostaglandin endoperoxide synthetase from sheep vesicular glands, *Biochim. Biophys. Acta* **487:**315–331.

van der Ouderaa, F. J., Buytenhek, M., Slikkerveer, F. J., and van Dorp, D. A., 1979, On the haemoprotein character of prostaglandin endoperoxide synthetase, *Biochim. Biophys. Acta* **572:**29–42.

Watanabe, K., Yamamoto, S., and Hayaishi, O., 1979, Reactions of prostaglandin endoperoxides with prostaglandin I synthetase solubilized from rabbit aorta microsomes, *Biochem. Biophys. Res. Commun.* **87:**192–199.

Yoshimoto, T., Yamamoto, S., Okuma, M., and Hayaishi, O., 1977, Solubilization and resolution of thromboxane synthesizing system from microsomes of bovine blood platelets, *J. Biol. Chem.* **252:**5871–5874.

4

A Two-Receptor Model for the Mechanism of Action of Prostaglandins in the Renal Collecting Tubule

WILLIAM L. SMITH and ARLYN GARCIA-PEREZ

1. INTRODUCTION

Studies on the mechanism of action of steroid hormones have led to the concept that different steroids act through different receptors present in different target cells to elicit a common biochemical response, namely, regulation of transcription. One might presume that different prostaglandins also operate through different receptors to cause some common response. However, this response is still not defined. In this chapter we summarize our recent studies on the metabolism and function of prostaglandins by the renal collecting tubule (Garcia-Perez and Smith, 1983, 1984). Our major focus is to describe the development of a two-receptor model for the actions of prostaglandins in these cells.

Briefly, we propose that prostaglandins operate in the collecting tubule through two distinct receptor populations having different specificities to elicit two different biochemical responses. Binding of a prostaglandin to one type of receptor (type I) leads to the activation of adenylate cyclase and rapid elevation of cAMP levels in collecting tubule cells. We suggest that this increase in cAMP leads to inhibition of the release of arachidonic acid from phosphoglycerides. Thus, occupancy of this type I prostaglandin receptor leads to feedback inhibition of prostaglandin formation. A second type of prostaglandin receptor (type II) in collecting tubule cells is coupled

WILLIAM L. SMITH and ARLYN GARCIA-PEREZ ● Department of Biochemistry, Michigan State University, East Lansing, Michigan 48824.

to the desensitization of adenylate cyclase to circulating hormones. The major circulating hormone to which collecting tubules respond is antidiuretic hormone (ADH).

2. RESULTS AND DISCUSSION

2.1. Renal Collecting Tubules as a Model for Studies of the Mechanism of Prostaglandin Action

We begin our discussion of the mechanism of action of prostaglandins in collecting tubules by indicating why these cells are an appropriate model for looking at a physiological action of prostaglandins. Diagrammed in Fig. 1 are the major functional units of the kidney and the cellular sites at which prostaglandins are formed by the kidney. The renal collecting tubule is the terminal part of the tubule. Immunocytochemical studies in our laboratory (Smith and Bell, 1978; Smith *et al.*, 1979) and microdissection studies by Currie and Needleman (1982) have demonstrated that the thin limb and the collecting tubule are the only sites of prostaglandin synthesis in the renal tubule. Thus, collecting tubules are clearly capable of producing prostaglandins.

Early studies by Grantham and Orloff (1968), which have since been confirmed by many other workers (Handler and Orloff, 1981), have established that addition of low concentrations of PGE_2 (*ca.* 10^{-9} M) inhibits the hydroosmotic effect of ADH on rabbit cortical collecting tubules *in vitro*. That is, treatment of isolated perfused collecting tubules with PGE_2 prevents the ADH-induced transcellular movement of water from the lumen of the tubule to the surrounding interstitium. This hydroosmotic effect of ADH is known to be mediated by cAMP (Handler and Orloff, 1981). However, Grantham and Orloff demonstrated that PGE_2 does not prevent the hydroosmotic effect of cAMP itself, suggesting that PGE_2 inhibits ADH-induced cAMP formation. Paradoxically, Grantham and Orloff found that high concentrations of prostaglandins (*ca.* $\geqslant 10^{-7}$ M), when used alone, actually cause a hydroosmotic effect in rabbit collecting tubules.

Inhibition of the hydroosmotic response to ADH by PGE_2 is not simply a pharmacological curiosity observed with isolated perfused tubules but also appears to occur *in vivo*. Animals treated with cyclooxygenase inhibitors or subjected to essential fatty acid deficiency produce a hyperosmotic urine (Anderson *et al.*, 1975; Fejes-Toth *et al.*, 1977; Hansen, 1981). Moreover, the medullas of rats fed with cyclooxygenase inhibitors contain elevated levels of cAMP (Lum *et al.*, 1977). One would expect these results if no prostaglandins were being produced by the collecting tubule to blunt the cAMP-elevating effect of ADH. The fact that collecting tubules form prostaglandins, coupled with evidence that prostaglandins inhibit the hyroosmotic response of ADH both in isolated tubules and in intact animals, argues strongly that prostaglandins normally play a physiological role in modulating the response of collecting tubule cells to ADH.

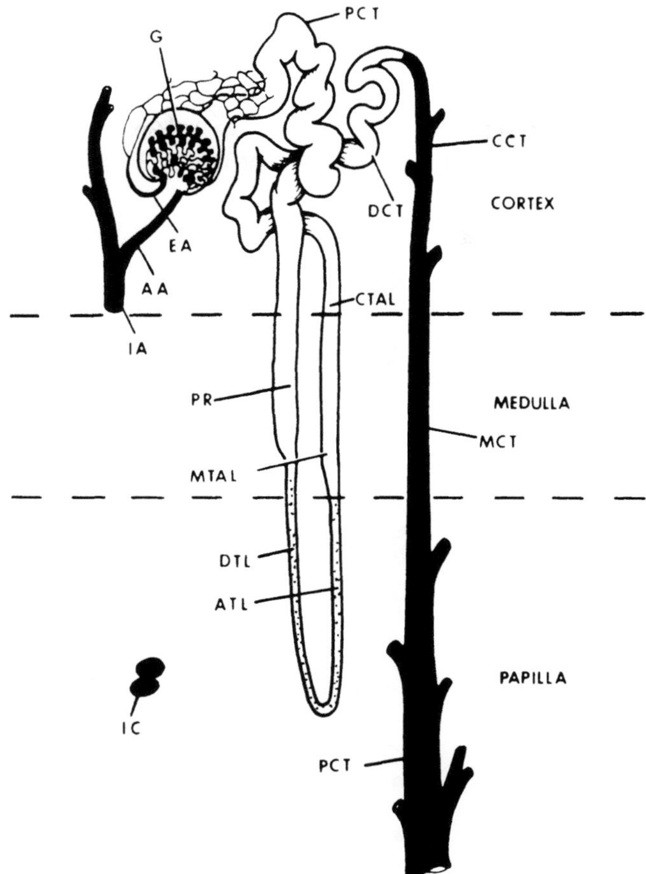

FIGURE 1. Renal sites of prostaglandin synthesis (colored in black). G, glomerulus; IA, interlobular artery; AA, afferent arteriole; EA, efferent arteriole; PCT, proximal convoluted tubule; PR, pars recta; DTL, descending thin limb; ATL, ascending thin limb; MTAL, medullary thick ascending limb; CTAL, cortical thick ascending limb; DCT, distal convoluted tubule; CCT, cortical collecting tubule; MCT, medullary collecting tubule; PCT, papillary collecting tubule; IC, interstitial cells. (From Smith *et al.*, 1983).

2.2. Isolation and Properties of Canine Cortical Collecting Tubule Cells

In order to study ADH–prostaglandin interactions at the biochemical level, it is necessary to have large numbers of collecting tubule cells. We have obtained large and homogeneous populations of canine cortical collecting tubule (CCCT) cells by immunodissection (Fig. 2; Garcia-Perez and Smith, 1983, 1984). We first developed a rat lymphocyte–mouse myeloma hybrid cell line (*cct*-1) that secretes a rat IgG$_{2c}$ that interacts with the cell surface of canine collecting tubule cells but

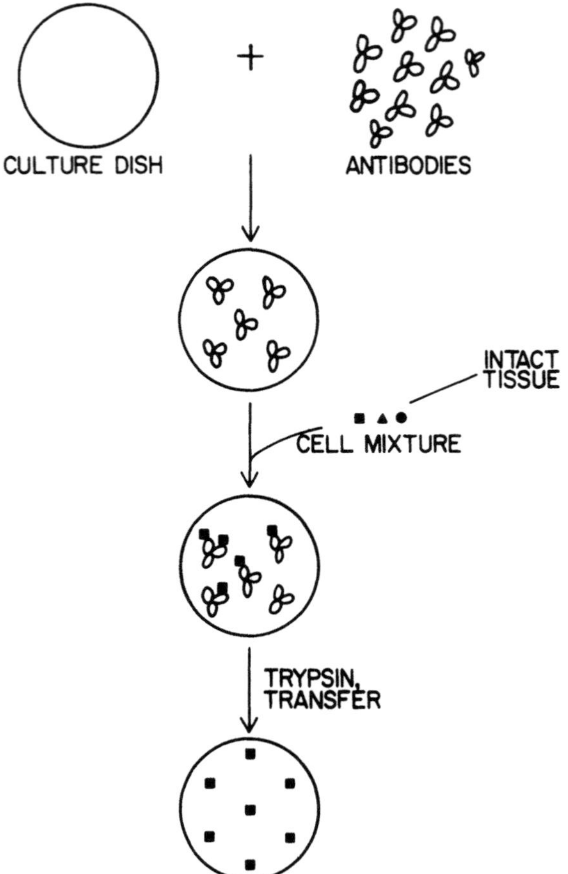

FIGURE 2. Procedure for the isolation of canine cortical collecting tubule cells using rat anti-canine collecting tubule monoclonal antibodies (secreted by *cct*-1). Intact cortical tissue is dispersed by treatment with collagenase to yield a mixture of cortical cells, which are fractionated by differential adherence to the antibody-coated culture dish. (From Smith *et al.*, 1983.)

not other renal cells. We then coated plastic tissue culture plates with the anti-collecting tubule antibodies secreted by *cct*-1, and we have used these antibody-coated dishes to selectively remove collecting tubule cells from heterogeneous single-cell dispersions prepared from the renal cortex (Fig. 2).

Isolated CCCT cells can be grown in culture and retain their major differentiated properties under culture conditions (Garcia-Perez and Smith, 1983). The CCCT cells exhibit a histochemical and immunohistochemical staining pattern character-istic of that seen with collecting tubules in cryotome sections of intact kidney. Moreover, CCCT cells form cAMP in response to ADH and isoproterenol but not other renal hormones, and CCCT cells form prostaglandins in response to both bradykinin and ADH.

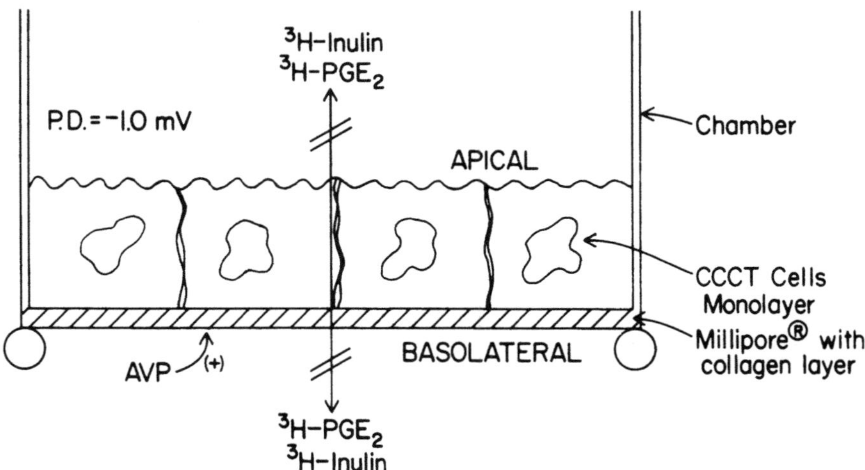

FIGURE 3. A CCCT cell monolayer system for studying functional asymmetry of the cortical collecting tubule. The CCCT cells seeded on a Millipore® filter bonded to a hollow polycarbonate cylinder form a confluent cell monolayer with distinct apical and basolateral surfaces. The chamber stands on three legs in a culture dish. The three criteria routinely used to establish the functional polarity of these chambers (i.e., transcellular potential differences, impermeability to inulin, and asymmetry of AVP-induced cAMP formation) are depicted. (From Garcia-Perez and Smith, 1984.)

When seeded at confluency on Millipore® filters (Fig. 3), CCCT cells develop a transcellular potential difference of 1 mV (lumen side negative) and are essentially impermeable to either inulin or PGE_2 (Garcia-Perez and Smith, 1984). When examined by transmission electron microscopy, CCCT cells on Millipore® filters are found to have microvilli on their apical (luminal) surface. In addition, CCCT cells form cAMP in response to ADH added to the basolateral but not the apical side of the cell. Moreover, CCCT cells form prostaglandins in response to bradykinin added to the apical but not the basolateral side of the cell. The major prostaglandin formed by these cells is PGE_2. Overall, then, this CCCT cell system retains the expected differentiated properties of the collecting tubule and is a reasonable model system for studying the function and metabolism of prostaglandins in the collecting tubule.

2.3. Prostaglandin–ADH Interactions in CCCT Cells

In the experiment depicted in Fig. 4, we measured the level of cAMP in the media bathing the apical and basolateral surfaces of CCCT cells on Millipore® filters following a 60-min incubation with 10^{-6} M AVP added to either the apical or basolateral surface of the monolayer. As mentioned earlier, the results indicate that AVP works only from the basolateral surface to cause cAMP release.

As shown in Fig. 5, PGE_2, unlike AVP, causes cAMP release from CCCT cells when added to either the apical or basolateral surface of the cells. Half-maximal

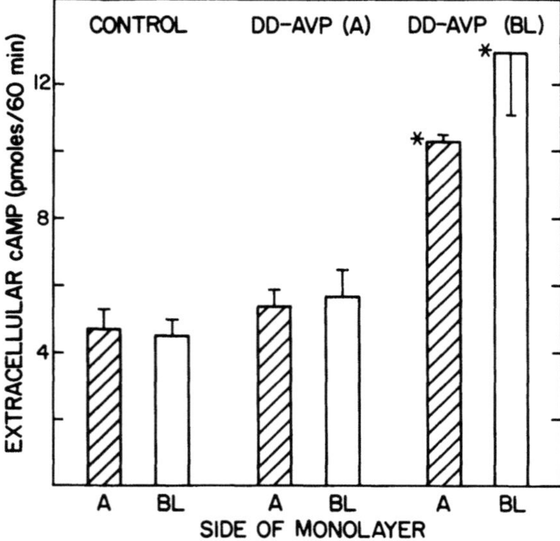

FIGURE 4. Sidedness of DDAVP effect on cAMP release from CCCT cell monolayers on Millipore® filters. Indicated on the horizontal axis is the side of release of cAMP. A, apical; BL, basolateral. The bars indicate the release in response to no effector or to 10^{-6} M DDAVP added to either the apical (A) or basolateral (BL) side of the monolayer. The data represent the mean of six chambers. All treatments were at 37° for 60 min in the presence of 10^{-4} M IBMX. Extracellular cAMP was measured by radioimmunoassay. *Significantly different from control values (i.e., absence of DDAVP) ($P < 0.05$). (From Garcia-Perez and Smith, 1984.)

increases in cAMP release occurred with 5×10^{-8} M PGE$_2$. We would like to emphasize that at concentrations of $\leq 10^{-10}$ M, PGE$_2$ had no detectable effect on cAMP release in this CCCT cell system.

Figure 6 illustrates what happens if we add both AVP and 10^{-10} M PGE$_2$ to the CCCT cell monolayer. Vasopressin, in this case at a concentration of 10^{-8} M, when added to the basolateral surface of the monolayer causes its usual two- to threefold increase in cAMP release. Although 10^{-10} M PGE$_2$ has no independent effect on cAMP release, 10^{-10} M PGE$_2$ completely inhibits AVP-induced cAMP release. In the experiment depicted in Fig. 6, we measured extracellular cAMP, but we have obtained essentially the same results when measuring intracellular cAMP levels. In addition, although PGE$_2$ was added to both sides of the CCCT cell monolayer in this experiment, PGE$_2$ will actually operate from either side of the monolayer to inhibit AVP-induced cAMP formation.

Several prostaglandins other than PGE$_2$ will cause inhibition of AVP-induced cAMP formation by CCCT cells. Among these is PGF$_{2\alpha}$ · PGF$_{2\alpha}$ at concentrations of 10^{-12} M or greater causes inhibition of AVP-induced cAMP formation. This turns out to be a very important observation. Although PGF$_{2\alpha}$ is as effective as PGE$_2$ in inhibiting AVP-induced cAMP formation, PGF$_{2\alpha}$, unlike PGE$_2$, does not activate adenylate cyclase even at high concentrations. We also mention that we have found in preliminary experiments that PGD$_2$ will activate the CCCT cell adenylate cyclase but that PGD$_2$ does not appear to inhibit AVP-induced cAMP formation.

Thus, an inhibitory response can be produced by comparable concentrations of PGE$_2$ or PGF$_{2\alpha}$ but not PGD$_2$, whereas PGE$_2$ and PGD$_2$ but not PGF$_{2\alpha}$ will

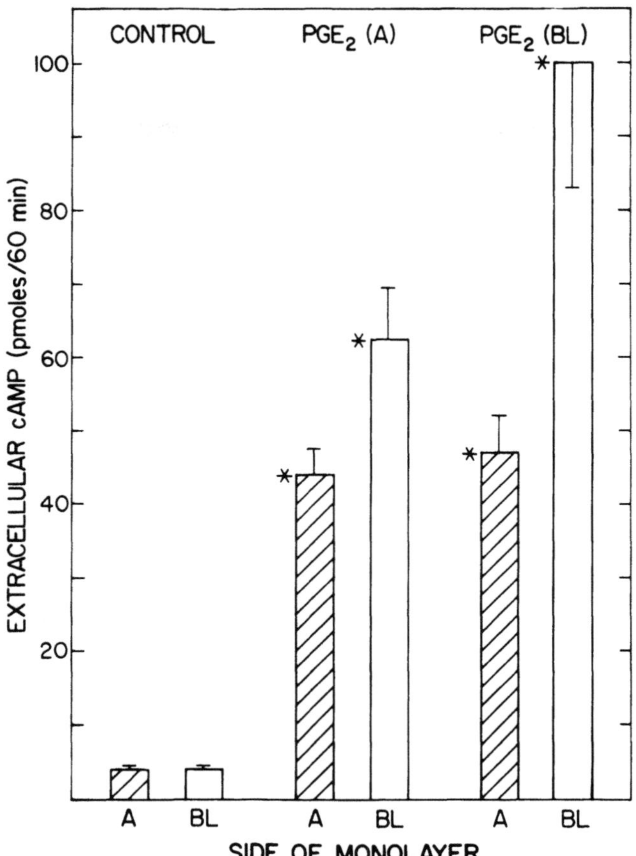

FIGURE 5. Sidedness of PGE_2 effect on cAMP release from CCCT cell monolayers on Millipore® filters. The bars indicate the amount of extracellular cAMP measured on the apical (A) or the basolateral (BL) side of cell monolayers following treatment with no effector or with 10^{-6} M PGE_2 added to the apical (A) or basolateral (BL) side. The data represent the mean of six chambers. All treatments were performed at 37° for 60 min in the presence of 10^{-4} M IBMX. Extracellular cAMP was measured by radioimmunoassay. *Significantly different from control values (i.e., absence of PGE_2) ($P < 0.05$). (From Garcia-Perez and Smith, 1984.)

activate the adenylate cyclase of CCCT cells. In essence, there are two·types of responses presumably coupled to two classes of prostaglandin receptors in CCCT cells. One receptor (type I) is coupled to the activation of the adenylate cyclase, and another (type II) is coupled to the desensitization of these cells to the cAMP-elevating effect of AVP.

In the inhibition experiment depicted in Fig. 6, the cells were actually preincubated with PGE_2 for 60 min prior to the addition of AVP. This preincubation is

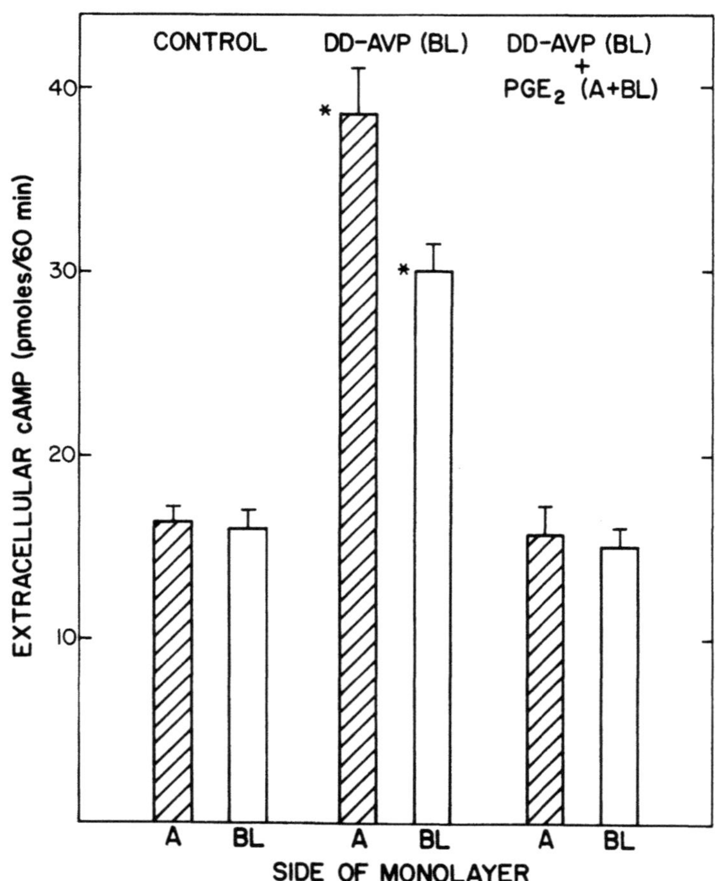

FIGURE 6. Inhibition by PGE$_2$ of DDAVP-induced cAMP release from CCCT cell monolayers on Millipore® filters. Each bar represents the amount of cAMP measured on the apical (A) or basolateral (BL) side of the CCCT cell monolayer following treatment with (a) 10^{-8} M DDAVP added to the basolateral side, (b) 10^{-8} M DDAVP (added to the basolateral side) plus 10^{-10} M PGE$_2$ (added to both sides), or (c) no effector. All samples were preincubated for 60 min at 37°; PGE$_2$ was added for the duration of the preincubation period to samples that were to be treated with PGE$_2$. All other samples were preincubated with buffer alone. At the end of the preincubation period, the medium was removed, and the monolayers were incubated for an additional 60 min at 37° with the effectors indicated. Following this incubation period, the media from the two sides were removed and assayed for cAMP by radioimmunoassay. All treatments were performed with triplicate chambers in the presence of 10^{-4} M IBMX. *Significantly different from control values ($P < 0.05$). (From Garcia-Perez and Smith, 1984.)

essential for the development of the inhibitory response. For example, CCCT cells must be preincubated for 10–20 min with 10^{-11} M PGE$_2$ before the inhibitory effect of PGE$_2$ on AVP-induced cAMP formation is expressed. In contrast, 10^{-6} M PGE$_2$ acts within 2 min to increase cAMP levels in CCCT cells.

Figure 7 summarizes our current concepts of prostaglandin–ADH interactions in CCCT cells. Vasopressin induces PGE$_2$ formation. Newly formed prostaglandins are released on both sides of the cell monolayer. Prostaglandin E$_2$ can then have two effects. One effect, caused by interaction with a type I prostaglandin receptor, is to activate adenylate cyclase. This effect occurs within 2 min and leads to the elevation of cAMP in CCCT cells. We suggest that the physiological function of this adenylate cyclase activation is to inhibit prostaglandin production. We have not tested this concept in CCCT cells *per se,* but the elevation of cellular cAMP levels has been shown to inhibit prostaglandin formation in platelets, foreskin fibroblasts, and MDCK cells (Minkes *et al.,* 1977; Hopkins and Gorman, 1981; Hassid, 1982). Moreover, the fact that collecting tubules and thin-limb cells are the only renal tubule cells that form prostaglandins (Smith and Bell, 1978; Smith *et al.,* 1979; Currie and Needleman, 1982) and also the only renal tubule cells that respond to prostaglandins to form cAMP (Torikai and Kurokawa, 1981) suggests that only cells that synthesize prostaglandins have a prostaglandin-responsive adenylate cylase.

A second response of CCCT cells to newly formed PGE$_2$ is inhibition of AVP-induced cAMP formation. We propose that this action results from binding of prostaglandins to a type II receptor. We have not yet investigated the biochemical mechanism for this inhibitory effect of prostaglandins. However, the net result is

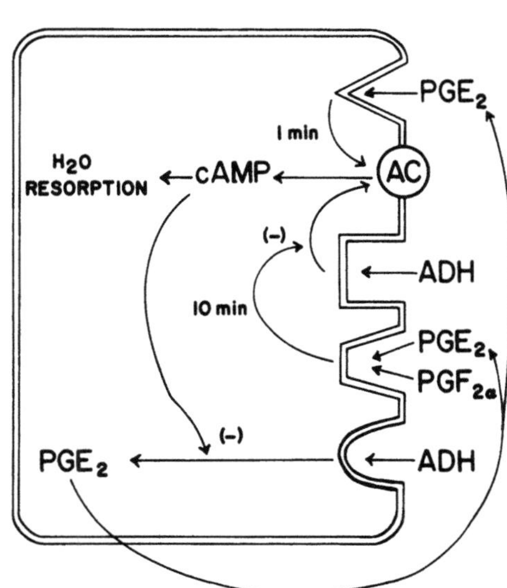

FIGURE 7. Two-receptor model for the action of PGE$_2$ on the cortical collecting tubule. ADH, antidiuretic hormone; AC, adenylate cyclase.

analogous to that seen in the heterologous desensitization of adenylate cyclase studied by Kassis and Fishman (1982) in human fibroblasts and by Garrity, Robertson, and co-workers (1983) in liver tissue. In these two instances, prostaglandins appear to modify a guanine-nucleotide-binding protein(s) in such a way that interactions between hormone–receptor complexes and the adenylate cyclase catalytic unit are impaired.

As mentioned earlier, the activation of adenylate cyclase by PGE_2 occurs within 2 min, when one can first consistently measure increases in cAMP. However, the desensitization process requires 10–20 min to develop in the presence of low PGE_2 concentrations.

Desensitization systems are not restricted to prostaglandin-forming cells. For example, PGE_2 inhibits ADH-induced cAMP synthesis by the medullary thick limb of Henle's loop (Torikai and Kurokawa, 1983), a region of the nephron that does not form prostaglandins. The thick limb is adjacent to the collecting tubule in the kidney and probably responds to prostaglandins formed by the collecting tubule.

3. SUMMARY

We have presented evidence suggesting that there are two functional classes of prostaglandin receptors (type I and type II) in the renal collecting tubule. Type I receptors are coupled to adenylate cyclase activation. We propose that this type of receptor is found only in prostaglandin-forming cells and that occupancy of this receptor leads to feedback regulation of prostaglandin synthesis. Type II receptors present in collecting tubule cells are somehow coupled to events that lead to the desensitization of these cells to ADH.

ACKNOWLEDGMENTS. This work was supported in part by U.S. Public Health Service Grant NIH AM22042, by U.S. Public Health Service Predoctoral Traineeship NIH HL07404, and by an Established Investigatorship from the American Heart Association (W.L.S.).

REFERENCES

Anderson, R. J., Berl, T., McDonald, K. M., and Schrier, R. W., 1975, Evidence for an *in vivo* antagonism between vasopressin and prostaglandin in the mammalian kidney, *J. Clin. Invest.* **56**:420–426.

Currie, M. G., and Needleman, P., 1982, Distribution of prostaglandin synthesis along the rabbit nephron, *Fed. Proc.* **41**:1545.

Fejes-Toth, G. A., Magyar, A., and Walter, J., 1977, Renal response to vasopressin after inhibition of prostaglandin synthesis, *Am. J. Physiol.* **232**:F416–F423.

Garcia-Perez, A., and Smith, W. L., 1983, Use of monoclonal antibodies to isolate cortical collecting tubule cells: AVP induces PGE release, *Am. J. Physiol.* **244**:C211–C220.

Garcia-Perez, A., and Smith, W. L., 1984, Apical–basolateral membrane asymmetry in canine cortical collecting tubule cells: Bradykinin, arginine vasopressin, prostaglandin E_2 interrelationships, *J. Clin. Invest.* **74**:63–74.

Garrity, M. J., Andreason, T. J., Storm, D. R., and Robertson, R. P., 1983, Prostaglandin E-induced heterologous desensitization of hepatic adenylate cyclase: Consequences on the guanyl nucleotide regulatory complex, *J. Biol. Chem.* **258**:8692–8697.

Grantham, J. J., and Orloff, J., 1968, Effect of prostaglandin E_1 on the permeability response of the isolated collecting tubule to vasopressin, adenosine 3′,5′-monophosphate and theophylline, *J. Clin. Invest.* **47**:1154–1161.

Handler, J. S., and Orloff, J., 1981, Antidiuretic hormone, *Annu. Rev. Physiol.* **43**:611–624.

Hansen, H. S., 1981, Essential fatty acid-supplemented diet increases renal excretion of prostaglandin E_2 and water in essential fatty acid-deficient rats, *Lipids* **16**:849–854.

Hassid, A., 1982, Regulation of prostaglandin biosynthesis in cultured cells, *Am. J. Physiol.* **243**:C205–C211.

Hopkins, N. K., and Gorman, R. R., 1981, Regulation of endothelial cell cyclic nucleotide metabolism by prostacyclin, *J. Clin. Invest.* **67**:540–546.

Kassis, S., and Fishman, P. H., 1982, Different mechanisms of desensitization of adenylate cyclase by isoproterenol and prostaglandin E_1 in human fibroblasts, *J. Biol. Chem.* **257**:5312–5381.

Lum, G. M., Aisenberg, G. A., Dunn, M. J., Berl, T., Schrier, R. W., and McDonald, K. M., 1977, *In vivo* effect of indomethacin to potentiate the renal medullary cyclic AMP response to vasopressin, *J. Clin. Invest.* **59**:8–13.

Minkes, M., Stanford, N., Chi. M.-Y., Roth, G. J., Raz, A., Needleman, P., and Majerus, P. W., 1977, Cyclic adenosine 3′,5′-monophosphate inhibits the availability of arachidonate to prostaglandin synthetase in human platelet suspensions, *J. Clin. Invest.* **59**:449–454.

Smith, W. L., and Bell, T. G., 1978, Immunohistochemical localization of the prostaglandin-forming cyclooxygenase in renal cortex, *Am. J. Physiol.* **235**:F451–F457.

Smith, W. L., Bell, T. G., and Needleman, P., 1979, Increased renal tubular synthesis of prostaglandins in the rabbit kidney in response to ureteral obstruction, *Prostaglandins* **18**:269–277.

Smith, W. L., Grenier, F. C., DeWitt, D. L., Garcia-Perez, A., and Bell, T. G., 1983, Cellular compartmentalization of the biosynthesis and function of PGE_2 and PGI_2 in the renal medulla, in: *Prostaglandins and the Kidney* (M. J. Dunn, C. Patrons, and G. A. Cinotti, eds.), Plenum Press, New York, pp. 27–39.

Torikai, S., and Kurokawa, K., 1981, Distribution of prostaglandin E_2-sensitive adenylate cyclase along the rat nephron, *Prostaglandins* **21**:427–438.

Torikai, S., and Kurokawa, K., 1983, Effect of PGE_2 on vasopressin-dependent cell cAMP in isolated single nephron segments, *Am. J. Physiol.* **245**:F58–F66.

5

Prostaglandin Endoperoxide Synthase from Human Cell Line Lu-65

NOREEN J. HICKOK, MARY ALOSIO, and RICHARD S. BOCKMAN

1. INTRODUCTION

In this chapter, we present our data on the isolation and properties of prostaglandin endoperoxide synthase (PES; E.C. 1.14.99.1) from a human cell line. This enzyme has been isolated and purified to homogeneity in other species; partially purified human platelet PES that retained activity has been prepared by DEAE-chromatography (Hammarström and Falardeau, 1977) and has been purified to near-homogeneity (Ho *et al.*, 1980). In other studies, an inactive human platelet PES has been identified by SDS-PAGE (Roth and Majerus, 1975). Preliminary data on the characterization of this PES from a human cell line, Lu-65, is presented.

The prostaglandins (PGs) are a family of C_{20} fatty acids formed through action of a membrane-bound enzyme complex, PG synthase (Samuelsson, 1972). In intact cells, the predominant precursor of PGs is arachidonic acid, which is found as a constituent of phospholipids. A heterogeneous group of enzymes known collectively as phospholipases cleaves arachidonic acid from phospholipids (Flower and Blackwell, 1976). The free arachidonate may then be cyclized by PES to form the short-lived intermediates PGG_2 and PGH_2. Specific isomerases are final-stage enzymes that determine the specific PG products.

Prostaglandin endoperoxide synthase was first purified from bovine seminal vesicles (Miyamoto *et al.*, 1976), closely followed by purification from ram seminal

NOREEN J. HICKOK, MARY ALOSIO, and RICHARD S. BOCKMAN ● Memorial Sloan-Kettering Cancer Center, New York, New York 10021.

vesicles (RSV) (Hemler *et al.*, 1976; van der Ouderaa *et al.*, 1977). Human PES was solubilized from platelets, and the enzymic activity recovered after DEAE-chromatography (Hammarström and Falardeau, 1977) or after isoelectric focusing, gel filtration, and hydrophobic chromatography (Ho *et al.*, 1980). The purified enzymes were found to have two activities that copurified, suggesting that both activities resided within one protein unit (Miyamoto *et al.*, 1976). The first activity was a cyclooxygenase, which formed the cyclopentane intermediate PGG_2. In this latter step, the enzyme also catalyzed the addition of two oxygen molecules to arachidonic acid, forming a cycloendoperoxide between C-9 and C-11 and a hydroperoxide at C-15. Through labeling studies, Samuelsson (1965) demonstrated that the cycloendoperoxide was formed from molecular oxygen. The cyclooxygenase required heme iron as a cofactor (Miyamoto *et al.*, 1976; Hemler *et al.*, 1976; van der Ouderaa *et al.*, 1977).

The second catalytic activity of PES was a hydroperoxidase. In this step, the hydroperoxy at C-15 of PGG_2 was reduced to a hydroxy, forming the cycloendoperoxide PGH_2. Heme iron was required for this activity. Radical scavengers such as hydroquinone and phenol were found to augment hydroperoxidase activity (Miyamoto *et al.*, 1976).

Incubation of the PES with aspirin acetylated the enzyme, irreversibly inhibiting the cyclooxygenase activity. Other nonsteroidal antiinflammatory agents such as indomethacin were reversible inhibitors of PES cyclooxygenase. None of these agents affected the hydroperoxidase activity (Miyamoto *et al.*, 1976). [^{3}H]Acetylaspirin treatment of PES resulted in covalent binding of the [^{3}H]acetyl moiety to the active site of the enzyme, providing a specific label for PES (Roth and Majerus, 1975). By this reaction, human platelets were labeled with [^{3}H]acetylaspirin, the human platelet PES was solubilized from the membrane by boiling in sodium dodecyl sulfate (SDS) and fractionated by SDS-polyacrylamide electrophoresis (PAGE), and the inactive, partially purified human platelet PES was visualized by autoradiography (Roth and Majerus, 1975). Subsequent structural studies using RSV-PES labeled with [^{3}H]acetylaspirin demonstrated a molecular weight of 70,000 for the monomer and identified the enzyme protein as a dimer of identical monomers in Tween-20 solution (Roth *et al.*, 1980). [^{3}H]Acetylated PES from RSV digested with pepsin have permitted the isolation of fragments containing the [^{3}H]acetyl moiety. This peptide has been purified by high-pressure liquid chromatography (HPLC), and the active site sequenced, revealing the primary structure of the critical region in RSV PES (Roth *et al.*, 1983).

Spin-resonance studies showed that a radical was formed by PES during arachidonic acid metabolism (Egan *et al.*, 1979). Radical formation during prostaglandin synthesis has several biological consequences, including cooxygenation of a host of endogenous and exogenous molecules that could act as radical acceptors. The PES may itself be affected, with the possibility that the observed autoinactivation of PES is a consequence of radical attack (Egan *et al.*, 1976). In fact, in the presence of heme alone, radical formation (Egan *et al.*, 1976) and rapid inactivation of PES activity are seen (Hemler and Lands, 1980). When phenol was added to the reaction, the EPR signal representing radical formation was greatly diminished. Furthermore, PES showed a time-dependent activation in the presence of phenol (Egan *et al.*, 1976), which also caused a substantial stabilization of the labile enzyme to the point that PES in 25 μM hematin and 1 mM phenol could be left on the benchtop overnight with little loss of activity (Hemler and Lands, 1980). Therefore, it was evident that PES activity could be regulated by the careful selection of cofactors.

2. METHODS

Several years ago, we had established a human lung tumor cell line, Lu-65, that was noted to produce large amounts of PGE_2 (Bockman *et al.*, 1983). We set about to purify the enzyme in order to elucidate the nature of the controls of PES in this system. Our purification scheme initially followed the procedure of van der Ouderaa *et al.* (1977). We have since modified the procedure to maximize the yield of active enzyme. To optimize this purification scheme, autoinactivation of PES had to be minimized. van der Ouderaa *et al.* (1977) had added Na_2EDTA to all buffers to inhibit calcium-dependent phospholipases and thereby block arachidonic acid release from phospholipids. They also used diethyldithiocarbamic acid (DDC) in millimolar concentrations throughout the solubilization steps in an effort to inhibit PES activity, which would ensure that radical formation did not occur. To stabilize the labile PES, we added DDC and EDTA as well as supplementing all buffers during the solubilization steps with 25 μM heme and 1 mM phenol (Hemler and Lands, 1980).

During the PES purification, enzyme activity was followed by measuring the hydroperoxidase-mediated reduction of tetramethyl-*p*-phenylenediamine in the presence of 1.7 mM heme and H_2O_2 (van der Ouderaa *et al.*, 1977). Indomethacin-inhibitable oxygen consumption (the utilization of molecular oxygen during cyclization of arachidonic acid by the cyclooxygenase activity) was used to confirm that the hydroperoxidase activity we were following also possessed a cyclooxygenase activity. The work discussed in this chapter represents the results of several purifications.

As PES is a membrane-bound enzyme, homogenization followed by differential centrifugation was used to remove the majority of contaminating protein. The enzyme was solubilized from the sedimented membrane with 1% Tween-20. In our early purifications, the solubilized PES was first applied to an AcA-34 gel filtration column equilibrated with the heme–phenol buffer, and the active fractions were pooled and then separated on a DE-52 anion exchange column. In this procedure, an unacceptable amount of peroxidase activity was lost during the AcA-34 column chromatography. Therefore, in the subsequent purifications, the solubilized enzyme was first fractionated by DE-52 column chromatography in the hope that absorption on the anion exchanger would stabilize the activity. This proved to be the case, with a twofold increase in the yield of hydroperoxidase activity. The elution profile of this DE-52 column is shown in Fig. 1. Two peaks (1 and 2) of peroxidase activity were eluted—peak 1 eluted at 600 mM Tris-HCl, and peak 2 at 1 M Tris-HCl. Both peaks possessed cyclooxygenase activity as determined by O_2 consumption measured with a Clark oxygen electrode. The two peaks were then fractionated on a calibrated AcA-34 column that was eluted with 50 mM Tris-HCl containing 0.1% β-octylglucoside. Peak 1 eluted as a single peak of approximately 115,000 daltons, and peak 2 eluted as a single protein band at approximately 70,000 daltons. Only the peak 1 material had cyclooxygenase activity that was inhibited by indomethacin. Peak 2 showed only slight inhibition by indomethacin and may not be a PES. Therefore, all further experiments were performed with peak 1 material.

To confirm that the peak 1 activity metabolized arachidonic acid to PGH_2 or

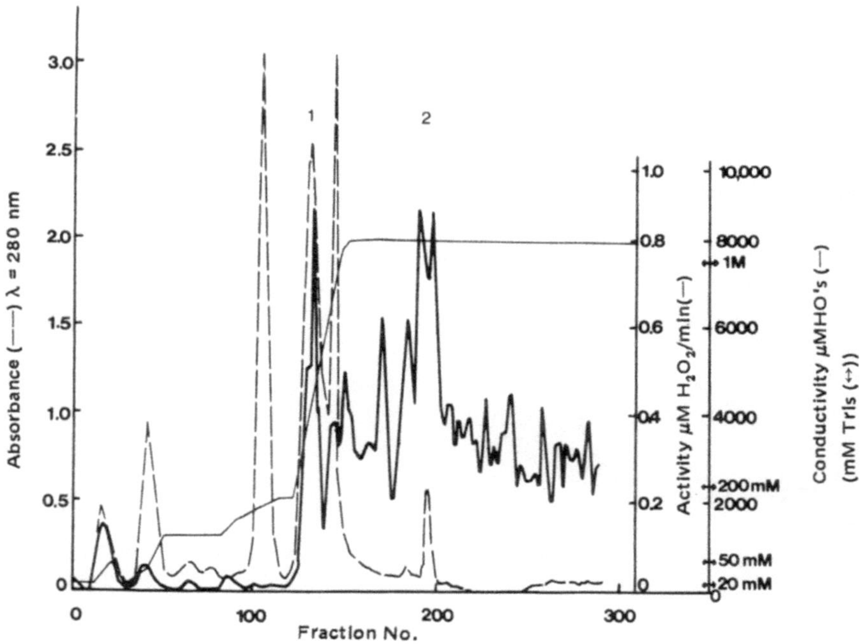

FIGURE 1. Fractionation of Lu-65 PES on DE52. Lu-65 PES was applied to a DE52 column (3 cm × 25 cm) in 50 mM Tris-HCl, 25 μM hematin, 1 mM phenol, pH 8.0, containing 0.1 mM Na_2EDTA, 0.1% Tween-20, and 0.01% NaN_3 and washed with the same buffer without hematin and phenol. The ionic strength of the column was raised by a linear gradient from 50 mM to 200 mM Tris-HCl, pH 8.0, containing the same additives as the washing buffer, and the PES activity eluted with a linear gradient from 200 mM Tris-HCl to 1 M Tris-HCl with the above additives. Absorbance at 280 nm (— —) was used to follow protein elution, hydroperoxidase activity (——) indicated the PES active fractions, and conductance (——) was used to monitor the progress of the gradient. Two peaks of peroxidase activity eluted (1 and 2) and were incubated with arachidonic acid and indomethacin to confirm cyclooxygenase as well as hydroperoxidase activity.

its hydrolysis products, the peak 1 material was incubated with [³H]arachidonic acid in the presence of 50 nM heme and 5.8 μM hydroquinone. The supernatants were extracted using C_{18} Sep-paks® (Powell, 1980), and those products eluting in the prostaglandin fraction were further separated by reverse-phase HPLC (Terragno *et al.*, 1981) (Fig. 2). Two peaks of radioactive metabolites were found. The first peak was formed under enzyme-free conditions and probably comprised reactive impurities in the arachidonic acid. The second peak encompassed the elution volumes of authentic PGD_2 and PGE_2 standards. Formation of these two prostaglandins would be in keeping with nonenzymatic breakdown of the primary PES product, PGH_2.

When we measured the pH dependence of the cyclooxygenase, the PES from Lu-65 showed no enzyme activity below pH 6, with peak activity at pH 8. No enzymatic activity above pH 9.5 was measurable. Therefore, in its pH profile, the Lu-65 human PES shows the same pH dependence as human platelet PES (Ho *et al.*, 1980) and is similar to PES from RSV.

Roth and Majerus (1975) fractionated their [³H]acetylated human platelet PES on SDS-PAGE, obtaining a molecular weight of 80,000; fractionation of the purified human platelet PES was in agreement with this molecular weight (Ho *et al.*, 1980). The isolated Lu-65 enzyme from the initial purification procedure (AcA-34 column

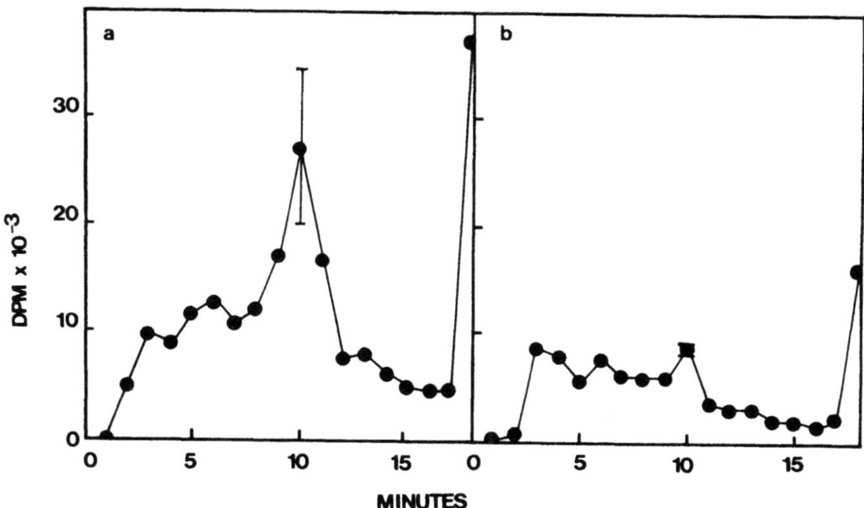

FIGURE 2. Elution of [³H]arachidonic acid metabolites by HPLC. A: [³H]Arachidonic acid metabolites formed in the presence of 38.5 units of peroxidase activity (1 unit = the consumption of 1×10^{-9} mol H_2O_2/min) and cofactors (100 nM hematin, 5.7 μM hydroquinone), followed by extraction of prostaglandin products by C_{18} Sep-Paks (Powell, 1980) and isocratic fractionation using reverse-phase HPLC (Terragno *et al.*, 1981). Authentic PGE_2 and PGD_2 standards elute between 9 and 11 min. B: Incubation of [³H]arachidonic acid under the same conditions in enzyme-free buffer.

chromatography followed by DEAE-chromatography) was fractionated on a 9% gel by SDS-PAGE under reducing conditions using the discontinuous buffer system of Laemmli (1970). Under these conditions, two major protein bands were visualized by Coomassie blue staining. Based on protein standards in adjacent lanes, the PES derived from Lu-65 had molecular weights of 68,000 and 45,000.

We iodinated peak 1 material and immunoprecipitated the iodinated protein with anti-RSV PES using precipitation with *Staph*. A (Kessler, 1981). The iodinated protein that was bound by the anti-RSV PES antibody was then fractionated on a 9% gel by SDS-PAGE under reducing conditions and the protein bands visualized by autoradiography (Bonner and Lasky, 1974) (Fig. 3). One specific band at 45,000 daltons was precipitated. This could correspond to the lower-molecular-weight band seen after Coomassie blue staining in the nonradiolabeled preparation. In other experiments, we metabolically labeled Lu-65 cells with [^{14}C]mixed amino acids (Amersham), solubilized the labeled enzyme using the procedure outlined for purification, and immunoprecipitated the [^{14}C]Lu-65 PES with anti-RSV PES followed by *Staph*. A treatment. The immunoprecipitated protein was fractionated under reducing conditions on 9% SDS-PAGE, and the protein visualized by autoradiography. Again, the band at 45,000 daltons showed greater density in the immune than nonimmune serum; however, a unique band was visualized at 68,000 daltons, corresponding to the higher-molecular-weight band seen after Coomassie blue staining. We are presently performing competition experiments with RSV PES and labeling Lu-65 PES with [^{3}H]acetylaspirin to determine whether the 45,000-dalton, the 68,000-dalton, or both bands are the human PES. We are also exploring the

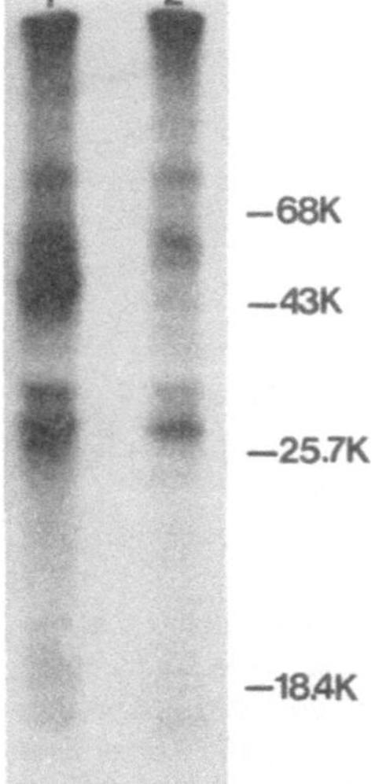

FIGURE 3. Fractionation of [^{125}I]Lu-65 PES by SDS-PAGE under reducing conditions. Lu-65 PES was iodinated, and the labeled PES immunoprecipitated with rabbit antiserum. Molecular weight markers are as indicated. Column 1: Immunoprecipitation of Lu-65 PES by rabbit anti-RSV PES. Column 2: Immunoprecipitation of Lu-65 PES by rabbit nonimmune serum. Only one specific band at approximately 45,000 daltons was immunoprecipitated, corresponding to the lower-molecular-weight band seen by Coomassie brilliant blue staining (data not shown).

possibility that this 45,000-dalton peak represents an active enzyme cleavage product of a larger molecule.

The visible spectrum of Lu-65 PES from peak 1 was compared with that of RSV PES. In the presence of 5 nM heme, the visible spectra for both enzymes were identical. Of particular note was the absorption band at 420 nm, which was consistent with PES being a heme-containing protein.

Kulmacz and Lands (1982) observed that RSV PES was inactivated by trypsin and that the addition of exogenous heme protected RSV PES from trypsin inactivation. The PES from Lu-65 appeared to differ from the RSV enzyme in that the hydroperoxidase activity was augmented by the presence of heme or trypsin and that maximal augmentation occurred in the presence of both agents. In our hands, the hydroperoxidase activity of RSV PES showed no change or was inhibited in the presence of heme or trypsin. No change in RSV PES activity was noted in the presence of both agents.

Another difference in the properties of the human enzyme was observed in its substrate affinity. We calculated the K_m for arachidonic acid of RSV cyclooxygenase and Lu-65 cyclooxygenase from the Lineweaver–Burk plots. In 0.1 M tris-HCl, pH 8.0, containing 100 nM heme and 5.8 μM hydroquinone, we measured the K_m for RSV cyclooxygenase to be in the low micromolar concentration. van der Ouderaa (1982) reported a similar value of 15 μM for the RSV cyclooxygenase. In contrast, the calculated K_m for Lu-65 cyclooxygenase fell in the millimolar range.

As this is clearly not a physiological K_m, two possibilities are being explored. The first possibility to explain the large K_m is that an effector molecule is present in the cell that causes lowering of the Lu-65 PES K_m. To test this hypothesis, we measured the K_m for cyclooxygenase activity in a crude membrane fraction from the cell homogenate. The K_m under these conditions still fell in the millimolar range; therefore, we could not detect a lower K_m for the PES in a crude microsomal membrane preparation. We are currently testing the crude homogenate. The second possibility to explain this large K_m is that through proteolysis part of the enzyme was cleaved, leaving a viable enzyme but one with a decreased efficiency metabolizing arachidonic acid. However, when deWitt et al. (1981) exposed membrane-bound RSV PES to different proteases, they found that proteolytic cleavage was accompanied by inactivation of the cyclooxygenase activity. It is possible that the human and the ovine PES could have different proteolytic sensitivities and that proteolytic cleavage of the Lu-65 PES could explain the mixed apparent molecular weights seen in the reducing SDS-PAGE gels. Through the use of protease inhibitors during homogenization and solubilization, we hope to determine whether the human PES molecule has been degraded by endogenous proteases.

3. SUMMARY

In summary, we report the purification of PES from a human cell line. This PES shows the same pH dependence as RSV PES and human platelet PES, has the same absorption spectrum as RSV PES, and could be shown to metabolize [^{3}H]arachidonic acid to appropriate [^{3}H]prostaglandin products. The PES from Lu-65 has a different apparent molecular weight by gel filtration, has a different sensitivity to trypsin, and has a K_m several orders of magnitude larger than that measured for RSV PES. Whether this human PES is an active proteolysis product or is a protein that requires an effector molecule to lower its K_m to a physiological range is not clear at this time. However, our data would suggest that purified Lu-65 PES differs in several of its properties from RSV PES.

REFERENCES

Bockman, R. S., Bellin, A., Repo, M. A., Hickok, N. J., and Kameya, T., 1983, *In vivo* and *in vitro* biologic activities of two human cell lines derived from anaplastic lung cancers, *Cancer Res.* **43**:4571–4576.

Bonner, W. M., and Laskey, R. A., 1974, A film detection method for tritium-labelled proteins and nucleic acids in polyacrylamide gels, *Eur. J. Biochem.* **46**:83–88.

Egan, R. W., Paxton, J., and Kuehl, F. A., Jr., 1976, Mechanism for irreversible self-deactivation of prostaglandin synthetase, *J. Biol. Chem.* **251**:7329–7335.

Egan, R. W., Gale, P. H., and Kuehl, F. A., Jr., 1979, Reduction of hydroperoxides in the prostaglandin synthetic pathway by a microsomal peroxidase, *J. Biol. Chem.* **254**:3295–3302.

Egan, R. W., Gale, P. H., Beveridge, G. C., Marnett, L. J., and Kuehl, F. A., Jr., 1980, Direct and indirect involvement of radical scavengers during prostaglandin biosynthesis, *Adv. Prostaglandin Thromboxane Res.* **6**:153–155.

Flower, R. J., and Blackwell, G. J., 1976, The importance of phospholipase A$_2$ in prostaglandin biosynthesis, *Biochem. Pharmacol.* **25**:285–291.

Hammarström A., and Falardeau, P., 1977, Resolution of prostaglandin endoperoxide synthase and thromboxane synthase of human platelets, *Proc. Natl. Acad. Sci. U.S.A.* **74**:3691–3695.

Hemler, M. E., and Lands, W. E. M., 1980, Protection of cyclooxygenase activity during heme-induced destabilization, *Arch. Biochem. Biophys.* **201**:586–593.

Hemler, M. E., Lands, W. E. M., and Smith, W. L., 1976, Purification of the cyclooxygenase that forms prostaglandins. Demonstration of two forms of iron in the holoenzyme, *J. Biol. Chem.* **251**:5575–5579.

Ho, P. P. K., Towner, R. D., and Esterman, M. A., 1980, Purification and characterization of fatty acid cyclooxygenase from human platelets, *Prep. Biochem.* **10**:597–613.

Kessler, S. W., 1981, Use of protein A-bearing staphylococci for the immunoprecipitation and isolation of antigens from cells, *Methods Enzymol.* **73**:442–459.

Kulmacz, R. J., and Lands, W. E. M., 1982, Protection of prostaglandin synthase from trypsin upon binding of heme, *Biochem. Biophys. Res. Commun.* **104**:758–764.

Laemmli, U. K., 1970, Cleavage of structural proteins during the assembly of the head of bacteriophage T4, *Nature* **227**:680–685.

Miyamoto, T., Ogino, N., Yamamoto, S., and Hayaishi, O., 1976, Purification of prostaglandin endoperoxide synthetase from bovine vesicular gland microsomes, *J. Biol. Chem.* **251**:2629–2636.

Powell, W. S., 1980, Rapid extraction of oxygenated metabolites of arachidonic acid from biological samples using octadecylsilyl silica, *Prostaglandins* **20**:947–957.

Roth, G. J., and Majerus, P. W., 1975, The mechanism of the effect of aspirin on human platelets. I. Acetylation of a particulate fraction protein, *J. Clin. Invest.* **56**:624–632.

Roth, G. J., Siok, C. J., and Ozols, J., 1980, Structural characteristics of prostaglandin synthetase from sheep vesicular gland, *J. Biol. Chem.* **255**:1301–1304.

Roth, G. J., Machuga, E. T., and Ozols, J., 1983, Isolation and covalent structure of the aspirin-modified, active-site region of prostaglandin synthetase, *Biochemistry* **22**:4672–4675.

Samuelsson, B., 1965, On the incorporation of oxygen in the conversion of 8,11,14-eicosatrienoic acid to prostaglandin E_1, *J. Am. Cancer Soc.* **87**:3011–3013.

Samuelsson, B., 1972, Biosynthesis of prostaglandins, *Fed. Proc.* **31**:1442–1450.

Terragno, A., Rydzik, R., and Terragno, N., 1981, High performance liquid chromatography and UV detection for the separation and quantitation of prostaglandins, *Prostaglandins* **21**:101–112.

van der Ouderaa, F. J., and Buytenhek, M., 1982, Purification of PGH synthase from sheep vesicular glands, *Methods Enzymol.* **86**:60–68.

van der Ouderaa, F. J., Buytenhek, M., Nugteren, D. H., and van Dorp, D. A., 1977, Purification and characterisation of prostaglandin endoperoxide synthetase from sheep vesicular glands, *Biochim. Biophys. Acta* **487**:315–331.

6

Coronary Arterial Prostacyclin Synthetase and Prostaglandin E_2 Isomerase Activities

DENNIS B. McNAMARA, MORRIS D. KERSTEIN, ALICE Z. LANDRY, JOHN L. HUSSEY, HOWARD L. LIPPTON, ROBERT S. ROSENSON, ALBERT L. HYMAN, and PHILIP J. KADOWITZ

1. INTRODUCTION

Prostaglandin E_2 (PGE_2) isomerase and prostacyclin (PGI_2) synthetase are the major endoperoxide (PGH_2)-metabolizing enzymes present in microsomes isolated from vascular tissue. The PGE_2 isomerase activity specifically requires the addition of GSH (Ogino *et al.*, 1977). Prostacyclin has potent vasodilator and platelet antiaggregatory activity (Vane, 1983). Therefore, modulation of PGI_2 synthetase activity is of prime physiological importance in the regulation of blood vessel function. Reducing agents such as GSH have been reported to augment prostaglandin production by protecting the cyclooxygenase enzyme system (arachidonic acid $\rightarrow$ PGH_2)

DENNIS B. McNAMARA, ALICE Z. LANDRY, and PHILIP J. KADOWITZ ● Department of Pharmacology, Tulane University School of Medicine, New Orleans, Louisiana 70112.　MORRIS D. KERSTEIN and JOHN L. HUSSEY ● Department of Surgery, Tulane University School of Medicine, New Orleans, Louisiana 70112.　ROBERT S. ROSENSON ● Department of Medicine, Tulane University School of Medicine, New Orleans, Louisiana 70112.　HOWARD L. LIPPTON ● Departments of Medicine and Pharmacology, Tulane University School of Medicine, New Orleans, Louisiana 70112.　ALBERT L. HYMAN ● Departments of Pharmacology, Surgery and Medicine, Tulane University School of Medicine, New Orleans, Louisiana 70112.

from self-catalyzed deactivation (Egan *et al.*, 1976). However, little evidence exists for the effects of these agents on augmenting prostacyclin synthetase activity (McNamara *et al.*, 1984). We report the concentration-dependent modulation of the formation of 6-keto-PGF$_{1\alpha}$, the stable breakdown product of PGI$_2$, by GSH and dithiothreitol (DTT) and unmasking of PGE$_2$ isomerase in microsomes isolated from bovine coronary artery.

2. MATERIALS AND METHODS

2.1. Isolation of Coronary Arterial Microsomes

Sheep and dog hearts were obtained from anesthetized animals; bovine hearts were obtained from a local slaughterhouse; the human heart was obtained from a 17-year-old male at the time of harvesting of the kidneys for renal transplantation. All hearts were immediately placed in ice-cold 0.1 M phosphate buffer, pH 7.4, for transport to the laboratory. The left anterior descending and left circumflex coronary arteries were dissected free, cleaned, pooled, minced, and homogenized in three volumes of the ice-cold phosphate buffer using a Polytron homogenizer. The homogenate was centrifuged at 10,000 $\times$ g for 15 min, and the supernatant was strained through cheesecloth and centrifuged at 105,000 $\times$ g for 60 min. All procedures were carried out at 0–4°C. The microsomal pellet obtained was suspended in phosphate buffer and stored at $-55°C$. Protein was assayed by the method of Lowry *et al.* (1951).

2.2. Assay of Microsomal Metabolites of PGH$_2$

The incubation mixture contained coronary arterial microsomes in 100 μl of 0.1 M potassium phosphate buffer, pH 7.4, and any additions, e.g., GSH. The reaction was initiated by the addition of the microsomal fraction to a 0°C tube containing PGH$_2$ (previously blown dry under a N$_2$ stream), vortexed, and incubated at 37°C for 2 min. The reaction was stopped, and the products were extracted by adding 400 μl of ethyl acetate : methanol : 0.2 M citric acid, pH 2.0 (15 : 2 : 1), vortexed, and centrifuged. The upper organic layer was spotted for TLC on Analtech silica gel GHL plates along with authentic prostaglandin standards (Upjohn Co.) and developed using the solvent system ethyl acetate : acetic acid : hexane : water (54 : 12 : 25 : 60, organic phase). The migration of authentic prostaglandin standards was located by exposing the plates to iodine vapor, and that of radiolabeled products by radiochromatogram scan. The [1-^{14}C]PGH$_2$ preparation from [1-^{14}C]arachidonic acid (50–60 mCi/mmol, Amersham) and all other procedures were as previously described in detail (Kerstein *et al.*, 1983; She *et al.*, 1981).

TABLE I. Effect of Microsomal Protein Concentration in Bovine Coronary Arterial
Prostacyclin Synthetase Activity

| Protein | Prostacyclin synthetase | | | |
| (μg) | 2.5 μM PGH_2 | | 10 μM PGH_2 | |
	$-GSH$	$+GSH^a$	$-GSH$	$+GSH^a$
10	8 ± 1^b	8 ± 3	30 ± 7	23 ± 5
25	10 ± 4	20 ± 10	38 ± 6	102 ± 36
50	27 ± 7	88 ± 15	141 ± 30	224 ± 39
100	96 ± 8	133 ± 9	289 ± 31	380 ± 71
200	138 ± 16	148 ± 15	352 ± 92	608 ± 40

[a] GSH = 2 mM; incubation volume = 0.1 ml.
[b] Data are picomoles of 6-keto-$PGF_{1\alpha}$/2 min and are expressed as mean $\pm$ S.E.M. of duplicate incubations of separate microsomal fractions isolated from four hearts.

3. RESULTS

3.1. Prostacyclin Synthetase Activity

The data in Table I show that the formation of 6-keto-$PGF_{1\alpha}$, the stable hydrolytic product of prostacyclin, is dependent on the concentration of microsomal protein and the concentration of PGH_2; in addition, these data indicate that 6-keto-

TABLE II. Effect of GSH Concentration
on Bovine Coronary Arterial Prostacyclin
Synthetase Activity^a

| GSH | Prostacyclin synthetase^b | |
(mM)	PGH_2 2.5 μM*	PGH_2 10 μM**
—	47 ± 8	129 ± 12
0.1	58 ± 3	127 ± 16
0.5	56 ± 4	191 ± 30
1	65 ± 8	167 ± 19
1.25	—	268 ± 65
1.5	80 ± 9	—
1.75	—	215 ± 41
2	97 ± 10	315 ± 37
5	103 ± 7	300 ± 51

[a] Microsomal protein = 50 μg; incubation volume = 0.1 ml.
[b] Data are picomoles of 6-keto-$PGF_{1\alpha}$/2 min per 50 μg protein expressed as mean $\pm$ S.E.M. of duplicate incubations of separate microsomal fractions isolated from 4* or 3** hearts.

$PGF_{1\alpha}$ formation is augmented in the presence of 2 mM GSH. The activity of the prostacyclin synthetase observed in the absence of GSH is similar to that previously reported for bovine coronary arterial microsomes (Gerritsen and Printz, 1981). Thromboxane B_2 (TxB_2), the stable hydrolytic product of TxA_2, formation was not observed, indicating the absence of detectable thromboxane synthetase activity. Tranylcypromine, an inhibitor of prostacyclin synthetase (She *et al.*, 1981), inhibited the formation of 6-keto-$PGF_{1\alpha}$ by 90% both in the presence and absence of 2 mM GSH (data not shown).

The data presented in Table II indicate that GSH exhibited a concentration-dependent modulation of prostacyclin synthetase activity. This modulation of the formation of 6-keto-$PGF_{1\alpha}$ occurred over a narrow GSH concentration range and was observed at two concentrations of PGH_2. The activity of prostacyclin synthetase was also modulated in a concentration-dependent manner by dithiothreitol (DTT) (Table III). The modulation of prostacyclin synthetase by GSH was not species specific, as it was observed in coronary arterial microsomes prepared from sheep, dog, and man (Table IV).

3.2. Prostaglandin E_2 Isomerase Activity

The data in Table V show an increase in the formation of PGE_2 by bovine coronary arterial microsomes in response to increasing concentrations of GSH at two concentrations of PGH_2. The formation of PGE_2 in the presence of GSH (at

Table III. Effect of DTT Concentration on Bovine Coronary Arterial Prostacyclin Synthetase Activity[a]

DTT (mM)	Prostacyclin synthetase[b]	
	PGH_2 2.5 μM*	PGH_2 10 μM**
—	43 ± 11	74 ± 24
0.1	83 ± 13	80 ± 15
0.5	100 ± 18	130 ± 16
1	108 ± 25	135 ± 14
1.25	124 ± 8	—
1.75	—	286 ± 23
2	117 ± 13	301 ± 64
5	135 ± 9	—

[a] Microsomal protein = 50 μg; incubation volume = 0.1 ml; DTT = dithiothreitol.
[b] Data are picomoles of 6-keto-$PGF_{1\alpha}$/2 min per 50 μg protein expressed as mean ± S.E.M. of duplicate incubations of separate microsomal fractions isolated from 4* or 3** hearts.

TABLE IV. Microsomal Prostacyclin Synthetase Activity from Coronary Arteries of Sheep, Dog, and Human

Animal	Additions	Prostacyclin synthetase[a]	
		PGH$_2$ 2.5 μM	PGH$_2$ 10 μM
Sheep (3)	No additions	24 ± 1	100 ± 2
	2 mM GSH	76 ± 3	158 ± 1
Dog (1)	No additions	30 ± 4	54 ± 1
	2 mM GSH	88 ± 8	114 ± 23
Human (1)	No additions	23	—
	2 mM GSH	70 ± 7	—

[a] Incubation volume = 0.1 ml. Data are picomoles of 6-keto-PGF$_{1\alpha}$/2 min per 50 μg protein and are expressed as mean ± S.E.M. of duplicate incubations of separate microsomal fractions isolated from the number of hearts shown in parentheses.

10 μM PGH$_2$) showed an initial increase that fell at higher GSH concentrations (Table V). The formation of PGE$_2$ was greater in the presence of GSH (at 10 μM PGH$_2$) for sheep and dog, also suggesting the presence of an active PGE$_2$ isomerase (Table VI). However, in the presence of DTT with no GSH in the incubation medium, the formation of PGE$_2$ decreased in a manner inversely related to the DTT concentration (Table VII).

TABLE V. Effect of GSH Concentration on Activities of Bovine Coronary Arterial PGE$_2$ Isomerase

GSH (mM)	PGE$_2$ isomerase[a]	
	PGH$_2$ 2.5 μM*	PGH$_2$ 10 μM**
—	56 ± 7	246 ± 9
0.1	55 ± 2	254 ± 24
0.5	73 ± 5	329 ± 34
1	74 ± 6	315 ± 11
1.25	—	308 ± 45
1.5	62 ± 3	—
1.75	—	354 ± 68
2	61 ± 6	234 ± 15
5	47 ± 3	237 ± 19

[a] Microsomal protein = 50 μg; incubation volume = 0.1 m. Data are picomoles of PGE$_2$/2 min per 50 μg protein expressed as mean ± S.E.M. of duplicate incubations of separate microsomal fractions isolated from 4* or 3** hearts.

**TABLE VI. Microsomal PGE$_2$ Isomerase Activities from
Coronary Arteries of Sheep, Dog, and Human**

		PGE$_2$ isomerase[a]	
		PGH$_2$	PGH$_2$
Animal	Additions	2.5 μM	10 μM
Sheep (3)	No additions	85 ± 1	343 ± 2
	2 mM GSH	66 ± 3	414 ± 3
Dog (1)	No additions	56 ± 10	282 ± 34
	2 mM GSH	84 ± 8	408 ± 40
Human (1)	No additions	72	—
	2 mM GSH	64 ± 1	—

[a] Incubation volume = 0.1 ml. Data are picomoles of PGE$_2$/2 min per 50 μg protein and are expressed as mean ± S.E.M. of duplicate incubations of separate microsomal fractions isolated from the number of hearts shown in parentheses.

4. DISCUSSION

4.1. Prostacyclin Synthetase Activity

These data indicate that coronary arterial microsomal prostacyclin synthetase activity can be modulated by the concentration of the sulfhydryl reducing agents GSH and DTT. As the modulation of 6-keto-PGF$_{1\alpha}$ formation occurs over a narrow

**TABLE VII. Effect of DTT Concentration
on Activities of Bovine Coronary Arterial
PGE$_2$ Isomerase**

	PGE$_2$ isomerase[a]	
DTT	PGH$_2$	PGH$_2$
(mM)	2.5 μM*	10 μM**
—	58 ± 5	252 ± 10
0.1	41 ± 4	236 ± 28
0.5	36 ± 7	219 ± 18
1	36 ± 9	226 ± 18
1.25	27 ± 3	166 ± 37
1.5	30 ± 5	—
1.75	—	192 ± 23
2	34 ± 6	162 ± 31
5	30 ± 10	—

[a] Microsomal protein = 50 μg; incubation volume = 0.1 ml; DTT = dithiothreitol. Data are picomoles of PGE$_2$/2 min per 50 μg protein expressed as mean ± S.E.M. of duplicate incubations of separate microsomal fractions isolated from 4* or 3** hearts.

concentration range of the reducing agent, the data suggest the possibility that physiological alterations in intracellular GSH may modulate prostacyclin formation. Whether the observed change in prostacyclin synthetase activity mediated by GSH and DTT represents changes in the redox state of the enzyme or more specifically sulfhydryl oxidation is not certain from these data. However, it should be noted that the concentrations of GSH employed in this investigation are consistent with intracellular GSH concentration (Meister and Anderson, 1983) and that alterations in tissue GSH levels have been reported under normal physiological conditions (Farooqui and Ahmed, 1984; Thomas *et al.*, 1984). We suggest that the higher activity of prostacyclin synthetase observed in the presence of GSH is representative of basal prostacyclin formation, whereas the lower activity observed below 1 mM GSH represents subbasal prostacyclin-generating activity.

Decreases in GSH concentration (antioxidant buffering capacity) may also occur in response to the transformation of molecular oxygen into highly reactive superoxide anion, hydrogen peroxide or hydroxyl radical. Formation of these reactive oxygen species could occur in polymorphonuclear leukocyte (PMN) infiltration during acute myocardial infarction (Romson *et al.*, 1982). Additionally, a reduction in prostacyclin formation could mediate or potentiate PMN infiltration, as prostacyclin has been associated with inhibition of neutrophil chemotaxis, aggregation, lysosomal enzyme release, and superoxide anion production (Fantone and Kinnes, 1983; Lefer *et al.*, 1978). These data may partially explain the effect of prostacyclin infusion on reducing the area of infarction (Lefer *et al.*, 1978).

Prostacyclin formation by aortas of rats fed a diet deficient in the antioxidant vitamin E has been correlated with an increase in aortic peroxide levels (Okuma *et al.*, 1980). Whether this is related to a decrease in prostacyclin synthetase activity subsequent to a decrease in the antioxidant buffering capacity as a result of vitamin E deficiency or to direct inhibition of this enzyme by peroxides or both requires further investigation. Similar decreases in prostacyclin formation by atherosclerotic coronary arterial tissue have been correlated with production of lipid peroxides (Fantone and Kinnes, 1983). Additionally, nitroglycerin is known to decrease GSH concentrations in various tissues including blood vessels (Needleman, 1976). If coronary arterial vasodilatation in response to nitroglycerin is mediated at least in part by prostacyclin (Schror *et al.*, 1981), tolerance to the vasodilator effect of nitroglycerin might be attributed in part to a parallel fall in prostacyclin formation and GSH concentration.

A fall in coronary artery prostacyclin generation could affect vascular tone and vessel–platelet interaction, which could lead to thromboembolic problems in large vessels or downstream in microvessels. The recent report (Beetens *et al.*, 1983) that the antioxidant ascorbate increases the release of 6-keto-PGF$_{1\alpha}$ from aortic rings lends support to our findings and raises the possibility that prostacyclin synthetase activity in other vascular tissue may be modulated in a similar manner to that reported here. However, we have not observed this effect of GSH on microsomal prostacyclin synthetase of human saphenous vein (Kerstein *et al.*, 1983) or human

(McMullen-Laird *et al.*, 1982) or bovine (unpublished results) intrapulmonary artery and vein.

In addition to producing a fall in the formation of the potent vasodilator prostacyclin, a decrease in coronary arterial prostacyclin synthetase activity might be expected to produce an increase in the concentration of unmetabolized PGH_2 and in the formation of other metabolites of PGH_2 such as PGE_2, PGD_2, or $PGF_{2\alpha}$. The vascular response to such an alteration in PGH_2 metabolism would depend on the metabolites formed, their concentrations, and their action on the coronary vasculature. We have previously shown that prostacyclin, a stable analogue of prostacyclin, PGE_2, and PGD_2 are vasodilators in the canine coronary vascular bed whereas $PGF_{2\alpha}$ was inactive (Hyman *et al.*, 1978). Unmetabolized PGH_2, however, might be expected to cause vasoconstriction. This effect could be mediated by a conversion of the unmetabolized PGH_2 to thromboxane A_2, a potent vasoconstrictor, by platelet thromboxane synthetase in a manner similar to (but in the reverse direction) the "steal hypothesis" of Vane (Bunting *et al.*, 1983). Thromboxane A_2 thus generated could also aggregate platelets, especially under these conditions in which prostacyclin synthesis was decreased, and initiate thrombogenesis. Alternatively, unmetabolized PGH_2 could directly cause vasoconstriction, as it has been suggested that thromboxane A_2 and PGH_2 interact with a single thromboxane receptor site (Carrier *et al.*, 1984). A stable analogue of PGH_2, U46619, which is thought to be a functional analogue of TxA_2, can release calcium from vascular mitochondrial storage sites (McNamara *et al.*, 1980). Such a release of bound calcium would increase intracellular free calcium, which could induce contraction of vascular smooth muscle. A similar sequence of events in the mediation of vascular smooth muscle contraction has been previously postulated by us (McNamara *et al.*, 1980) and by others (Greenburg, 1981). In addition, prostanoid metabolites of PGH_2 have also been shown to release calcium bound to vascular mitochondria (McNamara *et al.*, 1980) as well as that bound to smooth muscle sarcoplasmic reticulum (Carstein and Miller, 1977).

4.2. Prostaglandin E_2 Isomerase Activity

Prostaglandin E_2 isomerase is a GSH-requiring enzyme (Ogino *et al.*, 1977). The demonstration of the presence of an active PGE_2 isomerase is complicated, as in the absence of GSH, PGH_2 spontaneously breaks down to PGE_2. Thus, in the presence of GSH, PGE_2 formation is both enzymatic and nonenzymatic, and the nonenzymatic component cannot be distinguished from the enzymatic component. The presence of an active PGE_2 isomerase is suggested by an increased PGE_2 formation in the presence of GSH. However, as the PGH_2 concentration is metabolically depleted in the presence of GSH, nonenzymatic conversion of PGH_2 to PGE_2 becomes less significant, and it could be misleading to simply subtract the quantity of PGE_2 formed in the absence of GSH from that formed in the presence of GSH to obtain "net PGE_2" formation; such a figure is, however, useful and is employed (Gerritsen and Printz, 1981) as an index of PGE_2 isomerase activity.

Demonstration of an active PGE_2 isomerase in the coronary arterial microsomes is complicated by the fact that in the presence of GSH the enzymatic formation of both PGI_2 and PGE_2 would be augmented and the nonenzymatic component of total PGE_2 formation would decrease through a shunting of PGH_2 to PGI_2 formation. This is indicated by the fall in PGE_2 formation seen in the presence of DTT, which cannot serve as a cofactor for PGE_2 isomerase activity (Ogino *et al.*, 1977) but can modulate PGI_2 formation. Thus "net PGE_2" should not be employed as an index of PGE_2 isomerase activity for coronary arterial microsomes. However, as both GSH and DTT similarly affect PGI_2 formation, the amount of PGE_2 formed in the presence of DTT (nonenzymatic formation) is more representative of nonenzymatic PGE_2 formation than is that formed in the absence of GSH. Thus, the amount of PGE_2 formed in the presence of DTT (nonenzymatic), as presented on Table VII, should be subtracted from the amount of PGE_2 formed in the presence of GSH (enzymatic plus nonenzymatic), as presented in Table V. This difference more closely approximates PGE_2 isomerase activity. The data obtained in this study, when so interpreted, unmask the presence of an active PGE_2 isomerase in coronary arterial microsomes. This enzyme has been previously reported to be absent in the bovine coronary artery (Gerritsen and Printz, 1981).

5. SUMMARY AND CONCLUSIONS

5.1. Coronary Arterial Prostacyclin Synthetase

In this study, prostacyclin synthetase activity of bovine coronary arterial microsomes could be altered over a two- to threefold range by GSH or dithiothreitol in a concentration-dependent manner and over a microsomal protein range of 10–200 µg. Modulation of coronary artery prostacyclin synthetase activity was also seen in vessels from sheep, dog, and man. These data suggest that coronary artery prostacyclin synthetase activity is unusually sensitive to the redox state or sulfhydryl oxidation of the enzyme. We suggest that the higher prostacyclin synthetase activity is representative of basal prostacyclin formation and that the lower activity observed below 1 mM GSH represents subbasal prostacyclin activity. These data also suggest that modulation of coronary arterial prostacyclin synthetase activity under physiological and pathophysiological conditions could be mediated by those conditions or interventions that induce alterations in cellular antioxidant concentration.

5.2. Coronary Arterial Prostaglandin E_2 Isomerase

Contrary to a previous report, bovine coronary arterial microsomes contain an active GSH-dependent $PGH_2 \rightarrow PGE_2$ isomerase. In this respect, the coronary artery is similar to vascular segments found in other vascular beds.

ACKNOWLEDGEMENTS. The authors thank Ms. Jan Ignarro for her help in the preparation of the manuscript. This investigation was supported at Tulane University by National Institutes of Health grants HL11802, HL18070, and HL29456 and the American Heart Association-Louisiana, Inc.

REFERENCES

Beetens, J. R., Claeys, M., and Herman, A. G., 1983, Vitamin C increases the formation of prostacyclin by aortic rings from various species and neutralizes the inhibitory effect of the 15-hydroxy-arachidonic acid, *Br. J. Pharmacol.* **80**:249–254.

Bunting, S., Moncada, S., and Vane, J. R., 1983, The prostacyclin–thromboxane A_2 balance: Pathophysiological and therapeutic implications, *Br. Med. Bull.* **39**:271–276.

Carrier, R., Cragoe, E. J., Ethier, D., Ford-Hutchinson, A. W., Girard, Y., Hall, R. A., Hamel, P., Rokach, J., Share, N. N., Stone, C. A., and Yusko, P., 1984, Studies on L-640,035: A novel antagonist of contractile prostanoids in the lung, *Br. J. Pharmacol.* **82**:389–395.

Carstein, M. E., and Miller, J. D., 1977, Effects of prostaglandins and oxytocin on calcium release from a uterine microsomal fraction, *J. Biol. Chem.* **252**:1576–1581.

Egan, R. W., Paxton, J., and Kuehl, F. A., Jr., 1976, Mechanism for irreversible self-deactivation of prostaglandin synthetase, *J. Biol. Chem.* **251**:7329–7335.

Fantone, J. C., and Kinnes, D. A., 1983, Prostaglandin E_1 and prostaglandin I_2 modulation of superoxide production by human neutrophils, *Biochem. Biophys. Res. Commun.* **113**:506–512.

Farooqui, M. Y. H., and Ahmed, A., 1984, Circadian periodicity of tissue glutathione and its relationship with lipid peroxidation in rats, *Life Sci.* **34**:2413–2418.

Gerritsen, M. E., and Printz, M. P., 1981, Sites of prostaglandin synthesis in the bovine heart and isolated bovine coronary microvessels, *Circ. Res.* **49**:1152–1163.

Greenberg, S., 1981, Effect of prostacyclin and 9,11-epoxymethanoprostaglandin H_2 on calcium and magnesium fluxes and tension development in canine intralobar pulmonary arteries and veins, *J. Pharmacol. Exp. Ther.* **219**:326–337.

Hyman, A. L., Kadowitz, P. J., Lands, W. E. M., Crawford, C. G., Fried, J., and Barton, J., 1978, Coronary vasodilator activity of 13,14-dehydroprostacyclin methyl ester: Comparison with prostacyclin and other prostanoids, *Proc. Natl. Acad. Sci. U.S.A.* **75**:3522–3526.

Kerstein, M. D., Saroyan, M., McMullen-Laird, M., Hyman, A. L., Kadowitz, P. J., and McNamara, D. B., 1983, Metabolism of prostaglandins in human saphenous vein, *J. Surg. Res.* **35**:91–100.

Lefer, A. M., Ogletree, M. S., Smith, J. B., Silver, M. J., Nicolaou, K. C., and Gasic, G. P., 1978, Prostacyclin: A potentially valuable agent for preserving myocardial tissue in acute myocardial infarction, *Science* **200**:52–54.

Lowry, O. H., Rosebrough, N. J., Farr, A. L., and Randall, R. S., 1951, Protein measurement with folin phenol reagent, *J. Biol. Chem.* **193**:265–275.

McMullen-Laird, M., McNamara, D. B., Kerstein, M. D., Hyman, A. L., and Kadowitz, P. J., 1982, Human lung metabolism of prostaglandin endoperoxide, *Circulation* **66(II)**:166.

McNamara, D. B., Roulet, M. J., Gruetter, C. A., Hyman, A. L., and Kadowitz, P. J., 1980, Correlation of prostaglandin-induced mitochondrial calcium release with contraction in bovine intrapulmonary vein, *Prostaglandins* **20**:311–320.

McNamara, D. B., Hussey, J. L., Kerstein, M. D., Rosenson, R. J., Hyman, A. L., and Kadowitz, P. J., 1984, Modulation of prostacyclin synthetase and unmasking of PGE_2 isomerase in bovine coronary arterial microsomes, *Biochem. Biophys. Res. Commun.* **118**:33–39.

Meister, A., and Anderson, M. E., 1983, Glutathione, *Annu. Rev. Biochem.* **52**:711–760.

Needleman, P., 1976, Organic nitrate metabolism, *Annu. Rev. Pharmacol.* **16**:81–93.

Ogino, N., Miyamoto, T., Yamamoto, S., and Hayaishi, O., 1977, Prostaglandin endoperoxide E isomerase from bovine vesicular gland microsomes, a glutathione requiring enzyme, *J. Biol. Chem.* **252**:890–895.

Okuma, M., Takayma, H., and Uchino, H., 1980, Generation of prostacyclin-like substance and lipid peroxidation in vitamin E-deficient rats, *Prostaglandins* **19**:527–536.

Romson, J. L., Hook, B. G., Rigot, V. H., Schork, M. A., Swanson, D. P., and Lucchesi, B. R., 1982, The effect of ibuprofen on accumulation of Indium-111-labeled platelets and leukocytes in experimental myocardial infarction, *Circulation* **66**:1002–1011.

Schror, K., Grodzinska, L., and Darius, H., 1981, Stimulation of coronary vascular prostacyclin and inhibition of human platelet thromboxane A_2 after low dose nitroglycerin, *Thromb. Res.* **23**:59–67.

She, H. S., McNamara, D. B., Spannhake, E. W., Hyman, A. L., and Kadowitz, P. J., 1981, Metabolism of prostaglandin endoperoxide by microsomes from cat lung, *Prostaglandins* **21**:531–541.

Thomas, G., Skrinska, V., Lucas, F., and Butkus, A., 1984, Platelet glutathione and thromboxane synthesis in diabetes mellitus, *Fed. Proc.* **43**:1040.

Vane, J. R., 1983, Adventures and excursions in bioassay: The stepping stones to prostacyclin, *Br. J. Pharmacol.* **79**:821–838.

7

Glutathione Transferases Catalyzing Leukotriene C Synthesis and Metabolism of Leukotrienes C_4 and E_4 *in Vivo* and *in Vitro*

SVEN HAMMARSTRÖM, LARS ÖRNING, KERSTIN BERNSTRÖM, BENGT GUSTAFSSON, ELISABETH NORIN, BENGT MANNERVIK, HELGI JENSSON, and PER ÅLIN

1. GLUTATHIONE TRANSFERASES CATALYZING LEUKOTRIENE C SYNTHESIS

Glutathione transferases constitute a group of enzymes that catalyze several reactions involving glutathione (Mannervik, 1985). Their main function, according to current concepts, is to detoxify and accelerate the excretion of certain xenobiotic compounds (Chasseaud, 1979) by catalyzing the conjugation of glutathione with these electrophilic substrates. In addition, glutathione transferases also catalyze other reactions (e.g., peroxidase and isomerase reactions). We have recently reported (Mannervik *et al.*, 1984) that six basic glutathione transferases from rat liver cytosol (Mannervik

SVEN HAMMARSTRÖM, LARS ÖRNING, and KERSTIN BERNSTRÖM ● Department of Physiological Chemistry, Karolinska Institutet, S-104 01 Stockholm, Sweden. BENGT GUSTAFSSON and ELISABETH NORIN ● Department of Germfree Research, Karolinska Institutet, S-104 01 Stockholm, Sweden. BENGT MANNERVIK, HELGI JENSSON, and PER ÅLIN ● Department of Biochemistry, Arrhenius Laboratory, University of Stockholm, S-106 91 Stockholm, Sweden.

and Jensson, 1982) catalyze the conversion of LTA_4 or its methyl ester to LTC_4. One of the enzymes was a considerably more efficient catalyst of the reaction than the remaining five.

1.1. Formation of LTC_4 (C-1) Monomethyl ester by Cytosolic Rat Glutathione Transferases

Incubation of rat cytosolic glutathione transferase 4–4 with tritium-labeled LTA_4 methyl ester and glutathione yielded a tritium-labeled product that emerged later than LTC_4 on reverse-phase HPLC. The UV spectrum of this compound showed a λ_{max} at 280 nm and shoulders at 270 and 292 nm as previously reported for LTC_4 (Murphy *et al.*, 1979). Treatment of the product with soybean lipoxygenase shifted the λ_{max} to 308 nm. After mild alkaline hydrolysis, the product cochromatographed with authentic LTC_4 on reverse-phase HPLC. Incubation of the product with γ-glutamyl transpeptidase followed by mild alkaline hydrolysis gave rise to LTD_4 as judged by cochromatography on HPLC with the synthetic reference compound. The results showed that the product formed from LTA_4 methyl ester by transferase 4–4 is the methyl ester of the naturally occurring isomer of LTC_4 (cf. Hammarström *et al.*, 1980). The reaction involves a nucleophilic substitution of the oxygen at C-6 in LTA_4 methyl ester by the thiolate anion of glutathione and proceeds with inversion of the configuration at C-6.

Incubation of LTA_4 methyl ester with transferases 1–1, 1–2, 2–2, 3–3, or 3–4 in each case yielded tritium-labeled products with similar chromatographic and UV-absorbing properties as the product formed by transferase 4–4. After mild alkaline hydrolysis, these products cochromatographed with synthetic LTC_4 on HPLC.

1.2. Kinetics of the Reaction with LTA_4 Methyl Ester Catalyzed by Cytosolic Glutathione Transferases

Kinetic parameters for the conversion of leukotriene A_4 methyl ester to leukotriene C_4 methyl ester by rat liver cytosolic glutathione transferases are shown in Table I. The rate of formation of leukotriene C_4 methyl ester was constant for more than 1 min under the conditions used and was proportional to the enzyme concentration.

The kinetics of LTC_4 monomethyl ester formation were studied as a function of LTA_4 methyl ester concentration at 5 mM glutathione concentration. The K_m values for the different isoenzymes ranged from 2.3 to 15 μM, and the V_{max} values from 17 to 615 nmol/min per mg protein. Transferase 4–4 had the highest V_{max}. This enzyme is a homodimer of subunit "4." A heterodimeric transferase (3–4) containing one subunit "4" and one subunit "3" had a V_{max} slightly over 25% of that of transferase 4–4. The other transferases (3–3, 1–1, and 1–2) had V_{max} values of approximately 20 nmol/min per mg (3% of that of 4–4). Subunit "4" was thus a more effective catalyst than subunits "3," "2," and "1" with regard to the transformation of LTA_4 methyl ester to LTC_4 monomethyl ester.

TABLE I. Specific Activities and Kinetic Constants for the Conversion of LTA_4 Methyl Ester to LTC_4 Monomethyl Ester by Cytosolic Rat Glutathione Transferase[a]

Glutathione transferase	Specific activities (nmol/min per mg)		Kinetic constants (LTA_4Me)	
	LTA_4Me[b]	$CDNB$[c]	V_{max} (nmol/min per mg)	K_m (μM)
1–1	7.6 ± 2.7 [d]	33,000[e]	17	2.3
1–2	5.6 ± 2.8	28,000	17	3.3
2–2	0.83 ± 0.25	19,000	n.d.	n.d.
3–3	2.5 ± 1.1	38,000	22	9.2
3–4	32 ± 22	28,000	162 ± 20	15 ± 0.3 ($n = 2$)
4–4	102 ± 18	18,000	615 ± 15	11 ± 4 ($n = 2$)

[a] From Mannervik *et al* (1984).
[b] Leukotriene A_4 methyl ester
[c] 1-Chloro-2,4-dinitrobenzene.
[d] Mean values ± S.D. from three experiments with single or duplicate analyses.
[e] Data from Mannervik and Jensson (1982)

The pattern of the relative catalytic activities of the cytosolic glutathione transferases reflected their respective subunit composition as demonstrated with other substrates (Mannervik and Jensson, 1982). V_{max} values for the reactions with leukotriene A_4 methyl ester ranged from 17 to 615 nmol/min per mg. Transferase 4–4 had the highest (615) and transferase 3–4 the second highest V_{max} value (162). K_m values were near 10 μM for transferases containing subunits 3 and/or 4 and around 3 μM for transferases composed of subunits 1 and/or 2. Thus, the transferases with highest apparent "affinity" for LTA_4 methyl ester had the lowest capacity to transform it into LTC_4 monomethyl ester and vice versa.

Leukotriene A_4 free acid was converted by the same enzymes to LTC_4. Further studies regarding the kinetics of the latter reaction are needed to compare the catalytic efficiencies of cytosolic rat glutathione transferases with those of microsomal enzymes from murine mastocytoma cells and rat basophilic leukemia cells, which catalyze the LTC synthase reaction (Söderström *et al.*, 1985).

2. Metabolism of Leukotrienes C_4 and E_4

Previous investigations in our laboratories have shown that leukotriene C is metabolized *in vitro* (Örning and Hammarström, 1980; Hammarström, 1981; Bernström and Hammarström, 1981) and *in vivo* (Hammarström *et al.*, 1981; Appelgren and Hammarström, 1982) by modifications of the peptide part. Evidence indicating alterations of the fatty acid part has also been published (Goetzl, 1982; Hendersson and Klebanoff, 1982; Lee *et al.*, 1983), but the *in vivo* significance of the latter reactions has not been established. Whole-body autoradiographic distribution experiments have indicated that in mice the major excretion pathway for leukotriene C is via the bile into feces (Appelgren and Hammarström, 1982). Chromatographic analyses of the radioactivity present in different mouse organs showed that leukotrienes C_3, D_3, and E_3 could be detected in lung, liver, intestine, kidney, and blood for 1 hr or more after the administration of leukotriene C_3 (Appelgren and Hammarström, 1982). One to five unidentified metabolites were observed in the mouse experiments. A compound with analogous chromatographic behavior to one of these products has now been identified as a metabolite of leukotriene E_4 formed by rat liver homogenates. The same product was also found to be an excreted metabolite of leukotriene C_4 in the rat.

2.1. Metabolism of Leukotriene E_4 by Rat Liver Homogenates

[$^3H_8$]Leukotriene E_4 was converted by homogenates of rat liver to a product eluting before leukotriene E_4 on RP-HPLC (compound I; relative elution time, 0.85 ± 0.1 compared with leukotriene E_4) (Bernström and Hammarström, 1985). The ultraviolet spectrum of the product was identical to the spectrum of leukotriene E_4 (Bernström and Hammarström, 1981). Desulfurization with Raney nickel gave a product that cochromatographed by gas–liquid chromatography with the product obtained from [$^3H_8$]leukotriene C_4. A positive ion fast atom bombardment mass spectrum showed ions at m/z 482, 504, and 520. These ions most likely represent $[M + H]^+$, $[M + Na]^+$, and $[M + K]^+$, respectively, suggesting that the molecular weight of compound I is 481. Since the molecular weight of leukotriene E_4 is 439, this suggested the presence of an acetyl group in compound I. 5-Hydroxy-6-S-(2-acetamido-3-thiopropionyl)-7,9,11,14-eicosatetraenoic acid (N-acetylleukotriene E_4) was synthesized from leukotriene E_4 by treatment with acetic anhydride. The product gave the same fast atom bombardment mass spectrum, had the same ultraviolet spectrum, and cochromatographed by RP-HPLC with compound I, both as the free acid and as the dimethyl ester derivative.

[1-^{14}C]Acetylcoenzyme A and [$^3H_8$]leukotriene E_4 were incubated with dialyzed rat liver mitochondria. The N-acetylleukotriene E_4 formed was purifed by RP-HPLC, and the contents of tritium and ^{14}C were determined as 34 pmol of ^{14}C from [1-^{14}C]acetylcoenzyme A and 33 pmol of tritium from [5,6,8,9,11,12,14,15-$^3H_8$]leukotriene E_4 (Fig. 1).

The contractile effects of compound I, synthetic N-acetylleukotriene E_4, and leukotriene C_4 were determined on isolated segments of guinea pig ileum. Compound I and N-acetylleukotriene E_4 were equipotent, and their potency was 10–14 times lower than that of leukotriene C_4. FPL 55712 (40 ng/ml) antagonized the contractile effect of N-acetylleukotriene E_4. The onset of the contractions was similar for compound I and N-acetylleukotriene E_4 and resembled that of contractions elicited by leukotrienes D_4 and E_4.

2.2. *In Vivo* Metabolism of Leukotriene C_4, in the Rat

After subcutaneous administration of leukotriene C_4, labeled with tritium in the fatty acid part, the radioactivity was largely excreted over a 72-hr period (Örning *et al.*, 1984). Of the recovered tritium, 89–92% appeared in feces, and the rest in

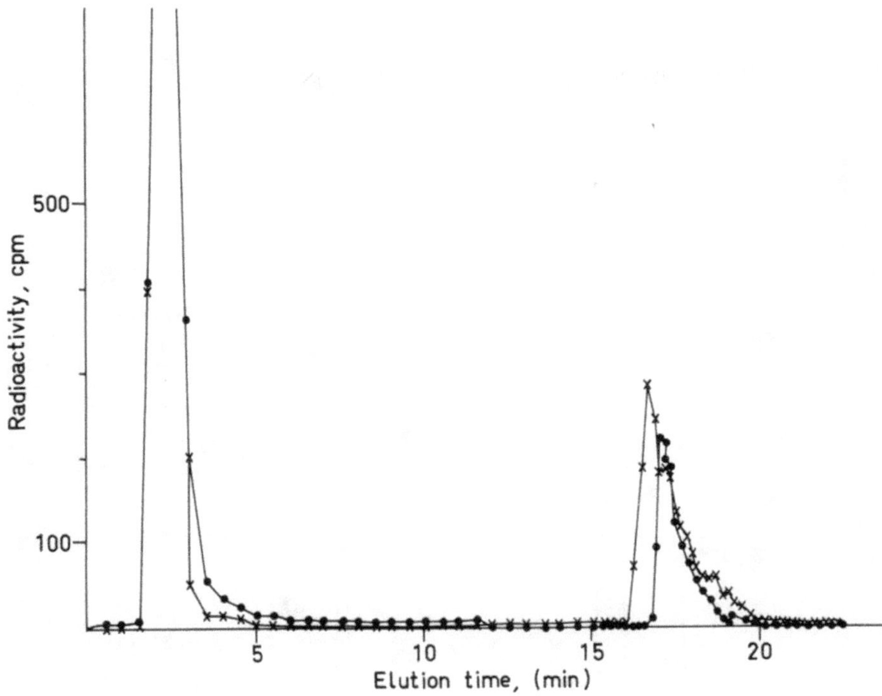

FIGURE 1. Conversion of [1-^{14}C]acetylcoenzyme A and [^{3}H$_8$]leukotriene E$_4$ to compound 1 using a pressure-dialyzed suspension of rat liver mitochondria. Compound 1 contained 33 pmol of leukotriene E$_4$ (from ^{3}H) and 34 pmol of acetate (from ^{14}C). Conditions for HPLC: A column of C$_{18}$ Nucleosil (4.6 × 250 mm; 5 μm particles) was eluted with methanol/water/acetic acid/o-phosphoric acid (65 : 35 : 0.07 : 0.03, v/v/v/v; pH 5.4) at a flow rate of 1 ml/min. x, ^{3}H; •, ^{14}C. (From Bernström and Hammarström, 1984).

urine from conventional and germ-free rats. Approximately 75% of the tritium-labeled leukotriene C$_4$ metabolites were recovered in the ethanolic extract of feces from germ-free rats, and the rest was released after refluxing with chloroform/methanol. The tritium-labeled material in the 80% ethanol eluate after chromatography on Amberlite XAD-8 was purified by silicic acid chromatography. Fifty-four and 82% of the radioactivity applied to the column was recovered in a methanol/ethyl acetate (7 : 3, v/v) eluate. This material was further fractionated by reverse-phase HPLC. The chromatogram of fecal metabolites from germ-free rats (Fig. 2A) revealed at least eight components, designated I, III, IV, VII, VIII, IX, X, and XI. Compound VIII was the predominant one, and compound IX was the second most abundant metabolite. They constituted 23% and 13% of the radioactivity that was subjected to HPLC, corresponding to 4.6% and 2.7%, respectively, of the radioactivity administered to the animals.

Figure 2B shows a chromatogram of fecal metabolites of leukotriene C$_4$ from conventional rats. Eight compounds, designated II, III, and IV–IX can be seen.

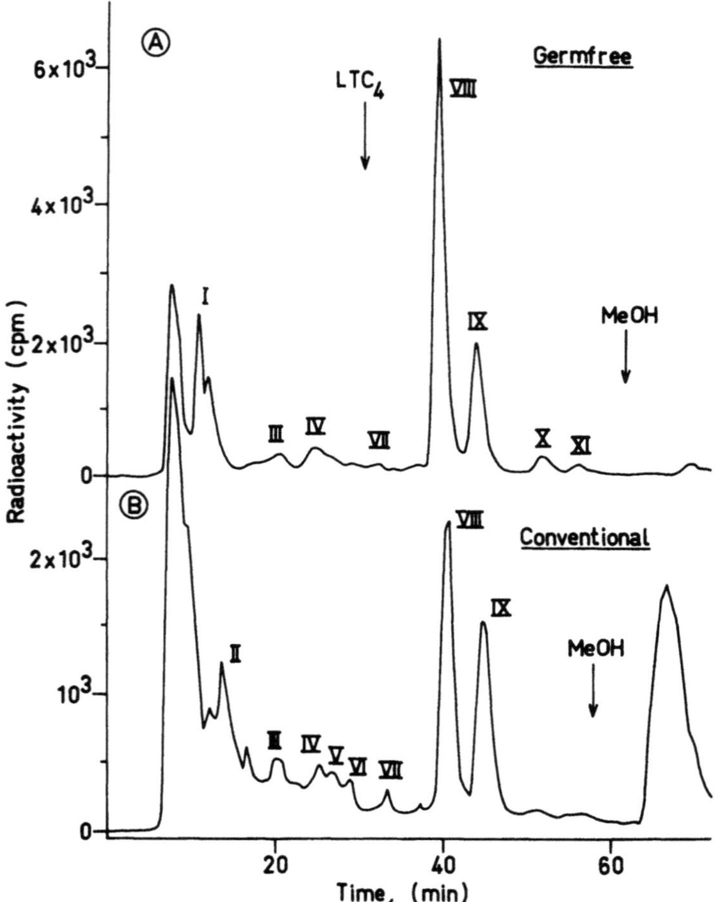

FIGURE 2. Reverse-phase HPLC of radioactive products excreted in feces from germ-free (A) and conventional (B) rats after administration of [³H₈]leukotriene C₄. Conditions for HPLC: A column of C₁₈ Polygosil (500 × 10 mm) was eluted with methanol/water/acetic acid 70/30/0.1 (v/v/v) adjusted to pH 5.4 with NH₄OH at a flow rate of 4.5 ml/min. (From Örning *et al.*, 1984.)

The radioactivity, eluted with the solvent front and with methanol at the end of the chromatography, is probably heterogeneous and has not been further analyzed. Compounds VIII and IX constituted 9% and 7% of the radioactivity analyzed by HPLC and 0.6% and 0.5%, respectively, of the radioactivity administered to the conventional rats.

The UV spectrum of compound VIII (Fig. 3) is similar to that of leukotriene C₄ (Murphy *et al.*, 1979), suggesting that the conjugated triene with an allylic thioether substituent had been retained in compound VIII. Treatment with soybean lipoxygenase gave a similar spectral shift to that previously reported for leukotriene C₄ (Murphy *et al.*, 1979). Thus, the original positions of the triene and the thioether

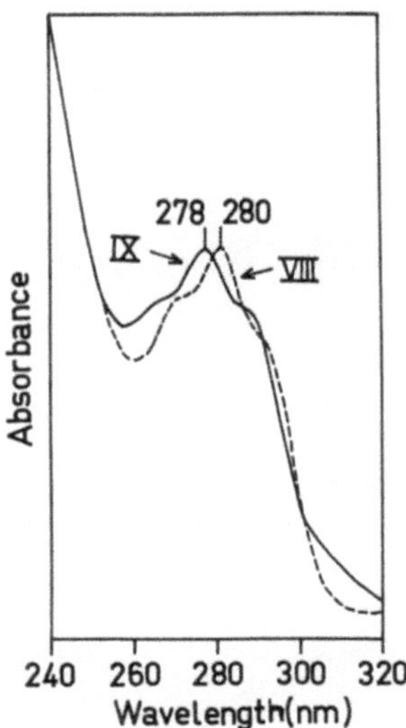

FIGURE 3. Ultraviolet spectrum of compounds VIII and IX from Fig. 1. (From Örning *et al.*, 1984.)

substituent were unchanged. Positive and negative ion fast atom bombardment mass spectra of compound VIII showed intense ions at the same *m/z* values as the corresponding spectra of synthetic N-acetyl leukotriene E_4. Compound VIII also cochromatographed by reverse-phase HPLC with synthetic N-acetyl leukotriene E_4 both as the free acid and as the methyl ester derivative. The contractile activity of compound VIII on guinea pig ileum was also identical to that of N-acetyl leukotriene E_4. Based on this, it was concluded that the structure of compound VIII was 5-hydroxy-6-*S*-(2-acetamido-3-thiopropionyl)-7,9-*trans*-11,14-*cis*-eicosatetraenoic acid (N-acetyl leukotriene E_4).

The UV spectrum of compound IX (Fig. 3) was similar to that previously reported for 11-*trans*-leukotriene C_4 (Clark *et al.*, 1980). Its elution time on RP-HPLC was 1.1 times that of compound VIII, which is similar to the relative elution time of 11-*trans*-leukotriene C_4 compared to leukotriene C_4 (Clark *et al.*, 1980). Desulfurization with Raney nickel yielded 5-hydroxyeicosanoic acid. N-Acetyl-11-*trans*-leukotriene E_4 was prepared from 11-*trans*-leukotriene E_4 in the same way as described above for N-acetylleukotriene E_4. It cochromatographed with compound IX. Compound IX was thus tentatively identified as the 11-*trans* isomer of compound VIII, i.e., N-acetyl 11-*trans*-leukotriene E_4.

The transformation of leukotriene C_4 to N-acetyl leukotriene E_4 (Fig. 4) is analogous to the pathway of mercapturic acid biosynthesis (Chasseaud, 1979), in which xenobiotics are transformed to N-acetylcysteine derivatives prior to excretion. It is conceivable that the original function of this pathway was to inactivate and facilitate the excretion of the endogenous hormonelike substance, leukotriene C_4.

On the basis of *in vitro* experiments, it has been suggested that cysteine-containing leukotrienes are metabolized to 6-*trans*-leukotriene B_4 and the 12-*epi*

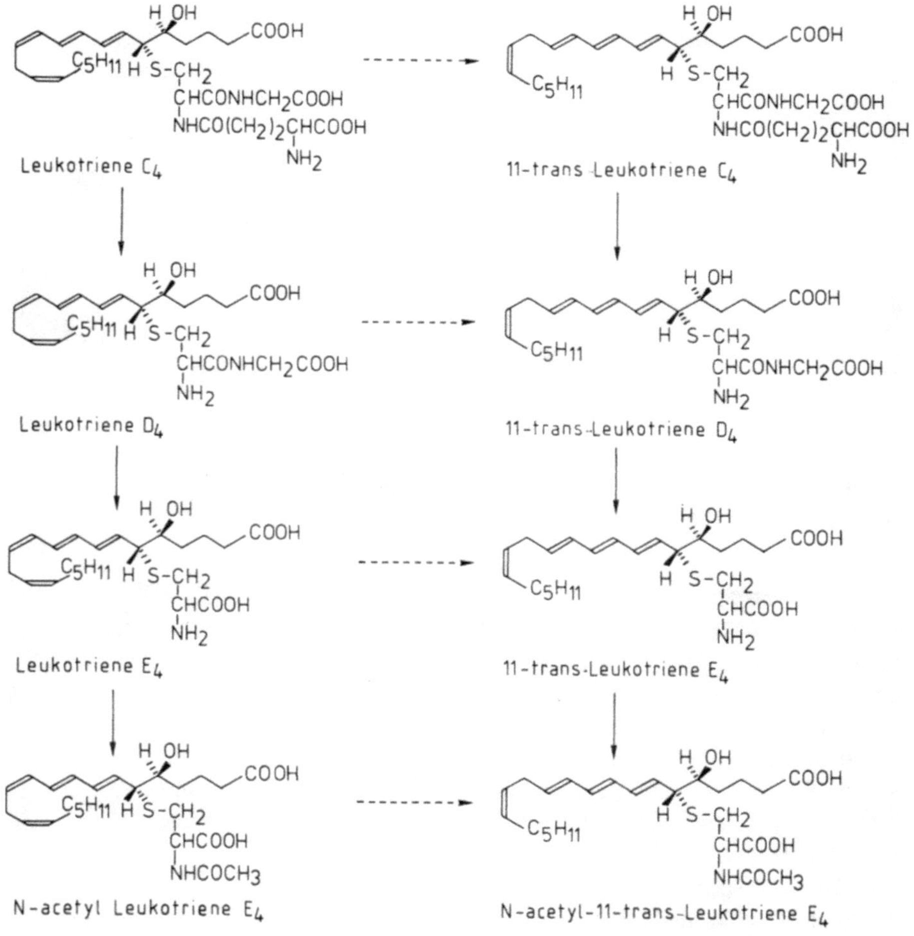

FIGURE 4. Scheme for the conversion of leukotriene C_4 to N-acetyl leukotriene E_4 and N-acetyl-11-*trans* leukotriene E_4.

isomer of this compound (Goetzl, 1982; Hendersson and Klenbanoff, 1983; Lee *et al.*, 1983). In the present experiments, these non-cysteine-containing metabolites were not detected in either feces or urine of rats. The high recoveries of tritium and the lack of excretion of leukotriene B_4 isomers indicate that in the rat, the major pathways of metabolism of systemically administered leukotriene C_4 do not involve cleavage of the C-6 to sulfur bond.

The characterization of excreted metabolites of leukotriene C_4 described above should facilitate the development of analytical methods for the quantitative determination of endogenous leukotriene formation under various physiological conditions in the rat.

REFERENCES

Appelgren, L. E., and Hammarström, S., 1982, Distribution and metabolism of ^{3}H-labeled leukotriene C$_3$ in mice, *J. Biol. Chem.* **257**:531–535.

Bernström, K., and Hammarström, S., 1981, Metabolism of leukotriene D by porcine kidney, *J. Biol. Chem.* **256**:9579–9582.

Bernström, K., and Hammarström, S. 1985, Metabolism of leukotriene E$_4$ by rat tissues, *Arch. Biochem. Biophys.* submitted.

Chasseaud, L. F., 1979, The role of glutathione and glutathione-S-transferases in the metabolism of chemical carcinogens, *Adv. Cancer Res.* **29**:175–274.

Clark, D. A., Goto, G., Marfat, A., Corey, E. J., Hammarström, S., and Samuelsson, B., 1980, 11-*trans*-Leukotriene C: A naturally occurring slow reacting substance, *Biochem. Biophys. Res. Commun.* **94**:1133–1139.

Goetzl, E. J., 1982, The conversion of leukotriene C$_4$ to isomers of leukotriene B$_4$ by human eosinophil peroxidase, *Biochem. Biophys. Res. Commun.* **106**:270–275.

Hammarström, S., 1981, Metabolism of leukotriene C$_3$ in the guinea pig. Identification of metabolites formed by lung, liver and kidney, *J. Biol. Chem.* **256**:9573–9578.

Hammarström, S., Samuelsson, B., Clark, D. A., Goto, G., Marfat, A., Mioskowski, C., and Corey, E. J., 1979, Stereochemistry of leukotriene C-1, *Biochem. Biophys. Res. Commun.* **92**:946–953.

Hammarström, S., Bernström, K., Örning, L., Dahlén, S. E., Hedqvist, P., Smedegård, G., and Revenäs, B., 1981, Rapid *in vivo* metabolism of leukotriene C$_3$ in the monkey *Macaca irus*, *Biochem. Biophys. Res. Commun.* **101**:1109–1115.

Hendersson, W. R., and Klebanoff, S. J., 1983, Leukotriene production and inactivation by normal, chronic granulomatous disease and myeloperoxidase deficient neutrophils, *J. Biol. Chem.* **258**:13522–13527.

Lee, C. W., Lewis, R. A., Tauber, A. I., Mehrotra, M., Corey, E. J., and Austen, K. F., 1983, The myeloperoxidase-dependent metabolism of leukotrienes C$_4$, D$_4$, and E$_4$ to 6-*trans*-leukotriene B$_4$ diastereoisomers and the subclass-specific S-diastereoisomeric sulfoxides, *J. Biol. Chem.* **258**:15004–15010.

Mannervik, B., 1985, The isoenzymes of glutathione transferase, *Adv. Enzymol.* **57**:357–417.

Mannervik, B., and Jensson, H., 1982, Binary combinations of four protein subunits with different catalytic specificities explain the relationship between six basic glutathione-S-transferases in rat liver cytosol, *J. Biol. Chem.* **257**:9909–9912.

Mannervik, B., Jensson, H., Ålin, P., Örning, L., and Hammarström, S., 1984, Transformation of leukotriene A$_4$ methyl ester to leukotriene C$_4$ monomethyl ester by cytosolic rat glutathione transferases, *FEBS Lett.* **175**:289–293.

Murphy, R. C., Hammarström, S., and Samuelsson, B., 1979, Leukotriene C: A slow reacting substance from murine mastocytoma cells, *Proc. Natl. Acad. Sci. U.S.A.* **76**:4275–4279.

Örning, L., and Hammarström, S., 1980, Inhibition of leukotriene C and leukotriene D biosynthesis, *J. Biol. Chem.* **255**:8023–8026.

Örning, L., Norin, E., Gustafsson, B., and Hammarström, S., 1984, On the metabolism of leukotriene C$_4$ in the rat: Identification of two fecal metabolites, *J. Biol. Chem.* **260**:(in press).

Söderström, M., Mannervik, B., Örning, L., and Hammerström, S., 1985, Leukotriene C$_4$ formation catalyzed by three distinct forms of human cytosolic glutathione transferase, *Biochem. Biophys. Res. Commun.* **228**:265–270.

8

Formation and Role of Lipoxygenase Products in Human Platelets

M. LAGARDE, M. CROSET, K. S. AUTHI,
M. DECHAVANNE, S. RENAUD, and
N. CRAWFORD

1. INTRODUCTION

Blood platelets are involved in primary hemostasis through their aggregation in response to different agents. When platelets are aggregated by some of these agents such as thrombin or collagen, polyunsaturated fatty acids (mainly arachidonic acid) are liberated from membrane phospholipids and then subsequently oxygenated via cyclooxygenase and/or lipoxygenase. The cyclooxygenase pathway has been well documented (Smith, 1980). From arachidonic acid (AA), it leads to proaggregatory prostanoids, prostaglandin endoperoxides (PGG_2/PGH_2) and thromboxane A_2 (TxA_2), which together with 12-hydroxyheptadecatrienoic acid represent the main compounds of the pathway, primary prostaglandins being minor products. In comparison, dihomo-γ-linolenic acid (DHLA) and 5,8,11,14,17-eicosapentaenoic acid (EPA), precursors of mono- and trienoic series prostanoids, respectively, are well known to be inhibitors of platelet functions (Willis, 1981). Less data about platelet lipoxygenase pathway, especially for its biological role, are available. This enzyme is assumed to be a 12-lipoxygenase (Hamberg et al., 1974; Nugteren, 1975) associated with a glutathione-dependent peroxidase (Bryant and Bailey, 1980). It has been

M. LAGARDE, M. CROSET, M. DECHAVANNE, and S. RENAUD • INSERM U 63, Institut Pasteur, Laboratoire d'Hémobiologie, Faculté Alexis Carrel, 69372 Lyon Cédex 08, France. K. S. AUTHI and N. CRAWFORD • Department of Biochemistry, Royal College of Surgeons, London, United Kingdom.

reported that 12-hydroperoxyeicosatetraenoic acid (12-HPETE), the lipoxygenase product of AA, is able to inhibit both thromboxane synthase (Hammarström and Falardeau, 1977) and platelet diglyceride lipase (Rittenhouse-Simmons, 1980). More recently, 12-HPETE was described as an inhibitor of PGH_2/TxA_2-induced platelet aggregation (Aharony *et al.*, 1982).

We report here the activity of human platelet lipoxygenase towards prostaglandin precursors (AA, DHLA, and EPA) and 5,8,11-eicosatrienoic acid (20:3n-9), a marker of linoleic acid deficiency (Holman, 1960) that appeared to potentiate platelet aggregation (Lagarde *et al.*, 1983). We also present evidence that their lipoxygenase products, especially 12-hydroxy derivatives, modulate prostanoid-induced platelet aggregation.

2. OXYGENATION OF EICOSAENOIC ACIDS BY LIPOXYGENASE OF INTACT PLATELETS

When incubated with prostaglandin precursors, human platelets synthesize lipoxygenase products from each of them that were identified as mainly 12-hydroxyeicosaenoic acids by high-performance liquid chromatography (HPLC). However, a small amount of a compound that comigrated with the 15-hydroxy derivative could also be detected, especially from DHLA (Lagarde *et al.*, 1984a). However, it averaged only one-tenth of the 12-hydroxy derivative according to the peak heights and might have been provided by leukocytes contaminating the platelet preparations. We may then confirm that human platelet lipoxygenase is essentially a 12-lipoxygenase.

Although prostaglandin precursors are qualitatively oxygenated by platelet lipoxygenase in a similar way, great quantitative differences can be observed. Studies on physiological concentrations of platelets (3×10^8/ml) isolated from their plasma and 10^{-5} M AA, DHLA, EPA, or 20:3n-9 for 4 min revealed that AA was the best substrate. In comparison, EPA appeared to be the poorest. In particular, increasing the substrate concentration of eicosaenoic acids except EPA potentiated their oxygenation (Lagarde *et al.*, 1984a). At the opposite extreme, the EPA oxygenation by lipoxygenase was markedly increased by AA or 12-HPETE (Boukhchache and Lagarde, 1982; Morita *et al.*, 1983). This potentiating effect seemed to be specific, since 12-HETE or 15-HPETE did not share the action of 12-HPETE (Lagarde *et al.*, 1984a). These results (Table I) suggest that oxygenation of EPA is a peroxide-requiring process, as has been reported previously (Culp *et al.*, 1979). In this context, the increased concentration of EPA would not be sufficient to generate enough peroxides for its enhanced conversion, in contrast to the other eicosaenoic acids. The absence of potentiation by 15-HPETE could be attributed to the counteracting effect of 15-lipozygenase products platelet lipoxygenase as described earlier (Vanderhoek *et al.*, 1980).

Finally, the oxygenation of DHLA and 20:3n-9 by platelet lipoxygenase was also potentiated by AA or 12-HPETE (Table I). This fact raises the possibility of

TABLE I. Lipoxygenation of Eicosaenoic Acids Alone or in the Presence of AA or 12-HPETE[a]

Concentration	AA	DHLA	EPA	20:3n-9
10^{-5} M ($n = 5$)	1.8	0.6	0.4	0.8
10^{-4} M ($n = 2$)	183.2	85.4	2.2	241.3
10^{-5} M + AA (10^{-5} M) ($n = 5$)	—	1.4	2.4	1.6
10^{-5} M + 12-HPETE (5×10^{-6} M) ($n = 5$)	—	2.4	3.6	2.5

[a] Results expressed in nanomoles/10^9 platelets per 4-min incubation

increasing the oxygenation of minor polyunsaturated fatty acids, which can be elevated in platelet phospholipids under certain dietary lipid modifications (Willis, 1981). After platelet activation, these minor polyunsaturated fatty acids would be then liberated together with AA, which might enhance their oxygenation.

3. SUBCELLULAR LOCALIZATION OF PLATELET LIPOXYGENASE

Controversial data have been reported concerning the localization of platelet lipoxygenase. In studies with bovine and rat platelets, the majority of the activity was found in the cytosol (Nugteren, 1975; Chang *et al.*, 1982), whereas others mainly localized the enzyme activity in the particulate fraction of human platelets (Ho *et al.*, 1977). Using AA as substrate, we have recently found (Lagarde *et al.*, 1984b) that the lipoxygenase activity of human platelets is bimodally distributed between cytosol and membranes, the cytosol activity being predominant. When the membrane fraction was separated into surface and intraplatelet membranes by free-flow electrophoresis as described previously (Menashi *et al.*, 1981), significantly higher activity could be measured in intraplatelet membranes as compared to surface membranes. Enzyme activities in both cytosol and membranes did not differ significantly in terms of pH dependence (neutral optimal pH), calcium requirement (no calcium or calmodulin dependency), and carbon specificity (the enzyme was essentially a 12-lipoxygenase in both cases). The only consistent difference was a higher rate of the membrane-bound lipoxygenase activity at the early stages (first minute), which is confirmed by a higher V_{max} for the oxygenation of AA, DHLA, EPA, or 20:3n-9. These data suggest that the membrane lipoxygenase activity could be of physiological relevance, at least at the early stages of platelet activation.

Taking into account the preferential location of such an activity in intraplatelet membranes, we may even hypothesize that this membrane subpopulation, which has been recognized as the exclusive location of prostanoid formation (Gerrard *et al.*, 1976; Carey *et al.*, 1982), could be the main site of polyunsaturated fatty acid oxygenation in platelets. This hypothesis is reinforced by our previous findings, which localized platelet phospholipase A_2 and diglyceride lipase activities, both enzymes involved in the liberation of polyunsaturated fatty acids from phospholipids (MacKean *et al.*, 1981; Bell *et al.*, 1979; Lagarde *et al.*, 1982), in this intracellular membrane system. This hypothesis leads to the concept that both the liberation of polyunsaturated fatty acids and their conversion by platelet oxygenases could be closely related in space, which fits well with results obtained from comparing the oxygenation of exogenous and endogenous AA into TXB_2 and 12-HETE. These results showed a preferential conversion of the endogenous fatty acid (Sautebin *et al.*, 1983).

4. MODULATION OF PLATELET AGGREGATION BY LIPOXYGENASE PRODUCTS

Although the biological activity of prostanoids has been well defined in platelets, the role of the lipoxygenase pathway remains to be specified. In our approach, we have prepared 12- and 15-hydroxy derivatives from prostaglandin presurors with platelet homogenates and soybean lipoxygenase used as enzymic sources, respectively. The products were then purified by HPLC as described (Croset and Lagarde, 1983) and tested on platelet aggregation induced by the 9-methano analogue of PGH_2 (a TxA_2 mimetic). Each product was able to counteract the aggregation with an IC_{50} in the micromolar range. However, the 12-hydroxy derivatives were slightly more potent than 15-hydroxy. Further investigations on the specificity of the inhibition were done with other lipoxygenase products of AA. We then found that 5-HETE was about threefold less potent than 12-HETE, suggesting that the presence of the hydroxide on carbon 12 would be optimal for the biological activity. Most relevant was the finding that leukotriene B_4 (5*S*,12*R*-di HETE) was completely devoid of any activity, whereas its isomer 5*S*,12*S*-diHETE, the double lipoxygenation derivative of AA (Borgeat *et al.*, 1981; Marcus *et al.*, 1982), shared the activity of 12-HETE (Croset and Lagarde, 1983). This fact strongly suggests that the inhibition of prostanoid-induced platelet aggregation by lipoxygenase products is quite stereospecific. The *S* configuration of the derivative would be required to counteract the aggregating activity of prostanoids, which are 15-*S* compounds. These data are summarized in Table II.

Compared to platelet lipoxygenase products of AA, DHLA, or EPA, the 12-hydroxy derivative of 20:3n-9 (12-OH-5,8,10-20:3 as checked by mass spectrometry) exhibited a different activity pattern. It potentiated platelet aggregation at low concentrations (below 5×10^{-7} M) but it was inhibiting over 10^{-6} M. This biphasic effect was quite similar to that obtained with PGE_2. Experiments performed with platelets enriched with 20:3n-9 revealed that when triggered with thrombin, the

TABLE II. Inhibition of Platelet
Aggregation Induced by the 9-Methoxy
Analogue of PGH$_2$[a]

Compound	Inhibition (%)[b]
12-HETE	100
15-HETE	97
12-OH-20:3(8,10,14)	86
15-OH-20:3(8,11,13)	44
12-OH-20:5(5,8,10,14,17)	135
15-OH-20:(5,8,11,13,17)	92
5-HETE	33
5S,12S-diHETE	46
LTB$_4$ (5S,12R-diHETE)	0

[a] The IC$_{50}$ for 12-HETE was 6.6×10^{-6} M when 8×10^{-8} M analogue of PGH$_2$ was used.

[b] Results, expressed in percentage of the activity exhibited by 12-HETE, are the means of at least 5 determinations.

amount of lipoxygenase product of this acid produced during the incubation compares with the potentiating concentrations of 12-OH-5,8,10-20:3 reported above (Lagarde *et al.*, 1983). This indicates that only the potentiating activity of the latter derivative would be of physiological relevance.

5. CONCLUSIONS

This chapter reports data that tend to demonstrate that the lipoxygenase pathway of human platelets could have a physiological role in modulating the proaggregatory activity of prostaglandin endoperoxides/thromboxane A$_2$. The inhibition of platelet aggregation by the lipoxygenase product of arachidonic acid might be shared by 12-hydroxy derivatives of other prostaglandin precursors, since their formation appears markedly enhanced by arachidonic acid through 12-HPETE. From the physiological point of view, the 12-lipoxygenase product of 20:3n-9 would have a potentiating effect on platelet aggregation. At this time, the mechanism of action of such modulations remains to be determined.

On the other hand, a common location (platelet intracellular membranes) for the formation of proaggregatory prostanoids and of a consistent part of 12-hydroxy-eicosaenoic acids would allow more efficiency in the regulation of prostanoid-induced aggregation by lipoxygenase products. Taking into account all our finds, we may summarize these molecular interactions in terms of platelet topography with a hypothetical scheme (Fig. 1).

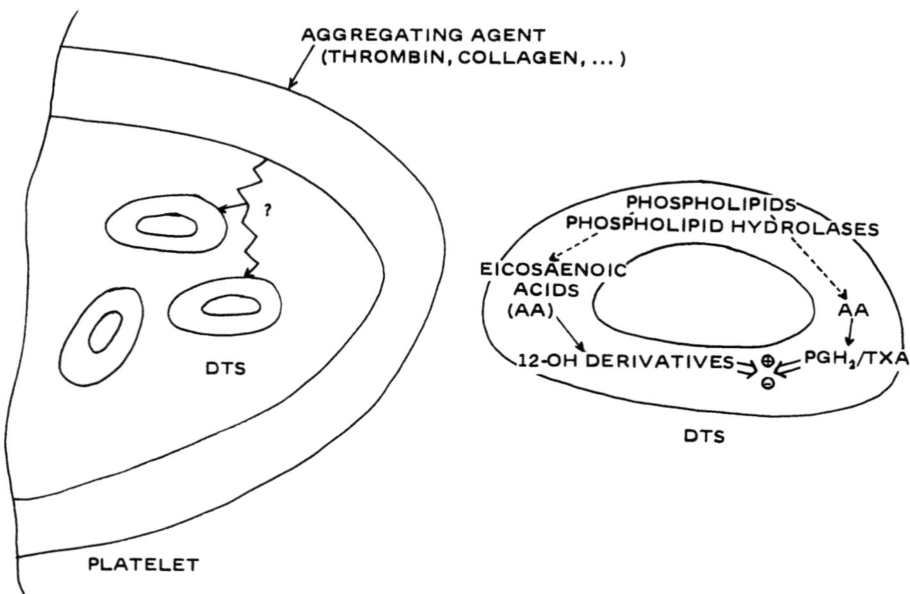

FIGURE 1. Hypothetical scheme suggeting that both the liberation of eicosaenoic acids (mainly arachidonic acid) and their oxygenation by cyclooxygenase and/or lipoxygenase could occur in platelet intracellular membranes (dense tubular system, DTS). This spatial relationship would allow lipoxygenase products to be more efficient in counteracting the aggregating activity of PGH_2/TXA_2.

REFERENCES

Aharony, D., Smith, J. B., and Silver, M. J., 1982, Regulation of arachidonate induced platelet aggregation by the lipoxygenase product, 12-hydroperoxyeicosatetraenoic acid, *Biochim. Biophys. Acta* **718**:193–200.

Bell, R. L., Kennerly, D. A., Stanford, N., and Majerus, P. W., 1979, Diglyceride lipase: A pathway for arachidonate release from human platelets, *Proc. Natl. Acad. Sci. U.S.A.* **76**:3238–3241.

Borgeat, P., Picard, S., and Vallerand, P., 1981, Transformation of arachidonic acid in leukocytes. Isolation and structural analysis of a novel dihydroxy derivative. *Prostaglandins Md.* **6**:557–570.

Boukhchache, D., and Lagarde, M., 1982, Interactions between prostaglandin precursors during their oxygenation by human platelets, *Biochim. Biophys. Acta* **713**:386–392.

Bryant, R. W., and Bailey, J. M., 1980, Altered lipoxygenase metabolism and decreased glutathione peroxidase activity in platelets from selenium-deficient rats, *Biochem. Biophys. Res. Commun.* **92**:268–276.

Carey, F., Menashi, S., and Crawford, N., 1982, Localization of cyclooxygenase and thromboxane synthetase in human platelet intracellular membranes, *Biochem. J.* **204**:847–851.

Chang, W. C., Nakao, J., Orimo, H., and Murota, S., 1982, Effects of reduced glutathione on the 12-lipoxygenase pathway in rat platelets, *Biochem. J.* **202**:771–776.

Croset, M., and Lagarde, M., 1983, Stereospecific inhibition of PGH_2-induced platelet aggregation by lipoxygenase products of icosaenoic acids, *Biochem. Biophys. Res. Commun.* **112**:878–883.

Culp, B. R., Titus, B. G., and Lands, W. E. M., 1979, Inhibition of prostaglandin biosynthesis by eicosapentaenoic acid, *Prostaglandins Med.* **3**:269–278.

Gerrard, J. M., White, J. G., Rao, G. H. R., and Townsend, D. W., 1976, Localization of platelet prostaglandin production in the platelet dense tubular system, *Am. J. Pathol.* **83**:283–298.

Hamberg, M., Svensson, J., and Samuelsson, B., 1974, Prostaglandin endoperoxides. A new concept concerning the mode of action and release of prostaglandins, *Proc. Natl. Acad. Sci. U.S.A.* **71**:3824–3828.

Hammarström, S., and Falardeau, P., 1977, Resolution of prostaglandin endoperoxide synthase and thromboxane synthase of human platelets, *Proc. Natl. Acad. Sci U.S.A.* **74**:3691–3695.

Ho, P. P. K., Walters, P., and Sullivan, H. R., 1977, A particulate arachidonate lipoxygenase in human blood platelets, *Biochem. Biophys. Res. Commun.* **76**:398–405.

Holman, R. T., 1960, The ratio of trienoic : tetraenoic acids in tissue lipids as a measure of essential fatty acid requirement, *J. Nutr.* **70**:405–411.

Lagarde, M., Authi, K. S., and Crawford, N., 1982, Fatty acid specificity of human platelet-membrane phospholipase A_2 and diacylglycerol lipase, *Biochem. Soc. Trans.* **10**:241–243.

Lagarde, M., Burtin, M., Sprecher, H., Dechavanne, M., and Renaud, S., 1983, Potentiating effect of 5,8,11-eicosatrienoic acid on human platelet aggregation, *Lipids* **18**:291–294.

Lagarde, M., Croset, M., Authi, K. J., and Crawford, N., 1984a, Subcellular localization and some properties of lipoxygenase activity in human blood platelets, *Biochem. J.* **222**:495–500.

Lagarde, M., Croset, M., Boukhchache, D., Greffe, A., Dechavanne, M., and Renaud, S., 1984b, Lipoxygenase activity of intact human platelets, *Prostaglandins Leukotrienes Med.* **13**:61–66.

Lagarde, M., Menashi, S., and Crawford, N., 1981, Localization of phospholipase A_2 and diglyceride lipase activities in human platelet intracellular membranes, *FEBS Lett.* **124**:23–26.

MacKean, M. L., Smith, J. B., and Silver, M. J., 1981, Formation of lysophosphatidylcholine by human platelets in response to thrombin, *J. Biol. Chem.* **256**:1522–1524.

Marcus, A. J., Broeckman, M. J., Safier, L. B., Ullman, H. L., Islam, N., Serham, C. N., Rutherford, L. E., Korchak, H. M., and Weissmann, G., 1982, Formation of leukoktrienes and other hydroxy acids during platelet–neutrophil interactions *in vivo, Biochem. Biophys. Res. Commun.* **109**:130–137.

Menashi, S., Weintroub, H., and Crawford, N., 1981, Characterization of human platelet surface and intracellular membranes isolated by free flow electrophoresis, *J. Biol. Chem.* **256**:4095–4101.

Morita, I., Takahashi, R., Saito, Y., and Murota, S., 1983, Stimulation of eicosapentaenoic acid metabolism in washed human platelets by 12-hydroperoxyeicosatetraenoic acid, *J. Biol. Chem.* **258**:10197–10199.

Nugteren, D. H., 1975, Arachidonate lipoxygenase in blood platelets, *Biochim. Biophys. Acta* **380**:299–307.

Rittenhouse-Simmons, S., 1980, Indomethacin-induced accumulation of diglyceride in activated human platelets, *J. Biol. Chem.* **225**:2259–2262.

Sautebin, L., Caruso, D., Galli, G., and Paoletti, R., 1983, Preferential utilization of endogenous arachidonate by cyclooxygenase in incubations of human platelets, *FEBS Lett.* **157**:173–178.

Smith, J. B., 1980, The prostanoids in hemostasis and thrombosis, *Am. J. Pathol* **99**:742–804.

Vanderhoek, J. Y., Bryant, R. W., and Bailey, J. M., 1980, 15-hydroxy-5,8,11,13-eicosatetraenoic acid. A potent and selective inhibitor of platelet lipoxygenase, *J. Biol. Chem.* **255**:5996–5998.

Willis, A. L., 1981, Nutritional and pharmacological factors in eicosanoid biology, *Nutr. Rev.* **39**:289–301.

9

Lipoxygenation of Mitochondrial Membranes by Reticulocyte Lipoxygenase

GARY FISKUM, ROBERT W. BRYANT, CHOW-ENG LOW, ANDREW PEASE, and J. MARTYN BAILEY

1. INTRODUCTION

There is considerable interest in the lipoxygenation of mitochondrial membranes because of the inhibitory influence this process can have on mitochondrial activities such as oxidative phosphorylation (Vladimirov *et al.*, 1980). In addition, mitochondrial lipid peroxidation has been associated with several important pathological conditions. These include hypoxia and postischemic reperfusion (Hillered and Ernster, 1983), aging (Nohl *et al.*, 1978), and the cytotoxicity of many chemicals as well as UV or ionizing radiation (Vladimirov *et al.*, 1980).

In general, mitochondrial lipid peroxidation has been attributed to nonenzymatic attack by superoxide and hydroxyl radicals (Vladimirov *et al.*, 1980). However, one system has been identified in which specific enzymatic peroxidation of mitochondrial membranes may occur. Professor Samuel Rapoport and co-workers have obtained evidence indicating that a lipoxygenase is involved in the physiological process whereby mitochondria are degraded and eliminated during the maturation of reticulocytes to erythrocytes (Rapoport *et al.*, 1979). Lipoxygenase activity appears in developing reticulocytes but is absent in mature erythrocytes (Thiele *et al.*, 1979). Furthermore, the purified enzyme can, under some conditions, cause

GARY FISKUM, ROBERT W. BRYANT, CHOW-ENG LOW, ANDREW PEASE and J. MARTYN BAILEY • Department of Biochemistry, George Washington University School of Medicine and Health Sciences, Washington, D.C. 20037. *Present address* of R.W.B.: Shering Corporation, Pharmaceutical Research Division, Bloomfield, New Jersey 07003.

gross structural and functional alterations of isolated mitochondria (Schewe *et al.*, 1975). The mitochondrial destruction observed *in vitro* is apparently associated with the peroxidation of membrane lipids and is not obtained when mitochondria are incubated with lipoxygenases from other tissues (Rapoport *et al.*, 1979). The reticulocyte enzyme also exhibits specificity with respect to its attack toward mitochondrial membranes, since it does not readily lipoxygenate erythrocyte ghosts (Rapoport *et al.*, 1979). The reticulocyte lipoxygenase exhibits chemical specificity toward polyunsaturated free fatty acids with K_m values decreasing as a function of the number of double bonds (Halangk *et al.*, 1977). Also, the V_{max} values for $C_{20:3}$ and $C_{18:2}$ are significantly greater than those for $C_{18:3}$ and $C_{20:4}$ fatty acids. The major product generated from arachidonic acid is 15-hydroperoxy-5,8,11,14-eicosatetraenoic acid (15-HPETE), although some 12-hydroperoxy derivatives are also observed as minor components (5–10%) (Bryant *et al.*, 1982). In either case, oxygen is inserted with S stereospecificity.

The precise role of lipoxygenase in the degradation of maturing reticulocyte mitochondria has recently been questioned by the findings of several laboratories that reticulocytes possess a very active mitochondria-specific protease (Boches and Goldberg, 1982; Muller *et al.*, 1980; Speiser and Etlinger, 1982). However, it is possible that the two enzymes may act synergistically, with either the lipoxygenase making the mitochondria susceptible to proteolysis (Schewe *et al.*, 1977) or the protease potentiating lipoxygenation of the mitochondria.

The conditions necessary for lipoxygenase-mediated mitochondrial destruction and the identity of the products formed during incubation of reticulocyte lipoxygenase with isolated mitochondria have recently been explored in our laboratories with the aid of various techniques commonly used in bioenergetics and analytical biochemistry (Fiskum *et al.*, 1983). The following is a brief description of our findings and how they relate to the current state of knowledge regarding mitochondrial lipoxygenation.

2. MITOCHONDRIAL LIPOXYGENATION AND RESPIRATORY INHIBITION

The model used for studying mitochondrial lipoxygenation was one in which intact or swollen rat liver mitochondria were incubated at 30°C in a medium containing rabbit reticulocyte lipoxygenase and concentrations of K^+, Mg^{2+}, and phosphate that are similar to those present in cytosol. The oxygen concentration of the mitochondrial suspension was monitored with an oxygen electrode and was used to follow the consumption of oxygen due to lipoxygenation or mitochondrial respiration.

Figure 1 describes typical O_2 electrode chart recordings from experiments performed with suspensions of rabbit reticulocyte lipoxygenase and either intact or swollen rat liver mitochondria. The addition of intact mitochondria to medium containing the NADH-linked respiratory inhibitor rotenone resulted in very little consumption of O_2 (Fig. 1A). Subsequent addition of the rotenone-insensitive ox-

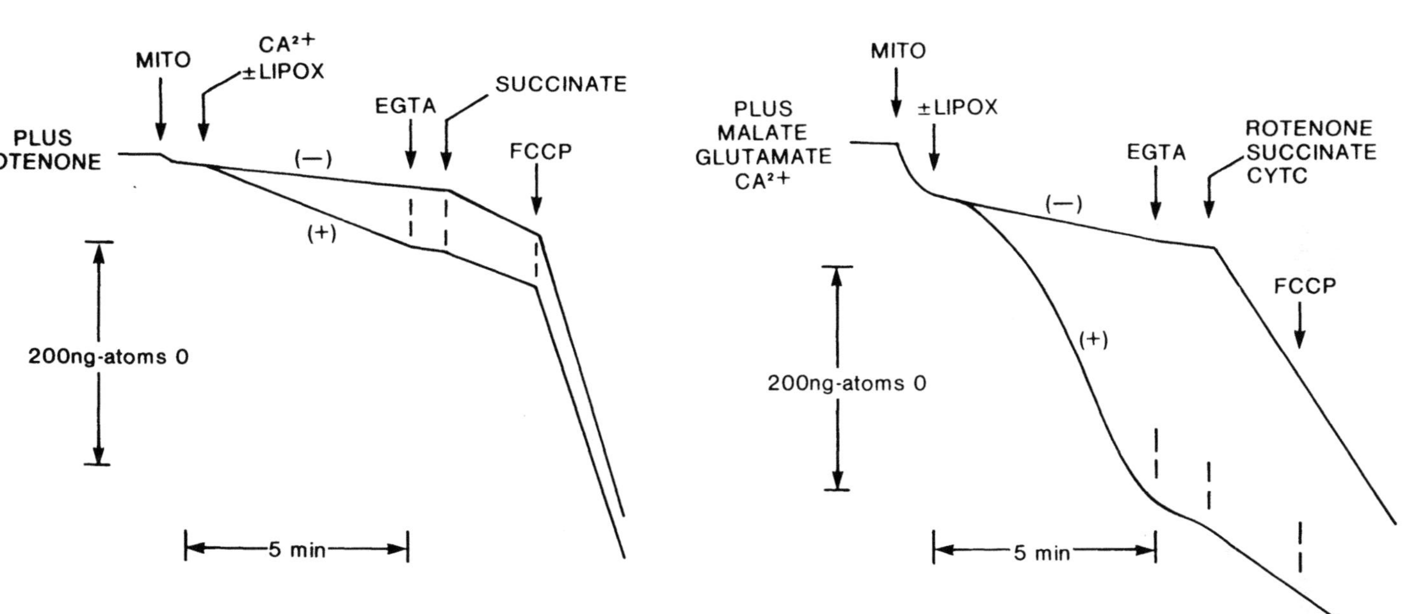

FIGURE 1. Oxygen electrode measurements of mitochondrial respiration and mitochondrial lipoxygenation by reticulocyte lipoxygenase. Rat liver mitochondria (MITO) were suspended at 0.25 mg protein/ml in medium containing 125 mM KCl, 2 mM K_2HPO_4, 1 mM $MgCl_2$, 1 mg/ml BSA, 5 mM HEPES (pH 7.0). The temperature was maintained at 30°C. A: Rotenone (4 μM) was present prior to the addition of mitochondria. $CaCl_2$ (80 μM) was added in the absence (−) or presence (+) of 20 μg/ml purified rabbit reticulocyte lipoxygenase (LIPOX). Subsequent additions were EGTA (0.25 mM), succinate (5 mM), and FCCP (0.4 μM). B: Malate (5 mM), glutamate (5 mM), and $CaCl_2$ (80 μM) were present prior to the addition of mitochondria. All other additions were the same as in A except for the addition of cytochrome c (CYTC) at a concentration of 5 μM.

idizable substrate succinate initiated mitochondrial respiration, which was further stimulated by addition of the respiratory uncoupler FCCP (carbonylcyanide *p*-trifluoromethoxyphenylhydrazone). When reticulocyte lipoxygenase was added along with Ca^{2+} to the mitochondrial suspension, a significant increase in the rate of rotenone-insensitive O_2 consumption was observed. Addition of the Ca^{2+} chelator EGTA eliminated this apparent lipoxygenase-mediated oxygenation. Subsequent addition of succinate and FCCP provided rates of mitochondrial respiration that were almost identical to those observed in the absence of pre-incubation with lipoxygenase.

The modest consumption of O_2 by the reticulocyte lipoxygenase and the lack of an effect of the enzyme on mitochondrial respiration were somewhat surprising in light of reports describing major alterations of mitochondrial structure and function by this enzyme (Schewe *et al.*, 1975; Krause and Halangk, 1977). Our results were obtained with both pure and partially purified preparations of lipoxygenase and did not significantly change when the temperature was raised to 37°C or when the incubation period was extended to 30 min. This indicated to us that either our experimental conditions were unsatisfactory or that intact mitochondria are poor substrates for the reticulocyte lipoxygenase. We therefore decided to test our lipoxygenase assay system with mitochondria that were swollen and partially disrupted by excessive respiration-dependent Ca^{2+} accumulation.

In the experiment shown in Fig. 1B, intact rat liver mitochondria were added to normal medium containing an abundance of Ca^{2+} (320 nmol/mg protein) and the NADH-linked oxidizable substrates malate and glutamate (minus rotenone). Under these conditions, mitochondrial respiration was initially rapid because of the active sequestration of Ca^{2+} (Fiskum, 1984). Soon thereafter, respiration decelerated to an abnormally slow pace because of Ca^{2+}-induced swelling, disruption of the outer membrane, and loss of cytochrome *c* into the surrounding medium (Chappell and Crofts, 1965). The subsequent addition of rotenone, succinate, and excess cytochrome *c* was then used to initiate rotenone-insensitive, $FADH_2$-linked respiration. At this point, the mitochondria were uncoupled (leaky to protons) and could not be further stimulated by FCCP. When rat liver mitochondria were incubated under these conditions, the addition of lipoxygenase was followed by a period of very active oxygen consumption. Moreover, the lipoxygenase-mediated oxygenation of swollen mitochondria resulted in a greater than 50% inhibition of succinate-dependent mitochondrial respiration (Fig. 1B) and complete release of intramitochondrial enzymes into the surrounding medium (data not shown). Since the normal Ca^{2+}, Mg^{2+} permeability barrier of the outer mitochondrial membrane to large molecules was lost during the Ca^{2+}-induced swelling, these results suggested that the inner membrane may be the primary site of attack by the reticulocyte lipoxygenase.

Experiments such as those described in Fig. 1 were also performed with mitochondria that had their outer membranes physically removed by digitonin solubilization and subsequent centrifugation (Schnaitman and Greenawalt, 1968). As expected, these "mitoplast" preparations were as good or better substrates for the reticulocyte lipoxygenase than the Ca^{2+}-swollen mitochondria were. For intact mitoplasts, the extent of lipoxygenase-dependent oxygen consumption over a 5-min period was 634 ± 127 ng-atoms O/min per mg mitochondrial protein ($n = 7$) versus 464 + 117 for swollen mitochondria ($n = 8$) and only 96 ± 50 for intact mitochondria ($n = 7$). The inhibition of mitochondrial electron transport was also greatest with the mitoplasts (86 ± 7%) compared to that for swollen mitochondria (66 ± 8%) or to the insignificant value obtained with intact mitochondria (7 ± 6%).

The large difference in the extent of lipoxygenation and subsequent respiratory inhibition between intact mitochondria and either swollen mitochondria or mitoplasts indicates that mitochondria are rather poor substrates for the reticulocyte lipoxygenase and that this is because of the inaccessibility of the enzyme to the mitochondrial inner membrane. The lipoxygenase-mediated destruction of mitochondria observed by other investigators may have occurred under conditions (e.g., high phosphate concentrations) that predispose mitochondria to spontaneous swelling.

One common means by which isolated mitochondria may undergo spontaneous degradation is the activation of endogenous phospholipase A_2 and the release of free fatty acids from membrane phospholipids (Scarpa and Lindsay, 1972). We took the precaution of including BSA (1 mg/ml) in our medium to protect against damage caused by the possible presence of free fatty acids. The observation that BSA made little difference in the activity of the lipoxygenase toward rat liver mitoplasts (Table I) but completely inhibited the lipoxygenation of 20 μM free arachidonate (data not shown) indicated that mitochondrial phospholipid fatty acyl groups rather than free fatty acids had been oxygenated. This was confirmed by other studies to be described below in which lipid analyses were performed on the mitochondrial membranes after they had been incubated with the reticulocyte lipoxygenase. No oxygenated fatty acids were detected by thin-layer chromatography or mass spectrometry unless the reacted mitochondrial lipids were first subjected to saponification.

Table I also describes the results of other experiments designed to determine the optimal conditions for lipoxygenation of the mitochondrial inner membrane. As

TABLE I. Lipoxygenation of Rat Liver Mitoplasts

Incubation conditions	Rate of oxygen consumption (% control)
Control ($+$BSA, CA^{2+}, MG^{2+})[a]	100
$-$BSA	80
$+$Soybean lipoxygenase[b]	0
$+$Boiled reticulocyte lipoxygenase	0
$+$EGTA	0
$-$Mg^{2+}	135
$+$PMSF[c]	115

[a] The conditions were the same as those used in the experiment described in Fig. 1A ($n = 2$–10).
[b] Purified soybean lipoxygenase was present at a concentration that gave the same rate of O_2 consumption in the presence of free arachidonic acid as that for reticulocyte lipoxygenase.
[c] Phenylmethyl sulfonylfluoride (PMSF) was present at a concentration of 0.3 mM.

noted earlier, lipoxygenation was dependent on the presence of the active reticulocyte lipoxygenase. Thus, no O_2 consumption was observed with soybean 15-lipoxygenase or the heat-inactivated reticulocyte enzyme. Lipoxygenation was also entirely dependent on the presence of Ca^{2+} (but not Mg^{2+}), as indicated by the 100% inhibition caused by the presence of EGTA. This is particularly intriguing since the presence or absence of Ca^{2+} or EGTA had no influence on the lipoxygenation of free fatty acids by either the reticulocyte or soybean enzymes. Preliminary results obtained with suspensions of rat liver mitoplasts and the reticulocyte lipoxygenase indicate that the K_A for free Ca^{2+} falls within the range of 10–20 μM. This suggests that Ca^{2+} may play a role in the triggering of mitochondrial lipoxygenation *in vivo* and adds the reticulocyte enzyme to the list of lipoxygenases that are regulated in one way or another by low concentrations of Ca^{2+} (Chapter 14, this volume).

The presence of the serine protease inhibitor phenylmethylsulfonylfluoride (PMSF) had no influence on the lipoxygenation of mitoplasts by the reticulocyte enzyme. Thus, proteolysis is probably not required for lipoxygenation when the inner membrane is accessible to the lipoxygenase. However, it remains to be determined whether proteolysis is necessary for lipoxygenation of mitrochondria present within the intact reticulocyte.

3. PRODUCT FORMATION DURING MITOCHONDRIAL LIPOXYGENATION

Very little information is available regarding the actual products formed during lipoxygenation of mitochondrial membranes by enzymatic or nonenzymatic mechanisms. The reaction products have typically only been identified via determination of malonyl dialdehyde or other thiobarbituric-acid-reactive substances (see, e.g., Rapoport *et al.*, 1979; Schewe *et al.*, 1975). Thus, only the presence of peroxidative products but not their identity has been determined.

In order to gain further knowledge along these lines, we decided to analyze the lipids present in lipoxygenase-treated Ca^{2+}-swollen mitochondria with the aid of gas chromatography and mass spectrometry. At the end of a 15-min incubation of swollen mitochondria in the presence or absence of reticulocyte lipoxygenase, an aliquot of the suspension was taken and diluted 1 : 2 with methanol. Thereafter, chloroform and NaOH (0.2 M final concentration) were added to the methanolic solution in order to saponify the membrane lipids (90 min at 37°C). After the saponification reaction was neutralized with formic acid, the fatty acids and non-saponifiables were extracted with chloroform. This extract was further separated by silicic acid column chromatography into two fractions that were enriched with either oxygenated or nonoxygenated fatty acids. The fractions were then methylated and analyzed by gas chromatography or methylated and trimethylsilylated and analyzed by GC–mass spectrometry.

Gas chromatography of the nonoxygenated fraction indicated that at least 50%

of 18:2, 20:4, and 22:6 fatty acyl groups had been utilized as substrates for the reticulocyte lipoxygenase. The wide variety of products that were formed during the incubation is evident from the total ion chromatogram of the saponifiable, oxygenated membrane lipids shown in Fig. 2. Mass spectra of several of the more prominent peaks allowed us to tentatively identify several lipoxygenase products. These included monohydroxy-18:2, -22:6, and -20:4, trihydroxy-18:1 and -20:3, and 11-OH-12,13-epoxy-18:1. Such products have also been observed after incubation of the reticulocyte lipoxygenase with polyunsaturated free fatty acids and

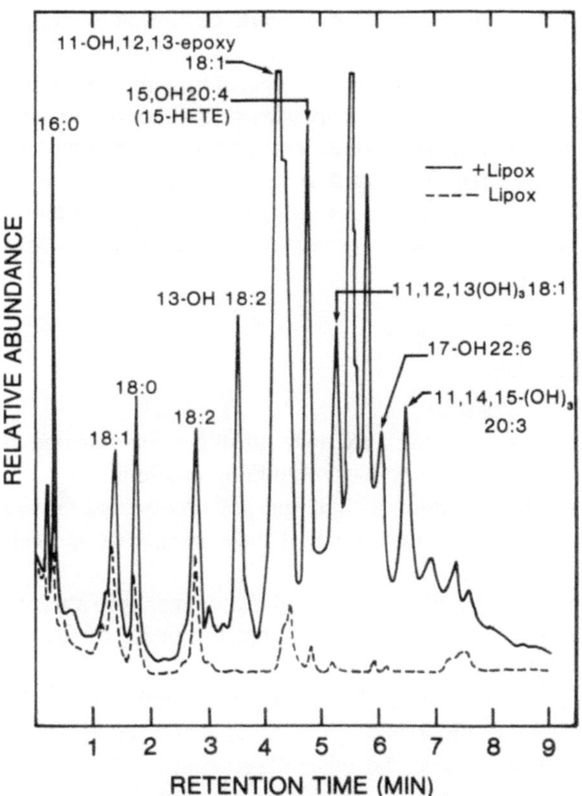

FIGURE 2. Gas chromatographic–MS analysis of saponified products (hydroxy-fatty-acid-enriched fraction) formed during incubation of reticulocyte lipoxygenase with swollen mitochondria: total ion chromatogram. Hydroxy fatty acid metabolites were isolated as described in the text. Methylated and trimethylsilylated products were analyzed on a Hewlett-Packard Model 5992 GC–MS fitted with a column (2 mm × 100 cm) containing 1% SE30. Samples were injected at a column temperature of 160°C and an injection port temperature of 220°C. After 1 min, a linear temperature program was initiated at 10°C/min and run to a maximum of 280°C. Mass spectra (60–700 amu) were acquired at approximately 1-sec intervals. Metabolite identification was based on comparison of experimental spectra of known compounds or based on known fragmentation patterns.

represent the metabolites of the originally formed hydroperoxides. The large amounts of monohydroxy compounds that were obtained in these experiments suggest that the hydroperoxy phospholipids may have been rapidly reduced to their respective hydroxides, possibly via the activity of mitochondrial glutathione peroxidase (Lawrence and Burk, 1976; Sies and Moss, 1978).

These preliminary results indicate that identification of lipoxygenated mitochondrial membrane lipids is feasible through the use of GC–mass spectrometry. Further progress along these lines will probably require separation of the lipoxygenated products by HPLC prior to analysis by GC–mass spectrometry. In fact, such a procedure has recently been applied toward identification of the peroxidative products formed in the whole liver of carbon tetrachloride-treated animals (Hughes *et al.*, 1983).

Further elucidation of the products formed during mitochondrial lipoxygenation should increase our understanding of the structure–function relationships that exist between the mitochondrial membranes and normal or abnormal mitochondrial activities. This, in turn, may be helpful in determining the precise sequence of biochemical events that is responsible for mitochondrial degradation under physiological conditions (e.g., in the maturing reticulocyte) as well as under many different pathological conditions.

4. CONCLUSIONS

Intact rat liver mitochondria are poor substrates for reticulocyte lipoxygenase when suspended at 30°C in a cytosol-resembling medium.

Greater than 50% of the mitochondrial polyunsaturated fatty acyl groups become oxygenated when the mitochondrial inner membrane is made accessible to the lipoxygenase enzyme.

Mitochondrial inner membrane lipoxygenation results in almost total inhibition of mitochondrial electron transport and complete release of matrix enzymes into the suspending medium.

Oxygenation of mitochondrial membranes by reticulocyte lipoxygenase requires micromolar concentrations of Ca^{2+} but is unaffected by the presence of high concentrations of BSA. These and other observations indicate that mitochondrial phospholipids are directly oxygenated by lipoxygenase in a Ca^{2+}-dependent and phospholipase-independent manner.

The major saponifiable products generated during incubation of reticulocyte lipoxygenase with mitoplasts or swollen mitochondria are monohydroxy, trihydroxy, and hydroxyepoxy derivatives of 18:2, 20:4, and 22:6 fatty acids.

REFERENCES

Boches, F. S., and Goldberg, A. L., 1982, Role for the adenosine triphosphate-dependent proteolytic pathway in reticulocyte maturation, *Science* **215**:978–980.

Bryant, R. W., Bailey, J. M., Schewe, T., and Rapoport, S. M., 1982, Positional specificity of a reticulocyte lipoxygenase: Conversion of arachidonic acid to 15S-hydroperoxy-eicosatetraenoic acid, *J. Biol. Chem.* **257**:6050–6055.

Chappell, J. B., and Crofts, A. R., 1965, Calcium accumulation and volume changes of isolated mitochondria, *Biochem. J.* **95**:378–392.

Fiskum, G., 1984, Physiological aspects of mitochondrial calcium transport, in: *Metal Ions in Biological Systems,* Vol. 17: *Calcium and Its Role in Biology* (H. Sigel, ed.), Marcel Dekker, New York, pp. 187–214.

Fiskum, G., Bryant, R. W., Pease, A., and Bailey, J. M., 1983, Influence of reticulocyte lipoxygenase and 15-HETE on mitochondrial energy coupling, *Fed. Proc.* **42**:2251.

Halangk, W., Schewe, T., Hiebsch, C., and Rapoport, S. M., 1977, Some properties of the lipoxygenase from rabbits reticulocytes, *Acta Biol. Med. Germ.* **36**:405–410.

Hillered, L., and Ernster, L., 1983, Respiratory activity of isolated rat brain mitochondria following *in vitro* exposure to oxygen radicals, *J. Cereb. Blood Flow Metab.* **3**:207–214.

Hughes, H., Smith, C. V., Horning, E. C., and Mitchell, J. R., 1983, High performance liquid chromatography and gas chromatography–mass spectrometry determination of specific lipid peroxidation products *in vivo, Anal. Biochem.* **130**:431–436.

Krause, W., and Halangk, W., 1977, Relationship between conformational state and susceptibility to reticulocyte lipoxygenase of rat liver mitochondria, *Acta Biol. Med. Germ.* **36**:381–387.

Lawrence, R. A., and Burk, R. F., 1976, Glutathione peroxidase activity in selenium-deficient rat liver, *Biochem. Biophys. Res. Commun.* **71**:952–958.

Muller, M., Dubiel, W., Rathmann, J., and Rapoport, S. M., 1980, Determination and characteristics of energy-dependent proteolysis in rabbit reticulocytes, *Eur. J. Biochem.* **109**:405–410.

Nohl, H., Breuninger, V., and Hegner, D., 1978, Influence of mitochondrial radical formation on energy-linked respiration, *Eur. J. Biochem.* **90**:385–390.

Rapoport, S. M., Schewe, T., Wiesner, R., Halangk, W., Ludwig, P., Janicke-Hohne, M., Tannert, C., Hiebsch, C., and Klatt, D., 1979, The lipoxygenase of reticulocytes: Purification, characterization and biological dynamics of the lipoxygenase; its identity with the respiratory inhibitors of the reticulocyte, *Eur. J. Biochem.* **96**:545–561.

Scrapa, A, and Lindsay, J. G., 1972, Maintenance of energy-linked functions in rat liver mitochondria aged in the presence of nupercaine, *Eur. J. Biochem.* **27**:401–407.

Schewe, T., Halangk, W., Hiebsch, C., and Rapoport, S. M., 1975, A lipoxygenase in rabbit reticulocytes which attacks phospholipids and intact mitochondria, *FEBS Lett.* **60**:149–152.

Schewe, T., Halangk, W., Hiebsch, C., and Rapoport, S., 1977, Degradation of mitochondria by cytosolic factors in reticulocytes, *Acta Biol. Med. Germ.* **36**:363–372.

Schnaitman, C. A., and Greenawalt, J. W., 1968, Enzymatic properties of the inner and outer membranes of rat liver mitochondria, *J. Cell Biol.* **38**:158–175.

Sies, H., and Moss, K. M., 1978, A role of mitochondrial glutathione peroxidase in modulating mitochondrial oxidations in liver, *Eur. J. Biochem.* **84**:377–383.

Speiser, S., and Etlinger, J. D., 1982, Loss of ATP-dependent proteolysis with maturation of reticulocytes and erythrocytes, *J. Biol. Chem.* **257**:14122–14127.

Thiele, B. J., Belkner, J., Andree H., Rapoport, T. A., and Rapoport, S. M., 1979, Synthesis of nonglobin proteins in rabbit-erythroid cells: Synthesis of a lipoxygenase in reticulocytes, *Eur. J. Biochem.* **96**:563–569.

Vladimirov, Y. A., Olenev, V. I., Suslova, T. B., and Cheremisina, Z. P., 1980, Lipid peroxidation in mitochondrial membrane, *Adv. Lipid Res.* **17**:173–249.

10

The Major Pathway for Leukotriene B₄ Catabolism in Human Polymorphonuclear Leukocytes Involves ω-Oxidation by a Cytochrome P-450 Enzyme

STEVEN SHAK and IRA M. GOLDSTEIN

1. INTRODUCTION

Leukotriene B_4 (LTB_4), or $5(S),12(R)$-dihydroxy-6,14-*cis*-8,10-*trans*-eicosatetraenoic acid, is a product of the 5-lipoxygenase pathway in human polymorphonuclear leukocytes (PMN) and has received a great deal of attention recently because of the role that it may play as a mediator of inflammation (Samuelsson, 1983). Leukotriene B_4 acts *in vitro* to provoke directed migration (chemotaxis) as well as enhanced random migration (chemokinesis) of human PMN (Goetzl and Pickett, 1980). It also provokes selective release of granule-associated (lysosomal) enzymes from cytochalasin-B-treated PMN (Goetzl and Pickett, 1980; Feinmark *et al.*, 1981) and causes these cells to aggregate (O'Flaherty *et al.*, 1981). *In vivo*, LTB_4 induces adherence of PMN to the walls of postcapillary venules in the hamster cheek pouch (Dahlen *et al.*, 1981), PMN-dependent increases in vascular permeability in rabbit skin (Wedmore and Williams, 1981), and exudation of PMN in human skin (Soter *et al.*, 1983).

STEVEN SHAK and IRA M. GOLDSTEIN ● Cardiovascular Research Institute and Rosalind Russell Arthritis Research Laboratory, Department of Medicine, University of California, San Francisco, California 94102, and Medical Service, San Francisco General Hospital, San Francisco, California 94110.

Results of recent studies indicate that LTB_4 can be converted by mixed human leukocytes to products that are less biologically active (Hansson *et al.*, 1981; Camp *et al.*, 1982; Jubiz *et al.*, 1982; Salmon *et al.*, 1982; Powell, 1984). One pathway by which LTB_4 can be catabolized appears to involve ω-oxidation and yields $5(S),12(R),20$-trihydroxy-6,14-*cis*-8,10-*trans*-eicosatetraenoic acid (20-OH-LTB_4) and $5(S),12(R)$-dihydroxy-6,14-*cis*-8,10-*trans*-eicosatetraen-1,20-dioic acid (20-COOH-LTB_4) (Hansson *et al.*, 1981; Powell, 1984).

Using high-performance liquid chromatography (HPLC), we have developed a sensitive, reproducible assay that permits quantification of LTB_4 and its ω-oxidation products. With this assay, we have found that human PMN (but not human monocytes, lymphocytes, or platelets) convert exogenous LTB_4 almost exclusively to 20-OH- and 20-COOH-LTB_4. We also found that the pathway for ω-oxidation in PMN is specific for LTB_4 and $5(S),12(S)$-dihydroxy-8,14-*cis*-6,10-*trans*-eicosatetraenoic acid (5,12-diHETE) and is sufficiently active to catabolize almost all of the LTB_4 that is generated when PMN are stimulated by the calcium ionophore A23187. Finally, we found that the rate of ω-oxidation of LTB_4 in human PMN is markedly reduced by carbon monoxide, a potent inhibitor of ferrous heme enzymes such as cytochrome P-450.

2. HPLC ASSAY FOR LTB_4, 20-OH-LTB_4, AND 20-COOH-LTB_4

In preparation for analysis by reverse-phase HPLC, LTB_4 and its ω-oxidation products were extracted and partially purified from cell-free supernatants by chromatography on Sep-Pak C_{18} cartridges (Waters Associates, Milford, MA) using minor modifications of the method described by Powell (1982). Individual cartridges were pretreated with 20 ml ethanol followed by 10 ml distilled water. Supernatants were diluted with water to bring the final concentration of ethanol to 10% (v/v), adjusted to pH 3.0 with HCl, and then passed through the cartridges. Cartridges were eluted sequentially with 5.0 ml 10% (v/v) ethanol, 5.0 ml chloroform : petroleum ether (35 : 65), and 4.0 ml methyl formate. The methyl formate fractions contained LTB_4, 20-OH-LTB_4, and 20-COOH-LTB_4 but were relatively free of polar phospholipids, arachidonic acid, and monohydroxy acids. Before analysis of HPLC, methyl formate was evaporated under nitrogen and replaced with 100 μl of solvent system A (see below). In experiments performed with synthetic 20-COOH-LTB_4, 20-OH-LTB_4, LTB_4 (generously provided by Dr. J. Rokach, Merck-Frosst, Montreal, Quebec), and prostaglandin B_2 (PGB_2) (Sigma Chemical Co., St. Louis, MO), recoveries after SEP-PAK chromatography were $81.0 \pm 1.8\%$, $83.9 \pm 2.5\%$, $79.1 \pm 3.1\%$, and $82.8 \pm 2.2\%$, respectively (mean $\pm$ S.D., $n = 7$). Since there were no significant differences among the recoveries of the leukotrienes, recoveries of the internal standard, PGB_2, were used to correct for losses during preparation of samples.

HPLC was performed using a Beckman Model 334 gradient system (Beckman Instruments, Palo Alto, CA), a Techsphere-Ultra-C_{18} column (5.0 μm, 5.0 mm × 25 cm) (Phenomenex/HPLC Technology, Palos Verdes Estates, CA), and a Brownlee precolumn cartridge (Aquapore RP-300, 2.1 mm × 3.0 cm) (Brownlee, Santa Clara, CA). Petroleum ether, ethanol, and methyl formate were obtained from Mallinkrodt (St. Louis, MO), Alltech Associates (Deerfield, IL), and Eastman Kodak (Rochester, NY), respectively. All other organic solvents were purchased from Burdick and Jackson (Muskegon, MI). The gradient program shown in Fig. 1 was devised to elute LTB_4 and its ω-oxidation products after the solvent front and to effectively wash and recycle the column in 45 min. Solvent systems A and B consisted of methanol : acetonitrile : water : acetic acid (35 : 25 : 45 : 0.02) and methanol : acetonitrile (75 : 25), respectively. Lipids were eluted at a rate of 1.0 ml/min with continuous monitoring for UV absorbance at 280 nm. Although some variability was noted from day to day, retention times in consecutive experiments were highly reproducible. For example, in one series of consecutive experiments, retention times for 20-COOH-LTB_4, 20-OH-LTB_4, PGB_2, and LTB_4 were 8.35 ± 0.05, 9.00 ± 0.04, 15.17 ± 0.04, and 17.06 ± 0.05 min, respectively (mean ± S.D., $n = 8$) (Fig. 1).

Integrated peak areas observed after repeated injections of a single amount (1.0–200 ng) of each lipid were highly reproducible, and there was a linear relationship between the amounts of each lipid injected and the observed peak areas

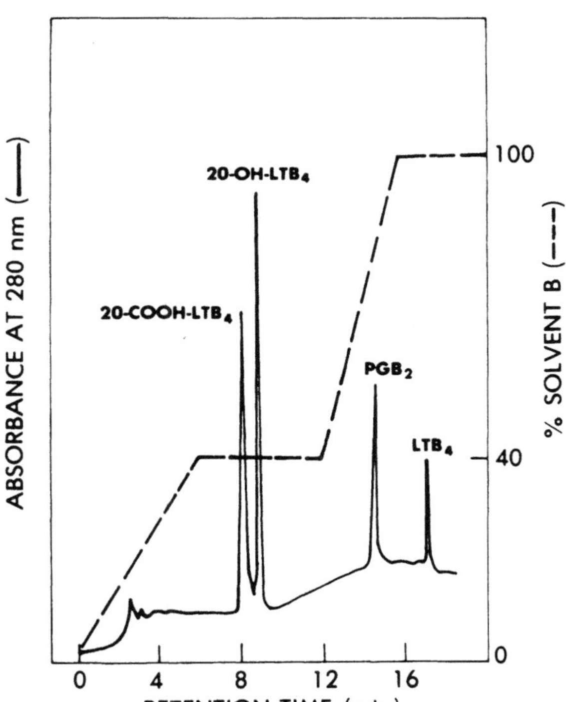

FIGURE 1. Reverse-phase HPLC assay for leukotriene B₄ and its ω-oxidation products.

(data not shown). Response factors for each lipid were calculated from the ratios of amounts to integrated peak areas. The 95% confidence limits for the response factors were determined to be ± 7.0%. Thus, integrated peak areas were used to quantify the leukotrienes.

Lipids that were isolated and purified by reverse-phase HPLC were methylated with ethereal diazomethane as described previously (Shak *et al.*, 1983) and subjected to normal-phase HPLC on a Techsphere-Ultra Silica column (5.0 μm, 5.0 mm × 25 cm). (Phenomenex/HPLC Technology) with a Brownlee precolumn cartridge (Si, 2.1 × 3.0 cm) Methyl esters were eluted isocratically at a flow rate of 2.0 ml/min with hexane : 2-propanol (100 : 12). Gas chromatography–mass spectrometry was performed as described previously (Shak *et al.*, 1983).

3. CATABOLISM OF EXOGENOUS LTB$_4$ BY HUMAN PMN

Polymorphonuclear leukocytes were isolated from venous blood of healthy adult donors by centrifugation on Hypaque–Ficoll, dextran sedimentation, and hypotonic lysis of erythrocytes (Boyum, 1968). Isolated cells (95–98% PMN, 2–4% eosinophils, <1% monocytes and lymphocytes) were suspended in Hanks' balanced salt solution (GIBCO, Grand Island, NY) containing 1.3 mM CaCl$_2$ and adjusted to pH 7.4 with NaHCO$_3$. This buffer was used throughout. Suspensions of mononuclear leukocytes (10–20% monocytes, 80–90% lymphocytes, 0–2% eosinophils, <1% PMN) also were prepared by Hypaque–Ficoll centrifugation (Boyum, 1968).

Cells were incubated with LTB$_4$ in polypropylene test tubes at 37°C with constant agitation. Incubations were terminated by the addition of 1.5 volumes of ice-cold ethanol that contained 334 ng/ml PGB$_2$ (internal standard), followed by centrifugation at 12,800 × g for 1 min. Supernatants were either analyzed immediately or stored under nitrogen at −20°C.

When LTB$_4$ (1.0 μM) was incubated at 37°C with highly purified human PMN (20 × 10^6 cells/ml), only 15.4 ± 11.6% (mean ± S.D.; range, 3.0–36%; n = 13) of the LTB$_4$ added initially was recovered intact after 15 min. In contrast, greater than 97% of LTB$_4$ was recovered intact after incubation at 37°C for 15 min in buffer alone. Disappearance of LTB$_4$ from reaction mixtures containing PMN was dependent on cell number. Recoveries of 63.8, 41.2, and 18.5% were observed after 15 min at 37°C when LTB$_4$ (1.0 μM) was incubated with 1.0, 7.0, and 20 × 10^6 PMN/ml, respectively. In addition, the disappearance of LTB$_4$ in the presence of human PMN was dependent on temperature. Greater than 95% of LTB$_4$ (1.0 μM) was recovered intact after incubation with PMN for 15 min at 0°C (n = 6).

In these experiments, two more polar peaks that retained UV absorption characteristics of conjugated trienes appeared on reverse-phase HPLC coincident with the disappearance of LTB$_4$ peaks (Fig. 2). These new peaks, which eluted with retention times of 8.1 and 8.9 min, were observed after 5 min of incubation. As

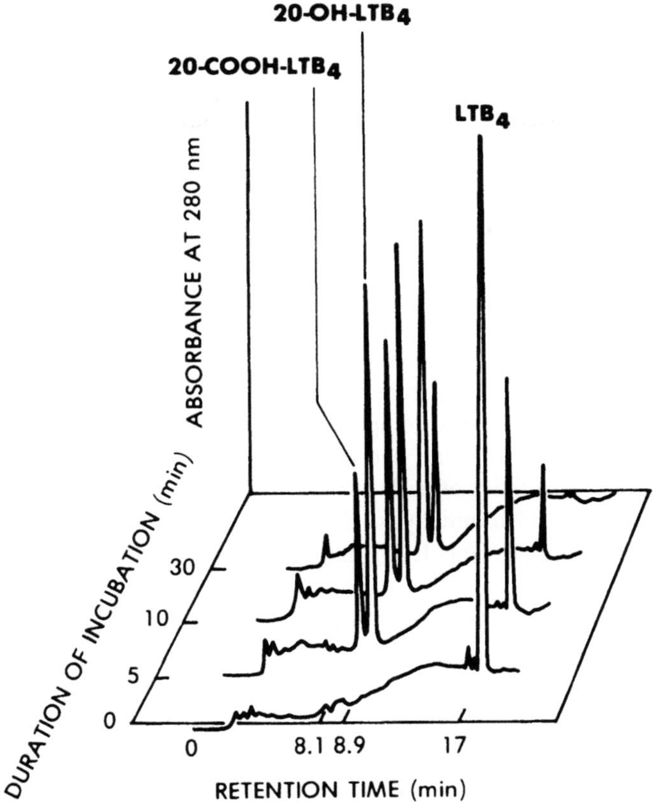

FIGURE 2. HPLC analysis of the catabolism of LTB₄ by human PMN. Shown are representative reverse-phase HPLC chromatograms obtained after incubating LTB₄ (1.0 μM) with PMN (20 × 10⁶/ml) at 37°C for 0, 5, 10, and 30 min.

the duration of incubation increased, the peak with a retention time of 8.9 min diminished in area, whereas the peak with a retention time of 8.1 min increased in area.

The products that eluted with retention times of 8.1 and 8.9 min were identified in three ways as 20-COOH-LTB₄ and 20-OH-LTB₄ respectively. First, the material in the 8.1- and 8.9-min peaks cochromatographed on reverse-phase HPLC with synthetic 20-COOH-LTB₄ and 20-OH-LTB₄. Second, when the material in each peak was collected separately, methylated, and subjected both to normal-phase and to reverse-phase HPLC, retention times were identical to those observed when methylated synthetic 20-COOH-LTB₄ and 20-OH-LTB₄ were chromatographed using both HPLC systems. Finally, the products in the 8.1- and 8.9-min peaks were identified as 20-COOH-LTB₄ and 20-OH-LTB₄ by gas chromatography–mass spectrometry.

3.1. Time Course of the Catabolism of Exogenous LTB$_4$ by Human PMN

The disappearance of LTB$_4$ (1.0μM) at 37°C from reaction mixtures containing 20 $\times$ 10^6 PMN/ml was rapid ($t_{1/2}$ of approximately 4 min). The accumulation of 20-OH-LTB$_4$ and 20-COOH-LTB$_4$ also was rapid, particularly during the first 5 min of incubation (data not shown). After 5 min of incubation, the concentration of 20-OH-LTB$_4$ exceeded the concentrations of both LTB$_4$ and 20-COOH-LTB$_4$. After 10 min of incubation, however, the concentration of 20-COOH-LTB$_4$ increased further, and the concentrations of 20-OH-LTB$_4$ and LTB$_4$ decreased proportionately. Greater than 90% of the LTB$_4$ added initially could be accounted for during the first 10 min of incubation.

Results of experiments using trace amounts of [^{3}H]LTB$_4$ (Amersham, Arlington Heights, IL) indicated that this compound was converted by human leukocytes almost exclusively to [^{3}H]20-OH- and [^{3}H]20-COOH-LTB$_4$ (data not shown). In addition, the specific activity of each of the ω-oxidation products was identical to that of the [^{3}H]LTB$_4$ added initially. ω-Oxidation, therefore, is the major pathway for the catabolism of exogenous LTB$_4$ in human PMN.

3.2. Comparison of PMN with Other Cell Types for the Ability to Catabolize Exogenous LTB$_4$

Suspensions of human PMN prepared by Hypaque–Ficoll centrifugation and dextran sedimentation are contaminated with small numbers of mononuclear leukocytes and larger numbers of platelets. To determine whether these other blood cell types generate ω-oxidation products from LTB$_4$, paired experiments were performed using PMN and mononuclear leukocytes from the same donor (Table I). Suspensions of mononuclear leukocytes containing large numbers of platelets converted only small amounts of LTB$_4$ to 20-OH- and 20-COOH-LTB$_4$ (very likely attributable to contaminating PMN). Approximately 12-fold greater amounts of ω-oxidation products were generated after 15 min in PMN-enriched suspensions. Although none of the cell types that we examined converted 20-OH- and 20-COOH-LTB$_4$ very rapidly to other compounds, suspensions of mononuclear leukocytes and platelets did convert significant amounts of LTB$_4$ to unidentified polar products. Nevertheless, it appears that among human peripheral blood cells, only PMN rapidly and efficiently catabolize LTB$_4$ by ω-oxidation.

3.3 Specificity of ω-Oxidation by Human PMN

The specificity of ω-oxidation was examined directly by incubating various dihydroxylated derivatives of arachidonic acid with human PMN. Leukotriene A$_4$ methyl ester (generously provided by Dr. J. Rokach) was hydrolyzed to generate all-*trans* conjugated isomers of LTB$_4$ (6-*trans*-LTB$_4$ and 12-epi-6-*trans*-LTB$_4$), which were purified by HPLC as described previously (Shak *et al.*, 1983). Soybean

TABLE I. Catabolism of LTB$_4$ by PMN, Mononuclear Leukocytes, and Platelets

	LTB$_4$[a]	20-OH $-$ + 20-COOH-LTB$_4$[a]
PMN-enriched suspension[b]		
0 min incubation	100	0
15 min incubation	7 ± 5	81 ± 8
Mononuclear leukocyte/platelet-enriched suspension[c]		
0 min incubation	100	0
15 min incubation	71 ± 19	7 ± 5

[a] Leukotriene B$_4$ (1.0 μM) was incubated at 37°C with paired cell suspensions prepared from the same donor. Results are expressed as a percentage of the initial concentration of LTB$_4$ added (mean ± S.D., n = 4).

[b] The PMN-enriched suspensions contained (per ml) 20 × 10^6 PMN, 6–8 × 10^4 eosinophils, <1 × 10^4 mononuclear leukocytes, and 20–80 × 10^6 platelets.

[c] Mononuclear leukocyte/platelet-enriched suspensions contained (per ml) 80–100 × 10^6 lymphocytes, 20 × 10^6 monocytes, <1 × 10^6 PMN, <2 × 10^6 eosinophils, and >1 × 10^{10} platelets.

lipoxygenase (Sigma) was used to generate 5(S),15(S)-dihydroxy-8,11-*cis*-6,13-*trans*-eicosatetraenoic acid (5,15-diHETE) and 8(S),15(S)-dihydroxy-5,11-*cis*-9,13-*trans*-eicosatetraenoic acid (8,15-diHETE) from arachidonic acid (Shak *et al.*, 1983). The doubly oxygenated compound 5,12-diHETE was prepared by adding A23187 (10μM) (Sigma) to suspensions of human PMN containing large numbers of platelets (Lindgren *et al.*, 1981; Borgeat *et al.*, 1981).

Whereas PMN converted large amounts of LTB$_4$ and 5,12-diHETE to ω-oxidation products, there was no catabolism by these cells of the 15-lipoxygenase products, 8,15-diHETE and 5,15-diHETE. The two all-*trans* conjugated isomers of LTB$_4$ underwent minimal ω-oxidation.

3.4. Catabolism of Exogenous LTB$_4$ by Human PMN Stimulated with Phorbol Myristate Acetate

Evidence has appeared recently suggesting that leukotrienes can be catabolized by human PMN as a consequence of the generation by these cells of oxygen-derived free radicals and H$_2$O$_2$ (Henderson *et al.*, 1982; Henderson and Klebanoff, 1983a,b). Henderson and Klebanoff (1983b), for example, found that more LTB$_4$ and leukotriene C$_4$ were recovered from supernatants of chronic granulomatous disease PMN (which are unable to generate superoxide anion radicals and H$_2$O$_2$) (Babior, 1978) than from supernatants of normal PMN after stimulation by A23187. Results of our studies suggest that oxygen-derived free radicals (e.g., superoxide) and/or H$_2$O$_2$ play only a minor role, if any, in the catabolism of LTB$_4$ by PMN.

To determine whether stimulation of PMN oxidative metabolism influences the catabolism of LTB$_4$, we compared "resting" PMN with PMN stimulated by phorbol myristate acetate (PMA) (Sigma) for their ability to convert exogenous LTB$_4$ to 20-OH- and 20-COOH-LTB$_4$ (Table II). The PMA (500 ng/ml) did not stimulate PMN to produce either LTB$_4$ or its ω-oxidation products but did cause

TABLE II. Catabolism of LTB_4 by PMN Stimulated with PMA

PMA (ng/ml)	20-COOH-LTB_4[a]	20-OH-LTB_4[a]	LTB_4[a]
0	37 ± 9	32 ± 13	15 ± 9
500	9 ± 4	26 ± 2	39 ± 8
	$P < 0.01$[b]	$P > 0.2$[b]	$P < 0.005$[b]

[a] Human PMN (20×10^6 cells/ml) were incubated with LTB_4 (1.0 μM) in the presence and absence of PMA. Results are expressed as a percentage of the initial concentration of LTB_4 (mean ± S.D.; $n = 3$).

[b] A paired Student's t test was used to compare results obtained in the presence and absence of PMA.

these cells to generate superoxide anion radicals (20–30 nmol cytochrome c reduced/10^6 PMN per 5 min) (measured as superoxide-dismutase-inhibitable ferricytochrome c reduction) (Goldstein *et al.*, 1975). Nevertheless, like "resting" PMN, PMA-stimulated PMN converted exogenous LTB_4 primarily to 20-OH- and 20-COOH-LTB_4. Interestingly, PMN stimulated by PMA converted less LTB_4 to its ω-oxidation products and less 20-OH-LTB_4 to 20-COOH-LTB_4 than did "resting" PMN. Thus, even in PMN stimulated to produce abundant amounts of superoxide, ω-oxidation is the major pathway for LTB_4 catabolism. This is not surprising, since oxidants produced by stimulated PMN might react with LTB_4 at or near sites of the double bonds (by abstraction of allylic hydrogen atoms or by direct addition) (Foote, 1968) but would not be expected to react with the relatively inert terminal methyl group to yield 20-OH- or 20-COOH- products.

4. LEUKOTRIENE B_4 GENERATION BY HUMAN PMN STIMULATED WITH THE CALCIUM IONOPHORE A23187

Having confirmed the findings of others (Hansson *et al.*, 1981; Camp *et al.*, 1982) that PMN stimulated by the calcium ionophore A23187 generate 20-OH-LTB_4 and 20-COOH-LTB_4, we also found that the pathway for ω-oxidation is sufficiently active to catabolize almost all of the LTB_4 that is generated by A23187-stimulated PMN. Although there was some variability from experiment to experiment, we consistently found that PMN incubated with A23187 (0.1–20 μM) for 60 min generated much larger amounts of 20-OH- and 20-COOH-LTB_4 than of LTB_4 itself (Fig. 3).

Results of a representative experiment demonstrating the time course of LTB_4 generation and ω-oxidation by human PMN stimulated by A23187 are shown in Fig. 4. After 60 min of incubation, much larger amounts of 20-OH- and 20-COOH-LTB_4 than of LTB_4 were detected in the medium surrounding PMN stimulated by 10 μM A23187. In fact, only at the earliest time point sampled (i.e., 2 min) was more LTB_4 detected than the total of its ω-oxidation products. Concentrations of LTB_4 reached a maximum after 5 min of incubation and then declined. In contrast,

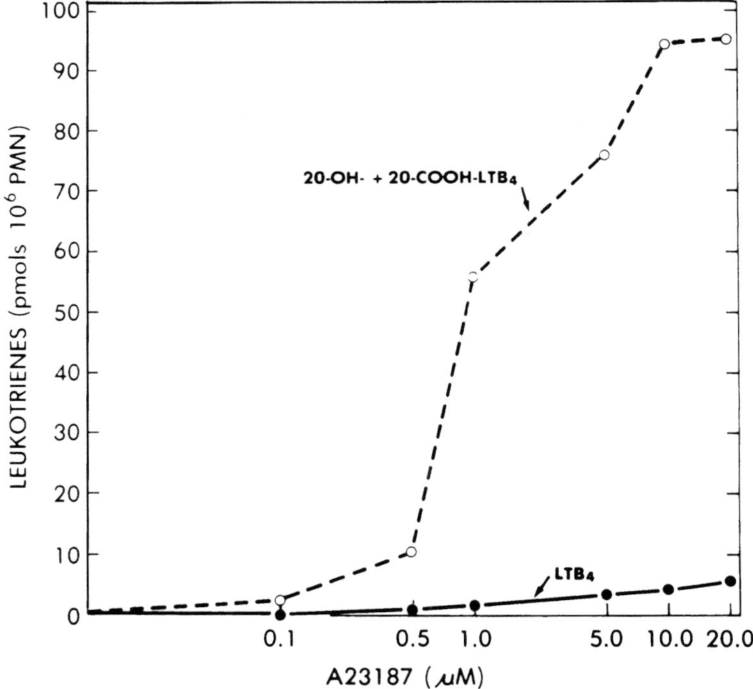

FIGURE 3. Generation of LTB₄ and its ω-oxidation products by human PMN stimulated with different concentrations of the calcium ionophore A23187. PMN (20 × 10⁶ cells/ml) were incubated at 37°C for 60 min with A23187, and the quantities of LTB₄ and total ω-oxidation products were determined.

concentrations of the ω-oxidation products continued to increase. When PMN were stimulated by a low concentration of A23187 (0.5 μM), only negligible amounts of LTB₄ were detected after 10 min of incubation (0.15 pmol/10⁶ PMN). In contrast, much larger amounts of 20-OH- and 20-COOH-LTB₄ accumulated in the reaction mixtures gradually over 60 min (data not shown). Thus, almost all of the LTB₄ generated by PMN in response to stimulation with A23187 can be catabolized by ω-oxidation.

Jubiz *et al.* (1982) reported that PMN stimulated by the synthetic chemotactic peptide N-formylmethionylleucylphenylalanine (FMLP) produced primarily small amounts of 20-COOH-LTB₄. They suggested that FMLP "more specifically" stimulates the biosynthesis of 20-COOH-LTB₄ by PMN than the biosynthesis of LTB₄. Results of our studies, however, suggest that it is more likely that FMLP stimulates LTB₄ synthesis in human PMN only weakly (resembling 0.5 μM A23187) and that the small amount of LTB₄ produced by the cells is converted rapidly and almost completely to its ω-oxidation products.

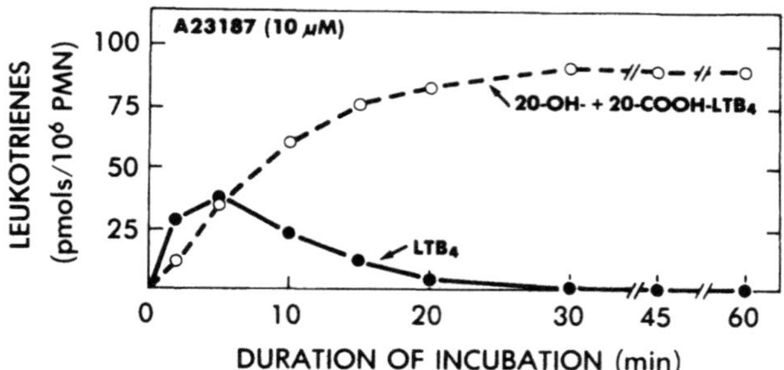

FIGURE 4. Time course of the generation of LTB$_4$ and its ω-oxidation products by stimulated PMN. Shown are the results of a representative experiment in which PMN (20 $\times$ 10^6 cells/ml) were incubated at 37°C with 10 μM A23187.

5. INHIBITION OF ω-OXIDATION OF LTB$_4$ BY CARBON MONOXIDE

ω-Oxidation of stable prostaglandins in liver, adrenal, lung, and kidney is mediated by cytochrome P-450 enzymes and is inhibited by carbon monoxide (Kupfer, 1980). To determine whether catabolism of LTB$_4$ by human PMN also is mediated by a cytochrome P-450 enzyme, experiments were performed to examine the effect of carbon monoxide on this process. The PMN were exposed to carbon monoxide using gas-saturated buffers (Omura and Sato, 1964). Buffer was first degassed under vacuum and then equilibrated with either carbon monoxide, nitrogen, or air for 30 min. The effect of carbon monoxide on ω-oxidation of LTB$_4$ was then examined by suspending PMN in desired mixtures of carbon-monoxide-saturated and air-saturated buffers. Control experiments were performed with PMN suspended in mixtures of nitrogen-saturated and air-saturated buffers.

ω-Oxidation of LTB$_4$ (1.0μM) by PMN (20 $\times$ 10^6 cells/ml) exposed to carbon monoxide (carbon monoxide : oxygen ratio = 11) was inhibited significantly as compared to that observed in experiments with PMN exposed to nitrogen and air (Fig. 5). Inhibition varied with the ratio of carbon monoxide to oxygen and was completely reversible (data not shown). In addition, the amount of LTB$_4$ detected in the medium surrounding PMN that had been stimulated with A23187 (10μM) varied depending on whether the buffer was equilibrated with carbon monoxide or air. In contrast to the results shown in Fig. 4, large amounts of LTB$_4$ accumulated in the medium during 60 min of incubation of PMN with A23187 in buffer containing carbon monoxide, and only small amounts of 20-OH- and 20-COOH-LTB$_4$ were detected (data not shown). Reexposure of the reaction mixtures to oxygen after 60 min of incubation resulted in the rapid disappearance of LTB$_4$ and the accumulation of large amounts of 20-OH- and 20-COOH-LTB$_4$. Thus, carbon monoxide reversibly inhibits ω-oxidation by PMN of both exogenous LTB$_4$ and LTB$_4$ generated after exposure of cells to A23187. ω-Oxidation of LTB$_4$ in PMN, therefore, resembles

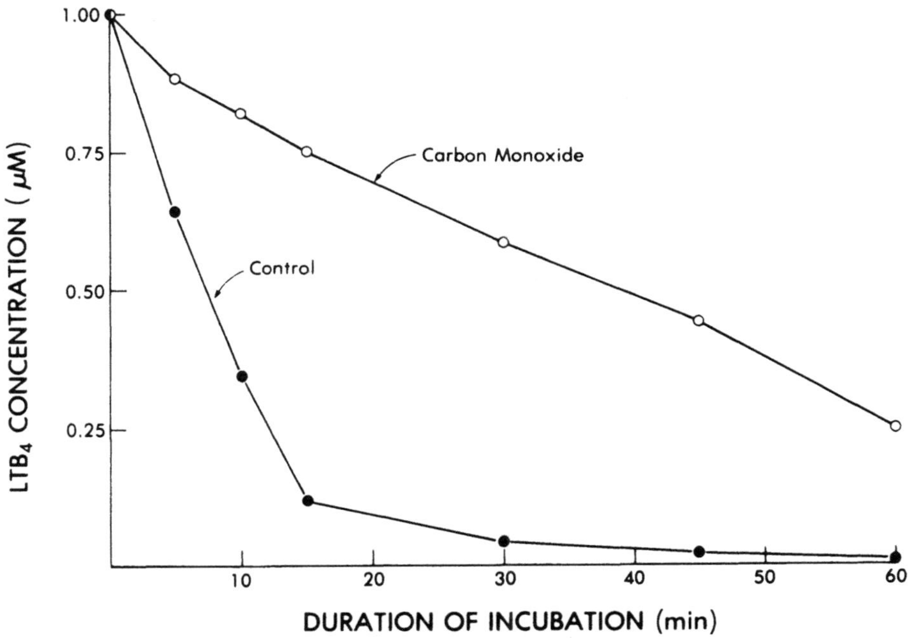

FIGURE 5. Inhibition of LTB₄ ω-oxidation by carbon monoxide.

ω-oxidation of prostaglandins in other cell types and probably is mediated by a cytochrome P-450 enzyme.

6. CONCLUSIONS

Polymorphonuclear leukocytes are the most prominent cell type observed at sites of acute inflammation. The precise role that LTB₄ plays in provoking inflammation has not been determined. Results of our studies indicate that human PMN not only generate and respond to LTB₄ but also are capable of catabolizing this mediator. They do so rapidly, specifically, and primarily by ω-oxidation (mediated by a cytochrome P-450 enzyme). Although generation of LTB₄ by PMN might promote inflammation, catabolism of LTB₄ by PMN may be a mechanism whereby inflammatory reactions are modulated.

ACKNOWEDGMENTS. This work was supported by grants from the National Institutes of Health (HL-28475, AI-14752, and HL-19155), the Treadwell Foundation, and the Strobel Medical Research Fund of the American Lung Association of San Francisco. Dr. Shak was supported by a National Pulmonary Faculty Training Center Grant (HL-07159) from the National Institutes of Health.

REFERENCES

Babior, B. M., 1978, Oxygen-dependent microbial killing by phagocytes, *N. Engl. J. Med.* **298**:659–668.

Borgeat, P., Picard, S., Vallerand, P., and Sirois, P. 1981, Transformation of arachidonic acid in leukocytes. Isolation and structural analysis of a novel dihydroxy derivative, *Prostaglandins Med.* **6**:557–570.

Boyum, A., 1968, Isolation of mononuclear cells and granulocytes from human blood. Isolation of monocuclear cells by one centrifugation and of granulocytes of combining centrifugation and sedimentation at 1 *g*, *Scand. J. Clin. Lab. Invest.* **21**(Suppl. 97):77–89.

Camp, R. D. R., Woollard, P. M., Mallet, A. I., Fincham, N. J., Ford-Hutchinson, A. W., and Bray, M. A., 1982, Neutrophil aggregatory and chemotactic properties of a 5,12,20-trihydroxy-eicosatetraenoic acid isolated from human leukocytes, *Prostaglandins* **23**:631–641.

Dahlén, S.-E., Bjork, J., Hedqvist, P., Arfors, K.-E., Hammarström, S., Lindgren, J. A., and Samuelsson, B., 1981, Leukotrienes promote plasma leakage and leukocyte adhesion in postcapillary venules: *In vivo* effect with relevance to the acute inflammatory response, *Proc. Natl. Acad. Sci. U.S.A.* **78**:3887–3891.

Feinmark, S. J., Lindgren, J. A., Claesson, H.-E., Malmsten, C., and Samuelsson, B., 1981, Stimulation of human leukocyte degranulation by leukotriene B_4 and its ω-oxidation metabolites, *FEBS Lett.* **136**:141–144.

Foote, C. S., 1968, Photosensitized oxygenations and the role of singlet oxygen, *Accounts Chem. Res.* **1**:104–110.

Goetzl, E. J., and Pickett, W. C., 1980, The human PMN leukocyte chemotactic activities of complex hydroxy-eicosatetraenoic acids (HETEs), *J. Immunol.* **125**:1789–1791.

Goldstein, I. M., Roos, D., Kaplan, H. B., and Weissmann, G., 1975, Complement and immunoglobulins stimulate superoxide production by human leukocytes independent of phagocytosis, *J. Clin Invest.* **56**:1155–1163.

Hansson, G., Lindgren, J. A., Dahlen, S.-E., Hedqvist, P., and Samuelsson, B., 1981, Identification and biological activities of novel ω-oxidized metabolities of leukotriene B_4 from human leukocytes, *FEBS Lett.* **130**:107–112.

Henderson, W. R., and Klebanoff, S. J., 1983a, Leukotriene B_4, C_4, D_4, and E_4 inactivation by hydroxyl radicals, *Biochem. Biophys. Res. Commun.* **110**: 266–272.

Henderson, W. R., and Klebanoff, S. J., 1983b, Leukotriene production and inactivation by normal, chronic granulomatous disease and myeloperoxidase-deficient neutrophils, *J. Biol. Chem.* **258**:13522–13527.

Henderson, W. R., Jorg, A., and Klebanoff, S. J., 1982, Eosinophil peroxidase-mediated inactivation of leukotrienes B_4, C_4, and D_4, *J. Immunol.* **128**:2609–2613.

Jubiz, W., Radmark, O., Malmsten, C., Hansson, G., Lindgren, J. A., Palmblad, J., Uden, A.-M., and Samuelsson, B., 1982, A novel leukotriene produced by stimulation of leukocytes with formylmethionylleucylphenylalanine, *J. Biol. Chem.* **257**:6106–6110.

Kupfer, D., 1980, Endogenous substrates of monooxygenases: Fatty acids and prostaglandins, *Pharmacol. Ther.* **11**:469–496.

Lindgren, J. A., Hansson, G., and Samuelsson, B., 1981, Formation of novel hydroxylated eicosatetraenoic acid metabolites in preparations of human polymorphonuclear leukocytes, *FEBS Lett.* **128**:329–335.

O'Flaherty, J. T., Hammett, M. J., Shewmake, T. B., Wykle, R. L., Love, S. H., McCall, C. E., and Thomas, M. J., 1981, Evidence for 5,12-dihydroxy-6,8,10,14-eicosatetraenoate as a mediator of human neutrophil aggregation, *Biochem. Biophys. Res. Commun.* **103**:552–558.

Omura, T., and Sato, R., 1964, The carbon monoxide-binding pigment of liver microsomes. I. Evidence for its hemoprotein nature, *J. Biol. Chem.* **239**:2370–2378.

Powell, W. S., 1982, Rapid extraction of arachidonic acid metabolites from biological samples using octadecylsilyl silica, *Methods Enzymol.* **86**:467–477.

Powell, W. S., 1984, Properties of leukotriene B_4 20-hydroxylase from polymorphonuclear leukocytes, *J. Biol. Chem.* **259**:3082–3089.

Salmon, J. A., Simmons, P. M., and Palmer, R. M. J., 1982, Synthesis and metabolism of leukotriene B_4 in human neutrophils measured by specific radioimmunoassay, *FEBS Lett.* **146**:18–22.

Samuelsson, B., 1983, Leukotrienes: Mediators of immediate hypersensitivity reactions and inflammation, *Science* **220**:568–575.

Shak, S., Perez, H. D., and Goldstein, I M., 1983, A novel dioxygenation product of arachidonic acid possesses potent chemotactic activity for human polymorphonuclear leukocytes, *J. Biol. Chem.* **258**:14948–14953.

Soter, N. A., Lewis, R. A., Corey, E. J., and Austen, K. F., 1983, Local effects of synthetic leukotrienes (LTC_4, LTD_4, LTE_4, and LTB_4) in human skin, *J. Invest. Dermatol.* **80**:115–119.

Wedmore, C. V., and Williams, T. J., 1981, Control of vascular permeability by polymorphonuclear leukocytes in inflammation, *Nature* **289**:646–650.

11

Hematin-Assisted Intramolecular Oxygen Transfer Mechanism Is Involved in the Formation of 8-Hydroxy-11,12-epoxyeicosa-5,9,14-trienoic Acid (8H-11, 12-EPETE) from 12-HPETE

C. R. PACE-ASCIAK

1. INTRODUCTION

We recently described the isolation and structure of two hydroxyepoxides formed from 12-HPETE by an enzyme system present in rat lung (Pace-Asciak *et al.*, 1983a). These were shown to be 8-hydroxy-11,12-epoxy (8H-11,12-EPETE) and 10-hydroxy-11,12-epoxy (10H-11,12-EPETE) eicosatrienoic acid. Evidence was presented to show that both oxygen atoms in the hydroxyepoxides were derived from molecular oxygen (Pace-Asciak *et al.*, 1983a). Also, when [^{18}O]oxygenated 12-HPETE was incubated with this enzyme preparation, ^{18}O atoms were found in both the hydroxyl and the epoxide groups, suggesting that the hydroxyl group at carbon 8 and carbon 10 was derived from the hydroperoxide of 12-HPETE (Pace-Asciak *et al.*, 1983a). However, whether the terminal hydroxyl group of the hydroperoxide of 12-HPETE was transferred via an inter- or intramolecular mechanism was not determined. The present chapter reports evidence with a mixture of [^{16}O]- and [^{18}O]-labeled 12-HPETE to show a unique intramolecular rearrangement

C. R. PACE-ASCIAK • Research Institute, The Hospital for Sick Children, Toronto, Ontario, and Department of Pharmacology, University of Toronto, Toronto, Ontario M5G 1X8, Canada.

of 12-HPETE into the hydroxyepoxides catalyzed by bovine hematin in a protein-free environment as well as by a rat lung cytosol fraction.

2. MATERIALS AND METHODS

2.1. Preparation of $[12\text{-}^{18}O^{18}O]$- and $[12\text{-}^{16}O^{16}O]\text{-}[^2H_8]HPETE$

$[^2H_8]12$-HPETE containing $^{18}O^{18}O$ or $^{16}O^{16}O$ was prepared as recently reported from rat platelets except that octadeuterated arachidonic acid was used as substrate (Pace-Asciak *et al.*, 1983a). The $[^{18}O]$- and $[^{16}O]$-labeled HPETE were purified on HPLC using hexane/isopropanol/acetic acid (98/2/0.1 v/v) and two Resolve columns (Waters) connected in series. The HPETEs were stable to storage in distilled diethyl ether at $-20°C$ for several months. A mixture of $[12\text{-}^{18}O^{18}O]$- and $[12\text{-}^{16}O^{16}O]\text{-}[^2H_8]HPETE$ was made up in a ratio of 4 : 5, and an aliquot was assayed by GC–MS after reduction with triphenyl phosphine and conversion into the methyl ester and trimethylsilyl ether.

2.2. Incubation of $[12\text{-}^{18}O^{18}O]$- and $[12\text{-}^{16}O^{16}O]\text{-}[^2H_8]HPETE$ (4 : 5)

2.2.1. Rat Lung Enzyme

The above oxygen- and deuterium-labeled 12-HPETE mixture was incubated for 20 min at 37° with 3 ml of a 0–30% ammonium sulfate fraction in phosphate buffer of a rat lung high-speed supernatant prepared as described previously (Pace-Asciak *et al.*, 1982, 1983a,b). The incubation was terminated by the addition of ice-cold methanol (3 ml) and ice-cold diethyl ether (12 ml). The contents were acidified to pH 3 with 0.5 N HCl and centrifuged at 2°C. The organic layer was removed, washed with water, centrifuged again, and transferred. The diethyl ether extract was taken to complete dryness *in vacuo*, redissolved in a small volume of ether, and subjected to derivatization for GC–MS analysis.

2.2.2. Hematin

The labeled 12-HPETE mixture (see above) was incubated for 20 min at 37°C in 1 ml phosphate buffer, pH 7.4, containing an equal amount of bovine hematin (Sigma). Workup of the incubation mixture was carried out as in Section 2.2.1 above.

2.2.3. Ferric and Ferrous Chloride

These were used in separate experiments at a concentration of 6×10^{-6} M in incubations as above but containing $[^{14}C]$-labeled 12-HPETE instead of the oxygen- and deuterium-labeled 12-HPETE. Analysis of the incubation products was achieved by HPLC as above, and radioactive products were monitored through an on-line Berthold radioactivity detector.

3. RESULTS AND DISCUSSION

On incubation of 12-HPETE with the 0–30% ammonium sulfate fraction of rat lung high-speed supernatant, several products are obtained. These are 12-HETE and two isomeric hydroxyepoxides, i.e., 8H-11,12-EPETE and 10H-11,12-EPETE (Pace-Asciak *et al.*, 1983a). It is possible to isolate these hydroxyepoxides because the 0–30% ammonium sulfate fraction lacks to a large extent epoxide hydratase, which hydrolyzes the hydroxyepoxides to triols (Pace-Asciak *et al.*, 1982, 1983a). In the present experiments, a mixture of [12-$^{18}O^{18}O$]- and [12-$^{16}O^{16}O$]octadeuterated HPETE (4 : 5) was incubated with the 0–30% ammonium sulfate fraction of rat lung and after extraction with diethyl ether and derivatization into the pentafluorobenzyl (PFB) ester trimethylsilyl (TMSi) ether, the products were analyzed by capillary column gas chromatography–mass spectrometry using negative ion chemical ionization detection (NICI). The use of NICI in the study was important because this type of GC–MS provides mass spectra for these compounds dominated by fragments (M-PFB) containing all oxygen and deuterium atoms of the substrate 12-HPETE (Pace-Asciak, 1984a).

The results for the extracted 12-HETE, 8H-11,12-EPETE, and 10H-11,12-EPETE are shown in Fig. 1A with hematin as catalyst and in Fig. 1B for the rat lung enzyme. The use of octadeuterated substrate was essential to distinguish products derived from the substrate 12-HPETE from products derived from endogenous precursors in the rat lung enzyme experiments. It is clear from these data that the ratio of $^{18}O : ^{16}O$ of the substrate 12-HPETE is recovered in 12-HETE as well as in both hydroxyepoxides. Furthermore, the latter contain either two ^{18}O atoms or two ^{16}O atoms with negligible amounts of the mixed species, i.e., $^{18}O^{16}O$. In both experiments, i.e., hematin and rat lung, some hydrolysis of 8H-11,12-EPETE took place as observed by the presence of the trihydroxy derivative 8,11,12-THETE. The present data show that this product contained two ^{18}O or two ^{16}O atoms, as did its precursor, 8H-11,12-EPETE; the third oxygen atom contains ^{16}O (derived from water), confirming our previous observations (Pace-Asciak *et al.*, 1983a). These experiments demonstrate that the terminal hydroxyl group of the hydroperoxide of 12-HPETE is transferred by the action of both hematin and the rat lung enzyme (1) to the C-8 alkyl position with double bond migration from $\Delta^8(cis)$ to $\Delta^9(trans;$ unpublished observations) and epoxide formation at C-11,12; or (2) to the C-10 alkyl position with epoxide formation at C-11,12.

In order to gain some information on the role of iron and possibly the porphyrin group on this unique intramolecular rearrangement, the effect of $FeCl_2$ and $FeCl_3$ was investigated. Figure 2 shows HPLC patterns comparing the effect of hematin and both iron chlorides on the production of the hydroxyepoxides. It is clear from

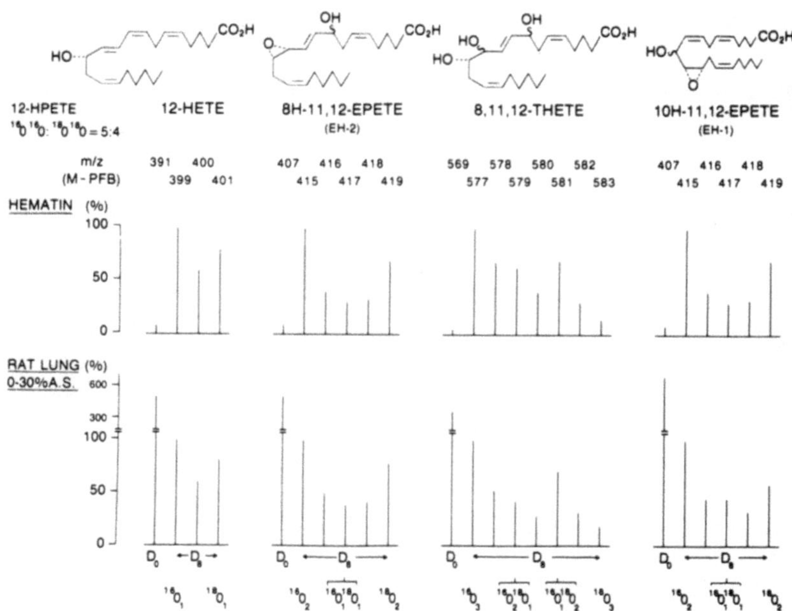

FIGURE 1. Results of GC–MS–NICI analyses of products resulting from incubation in phosphate buffer of a mixture of [12-$^{16}O^{16}O$]- and [12-$^{18}O^{18}O$]- (5 : 4) [5,6,8,9,11,12,14,15-$^2H_8$]HPETE with (A) Hematin and (B) rat lung 0–30% ammonium sulfate fraction. Products were analyzed as PFB·TMSi derivatives using the SIM mode (see Section 2.2 for details).

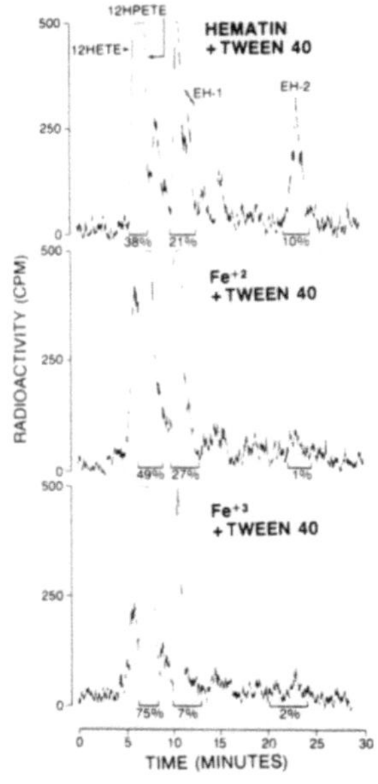

FIGURE 2. High-performance LC profiles of extracts derived from incubation of [1-^{14}C]12-HPETE in phosphate buffer containing hematin (0.6 μM), $FeCl_2$ (6 μM), and $FeCl_3$ (6 μM). EH-1 = 10H-11,12-EPETE; EH-2 = 8H-11,12-EPETE.

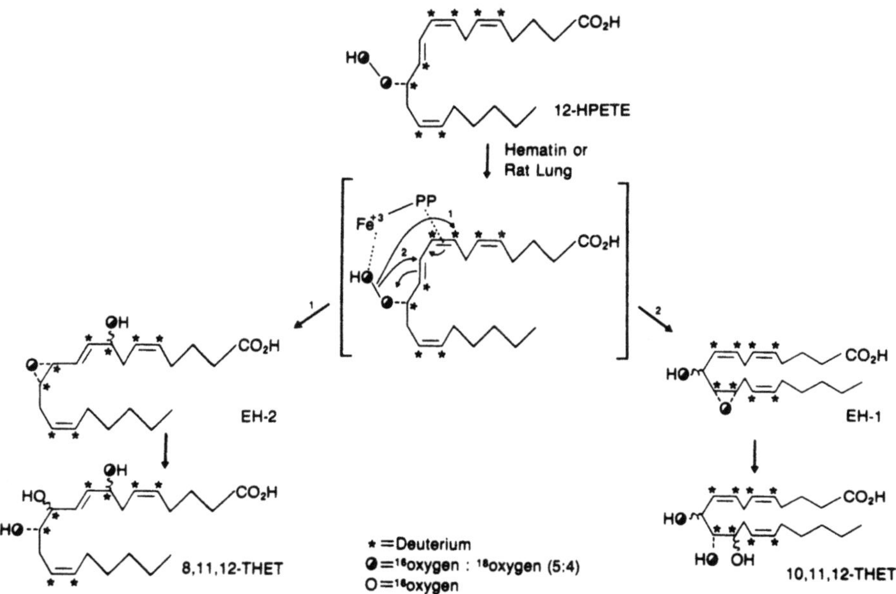

FIGURE 3. Scheme depicting a hematin-assisted intramolecular transfer of the terminal hydroxyl group from 12-HPETE to form two hydroxy epoxides as shown in this study. Reaction also proceeds with a rat lung preparation. Products shown have recently been renamed; EH-2 = hepoxilin A3; EH-1 = hepoxilin B3; 8,11,12-THET = trioxilin A3; 10,11,12-THET = trioxilin B3. (From Pace-Asciak and Martin, 1984b).

these data that 8H-11,12-EPETE was only formed with hematin as catalyst, although 10H-11,12-EPETE could be formed in all three experiments. These data reveal a requirement for the porphyrin moiety in the transfer of the OH group from 12-HPETE to the C-8 position, although its transfer to the proximal C-10 position is not as greatly affected by the absence of the porphyrin moiety.

These results favor a molecular complex in which the porphyryl moiety of hematin is associated with the $\Delta^{8,10}$ conjugated diene system of 12-HPETE possibly aligning the diene through some form of van der Waals forces for intramolecular transfer of the hydroxyl group from the hydroxperoxide (Fig. 3). Molecular complexes between porphyrins and aromatic compounds have previously been described (Hill *et al.*, 1973; Konishi *et al.*, 1980), making possible a four-way interaction in a cage system composed of iron–hydroperoxide–diene–porphyrin (see scheme in Fig. 3). Our recent observation of a similar transformation by hemoglobin indicates that the protein part of hemoglobin does not interfere with or participate in this intramolecular rearrangement of 12-HPETE (Pace-Asciak, 1984b).

Since 12-HPETE is abundantly formed by activated platelets, formation of the hydroxyepoxides studied in this chapter could be envisaged *in vivo* through hemoglobin catalysis. In view of our recent demonstration that 8H-11,12-EPETE is quite potent in enhancing glucose-stimulated insulin release in isolated pancreatic islets (Pace-Asciak and Martin, 1984a), formation of these hydroxyepoxides could have profound physiological or pathophysiological significance. Similar transformations in the 15-HPETE series (Sok *et al.*, 1983) and the C18 series (Dix and Marnett, 1983; Garssen *et al.*, 1976; Hamberg, 1975) have recently been reported, making catalysis of fatty hydroperoxides by hemoglobin and related substances a general phenomenon.

ACKNOWLEDGMENTS. This study was supported by a grant (MT-4181) from the Medical Research Council of Canada. The author thanks Mr. I. Johnson and Dr. L. Diosady for the use of the mass spectrometer facility.

REFERENCES

Dix, T. A., and Marnett, L. J., 1983, Hematin-catalyzed rearrangement of hydroperoxylinoleic acid to epoxy alcohols via an oxygen rebound, *J. Am. Chem. Soc.* **105**:7001–7002.

Garssen, G. J., Veldink, G. A., Vliegenthart, J. F. G., and Boldingh, J., 1976, The formation of *threo*-11-hydroxy-*trans*-12:13-epoxy-9-*cis*-octadecenoic acid by enzymic isomerisation of 13-L-hydroperoxy-9-*cis*-11-*trans*-octadecadienoic acid by soybean lipoxygenase-1, *Eur. J. Biochem.* **62**:33–36.

Hamberg, M., 1975, Decomposition of unsaturated fatty acid hydroperoxides by hemoglobin: Structures of major products of 13L-hydroperoxy-9,11-octadecadienoic acid, *Lipids* **10**:87–92.

Hill, H. A. O., Salder, P. J., and Williams, R. J. P., 1973, The effect of 1,3,5-trinitrobenzene on ^{1}H nuclear magnetic resonance and electron paramagnetic resonance spectra of some cobalt (II) porphyrins, *J. Chem. Soc. Dalton Trans.* 1663–1667.

Konishi, S., Hoshino, M., and Imamura, M., 1980, Formation of the charge-transfer and constrained complexes of cobalt (II) tetraphenylporphyrin in rigid solution, *J. Phys. Chem.* **84**:3437–3440.

Pace-Asciak, C. R., 1984a, Arachidonic acid epoxides: Demonstration through ^{18}oxygen studies of an *intra*molecular transfer of the terminal hydroxyl group of 12(*S*)-hydroperoxyeicosa-5,8,10,14-tetraenoic acid to form hydroxy epoxides, *J. Biol. Chem.* **259**:8332–8337.

Pace-Asciak, C. R., 1984b, Hemoglobin- and hemin-catalyzed transformation of 12L-hydroperoxy-5,8,10,14-eicosatetraenoic acid, *Biochim. Biophys. Acta* **793**:485–488.

Pace-Asciak, C. R., and Martin, J., 1984a, 8-Hydroxy-11,12-epoxyeicosatrienoic acid (8H-11,12-EPETE) enhances insulin secretion by isolated perfused pancreatic islets, in: *Proceedings of the IVth International Washington Spring Symposium on Prostaglandins and Leukotrienes*, Academic Press, Orlando, FL, Abstr. 88.

Pace-Asciak, C. R., and Martin, J., 1984b, Hepoxilin, a new family of insulin secretagogues formed by intact rat pancreatic islets, *J. Prostagl. Leuk. Med.* **16**:173–180.

Pace-Asciak, C. R., Mizuno, K., and Yamamoto, S., 1982, The enzymatic conversion of arachidonic acid into 8,11,12-trihydroxyeicosatrienoic acid: Resolution of rat lung enzyme into two active fractions, *Biochim. Biophys. Acta* **712**:142–145.

Pace-Asciak, C. R., Granström, E., and Samuelsson, B., 1983a, Arachidonic acid epoxides: Isolation and structure of two hydroxy epoxide intermediates in the formation of 8,11,12- and 10,11,12-trihydroxyeicosatrienoic acids, *J. Biol. Chem.* **258**:6835–6840.

Pace-Asciak, C. R., Mizuno, K., and Yamamoto, S., 1983b, Resolution by DEAE-cellulose chromatography of the enzymatic steps in the transformation of arachidonic acid into 8,11,12- and 10,11,12-trihydroxyeicosatrienoic acid by the rat lung, *Prostaglandins* **25**:79–84.

Sok, D. E., Chung, T., and Sih, C. J., 1983, Mechanisms of leukotriene formation: Hemoglobin-catalyzed transformation of 15-HPETE into 8,15-diHETE and 14,15-diHETE isomers, *Biochem. Biophys. Res. Commun.* **110**:273–279.

Regulatory Mechanisms in Eicosanoid Release and Metabolism

Molecular Mechanisms of the Modulation of Phospholipid Metabolism by Glucocorticoids

FUSAO HIRATA

1. INTRODUCTION

Glucocorticoids are hormones from the adrenal cortex and have diverse effects on the metabolisms of purines, fats, amino acids, and carbohydrates in a variety of tissues and organs. The actions of glucocorticoids have been well documented with respect to the capacity of organisms to resist environmental changes and noxious stimuli, especially stress (Axelrod and Reissin, 1984). These hormones are reported generally to act in concert with a variety of other hormones in a permissive manner. Many hormones, neurotransmitters, and drugs cause release of arachidonic acid, a precursor of leukotrienes and prostaglandins, from target tissues or organs, and glucocorticoids suppress such receptor-mediated release of arachidonic acid (Kuehl and Egan, 1980). Since prostaglandins and leukotrienes are now believed to be inflammatory mediators, the antiinflammatory activity, a major action, of glucocorticoids has been proposed to be associated with inhibition of this arachidonic acid release (Kuehl and Egan, 1980). In the present chapter, I describe the molecular mechanism of actions of glucocorticoids whereby these hormones block the release of arachidonic acid stimulated by hormones, neurotransmitters, and drugs.

2. PHOSPHOLIPASE INHIBITORY PROTEIN(S)

Arachidonic acid is present in an esterified form rather than as a free fatty acid. Therefore, the stimulation of receptors by various ligands appears to result

---

FUSAO HIRATA ● Laboratory of Cell Biology, National Institute of Mental Health, Bethesda, Maryland 20205.

in activation of cellular phospholipase(s), enzymes that cleave fatty acids from the glycerol moiety of phospholipids (Kuehl and Egan, 1980). When neutrophils are stimulated by synthetic chemoattractants such as fMetLeuPhe, arachidonic acid is released mainly from the phosphatidylcholine fraction, suggesting that phospholipase A_2 is activated by the stimulation of receptors for chemoattractants (Hirata *et al.*, 1979; Borman *et al.*, 1984). When these cells are treated with glucocorticoids, the f-Met-Leu-Phe-induced release of arachidonic acid is markedly suppressed (Hirata *et al.*, 1981). However, the capacity to release this fatty acid after treatment with proteases or excess Ca^{2+} ionophores is still retained by glucocorticoid-treated neutrophils. These observations suggest that the treatment of cells with glucocorticoids results in induction of the synthesis of a phospholipase inhibitory protein(s) rather than in alterations in phospholipase A_2 itself.

By conventional methods, we purified this phospholipase inhibitory protein to near homogeneity from the culture media of rabbit neutrophils treated with glucocorticoids and named it "lipomodulin" (Hirata, 1981). On sodium dodecylsulfate electrophoresis, the purified protein has a molecular weight of 40,000 (Hattori *et al.*, 1983). In addition to this species, proteins with apparent molecular weights of 30,000 and 15,000 are also found to be immunoreactive with antilipomodulin antibody and to have anti-phospholipase-A_2 activity *in vitro* (Hirata *et al.*, 1982). Thus, it is tentatively thought that the smaller peptide might be derived from the species with the $M_r = 40,000$ (Hirata *et al.*, 1982). Purified lipomodulin forms a stoichiometric complex with phospholipase A_2. The extent of phospholipid hydrolysis by porcine pancreatic phospholipase is not changed in the absence and presence of lipomodulin, but the initial rate of the phospholipase A_2 reaction is reduced by lipomodulin in a dose-dependent manner. Furthermore, lipomodulin affects the apparent K_a of phospholipase A_2 for Ca^{2+}, an activator, but not the apparent K_m of phospholipase A_2 for phospholipid substrates. Since certain types of detergents such as dodecylsulfate and deoxycholate can counteract the inhibition of phospholipase A by lipomodulin, it is likely that lipomodulin interacts with the hydrophobic region of phospholipases near the active site.

3. MECHANISM OF INDUCTION OF THE SYNTHESIS OF LIPOMODULIN BY GLUCOCORTICOIDS

Glucocorticoids are proposed first to bind to intracellular receptor proteins. The steroid–receptor complex is then transfered into nuclei, where it regulates the expression of genes and consequently the synthesis of proteins (Thompson and Lippman, 1974). The increased synthesis of lipomodulin in rabbit neutrophils is dependent on the affinities of glucocorticoids for the receptors. The action of glucocorticoids on the suppression of arachidonate release is blocked by cycloheximide or actinomycin D, inhibitors of protein synthesis (Borman *et al.*, 1984). In addition, dexamethasone induces the synthesis of lipomodulin in U937 cells, a histiocytic leukemia cell line that has intracellular steroid receptors, but not in HL60 cells, a promyeloleukemia cell line that has no steroid receptors (Table I). These observations support the proposed mechanism whereby glucocorticoids exert their actions. Further concrete evidence is that the level of mRNA in dexamethasone-treated U937 cells is three fold higher than that in the nontreated cells, as measured by the incorporation of [^{3}H]leucine into lipomodulin after the translation of mRNA with reticulocyte lysates. How the transcription of the lipomodulin gene is regulated by the steroid–receptor complex remains to be elucidated. Probably, the complex binds to a specific site on the gene, as reported in the case of mammary tumor virus (Payver *et al.*, 1983). Since cloning of the lipomodulin gene is currently being carried out in my laboratory, the questions will be solved in the near future.

4. REGULATION OF CELLULAR PHOSPHOLIPASE A ACTIVITY BY LIPOMODULIN

How phospholipase A_2 is activated in a receptor-mediated fashion is a clue to understanding the mechanism of signal transduction across biomembranes. Since antilipomodulin antibody can bind to intact cells (Hirata *et al.*, 1982), some parts of lipomodulin appear to reside on the cell surface. Lipomodulin can be cleaved by proteases such as trypsin and chymotrypsin. Under conditions in which the cells are not killed by proteases (as determined by the trypan blue exclusion test), the rates of arachidonic acid release are inversedly related to the amounts of antilipomodulin antibody bound to the cells. Since some receptors are well documented to require activation of serine proteases for initiation of signaling (Glenn and Runningham, 1979), such cleavage of lipomodulin by membrane-bound serine proteases whose activities are activated in a receptor-mediated fashion might be one of the mechanisms to release arachidonic acid.

On the other hand, protein phosphorylation is widely believed to control various intracellular events that occur in many cells after physiological stimuli. When purified lipomodulin is phosphorylated *in vitro* by various kinases (Hirata *et al.*, 1981, 1984), it loses the ability to inhibit phospholipase A_2. However, its dephos-

TABLE I. Increased mRNA Level by Glucocorticoids

	HL 60 cells	U937 cells
Glucocorticoid receptor		
sites per cell	100	2,900
Km for dexamethasone (nM)		18
Lipomodulin synthesized (cpm/10^4 cells)^a		
minus dexamethasone	640	430
plus dexamethasone	640	1,160
mRNA for lipomodulin (%)^a		
minus dexamethasone	1.00	0.6
plus dexamethasone	1.00	1.8

^{a} Measured by incorporation of [H]-leucine into the immunoprecipitable protein.

phorylation by alkaline phosphatase restores its inhibitory activity, suggesting that the phosphorylation–dephosphorylation of lipomodulin is one of the mechanisms that regulate cellular phospholipase activity in a reversible way. In fact, the stimulation of neutrophils with fMetLeuPhe results in a transient increase in the phosphorylation of lipomodulin followed by its dephosphorylation. Rates of the release of arachidonic acid follow an essentially identical time course (Hirata *et al.*, 1981). Such a transient increase in phosphorylation followed by dephosphorylation is also observed in mitogen-stimulated lymphocytes (Hirata *et al.*, 1984).

5. SUMMARY

Release of arachidonic acid requires activation of phospholipases. Such phospholipase activation is often observed in cells responding to various stimuli. Glucocorticoids induce the synthesis of a naturally occurring phospholipase inhibitory protein, lipomodulin, thus leading to inhibition of arachidonic acid release. This regulation appears to occur in the transcriptional stage rather than in the translational stage. Cleavage by proteases or phosphorylation by protein kinases inactivates lipomodulin and enhances the action of cellular phospholipases to release more arachidonic acid. Since the phosphorylated lipomodulin is rapidly dephosphorylated to restore its inhibitory activity on phospholipase A_2, the phosphorylation–dephosphorylation process appears to be a mechanism in short-term regulation of phospholipid metabolism. On the other hand, the resynthesis of lipomodulin obviously requires a longer time, which suggests that the degradation and resynthesis of lipomodulin might be a mechanism in the longer-term regulation of phospholipid metabolism (see Fig. 1).

GLUCOCORTICOID-RESPONSIVE CELL

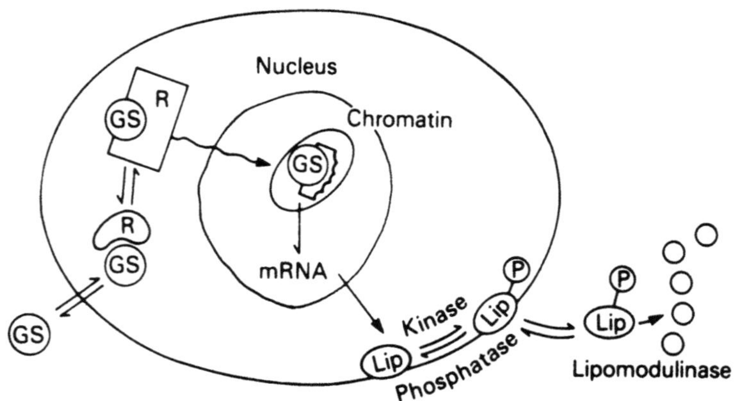

Glucocorticoid Responses

FIGURE 1. Hypothetical mechanism of glucocorticoid modulation of arachidonic acid release.

REFERENCES

Axelrod, J., and Reissin, T., 1984, Stress hormones: Their interaction and regulation, *Science* **224**:452–459.

Borman, B. J., Huag, C. K., MacKin, W. M., and Becker, E. L., 1984, Receptor mediated activation of a phospholipase A in rabbit neutrophil membrane, *Proc. Natl. Acad. Sci. U.S.A.* **81**:767–781.

Glenn, K. D., and Runningham, D. D., 1979, Thrombin-stimulated cell division involves proteolysis of its cell surface receptor, *Nature* **278**:711–714.

Hattori, T., Hirata, F., Hoffmann, T., Hizuta, A., and Herberman, R. B., 1983, Inhibition of human natural killer activity and antibody dependent cellular cytotoxicity by lipomodulin, a phospholipase inhibitory protein, *J. Immunol.* **131**:662–665.

Hirata, F., 1981, The regulation of lipomodulin, a phospholipase inhibitory protein, in rabbit neutrophils by phosphorylation, *J. Biol. Chem.* **256**:7730–7733.

Hirata, F., Concoran, B. A., Venkatasubramanian, K., Schiffmann E., and Axelrod, J., 1979, Chemo-attractants stimulate degradation of methylated phospholipid and release of arachidonic acid in rabbit leukocytes, *Proc. Natl. Acad. Sci. U.S.A.* **76**:2640–2643.

Hirata, F., Schiffmann, E., Venkatasubramanian, K., Salomon D., and Axelrod J., 1981, A phospholipase A inhibitory protein in rabbit neutrophils induced by glucocorticoids, *Proc. Natl. Acad. Sci. U.S.A.* **77**:2533–2536.

Hirata, F., Notsu, Y., Iwata, M., Parente L., DiRosa, M., and Flower, R. J., 1982, Identification of several species of phospholipase inhibitory protein(s) by radioimmunoassay for lipomodulin, *Biochem. Biophys. Res. Comun.* **109**:223–230.

Hirata, F., Notsu, Y., Matsuda, K., Hattori, T., and Del Carmine, R., 1984, Phosphorylation at tyrosine site of lipomodulin, a phospholipase inhibitory protein, in mitogen stimulated murine thymocytes, *Proc. Natl. Acad. Sci. U.S.A.* **81**:4717–4721.

Kuehl, F. A., Jr., and Egan, R. W., 1980, Prostaglandins, arachidonic acid and inflammation, *Science* **210**:978–984.

Payver, F., DeFranco, D., Firestone, G. L., Edgar, B., Wrange, O., Okret, S., Gustafsson, J. A., and Yamamoto, K. P., 1983, Sequence-specific binding of glucocorticoid receptor to MTV DNA at sites within and upstream of the transcribed region, *Cell* **35**:381–392.

Thompson, E. B., and Lippman M. E., 1974, Mechanisms and action of glucocorticoids, *Metabolism* **23**:159–169.

13

Transmembrane Signal in the Secretory Process of A23187-Stimulated Human Leukocytes

A Study of Eicosanoid Release

MONIQUE BRAQUET, ROGER DUCOUSSO, MARIE-YVONNE CHAPELAT, PIERRE BORGEAT, and PIERRE BRAQUET

1. INTRODUCTION

Leukocytes play key roles in mechanisms of body defense against both endotoxins and exotoxins. The factors involved in triggering such defense mechanisms involve an activation of cellular membrane processes, among which is the release from phospholipids of arachidonic acid (AA) and its subsequent conversion into various eicosanoids. The liberation of such eicosanoids [e.g., leukotrienes, prostaglandins (PGs), HPETEs], in addition to other processes, is involved not only in the destruction of ingestible particles but also in the recruitment (chemotactic effect) of other cells for this process. All of these events depend on the generation and transmission of a membrane signal that involves mainly membrane metabolism and ion transport (Smolen *et al.*, 1984).

Recent studies have revealed that K^+ might play a role in modulating the AA

MONIQUE BRAQUET and ROGER DUCOUSSO ● Centre de Recherches du Service de Santé des Armées, F-92141 Clamart, France.   MARIE-YVONNE CHAPELAT, and PIERRE BRAQUET ● I. H. B. Research Laboratories, F-92350 Le Plessis Robinson, France.   PIERRE BORGEAT ● Groupe de Recherches sur les Leucotriènes, Laboratoire d'Endocrinologie Moléculaire Centre Hospitalier de l'Université Laval, Québec G1V 4G2, Canada.

cascade (Oelz *et al.*, 1978; Dusing *et al.*, 1982; Gill, 1980; Skrabal *et al.*, 1981). Since Ca^{2+} also plays a fundamental role in triggering the AA cascade (Oelz *et al.*, 1978), Ca^{2+}-dependent K^+ permeability ($[Ca^{2+}]_i$-dep P_K) (Gardos, 1958; Kregenow and Hoffman, 1972; Lew and Ferreira, 1977, 1978; Hoffman *et al.*, 1980; Schartz and Passow, 1983) could serve as the mechanism that links $[Ca^{2+}]_i$, K^+ efflux, and AA metabolism. As we have shown that $[Ca^{2+}]_i$-dep P_K could be a part of the membrane signal of the A23187-stimulated rabbit platelet (Braquet *et al.*, 1985), we decided to study the early events involved in such secretory phenomena using the A23187-stimulated human leukocyte.

2. MATERIALS AND METHODS

Blood samples (400 ml) of normal male volunteers, 20–30 years of age, were collected in the presence of CDP (citrate 0.11 M; phosphate, 16 mM; dextrose, 0.13 M) and were centrifuged at 1000 g for 25 min. The leukocyte fraction was obtained following 6% dextran-500 sedimentation, centrifugation over Ficoll Paque (Pharmacia), and ammonium chloride (160 mM) + tris (170 mM) treatment. Then, the cells were suspended in the various incubation media (see below); the platelet–leukocyte ratio was ≤ 8.0.

Two different buffers were used to prepare the incubation media. The standard phosphate-buffered medium contained (in mM): Na^+, 146; K^+, 4; Mg^{2+}, 1: Cl^-, 142; HPO_4^{2-}, 2.5; $H_2PO_4^-$, 0.5; glucose, 10 (pH $\cong$ 7.4; determined osmolarity $\cong$ 284 mOsM). To change the Na^+/K^+ ratio, the concentrations of Na^+ and K^+ were varied inversely over the range 0–150 mM. The media were buffered with both tris(hydroxymethyl)aminomethane (tris, 10 mM; Sigma) and 3-(N-morpholino)propanesulfonic acid (MOPS, 10 mM, Sigma) when Rb^+ or Cs^+ was used to replace K^+ or when the pH was varied.

Both $CaCl_2$ and $MgCl_2$ were added to the leukocyte suspension to obtain 2 mM and 0.5 mM final concentrations, respectively, and the suspensions were preincubated for 5 min at 37°C. Cells were then stimulated with the Ca^{2+} ionophore A23187 (1.5μM) and incubated for a further 5-min period. In the time-course study of the action of A23187, incubation periods were varied from 15 sec to 15 min. Incubations were stopped by addition of one volume of a mixture of methanol/acetonitrile (1/1, v/v) containing 200 ng of PGB_2. In some cases, valinomycin was used to replace A23187 or was added together with A23187, either simultaneously or 5 min before the ionophore. The possible effects on eicosanoid release of the following ion-transport-modifying agents were determined: ouabain, quinine, tetraethylammonium, tetrodotoxin (TTX), 4,4'-diisothiocyano-2,2'-disulfonic acid stilbene (DIDS), the cyanine dye di-S-C_3(5), amiloride, apamin, bumetanide, piretanide, furosemide, and 4-aminopyridine.

Various cyclooxygenase (CO) and lipoxygenase (LO) metabolites were determined using reversed-phase high-performance liquid chromatography (RP-HPLC) or radioimmunoassay (RIA; Borgeat *et al.*, 1984).

3. RESULTS

The major metabolites that were released in the platelet leukocyte preparation (at pH 7.4 in standard medium) stimulated with the Ca^{2+} ionophore A23187 were 5-HETE, LTB_4, $5(S,12(S)$-diHETE, and the isomers of LTB_4 (Δ^6-*trans* LTB_4 and 12-epi-Δ^6-*trans* LTB_4). Metabolites of LTB_4 (i.e., 20-OH LTB_4 or 20-COOH-LTB_4) as well as LTC_4 and LTD_4 were also detected (CO products were essentially due to the platelet contamination).

A time-course study (also in physiological medium) revealed that A23187 did not induce any significant release of eicosanoids during the first 30 sec after its addition to the medium. However, eicosanoid release began 30–45 sec after addition of A23187, and maximal stimulated release usually occurred at 3–5 min.

As $[K^+]_o$ was increased from 10 to 150 mM at the expense of $[Na^+]_o$, there occurred a progressive decrease in the release of all CO and LO metabolites that were monitored, the release being nearly abolished at 150 mM $[K^+]_o$. At $[K^+]_o = 0$ mM, the release of all eicosanoids monitored also was markedly reduced. Rubidium or Cs^+, used in place of $[K^+]_o$, had essentially the same effect.

Replacement of Na^+ by choline, in MOPS–tris-buffered medium, abolished the release of all eicosanoids monitored independently of $[K^+]_o$. Ouabain (3×10^{-3} M), which inhibits the electrogenic Na^+ pump (Na^+, K^+-ATPase), produced a progressive, time-dependent partial inhibition of the release of both LO and CO metabolites in the standard medium. No significant change in the release of AA metabolites was observed in the presence of furosemide, bumetanide, or piretanide (all at 10^{-7}–10^{-4} M) in the standard medium.

In the standard medium, apamin produced a slight inhibition of the CO pathway, the maximal effect usually being observed at 10^{-6} M or 10^{-7} M; quinine induced a strong inhibition of both CO and LO pathways but only at the highest concentration used (10^{-3} M); di-S-C_3(5) at 10^{-4} or 10^{-3} M markedly inhibited both CO and LO pathways (Table I).

In the standard medium, 4-aminopyridine (10^{-4} M) did not influence LO or CO pathways, but at 10^{-3} M, it inhibited the release of TxB_2 (46% decrease) and LTB_4 (30% decrease); TEA, even at 25 mM, did not influence LO or CO pathways;

TABLE I. Effect of the Cyanine Dye Di-S-C_3(5)
on Eicosanoid Release in A23187-
Stimulated Human Leukocytes plus Platelets

Concentration (M) of di-S-C_3(5)	Eicosanoid release (percent of control)[a]		
	TxB_2	PGE_2	LTB_4
10^{-3}	16.1	25.2	17.4
10^{-4}	14.7	18.4	12.1
10^{-5}	40.0	57.1	37.1
10^{-6}	93.6	102.0	106.4

[a] Means of three samples in all cases.

TTX induced a moderate effect on eicosanoid release, but only at the highest concentration used (10^{-4} M): LTB$_4$, -32%, 5-HETE, -42%, 20-OH-LTB$_4$, -51%.

When valinomycin and A23187 were added to the human leukocyte plus platelet preparation simultaneously, a significant synergistic effect appeared after 5 min of incubation for the lowest concentrations of Ca^{2+} ionophore used ($\leqslant 10^{-6}$ M) (Fig. 1). This effect seemed to be maximal when the concentration of A23187 was about 5×10^{-7} M and when the concentration of valinomycin was 10^{-5} M. A time-course study (0 to 180 sec) revealed that valinomycin significantly enhanced the initial rate of AA metabolite release. Conversely, when valinomycin was added 10 min before A23187, inhibitory effects were usually observed for all concentrations of K$^+$ ionophore that were used.

Variations of external pH (pH$_o$) in the MOPS–tris-buffered medium produced profound modifications of the release of both CO and LO metabolites. Thus, at all values of pH$_o$ $\leqslant$ 6.0, eicosanoid release was abolished. This effect was particularly evident for LTB$_4$: at pH$_o$ = 6.5, LTB$_4$ release was about 50% of control, whereas at physiological pH$_o$ (i.e., 7.4) release was maximal (Fig. 2). Alkaline conditions appeared to modify the profile of metabolites released in favor of nonenzymatic derivatives of LTA$_4$ (Δ^6-*trans*-LTB$_4$, 12-epi-LTB$_4$). In MOPS–tris-buffered medium, DIDS (10^{-5} M) profoundly inhibited eicosanoid release over the whole range of pH$_o$ studied (i.e., 5.0–8.0). Amiloride (10^{-4} M) in MOPS–tris-buffered medium, like DIDS, inhibited the release of all eicosanoids that were monitored; inhibition was nearly complete at all pH$_o$ values $\leqslant$7.0.

4. DISCUSSION

The results presented herein have confirmed and extended those of previous workers concerning the characterization of early events involved in secretory phenomena associated with leukocyte activation. Concerning the finding that A23187

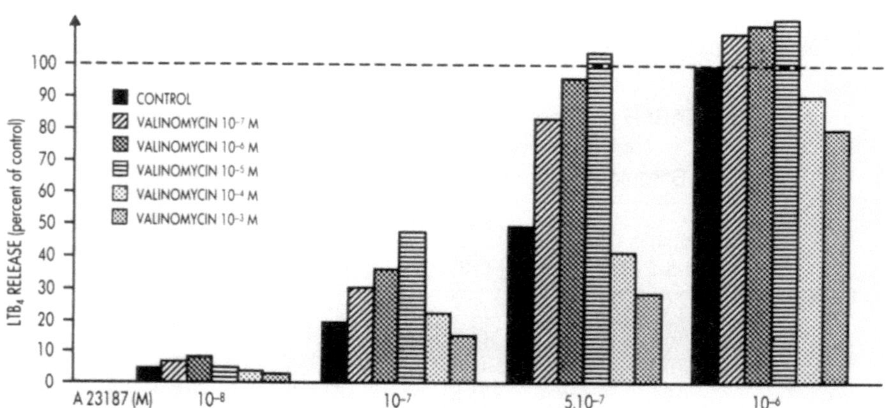

FIGURE 1. Dose-dependent effect of the mixture A23187 plus valinomycin versus A23187 alone on LTB$_4$ release in human leukocytes plus platelets.

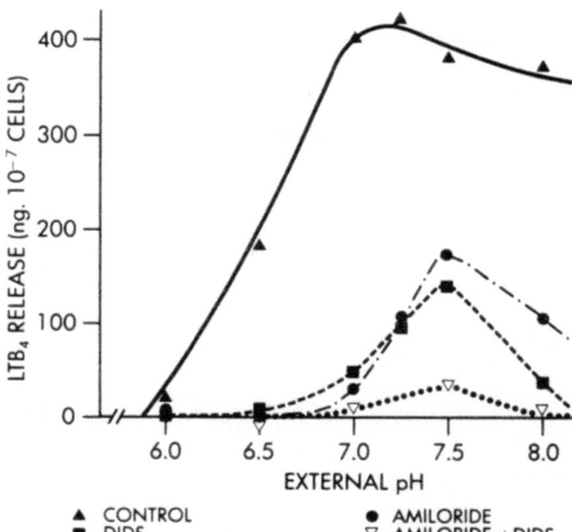

FIGURE 2. Effects of DIDS, of amiloride, and of amiloride plus DIDS on LTB$_4$ release in human leukocytes plus platelets at different external pH values.

initiated eicosanoid release 30 sec after its addition to the medium, it is noteworthy that the generation of superoxide ($O_2{\cdot}^-$) by immune complex or by concanavalin A also occurred with a lag period of 30–42 sec (Korchak and Weissman, 1978). In support of results that have shown that modifications of $[K^+]_o$ led to changes in eicosanoid release *in vitro* (Oelz *et al.*, 1978; Dusing *et al.*, 1982) and *in vivo* (Gill, 1980; Skrabal *et al.*, 1981), it is shown herein that eicosanoid release was also decreased in A23187-stimulated leukocytes at various $[K^+]_o$. The decreased eicosanoid release at $[K^+]_o = 0$ was likely related to an inhibition of the electrogenic Na^+ pump and the subsequent membrane depolarization, whereas the decreased eicosanoid release by high $[K^+]_o$ was probably related to an inhibition of the K^+ efflux that is associated with A23187 action in producing a transient hyperpolarization. Eicosanoid release was also decreased when Na^+ was replaced by choline. These results, taken together with others (Showell *et al.*, 1977; Korchak and Weissman, 1980), indicate that at least three events involved in leukocyte activation (transmembrane signaling) are dependent on Na^+ influx.

In line with a mechanism by which stimulation of K^+ efflux could be caused by an increased $[Ca^{2+}]_i$, which would favor opening of the K^+ channels ($[Ca^{2+}]_i$-dep P_K), Gallin *et al.* (1975) have shown that $[Ca^{2+}]_i$-dep P_K is involved in macrophage activation (see also Oliveira-Castro and Dos Reis, 1981), and it is shown herein that the inhibitors of $[Ca^{2+}]_i$-dep P_K, quinine and di-S-C$_3$(5), markedly inhibited eicosanoid release. With further regard to the possible role of K^+ fluxes on eicosanoid release, it is shown herein that K^+ ionophore (valinomycin) does not trigger the AA cascade in leukocytes but that addition of valinomycin simultaneously with A23187 had a synergistic effect on eicosanoid release. Thus, although K^+ efflux *per se* does not trigger the AA cascade, K^+ efflux coupled with Ca^{2+} mobilization appears to be important for this process.

The results discussed above, together with those obtained by previous workers

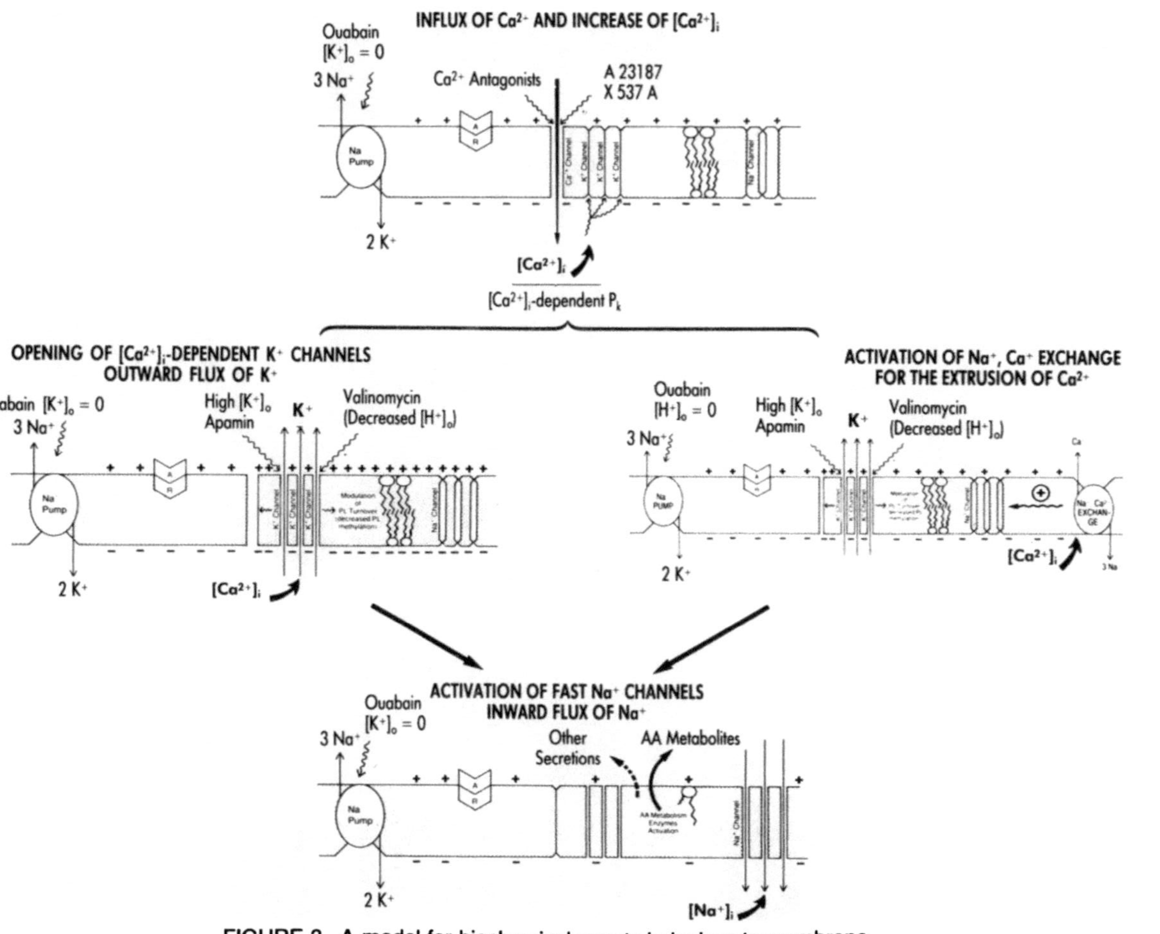

FIGURE 3. A model for biochemical events in leukocyte membrane.

on changes in membrane potential and Ca^{2+} efflux that occur during leukocyte activation, can be used to formulate the following sequence of events involved in leukocyte activation (see Fig. 3): Ca^{2+} mobilization and/or Ca^{2+} and Na^{+} influxes $\rightarrow$ increased $[Ca^{2+}]_i$ $\rightarrow$ transient hyperpolarization via K^{+} efflux ($[Ca^{2+}]_i$-dep P_K) and simultaneous commencement of Ca^{2+} extrusion or sequestering mechanism $\rightarrow$ changes in Na^{+} channels $\rightarrow$ Na^{+} influx $\rightarrow$ membrane depolarization.

REFERENCES

Borgeat, P., Fruteau de Laclos, B., Rabinovitch, S., Picard, S., Braquet, P., Hebert, J., and Laviolette, J., 1984, Eosinophil-rich human polymorphonuclear leukocyte preparations characteristically release leukotriene C₄ upon ionophore A 23187 challenge, *Allergy Clin. Immunol.* **2:**310–317.

Braquet, P., Spinnewyn, B., Lehuu, B., Braquet, M., Chabrier, E., Dray, F., and DeFeudis, F. V., 1985, Is Ca^{2+}-dependent K^{+} permeability involved in triggering the arachidonic acid cascade?— A study with rabbit platelets, *Prostaglandins Leukotrienes Med.* (in press).

Dusing, R., Scherhag, R., Tippelman, R., Udde, U., Glanzer, K., and Kramer, H. J., 1982, Arachidonic acid metabolism in isolated rat aorta. Dependence of prostacyclin biosynthesis on extracellular potassium, *J. Biol. Chem.* **257:**1993–1996.

Gallin, E. K., Wiederhold, M. L., Lipsky, P. E., and Rosenthal, A. S., 1975, Spontaneous and induced membrane hyperpolarizations in macrophages, *J. Cell. Physiol.* **86:**653–661.

Gardos, G., 1958, The function of calcium in the potassium permeability of human erythrocytes, *Biochim. Biophys. Acta* **30:**653–654.

Gill, J. R., 1980, Bartter's syndrome, *Annu. Rev. Med.* **31:**405–419.

Hoffman, J. F., Yingst, D. R., Goldinger, J. M., Blum, R. M., and Knauf, P. A., 1980, On the mechanism of Ca-dependent K transport in human red blood cells, in: *Membrane Transport in Erythrocytes* (U. V. Lassen, H. H. Ussing, and J. O. Wieth, eds.), Munksgaard, Copenhagen, pp. 178–192.

Korchak, H. M., and Weissman, G., 1978, Changes in membrane potential of human granulocytes antecede the metabolic responses to surface stimulation, *Proc. Natl. Acad. Sci. U.S.A.* **75:**3818–3822.

Korchak, H. M., and Weissman, G., 1980, Stimulus–response coupling in the human neutrophil. Membrane potential changes and the role of extracellular Na^{+}, *Biochim. Biophys. Acta* **601:**180–194.

Kregenow, F. M., and Hoffman, J. F., 1972, Some kinetic and metabolic characteristics of calcium-induced potassium transport in human red cells, *J. Gen. Physiol.* **60:**406–429.

Lew, V. L., and Ferreira, H. G., 1977, The effect of Ca on the K permeability of red cells, in: *Membrane Transport in Red Cells* (J. C. Ellory and V. L. Lew, eds.), Academic Press, New York, pp. 93–100.

Lew, V. L., and Ferreira, H. G., 1978, Calcium transport and the properties of a calcium activated potassium channel in red cell membranes. *Curr. Top. Membr. Transp.* **10:**217–277.

Oelz, O., Knapp, H. R., Roberts, L. J., Oelz, R., Sweetman, B. J., Oates, J. A., and Reed, P. W., 1978, Calcium-dependent stimulation of thromboxane and prostaglandin biosynthesis by ionophores, in: *Advances in Prostaglandin and Thromboxane Research*, Vol. 3 (C. Galli, ed.), Raven Press, New York, pp. 148–158.

Oliveira-Castro, G. M., and Dos Reis, G. A., 1981, Electrophysiology of phagocytic membranes. III. Evidence for a calcium-dependent potassium permeability change during slow hyperpolarizations of activated macrophages, *Biochim. Biophys. Acta* **640:**500–511.

Schartz, W., and Passow, H., 1983, Ca^{2+}-activated K^{+} channels in erythrocytes and excitable cells, *Annu. Rev. Physiol.* **45:**359–374.

Showel, H. J., Naccache, P. H., Sha'afi, R. I., and Becker, E. L., 1977, The effects of extra-cellular K^{+}, Na^{+}, and Ca^{++} on lysosomal enzyme secretion from polymorphonuclear leukocytes, *J. Immunol.* **119:**804–811.

Skrabal, F., Aubock, J., and Hortnagl, H., 1981, Low sodium/high potassium diet for prevention of hypertension: Probable mechanism of action, *Lancet* **2**:895–900.

Smolen, J. E., Korchak, H. M., and Weissman, G., 1984, Stimulus–secretion coupling in human polymorphonuclear leukocytes, in: *Cell Biology of the Secretory Process* (M. Cantin, ed.), S. Karger, Basel, pp. 517–545.

Postphospholipase Activation of Lipoxygenase/ Leukotriene Systems

JACK Y. VANDERHOEK and J. MARTYN BAILEY

1. INTRODUCTION

Once arachidonic acid is liberated by the action of acylhydrolases on cellular phospholipids or other esterified precursors, it can be metabolized either by the cyclooxygenase or lipoxygenase pathways to a variety of biologically active compounds (Kuehl and Egan, 1980). Lipoxygenase metabolites including hydroperoxyeicosatetraenoic acids (HPETEs), hydroxyeicosatetraenoic acids (HETEs), and leukotrienes appear to be involved in inflammatory processes (Samuelsson, 1983). Lipoxygenases in different cell types appear to exist in a relatively inactive site. An increasing number of studies indicate that the activation of different lipoxygenases can be accomplished by a diverse group of stimuli. It is the purpose of this chapter to discuss the research carried out in our laboratory on stimulators of mammalian lipoxygenases as well as to review the work of other investigators.

2. 5-LIPOXYGENASE

The 5-lipoxygenase initiates the conversion of arachidonic acid to leukotrienes via the formation of the 5-HPETE precursor. There are several reports that this enzyme can be activated by metabolites from other lipoxygenases. In the PT-18 mast/basophil cell line, the 5-lipoxygenase oxidizes exogenous arachidonic acid to only a very small extent (Vanderhoek et al., 1982). However, in the presence of

JACK Y. VANDERHOEK and J. MARTYN BAILEY ● Department of Biochemistry, George Washington University School of Medicine and Health Sciences, Washington, D.C. 20037.

the 15-lipoxygenase reduction product 15-HETE, the production of 5-HETE and 5,12-diHETE was stimulated in a dose-dependent manner by as much as 25- and 60-fold, respectively (Figs. 1, 2). This was the first demonstration that a cryptic lipoxygenase could be activated by a nonperoxidic lipoxygenase metabolite. The activation process occurs rapidly and is reversible.

There appear to be definite structural requirements in the HETE molecule to activate this 5-lipoxygenase (Fig. 3): 15-HPETE is less effective, the 15-acetate derivative of 15-HETE is comparable to 15-HETE, whereas 15-HETE methyl ester and 12-HETE are ineffective. These results show that a free carboxyl group is essential. However, the oxygen functionality at C-15 does not have to be a free hydroxyl group, since both acetate and hydroperoxy groups are capable of stimulating the 5-lipoxygenase. Different results were observed with the human peripheral blood PMN 5-lipoxygenase. A four- to eightfold stimulation of this enzyme (as

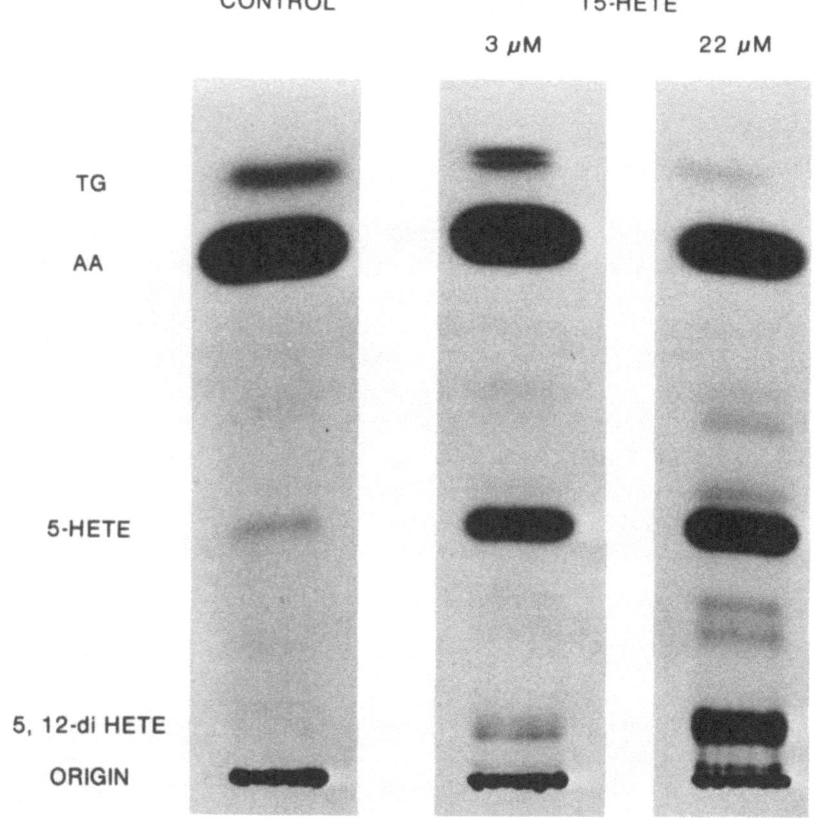

FIGURE 1. Activation of 5-lipoxygenase-catalyzed oxidation of arachidonic acid in PT-18 mast/basophil cells by 15-HETE. Autoradiograph of a thin-layer chromatogram of [^{14}C]arachidonic acid metabolites formed by PT-18 cells in absence (left lane) or presence (center lane, 3 μM; right lane, 22 μM) of 15-HETE. AA, arachidonic acid; TG, triglycerides.

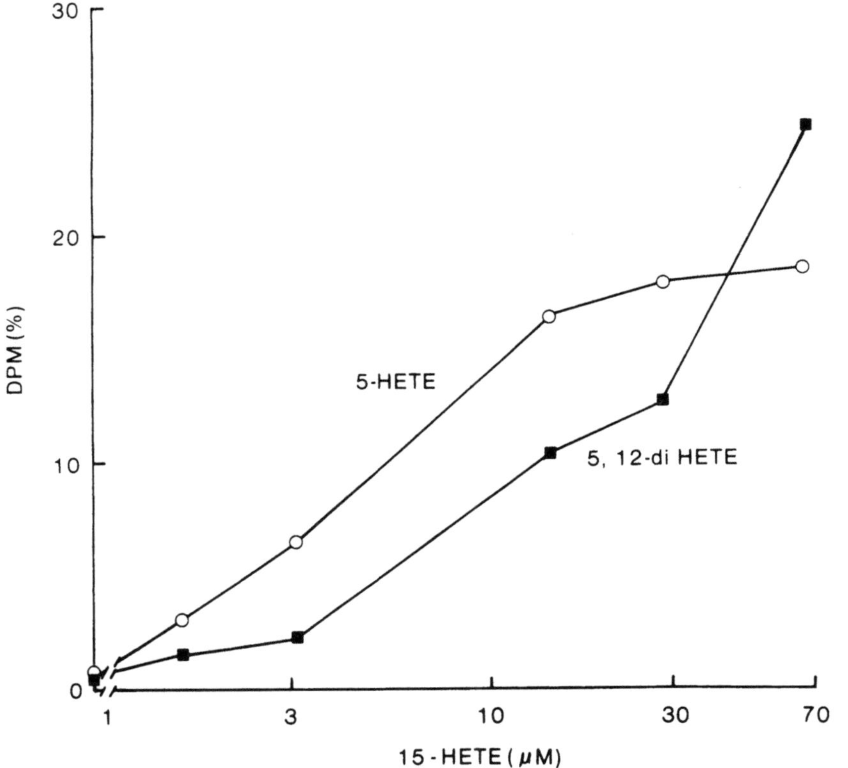

FIGURE 2. Effect of 15-HETE concentration on leukotriene B_4 and 5-HETE formation in PT-18 cells. PT-18 cells were preincubated with different concentrations of 15-HETE for 10 min at 37° prior to the addition of [^{14}C]arachidonic acid; [^{14}C]5-HETE and [^{14}C]5,12-diHETE (including leukotriene B_4) formation were analyzed by thin-layer chromatography; products, located by autoradiography, were scraped and counted in a liquid scintillation counter.

evidenced by 5-HETE and 5,12-diHETE production) was observed when these cells were coincubated with platelets, exogenous arachidonic acid, and calcium (Maclouf *et al.*, 1982a). The optimum platelet-to-leukocyte ratio was 25 : 1. It was found that the activation was caused by the platelet 12-lipoxygenase product 12-HPETE, and maximal activation was observed at 3–4 μM 12-HPETE. The reduction product 12-HETE was ineffective. It is possible that the stimulating effect of 12-HPETE may be a somewhat nonspecific hydroperoxide effect, since low concentrations of fatty acid hydroperoxides have been reported to be essential in reducing the initial lag phase of other lipoxygenation reactions (Smith and Lands, 1972; Siegel *et al.*, 1979). These findings contrast with those reported above for the PT-18 cells, where the 15-HETE was a more effective stimulator than the corresponding hydroperoxy 15-HPETE. Furthermore, the observed activation of the PT-18 5-lipoxygenase appears to be highly cell specific in view of our earlier report that 15-HETE inhibits

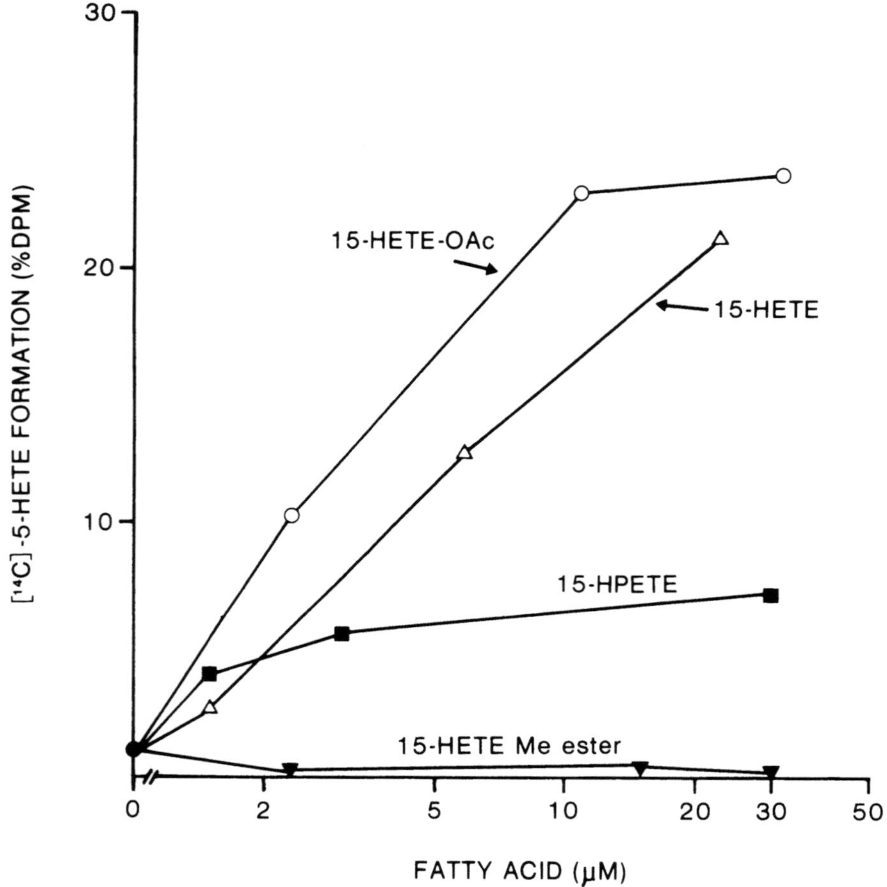

FIGURE 3. Effect of various 15-HETE derivatives on the 5-lipoxygenase activity in PT-18 cells. The 5-HETE formation from [¹⁴C]arachidonic acid was measured as described in the legend of Fig. 2.

the 5-lipoxygenase in rabbit peritoneal PMNs (Vanderhoek *et al.*, 1980). It appears that the activation mechanisms of the 5-lipoxygenase in the PT-18 and leukocyte systems are probably quite different.

A variety of other substances have also been reported to stimulate mammalian lipoxygenases. The 5-lipoxygenase in human PMNs is normally in a relatively inactive state (Borgeat and Samuelsson, 1979). However, in the presence of both arachidonic acid and the calcium ionophore A23187, the production of 5-HETE and 5,12-diHETE was stimulated by as much as 20- and 50-fold, respectively. This result suggested that the PMN 5-lipoxygenase was Ca^{2+} dependent. This was confirmed in RBL-1 cells, where it was established that the 5-lipoxygenase present in the 10,000 $\times$ g supernatant of RBL-1 cells was stimulated by Ca^{2+} in a dose-dependent manner (Jakschik *et al.*, 1980). Thus, 0.4 mM Ca^{2+} increased 5-HETE formation from exogenous [^{14}C]arachidonic acid fivefold. Subsequent work showed that Ca^{2+} induced a dimerization from an inactive 90,000-molecular-weight monomer to an active 189–190,000 dimeric species (Parker and Aykent, 1982).

Human PMNs metabolized exogenous arachidonic acid (250 μM) complexed with albumin predominantly to 5-HETE and LTB_4 on stimulation with the inflammatory agent N-formyl-Met-Leu-Phe or the complement component C5a (Clancy *et al.*, 1983). No lipoxygenase metabolites were formed in the absence of these chemotactic factors, nor do these factors stimulate membrane phospholipases to release endogenous arachidonic acid under the experimental conditions used. Thus, C5a (0.1 μM) increased 5-HETE production in 10^7 cells from <7 pmol to 440 pmol, whereas LTB_4 formation was increased from <2 pmol to 66 pmol. The 5-lipoxygenase from guinea pig peritoneal PMNs also exhibited a Ca^{2+} requirement (Ochi *et al.*, 1983). Other divalent cations such as Cu^{2+} or Fe^{2+} were ineffective. The enzyme activity was further increased by the addition of ATP. A maximum fourfold stimulation of 5-HETE and LTB_4 production was observed with about 2 mM ATP. Other nucleotides were less effective.

The human myeloid cell line K562 produces 5-HETE as well as 11/12-HETE and 15-HETE from exogenous arachidonic acid. The tumor promoter 12-O-tetradecanoylphorbol-13-acetate was reported to increase 5-HETE formation less than twofold (Valone *et al.*, 1983). The nonionic detergent BRIJ56 was recently reported to enhance the 5-lipoxygenase in rat pleural neutrophils (Myers and Siegel, 1983). Although maximal production of metabolites could be induced by either 0.1% detergent or 10 μM A23187, a further augmentation of product formation was observed when both of these agents were present. A nearly sevenfold stimulation of 5-HETE was observed with 0.1% BRIJ56 and 10 μM A23187 compared to A23187 alone. These workers also found that the nonsteroidal antiinflammatory drugs indomethacin, ibuprofen, and naproxen stimulated the 5-lipoxygenase in the rat neutrophils at concentrations that inhibited the cellular cyclooxygenase activities. A maximal 50% enhancement was observed with 1 μM indomethacin. Table I summarizes the effects of different stimulators on various mammalian 5-lipoxygenases.

3. 12-LIPOXYGENASE

The platelet was the most commonly used cell source in the various reports on the activation of mammalian 12-lipoxygenases. The human platelet 12-lipoxygenase was shown to be stimulated by its own arachidonic acid metabolite, 12-HPETE (Siegel *et al.*, 1979). A twofold stimulation was observed in platelet homogenates, and half-maximal activation occurred at about 0.33 μM 12-HPETE. The corresponding reduction product, 12-HETE, was inactive. Since 8-HPETE, 9-HPETE, and 11-HPETE were less effective than 12-HPETE, there is an isomeric specificity for this activation process. Inhibition of the platelet cyclooxygenase

TABLE I. Stimulators of Mammalian 5-Lipoxygenases

Cellular source		Analogue effects			
Cell	Species	Activator	Less active	Inactive	References
PT-18	Mouse	15-HETE 15-HETE-acetate	15-HPETE	12-HETE 15-HETE Me ester	Vanderhoek $et\ al.$, 1982 Unpublished data
Mastocytoma		Ethanol			Westcott and Murphy, 1983
RBL-1	Rat	A23187 Ca^{2+}		Mg^{2+}, Mn^{2+}	Jakschik $et\ al.$, 1978 Jakschik $et\ al.$, 1980 Parker and Aykent, 1982
Neutrophils (pleural)		A23187 BRIJ56 Indomethacin	Ibuprofen, naproxen		Myers and Siegel, 1983
PMN (peritoneal)	Guinea pig	Ca^{2+}		Co^{2+}, Cu^{2+}, Ba^{2+}, Fe^{2+}, Mg^{2+}, Zn^{2+}	Ochi $et\ al.$, 1983
		ATP	ADB, AMP, CTP, UTP, cAMP	Adenine, GTP, TTP	Ochi $et\ al.$, 1983
PMN (peripheral blood)	Human	A23187			Borgeat and Samuelsson, 1979
		12-HPETE N-Formyl-Met- Leu-Phe		12-HETE	Maclouf $et\ al.$, 1982a Clancy $et\ al.$, 1983
		C5a	des-Arg-C5a		
K562	Human	12-O-Tetra decanoylphorbol- 13-acetate	Phorbol-13,20- diacetate	4α-Phorbol, phor- bol-12,13-diacetate	Valone $et\ al.$, 1983

pathway by aspirin or indomethacin resulted in a lag phase in 12-HETE production from arachidonic acid (Hamberg and Hamberg, 1980). These results suggested that (endo)peroxide intermediates formed by the fatty acid cyclooxygenase pathway can also activate the platelet 12-lipoxygenase.

Although the platelet 12-lipoxygenase present in the $8000 \times g$ supernatant was shown to have no Ca^{2+} dependence (Jakschik *et al.*, 1980), two drugs known to interfere with intracellular Ca^{2+}, the membrane drug chlorpromazine and the intracellular Ca^{2+} antagonist 8-(N,N-diethylamino)-octyl-3,4,5-trimethoxybenzoate, selectively stimulated exogenous arachidonic acid oxidation via the 12-lipoxygenase pathway in intact human platelets by a factor of two to three (Maclouf *et al.*, 1982b). A coronary vasoconstrictor factor isolated from human plasma was reported to stimulate 12-HETE formation from exogenous arachidonic acid in human platelet preparations about sixfold (Moretti and Lin, 1980). Factors present in serum from adjuvant arthritic rats were also reported to specifically enhance lipoxygenase activities in rabbit and horse platelet preparations and rat peritoneal macrophages (Ahnfelt-Ronne and Arrigoni-Martelli, 1980). These factors were partially purified and had a molecular weight range of 100,000. Eosinophil stimulation promoter, a murine lymphokine that enhances eosinophil migration, was recently reported to induce a two- to threefold increase in 12-HETE production in murine eosinophils (Turk *et al.*, 1983). 12-O-Tetradecanoylphorbol-13-acetate doubled 11/12-HETE production in human K562 cells (Valone *et al.*, 1983). The data on the stimulation of the 12-lipoxygenase are summarized in Table II.

4. 15-LIPOXYGENASE

The 15-lipoxygenase in peripheral blood PMNs is in a relatively inactive state since these cells oxygenate only small amounts of exogenous arachidonic acid to 15-HETE (Borgeat and Samuelsson, 1979). We have recently found that the nonsteroidal antiinflammatory agent ibuprofen selectively activates the human PMN 15-lipoxygenase (Vanderhoek and Bailey, 1984). Pretreatment of the PMNs with 1–5 mM ibuprofen prior to the addition of exogenous [^{14}C]arachidonic acid resulted in the stimulation of 15-HETE formation (up to 20-fold), whereas no enhancement of the 5-lipoxygenase activity was observed under these conditions (Fig. 4). The activation was readily reversible (Fig. 5), and maximum stimulation of the 15-lipoxygenase was observed when the time interval between the addition of ibuprofen and arachidonic acid was less than 1 min. When the effects of aspirin, indomethacin, and ibuprofen on the 15-lipoxygenase were compared in six donors, ibuprofen produced an average ninefold increase, whereas aspirin and indomethacin produced only 1.5- and twofold enhancements, respectively (Table III). Indomethacin also has been reported to stimulate the partially purified 15-lipoxygenase in rabbit peritoneal leukocytes, although only a 65% enhancement in 15-lipoxygenase metabolite formation was observed (Narumiya *et al.*, 1981). These workers also reported that this enzyme lacked any divalent cation dependence. The tumor promoter 12-O-

TABLE II. Stimulators of Mammalian 12-Lipoxygenases

| Cellular source | | | Analog effects | | |
Cell	Species	Activator	Less active	Inactive	References
Platelet	Human	12-HPETE (Endo)peroxides	8-HPETE, 9-HPETE	12-HETE	Siegel *et al.*, 1979 Hamberg and Hamberg, 1980
		Vasoconstrictor factor			Moretti and Lin, 1980
		TMB-8[a] Chlorpromazine			Maclouf *et al.*, 1982b
	Horse Rabbit	Arthritic rat serum factors			Ahnfelt-Ronne and Arrigoni-Martelli, 1980
Macrophage	Rat	Arthritic rat serum factors			
Eosinophil	Mouse	Eosinophil stimulation promoter			Turk *et al.*, 1983
K562	Human	12-O-Tetradecanoyl-phorbol-13-acetate	Phorbol-13,20-diacetate, 4α-phorbol, phorbol-12,13-diacetate		Valone *et al.*, 1983

[a] TMB-8: 8-(N,N-diethylamino)-octyl-3,4,5-trimethoxybenzoate.

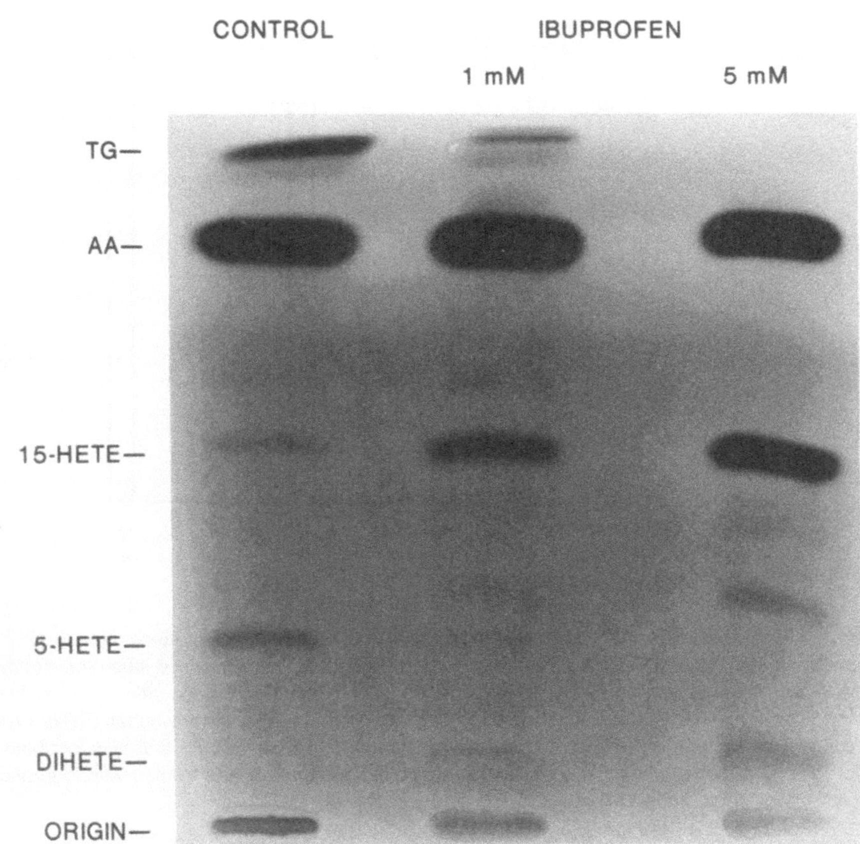

FIGURE 4. Activation of human PMN 15-lipoxygenase by ibuprofen. Autoradiograph of [¹⁴C]arachidonic acid metabolites formed by human peripheral blood PMN leukocytes in the absence (left lane) or presence (middle, 1 mM; right, 5 mM) of ibuprofen. Products were extracted and separated by thin-layer chromatography; 15-HETE was identified by ultraviolet spectroscopy, radioimmunoassay, and HPLC retention times in comparison with authentic 15-HETE.

tetradecanoylphorbol-13-acetate increased 15-HETE production in human K562 cells about 60% (Valone *et al.*, 1983). The effects of different activators on the 15-lipoxygenase are summarized in Table IV.

5. NEW REGULATORY SITE OF LIPOXYGENASE-CATALYZED ARACHIDONIC ACID METABOLISM

It is generally accepted that formation of the different classes of biologically active arachidonic acid metabolites is initially controlled by cellular lipase activity, which releases arachidonic acid from cellular lipid pools. As shown in Fig. 6, the

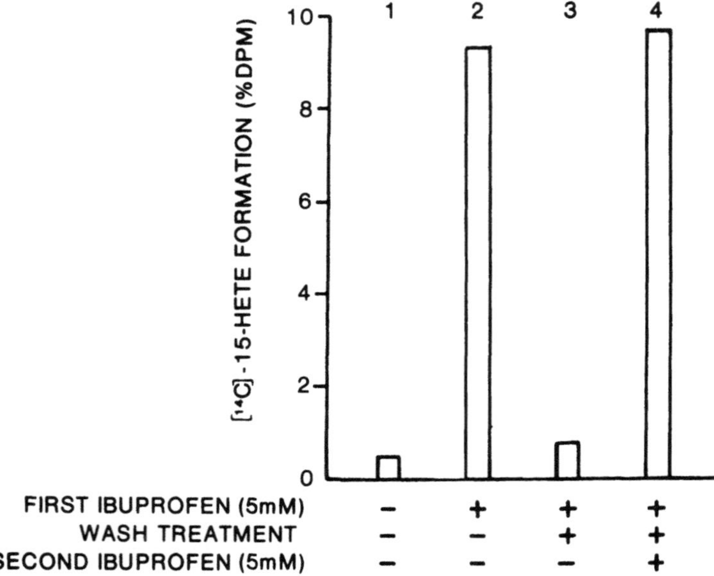

FIGURE 5. Reversibility of ibuprofen activation of human PMN 15-lipoxygenase. The 15-HETE formation from [^{14}C]arachidonic acid with untreated PMNs and that with 5 mM ibuprofen-treated PMNs are shown in columns 1 and 2, respectively. Other ibuprofen-treated cells were washed with 1% bovine serum albumin to remove the original ibuprofen. After these washed PMNs were resuspended in buffer, 15-HETE formation was measured in one aliquot that was incubated with [^{14}C]arachidonic acid (column 3) and in a second aliquot that was treated with ibuprofen (5 mM) prior to the addition of arachidonic acid (column 4).

TABLE III. Relative Potencies of Ibuprofen, Aspirin, and Indomethacin in Stimulating the PMN 15-Lipoxygenase in Different Human Subjects

	Stimulation of [^{14}C]15-HETE[a] formation (times control)		
Subject	Ibuprofen[b] 5 mM	Aspirin[c] 5 mM	Indomethacin[d] 0.7 mM
W.M.	12.4	1.1	1.1
G.K.	10.4	2.0[c]	1.8
T.H.	13.6	3.4	2.2
T.S.	6.4	1.0	0.79
J.S.	6.6	1.3	4.7
J.V.	5.4	0.58	1.1[d]

[a] Stimulation of [^{14}C]15-HETE formation from [^{14}C]arachidonic acid. Values obtained are relative to controls without the drug.

[b] Concentration range tested was 0.2–5 mM. At 5 mM, ibuprofen exhibited highest stimulation of the 15-lipoxygenase.

[c] Concentration range tested was 0.2–8 mM. At 5 mM, aspirin exhibited highest stimulation except in subjects G.K. and J.V., where 0.4 and 1 mM, respectively, were found to be the optimum concentrations.

[d] Concentration range tested was 0.05–0.7 mM and was not further extended because of insolubility of indomethacin at higher concentrations. At 0.7 mM, indomethacin exhibited highest stimulation of the 15-lipoxygenase except in subject J.V., where 0.05 mM was found to be the optimum concentration.

TABLE IV. Stimulators of Mammalian 15-Lipoxygenases

Cellular source			Analog effects		
Cell	Species	Activator	Less active	Inactive	References
PMN (peripheral blood)	Human	Ibuprofen	Aspirin Indomethacin		Vanderhoek and Bailey, 1984
K562		12-O-Tetradecanoyl-phorbol-13-acetate	Phorbol-13,20-diacetate	4α-Phorbol, phorbol-12,13-diacetate	Valone et al., 1983
PMN (peritoneal)	Rabbit	Indomethacin			Narumiya et al., 1981

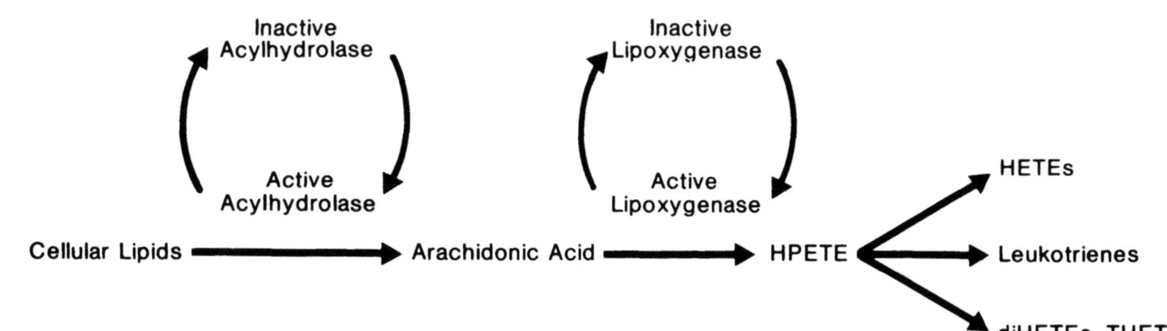

FIGURE 6. Regulatory sites of lipoxygenase-catalyzed arachidonic acid metabolism. Proposed scheme indicating the new, second site of regulation whereby an inactive lipoxygenase is converted to an active lipoxygenase species. The first site of regulation is the previously established activation of the cellular acylhydrolase to release free arachidonic acid.

activation of the inactive acylhydrolase results in the hydrolysis of arachidonic acid from membrane lipids and is the first site of regulation of arachidonic acid metabolism. Once free arachidonic acid has become available, it can be metabolized by either the cyclooxygenase pathway or various lipoxygenase routes. The studies of Lands and co-workers have shown that the activation of the cyclooxygenase enzyme constitutes a secondary control point in the conversion of arachidonic acid to the different prostanoids (Lands, 1984). In view of our data as well as those of other investigators, which have been discussed above, we now propose that a second postlipase control point also exists in the formation of lipoxygenase metabolites. This second regulatory feature involves the conversion of inactive lipoxygenase to an active lipoxygenase species, which is the necessary prerequisite to the formation of lipoxygenase metabolites such as leukotrienes (see Fig. 6). The structural diversity of the various lipoxygenase activators indicates that they probably act via different mechanisms, although it is possible that they may act via a common final step that results in the active lipoxygenase. Delineation of these mechanisms should suggest new hypotheses for potential modulation of inflammatory reactions.

REFERENCES

Ahnfelt-Ronne, I., and Arrigoni-Martelli, E., 1980, Adjuvant arthritic rat serum enhances HETE biosynthesis, *Agents Actions* **10**:550–553.

Borgeat, P., and Samuelsson, B., 1979, Arachidonic acid metabolism in polymorphonuclear leukocytes: Effects of ionophore A23187, *Proc. Natl. Acad. Sci. U.S.A.* **76**:2148–2152.

Clancy, R. M., Dahinden, C. A., and Hugli, T. E., 1983, Arachidonate metabolism by human polymorphonuclear leukocytes stimulated by N-formyl-Met-Leu-Phe or complement component C5a is independent of phospholipase activation, *Proc. Natl. Acad. Sci. U.S.A.* **80**:7200–7204.

Hamberg, M., and Hamberg, G., 1980, On the mechanism of the oxygenation of arachidonic acid by human platelet lipoxygenase, *Biochem. Biophys. Res. Commun.* **95**:1090–1097.

Jakschik, B. A., Lee, L. H., Shuffer, G., and Parker, C. W., 1978, Arachidonic acid metabolism in rat basophilic leukemia (RBL-1) cells, *Prostaglandins* **16**:733–748.

Jakschik, B. A., Sun, F. F., Lee, L. H., and Steinhoff, M. M., 1980, Calcium stimulation of a novel lipoxygenase, *Biochem. Biophys. Res. Commun.* **95**:103–110.

Kuehl, F. A., and Egan, R. W., 1980, Prostaglandins, arachidonic acid and inflammation, *Science* **210**:978–984.

Lands, W. E. M., 1984, Biological consequences of fatty acid oxygenase reaction mechanisms, *Prostaglandins Leukotrienes Med.* **13**:35–46.

Maclouf, J., DeLaclos, B. F., and Borgeat, P., 1982a, Stimulation of leukotriene biosynthesis in human blood leukocytes by platelet-derived 12-hydroperoxy-eicosatetraenoic acid, *Proc. Natl. Acad. Sci. U.S.A.* **79**:6042–6046.

Maclouf, J., De la Baume, H., Levy-Toledano, S., and Caen, J. P., 1982b, Selective stimulation of human platelet lipoxygenase product 12-hydroxy-5,8,10,14-eicosatetraenoic acid by chlorpromazine and 8-(N,N-diethylamino)-octyl-3,4,5-dimethoxybenzoate, *Biochim. Biophys. Acta* **711**:377–385.

Moretti, R. L., and Lin, C. Y., 1980, Coronary vasoconstrictor extracted from blood plasma stimulates platelet lipoxidase activity, *Prostaglandins* **19**:99–108.

Myers, R. F., and Siegel, M. I., 1983, Differential effects of anti-inflammatory drugs on lipoxygenase and cyclooxygenase activities of neutrophils from a reverse passive Arthus reaction, *Biochem. Biophys. Res. Commun.* **112**:586–594.

Narumiya, S., Salmon, J. A., Cottee, F. H., Weatherley, B. C., and Flower, R. J., 1981, Arachidonic acid 15-lipoxygenase from rabbit peritoneal polymorphonuclear leukocytes. Partial purification and properties, *J. Biol. Chem.* **256**:9583–9592.

Ochi, K., Yshimoto, T., Yamamoto, S., Taniguchi, K., and Miyamoto, T., 1983, Arachidonate 5-lipoxygenase of guinea pig peritoneal polymorphonuclear leukocytes. Activation by adenosine 5'-triphosphate, *J. Biol. Chem.* **258**:5754–5758.

Parker, C. W., and Aykent, S., 1982, Calcium stimulation of the 5-lipoxygenase from RBL-1 cells, *Biochem. Biophys. Res. Commun.* **109**:1011–1016.

Samuelsson, B., 1983, Leukotrienes: Mediators of immediate hypersensitivity reactions and inflammation, *Science* **220**:568–575.

Siegel, M. I., McConnell, R. T., Abrahams, S. L., Porter, N. A., and Cuatrecasas, P., 1979, Regulation of arachidonate metabolism via lipoxygenase and cyclooxygenase by 12-HPETE, the product of human platelet lipoxygenase, *Biochem. Biophys. Res. Commun.* **89**:1273–1280.

Smith, W. L., and Lands, W. E. M., 1972, Oxygenation of unsaturated fatty acids by soybean lipoxygenase, *J. Biol. Chem.* **247**:1038–1048.

Turk, J., Rand, T. H., Maas, R. L., Lawson, J. A., Brash, A. R. Roberts, L. J. Colley, D. G., and Oates, J. A., 1983, Identification of lipoxygenase products from arachidonic acid metabolism in stimulated murine eosinophils, *Biochim. Biophys. Acta* **750**:78–90.

Valone, F. H., Obrist, R., Tarlin, N., and Bast, R. C., 1983, Enhanced arachidonic acid lipoxygenation by K562 cells stimulated with 12-O-tetradecanoylphorbol-13-acetate, *Cancer Res.* **43**:197–201.

Vanderhoek, J. Y., and Bailey, J. M., 1984, Activation of a 15-lipoxygenase/leukotriene pathway in human polymorphonuclear leukocytes by the anti-inflammatory agent ibuprofen, *J. Biol. Chem.* **259**:6752–6756.

Vanderhoek, J. Y., Bryant, R. W., and Bailey, J. M., 1980, Inhibition of leukotriene biosynthesis by the leukocyte product 15-hydroxy-5,8,11,13-eicosatetraenoic acid, *J. Biol. Chem.* **255**:10064–10065.

Vanderhoek, J. Y., Tare, N. S., Bailey, J. M. Goldstein, A. L., and Pluznik, D. H., 1982, New role for 15-hydroxyeicosatetraenoic acid: Activator of leukotriene biosynthesis in PT-18 mast/basophil cells, *J. Biol. Chem.* **257**:12191–12195.

Westcott, J. Y., and Murphy, R. C., 1983, The interaction of ethanol and exogenous arachidonic acid in the formation of leukotrienes and prostaglandin D_2 in mastocytoma cells, *Prostaglandins* **26**:223–239.

15

Regulation of Cyclooxygenase Synthesis in Vascular Smooth Muscle Cells by Epidermal Growth Factor

J. MARTYN BAILEY, TIMOTHY HLA, BARBARA MUZA, and JAMES PASH

1. INTRODUCTION

The prostaglandin synthases, which are widely distributed throughout the various organ systems of the body, are activated primarily by release of substrate arachidonic acid from cellular phospholipids (Hamberg et al., 1975). These prostaglandin-synthesizing systems may become refractory to further challenge by arachidonic acid substrate because of a self-inactivating feature of the cyclooxygenase enzyme first characterized by Smith and Lands (1972). This is mediated by free-radical intermediates of the reaction, which permanently inactivate the enzyme. An analogous refractoriness also develops following exposure to aspirin (Vane, 1971) and is caused by acetylation of a serine residue in the active site (Roth et al., 1975).

Recovery from the refractory state depends on the ability of the affected tissue to synthesize new cyclooxygenase enzyme and to integrate it into the phospholipase-activating systems of the cell (Whiting et al., 1980). Thromboxane (TxA_2) synthesis in blood platelets, which have essentially no protein-synthesizing machinery, is permanently inactivated following a single exposure to aspirin. Platelet aggregation is thus permanently impaired for the remainder of the 9- to 10-day circulating life span of the platelet. These observations have led to a number of clinical trials to

J. MARTYN BAILEY, TIMOTHY HLA, BARBARA MUZA, and JAMES PASH • Department of Biochemistry, George Washington University School of Medicine and Health Sciences, Washington, D.C. 20037.

test the effectiveness of aspirin in preventing heart attacks, since platelets are believed to be involved at several stages in the events that culminate in coronary thrombosis (Ross and Glomset, 1976). (For a review of these trials, see Chapter 39 of this volume.)

Prostacyclin (PGI_2) synthesized in blood vessels via the cyclooxygenase enzyme is a potent inhibitor of platelet aggregation, and PGI_2 release is believed to underlie the well-known property of vascular tissues to resist platelet adhesion and aggregation (Moncada *et al.*, 1977). It is apparent that inactivation of vascular prostacyclin synthesis by aspirin could oppose the beneficial antiaggregatory effects of aspirin on the platelets *per se*. There is evidence that these opposing activities may account for the equivocal results observed in clinical trials using high-dose aspirin (Weksler, 1983).

It has been shown that prostacyclin synthesis by cultured vascular smooth muscle cells can become refractory to further stimuli following a single exposure to either arachidonic acid or the physiological releasing agent thrombin or following brief exposure to low levels of aspirin (Bailey, 1980). This chapter describes experiments that characterize the recovery of prostacyclin synthesis in refractory vascular smooth muscle cells following either self-inactivation or inactivation by aspirin. It was found that recovery takes place relatively rapidly (within 1–2 hr) by a process that is sensitive to the protein synthesis inhibitor cycloheximide but that is not suppressed by actinomycin D. The recovery process showed an absolute requirement for a macromolecular component of serum, which has been identified as epidermal growth factor (EGF).

2. METHODS

2.1. Cell Culture Procedures

Confluent cultures of rat aorta smooth muscle cells were isolated from the aortas of Wistar rats by sequential elution of the cannulated vessel with collagenase and trypsin (Bailey, 1980). Cells were maintained in 25-cm^2 or 75-cm^2 flasks (Costar). Cells were grown in NCTC-135 or Ham's F 12 (Flow Laboratories) medium buffered with HEPES (Fisher) (hereinafter referred to as basal medium) and supplemented with 10% fetal bovine serum (M.A. Bioproducts). Antibiotics (Gibco) gentamycin (50 μg/ml), penicillin (50 units/ml), and streptomycin (50 μg/ml) were added to all cultures. Cells were harvested for subculturing using 0.25% trypsin (Gibco) in calcium- and magnesium-free (CMF) Hanks medium plus EDTA (0.54 mM) and incubated at 37° for 2 to 4 min to release the cells. Cell populations were determined using a Coulter counter and cell protein content by the Lowry procedure. Serum-containing growth medium was then added, and the resulting suspension aliquoted into culture flasks or wells at a subculturing ratio of 1 : 4. Cultures became confluent within 3–4 days, at which time they were used for assay of prostacyclin formation as described below.

Inactivation of cells by aspirin was carried out by replacing the growth medium with basal medium containing a previously prepared solution of aspirin (USP) in a final concentration of 200 μM. The cells were incubated for 30 min at 37°, following which the aspirin-containing medium was removed. Self-inactivation was produced by exposing the cells to arachidonic acid (2.5 μg/ml) for 10 min. Following either inactivation procedure, the cells were washed twice before the test medium was added. Cultures were harvested at intervals and tested for recovery of ability to synthesize prostacyclin as described below. Test compounds cycloheximide, actinomycin D (Boerhinger Mannheim), EGF, or FGF (Collaborative Research Inc., Lexington, MA) and apolipoproteins were dissolved in basal medium before addition to the medium. α-Tocopherol and serum lipids were dissolved in ethanol, 1–5 μl per culture, and appropriate additions of alcohol were made to control cultures. The EGF sample used was a highly purified preparation from mouse submaxillary glands showing a single major band on gel electrophoresis.

2.2. Incubation of Cells with [^{14}C]Arachidonic Acid and Product Extraction

Medium was removed from confluent cultures and the cells were washed twice with basal medium (pH 7.4) at 37° using two 1-ml portions for 25-cm^2 flasks and two 0.5ml portions for well cultures. [1-^{14}C]Arachidonic acid (0.75 μCi, 4 μg/ml) was added as follows: 1 ml per 25-cm^2 flask, 0.5 ml per 10-cm^2 well, and 0.25 ml per 2-cm^2 well. All cultures were incubated at 37° for 5 min, and the medium was collected and added to tubes containing 5 μg each of 6-keto-PGF$_{1\alpha}$, PGF$_{2\alpha}$, PGE$_2$, PGD$_2$, and 15-HETE as carriers. Samples were acidified to pH 3 with 1 N HCl and extracted three times with two volumes of ethyl acetate or once with six volumes of chloroform–methanol (2 : 1). Ethyl acetate layers were collected and backwashed with one volume of distilled water. Medium samples that contained serum were extracted with six volumes of chloroform–methanol, the methanol–aqueous layer was extracted twice more with chloroform, and the organic layers were separated and stored at $-20°$ until analyzed.

2.3. Thin-Layer Chromatography–Autoradiography Procedures

Silica gel G TLC plates (Analtech) were used to separate prostaglandins and hydroxy fatty acids using solvent system Iw (Bailey *et al.*, 1983), which consisted of the organic layer of a mixture of ethyl acetate : isooctane : water and acetic acid in the proportions 11 : 5 : 10 : 2. After sample application, plates were equilibrated in water vapor for 30 min before development. A separate lane of TLC standards was included with each plate, and the compounds in this lane only were visualized by spraying with 10% phosphomolybdic acid in methanol followed by heating at 110°C for 5 min. Radiocative compounds were analyzed by three different procedures. The plates were first scanned using a Vanguard radioactivity scanner (Model 930 Autoscanner, scanner gas 1.3% butane in helium) with an efficiency for ^{14}C of approximately 20%. The plates were then wrapped in plastic wrap (Union Carbide) and placed on $8'' \times 10''$ sheets of XAR X-OMat X-Ray film (Kodak) in the dark for 5–7 days. The exposed film was developed for 5 min in Kodak X-ray developer at 25°C and fixed in Kodak Rapid Fix for 3 min. After separation by TLC and visualization by autoradiography, the radioactive bands were tentatively identified by comparison with the authentic standards and scraped from the plates into 4 ml of Aquasol (New England Nuclear) for quantitative determination by liquid scintillation counting. All data were converted to d.p.m. using a quench curve.

3. RESULTS AND DISCUSSION

The measured doubling time for the rat aorta smooth muscle cells in the 10% serum growth medium used was 23.7 ± 1.6 hr, and cells usually became confluent at about 4 days. Confluent cultures superfused with $[^{14}C]$arachidonic acid synthesized primarily prostacyclin (identified as its stable breakdown product 6-keto-$PGF_{1\alpha}$) as the major component together with lesser quantities of prostaglandins E_2, $F_{2\alpha}$, and D_2 (Fig. 1). Prostacyclin production averaged approximately 90 nanomol/mg of cell protein and was confirmed by bioassay of the supernatant medium at varying times following addition of arachidonic acid to the cultures and by GC/MS of appropriate derivatives as described previously (Bailey *et al.*, 1983).

Release of prostacyclin was directly related to the concentration of arachidonic acid with which the cultures were superfused in the range from 3 μM to 25 μM. The release of prostacyclin was extremely rapid, occurring in a pulse lasting approximately 1 to 2 min following exposure to arachidonic acid. That the reaction was not substrate limited can be readily ascertained from the profile shown in Fig. 1, which indicates that the reaction typically became self-limiting when 50–75% of the arachidonic acid substrate still remained in the medium.

Following treatment with aspirin (0.2 mM) for 30 min, the ability to synthesize prostacyclin and other prostaglandins was completely destroyed (Fig. 1). When the aspirin-containing medium was removed and replaced with fresh medium, the prostacyclin-synthesizing ability of the cells recovered rapidly, reaching a level within 2 hr that was approximately equal to that of the cells immediately before aspirin treatment (Fig. 2, Control). In cultures that were followed for longer periods, prostacyclin synthesis then declined to about 50% of control levels at 24 hr. Recovery of the minor products PGE_2, PGD_2, and $PGF_{2\alpha}$ paralleled that of prostacyclin, confirming that the cyclooxygenase (rather than the prostacyclin endoperoxide isomerase) was the aspirin-labile component being measured.

In order to determine the mechanism underlying this recovery, aspirin-inactivated cultures were incubated with fresh medium in the presence and absence of cycloheximide (20 $\mu g/ml$) or actinomycin D (2 $\mu g/ml$) as inhibitors of protein and RNA snythesis, respectively. Cycloheximide completely blocked the recovery, whereas actinomycin D did not prevent restoration of full activity (Fig. 2).

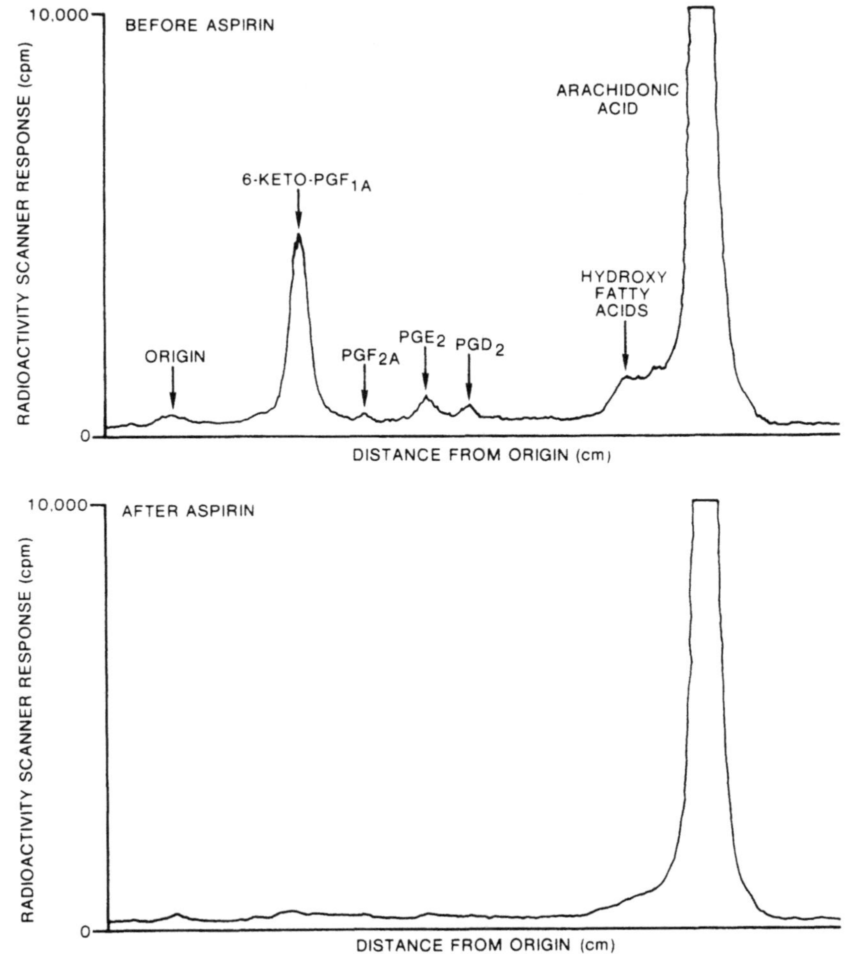

FIGURE 1. Prostacyclin synthesis in cultured rat aorta smooth muscle cells before and after aspirin treatment. Confluent cultures of rat aorta smooth muscle cells were incubated with basal growth medium containing aspirin [1-^{14}C]arachidonic acid (0.75 μCi, 4 μg/ml) at 37° from 5 min. The products were extracted from the medium and analyzed by TLC as described in Sections 2.2 and 2.3. The developed TLC plates were scanned in a Vanguard scanner at an attenuation of 10K. Upper curve: control cells. Lower curve: cells pretreated with 200 μM aspirin for 30 min.

In parallel experiments, cultures were exposed to graded concentrations of arachidonic acid in the range of 1 to 15 μg/ml for 15 min. Essentially complete inactivation of PGI$_2$ synthesis to a second challenge of [^{14}C]arachidonic acid was induced by prior exposure to all concentrations of AA in excess of 2.5 μg/ml. Following removal of the inactivating dose of arachidonic acid and addition of fresh serum-containing medium, self-inactivated cells recovered full prostacyclin synthesizing capacity within 1½ hr.

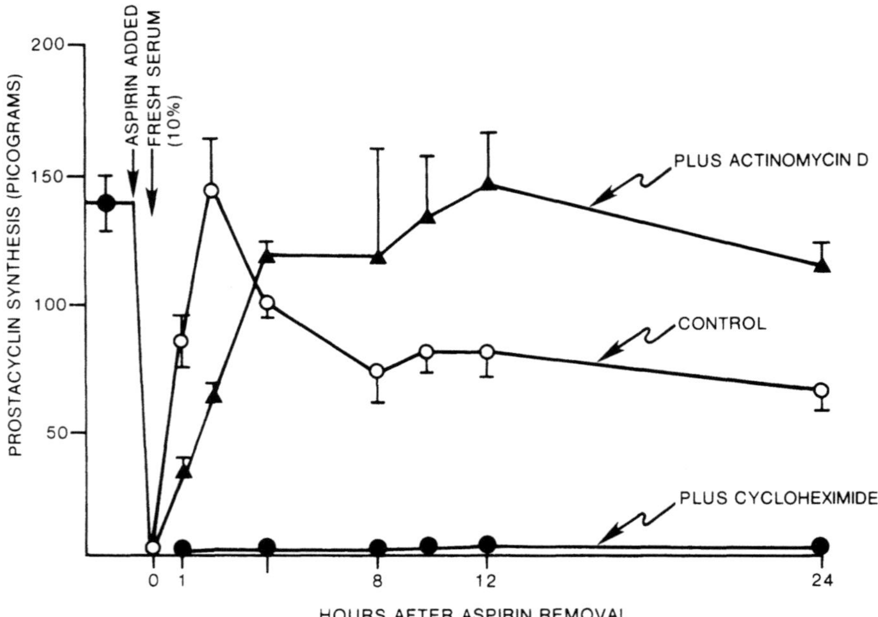

FIGURE 2. Recovery of prostacyclin synthesis in aspirin-treated rat aorta smooth muscle cells: inhibition by cycloheximide. Confluent cultures were treated with aspirin (200 µM) for 30 min as described in Section 2. The washed cells were incubated with fresh growth medium for varying intervals, and recovery of prostacyclin synthesis was tested. Medium contained 10% serum in basal medium alone (control) or supplemented with cycloheximide (20 µg/ml) or actinomycin D (2 µg/ml).

It was found that the recovery of the enzyme following either aspirin treatment or self-inactivation was dependent on a factor present in the calf serum component of the fresh growth medium. Cells incubated either in basal medium alone ("plain" medium) or in serum medium that was removed from confluent cell cultures ("spent" medium) did not support recovery, whereas fresh fetal bovine serum (10%) stimulated full recovery (Fig. 3). The factor was not related to platelet-derived growth factor since equal activities were found in serum prepared from whole blood and both platelet-rich and platelet-poor plasma. Following addition of fresh serum-containing medium, synthesis of all prostaglandin classes was increased proportionately, indicating that the effects observed were directly on the cyclooxygenase enzyme. Serum samples from other species including adult human and guinea pig were also tested and gave similar enhancement of prostacyclin synthesis.

Several experiments were carried out to determine if the cyclooxygenase recovery factor activity in serum was associated with a low-molecular-weight substance bound to the serum proteins. Activity was recovered substantially in the upper layer following filtration of serum through a Diaflo filter with a molecular weight cut-off of approximately 30,000. In addition, serum samples were delipidized by treatment with ethanol at $-70°$. Both the lipid-extractable and the reconstituted delipidized serum protein fractions were tested for recovery activity on aspirin-inactivated cells. Although this treatment resulted in about 50% inactivation of the recovery factor, the residual activity was found to be associated with the protein and not the lipid fractions. In further experiments, serum was separated by ultracentrifugation into the VLDL + LDL and HDL subfractions, and each was tested individually. Essentially no activity was found in any of the lipoproteins, and the activity that was recovered remained in the lipoprotein-free supernatant.

A number of serum constituents were tested for their ability to replace the activity of 10% whole serum, including α-tocopherol (1 to 10 μg/ml), insulin (10 μg/ml), estrogen (1 μg/ml), HDL (50 μg/ml), platelet-derived growth factor, fibroblast growth factor (FGF) (2 μg/ml), and endothelial cell growth factor (150 μg/ml). None gave significant stimulation of recovery over the basal serum-free medium alone. However, addition of purified epidermal growth factor (EGF) at concentrations of 10 and 20 ng/ml to serum-free medium progressively restored prostacyclin synthesis after aspirin inactivation (Fig. 4).

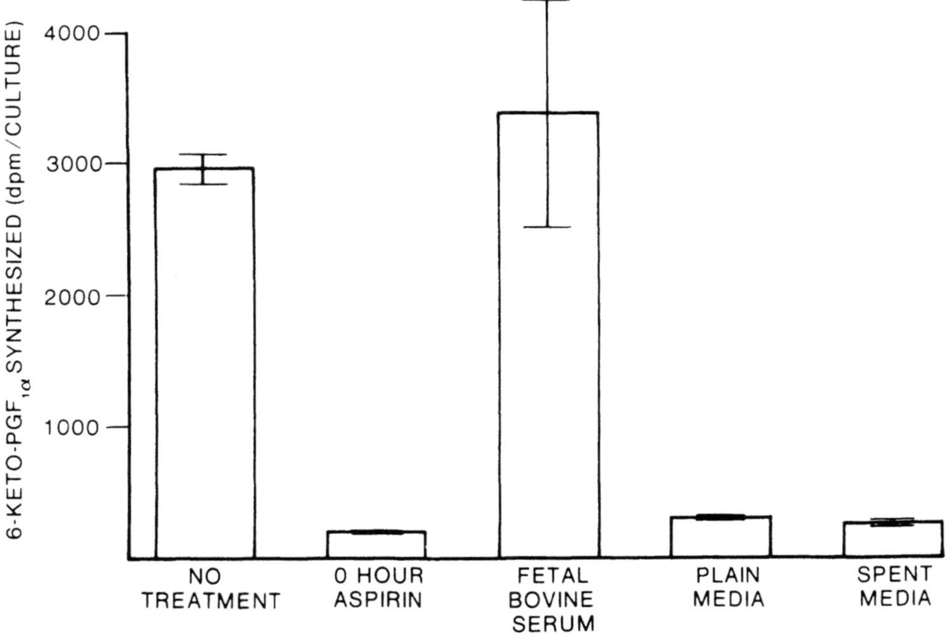

FIGURE 3. Recovery of prostacyclin synthesis after aspirin treatment: requirement for a serum factor. Monolayers of rat aorta smooth muscle cells were treated with aspirin for 30 min. The aspirin-containing medium was removed and replaced with the additions as indicated. Medium was basal medium alone (plain media) or supplemented with 10% fetal bovine serum or spent media. Spent medium consisted of 10% fetal bovine serum in basal medium that was removed from confluent cultures following 4 days of growth. Recovery of prostacyclin synthesis was measured after 3 hr as described in Section 2 using [1-¹⁴C]arachidonic acid as substrate.

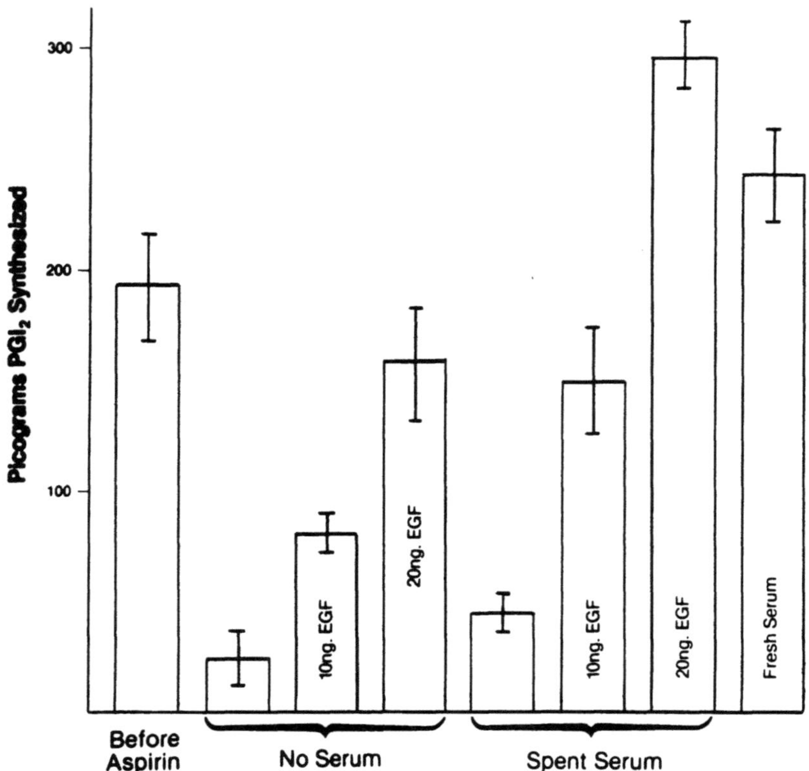

FIGURE 4. Restoration of prostacyclin synthase in aspirin-inactivated vascular smooth muscle cells: requirement for epidermal growth factor. Confluent cultures of rat aorta smooth muscle cells were inactivated by incubation with aspirin (200 μM) for 30 min. Following aspirin removal, recovery of prostacyclin synthesis was measured after 3 hr of incubation in the indicated media as described in Section 2.

As noted above, the stimulating factor is rapidly removed from serum by growing cells, and "spent" medium does not support recovery. Addition of small amounts of EGF (20 ng/ml) to spent medium fully restored the activity to a level equal to that of fresh serum medium (Fig. 4).

Similar observations were made for restoration of prostacyclin synthase activity in cells that had been self-inactivated following a single prior exposure to arachidonic acid. Addition of EGF in the range of 0.1 to 10 ng/ml to serum-free medium progressively restored prostacyclin synthesis in these refractory cells (Fig. 5).

Since the measured doubling time of the vascular smooth muscle cells used in these experiments is 24–30 hr, the rapid recovery of cyclooxygenase in self-inactivated or aspirin-treated cells within 2 hr following addition of EGF represents a relatively selective activation of synthesis of the cyclooxygenase protein compared to that of the average cell protein. Since recovery is blocked by cycloheximide but

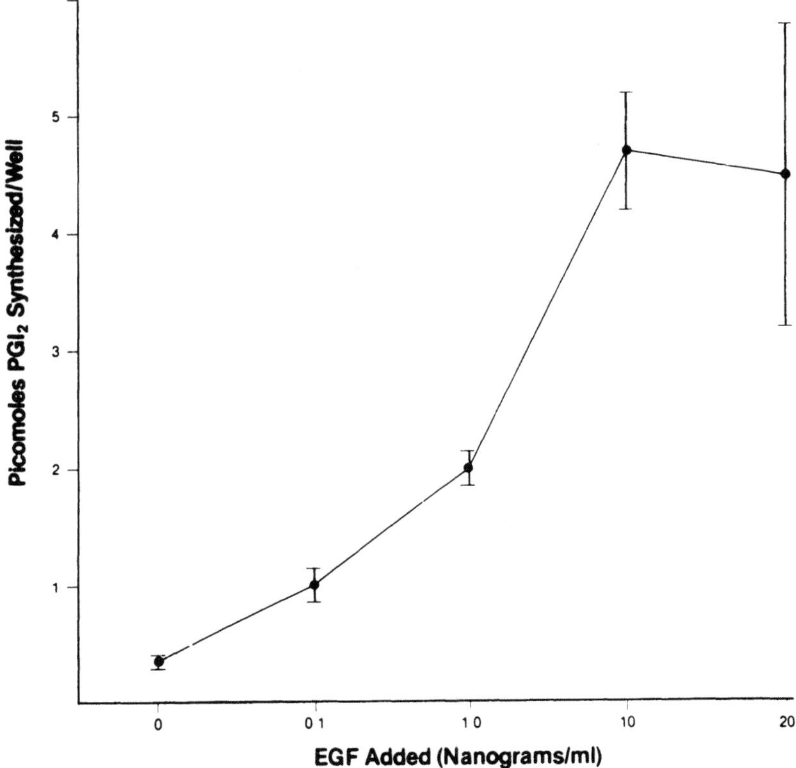

FIGURE 5. The EGF-dependent restoration of prostacyclin synthase activity in refractory (AA-inactivated) vascular smooth muscle cells. Confluent cultures were inactivated by exposure to arachidonic acid (2.5 μg/ml) in serum-free medium for 30 min as described in Section 2. Recovery of prostacyclin synthesis was measured after 1½ hr of incubation in serum-free medium containing the indicated additions of purified epidermal growth factor.

not by actinomycin D, it may be concluded that in rat vascular smooth muscle cells, EGF probably stimulates recovery of refractory cells by enhancing the translation rate of preexisting messenger RNA for the cyclooxygenase.

4. SUMMARY

Cyclooxygenase is a widely distributed enzyme that is regulated in part by a self-catalyzed inactivation reaction. Aspirin produces an analogous refractory state by irreversibly acetylating the enzyme. The mechanisms of recovery of prostacyclin synthesis were studied in cultured rat aorta smooth muscle cells rendered refractory to prostacyclin synthesis by self-inactivation or by prior aspirin exposure.

Confluent cultures, when superfused with [^{14}C]arachidonic acid, synthesized prostacyclin (PGI_2) together with prostaglandins E_2, D_2, and $F_{2\alpha}$. Brief treatment with physiological levels of aspirin (0.2 mM) or arachidonic acid (2.5 μg/ml) completely inactivated prostacyclin synthesis. Following removal of inhibitor and addition of fresh growth medium, PGI_2 synthesis recoverd rapidly with a $t_{1/2}$ of only 30–40 min compared to a doubling time of 24–30 hr for the cells. Recovery of PGE_2, PGD_2, and $PGF_{2\alpha}$ synthesis paralleled that of PGI_2, confirming that cyclooxygenase rather than endoperoxide–prostacyclin isomerase was the labile component. Recovery of PGI_2 synthesis after aspirin was blocked by cycloheximide but not by actinomycin D.

Recovery of inactivated cells required a nondialyzable serum component. The activity was completely removed from the medium by growing cells. It was recovered in protein following delipidation of serum and also in the infranatant after removal of lipoproteins by ultracentrifugation. Activity was not duplicated by α-tocopherol, estradiol, or insulin or by platelet (PDGF), fibroblast (FGF), or endothelial cell (ECGF) growth factors. However, nanogram quantities of purified mouse epidermal growth factor (EGF) in the range 0.1 to 20 ng/ml progressively replaced the biological activity of serum or restored the activity of "spent" serum medium.

These studies indicate that synthesis of cyclooxygenase in rat vascular smooth muscle cells requires a serum factor that can be replaced by EGF. In refractory cells, this induces rapid restoration of the enzyme by a cycloheximide-sensitive process not inhibited by actinomycin D.

REFERENCES

Bailey, J. M., Bryant, R. W., Whiting, J., and Salata, K., 1983, Characterization of 11-HETE and 15-HETE, together with prostacyclin, as major products of the cyclooxygenase pathway in cultured rat aorta smooth muscle cells, *J. Lipid Res.* **24:**1419–1428.

Bailey, J. M., 1980, Regeneration of prostacyclin synthase activity in cultured vascular cells following aspirin treatment, *Adv. Prostaglandin Thromboxane Res.* **6:**537–541.

Hamberg, M., Svensson, J., and Samuelsson, B., 1975, Thromboxanes: A new group of biologically active compounds derived from prostaglandin endoperoxides, *Proc. Natl. Acad. Sci. U.S.A.* **72:**2994–2998.

Moncada, S., Herman, A. G., Higgs, E. A., and Vane, J. R., 1977, Differential formation of prostacyclin (PGX or PGI_2) by layers of the arterial wall. An explanation for the antithrombotic properties of vascular endothelium, *Thromb. Res.* **11:**323–344.

Ross, R., and Glomset, J. A., 1976, The pathogenesis of atherosclerosis, *N. Engl. J. Med.* **295:**369–377.

Roth, G. J., Stanford, N., and Majerus, P. W., 1975, Acetylation of prostaglandin synthase by aspirin, *Proc. Natl. Acad. Sci. U.S.A.* **72:**3073–3076.

Smith, W. L., and Lands, W. E. M., 1972, Oxygenation of polyunsaturated fatty acids during prostaglandin biosynthesis by sheep vesicular gland, *Biochemistry* **11:**3276–3285.

Vane, J. R., 1971, Inhibition of prostaglandin synthesis as a mechanism of action for aspirin-like drugs, *Nature (New Biol.)* **231:**232–235.

Weksler, B. B., 1983, Differential inhibition by aspirin or vascular and platelet prostaglandin synthesis in atherosclerotic patients, *N. Engl. J. Med.* **308:**800–805.

Whiting, J., Salata, K., and Bailey, J. M., 1980, Aspirin: An unexpected side effect on prostaglandin synthesis in cultured vascular smooth muscle cells, *Science* **210:**663–665.

16

Fuel Metabolism as a Determinant of Arachidonic Acid Release and Oxygenation

Studies with Intact Rat Islets of Langerhans

STEWART A. METZ

1. INTRODUCTION

A variety of stimuli (such as bradykinin, angiotensin, and vasopressin) are believed to interact with cell-surface receptors and evoke a release of arachidonic acid (AA) and its oxygenated metabolites in many cells. One link between the two events is presumed to be the activation of calcium-dependent phospholipases (or possibly calcium-dependent lipoxygenases; Ochi *et al.*, 1983), possibly in some cases via mediation by calmodulin. This Ca^{2+} might arise from extracellular sources (for example, the calcium influx induced via cell depolarization), in which case the release of the oxygenated products of AA can be blocked by Ca^{2+} channel blockers (Levine, 1983), or from intracellular stores (blockable by TMB-8, which putatively inhibits release of Ca^{2+} from intracellular sources; cf. Rittenhouse-Simmons and Deykin, 1978). Alternatively, it has been suggested that receptor occupation can directly activate a phospholipase C, leading to degradation of acidic polyphosphoinositides. Such breakdown could release membrane-bound Ca^{2+} (Broekman, 1984), and the concomitant release of inositol phosphates could promote release of Ca^{2+} from intracellular stores (Streb *et al.*, 1983). Such accumulation of Ca^{2+} could then potentiate AA release. Whatever the sequence of events leading to

STEWART A. METZ • Division of Clinical Pharmacology, University of Colorado Health Sciences Center, Denver, Colorado 80262, and Denver VA Medical Center, Denver, Colorado 80220.

arachidonate release in a given cell, the initiating event has usually been considered to take place at the cell surface. In contrast, a possible role for fuel metabolism and related intracellular events in AA release and/or metabolism has received almost no attention. To examine the possibility of such a role, we studied the effect of one fuel (glucose) on the release of the lipoxygenase product 12-hydroxy-eicosatetraenoic acid (12-HETE) from a glucose-sensitive organ, intact rat islets of Langerhans.

2. LIPOXYGENASE-MEDIATED METABOLISM OF AA AND INSULIN RELEASE

We chose to study pancreatic islets since our previously reported data (Metz *et al.*, 1982, 1983a,b, 1984a,b) suggested a critical role for phospholipase activation coupled to a 12-lipoxygenase in stimulus–secretion coupling and insulin release, especially that induced by glucose. We found that in addition to synthesizing several cyclooxygenase products, intact adult rat islets and overnight-cultured, dispersed cells of the neonatal rat pancreas synthesize 12-HETE (identifed by multiple high-performance liquid chromatographic systems and confirmed by gas chromatography/ mass spectrometry). Several other degradation products of 12-hydroperoxyeicosa-tetraenoic acid (12-HPETE) were also tentatively identified, namely, 11,12-epoxy-8- or -10-hydroxyeicosatrienoic acids and trihydroxyeicosatrienoic acids (Metz *et al.*, 1983b). Inhibition of lipoxygenase by any of four structurally dissimilar enzyme blockers (nordihydroguaiaretic acid, BW755c, 15-HETE, ETYA) markedly impeded or abrogated glucose-induced insulin release (Metz *et al.*, 1982, 1983a,b, 1984a,b). These findings were supported by the studies of Turk and colleagues (1983), Yamamoto *et al.* (1983), Laychock (1983a), and Evans *et al.* (1983), who dem-onstrated, in addition, inhibition of glucose-induced insulin release by several phos-pholipase inhibitors and by two other lipoxygenase blockers. Our studies (employing direct application of lipoxygenase- and monooxygenase-derived metabolites of AA and use of pharmacological agents perturbing AA metabolism at several steps distal to lipoxygenase action) implicated the labile hydroperoxide 12-HPETE and possibly epoxides of AA as possible mediators of AA's stimulatory effect on insulin release (Metz *et al.*, 1983b, 1984a,b). A unitary hypothesis was consequently developed (Metz, 1984c) relating the phasic release of AA and/or its metabolites in response to a constant stimulation in many cell types including the islet (Evans *et al.*, 1983) to the biphasic pattern of insulin release induced by glucose.

3. EFFECT OF GLUCOSE ON THE SYNTHESIS OF LIPOXYGENASE-DERIVED METABOLITES OF ARACHIDONIC ACID

This hypothesis would be greatly strengthened if glucose were indeed found to stimulate the accumulation of a presumptive mediator of its effect on insulin release. Therefore, production of 12-HPETE (as assessed by its stable reduction product, 12-HETE) was monitored using intact rat islets. Other positional isomers of HETE and classical leukotrienes derived by the 5-lipoxygenase pathway were not found to be synthesized by the islets of Sprague–Dawley rats; therefore, we assume that the measurements of 12-HETE are a good reflection of net islet lipoxygenase activity. Indeed, in preliminary studies, inhibition of 12-HETE formation by BW755c and nordihydroguaiaretic acid was found to correlate very well ($r = 0.98$) with the blockade by these agents of glucose-induced insulin release (Metz *et al.*, 1984a).

These studies were designed not only to prove insights into the control of AA metabolism by the pancreatic islet but also to shed light on an unresolved controversy (summarized by Hedeskov, 1980, and Ashcroft, 1981) in islet physiology. Proponents of the "regulator-site" hypothesis propose that glucose promotes insulin release by interacting with a stereospecific and anomerically preferential "glucose receptor" on the β-cell membrane; other investigators, espousing the "substrate-site" hypothesis, attribute glucose's action as secretagogue to its stimulation of fuel flux (chiefly along the glycolytic pathway) in the β cell. Studies of AA metabolism could provide new insights into this controversy.

Intact islets were studied not only to preserve normal paracrine intercellular relationships but also to leave undisturbed a normal deacylation–reacylation process for AA and to preserve its normal coupling to the lipoxygenase. Towards this end, the lipids of intact islets were prelabeled for 90 min with tritiated arachidonic acid, washed, and then exposed to various concentrations of hexoses or drugs during static 30-min incubations (Fig. 1). Paired comparisons were only performed between equal numbers of size-matched islets studied on the same day from the same pool of islets in order to eliminate variations between different pools of islets. Medium samples were extracted on ODS minicolumns after addition of UV-detectable amounts of 12-HETE and 15-HETE to serve as internal standards and were chromatographed on a 5-μm reverse-phase ODS column as described elsewhere (Metz *et al.*, 1983b, Metz, 1984a; Metz, 1985). The very small amount of intact cellular material studied per tube (70–100 islets/tube equals about 700–900 μg wet weight of tissue) prevented determination of 12-HETE formation from endogenous AA but offered the advantage of precise calculation of recoveries and monitoring of peak elution through the use of HETE internal standards.

The major findings are reported elsewhere (Metz, 1984a,b; Metz, 1985) and are reviewed here to provide a framework with which to discuss the fuel hypothesis. D-Glucose (16.7mM) stimulated enzymatic 12-HPETE synthesis by a mean of 271% compared to 0–1.7mM glucose. The increment induced by glucose was abolished by the lipoxygenase inhibitors nordihydroguaiaretic acid and phenidone and by the phospholipase inhibitor bromphenacyl bromide. In preliminary studies, D-glucose also seemed to augment both the incorporation and release of AA, suggesting a mechanism of action (at least in part) at the level of phospholipase(s) and/or the reacylation steps. Indeed, a glucose-sensitive phospholipase A_2 has been identified in the pancreatic islet by Laychock (1982). Appropriate controls excluded the possibility that the effect of D-glucose was a "nonspecific" effect of osmolarity,

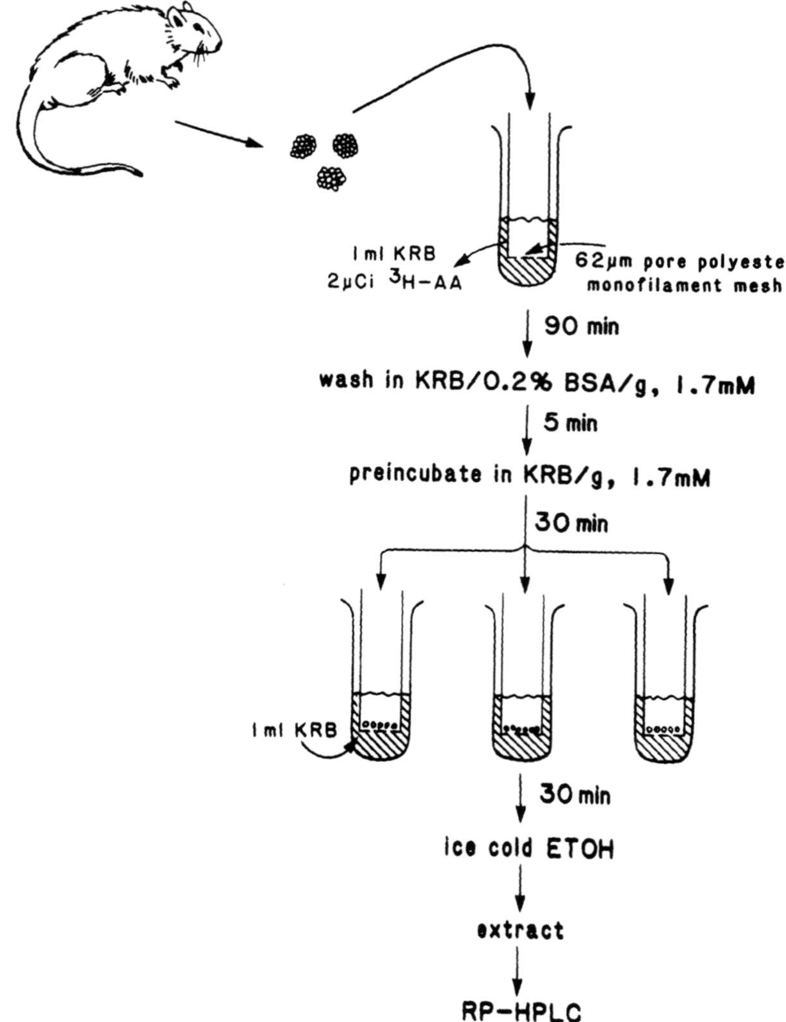

FIGURE 1. Protocol used to study metabolism of arachidonic acid (³H-AA) by intact Sprague–Dawley rat islets. KRB, Krebs–Ringer bicarbonate buffer; g, glucose concentration. Although depicted as being labeled in a single tube, islets were aliquoted into their individual tubes prior to labelling with ³H-AA.

hexose transport, or phosphorylation alone, generation of reduced glutathione, scavenging of hydroxyl radicals (Sagone *et al.*, 1983), or alterations of 12-HETE reuptake into cellular lipids. The effect was stereospecific (L-glucose being ineffective) and apparently was absolutely dependent on the metabolism of glucose, since the glucose effect was largely abrogated by mannoheptulose (which inhibits the rate-limiting step converting glucose to glucose-6-phosphate). Furthermore, either dihydroxyacetone or glyceraldehyde (trioses generated by the glycolytic metabolism of glucose) could substitute effectively for glucose, and the α anomer of D-glucose (which is a preferential substrate for glycolytic metabolism) was more potent than β-D-glucose.

The release of AA and 12-HETE might conceivably be consequent to membrane disruption induced by the emiocytotic release of insulin-containing granules rather than initiating such release. However, mannose (in concentrations nearly equipotent to 16.7 mM glucose in inducing insulin secretion) had no effect. Conversely, increasing glucose from 0 to 1.7 mM caused a small but detectable (and mannoheptulose-inhibitable) increase in 12-HETE release, although no insulin release occurred at such a low glucose concentration. Finally, 2-deoxyglucose (which primarily inhibits the glycolytic metabolism of glucose distal to the formation of glucose-6-phosphate and usually thereby inhibits glucose-induced insulin release; Zawalich, 1979) failed to alter glucose-induced release of 12-HETE. We interpret these data dissociating 12-HETE accumulation from the release of insulin to indicate that activation of the phospholipase–lipoxygenase cascade is a primary action of glucose (and not secondary to hormone release). Furthermore, detectable increases in 12-HPETE synthesis may not be entirely sufficient (nor even required) to mediate insulin release. However, the latter formulation does not exclude a critical, permissive role for "basal" AA and 12-HPETE release in insulin secretion; such activity could reflect the ability of the islet to utilize endogenous fuels in the face of extracellulary glycopenia (Malaisse *et al.*, 1983). It also does not negate an amplifier role for additional release of 12-HPETE above a threshold level. Along this line of reasoning, very low concentrations of glucose are sufficient to promote several other activities in the islet (glucose utilization and the concomitant changes in adenylate charge, lactate production, and cation flux), although at a rate below the thresholds needed to initiate insulin release (Zawalich, 1979; Malaisse, 1979b). In fact, our most recent data (S. Metz, unpublished data) have suggested that lipoxygenase-mediated metabolites of AA serve primarily a *potentiator* role in the islet, whereas, in contrast, lysophospholipids generated by the action of phospholipase A$_2$ may actually be able to *initiate* insulin secretion.

The failure of 2-deoxyglucose to impair glucose-induced 12-HETE accumulation suggests that glucose-6-phosphate (accumulating prior to the deoxyglucose-induced block in its usage) can be metabolized by a pathway other than glycolysis to stimulate AA release (Fig. 2). We speculate that this may be the hexose monophosphate shunt. Assuming that the necessary nonoxidative enzymes of the hexose monophosphate shunt (such as transketolase and transaldolase) are present in the islet, this pathway could yield the triose phosphate glyceraldehyde-3-phosphate (Brolin and Berne, 1967) (and possibly fructose-6-phosphate), thus, in essence, bypassing the glycolytic block between glucose-6-phosphate and fructose-6-phosphate (Fig. 2) and circumventing the consequent reduction in the generation of triose phosphates that would otherwise be induced by 2-deoxyglucose. The pivotal role for glucose-6-phosphate might also be supported by the failure of mannose to stimulate the generation of 12-HPETE, since this hexose yields little or no glucose-6-phosphate (Ashcroft *et al.*, 1970).

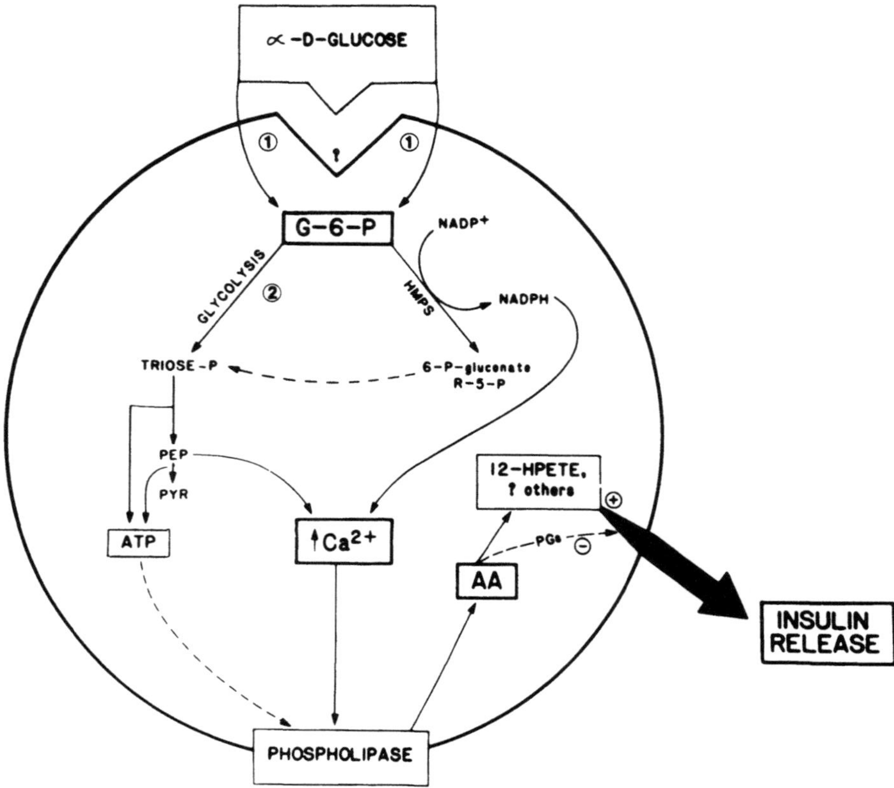

FIGURE 2. Schematization based on current or previously published data depicting the effect of D-glucose on the mobilization of arachidonic acid and its consequent conversion to 12-HPETE (and possibly other labile mediators of insulin release) in rat islets of Langerhans. Mannoheptulose inhibits at site 1, thereby impeding all synthesis of glucose-6-phosphate (G-6-P) from glucose. 2-Deoxyglucose inhibits at site 2, thereby inhibiting conversion of G-6-P to fructose-6-P and subsequently to the triose phosphates but not inhibiting the metabolism of G-6-P by other pathways such as the hexose monophosphate shunt (HMPS). PEP, phosphoenolpyruvate; PYR, pyruvate; R-5-P, ribulose-5-phosphate. Solid lines indicate established or likely pathways. Dashed lines indicate less well established, or secondary, pathways in the islet. The "?" at the cell surface is meant to indicate that it is conceivable that, in addition to the demonstrated role of glucose metabolism, a "glucoreceptor" mechanism could contribute to the stimulation of 12-HPETE synthesis by glucose.

Alternatively, one could interpret the failure of glucose's epimer mannose to support 12-HPETE synthesis as compatible with a very strict structural requirement for hexose-induced stimulation of lipoxygenase pathway products, i.e., a glucoreceptor mechanism. Indeed, we found that the α anomer of D-glucose was more potent at submaximal concentrations in inducing 12-HETE synthesis than β-D-glucose (Metz, 1984b); α-anomeric preference with regard to glucose-induced insulin release has been cited as support for a structurally specific membrane glucose receptor. However, α-D-glucose also preferentially promotes glycolytic flux and cAMP and Ca^{2+} accumulation; therefore, more recently the anomeric preference on insulin release has been used to support the fuel hypothesis (Malaisse *et al.*, 1976; Meglasson and Matschinsky, 1983) of insulin release. Thus, although we cannot absolutely exclude the possibility that glucose metabolism must work in concert with a membrane "glucoreceptor" to maximally promote AA release and oxygenation (Fig. 2), the data using mannoheptulose and the trioses clearly support the primacy of glucose's intracellular metabolism.

4. THE MECHANISM OF THE STIMULATION OF AA METABOLISM BY GLUCOSE

Glucose, by its metabolism is known to promote the accumulation of Ca^{2+} in critical pools in the β cell. Such actions on the availability of calcium are critical to the ability of glucose to trigger insulin release and probably are also critical to the release of arachidonic acid (Laychock, 1982). It has been speculated that the source of some of this calcium is the release of organelle-bound, intracellular Ca^{2+} by phosphoenolpyruvate (Sugden and Ashcroft, 1978), which is generated by the further glycolytic metabolism of the trioses (Fig. 2). Alternatively, calcium fluxes have been correlated (Malaisse *et al.*, 1979a,b) with the accumulation of NADPH, a by-product primarily of the hexose monophosphate shunt (Fig. 2). These two alternative putative sources of Ca^{2+} mobilization could explain the failure of the 2-deoxyglucose-induced block in the glycolytic metabolism of glucose-6-phosphate to reduce 12-HETE synthesis, since an alternate pathway would still be largely available for the further metabolism of glucose-6-phosphate (Fig. 2). (It is interesting to note that organic hydroperoxides themselves could also mobilize Ca^{2+} from intracellular organelles by altering the redox state of the NADPH and glutathione; Bellomo *et al.*, 1982).

Glucose also promotes the influx of extracellular Ca^{2+}. It is tempting to tie such Ca^{2+} fluxes to the breakdown of acidic phospholipids, which have a high affinity for Ca^{2+} prior to hydrolysis. Thus, metabolizable carbohydrates (although not amino acids) stimulated islet phosphatidylinositol hydrolysis in one study (Clements *et al.*, 1981). More recently, it has been reported that glucose promoted the labeling (Best and Malaisse, 1983) and degradation (Laychock, 1983b) of poly-phosphoinositides. These effects were both blocked by mannoheptulose and therefore required glucose metabolism. Furthermore, the former effect on polyphos-phoinositides was reproduced by other islet fuels and was inhibited by menadione, an oxidant of NADPH (Best and Malaisse, 1983). Such effects of a fuel (glucose) could mimic the effect of receptor stimulation in other cells in promoting the gating of Ca^{2+} via the breakdown of acidic phospholipids and the release of inositol phosphates (Broekman, 1984). Alternatively, the depolarization of the β cell in-duced by the metabolism of glucose could be central to the release of arachidonic acid, as it is in other cells (Lazarewicz *et al.*, 1983), by promoting Ca^{2+} influx through calcium channels.

Another effect of glucose metabolism is to yield ATP (Fig. 2). ATP synthesis, like arachidonic acid release, may be a necessary but not sufficient condition to permit insulin secretion (Hedeskov, 1980; Ashcroft, 1981; Malaisse *et al.*, 1979b). Provision of micromolar amounts of ATP (at least extracellularly) can increase arachidonic acid release (Schwartzman and Raz, 1982) and its oxygenation (Ochi *et al.*, 1983). Pong and Levine have provided data suggesting that phospholipase activation in BALB/3T3 cells requires intact oxidative phosphorylation and energy flux (Pong and Levine, 1977). The potential contributions of the metabolism of glucose as substrate in the synthesis of *sn*-glycerol-3-phosphate, inositol, phosphatidic acid, and perhaps other species that may play critical roles in islet phospholipid economy also merit investigation with regard to the release of AA. Fuels other than glucose also merit study, especially since fats such as octanoate augment islet accumulation of glucose-6-phosphate (Montague and Taylor, 1969) and amino acids augment accumulation of Ca^{2+} and NADPH in the islet (Hutton *et al.*, 1980). By these actions, other fuels could mimic glucose with respect to the potentiation of AA release. We have, in fact, recently observed that α-ketoisocaproic acid (a metabolite of the amino acid leucine) augments accumulation of 12-HPETE in intact islets (S. Metz, unpublished data), and that insulin release induced by this fuel is blocked by lipoxygenase inhibitors.

It should be emphasized that not all AA release from the islet is energy-dependent. Recently we have observed (S. Metz, unpublished data) that there is a "basal" release of AA (which may correspond to basal insulin secretion); this release (unlike that induced by glucose) is refractory to mannoheptulose, antimycin A (an inhibitor of electron transport and ATP formation) or blockade of Ca^{2+} availability using cobalt. Thus, the nutrient-stimulated islet may generate energy, which permits Ca^{2+} mobilization and consequent activation of typical phospholipases. The basal (or fuel-deprived) islet may depend on a second, Ca^{2+}- and energy-independent "pool" of arachidonate. Candidate enzyme systems activating such a pool might be ATP-independent transacylation mechanisms such as the "acyl CoA transferase in reverse" described by Trotter *et al.* (1982) coupled to an acyl CoA hydrolase, or a phospholipase A_1 or transacylase coupled with lysophospholipase activity (Abe *et al.*, 1974). Other atypical lipases (triglyceride lipase or a lysosomal phospholipase A_2) could also be candidates worthy of investigation.

5. RELEVANCE OF FUEL METABOLISM TO ARACHIDONIC ACID PHYSIOLOGY IN NONENDOCRINE CELLS

The available literature, though sparse, does suggest that fuel metabolism may indeed be an important determinant of arachidonic acid release and/or oxygenation in several nonendocrine cell types (see, for example, Pong and Levine, 1977). In sensitized human lung fragments (Kaliner and Austen, 1973), rabbit polymorphonuclear leukocytes (Walker and Parish, 1981), and horse eosinophils (Ziltener *et al.*, 1983), the release of lipoxygenase products can be reduced by the metabolic inhibitors 2-deoxyglucose or iodoacetate. In the latter study, glucose increased LTB_4 (but decreased LTC_4) production. In the pig thyroid, provision of phosphoenol-pyruvate and pyruvate kinase promotes arachidonic acid release (Haye *et al.*, 1984), presumably via the generation of ATP. In platelets, 12-HETE synthesis from exogenous arachidonate is augmented by glucose, an action attributed to its ability to scavenge hydroxyl radicals (Sagone *et al.*, 1983), whereas metabolic inhibitors (2-deoxyglucose plus antimycin A) reduce arachidonate release from thrombin-stimulated platelets, apparently by reducing ATP availability (Rittenhouse-Simmons and Deykin, 1977). The weak antioxidant effect of glucose (Sagone *et al.*, 1983) could potentiate a primary effect of glucose to augment 12-HPETE release under some circumstances by preventing the suicide inactivation of lipoxygenase by hydroxyl radicals. (Indeed, in our studies, mannitol had no effect in the presence of low glucose concentrations but tended to augment the 12-HETE response to high glucose concentrations.) However, high levels of glucose apparently can directly inhibit lipoxygenase (Morrison *et al.*, 1982).

Conversely, severe intracellular glycopenia (like ischemia) can, under some circumstances, paradoxically promote release of AA (Agardh *et al.*, 1980; Tannenbaum *et al.*, 1979). This may represent a specific effect of impeding the supply of energy and especially ATP needed for reesterification of AA, or it may represent a nonspecific noxious stimulus to AA release. Thus, the effects of fuel metabolism on AA release are complex and may vary from cell to cell. Nonetheless, intracellular events, especially those initiated by fuel fluxes, appear to merit further investigation as determinants of AA release and metabolism.

ACKNOWLEDGMENTS. These studies were supported by the Veterans Administration, and by grants from the Juvenile Diabetes Foundation, the Kroc Foundation, and NIH Grant AM 31112. The technical assistance of Ms. Nancy Mahr and Mr. Doug Holmes is gratefully acknowledged.

REFERENCES

Abe. M., Ohno. K., Sato. R., 1984, Possible identity of lysolecithin acyl-hydrolase with lysolecithin, lysolecithin acyl-transferase in rat-lung soluble fraction, *Biochem. Biophys. Acta.* **369**:361–70.

Agardh, C.-D., Westerberg, E., and Siesjö, B K., 1980, Severe hypoglycemia leads to accumulation of arachidonic acid in brain tissue, *Acta Physiol. Scand.* **109**:115–116.

Ashcroft, S. J. H., 1981, Metabolic control of insulin secretion, in: *The Islets of Langerhans: Biochemistry, Physiology and Pathology* (S. J. Cooperstein and D. Watkins, eds.), Academic Press, New York, pp. 117–148.

Ashcroft, S. J. H., Hedeskov, C. J., and Randle, P. J., 1970, Glucose metabolism in mouse pancreatic islets, *Biochem. J.* **118**:153–154.

Bellomo, G., Jewell, S. A., Thor, H., and Orrhenius, S., 1982, Regulation of intracellular calcium compartments: Studies with isolated hepatocytes and *t*-butyl hydroperoxide, *Proc. Natl. Acad. Sci. U.S.A.* **79**:6842–6846.

Best, L., and Malaisse, W. J., 1983, Effects of nutrient secretagogues upon phospholipid metabolism in rat pancreatic islets, *Mol. Cell Endocrinol.* **32**:205–214.

Broekman, M. J., 1984, Phosphatidylinositol 4,5-bisphosphate may represent the site of release of plasma membrane-bound calcium upon stimulation of human platelets, *Biochem. Biophys. Res. Commun.* **120**:226–231.

Brolin, S. E., and Berne, C., 1967, The enzymatic activities of the initial glycolytic steps in pancreatic islets and acini, *Metabolism* **16**:1024–1028.

Clements, R. S., Jr., Evans, M. H., and Pace, C. S., 1981, Substrate requirements for the phosphoinositide response in rat pancreatic islets, *Biochim. Biophys. Acta* **674**:1–9.

Evans, M. H., Pace, C. S., and Clements, R. J., Jr., 1983, Endogenous prostaglandin synthesis and glucose-induced secretion from the adult rat pancreatic islet, *Diabetes* **32**:509–515.

Haye, B., Champion, S., and Jacquemin, C., 1984, Existence of two pools of prostaglandins during stimulation of the thyroid by TSH, *FEBS Lett.* **41**:89–93.

Hedeskov, C. J., 1980, Mechanism of glucose-induced insulin secretion, *Physiol. Rev.* **60**:442–509.

Hutton, J. C., Senar, A., Herchuelz, A., Atwater, I., Kawazu, S., Boschero, A. C., Somera, G., Davis, G., and Malaisse, W. J., 1980, Similarities in the stimulus–secretion coupling mechanisms of glucose- and 2-keto acid-induced insulin release, *Endocrinology* **106**:203–219.

Kaliner, M., and Austen, K. F., 1973, A sequence of biochemical events in the antigen-induced release of chemical mediators from sensitized human lung tissue, *J. Exp. Med.* **183**:1077–1094.

Laychock, S. G., 1982, Phospholipase A_2 activity in pancreatic islets is calcium-dependent and stimulated by glucose, *Cell Calcium* **3**:43–54.

Laychock, S. G., 1983a, Fatty acid incorporation into phospholipids of isolated pancreatic islets of the rat, *Diabetes* **32**:6–13.

Laychock, S. G., 1983b, Identification and metabolism of polyphosphoinositides in isolated islets of Langerhans, *Biochem. J.* **216**:101–106.

Lazarewicz, J. W., Leu, V., Sun, G. Y., and Sun. A. Y., 1983, Arachidonic acid release from K^+-evoked depolarization of brain synaptosomes, *Neurochem. Int.* **5**:471–478.

Levine, L., 1983, Inhibition of the A23187-stimulated leukotriene and prostaglandin biosynthesis of rat basophil leukemia (RBL-1) cells by non-steroidal antiinflammatory drugs, anti-oxidants and calcium channel blockers, *Biochem. Pharmacol.* **32**:3023–3026.

Malaisse, W. J., Sener, A., Koser, M., and Herchuelz, A., 1976, Stimulus–secretion coupling of glucose-induced insulin release, *J. Biol. Chem.* **251**:5936–5943.

Malaisse, W. J., Hutton, J. C., Kawazu, S., Herchuelz, A., Valverde, I., and Sener, A., 1979a, The stimulus–secretion coupling of glucose-induced insulin release. XXXV. The links between metabolic and cationic event, *Diabetologia* **16**:331–341.

Malaisse, W. J., Sener, A., Herchuelz, A., and Hutton, J. C., 1979b, Insulin release: The fuel hypothesis, *Metabolism* **28**:373–386.

Malaisse, W. J., Best, L., Kawazu, S., Malaisse-Lagae, F., and Sener, A., 1983, The stimulus–secretion coupling of glucose-induced insulin release: Fuel metabolism in islets deprived of exogenous nutrient, *Arch. Biochem. Biophys.* **224**:102–110.

Meglasson, M. D., and Matschinsky, F. M., 1983, Discrimination of glucose anomers by glucokinase from liver and transplantable insulinoma, *J. Biol. Chem.* **258**:6705–6708.

Metz, S. A., 1984a, Glucose promotes lipoxygenase-mediated metabolism of arachidonic acid (AA) in rat islets, *Clin. Res.* **32**:51a, 403a.

Metz, S. A., 1984b, Glucose stimulation of arachidonic acid (AA)-derived mediators of insulin release shows anomeric preference and requires hexose metabolism, *Diabetes* **33**:(**Suppl. 1**):42a.

Metz, S., 1984c, Is phospholipase A_2 a "glucose sensor" responsible for the phasic pattern of insulin release? *Prostaglandins* **27**:147–158.

Metz, S. A., 1985, Glucose increases the synthesis of lipoxygenase-mediated metabolites of arachidonic acid in intact rat islets, *Proc. Natl. Acad. Sci. USA.* **82**:198–202.

Metz, S. A., Fujimoto, W. Y., and Robertson, R. P., 1982, Lipoxygenation of arachidonic acid: A pivotal step in stimulus-secretion coupling in the pancreatic beta cell, *Endocrinology* **111**:2141–2143.

Metz, S. A., Fujimoto, W. Y., and Robertson, R. P., 1983a, A role for the lipoxygenase pathway of arachidonic acid metabolism in glucose- and glucagon-induced insulin secretion, *Life Sci.* **32**:903–910.

Metz, S., VanRollins, M., Strife, R., Fujimoto, W., and Robertson, R. P., 1983b, Lipoxygenase pathway in islet endocrine cells: Oxidative metabolism of arachidonic acid promotes insulin release, *J. Clin. Invest.* **71**:1191–1205.

Metz, S. A., Fujimoto, W., and Robertson, R. P., 1984a, Oxygenation products of arachidonic acid: Third messengers for insulin release, *J. Allergy Clin. Immunol.* **74**:391–402.

Metz, S. A., Murphy, R. C., and Fujimoto, W. Y., 1984b, Effects on glucose-induced insulin release of lipoxygenase-mediated metabolites of arachidonic acid, *Diabetes* **33**:119–124.

Montague, W., and Taylor, K. W., 1969, Islet cell metabolism during insulin release, *Biochem. J.* **115**:257–262.

Morrison, A. R., Winokur, T. S., and Brown, W. A., 1982, Inhibition of soybean lipoxygenase by mannitol, *Biochem. Biophys. Res. Commun.* **108**:1757–1762.

Ochi, Y., Yoshimoto, T., Yamamoto, S., Taniguchi, K., and Miyamoto, T., 1983, Arachidonate 5-lipoxygenase of guinea pig peritoneal polymorphonuclear leukocytes. Activation by adenosine 5'-triphosphate, *J. Biol. Chem.* **258**:5754–5758.

Pong, S.-S., Levine, L., 1977, Prostaglandin production by methylcholanthrene-transformed mouse BALB/3T3: Effect of oxidative phosphorylation inhibitors, *Protaglandins* **13**:65–71.

Rittenhouse-Simmons, S., and Deykin, D., 1977, The mobilization of arachidonic acid in platelets exposed to thrombin or ionophore A23187, *J. Clin. Invest.* **60**:495–498.

Rittenhouse-Simmons, S., and Deykin, D., 1978, The activation by Ca^{2+} of platelet phospholipase A_2. Effects of dibutyryl cyclic adenosine monophosphate and 8(-N,N-diethylamino)-octyl-3,4,5-trimethoxybenzoate, *Biochim. Biophys. Acta* **543**:409–422.

Sagone, A. L., Jr., Greenwald, J., Kraut, E. H., Bianchine, J., and Singh, D., 1983, Glucose: A role as a free radical scavenger in biological systems, *J. Lab. Clin. Med.* **101**:97–104.

Schwartzman, M., and Raz, A., 1982, Purinergic vs peptidergic stimulation of lipolysis and prostaglandin generation in the perfused rabbit kidney, *Biochem. Pharmacol.* **31**:2453–2458.

Streb, H., Irvine, R. F., Berridge, M. J., and Schulz, I., 1983, Release of Ca^{2+} from a nonmitochondrial intracellular store in pancreatic acinar cells by inositol-1,4,5-trisphosphate, *Nature* **306**:67–69.

Sugden, M. D., and Ashcroft, S. J. H., 1978, Effects of phosphoenolpyruvate, other glycolytic intermediates and methylxanthine on calcium uptake by a mitochondrial fraction from rat pancreatic islets, *Diabetologica* **15**:173–180.

Tannenbaum, J., Sweetman, B. J., Nies, A. S., Aulsebrook, K., and Oates, J. A., 1979, The effect of glucose on the synthesis of protaglandins by the renal papilla of the rat *in vitro, Prostaglandins* **17**:337–350.

Trotter, J., Flesch, I., Schmidt, B., Ferber, E., 1982, Acyltransferase-catalyzed cleavage of arachidonic acid from phospholipids and transfer to lysophosphatides in lymphocytes and macrophages, *J. Biol. Chem.* **257**:1816–1823.

Turk, J., Kotagal, N., and Greider, M., 1983, Arachidonic acid metabolism in isolated pancreatic islets from the rat, *Diabetes (Suppl.)* **32**:139a.

Walker, J. R., and Parish, H. A., 1981, Metabolic requirements for rabbit polymorphonuclear leucocyte lipoxygenase activity, *Int. Arch. Allergy Appl. Immunol.* **66**:83–90.

Yamamoto, S., Ishii, M., Nakadate, T., Nakadi, T., and Kato, R., 1983, Modulation of insulin secretion by lipoxygenase products of arachidonic acid, *J. Biol. Chem.* **258**:12149–12152.

Zawalich, W. S., 1979, Intermediary metabolism and insulin secretion from isolated rat islets of Langerhans, *Diabetes* **28**:252–260.

Ziltener, H. J., Chavaillaz, P.-A., and Jörg, A., 1983, Leukotriene formation by eosinophil leukocytes, *Hoppe Seylers J. Physiol. Chem.* **364**:1029–1037.

17

Regulation of Differentiation of Canine Kidney (MDCK) Cells by Prostaglandins

SUZANNE K. BECKNER and MICHAEL C. LIN

1. INTRODUCTION

Very little is known about the complex process of cellular differentiation. Clearly, the appearance of differentiated characteristics must be sequentially organized during normal development. It is likely that such processes involve a complex interplay of factors that regulate gene activity as well as biochemical signals that regulate the functions of newly expressed gene products.

Carcinogenesis, which is characterized by dedifferentiation, would seem to represent the opposite of differentiation. It is likely that many of the same signals that regulate differentiation during normal development play some role in regulating the transformed state. Better understanding of such regulation would contribute to our knowledge of normal as well as abnormal cellular differentiation.

Prostaglandins have currently received a great deal of attention for their role in cellular processes. Evidence is accumulating for an important regulatory role of prostaglandins and leukotrienes in cellular proliferation, differentiation, and carcinogenesis, the topic of a 1981 symposium (Ramwell, 1982). There is evidence that prostaglandin-synthetase-dependent cooxidation plays a role in tumor initiation and promotion (Marnett, 1981; Fischer and Slaga, 1982). Additionally, thromboxanes and prostacyclin have been shown to directly modulate tumor growth (Honn and Meyer, 1981). A variety of growth factors, including epidermal, platelet-derived, and tumor growth factors, increase endogenous prostaglandin production

SUZANNE K. BECKNER and MICHAEL C. LIN ● Laboratory of Cellular and Developmental Biology, National Institute of Arthritis, Diabetes, Digestive and Kidney Disease, National Institutes of Health, Bethesda, Maryland 20205.

in a number of cells (Levine, 1982). In fact, prostaglandins of the F series act as potent growth factors for primary neonatal rat hepatocytes (Armato and Andreis, 1983). The role of prostaglandins in differentiation is less clear (Honn *et al.*, 1981). Prostaglandins both stimulate and inhibit differentiation depending on the cell examined. In the hematopoietic cell system, prostaglandins are certainly physiologically important regulators of stem cell proliferation and differentiation (Honn *et al.*, 1981).

This chapter discusses current findings for a stimulatory role of prostaglandins in the regulation of expression of a differentiated function, in this case, glucagon receptors in a transformed line of kidney cells.

1.1. Model Systems of Cellular Differentiation

Epithelial cells are the most abundant type of cell in the body, but very little is known about the development and maintainence of differentiated functions by such cells. It is likely that cell–cell contact as well as biochemical factors including local developmental hormones are involved in these processes.

A number of good epithelial cell model systems of differentiation exist and have recently been reviewed (Lin and Beckner, 1983). Additionally, hematopoietic stem cell systems provide good model systems for complex biochemical differentiation, as a number of specific growth and differentiation factors have been identified.

Even when a single differentiated function is studied, the cellular environment must be controlled. Cells in culture offer a number of advantages over *in vivo* studies, where numerous factors are at work. Although a primary culture represents a more physiological system to examine differentiation, in general, such preparations do not survive long enough to be useful for long-term studies. Additionally, primary cultures tend to lose their differentiated functions after several days in culture. On the other hand, established cell lines provide a continuous supply of homogeneous cell populations. Furthermore, it is possible to grow a number of cell lines in totally defined media, thus eliminating effects of undefined serum factors. Such a model system allows manipulation of the cellular environment, permitting examination of one or more cellular differentiated functions.

1.2. Hormone Responsiveness as a Differentiated Function

The ability of a cell to respond to hormones represents a highly differentiated function. There is ample evidence to indicate that hormone responsiveness develops in an orderly fashion during normal development. During fetal development, rat hepatocytes become increasingly responsive to glucagon, whereas responsiveness to insulin appears much earlier (Blazquez *et al.*, 1976). The ability of rat cerebral cortex to respond to catecholamines and adenosine during neonatal development also develops in an orderly fashion (Perkins and Moore, 1973).

It has also been demonstrated that the ability to respond to one hormone is required for the development of responsiveness to another hormone. In erythroleu-kemic cells, the appearance of β-adrenergic receptors precedes the induction of other functions (Lin and Lin, 1979). The synergistic effects of β-agonists and erythropoietin on the maturation of bone marrow cells (Brown and Adamson, 1977) support the significance of the early appearance of β-adrenergic receptors in erythroleukemic cells.

In the granulosa cell model system, treatment with follicle-stimulating hormone induces the appearance of receptors for luteinizing hormone. Similarly, insulin regulates the appearance of ACTH receptors in 3T3-L1 cells (Rubin *et al.*, 1977) and vasopressin receptors in LLC-PK kidney cells (Roy *et al.*, 1980). The appearance of β-adrenergic receptors in rat lung is modulated by T_4 (Whitsett *et al.*, 1980). Such studies suggest that responsiveness to these hormones appears sequentially during development. The appearance of responsiveness to a particular hormone also seems to be dependent on the cell type. Whereas receptors for β-adrenergic agonists are well developed in rat liver during the fetal stage, these same receptors in the cerebral cortex are not fully expressed until well after birth (Perkins and Moore, 1973).

Hormone responsiveness can be regulated at the receptor level or beyond, at the level of signal transmission. Hormone-responsive systems that are coupled to adenylate cyclase have been extensively studied and reviewed (Ross *et al.*, 1983). There are data to suggest that adenylate cyclase activity is altered following transformation (Anderson and Pastan, 1975; Beckner, 1984) and in various disease states (Levine *et al.*, 1980) by changes in one or more of its components. Additionally, activity can be altered by changes in the number of hormone receptors.

Because of the great deal of information available about the adenylate cyclase system and the probable role of hormone responsiveness in regulating subsequent differentiated functions, a model system in which hormone responsiveness can be regulated offers a good system in which to examine various questions concerning the biochemical regulation of differentiation.

1.3. The MDCK Cell Model System

The MDCK cells were established in 1958 by Madin and Darby (Gaush *et al.*, 1966) from the kidney of a normal cocker spaniel. These cells maintain microvillus projections and tight junctions on the surface, suggesting that they retain polarity in culture (Leighton *et al.*, 1969). The MDCK cells also retain a number of differentiated functions characteristic of kidney distal tubule in culture including fluid and electrolyte transport, a differentiated function regulated by cAMP (Rindler *et al.*, 1979).

The adenylate cyclase of MDCK cells is responsive to a number of hormones (Rindler *et al.*, 1979), including glucagon, vasopressin, prostaglandin E_1, and β-adrenergic agonists. Hormone responsiveness is conveniently monitored by measurement of intracellular cAMP generated during acute (3-min) exposure to the hormone, as described (Lin *et al.*, 1982a). Responsiveness is expressed as fold activation or cAMP produced in the presence of hormone relative to control, unstimulated cells. It is also possible to measure hormonal activation of adenylate cyclase in purified membrane preparations derived from MDCK cells. As can be seen in Table I, normal or parental MDCK cells respond to glucagon and vasopressin with six- and eightfold increases in intracellular cAMP, respectively. The response to isoproterenol and PGE_1 is much greater, 15- and 25-fold, respectively.

A cloned line of MDCK cells transformed with Harvey sarcoma virus (Scolnick *et al.*, 1976) has also been established. Transformed MDCK cells have a more fibroblastic morphology, and p21, the *ras* oncogene product (Shih *et al.*, 1980), has been demonstrated to be associated with the inner plasma membrane of MDCK cells (Willingham *et al.*, 1980). An additional consequence of viral transformation is the selective loss of glucagon responsiveness. As seen in Table I, transformed MDCK cells, in contrast to the parental line, do not respond to glucagon, although cAMP responsiveness to vasopressin, PGE_1 and isoproterenol are still evident.

It is possible to grow both normal and transformed MDCK cells in a totally defined medium (Taub *et al.*, 1979) consisting of 50% Dulbecco's minimum essential medium, 50% Hams F12 medium, 10 mM HEPES, 5 μg/ml insulin, 5 μg/ml transferrin, 50 nM hydrocortisone, 5 pM T3, and 10 nM selenium. Therefore, except where indicated, all experiments were performed in defined medium to eliminate serum effects. Thus, normal and transformed MDCK cells provide a good model system to compare biochemical factors that regulate the expression of a differentiated function, i.e., glucagon responsiveness, of an established cell line under defined conditions.

2. REGULATION OF MDCK CELL DIFFERENTIATION

It is possible to induce transformed MDCK cells to respond to glucagon (Table I) by the addition of sodium butyrate (Lin *et al.*, 1982a), which is known to induce differentiation in a number of systems (Leder and Leder, 1975). The induction of glucagon sensitivity by butyrate is time and concentration dependent (Lin *et al.*, 1982a). Glucagon sensitivity can be induced with 0.5–2 mM sodium butyrate, is detectable within 8 hr, and is maximal by 72 hr. The glucagon sensitivity induced

TABLE I. Hormone Responsiveness of MDCK Cells[a]

	Glucagon	Vasopressin	Isoproterenol	PGE_1
Normal	6	8	15	29
Transformed	1	3	16	37
Induced	5	4	10	18

[a] Normal or transformed MDCK cells were incubated in defined medium with (induced) or without 1 mM butyrate for 72 hr. Monolayers were washed, and cAMP generated over a 3 min period measured as described (Lin *et al.*, 1982a) in the presence or absence (basal) of saturating concentrations of the indicated hormone. Data are expressed as cAMP formed in the presence of hormone divided by that in its absence (basal). Data represent the average of values from triplicate dishes.

in transformed MDCK cells by butyrate differs from that of the parental line (Beckner *et al.*, 1985) in that the K_{act} for glucagon activation of adenylate cyclase is tenfold higher than that of the parental line (100 nM vs. 10 nM). Whether this represents a receptor defect or an alteration in the adenylate cyclase system is not known, but the fact that the adenylate cyclase sensitivity to PGE and isoproterenol is unchanged after induction suggests that the induced glucagon receptor may be less efficient in its interaction with adenylate cyclase.

That glucagon sensitivity can be induced in transformed MDCK cells permits manipulation and regulation of the expression of a differentiated function.

2.1. Induction by Endogenous Prostaglandins

During the course of studies of butyrate induction of glucagon sensitivity, it was noted that occasionally control, or uninduced cells, exhibited glucagon sensitivity. This was not a reversion to the parental phenotype, since the cells still expressed viral protein. This spontaneous glucagon sensitivity was further characterized and found to be dependent on culture conditions.

2.1.1. Development of Spontaneous Glucagon Sensitivity

Unlike butyrate induction of glucagon sensitivity, which was maximal after 72 hr, spontaneous glucagon sensitivity required 7–14 days in culture to develop. As can be seen in Table II, activation by glucagon was barely detectable after 7 days of culture (1.8-fold), but after 14 days, the activation by glucagon was 4.2-fold compared to the control. This result suggested that glucagon sensitivity appears as a function of culture age. Further studies revealed that culture density as well as culture age contributes to the appearance of glucagon sensitivity.

The glucagon sensitivity that appeared spontaneously resembled that induced by butyrate in that the K_{act} for adenylate cyclase activation by glucagon was tenfold higher (Fig. 1). In other words, the spontaneous appearance of glucagon sensitivity was identical to that induced by butyrate, thus providing more physiological conditions in which to examine the biochemical events involved in the expression of glucagon sensitivity. The increased K_{act} for glucagon activation of adenylate cyclase in the spontaneously sensitive cells also confirms that such hormone responsiveness does not reflect a reversion to the normal phenotype.

2.1.2. Prostaglandin Production by MDCK Cells

In addition to responding to a variety of hormones, the kidney itself is an endocrine organ and secretes renin and prostaglandins (Katz and Lindheimer, 1977). Like the kidney, MDCK cells have been shown to produce prostaglandins (Hassid, 1981). We had previously demonstrated that transformed MDCK cells produce only 1–2% of the amount of PGE_2 and $PGF_{2\alpha}$ as the parental line under identical culture conditions (Lin *et al.*, 1982b) because of decreased activity of phospholipase and

TABLE II. Time Course Development of Spontaneous
Glucagon Sensitivity[a]

Day of culture	Cell density ($\times 10^6$/dish)	Glucagon sensitivity
3	0.34	1.2
5	0.99	1.8
7	1.2	2.2
10	2.09	4.0
14	2.64	4.2

[a] Transformed MDCK cells were cultured in defined medium. At the indicated day of culture, cell number was determined, and glucagon responsiveness determined as in Table I.

PGE_2 synthetase. This suggested that endogenous prostaglandin production might be involved in the expression of glucagon sensitivity by MDCK cells. In three separate experiments (Table III), spontaneous glucagon sensitivity of transformed cells (up to fivefold activation) increased as the passage number of the cells increased (Table III). The correlation between glucagon sensitivity and the production of PGE_2, but not $PGF_{2\alpha}$, was evident. The production of PGE_2 varied from 0.3 to 19.8 pmol/dish per 3 days, and in every case, as the production of PGE_2 increased, so did the glucagon sensitivity.

2.1.3. Regulation of Endogenous Prostaglandin Production

Numerous attempts were made to regulate endogenous prostaglandin production in order to correlate this with enhanced glucagon sensitivity. Although indomethacin, known to inhibit endogenous prostaglandin production, did decrease the production of PGE_2 in transformed MDCK cells, it did not consistently inhibit the appearance of glucagon sensitivity. Similarly, bradykinin was shown to enhance the production of PGE_2 but did not induce glucagon sensitivity. For example, bradykinin caused only a twofold increase in endogenous PGE_2 production of transformed MDCK cells, which is still well below the amount of PGE_2 produced by parental MDCK cells, under comparable culture conditions. Similarly, there was no induction of glucagon sensitivity by phorbol esters, which are also known to enhance endogenous prostaglandin production in MDCK cells (Ohuchi and Levine, 1978).

Since the spontaneous appearance of glucagon sensitivity requires 7–14 days, it was necessary to include these regulators of endogenous prostaglandin production for this long period of time. Such experimental conditions as well as the toxicity and lability of these compounds make clear interpretation much more difficult than would be the case in a more acute experiment. Therefore, whether endogenous prostaglandin production is directly responsible for the appearance of glucagon sensitivity in transformed MDCK cells remains to be established.

2.2. Regulation of Induction by Exogenous Prostaglandins

Attempts to elevate prostaglandin levels in a more physiological manner were also unsuccessful. Normal rat kidney (NRK) cells produce levels of PGE_2 and $PGF_{2\alpha}$ comparable to normal MDCK cells, although these cells do not have glucagon receptors. It is therefore possible to coculture NRK cells with transformed MDCK cells to determine if exposure to prostaglandins endogenously produced by NRK

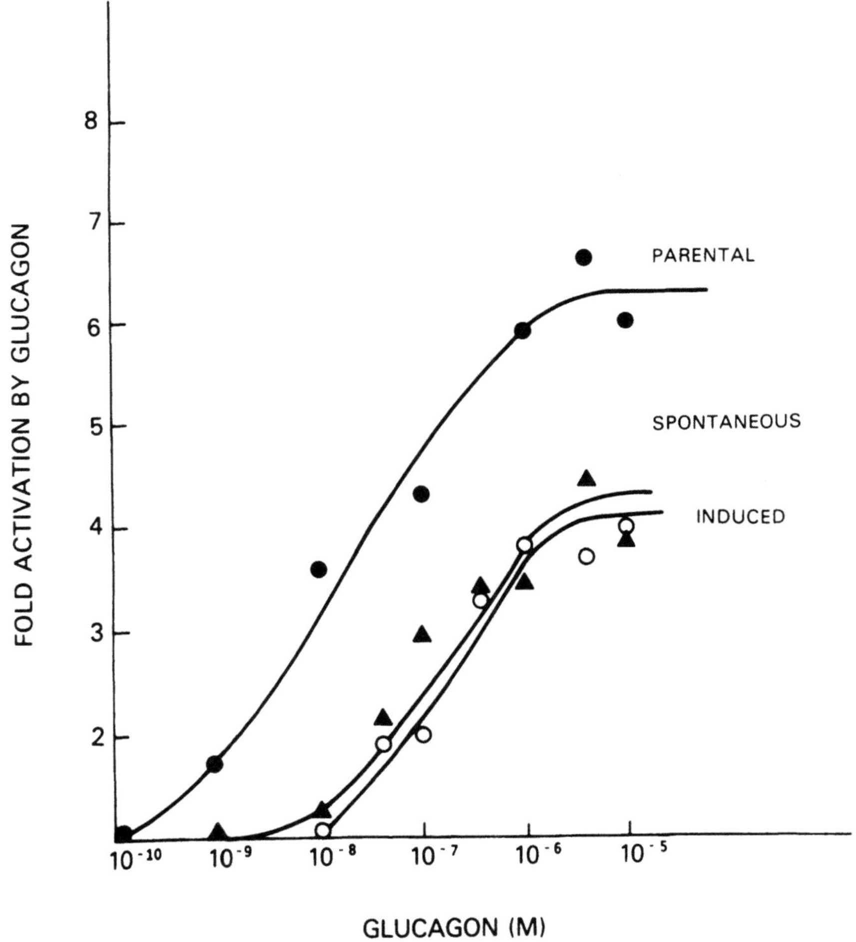

FIGURE 1. Dose–response curve to glucagon by normal, spontaneously glucagon sensitive transformed, and butyrate induced transformed MDCK cells. Normal cells were cultured in defined medium for 3 days. Transformed MDCK cells were cultured in defined medium for 14 days (spontaneous) or defined medium with 2 mM butyrate for 3 days (induced). The stimulation of cAMP production by the indicated concentration of glucagon was determined as described in Table I.

TABLE III. Correlation of Glucagon Sensitivity with Endogenous
Prostaglandin Production[a]

Expt. no.	Pass. no.	Glucagon sensitivity	PGE_2 (pmol/dish)	$PGF_{2\alpha}$ (pmol/dish)
1	17	1.2	1.9	4.4
	36	2.5	8.7	3.8
2	20	2.0	0.3	6.3
	36	5.1	9.6	4.5
	39	4.0	14.0	5.3
3	21	1.0	0.3	4.5
	36	5.8	19.8	4.1

[a] Transformed MDCK cell stocks were maintained in DMEM with 5% fetal bovine serum. At various passages indicated, cells were subcultured into the same medium and grown for 14 days with fresh medium added every 3 days. Glucagon sensitivity was determined as described in Table I. The amount of PGE_2 and $PGF_{2\alpha}$ in the medium at the time of measurement of glucagon sensitivity was determined by radioimmunoassay.

cells is sufficient to induce glucagon responsiveness in transformed MDCK cells. There was no increase in the spontaneous glucagon sensitivity of transformed MDCK cells that were cocultured with NRK cells compared to transformed MDCK cells cultured alone. The maximal final concentration of PGE_2 in the medium of either normal MDCK or NRK cells under defined medium conditions was less than 1 nM, a concentration less than that required exogenously to induce glucagon sensitivity in transformed MDCK cells (see below), which may explain this negative result.

As another approach to examine the role of prostaglandins in the expression of glucagon sensitivity, the ability of exogenous prostaglandins to induce glucagon sensitivity in transformed MDCK cells was examined. Such an approach eliminates the problems associated with long-term culture described above.

2.2.1. Characterization of Induction by Prostaglandins

Induction of glucagon sensitivity of transformed MDCK cells was most effective with PGE_1 and PGE_2, whose effects on transformed MDCK cell cAMP and induction were identical. The induction was maximal by 4–6 days and dependent on the concentration of PGE_1 (Fig. 2). Induction was maximal with 10–100 nM PGE_1, whereas higher concentrations were less effective. Whether this lack of induction with higher concentrations reflects a sort of desensitization is unknown. As before, the K_{act} of glucagon activation of adenylate cyclase induced by PGE was identical to that induced by butyrate and that which occurred spontaneously, that is, tenfold higher than that of the parental line. No other prostaglandins were as effective as the PGE series in inducing glucagon sensitivity, although some induction was consistently observed with PGI_2 (Table IV).

2.2.2. Role of cAMP

Prostaglandin E efficiently elevates cAMP (over a 3-min period) in transformed MDCK cells to a much greater extent (11- to 12-fold activation) than any other prostaglandin (Table IV). Prostaglandin B_2 (sixfold) and PGI_2 (threefold) also modestly increased cAMP production. The concentration range over which prostaglandins of the E series increase intracellular cAMP is 10 nM to 1 μM (Fig. 2), suggesting a role for cAMP in the induction process. Although PGI_2 has a modest effect on cAMP levels as well as induction, PGB_2 has a greater effect on cAMP levels but does not cause induction of glucagon sensitivity. However, the elevation of cAMP by PGE is rapid (within minutes) and disappears, whereas induction requires 3 to 5 days. Furthermore, isoproterenol, like PGE, efficiently stimulates cAMP production but is ineffective in inducing glucagon sensitivity. The data suggest that an increase in cAMP levels may be involved in the induction process but that increased cAMP is not sufficient to induce glucagon sensitivity. Whether an early, specific phosphorylation event mediated by PGE but not PGB_2 or isoproterenol is involved in the induction is under investigation.

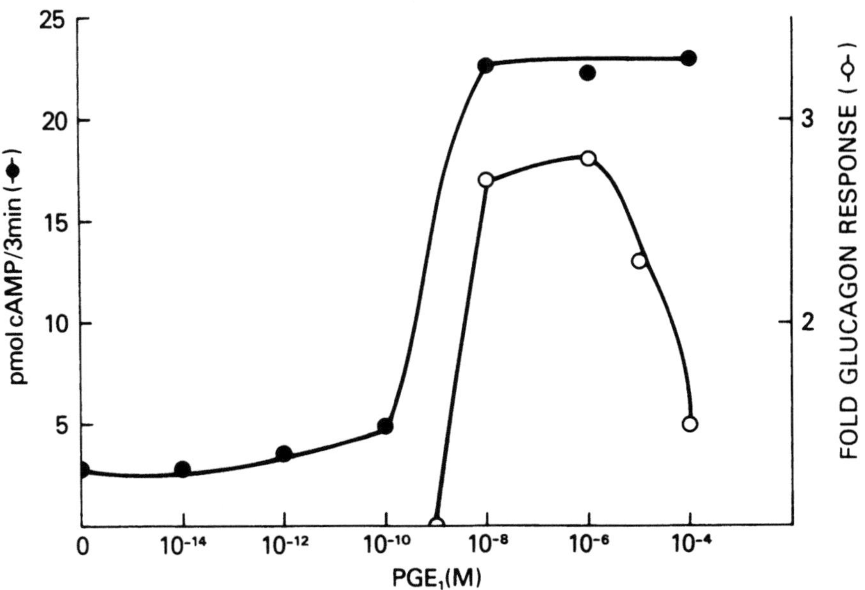

FIGURE 2. Concentration dependence of induction and cAMP stimulation by PGE_1. Transformed MDCK cells were incubated in defined medium with the indicated concentration of PGE_1 for 3 days, and glucagon sensitivity was determined as in Table I (o – – o). The cAMP responsiveness of parallel uninduced cells to the indicated concentration of PGE_1 was measured as in Table I (● – – ●).

TABLE IV. Effect of PGs on Induction and Cyclic AMP levels[a]

	Glucagon sensitivity	Cyclic AMP (pmol/10^6 cells per 3 min)
Control	1.0	9.0
PGE_1	2.2	101.1
PGE_2	4.3	109.8
$PGF_{2\alpha}$	1.4	9.0
PGA_2	1.5	28.8
PGB_1	1.0	14.4
PGB_2	1.0	54.0
PGI_2	2.5	27.0

[a] Transformed MDCK cells were induced by treatment with the indicated prostaglandin (100 nM) in defined medium for 5 days, and glucagon sensitivity was measured as described in Table I. The cAMP responsiveness to the various prostaglandins was determined in uninduced cultures as in Table I.

2.3. Inhibition of Induction

The spontaneous appearance of glucagon sensitivity as well as that induced by butyrate and PGE could be inhibited by cycloheximide, α-amanitine, and tunicamycin (Beckner *et al.*, 1985), suggesting that *de novo* protein synthesis is required for the expression of glucagon sensitivity. The spontaneous appearance of glucagon sensitivity occurred more readily in defined medium than in serum-containing medium, suggesting an inhibitory effect of serum. As expected, serum was found to inhibit PGE induction of glucagon sensitivity (Table V). When PGE induction was carried out in as little as 0.2% fetal bovine serum, there was no development of glucagon sensitivity. Further studies determined that the presence of serum was only required during the first 24 hr of induction, both that induced by PGE as well as that which appears spontaneously. This result suggests either that the trigger for the differentiation occurs early during the induction process or that the effect of serum is a long-lasting one. This serum effect is not mediated by the differentiation inhibitor described by Evinger-Hodges *et al.* (1982), although the identity of a specific inhibitor of MDCK cell differentiation remains to be determined.

Induction of glucagon sensitivity by any agent can also be inhibited by high concentrations of cAMP. For example, as seen in Table VI, 600 μM 8-Br-cAMP decreases the glucagon sensitivity induced by butyrate from 2.4- to 1.5-fold. This result may explain why concentrations of PGE_1 higher than that required to maximally elevate intracellular cAMP levels are not as efficient as lower concentrations in inducing glucagon sensitivity (Fig. 2) when included in the incubation for the 3–5 days required for induction.

Phorbol esters are well known to enhance prostaglandin production in MDCK cells (Ohuchi and Levine, 1978). In the experiments noted above, however, there was no effect of phorbol esters on the induction of glucagon sensitivity.

Surprisingly, TPA (12-*o*-tetradecanoylphorbol-13-acetate) was a potent inhibitor of PGE induction of glucagon sensitivity (Table VII). In the presence of 0.1–50

TABLE V. Inhibition of PGE Induction
by Serum[a]

	Glucagon sensitivity (fold activation)
Control	1.0
PGE_2	9.5
+ 0.05% FBS	2.5
+ 0.20% FBS	1.0

[a] Transformed MDCK cells were induced for 5 days in defined medium with or without (control) 100 nM PGE_2 and fetal bovine serum, and glucagon sensitivity was determined as in Table I. There was no effect of FBS alone on glucagon sensitivity.

nM TPA, the induction of glucagon sensitivity by PGE was completely inhibited. There was no effect of the inactive phorbol ester (4α-phorbol-12,13-didecanoate). In some systems, TPA has been shown to uncouple hormone receptors from biological responses, which might explain the inhibition of PGE induction by TPA. However, TPA had no effect on the ability of transformed cells to respond to isoproterenol or PGE, ruling out this explanation of the results. Whether the effect of TPA is related to its recently discovered regulation of C kinase activity is currently under investigation.

3. CONCLUDING REMARKS

As stated at the outset, evidence is emerging for a role of prostaglandins in carcinogenesis and differentiation. However, to date, no clear-cut relationships exist. Prostaglandins seem to either inhibit or facilitate cell proliferation and differentiation depending on the cell examined. Prostaglandins of the E series inhibit

TABLE VI. Inhibitory Effect of 8-Br-cAMP
on Butyrate Induction[a]

	Glucagon sensitivity
Control	1.1
Butyrate	2.4
+ 50 μM cAMP	2.3
+ 150 μM cAMP	2.1
+ 600 μM cAMP	1.5

[a] Transformed MDCK cells were induced for 72 hr in defined medium with 2 mM butyrate in the presence or absence of the indicated concentrations of 8-Br-cAMP. Glucagon sensitivity was determined as in Table I.

TABLE VII. Inhibition of PGE$_2$ Induction
by TPA

	Glucagon sensitivity
Control	1.0
PGE$_2$	6.5
+ 0.1 nM TPA	1.2
+ 2.0 nM TPA	1.3
+ 50 nM TPA	1.1
+ 50 nM 4-α-PDD	4.3

[a] Transformed MDCK cells were induced with PGE$_2$ in defined medium (100 nM) for 5 days in the presence or absence of the indicated concentration of TPA or 4α-PDD. Glucagon sensitivity was determined as described.

the differentiation of a granulocyte/macrophage stem cell system (Kurland and Moore, 1977) but stimulate differentiation in a myeloid cell line (M1) established from a mouse myelogenous leukemia (Ichikawa, 1969). Similarly, PGE$_2$ inhibited differentiation of monocyte and T lymphocyte precursor cell systems (Bockman and Rothschild, 1979), whereas PGA$_1$ enhanced erythropoiesis in the well-characterized Friend erythroleukemic cell system (Santoro *et al.*, 1979). Enhanced phospholipase and cyclooxygenase activity were found to accompany dimethyl-sulfoxide-induced differentiation in HL60 cells, a human promyelocytic leukemia cell line (Bonser *et al.*, 1981).

Variable effects of prostaglandins on the differentiation of nonhematopoietic model systems have been found. Prostaglandins induce differentiation of neuroblastoma cells (Prasad, 1982), presumably by elevation of cAMP levels. In the 3T3-L1 system, which differentiates into adipocytes, inhibitors of prostaglandin synthesis accelerate adipose conversion (Williams and Polakis, 1977), specifically by reducing levels of PGI$_2$ (Hopkins and Gorman, 1981). In a similar differentiation model system, ob 17 preadipocytes, differentiation can also be accelerated by inhibition of prostaglandin synthesis (Verrando *et al.*, 1981), but in this instance by reducing levels of PGF$_{2\alpha}$ (Negrel *et al.*, 1981). In summary, no clear correlations exist.

The present findings suggest that prostaglandins of the E series regulate differentiation in this model system by inducing the expression of glucagon sensitivity in transformed MDCK cells that have lost the capacity to respond to the hormone. Unfortunately, it was not possible to manipulate the degree of glucagon sensitivity by regulating endogenous prostaglandin production. This failure may reflect the greatly reduced capacity of transformed MDCK cells to produce prostaglandins under conditions favorable for induction. The basal production of prostaglandins by transformed MDCK cells is 1–2% that of normal MDCK cells. At best, this could only be increased twofold. Although endogenously produced prostaglandins seem more potent than exogenous prostaglandins (Hassid, 1983), it is unlikely that maximal concentrations of PGE_2 produced by transformed MDCK cells would exceed 1 nM, which is less than 10% of the concentration of PGE_2 required exogenously to induce glucagon sensitivity (Fig. 2).

Although serum is commonly observed to stimulate proliferation, a serum factor has also been found to inhibit differentiation in a muscle model system (Evinger-Hodges *et al.*, 1982). In the MDCK cell model system, serum also acts as an inhibitor of differentiation (Table V). However, in normal MDCK cells, dome or blister formation is enhanced by inducers of differentiation and also by serum (Lever, 1981). Certainly, our understanding of how different cells respond to undefined serum factors is incomplete. The opposite effects of serum on the induction of glucagon sensitivity in transformed MDCK cells and dome formation in normal MDCK cells, both differentiated functions, suggest that differentiation represents a complex interplay of biochemical signals.

Although the biochemical mechanisms underlying the induction of glucagon sensitivity remain to be elucidated, the MDCK cell system represents a good model system to examine these questions. That glucagon sensitivity can be induced with PGE and this effect blocked by phorbol esters suggests that protein phosphorylation may be involved in these processes. Further understanding of the biochemical mechanisms of prostaglandin regulation of differentiation in this and other model systems should provide insight into processes at work during abnormal differentiation and carcinogenesis.

REFERENCES

Armato, U., and Andreis, P. G., 1983, Prostaglandins of the F series are extremely powerful growth factors for primary neonatal rat hepatocytes, *Life Sci.* **33**:1745–1755.

Anderson, W. B., and Pastan, I., 1975, Altered adenylate cyclase activity: Its role in growth regulation and malignant transformation of fibroblasts, *Adv. Cyclic Nucleotide Res.* **5**:681–698.

Beckner, S. K., 1984, Decreased adenylate cyclase responsiveness of transformed cells correlates with the presence of a viral transforming protein, *FEBS Lett.* **166**:170–174.

Beckner, S. K., Wright, D. E., and Lin, M. C., 1985, Glucagon responsiveness induced in transformed MDCK cells differs from that of the parental cell line, *J. Biol. Chem.* submitted.

Blazquez, E., Rubalcava, B., Montesano, R., Orci, L., and Unger, R. H., 1976, Development of insulin and glucagon binding and the adenylate cyclase response in liver membranes of the prenatal, postnatal and adult rat: Evidence of glucagon resistance. *Endocrinology* **98**:1014–1023.

Bockman, R. S., and Rothschild, M., 1979, Prostaglandin E inhibition of T-lymphocyte colony formation: A possible mechanism of monocyte modulation of clonal expansion, *J. Clin. Invest.* **64**:812–819.

Bonser, R. W., Siegel, M. I., McConnell, R. T., and Cuatrecasas, P., 1981, The appearance of phospholipase and cyclo-oxygenase activities in the human promyelocytic leukemia cell line HL60 during dimethyl sulfoxide-induced differentiation, *Biochem. Biophys. Res. Commun.* **98**:614–620.

Brown, J. E., and Adamson, J. W., 1977, Modulation of *in vitro* erythropoiesis, *J. Clin. Invest.* **60**:70–77.

Evinger-Hodges, M. J., Ewton, D. Z., Seifert, S. C., and Florini, J. R., 1982, Inhibition of myoblast differentiation *in vitro* by a protein isolated from liver cell media, *J. Cell Biol.* **93**:395–401.

Fischer, S. M., and Slaga, T. J., 1982, Modulation of prostaglandin synthesis and tumor promotion, in: *Prostaglandins and Cancer: First International Conference* (T. J. Powels, R. S. Bockman, K. V. Honn, and P. Ramwell, eds.), Alan R. Liss, New York, pp. 255–264.

Gaush, C. R., Hard, W. L., and Smith, T. F., 1966, Characterization of an established line of canine kidney cells (MDCK), *Proc. Soc. Exp. Biol. Med.* **122**:931–935.

Hassid, A., 1983, Modulation of cAMP in cultured renal cells (MDCK) by endogenous prostaglandins, *J. Cell Physiol.* **116**:297–302.

Honn, K. V., and Meyer, J., 1981, Thromboxanes and prostacyclin: positive and negative modulators of tumor growth, *Biochem. Biophys. Res. Commun.* **102**:1122–1129.

Honn, K. V., Bockman, R. S., and Marnett, L. J., 1981, Prostaglandins and cancer: A review of tumor initiation through tumor metastasis, *Prostaglandins* **21**:833–864.

Hopkins, N. K., and Gorman, R. R., 1981, Regulation of 3T3-L1 fibroblasts differentiation by prostacyclin, *Biochim. Biophys. Acta* **663**:457–466.

Ichikawa, Y., 1969, Differentiation of a cell line in myeloid leukemia, *J. Cell Physiol.* **74**:223–234.

Katz, A. I., and Lindheimer, M. D., 1977, Actions of hormones on the kidney, *Annu. Rev. Physiol.* **39**:97–134.

Kurland, J., and Moore, M. A. S., 1977, Modulation of hemopoiesis by prostaglandins, *Exp. Haematol.* **5**:357–373.

Leighton, J., Brada, Z., Estes, L. W., and Justh, G., 1969, Secretory activity and oncogenicity of a cell line derived from canine kidney, *Science* **163**:472–473.

Lever, J. E., 1981, Regulation of dome formation in kidney epithelial cultures, *Ann. N.Y. Acad. Sci.* **372**:371–383.

Levine, L., 1982, Stimulation of cellular prostaglandin production by phorbol esters and growth factors and inhibition by cancer chemopreventive agents, in: *Prostaglandins and Cancer: First International Conference* (T. J. Powles, R. S. Bockman, K. V. Honn, and P. Ramwell, eds.), Alan R. Liss, New York, pp. 189–204.

Levine, M. A., Downs, R. W., Singer, M., Marx, S. J., Aurbach, G. D., and Spiegel, A. M., 1980, Deficient activity of guanine nucleotide regulatory protein in erythrocytes from patients with pseudohypoparathyroidism, *Biochem. Biophys. Res. Commun.* **94**:1319–1324.

Lin, C., and Lin, M. C., 1979, Appearance of β-adrenergic response of adenylate cyclase during the induction of differentiation of cell cultures, *Exp. Cell Res.* **122**:399–402.

Lin, M. C., and Beckner, S. K., 1983, Induction of Hormone Receptors and Responsiveness during cellular differentiation, in: *Current Topics in Membranes and Transport,* Vol. 18, (A. Kleinzeller and B. R. Martin, eds.), New York, Academic Press, pp. 287–315.

Lin, M. C., Koh, S. M., Dykman, D. D., Beckner, S. K., and Shih, T. Y., 1982a, Loss and restoration of glucagon receptors and responsiveness in a transformed kidney cell line, *Exp. Cell Res.* **142**:181–189.

Lin, M. C., Wang, S., and Beckner, S. K., 1982b, Induction of glucagon receptors and responsiveness in transformed kidney cells by prostaglandins, in: *Prostaglandins and Cancer: First International Conference* (T. J. Powels, R. S. Bockman, K. V. Honn, and P. Ramwell, eds.), Alan R. Liss, New York, pp. 493–497.

Marnett, L. J., 1981, Polycyclic aromatic hydrocarbon oxidation during prostaglandin biosynthesis, *Life Sci.* **29**:531–546.

Negrel, R., Grimaldi, P., and Ailhaud, G., 1981, Differentiation of ob 17 preadipocytes to adipocytes: Effects of $PGF_{2\alpha}$ and relationship to prostaglandin synthesis, *Biochem. Biophys. Acta.* **666**:15–24.

Ohuchi, K., and Levine, L., 1978, Stimulation of prostaglandin synthesis by tumor-promoting 12,13-diesters in canine kidney cells (MDCK), *J. Biol. Chem.* **253**:4783–4789.

Perkins, J. P., and Moore, M. M., 1973, Regulation of adenosine cyclic 3′,5′-monophosphate content of rat cerebral cortex: Ontogenic development of the responsiveness to catecholamines and adenosine, *Mol. Pharmacol.* **9**:774–782.

Prasad, K. N., 1982, Role of prostaglandins in differentiation of neuroblastoma cells in culture, in: *Prostaglandins and Cancer: First International Conference* (T. J. Powles, R. S. Bockman, K. V. Honn, and P. Ramwell, eds.), Alan R. Liss, New York, pp. 437–452.

Ramwell, P., 1982, Prostaglandins and Cancer: First International Conference, in: *Prostaglandins and Related Lipids* (T. J. Powels, R. S. Bockman, K. V. Honn, and P. Ramwell, eds.), Alan R. Liss, New York.

Rindler, M. J., Chuman, L. M., Shaffer, L., and Saier, M. H., 1979, Retention of differentiated properties in an established dog kidney epithelial cell line (MDCK), *J. Cell Biol.* **81**:635–648.

Ross, E. M., Pedersen, S. E., and Florio, V. A., 1983, Hormone sensitive adenylate cyclase: Identity, function and regulation of the protein components, in *Current Topics in Membranes and Transport*, Vol. 18 (A. Kleinzeller and B. R. Martin, eds.), 142, Academic Press, New York, pp. 109–142.

Roy, C., Preston, A. S., and Handler, J. S., 1980, Insulin and serum increase the number of receptors for vasopressin in a kidney-derived line of cells grown in a defined medium, *Proc. Natl. Acad. Sci. U.S.A.* **77**:5979–5983.

Rubin, C. S., Lai, E., and Rosen, O. M., 1977, Acquisition of increased hormone sensitivity during in vitro adipocyte development, *J. Biol. Chem.* **252**:3554–3557.

Santoro, M. G., Benedetto, A., and Jaffe, B. M., 1979, Prostaglandin A_1 induces differentiation in Friend erythroleukemic cells, *Prostaglandins* **17**:719–723.

Scolnick, E. M., Williams, D., Maryak, J., Vass, W., Goldberg, R. J., and Parks, W. P., 1976, Type C particle positive and type C particle negative rat cell lines: Characterization of the coding capacity of endogenous sarcoma virus-specific RNA, *J. Virol.* **20**:570–582.

Shih, T. Y., Papageorge, A. G., Stokes, P. E., Weeks, M. O., and Scolnick, E. M., 1980, Guanine nucleotide binding and autophosphorylating activities associated with the p21 protein of Harvey murine sarcoma virus, *Nature* **287**:686–691.

Taub, M., Chuman, L., Saier, M. H., and Sato, G., 1979, Growth of Madin–Darby canine kidney epithelial cell (MDCK) line in hormone supplemented, serum free medium, *Proc. Natl. Acad. Sci. U.S.A.* **76**:3338–3342.

Verrando, P., Negrel, R., Grimaldi, P., Murphy, M., and Ailbaud, G., 1981, Differentiation of 17 preadipocytes to adipocytes: Triggering effects of clofenapate and indomethacin, *Biochim. Biophys. Acta* **663**:255–265.

Whitsett, J. A., Darovec-Beckerman, C., Adams, K., Pollinger, J., and Needelman, H., 1980, Thyroid dependent maturation of β-adrenergic receptors in the rat lung, *Biochem. Biophys. Res. Commun.* **97**:913–917.

Williams, I. H., and Polakis, S. E., 1977, Differentiation of 3T3-L1 fibroblasts to adipocytes. The effect of indomethacin, prostaglandin E_1 and cyclic AMP on the process of differentiation, *Biochem. Biophys. Res. Commun.* **77**:175–186.

Willingham, M. C., Pastan, I., Shih, T. Y., and Scolnick, E. M., 1980, Localization of the *src* gene product of the Harvey strain of MSV to plasma membrane of transformed cells by electron microscopic immunocytochemistry, *Cell* **19**:1005–1014.

18

Human Trabecular Meshwork Cells

Arachidonic Acid Metabolism and Prostaglandin Release

MARY E. GERRITSEN, BERNARD I. WEINSTEIN, GARY G. GORDON, and A. LOUIS SOUTHREN

1. SUMMARY

In man, most of the aqueous humor drainage and the major resistance to outflow occur across the trabecular meshwork and inner wall of Schlemm's canal. Prostaglandins are known to alter intraocular presure (IOP) and have been implicated in both physiological and pathophysiological aspects of IOP regulation. In the present study, we have characterized pathways of arachidonic acid metabolism in human trabecular meshwork (HTM) cells and measured basal and hormone-stimulated PGE_2 release. The principal product of $[1-{}^{14}C]PGH_2$ metabolism and $[1-{}^{14}C]$arachidonic acid metabolism was PGE_2. Immunoreactive PGE_2 was released in excess of immunoreactive 6-keto-$PGF_{1\alpha}$ during 15-min incubations in phosphate-buffered saline. Incubation wth indomethacin (5 μM) for 20 min substantially reduced (80–85% inhibition) PGE_2 formation from AA. Overnight incubation with dexamethasone ($ID_{50} = 3.3 \times 10^{-9}$ M) inhibited basal and A23187-stimulated PGE_2 release. The effects of dexamethasone on PGE_2 synthesis and release could be prevented by pretreatment of the HTM cells with 2 μg/ml cycloheximide. A23187 (10^{-6}–10^{-5} M) elicited a significant ($P < 0.05$ compared to control) increase in PGE_2 release (RIA determinations).

MARY E. GERRITSEN • Department of Physiology, New York Medical College, Valhalla, New York 10595.   BERNARD I. WEINSTEIN, GARY G. GORDON, and A. LOUIS SOUTHREN • Departments of Ophthalmology and Medicine, New York Medical College, Valhalla, New York 10595.

2. INTRODUCTION

The aqueous humor is continuously elaborated by the cells of the ciliary epithelium into the posterior chamber of the eye and from there flows into the anterior chamber. The continuous production of aqueous humor also requires continuous drainage of the fluid, which occurs mainly through the trabecular meshwork into the canal of Schlemm and then into veins leaving the eye. Primary open angle glaucoma (POAG), which is associated with chronically elevated intraocular pressure (IOP), is a result of a decreased facility of outflow.

Various theories have been proposed to account for the defect in primary open angle glaucoma (POAG) leading up to increased IOP. Anatomic studies of the outflow channels in human eyes have suggested increased accumulation of extracellular material near the inner wall of Schlemm's canal (Rohen, 1973; Segawa, 1975; Rodrigues *et al.*, 1976), indicating that the outflow facility might be compromised. Glycosaminoglycans, including chondroitin sulfate, dermatin sulfate, and hyaluronic acid, are synthesized by cultured meshwork cells (Schachtschabel *et al.*, 1977, 1982; Polansky *et al.*, 1978) and may play a role in increased IOP. However, a direct relationship between glycosaminoglycans and the metabolism and function of trabecular meshwork cells has not been established. An alternative theory suggests some role of glucocorticoids in POAG. These steroids are contraindicated in these patients, as they frequently elicit a further pronounced rise in IOP and can lead to blindness. Studies of cortisol metabolism in trabecular meshwork cells obtained from normal and POAG eyes revealed a fundamental difference in the metabolism of glucocorticoids (Southren *et al.*, 1983).

Trabecular meshwork cells from POAG patients demonstrated significantly greater Δ^4-reductase and a significantly decreased 3-oxidoreductase activity compared to cells from normal patients. The net effect of these two actions would be an increased level of the intermediates 5β- and 5α-dihydrocortisol. Recent studies (Weinstein *et al.*, 1983; Southren *et al.*, 1984) have demonstrated a potentiation of the glucocorticoid effects in the rabbit by 5β-dihydrocortisol and suggest that this metabolite may be important in the etiology of POAG.

The precise mechanism of the glucocorticoid effects on IOP in POAG patients is unknown. Glucocorticoids have been demonstrated to inhibit prostaglandin (and other eicosanoids) production in many cells and tissues. This action appears to require the synthesis of a phospholipase inhibitory protein(s) called macrocortin (Blackwell *et al.*, 1980), lipomodulin (Hirata *et al.*, 1980), or renocortin (Russo-Marie and Duval, 1982). Prostaglandins introduced topically on the rabbit eye will elicit dose-dependent increases in IOP (Waitzman and King, 1967), although use of lower doses in other species (cat and monkey) can elicit a reduction in IOP (Stern and Bito, 1982; Camras *et al.*, 1979). Increased levels of prostaglandins in the anterior chamber have been observed following surgery, and cyclooxygenase inhibitors (aspirin, indomethacin) will reduce ocular inflammation following surgery (Miller *et al.*, 1973; Peyman *et al.*, 1979). Weinreb *et al.*, (1983) recently reported release of immunoreactive PGE_2 by cultured human trabecular meshwork cells and the inhibition of PGE_2 release by dexamethasone.

In the present study we examined (1) the metabolism of $[1 - {}^{14}C]$arachidonic acid by cultured human trabecular meshwork (HTM) cells, (2) release of PGE_2 and 6-keto $PGF_{1\alpha}$ by HTM cells, and (3) the effects of cortisol and dexamethasone on basal and A23187-stimulated PGE_2 synthesis by HTM cells.

3. METHODS

3.1. Cell Culture

Human trabecular meshwork (HTM) tissues were obtained from normal eye bank eyes post-mortem. Histological examination of the eyes used in this study confirmed the absence of POAG. The HTM cells were cultured as described previously (Polansky *et al.*, 1978, 1979; Worthen and Cleveland, 1982; Southren *et al.*, 1983) in Dulbecco's modified Eagle's medium (DME) supplemented with 10% fetal calf serum (FCS). Cells used in this study were from the third through sixth passages.

3.2. [¹⁴C]Arachidonic Acid Metabolism

The HTM cells were cultivated in six-well Linbro culture dishes. At confluence, medium was gently aspirated, and the cells were washed twice with 1 ml of 37°C phosphate-buffered saline (PBS), pH 7.4. The cells were incubated 2 hr with 4 μCi of $[1 - {}^{14}C]$arachidonic acid (AA) in DME without FCS. At the end of the incubation, the medium was removed and extracted with four volumes of ethylacetate : methanol : 0.2 M citric acid (15 : 2 : 1), and the organic phase was evaporated *in vacuo*. The residue was resuspended in 50 μl of ethyl acetate and applied to silica gel G GHL (Analtech) TLC plates. Chromatograms were developed in the organic phase of ethylacetate : hexane : acetic acid : water (56 : 24 : 12 : 60). Products were identified by comigration with authentic standards.

3.3. Prostaglandin Release

Confluent HTM cells were fed 24 hr prior to experiments. Cells were washed twice with 1 ml of PBS (23°C) and preincubated 15 min with PBS followed by a 15-min incubation with or without the Ca^{2+} ionophore A23187. Release of PGE_2 and 6-keto $PGF_{1\alpha}$ was quantified by radioimmunoassay using Seragen antisera against PGE_2 and Cappel Industries anti-6-keto-$PFG_{1\alpha}$, as described in earlier publications (Gerritsen and Cheli, 1983; Rodrigues and Gerritsen, 1984).

3.4. Steroid Treatment

Confluent HTM cells were fed 24 hr prior to experiments. Cells were washed twice with 1 ml of PBS and then incubated 12–14 hr with DMEM containing 1% dialyzed FCS with or without steroids at 37°C, 5% CO_2 in air. After this culture medium was removed, the cells were washed with 1 ml of PBS, preincubated 15 min with PBS (23°C), and then challenged with 5 μg/ml A23187. The PGE_2 release was determined by radioimmunoassay and standardized with respect to protein concentration determined by the method of Lowry *et al.* (1951).

4. RESULTS

The major product of [1-^{14}C]arachidonic acid metabolism by intact HTM cells comigrated with authentic PGE_2, with minor amounts of 6-keto-$PGF_{1\alpha}$, $PGF_{2\alpha}$, and PGD_2 formed (Fig. 1). Formation of all products was substantially reduced (80–85% inhibition) by preincubation with 5 μM indomethacin prior to addition of arachidonic acid. The HTM cells exhibited comparatively high conversion of $[1-^{14}C]AA$ to products compared to neighboring scleral fibroblasts isolated from the same eyes (Fig. 1) and to other cultured cells such as endothelial cells (Baenziger *et al.*, 1979). The radioimmunassay data (Table I) indicated that PGE_2 was released in excess of 6-keto-$PGF_{1\alpha}$ during a 15-min incubation with PBS. Similar results were obtained with two other HTM cell lines derived from different primary cultures.

The Ca^{2+} ionophore A23187 elicited a dose-dependent stimulation of PGE_2 release (Table II). A 12-hr pretreatment with dexamethasone inhibited total PGE_2 release into the growth medium (Fig. 2a), as well as inhibiting basal (PBS) (Fig. 2b) and A23187 (Fig. 2c) -stimulated PGE_2 release. These effects of dexamethasone were not observed in cells pretreated with 2 μg/ml cycloheximide (data not shown). In contrast, pretreatment with cortisol under the same conditions indicated a biphasic effect on A23187-stimulated release at concentrations of 10^{-9}–10^{-8} M and inhibited A23187-stimulated release at higher concentrations of cortisol (Table III).

5. DISCUSSION

This study demonstrates that the principal prostaglandin produced by cultured HTM cells is PGE_2. The release of PGE_2 can be augmented by addition of exogenous arachidonic acid or by incubation with A23187. Much smaller amounts of 6-keto-$PGF_{1\alpha}$, the stable hydrolysis product of PGI_2, are also elaborated by the HTM cells.

Synthesis of PGE_2 by the HTM cells was greatly reduced by preincubation with the cyclooxygenase inhibitor indomethacin. Overnight incubation with the nonmetabolizable glucocorticoid agonist dexamethasone elicited a dose-dependent reduction in basal and stimulated PGE_2 release. In contrast, the effect of cortisol was mixed, enhancing basal and stimulated PGE_2 release at low concentrations

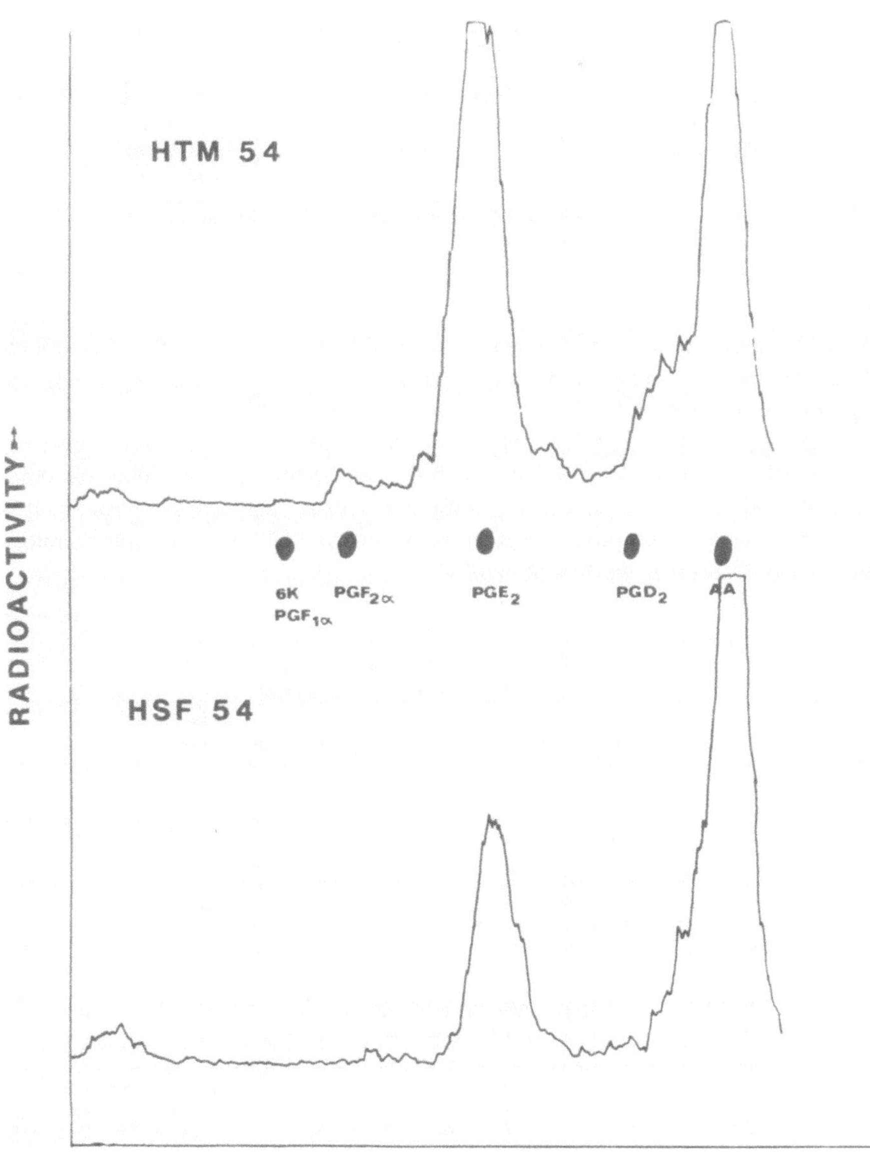

FIGURE 1. [^{14}C]Arachidonic acid metabolism by human trabecular meshwork (HTM) cells (top) and scleral fibroblasts (HSF) (bottom). The major product produced by both cell types was PGE$_2$. Representative radiochromatograms are drawn on the same scale.

TABLE I. Release of Immunoreactive PGE_2 and 6-Keto-$PGF_{1\alpha}$ by Human Trabecular Meshwork (HTM) and Scleral Fibroblast (HSN) Cells Obtained from Different Patients[a]

Cell line	6-Keto-$PGF_{1\alpha}$	PGE_2
HTM 54	0.05 ± 0.01	7.2 ± 3.4
HTM 61A	0.08 ± 0.02	5.4 ± 1.2
HTM 61A	< 0.01	0.6 ± 0.2
HSN 54	0.04 ± 0.02	0.74 ± 0.3

[a] Incubation was for 15 min in PBS at 23°C in air. Values (ng/well) are expressed as the means of six observations, each determined in duplicate, ± standard error of the mean.

($<10^{-8}$ M) and inhibiting PGE_2 release at higher doses (i.e., $>10^{-8}$ M). The effects of dexamethasone were prevented by treatment of cycloheximide, suggesting the requirement for protein synthesis in the dexamethasone effect.

The significance of PGE_2 synthesis in the HTM cell is unknown. However, in view of the stimulatory actions of PGE_2 and other prostaglandins on other secretory and absorptive cells (e.g., in the gut, kidney, and seminal vesicle), it is not unreasonable to propose a regulatory action of PGE_2 on transport/secretory functions of trabecular meshwork cells.

TABLE II. Release of Immunoreactive PGE_2 by HTM-54 Cells: Effects of A23187

Dose of A23187 (μM)	PGE_2 (ng/well per 15 min)[a]
1	2.6 ± 0.3
5	6.6 ± 1.1[b]
10	23.9 ± 6.5[b]

[a] Basal (PBS) levels of PGE_2 were 3.2 ± 1.5 ng/well per 15 min.
[b] Significantly different ($P < 0.05$ ANOVA; $n = 4$) compared to control. Each well contains cells equivalent to 350–410 μg protein.

TABLE III. Effects of Overnight Incubation of HTM cells (HTM-54) in DME plus 1% Dialyzed Serum with Cortisol or Without (Control)

Cortisol (M)	n	Percent of control[a]
10^{-12}	6	93.8 ± 4.6
10^{-11}	6	94.0 ± 2.8
10^{-10}	8	94.1 ± 9.2
10^{-9}	8	119.0 ± 11.6
10^{-8}	17	163.2 ± 14.3[b]
10^{-7}	15	83.6 ± 9.4
10^{-6}	9	41.0 ± 8.8[b]

[a] Data are expressed as percent of control (no cortisol) for each experiment ± S.E.M.
[b] Significant increase or decrease over control (100%) ($P < 0.05$, ANOVA).

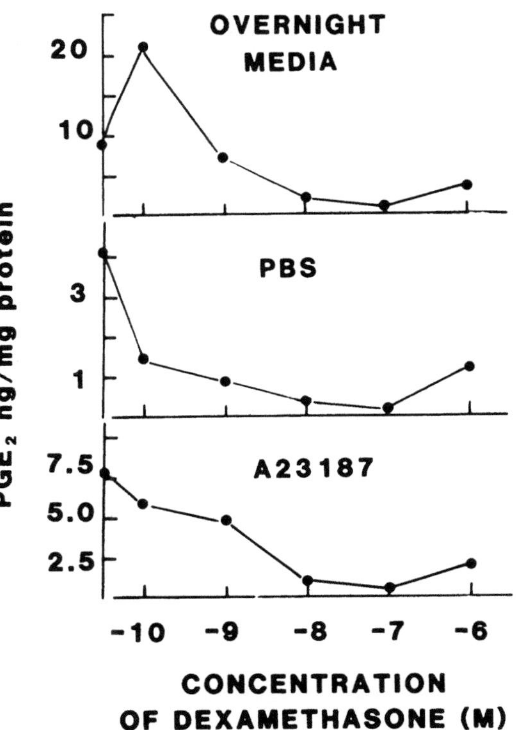

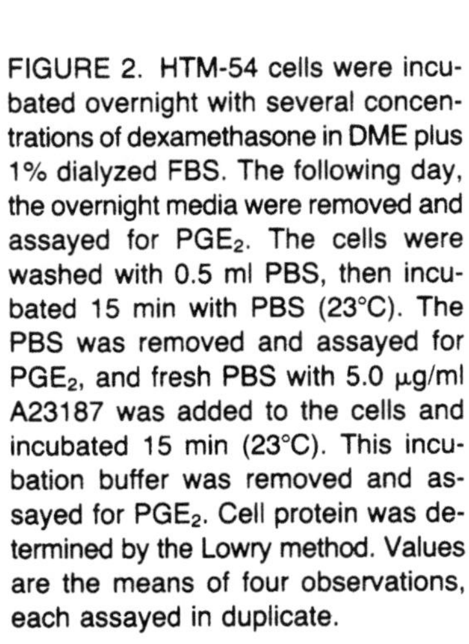

FIGURE 2. HTM-54 cells were incubated overnight with several concentrations of dexamethasone in DME plus 1% dialyzed FBS. The following day, the overnight media were removed and assayed for PGE_2. The cells were washed with 0.5 ml PBS, then incubated 15 min with PBS (23°C). The PBS was removed and assayed for PGE_2, and fresh PBS with 5.0 µg/ml A23187 was added to the cells and incubated 15 min (23°C). This incubation buffer was removed and assayed for PGE_2. Cell protein was determined by the Lowry method. Values are the means of four observations, each assayed in duplicate.

The biphasic effect of cortisol on PGE_2 synthesis and release by the HTM cells is an intriguing observation. The role of this glucocorticoid "dual response" to HTM cell function is a subject for future investigation.

ACKNOWLEDGMENTS. The authors express their sincere appreciation of the technical assistance provided by Astride M. Rodrigues and Carol D. Cheli.

REFERENCES

Baenziger, N., Becherer, P. R., and Majerus, P. W., 1979, Characterization of prostacyclin synthesis in cultured human arterial smooth muscle cells, venous endothelial cells and skin fibroblasts, *Cell* **16**:967–974.

Blackwell, G. J., Carnuccio, R., DiRosa, M., Flower, R. J., Parente, L., and Persico, P., 1980, Macrocortin: A polypeptide causing the antiphospholipase effect of glucocorticoids, *Nature* **287**:147–149.

Camras, C. B., Bito, L. Z., and Eakins, K. E., 1979, Reduction of intraocular pressure by prostaglandins applied topically to the eyes of conscious rabbits, *Invest. Ophthalmol. Vis. Sci.* **16**:1125–1134.

Gerritsen, M. E., and Cheli, C. D., 1983, Arachidonic acid and prostaglandin endoperoxide metabolism in isolated rabbit coronary microvessels and isolated and cultivated coronary microvessel endothelial cells, *J. Clin. Invest.* **72**:1658–1671.

Hirata, F., Scheffman, D., Venkatasubramanian, K., Salomon, D., and Axelrod, J., 1980, A phospholipase A_2 inhibitory protein in rabbit neutrophils induced by glucocorticoids, *Proc. Natl. Acad. Sci. U.S.A.* **77:**2533–2536.

Lowry, O. H., Rosebrough, N. J., Farr, A. L., and Randall, R. J., 1951, Protein measurement with the folin phenol reagent, *J. Biol. Chem.* **193:**265.

Miller, J. D., Eakins, K. E., and Atwal, M., 1973, The release of PGE_2-like activity into aqueous humor after paracentesis and its prevention by aspirin, *Invest. Ophthalmol. Vis. Sci.* **12:**939–942.

Peyman, K., Peyman, G. A., McGetrick, J., and Janevicius, R., 1979, Indomethacin inhibition of prostaglandin-mediated inflammation following intraocular surgery, *Invest. Ophthalmol. Vis. Sci.* **16:**760–762.

Polansky, J. R., Gospodarowicz, D., Weinreb, R., and Alvarado, J., 1978, Human trabecular meshwork cell culture and glycosaminoglycan synthesis, *Invest. Ophthalmol. Vis. Sci.* **17**(ARVO Suppl.):207.

Polansky, J. R., Weinreb, R. N., Baxter, J. D., and Alvarado, J., 1979, Human trabecular meshwork cells. I. Establishment in tissue culture and growth characteristics, *Invest. Ophthalmol. Vis. Sci.* **18:**1043–1049.

Rodrigues, A. M., and Gerritsen, M. E., 1984, Prostaglandin release from isolated rat cerebral cortex microvessels—comparison of 6-keto-$PGF_{1\alpha}$ and PGE_2 release from microvessels incubated in 100% O_2, room air and 95% N_2, 5% CO_2, *Stroke* **15(4):**717–722.

Rodrigues, M. M., Spaeth, G. L., Sivalingam, E., and Weinreb, S., 1976, Histopathology of 150 trabeculectomy specimens in glaucoma, *Trans. Ophthalmol. Soc. U.K.* **96:**245.

Rohen, M. W., 1973, Reinstrukturelle Veranderungen im Trabekelwerk des menschlichen Auges bei verschiedenen Glaukomformen, *Klin. Monatsbl. Augenheilkd.* **163:**401.

Russo-Marie, F., and Duval, D., 1982, Dexamethasone-induced inhibition of prostaglandin production does not result from a direct action on phospholipase activities but is mediated through a steroid-inducible factor, *Biochim. Biophys. Acta* **712:**177–185.

Schachtschabel, D. O., Bigalke, B., and Rohen, J. W., 1977, Production of glycosaminoglycans by cultures of the trabecular meshwork of the primate eye, *Exp. Eye Res.* **24:**71–80.

Schachtschabel, D. O., Rohen, J. W., Waver, J., and Sames, K., 1982, Synthesis and composition of glycosaminoglycans by cultured human trabecular meshwork cells, *Albrecht von Graefes Arch. Klin. Exp. Ophthalmol.* **218:**113–117.

Segawa, K., 1975, Ultrastructural changes of the trabecular tissue in primary open angle glaucoma, *Jpn. J. Ophthalmol.* **19:**317.

Southren, A. L., Gordon, G. G., Munnangi, P. R., Vittek, J., Schwartz, J., Monder, C., Dunn, M. W., and Weinstein, B. I., 1983, Altered cortisol metabolism on cells cultured from trabecular meshwork specimens obtained from patients with primary open-angle glaucoma, *Invest. Ophthalmol. Vis. Sci.* **24:**1413–1417.

Southren, A. L., l'Hommedieu, D., Ravikumar, S., Gordon, G. G., Dunn, M. W., and Weinstein, B. I., 1984, Potentiation of the ocular hypertensive effect of dexamethasone by 5β-dihydrocortisol, *Invest. Ophthalmol. Vis. Sci.* **25** (ARVO Suppl.):304.

Stern, F. A., and Bito, L. Z., 1982, Comparison of the hypotensive and other ocular effects of prostaglandins and $F_{2\alpha}$ on cat and rhesus monkey eyes, *Invest. Ophthalmol. Vis. Sci.* **22:**588–598.

Waitzman, M. B., and King, C. D., 1967, Prostaglandin influences on intraocular pressure and pupil size, *Am. J. Physiol.* **212:**329–334.

Weinreb, R. N., Mitchell, M. D., and Polansky, J. R., 1983, Prostaglandin production by human trabecular meshwork cells: *In vitro* inhibition by dexamethasone, *Invest. Ophthalmol. Vis. Sci.* **24:**1541–1545.

Weinstein, B. I., Gordon, G. G., and Southren, A. L., 1983, Potentiation of glucocorticoid activity by 5β-dihydrocortisol, its role in glaucoma, *Science* **222:**172–173.

Worthen, D. M., and Cleveland, P. H., 1982, Fibronectin production by cultured human trabecular meshwork cells, *Invest. Ophthalmol. Vis. Sci.* **22:**265–269.

19

Prostacyclin Can Inhibit DNA Synthesis in Vascular Smooth Muscle Cells

NANCY E. OWEN

1. INTRODUCTION

Vascular smooth muscle cell proliferation plays a pivotal role in physiological processes such as wound repair and in pathological processes such as atheromatous plaque formation. However, little is known about the mechanism by which DNA synthesis and cell growth are stimulated or regulated in vascular smooth muscle cells. In part, this lack of information is because of the lack of a suitable model system. In this regard, Ross (1971) was among the first to grow vascular smooth muscle cells in culture. It was observed that vascular smooth muscle cells, like other cells in culture, growth arrest at the G_1/G_0 interphase when deprived of serum or essential nutrients. It was subsequently determined that the vascular smooth muscle cell mitogenic factor in serum originated from platelets and has been appropriately termed platelet-derived growth factor (PDGF) (Ross et al., 1974).

In addition to its mitogenic properties, PDGF has also been shown to stimulate prostacyclin synthesis in both vascular smooth muscle and endothelial cells (Coughlin et al., 1980). Studies in 3T3 cells suggest that PDGF stimulated PGI_2 synthesis via effects on a Ca^{2+}-sensitive phospholipase that liberates arachidonic acid from the membrane phospholipids (Shier and Durkin, 1982). Prostacyclin (PGI_2) is the most potent vasodilator and platelet inhibitor thus far identified. The biological activity of PGI_2 has been suggested to be mediated through its effects on adenylate cyclase and elevation of cyclic AMP (cAMP) levels (Hopkins and Gorman, 1980). Cyclic AMP has been demonstrated to be a negative regulator of cell proliferation

---

NANCY E. OWEN ● Department of Biological Chemistry and Structure, University of Health Sciences, The Chicago Medical School, North Chicago, Illinois 60064.

(for reviews, see Berridge, 1975; Chalapowski *et al.*, 1975; Pastan *et al.*, 1975). In this regard, Huttner *et al.* (1977) have provided evidence to support the hypothesis that elevations in cAMP levels inhibit vascular smooth muscle cell proliferation. They demonstrated that agents such as PGE_1 that are known to elevate cAMP content inhibit cellular proliferation. A similar inhibition of cell proliferation was found using cyclic nucleotide phosphodiesterase inhibitors (e.g., caffeine or papaverine).

Based on these results, one could envision vascular smooth muscle cells as a finely tuned system wherein proliferation is both stimulated and attenuated indirectly by PDGF. Thus, PDGF, which is a potent mitogen for vascular smooth muscle cells, may regulate its own proliferative effects by initiating PGI_2 synthesis. The PGI_2, through elevations in cAMP levels, could, in turn, inhibit DNA synthesis. In the present studies, we investigated the effects of PGI_2 and elevations in cAMP on DNA synthesis in vascular smooth muscle cells.

2. METHODS

2.1. Reagents

Prostacyclin and prostaglandin E_1 were generously provided by the Upjohn Co., Kalamazoo, MI. 8-Bromo-cAMP, dibutyryl cAMP, 1-methyl-3-isobutylxanthine, and 2′,5′-dideoxyadenosine were purchased from the Sigma Chemical Co., St. Louis, MO. Forskolin was purchased from Calbiochem, La Jolla, CA.

2.2. Procedures

A-10 vascular smooth muscle cells were derived from embryonic rat thoracic aorta (Kimes and Brandt, 1976) and were purchased from American Type Culture Collection, Bethesda, MD. Cells were originally identified as vascular smooth muscle by measuring enzyme activity (myokinase and creatine phosphokinase), by electrophysiological measurements, and by electron microscopy. Cells were grown in Dulbecco's modified essential medium (DMEM) (KC Biologicals, Lenexa, KS) containing 10% fetal bovine serum (FBS) (KC Biologicals, Lenexa KS) in an atmosphere of 5% Co_2 : 95% air. They were used between passages 10 and 27.

For DNA synthesis studies, cells were removed from stock flasks by trypsinization and subcultured onto 24 cluster plates at a sparse density of 5000 cells/well in DMEM plus 10% FBS. Serum-stimulated DNA synthesis was tested since commercial serum presumably contains not only PDGF but also the requisite progression factors. Twenty-four hours following subculture, varying concentrations of inhibitors were added to the cells and allowed to preincubate for 24 hr. Following the 24-hr preincubation with inhibitor [³H]thymidine (1 μCi/ml) (1 mCi/ml, Amersham Searle, Arlington Heights, IL) was supplemented to the cells for an additional 24-

hr labeling period. In order to assess [3H]thymidine incorporation, the medium was aspirated, and cells were washed with tris-buffered saline (three times), fixed with 10% TCA (twice), washed with 95% ethanol (twice), and finally extracted with 0.1 N NaOH. Aliquots were counted in a liquid scintillation counter to determine [3H]thymidine incorporation.

For measurement of cellular protein, cells were extracted with 0.2% SDS following the tris-buffered saline wash step. Protein was determined by fluorometric assay as previously described (Owen and Villereal, 1982). Data are calculated as counts per minute per gram protein and expressed as percentage of control. The concentration of drug that yielded half-maximal inhibition (K_i) was calculated by fractional inhibition analyses of dose–response curves as described by Owen and Villereal (1981).

For cAMP measurements, cells were subcultured onto six-well cluster plates (35 mm) at a density of 1×10^5 cells/well in DMEM plus 10% FBS. Twenty-four hours following subculture, each well was washed with HEPES-buffered mininal essential medium (MEM). The cells were then incubated with 1-methyl-3-isobutylxanthine and PGI_2 in DMEM plus 10% FBS for 24 hr. The assay was terminated by the addition of 0.5 ml of 10% TCA, and the aqueous samples were extracted with ten volumes of diethylether. Cyclic AMP was assayed by radioimmunoassay (New England Nuclear, Boston, MA) as described by Steiner *et al.* (1972) with the inclusion of the acetylation modification as described by Harper and Brooker (1975).

3. RESULTS

3.1. Effect of Exogenous PGI_2 on DNA Synthesis

In order to evaluate whether PGI_2 could play a physiological role in regulation of DNA synthesis in vascular smooth muscle cells, the effect of adding PGI_2 to growing cells was determined. Cells were sparsely plated as described in Section 2.2, and PGI_2 was added from a 50 mM tris-HCl, pH 9.0, buffer in the presence of 100 μM 1-methyl-3-isobutylxanthine. As shown in Fig. 1, exogenous PGI_2 caused inhibition of DNA synthesis. The maximal effect was to depress DNA synthesis to approximately 50% of the control, and the half-maximal inhibitory concentration (K_i) of PGI_2 was calculated to be 10 ng/ml (28 nM). The actual values for [3H]thymidine incorporation are shown in Table I.

3.2. Effect of Exogenous PGE_1 on DNA Synthesis

Prostacyclin has been shown to exert its biological activity through its ability to elevate cAMP levels. As PGE_1 also elevates cAMP levels, it was of interest to test the effects of PGE_1 on DNA synthesis. As shown in Fig. 2 and Table I, when

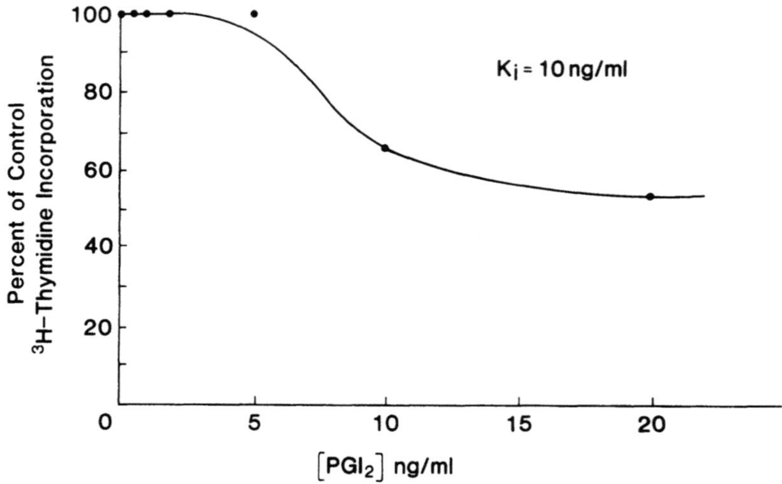

FIGURE 1. DNA synthesis in A-10 vascular smooth muscle cells; effect of prostacyclin. A-10 cells were subcultured onto 24-well cluster plates at a density of 5000 cells/well in DMEM plus 10% FBS. Twenty-four hours later varying concentrations of PGI_2 were added in the presence of 100 μM IBMX. After a 24-hr incubation, [^{3}H]thymidine (1 μCi/ml) was added. [^{3}H]Thymidine incorporation was assayed as described in the text. Data are expressed as percentages of control, where control is serum-stimulated DNA synthesis in the presence of 100 μM IBMX. Values are from a representative experiment; the K_i value represents the mean $\pm$ S.E.M. from six separate determinations.

PGE_1 was added to growing cells in the presence of 100 μM IBMX, it also blocked serum-stimulated DNA synthesis. The maximal effect of PGE_1 was to completely inhibit DNA synthesis, and the K_i was calculated to be 20 μM.

3.3. Effect of cAMP Analogues, Forskolin, and Phosphodiesterase Inhibitors on DNA Synthesis

It was of interest to evaluate the effects of stable cAMP analogues on serum-stimulated DNA synthesis. In this connection, both 8-bromo-cAMP and dibutyryl

TABLE I. Effect of PGI_2 and PGE_1 on DNA Synthesis[a]

Additions	[^{3}H]Thymidine incorporation (cpm/g protein)	Inhibition (%)
None	1.6×10^9	—
IBMX	1.1×10^9	30
IBMX + PGI_2	5.6×10^8	50
IBMX + PGI_1	3.3×10^8	70

[a] DNA synthesis was assayed as described in Fig. 1. The concentration of IBMX was 100 μM, PGI_2 20 ng/ml, and PGE_1 100 μM. Percent inhibition refers to percent of control, where control is 10% FBS (IBMX) or 10% FBS plus IBMX (IBMX + PGI_2 and IBMX + PGE_1). Values are from representative experiments.

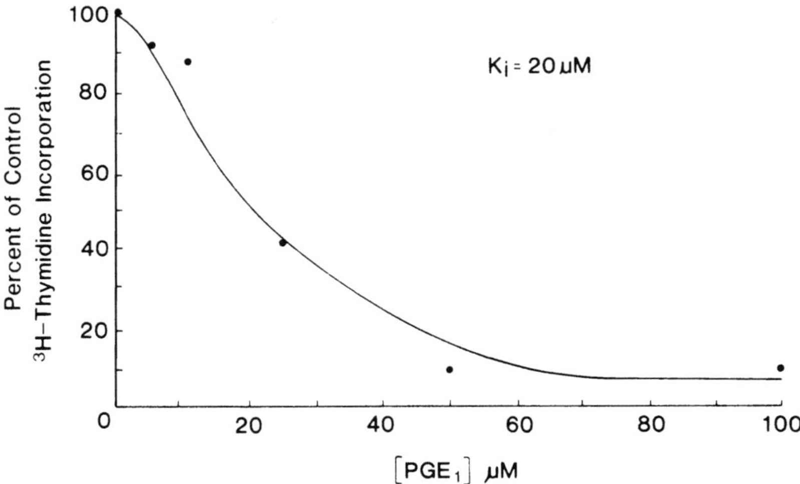

FIGURE 2. DNA synthesis in A-10 vascular smooth muscle cells: effect of prostaglandin E_1. Cells were subcultured, and DNA synthesis was assayed as described in Fig. 1 except that varying concentrations of PGE_1 were added in the presence of 100 μM IBMX. Data are expressed as percentages of control, where control is serum-stimulated DNA synthesis in the presence of 100 μM IBMX. Values are from a representative experiment: the K_i value represents the mean ± S.E.M. from four separate determinations.

cAMP were added to growing vascular smooth muscle cells. It can be seen in Fig. 3a,b that both analogues inhibited DNA synthesis. The K_i for 8-bromo-cAMP was calculated to be 40 μM, whereas the K_i for dibutyryl cAMP was calculated to be 50 μM. Both agents reduced DNA synthesis to approximately 25% of control values. When forskolin was added to cells, it also decreased DNA synthesis to approximately 30% of control, and the K_i was calculated to be 20 μM (Fig. 3c). Finally, the effect of the cyclic nucleotide phosphodiesterase inhibitor IBMX was tested. As demonstrated in Fig. 3d, IBMX also inhibited DNA synthesis, with a maximal reduction of 70% of control. The K_i for IBMX was found to be 30 μM. The studies described above indicated that elevating cAMP by three mechanistically separate means yielded a similar result (i.e., inhibition of serum-stimulated DNA synthesis).

3.4. Effect of 2′,5′-Dideoxyadenosine on PGI_2 Inhibition of DNA Synthesis

The above findings support the contention that PGI_2 inhibited DNA synthesis in VSMC. The data also indicated that sustained elevations in cAMP levels block DNA synthesis. If these two findings are causally related, i.e., if PGI_2 has its effects on DNA synthesis via effects on cAMP, then if the ability of PGI_2 to elevate cAMP levels is blocked, its ability to inhibit DNA synthesis should be abolished. This

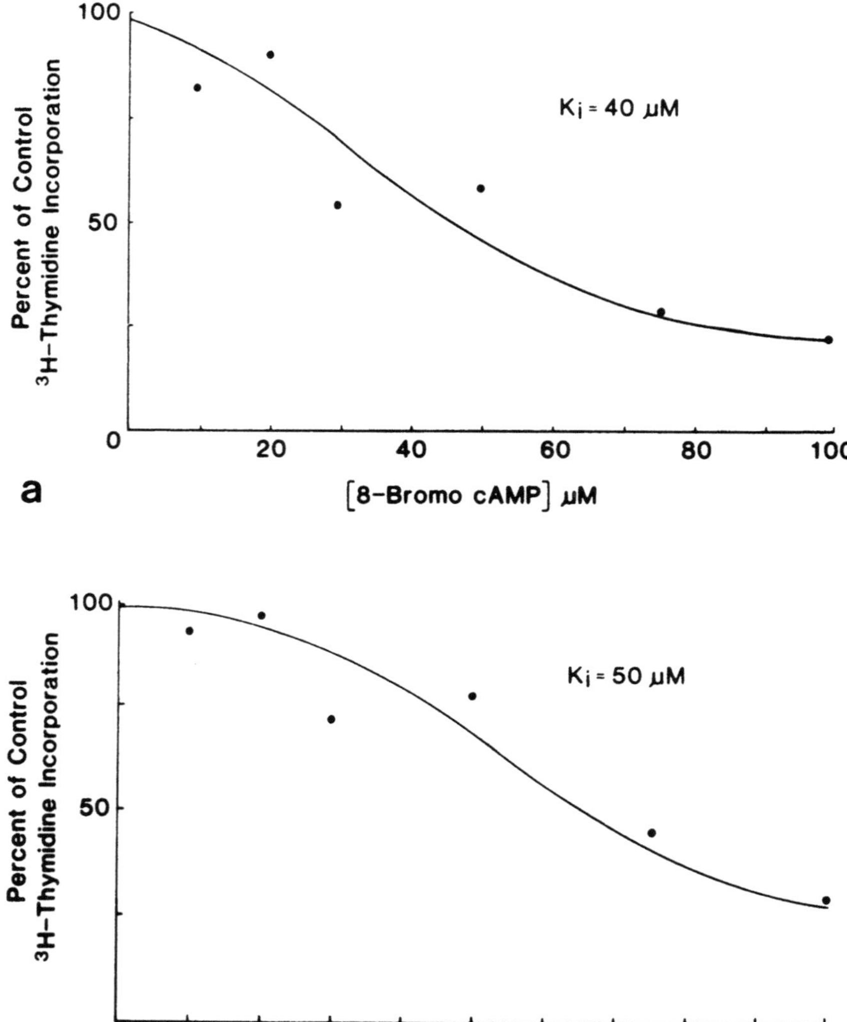

FIGURE 3. DNA synthesis in A-10 vascular smooth muscle cells: effect of cAMP analogues, forskolin, and IBMX. Cells were subcultured, and DNA synthesis was assayed as described in Fig. 1 except that varying concentrations of dibutyryl cAMP, 8-bromo-cAMP, forskolin, or IBMX were added. Data are expressed as percentages of control, where control is serum-stimulated DNA synthesis. Values are from representative experiments. The K_i value represents the owen ±S.E.M. from six separate determinations (a,b), from four separate experiments (c), or from five separate experiments (d).

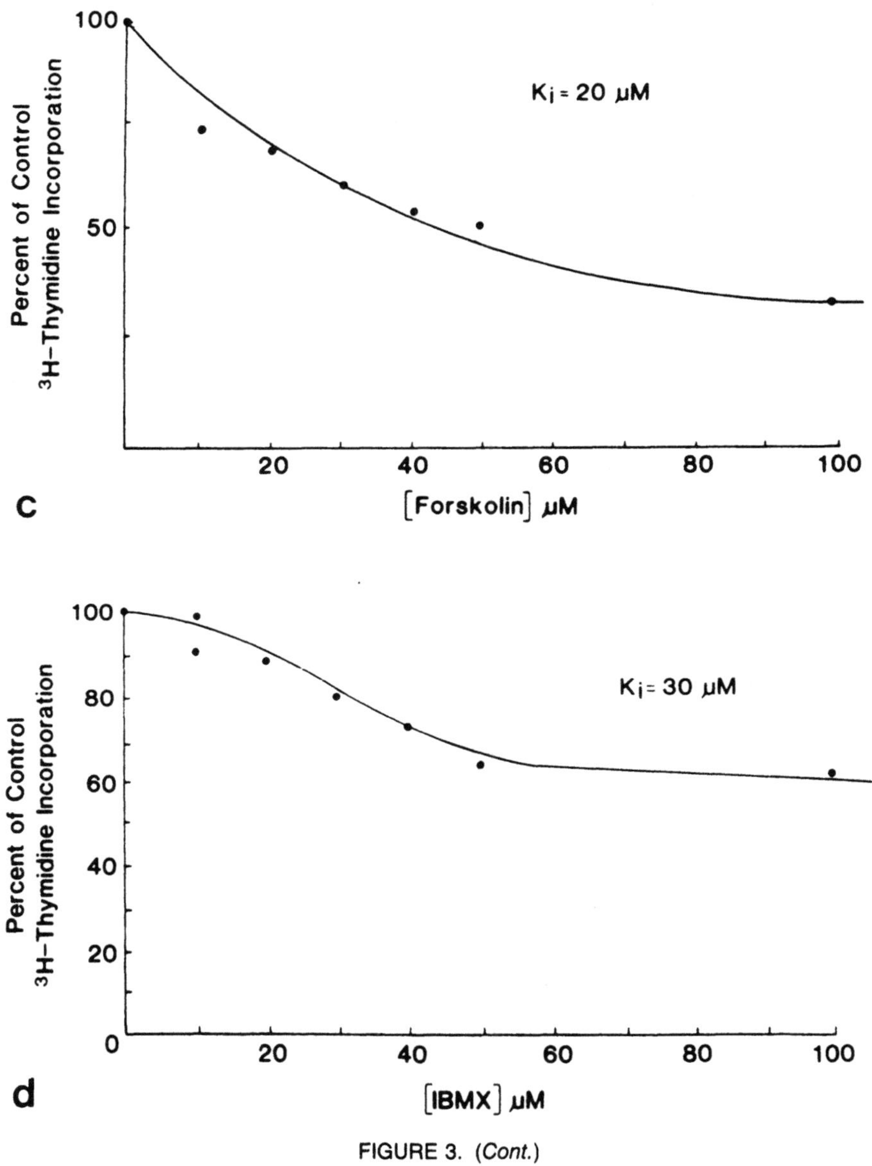

FIGURE 3. (*Cont.*)

was tested using 2',5'-dideoxyadenosine (DDA), which has been shown by Fain *et al.* (1972) to inhibit adenylate cyclase activation. Figure 4 depicts the results of studies in which the dose dependence of PGI_2 inhibition of DNA synthesis was measured in the presence and absence of DDA. Prostacyclin did not cause an inhibition of DNA synthesis in the presence of DDA.

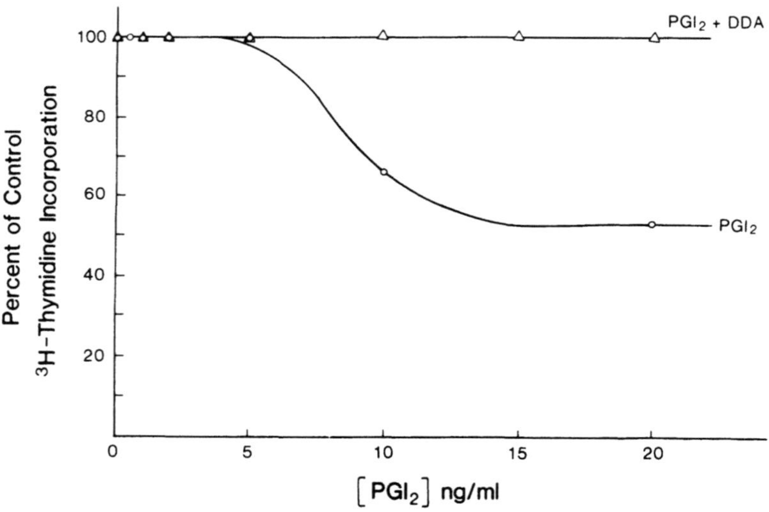

FIGURE 4. DNA synthesis in A-10 vascular smooth muscle cells: effect of prostacyclin $\pm$ 2′,5′-dideoxyadenosine. Cells were subcultured, and DNA synthesis was assayed as described in Fig. 1 except that varying concentrations of PGI_2 were added in the presence of 100 μM IBMX or in the presence of 100 μM IBMX plus 100 μM 2′,5′-dideoxyadenosine. Data are expresed as percentages of control, where control is serum-stimulated DNA synthesis in the presence of 100 μM IBMX. Values are from a representative experiment; the K_i value represents the mean $\pm$ S.E.M. from three separate determinations.

3.5. Relationship between cAMP levels and DNA Synthesis

Levels of cAMP were determined under the same conditions as used for the measurement of [³H]thymidine incorporation (i.e., cells were subcultured at the same density and were incubated with PGI_2 plus IBMX for 24 hr prior to measurement). As demonstrated in Fig. 5, control cells have approximately 5.5 pmol cAMP per g protein. This level is elevated threefold (i.e., to 13.7 pmol/g protein) in response to a maximal concentration of PGI_2 (30 nM plus IBMX). The data in Fig. 5 suggest that there may be an inverse relationship between the effects of PGI_2 on cAMP levels and on DNA synthesis in VSMC.

4. DISCUSSION

The present studies demonstrate that prostacyclin inhibits DNA synthesis in cultured vascular smooth muscle cells. The mechanism by which PGI_2 caused inhibition of DNA synthesis may be through its effects on adenylate cyclase and cellular cAMP levels. This contention is supported by the finding that PGE_1, another agent that activates adenylate cyclase, also inhibited DNA synthesis in VSMC. Furthermore, when cAMP levels were elevated using stable cAMP analogues, by

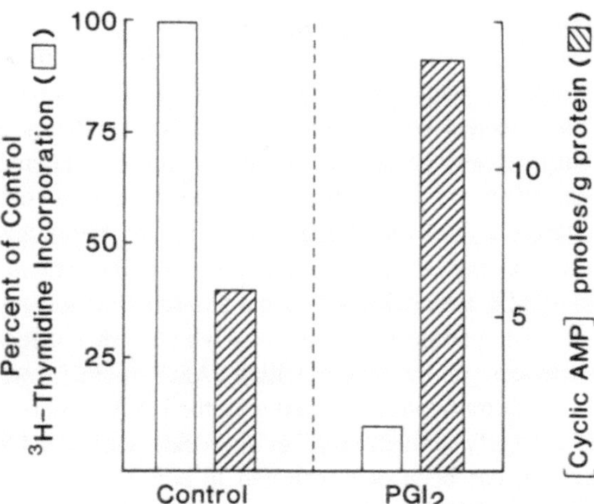

FIGURE 5. Relationship between DNA synthesis and cyclic AMP levels in A-10 vascular smooth muscle cells. DNA synthesis (□) was assayed as described in Fig. 1. Data are expressed as percentages of control. Control is serum-stimulated DNA synthesis in the presence of 100 μM IBMX; PGI₂ (20 ng/ml) was also added in the presence of 100 μM IBMX. Values represent mean ± S.E.M. from four separate determinations. Cyclic AMP (▨) was determined by radioimmunoassay as described in the text.

forskolin, or by the potent phosphodiesterase inhibitor 1-methyl-3-isobutylxanthine, DNA synthesis was always inhibited. These studies are the first demonstration that the potent vasodilatory and antiplatelet agent PGI₂ may also regulate vascular cell proliferation.

Further evidence that PGI₂ exerts its effects on DNA synthesis via an elevation in cAMP was provided by experiments using 2′,5′-dideoxyadenosine. These studies attempted to dissociate PGI₂ effects on DNA synthesis from PGI₂ effects on cAMP elevation. It appeared that the two could not be separated, since PGI₂ had no effect on DNA synthesis when added to cells in the presence of DDA. The final evidence that PGI₂ inhibited DNA synthesis via effects on cAMP was provided by actual measurements of cAMP. These data indicated that PGI₂ effects on cAMP levels and on DNA synthesis were inversely related.

The interpretation of these data is complicated by the extremely short biological half-life of PGI₂ (i.e., approximately 5 min at pH 7.45). However, all experiments with PGI₂ were done in the presence of the phosphodiesterase inhibitor IBMX. Thus, although PGI₂ was rapidly degraded and had its effects on adenylate cyclase very rapidly (within seconds), the effects were maintained over a longer period of time. This contention is substantiated by the finding that the elevation in cAMP levels was still threefold greater than control after a 24-hr incubation. Presumably, the reason that PGI₂ was found to be less potent than PGE₁ in inhibiting DNA synthesis (i.e., 30% inhibition versus 50% inhibition) is its short biological half-life.

Regardless of the absolute magnitude of inhibition, PGI₂ was found to cause a long-term elevation in cAMP levels (i.e., a threefold increase over control after 24 hr) and to inhibit DNA synthesis in vascular smooth muscle cells. These data are in agreement with those of Huttner *et al.* (1977) as well as with the large body of evidence suggesting that cAMP is a negative modulator of cellular proliferation (Berridge, 1975; Chalapowski *et al.*, 1975; Pastan *et al.*, 1975). On the other hand, recent studies by Franks *et al.* (1984) demonstrate that an increase in adenylate cyclase activity precedes DNA synthesis in cultured vascular smooth muscle cells. However, their studies also indicated that when cellular adenylate cyclase activity was elevated with cholera toxin, DNA synthesis appeared to be inhibited. On this basis, it was suggested that the increase in adenylate cyclase activity seen was a discrete event (i.e., a brief elevation and a return to basal levels) and that if adenylate cyclase is prevented from returning to basal levels the cells are blocked from synthesizing DNA. Thus, it appears that the present results with PGI_2 in the presence of IBMX are analogous to the abovementioned findings with cholera toxin.

The precise mechanism by which cAMP affects DNA synthesis may be highly dependent on the time at which cAMP exerts its effects. Nevertheless, certain steps in the stimulus-transfer pathway for DNA synthesis have been described. One step is the early activation of an amiloride-sensitive Na^+/H^+ exchange pathway. This system has been demonstrated to occur in many cell types (Schuldiner and Rozengurt, 1982; Moolenaar *et al.*, 1983; Mix *et al.*, 1984), has been shown to be activated by many mitogens [e.g., PDGF (Rozengurt and Mendoza, 1980; Cassel *et al.*, 1983; Owen, 1984)], and, finally, has been suggested to be a trigger for DNA synthesis (Rozengurt and Mendoza, 1980; Schuldiner and Rozengurt, 1982; Owen and Villereal, 1982). This Na^+/H^+ exchange system is present in A-10 vascular smooth muscle cells and can be stimulated by PDGF (Owen, 1984). Furthermore, the mechanism by which the system is stimulated is thought to involve an elevation in intracellular Ca^{2+} activity (Villereal and Owen, 1982). Most if not all effects of cAMP in cells are associated with activation of a protein kinase and subsequent protein phosphorylation. Therefore, it is likely that cAMP is initiating the phosphorylation of a protein that, in turn, interferes with a step(s) in the stimulus-transfer pathway for DNA synthesis. It is tempting to speculate that cAMP is causing the phosphorylation of a modulator of the Ca^{2+} sequestration system as has been proposed for other cell types (Haslam *et al.*, 1979; Owen and Le Breton, 1981). If this were the case, cAMP would effectively decrease intracellular Ca^{2+} levels and inhibit Na^+/H^+ exchange. Certainly, cAMP could have its effects entirely independent of Ca^{2+} as well.

Although the temporal and molecular mechanism of PGI_2 inhibition of DNA synthesis is not known, the phenomenon is likely to be important physiologically. In this regard, vascular smooth muscle cell proliferation is an integral aspect of atheromatous plaque formation (Ross and Glomset, 1976). Interestingly, vascular smooth muscle cells obtained from atherosclerotic lesions produce much less prostacyclin than do normal vascular smooth muscle cells (Larrue *et al.*, 1980). The present studies suggest that these two observations may be linked and that vascular smooth muscle cells may proliferate in atheromatous plaques in part because of depressed levels of prostacyclin production.

Thus, the present studies suggest that vascular smooth muscle cell proliferation may represent a finely tuned system: PDGF may regulate its own proliferative potential indirectly via effects on PGI_2 and cAMP.

ACKNOWLEDGMENTS. Special thanks to Koshy Chacko, Marsha Prastein, and Irene Stavros for their expert technical assistance. Thanks also to Cheryll Johnson for her help in manuscript preparation. This work was supported by a Chicago Heart Association Senior Fellowship and a National Institutes of Health Grant HL 35159.

REFERENCES

Berridge, M., 1975, The interaction of cyclic nucleotides and calcium in control of cellular activity, *Adv. Cyclic Nucleotide Res.* **6**:1–98.

Cassel, D., Rothenberg, P., Zhuang, Y., Deuel, T., and Glaser, L., 1983. Platelet-derived growth factor stimulates Na^+/H^+ exchange and induces cytoplasmic alkalinization in NR6 cells, *Proc. Natl. Acad. Sci. U.S.A.* **80**:6224–6228.

Chalapowski, F. J., Kelly, L. A., and Butcher, R. W., 1975, Cyclic nucleotides in cultured cells, *Adv. Cyclic Nucleotide Res.* **6**:245–338.

Coughlin, S. R., Moskowitz, U. A., Zetter, B. R., Antonaides, H. N., and Levine, L., 1980, Platelet-dependent stimulation of prostacyclin synthesis by platelet-derived growth factor, *Nature* **288**:600–602.

Epel, D., 1980, Ionic triggers in the fertilization of sea urchin eggs, *Ann. N.Y. Acad. Sci.* **339**:74–85.

Fain, J. N., Pointer, R. H., and Ward, W. F., 1972, Effects of adenosine nucleotides on adenylate cyclase, phosphodiesterase, cyclic adenosine monophosphate accumulation and lipolysis in fat cells, *J. Biol. Chem.* **247**:6866–6872.

Franks, D. J., Plamondon, J., and Hamet, P., 1984, An increase in adenylate cyclase activity precedes DNA synthesis in cultured vascular smooth muscle cells, *J. Cell. Physiol.* **119**:41–45.

Harper, J. F., and Brooker, G., 1975, Femtomole sensitive radioimmunoassay for cyclic AMP and cyclic GMP after 2′O-acetylation by acetic anhydride in aqueous solution, *J. Cyclic Nucleotide Res.* **1**:207–218.

Haslam, R. J., Lynham, J. A., and Fox, J. E. B., 1979, Effects of collagen, ionophore A23187 and prostaglandin E_1 on the phosphorylation of specific proteins in blood platelets, *Biochem. J.* **178**:397–406.

Hopkins, N. K., and Gorman, R. R., 1980, Regulation of endothelial cell nucleotide metabolism by prostacyclin, *J. Clin. Invest.* **67**:540–546.

Huttner, J. J., Gwebu, E. T., Panganamela, R. V., Milo, G. E., and Cornwell, D. G., 1977, Fatty acids and their prostaglandin derivatives: Inhibitors of proliferation in aortic smooth muscle cells, *Science* **197**:289–291.

Kimes, B. W., and Brandt, B. L., 1976, Characterization of two putative smooth muscle cell lines from rat thoracic aorta, *Exp. Cell. Res.* **98**:349–366.

Koch, K. S., and Leffert, H. L., 1979, Increased sodium ion influx is necessary to initiate rat hepatocyte proliferation, *Cell* **18**:153–163.

Larrue, J., Rigand, U., Daret, D., Demond, J., Durand, J., and Bricaud, H., 1980, Prostacyclin production by cultured smooth muscle cells from atherosclerotic rabbit aorta, *Nature* **285**:480–482.

Mix, L. L., Dinerstein, R. J., and Villereal, M. L., 1984, Mitogens increase intracellular pH and Ca^{2+} in human fibroblasts, *Biophys. J.* **45**:86a.

Moolenaar, W. H., Tsien, R. Y., van der Saag, P. T., and de Laat, S. W., 1983, Na^+/H^+ exchange and cytoplasmic pH in the action of growth factors in human fibroblasts, *Nature* **304**:645–648.

Owen, N. E., 1984, Platelet-derived growth factor stimulates Na influx in vascular smooth muscle cells, *Am. J. Physiol.* **247**:C501–C505.

Owen, N. E., and Le Breton, G. C., 1981, Calcium mobilization in intact human blood platelets as visualized by chlortetracycline fluorescence, *Am. J. Physiol.* **241**:H613–H619.

Owen, N. E., and Villereal, M. L., 1982, Evidence for a role of calmodulin in the serum-stimulation of Na^+ influx in human fibroblasts, *Proc. Natl. Acad. Sci. U.S.A.* **79**:3537–3541.

Owen, N. E., and Villereal, M. L., 1983, Lys-Bradykinin stimulates Na$^+$ influx and DNA synthesis in cultured human fibroblasts, *Cell* **32:**979–985.

Pastan, I., Johnson, G. S., and Anderson, W. B., 1975, Role of cyclic nucleotides in growth control, *Annu. Rev. Biochem.* **44:**491–522.

Ross, R., 1971, The smooth muscle cell: Growth of smooth muscle in culture and formation of elastic fibers, *J. Cell Biol.* **50:**172–185.

Ross, R. and Glomset, J., 1976, The pathogenesis of atherosclerosis, *N. Engl. J. Med.* **295:**420–425.

Ross, R., Glomset, J., Kareya, B., and Harker, L., 1974, A platelet-dependent serum factor that stimulates the proliferation of arterial smooth muscle cells *in vitro*, *Proc. Natl. Acad. Sci. U.S.A.* **71:**1207–1210.

Rozengurt, E., and Mendoza, S., 1980, Monovalent ion fluxes and the control of cell proliferation, *Ann. N.Y. Acad. Sci.* **339:**175–189.

Schuldiner, S., and Rozengurt, E., 1982, Na$^+$/H$^+$ antiport in Swiss 3T3 cells: Mitogenic stimulation leads to cytoplasmic alkalinization, *Proc. Natl. Acad. Sci. U.S.A.* **79:**7778–7782.

Shier, W. T., and Durkin, J. P., 1982, Role of stimulation of arachidonic acid release in the proliferative response of 3T3 mouse fibroblasts to platelet-derived growth factor, *J. Cell Physiol.* **112:**171–181.

Steiner, A. L., Parker, C. W., and Kipnis, D. M., 1972, Radioimmunoassay of cyclic nucleotides, *J. Biol. Chem.* **247:**1106–1113.

Villereal, M. L., and Owen, N. E., 1982, The involvement of Ca^{2+} in the serum-stimulation of Na$^+$ influx in human fibroblasts, in: *Ions, Cell Proliferation and Cancer* (A. L. Boynton, W. L. McKeehan, and J. F. Whitfield, eds.), Academic Press, New York, pp. 245–257.

20

Quantitation of Arachidonate Released during the Platelet Phosphatidylinositol Response to Thrombin

J. BRYAN SMITH, CAROL DANGELMAIER, and GERARD MAUCO

1. INTRODUCTION

Several workers have shown that decreases in the radioactivity of phosphatidylinositol (PI) and phosphatidylcholine (PC) occur when platelets prelabeled with radioactive arachidonic acid are treated with thrombin (see Rittenhouse-Simmons and Deykin, 1981, for review). These decreases have been attributed to arachidonate release by one of two mechanisms (Fig. 1). In one mechanism, PI is hydrolyzed by a PI-specific phospholipase C, yielding diacylglycerol and inositol phosphate. The diacylglycerol is further hydrolyzed by a lipase or lipases, releasing arachidonic acid and other fatty acids. In the other mechanism, PC is hydrolyzed by a phospholipase A_2, releasing arachidonic acid and other fatty acids from the 2-position and producing lysophosphatidylcholine (LPC). There is also some evidence that arachidonic acid is made available by the action of phospholipase A_2 on phosphatidylethanolamine.

There is presently some controversy over the relative importance of the above two pathways in contributing arachidonic acid for thromboxane A_2 formation by platelets. Some workers argue that the degradation of PI is more important since

J. BRYAN SMITH, CAROL DANGELMAIER, and GERARD MAUCO • Center for Thrombosis Research, Temple Medical School, Philadelphia, Pennsylvania 19140. G.M. was on academic leave from Université P. Sabatier, Faculté de Médecine Purpan, Toulouse, France; his present address is INSERM U101, Hôpital Purpan, F-31059 Toulouse, France.

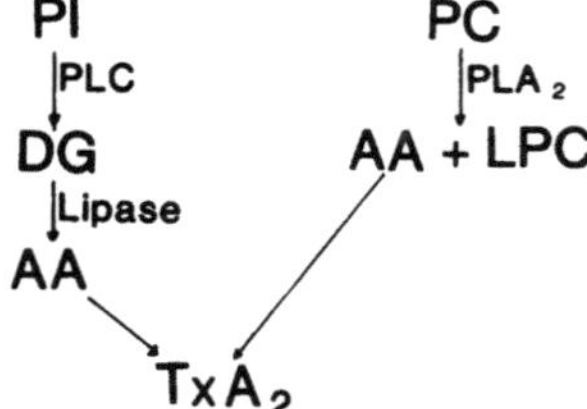

FIGURE 1. Schematic representation of the two major pathways proposed for arachidonic acid liberation in platelets.

PI contains almost exclusively arachidonic acid at the 2-position and the presence of phospholipase C and diglyceride lipases can readily be demonstrated in disrupted platelets. By contrast, studies of phospholipase A_2 in platelet membranes have so far only resulted in the detection of rather insubstantial amounts of activity. Furthermore, radiolabeled PI decreases more rapidly than radiolabeled PC when platelets prelabeled with radioactive arachidonate are treated with thrombin. On the other hand, some investigators argue that the amounts of PI in platelets are too small to provide enough of the arachidonic acid required for thromboxane formation and, besides, much of the diacylglycerol that is produced by the action of phospholipase C will be phosphorylated by diacylglycerol kinase to phosphatidic acid (PA) rather than being hydrolyzed by lipases. Other evidence in favor of the phospholipase A_2 pathway is the fact that increases in LPC and LPE have been detected in platelets within 10 sec of the addition of thrombin (McKean *et al.*, 1981; Broekman *et al.*, 1980).

One of the difficulties encountered in comparing the relative importance of PI and PC in providing arachidonic acid for thromboxane formation using platelets prelabeled with radioactive arachidonic acid is that PI has a much higher specific activity than PC. Thus, a decrease of 100 d.p.m. in PC is associated with the release of a much greater mass of arachidonic acid than a decrease of 100 d.p.m. of PI. Broekman *et al.* (1981) attempted to overcome this problem by studying the changes in endogenous phospholipids by phosphorous assay. They found that thrombin addition to platelets was followed by rapid alterations in the amounts of endogenous PI and PA and that the decrease in PI was not precisely reciprocated by an increase in PA. They postulated that this apparent discrepancy could represent the removal of diacylglycerol by diglyceride lipase.

Although a phosphorus assay enables the quantitative determination of decreases in the amounts of phospholipids, it does not permit the measurement of increases in the amounts of diacylglycerol and arachidonic acid. In the present work, we have compared decreases in PI, PI-4′-phosphate (PI-P), and PI-4′,5″-diphosphate (PI-P_2) with increases in PA, diacylglycerol, and arachidonic acid by measurement of their masses by gas chromatography–flame ionization detection of constituent fatty acids.

2. METHODS

A flow chart of the methods used in these studies is shown in Fig. 2. Human blood was obtained from healthy volunteers who had not taken any drug during the previous week. The blood was anticoagulated with acid-citrate-dextrose (ACD), and platelet-rich plasma was prepared by centrifugation at 180 g for 15 min at room temperature. The pH was adjusted to 6.5 with ACD, and a platelet pellet prepared by centrifugation at 1500 g for 15 min at room temperature. The pellet was resuspended in 0.1 volume of autologous plasma and subjected to gel filtration on a column of Sepharose 2B using a Tyrode's buffer without phosphate and containing 0.25% bovine serum albumin and 5 mM glucose. In some control experiments, not described here, the platelets were prelabeled with [^{3}H]arachidonic acid and [^{32}P]orthophosphate to determine the migration of radioactive compounds in the solvent systems described below.

The gel-filtered platelets (1 $\times$ 10^9 cells/ml) were warmed at 37°C for 3 min and either extracted directly or stimulated with 5 U/ml thrombin for times up to 1 min without stirring. In some experiments the cells were preincubated with 36 μM BW755C before thrombin addition to block cyclooxygenase and lipoxygenase activities. For lipid extraction, sodium EDTA was added (20 mM final concentration) plus 3.75 volumes of ice-cold chloroform/methanol, 1 : 2 (by volume). The extract was partitioned into two phases by addition of 1.25 volumes of chloroform and 1.25 volumes of 2.4 N hydrochloric acid. The lower phase was removed, and the

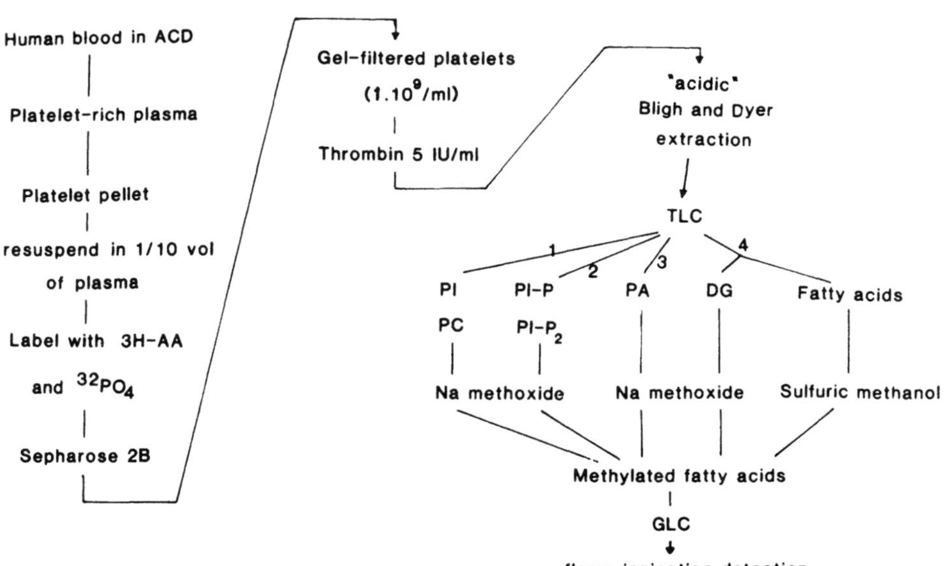

FIGURE 2. Methods used to study platelet phosphatidylinositol metabolism in the present study. For description see text.

upper phase was washed with 2.5 volumes of chloroform. The pooled organic extracts were evaporated under N_2 at 37°C and then stored in chloroform/methanol at -18°C. All of the organic solvents contained butylated hydroxytoluene (50 μg/ml) as an antioxidant.

Individual lipids were isolated by thin-layer chromatography using one of the following solvent systems: (1) chloroform/methanol/acetic acid/water, 81/10/45/1 by volume, for the resolution of PI; (2) chloroform/methanol/4 N ammonia, 9/7/2 by volume, for the resolution of PI-P and PI-P_2; (3) the upper phase of ethyl acetate/isooctane/acetic acid/water, 90/50/20/100 by volume, for PA; and (4) chloroform/methanol/hydrochloric acid, 87/13/0.5 by volume, for diacylglycerol and arachidonic acid. After development, the acids contained in the solvents were neutralized by exposure to ammonia vapor. The resolved compounds were detected using a primulin spray.

Except for the case of arachidonic acid, the silica gel scrapings were reacted for 15 min with 2 N sodium methoxide at 50°C in the presence of 5 μg of methyl heptadecanoate as an internal standard. In the case of arachidonic acid, methylation was performed with sulfuric acid in methanol at 80°C for 12 hr, and heptadecanoic acid was used as the internal standard. After neutralization, the methyl esters were extracted into hexane, taken to dryness under N_2, and redissolved in 50 μl of carbon disulfide.

Gas–liquid chromatography was performed in a Hewlett-Packard chromatograph model 5730A fitted with 6-foot by 1/8-inch glass columns containing 10% SP-2330 on 100/200 Chromosorb W AW (Supelco, Bellafonte, PA). The chromatograph was operated in a differential mode with a termperature gradient as follows: 2 min at 170°C, increasing 4°C/min to 220°C, and holding for 8 min. Injection and flame ionization ports were at 250°C. The carrier gas was nitrogen (20 ml/min). Quantitation of the different fatty acid methyl esters was performed with an automatic integrator (Hewlett-Packard, model 3390A).

3. RESULTS

The averaged data of four experiments indicated that the mass of platelet PI was 17 nmol/10^9 cells and consisted almost entirely (86.6%) of a species containing both stearic and arachidonic acids. On addition of thrombin, a decrease in the mass of PI was detected at 5 sec and continued during the minute of observation. The decrease in the mass of PI during this time was accounted for by the loss of equimolar amounts of stearic and arachidonic acids and accounted for about 7 nmol/10^9 cells.

We also observed that platelets contain small amounts of both PI-P and PI-P_2, amounting to 3.0 and 1.0 nmol/10^9 cells, respectively. These lipids also were enriched in stearic and arachidonic acids (76.9% of PI-P and 85.4% of PI-P_2 contained both arachidonic and stearic acids), and both lipids decreased in mass 5 sec after thrombin addition. However, the decrease in mass of PI-P was only 0.4

nmol/10^9 platelets, and although there was a similar decrease in PI-P_2 at 5 sec, the mass of this lipid actually increased by 0.4 nmol/10^9 cells at 1 min.

There was a dramatic increase in the amount of diacylglycerol present in platelets that peaked at 5 sec after the addition of thrombin. Unstimulated platelets contained 0.3 nmol/10^9 cells, whereas stimulated platelets contained 1.2 nmol/10^9 cells at 5 sec. The increase was attributable to the appearance of stearoyl, arachidolyglycerol that was undetectable prior to stimulation. By 1 min the diacylglycerol level had declined to 0.6 nmol/10^9 cells.

Phosphatidic acid increased progressively from a basal level of about 1 nmol/10^9 cells to about 4 nmol/10^9 cell at 1 min. This was mainly a result of the accumulation of the stearoyl, arachidonyl species of phosphatidic acid.

Thus, overall, the changes in PI-P and PI-P_2 are small, and the major changes observed are a decrease in PI and increases in PA and diacylglycerol. Moreover, the decrease in PI is not exactly compensated for by the latter increases, indicating that as much as 60% of the decrease in PI could occur by its being converted into arachidonic acid by the phospholipase C–diglyceride lipase pathway.

To examine the contribution of the above pathway to the total amount of arachidonic acid that is made available during thrombin stimulation of platelets, we pretreated platelets with BW755C and then incubated them for 1 min at 37°C with either saline or thrombin (5 U/ml). Figure 3 shows the GLC–flame ionization detector profiles of the free fatty acids present in these cells. The peak at 4.88 min is the heptadecanoic acid added as an internal standard to enable the calculation of mass. The major difference between the saline- and thrombin-treated platelets is the appearance of a large peak corresponding to arachidonic acid (20 nmol/10^9 cells) at 11.72 min. There also are small increases (of the order of 2 to 4 nmol/10^9 cells) in palmitic (16:0), stearic (18:0), and oleic (18:1) acids at 1 min after thrombin stimulation.

Table I shows the results of a typical experiment, performed in triplicate, in which the decrease in PI was measured in the same samples in which increases in diacylglycerol, PA, and arachidonic acid were determined. There was a decrease of 6.35 nmol/10^9 platelets of PI, and this was compensated for only by an increase of 2.5 nmol in diacylglycerol and phosphatidic acid combined. Thus, 3.85 nmol/10^9 platelets of the decrease in PI is unexplained and could provide arachidonic acid for thromboxane formation. On the other hand, 20–30 nmol/10^9 platelets of arachidonic acid became available 1 min after thrombin stimulation. Thus, at best, 19% of the arachidonic acid was made available by the decrease in PI.

4. DISCUSSION

Our results confirm and extend the observations of Broekman *et al.* (1981) that the decrease in PI is not precisely reciprocated by an increase in PA when thrombin is the stimulus. We show here that the decrease in PI is not compensated

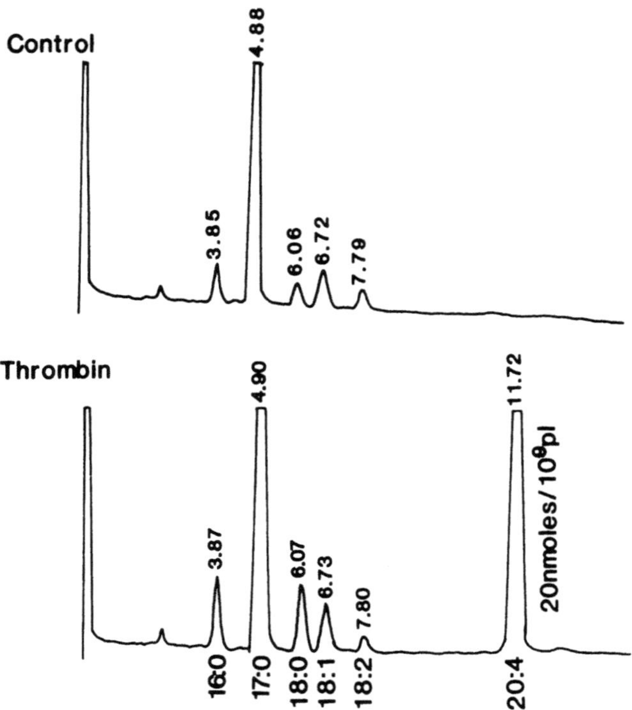

FIGURE 3. Thrombin-induced increase in free arachidonic acid in BW755C-treated platelets. Shown are representative recordings of fatty acid methyl esters obtained from saline- or thrombin-treated platelets analyzed by GLC with flame ionization detection.

fully even when the increases in both diacylglycerol and PA are taken into account. This apparent discrepancy could be explained by removal of diacylglycerol by diglyceride lipases or, alternatively, by the action of phospholipase A_2 on PI, thereby providing arachidonic acid for thromboxane formation. By making quantitative measurements of the decrease in PI at the same time as the increase in arachidonic

TABLE I. Decreases in PI and Increases in
PA and Diacylglycerol (Recycled) Measured
in a Single Triplicate Determination

	Change (nmol/10^9 platelets)	
PI decrease	6.35	
Recycled	2.5	
Released AA	3.8	(12.7–19.0%)
Total AA production	20–30	(100%)

acid, we can conclude that no more than 20% of the arachidonic acid is derived from PI. The remainder is presumably derived from phosphatidylcholine and/or phosphatidylethanolamine by the action of phospholipase A_2.

ACKNOWLEDGMENTS. This work was supported in part by Fondation pour la Recherche Medicale Française and NIH grants HL 14217 and HL 30783.

REFERENCES

Broekman, M. J., Ward, J. W., and Marcus, A. J., 1980, Phospholipid metabolism in stimulated human platelets. Changes in phosphatidylinositol, phosphatidic acid and lysophospholipids, *J. Clin. Invest.* **66**:275–283.

Broekman, M. J., Ward, J. W., and Marcus, A. J., 1981, Fatty acid composition of phosphatidylinositol and phosphatidic acid in stimulated platelets. Persistence of arachidonyl–stearyl structure, *J. Biol. Chem.* **256**:8271–8274.

McKean, M. L., Smith, J. B., and Silver, M. J., 1981, Formation of lysophosphatidylcholine by human platelets in response to thrombin, *J. Biol. Chem.* **256**:1522–1524.

Rittenhouse-Simmons, S., and Deykin, D., 1981, Release and metabolism of arachidonate in human platelets, in: *Platelets in Biology and Pathology* (J. Gordon, ed.), Elsevier/North-Holland Biomedical Press, Amsterdam, pp. 349–372.

21

Phospholipases in the Macrophage

EDWARD A. DENNIS, THEODORE L. HAZLETT, RAYMOND A. DEEMS, MERRICK I. ROSS, and RICHARD J. ULEVITCH

1. CONTROL OF ARACHIDONIC ACID PRODUCTION

It is well established that the synthesis of prostaglandins and leukotrienes is dependent on the availability of free arachidonic acid, which is normally found esterified in the *sn*-2 position of phospholipids. The controlling step of this production has been suggested to be at the level of arachidonic acid release from membrane phospholipids. A likely mechanism for this control is the activation of a phospholipase, probably one localized in the membranes of cells. The phospholipases comprise a widespread and abundant class of enzymes in biological systems (Dennis, 1983). They are defined by their positional specificity on the phospholipid backbone, as shown in Fig. 1. Many phospholipases have been suggested as possible modulators of arachidonic acid release. Figure 2 summarizes the various possible routes by which arachidonic acid can be released. The numerous enzyme sequences that lead to arachidonic acid release begin with one of three enzymes: phospholipase A_1, phospholipase A_2, or phospholipase C. The phospholipase A_2 route would be the simplest and most direct source of arachidonic acid. The second route would begin with a phospholipase A_1 followed by a lysophospholipase. The last begins with a phospholipase C followed by any number of other enzymes. A number of these enzymatic systems have been detected in a variety of circulating cell types including the platelet, polymorphonuclear leukocyte, and macrophage (Bell *et al.*,

EDWARD A. DENNIS, THEODORE L. HAZLETT, and RAYMOND A. DEEMS ● Department of Chemistry, University of California at San Diego, La Jolla, California 92093.　MERRICK I. ROSS and RICHARD J. ULEVITCH ● Department of Immunology, Research Institute of Scripps Clinic, La Jolla, California 92037.

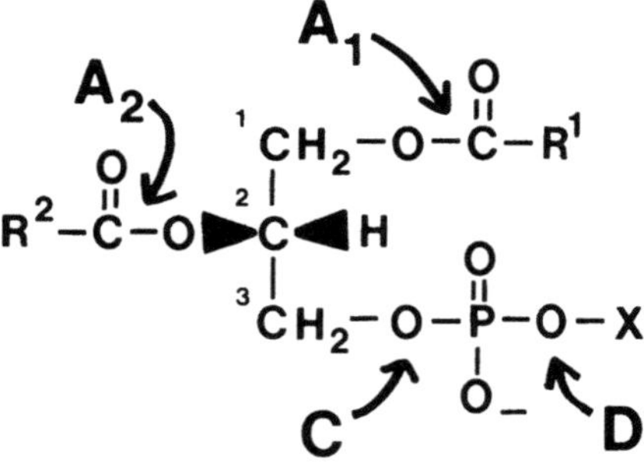

FIGURE 1. Site of action of the phospholipases. R^1 and R^2 refer to the fatty acids on the *sn*-1 and *sn*-2 positions of the stereospecifically numbered phospholipid, and X refers to the polar group—the most common are choline, ethanolamine, serine, inositol, and glycerol. (Reproduced with permission from Dennis, 1983.)

1979; Billah *et al.*, 1981; Kroner *et al.*, 1982; Lanni and Franson, 1981; Wightman *et al.*, 1981a,b, 1982).

Which of these pathways is responsible for the control of arachidonic acid production? Answering this question has proven far more difficult than simply determining the presence or absence of the various enzymes in a given cell type, for almost all of the enzymes shown in Fig. 2 have been found in each of the cell types mentioned. Because of this multiplicity of phospholipases, identifying the controlling enzymes becomes a problem of determining the relative levels of the various enzymes or enzyme systems *in vivo* in stimulated cells, a decidedly difficult task. This problem is compounded by the presence of other enzymes of lipid metabolism, e.g., lysophospholipases, lipases, acyltransferases, kinases. These enzymes not only modulate the levels of all phospholipid pools *in vivo* but also complicate experimental design and interpretation. The latter is especially true for the effects of lysophospholipase on *in vitro* assays.

In an attempt to overcome these obstacles, we are attempting to separate,

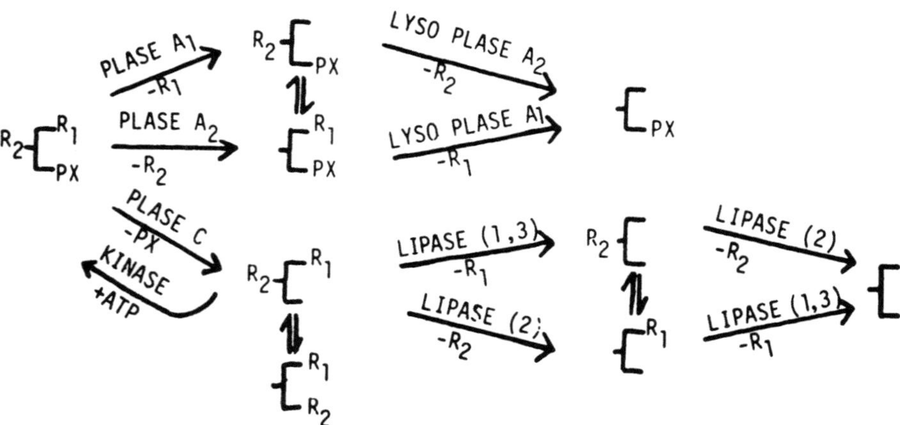

FIGURE 2. Possible pathways for arachidonic acid release from the *sn*-2 position of phospholipids including possible nonenzymatic migrations during isolation.

purify, and characterize the various phospholipases and lysophospholipases from the P388D$_1$ murine macrophage cell line. It is our hope that separation of the activities will allow the unambiguous determination of the enzyme characteristics that would, in turn, open the way to understanding their interactions *in vivo*. To date we have found at least three different phospholipase A activities toward phosphatidylcholine. We have also found a more active lysophospholipase. We have used sucrose density centrifugation to assign these activities to specific subcellular compartments. The data are presented here in terms of the general considerations necessary for evaluating the phospholipase activities and specificities involved in arachidonic acid release in cells.

2. Macrophage Cell Line Phospholipases

In order to insure a reproducible and abundant source of macrophage phospholipases, we have initiated studies on the P388D$_1$ cell line, which has been shown to have the usual characteristics of a macrophage (Koren *et al.*, 1975). We have developed a subcellular fractionation scheme, as shown in Fig. 3 and discussed in

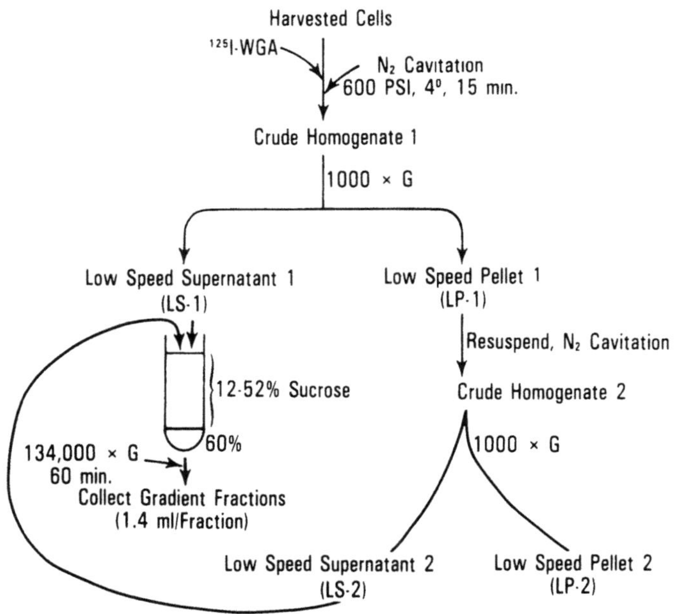

FIGURE 3. Subcellular fractionation of P388D$_1$ macrophages. The cells were nitrogen cavitated at 600 p.s.i. following surface labeling with [^{125}I]WGA. The resultant crude homogenate was centrifuged at 1000 × *g* to produce LS-2 and LP-1. LP-1 was resuspended, N₂ cavitated under the same conditions, and centrifuged at 1000 × *g* to produce LS-2 and LP-2. LS-1 and LS-2 were layered onto linear sucrose density gradients and fractionated by centriguation at 134,000 × *g* for 60 min.

detail elsewhere (M. I. Ross, R. A. Deems, A. J. Jesaitis, E. A. Dennis, and R. J. Ulevitch, unpublished data). It is based on nitrogen cavitation to break the cells followed by centrifugation at $100 \times g$. This produces a low-speed pellet (LP-1) and a supernatant fraction (LS-1). LP-1 was subjected to further nitrogen cavitation to produce a supernatant fraction (LS-2). Both LS-1 and LS-2 were subjected to subcellular fractionation on sucrose gradients.

LS-1 contains a Ca^{2+}-independent phospholipase that exhibits much less activity in the presence of Ca^{2+}. LP-1 contains a very active Ca^{2+}-dependent phospholipase that exhibits very little activity in the absence of Ca^{2+}. The Ca^{2+} dependence of the phospholipase in LP-1 is absolute, whereas the Ca^{2+}-independent enzyme that predominates in LS-1 may be slightly inhibited by high Ca^{2+} concentrations. The Ca^{2+}-dependent phospholipase has a pH optimum of 8.8 and appears to be membrane bound. We found that it could be released from the $1000 \times g$ pellet by further nitrogen cavitation and that the majority of this activity then appeared in the low-speed supernatant, which is designated LS-2. The Ca^{2+} dependence of the LS-2 enzyme is similar to that assayed in the LP-1 fraction and appears to be specific for Ca^{2+}. For example, Mg^{2+} will not replace it. The pH–rate profiles of LS-1 and LS-2 reveal that LS-1 contains a Ca^{2+}-independent pH 7.5 optimum phospholipase and also contains a pH 4.2 optimum Ca^{2+}-independent enzyme. LS-2, on the other hand, contains a pH 8.8 optimum Ca^{2+}-dependent enzyme and some of the pH 4.2 optimum phospholipase.

In summary, we have developed a scheme that produces a low-speed supernatant LS-1, which contains the majority of the Ca^{2+}-independent pH 7.5 enzyme, and a low-speed supernatant 2 (LS-2), which contains the majority of the Ca^{2+}-dependent pH 8.8 optimum enzymes. Both fractions contain the pH 4.2 optimum Ca^{2+}-independent enzyme. Both of these supernatants were subjected to sucrose gradient centrifugation in an attempt to determine the subcellular localization of these activities.

Using standard markers, we were able to localize the major subcellular organelles. The majority of the Ca^{2+}-independent pH 7.5 phospholipase was found in the cytoplasmic fraction along with LDH. The pH 4.2 optimum enzyme appears to be associated with the lysosomal fraction. The Ca^{2+}-dependent pH 8.8 enzyme is associated with the membrane fraction, most heavily with endoplasmic reticulum and mitochondria and perhaps some with the plasma membrane. We are currently pursuing the purification of this latter enzyme.

3. ROLE OF LYSOPHOSPHOLIPASES

The determination of the specificity of the separated phospholipases is a difficult undertaking. The pH 7.5 and pH 8.8 enzymes were identified by the release of free fatty acid from specifically labeled phospholipids in the *sn*-2 position, suggesting the presence of phospholipase A_2. However, fatty acid could also be released by the combined action of a phospholipase A_1 and a lysophospholipase, as shown in Fig. 4. Furthermore, we have identified a very active lysophospholipase in the cell

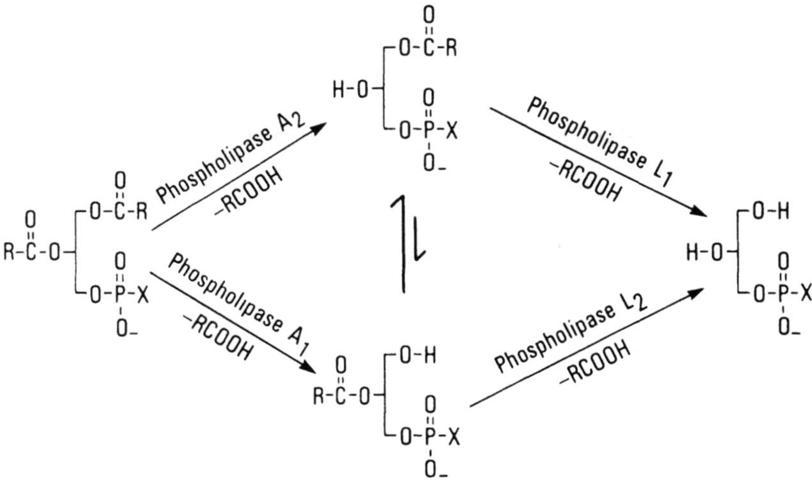

FIGURE 4. Products of phospholipases A_1 and A_2 and phospholipases L_1 and L_2 (also called lysophospholipase A_1 and A_2, respectively). Possible nonenzymatic migration between the 1-acyl and 2-acyl lysophospholipids is shown. (Reproduced with permission from Dennis, 1983.)

homogenate, and this lends credence to this possibility. The lysophospholipase enzyme is distributed principally in the 1000 × g supernatant (LS-1). It also appears to be soluble after centrifugation at 10,000 × g as well as 130,000 × g, as shown in Table I. The lysophospholipase in LS-1 has a broad pH optimum between pH 6 and pH 9, as shown in Fig. 5. The lysophospholipase apparently does not require metal ion.

The lysophospholipase in LS-1 is 20 to 40 times more active than the Ca^{2+}-independent pH 7.5 optimum phospholipase of LS-1 and the Ca^{2+}-dependent pH

TABLE I. Localization of Lysophospholipase Activity in P388D$_1$ Cells

	Total pellet activity[a]		Total supernatant activity	
	μunits/10^9 cells	%[b]	μunits/10^9 cells	%
Whole homogenate			37,444	100
1000 × g	14,487	39	24,333	65
10,000 × g	3,227	9	21,333	57
134,000 × g	2,667	7	18,500	49

[a] A radioactive assay procedure was employed to determine the lysophospholipase activity. The assay mixture contained 50 mM tris buffer at pH 8.0 and 100 μM 1-palmitoyl-*sn*-glycerol-3-phosphorylcholine (Sigma) containing a sufficient quantity of 1-palmitoyl [1-^{14}C]-*sn*-glycerol-3-phosphorylcholine (Amersham). The total assay volume was 1.0 ml, and the assay was carried out at 40°C for 45 to 180 min. The reaction was quenched by the addition of 1.0 ml of chloroform/methanol/acetic acid (2/4/1). An additional 0.5 ml of chloroform was then added to aid in the extraction. The sample was then vortexed and centrifuged. The chloroform layer of the resulting two-phase solution was removed, dried, resolubilized in 20 μl of chloroform, and spotted on Brinkmann Sil G-25 thin-layer chromatography plates. The plates were run in chloroform/methanol/water (65/25/4). The lysophospholipid and fatty acid spots were scraped into scintillation vials and counted to determine the amount of lysophospholipid hydrolyzed. One microunit of activity is defined as the amount of protein required to hydrolyze 1 pmol of substrate per minute.
[b] Percent of activity found in the whole homogenate.

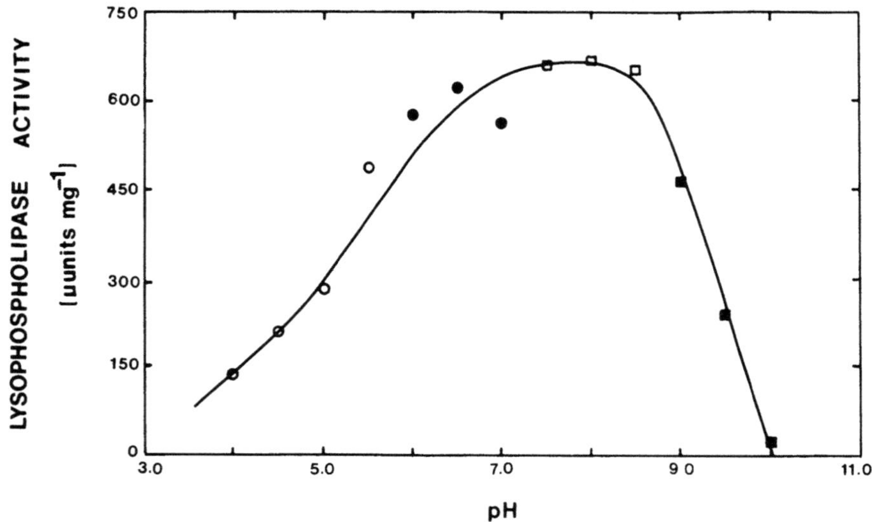

FIGURE 5. Rate of lysophospholipid hydrolysis by lysophospholipase in LS-1 as a function of pH. The buffers used were (○) acetate, (●) imidazole, (□) tris-HCl, and (■) glycine. Assay conditions are the same as those employed in the experiment described in Table I. Each assay contained 1.0 mg/ml of LS-1.

8.8 optimum enzyme of LS-2. This is true both in terms of total activity (Table I) and specific activity (Fig. 5). Because of the active lysophospholipase, any lysophospholipid product of phospholipase A_1 or A_2 would be rapidly converted to free fatty acid and glycerolphosphorylcholine. Thus, the release of $[^{14}C]$-labeled fatty acid at pH 7.5 and 8.8 could result from either a phospholipase A_2 or the combined action of a phospholipase A_1 and lysophospholipase. At pH 4.2, the phospholipase was identified by the production of lysophospholipids labeled in the palmitic acid with very little free $[^{14}C]$-labeled fatty acid being produced. This is indicative of a phospholipase A_1 as the predominant enzyme at pH 4.2.

4. CONCLUSION

We have begun the separation and characterization of the various phospholipases that could play a role in arachidonic acid release in the $P388D_1$ macrophage cell line. To date we have found a very active lysophospholipase that is localized in the cytoplasmic fraction with a pH optimum of 7–8. These results are identical to our previously reported findings regarding the lysophospholipase activity in human amnionic membranes (Dennis *et al.*, 1983; Jarvis *et al.*, 1984). These membranes also contain a lysophospholipase that is 20–40 times as active as any phospholipase A present and whose activity precludes the determination of the specificities of the other phospholipases.

We have also found that the P388D$_1$ cells contain a number of phospholipases: a pH 4.2 optimal Ca^{+2}-independent enzyme, which is probably a phospholipase A$_1$, a pH 7.5 optimal Ca^{+2}-independent enzyme, and a pH 8.8 Ca^{2+}-dependent enzyme. The specificity of the latter two enzymes could not be determined because of the high lysophospholipase activity. Phospholipase specificities are determined by the analysis of the catalytic products of hydrolysis of a specifically labeled phospholipid. For example, if the phospholipid is labeled in the *sn*-2 position, a phospholipase A$_1$ would produce labeled lysophospholipid, whereas a phospholipase A$_2$ would produce labeled fatty acid. The presence of a lysophospholipase has the potential to convert all of the lysophospholipid to fatty acid, and the assay thus loses its ability to distinguish specificities. Phospholipase C activity toward phosphatidylcholine is negligible under our experimental conditions. Of course, this does not preclude the presence of an inositol-specific phospholipase C in the macrophage cell line. Further work is in progress to purify and characterize these enzymes.

ACKNOWLEDGMENTS. These studies were supported by grants from the National Institutes of Health (GM-20,501 and AI 15136) and from the Lilly Research Laboratories. Edward A. Dennis wishes to thank Professor Manfred Karnovsky at Harvard Medical School and Professor Lawrence Levine at Brandeis University for hospitality and stimulating discussions on macrophages and prostaglandins as a Guggenheim Fellow, 1983–1984. Merrick I. Ross is a fellow of the Eleanor B. Pillsbury Residents Trust Fund of the University of Illinois, College of Medicine, Department of Surgery. Richard J. Ulevitch is the recipient of U.S.P.H.S. R.C.D.A. grant AI 00391.

REFERENCES

Bell, R. L., Kennerly, D. A., Stanford, N., and Majerus, P. W., 1979, Diglyceride lipase: A pathway for arachidonate release from human platelets, *Proc. Natl. Acad. Sci. U.S.A.* **76**:3238–3241.

Billah, M. M., Lapetina, E. G., and Cuatrecasas, P., 1981, Phospholipase A$_2$ activity specific for phosphatidic acid, *J. Biol. Chem.* **256**:5399–5403.

Dennis, E. A., 1983, Phospholipases, in: *The Enzymes,* ed. 3. Vol. 16 (P. D. Boyer, ed.), Academic Press, New York, pp. 307–353.

Dennis, E. A., Jarvis, A. A., and Hazlett, T. L., 1983, Lysophospholipases: Role in metabolism of phospholipids, *Fed. Proc.* **42**:2055.

Jarvis, A. A., Cain, C., and Dennis, E. A., 1984, Purification and characterization of a lysophospholipase from human amnionic membranes, *J. Biol. Chem.* **259**:15188–15195.

Koren, H. S., Handwerger, B. S., and Wunderlich, J. R., 1975, Identification of macrophage-like characteristics in a cultured murine tumor line, *J. Immunol.* **114**:894–897.

Kroner, E., Fischer, H., and Ferber, E., 1982, Sub-cellular fractionation of bone marrow-derived macrophages—localization of phospholipase A$_1$ and phospholipase A$_2$ and acyl-CoA-1-acylglycero-3-phosphorylcholine-O-acryltransferase, *Z. Naturforsch.* **37c**:314–420.

Lanni, C., and Franson, R. C., 1981, Localization and partial purification of a neutral-active phospholipase A$_2$ from BCG-induced rabbit alveolar macrophages, *Biochim. Biophys. Acta* **658**:54–63.

Wightman, P. D., Dahlgren, M. E., Davies, P., and Bonney, R. J., 1981a, The selective release of phospholipase A_2 by resident mouse peritoneal macrophages, *Biochem. J.* **200**:441–444.

Wightman, P. D., Humes, J. L., Davies, P., and Bonney, R. J., 1981b, Identification and characterization of two phospholipase A_2 activities in resident mouse peritoneal macrophages, *Biochem. J.* **195**:427–433.

Wightman, P. D., Dahlgren, M. E., and Bonney, R. J., 1982, Protein kinase activation of phospholipase A_2 in sonicates of mouse peritoneal macrophages, *J. Biol. Chem.* **257**:6650–6652.

22

Lipid Bodies

Widely Distributed Cytoplasmic Structures That Represent Preferential Nonmembrane Repositories of Exogenous [³H]Arachidonic Acid Incorporated by Mast Cells, Macrophages, and Other Cell Types

STEPHEN J. GALLI, ANN M. DVORAK,
STEPHEN P. PETERS, EDWARD S. SCHULMAN,
DONALD W. MacGLASHAN, Jr.,
TADASHI ISOMURA, KATHRYN PYNE,
V. SUSAN HARVEY, ILAN HAMMEL,
LAWRENCE M. LICHTENSTEIN,
and HAROLD F. DVORAK

1. INTRODUCTION

Although the biochemistry and pharmacology of the products of arachidonic acid (AA) oxidation have been described in detail (Samuelsson *et al.*, 1975; Moncada and Vane, 1978; Samuelsson, 1983), questions remain about the early events leading

STEPHEN J. GALLI, ANN M. DVORAK, KATHRYN PYNE, V. SUSAN HARVEY, and HAROLD F. DVORAK ● Department of Pathology, Beth Israel Hospital, and Harvard Medical School, and The Charles A. Dana Research Institute, Beth Israel Hospital, Boston, Massachusetts 02215. STEPHEN P. PETERS, EDWARD S. SCHULMAN, DONALD W. MacGLASHAN, Jr., and LAWRENCE M. LICHTENSTEIN ● Department of Medicine, Division of Clinical Immunology, Johns Hopkins University School of Medicine, The Good Samaritan Hospital, Baltimore, Maryland 21239.   TADASHI ISOMURA ● Second Department of Surgery, Kurume University School of Medicine, Kurume 830, Japan.   ILAN HAMMEL ● Department of Pathology, Sackler School of Medicine, Tel Aviv University, Ramat Aviv, Tel Aviv 69978, Israel.

to the formation of these molecules and particularly the sites within the cell from which AA can be initially mobilized. For example, it is widely held that phospholipids of cell membranes represent the major source of AA entering the cyclooxygenase or lipoxygenase pathways of oxidation. However, alternative routes of AA oxidation have been proposed. Thus, rabbits fed diets deficient in essential fatty acids exhibited markedly diminished content of PGE_1, PGE_2, and $PGF_{2\alpha}$ in several different organs despite maintaining stable or even supranormal levels of AA and dihomo-γ-linoleic acid in liver or red blood cell membrane phospholipids (Willis *et al.*, 1981). This result suggested that some prostaglandins ("basal prostaglandins") may be derived from a metabolic pool of precursors distinct from cellular membranes, although the localization of this metabolic pool within the cell has not been defined (Crawford, 1983).

In this chapter, we review evidence pointing to a potentially important nonmembrane source of AA products, cytoplasmic lipid bodies. Lipid bodies (or "lipid droplets") are structures of roughly spherical shape and inconstant size that are identifiable in transmission electron micrographs by their shape and variable osmiophilia and also because they lack a limiting membrane (Fawcett, 1981). In mature adipocytes, lipid bodies store metabolic energy in the form of triglycerides (Fawcett, 1981). However, lipid bodies may also be observed in a wide variety of other cell types. Table I presents a list of cells in which lipid bodies have been observed by one of us (A.M.D.), either in specimens submitted for diagnostic electron microscopy or during the course of research investigations. Although this cannot be considered a complete compilation, as it is likely that most if not all nucleated cells can exhibit at least a few lipid bodies under certain circumstances (Fawcett, 1981), it does illustrate the diversity of normal and neoplastic cells that can exhibit these structures *in situ*.

Although the factors governing the occurrence, size, and number of lipid bodies in cells other than adipocytes remain to be fully defined, lipid bodies can represent an early manifestation of cellular injury (e.g., Uchida *et al.*, 1983). They apparently

TABLE I. Human Cell Types That Can Contain
Cytoplasmic Lipid Bodies[a]

Mast cells[b,c]	Pericytes[b]
Macrophages[b,c]	Gastrointestinal Smooth muscle cells[b]
Neutrophils[b,c]	Cardiac muscle cells[b]
Eosinophils[b,c]	Chondroblasts[b]
Basophils[b,c]	Hepatocytes[b]
Lymphocytes[b,c]	Adrenal cortical cells[b]
Plasma cells[b]	Ciliated epithelial cells of bronchioles[b]
Platelets[b]	Absorptive epithelial cells of ileum[b]
Vascular endothelial cells[b]	Many carcinomas[b,c]
Vascular smooth muscle cells[b,c]	Many sarcomas[b]
Fibroblasts[b,c]	Many lymphomas[b]

[a] A partial list based on the personal experience of A.M.D.
[b] Observed in material obtained at surgery and immediately fixed for electron microscopy.
[c] Observed in cells maintained for variable intervals *in vitro* before fixation for electron microscopy.

may also develop in the absence of toxic stimuli, as judged by their occurrence in cells exhibiting no other evidence of injury. For example, neutrophils incubated with albumin-bound oleic acid develop lipid bodies but exhibit little or no alteration of bactericidal activity, phagocytosis, or chemotaxis (Hawley and Gordon, 1976). In addition, lipid bodies commonly occur *in vivo* in leukocytes and mast cells participating in a variety of inflammatory, immunologic, or pathological processes (Dvorak *et al.*, 1983a; Dvorak and Monahan, 1984b; Figs. 1, 2). Lipid bodies can also occur in platelets *in vivo* (Fig. 3) as well as in a variety of cell types *in vitro* (Hawley and Gordon, 1976; Dvorak *et al.*, 1983b; unpublished data).

2. COMPOSITION OF LIPID BODIES

For most cell types, the biochemical content of lipid bodies and their precise role in cellular metabolism are unknown. However, in those cases in which they have been investigated, lipid bodies appear to represent major repositories of prod-

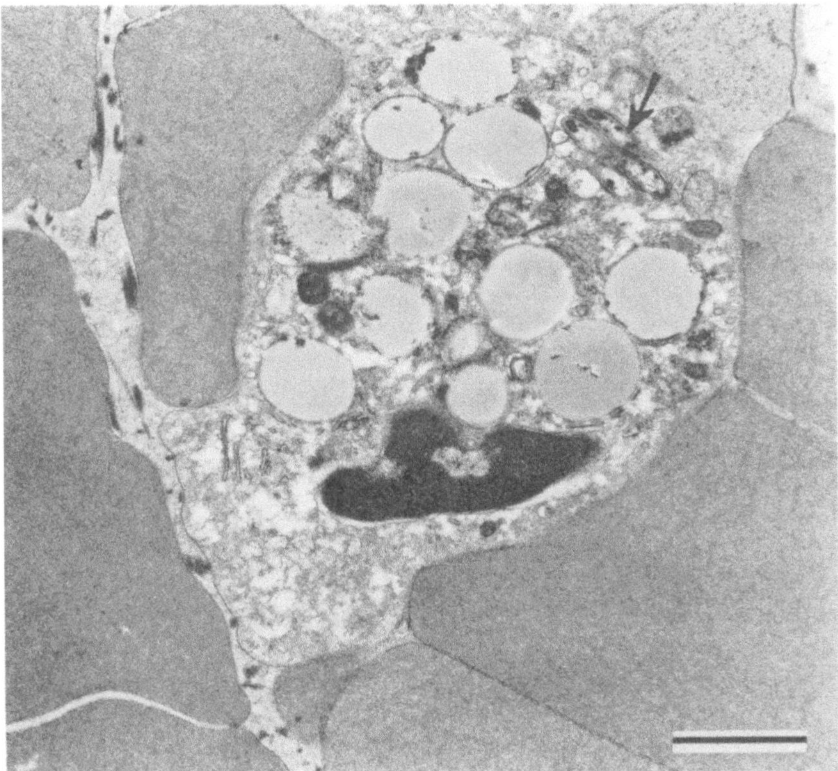

FIGURE 1. A neutrophil containing many cytoplasmic lipid bodies is shown surrounded by erythrocytes in a blood vessel of a patient with Whipple's disease. The neutrophil also contains a few Whipple's bacilli (arrow). Scale bar = 1.0 μm. (Reprinted with permission from Dvorak and Monahan, 1984b).

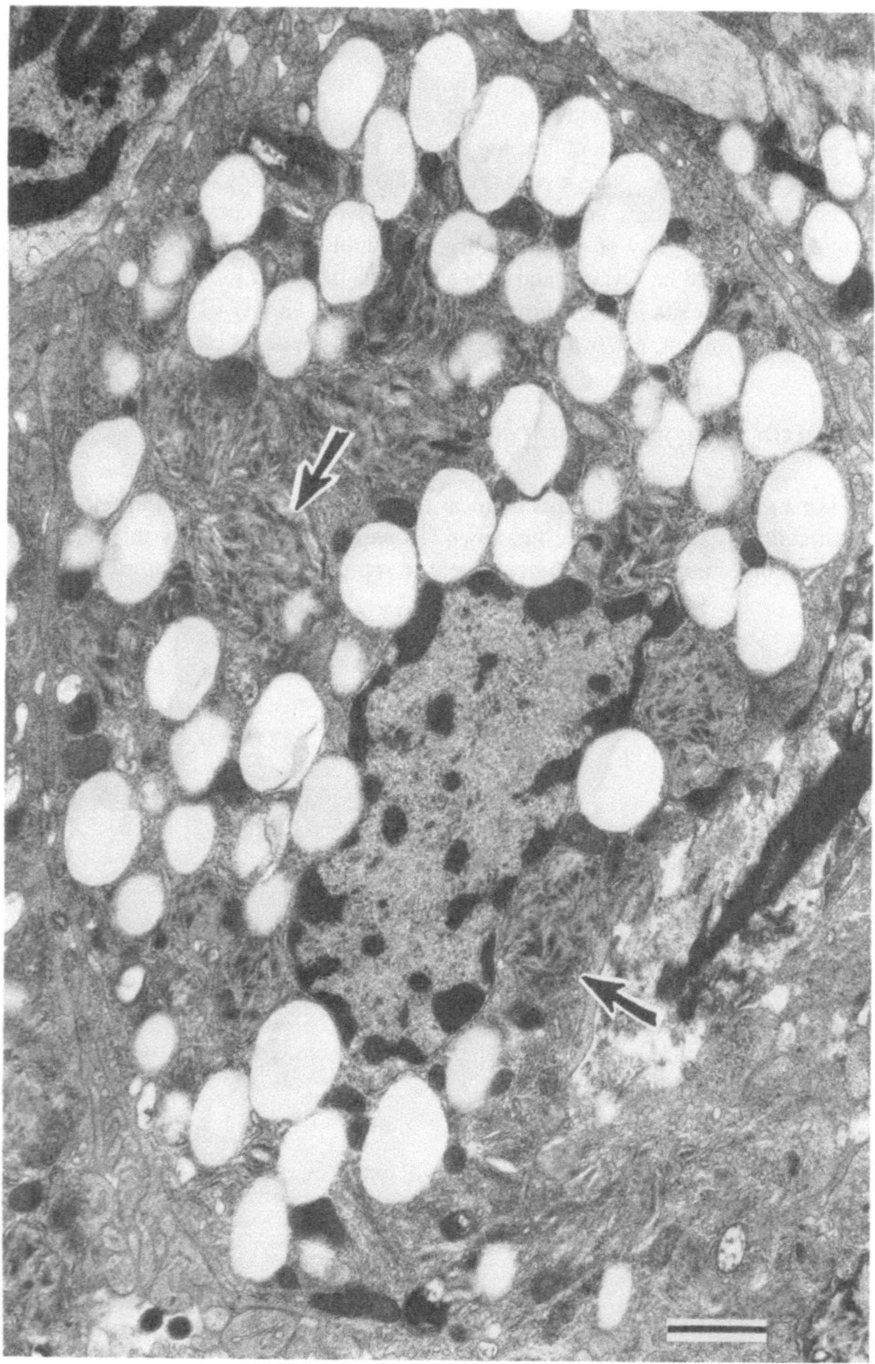

FIGURE 2. A macrophage in the small intestinal submucosa of a patient with Whipple's disease. There are dozens of cytoplasmic lipid bodies, which in this photomicrograph appear very pale. The cytoplasm also contains phagolysosomes with bacterial membranes (arrows). Scale bar = 1.0 μm. (Reprinted with permission from Dvorak and Monahan, 1984b).

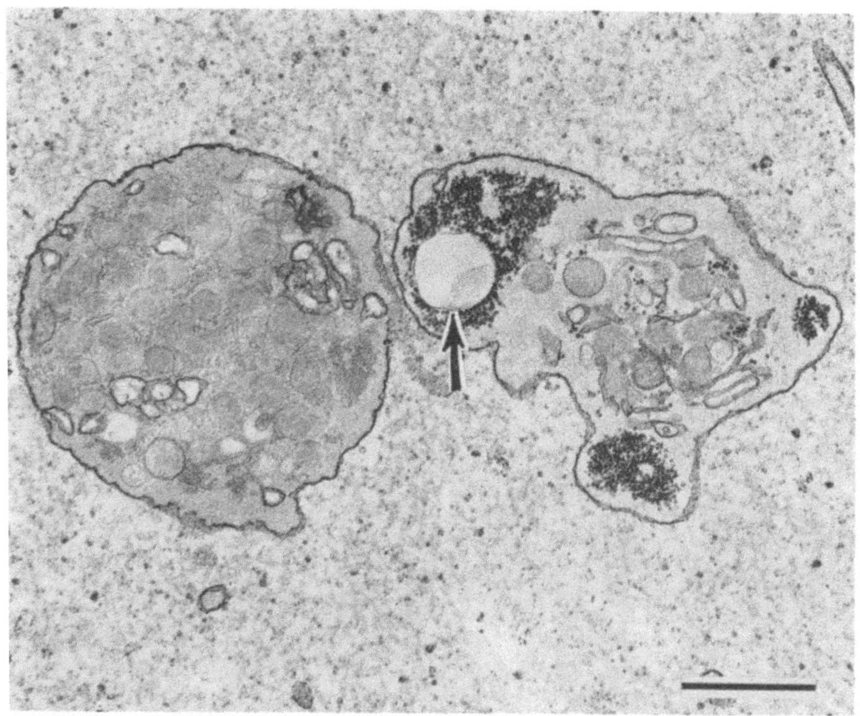

FIGURE 3. Two human platelets, one exhibiting a lipid body (arrow) surrounded by electron-dense glycogen. Both platelets contain multiple round α granules. Scale bar = 1.0 μm.

ucts derived from exogenous [³H]lipid precursors. For example, Dvorak *et al.* (1980) showed by ultrastructural autoradiography that guinea pig peritoneal macrophages incorporated [³H]palmitic acid, [³H]methylcholine, or [³H]myoinositol predominantly into cytoplasmic lipid bodies. More recently, we have used a similar approach to define the subcellular distribution of [³H]species derived from exogenous [³H]AA in a variety of leukocytes, mast cells and eosinophils (Dvorak *et al.*, 1983a; Weller and Dvorak, 1985).

Briefly, mast cells and macrophages were dispersed from human lung fragments by enzymatic digestion followed by countercurrent centrifugation elutriation and, in some cases, centrifugation over discontinuous Percoll gradients. The cells were then allowed to incorporate [³H]AA or other [³H]fatty acids *in vitro* and were examined by electron microscopic autoradiography. The mast cells contained numerous lipid bodies that were distributed predominantly in the perinuclear region of the cytoplasm amidst large numbers of intermediate filaments. These cells incorporated substantial amounts of [³H]AA, [³H]palmitic acid, or [³H]oleic acid *in vitro* at 37°C in overnight culture, and electron microscopic autoradiographs showed that cytoplasmic lipid bodies were much more heavily labeled than any other mast cell structure or compartment (Fig. 4). Thus, numerous silver grains appeared over

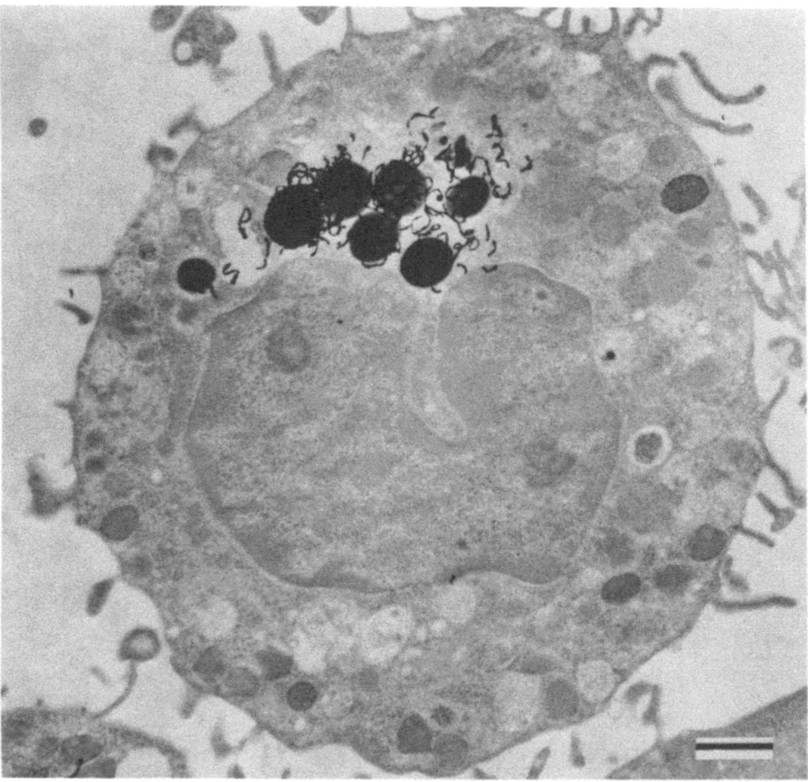

FIGURE 4. Electron microscopic autoradiograph of an unstimulated mast cell purified from human lung and incubated with [³H]arachidonic acid *in vitro*. The incorporated ³H label is localized predominantly to lipid bodies, some of which exhibit many silver grains. By contrast, there is little or no labeling of the plasma membrane, the cytoplasmic granules, or the nucleus. Scale bar = 1.0 μm.

lipid bodies as early as 12 hr after exposure to emulsion, and the number of grains increased progressively with longer exposure times. In contrast, plasma membranes were labeled weakly or not at all, with grains evident only in exposures of 7 days or longer. Lipid body labeling was not specific for [³H]AA in that [³H]palmitic acid or [³H]oleic acid also heav.ly labeled these structures. The intensity of overall labeling as judged by autoradiography was [³H]AA ≅ [³H]palmitic acid > [³H]oleic acid. Labeling of lipid bodies did not occur if cells were fixed in glutaraldehyde–paraformaldehyde before exposure to [³H]AA, and only slight labeling was observed in cells cultured with [³H]AA in 4°C. These findings paralleled isotope incorporation data and suggest that the [³H]AA uptake observed at 37°C under standard culture conditions requires cellular metabolic activity.

Human macrophages also incorporated substantial [³H]AA during overnight culture. As in human mast cells, ³H label was concentrated predominatly in cytoplasmic lipid bodies (Fig. 5), which were intimately associated with numerous intermediate filaments (Fig. 2, top insert). Unlike mast cells, macrophages also appeared to incorporate ^{3}H label into the plasma membrane. Although all macrophages contained heavily labeled lipid bodies, the greatest number of lipid bodies generally were observed in macrophages without phagolysosomes. In macrophages containing phagolysosomes, lipid bodies were preferentially located adjacent to phagosomes and cytoplasmic vacuoles. Some of these lipid bodies appeared to be discharging into phagosomes that exhibited autoradiographic grains as well as phagocytosed carbon particles. Occasional neutrophils and ciliated respiratory epithelial cells (Fig. 6) contaminating these preparations also contained heavily labeled lipid bodies.

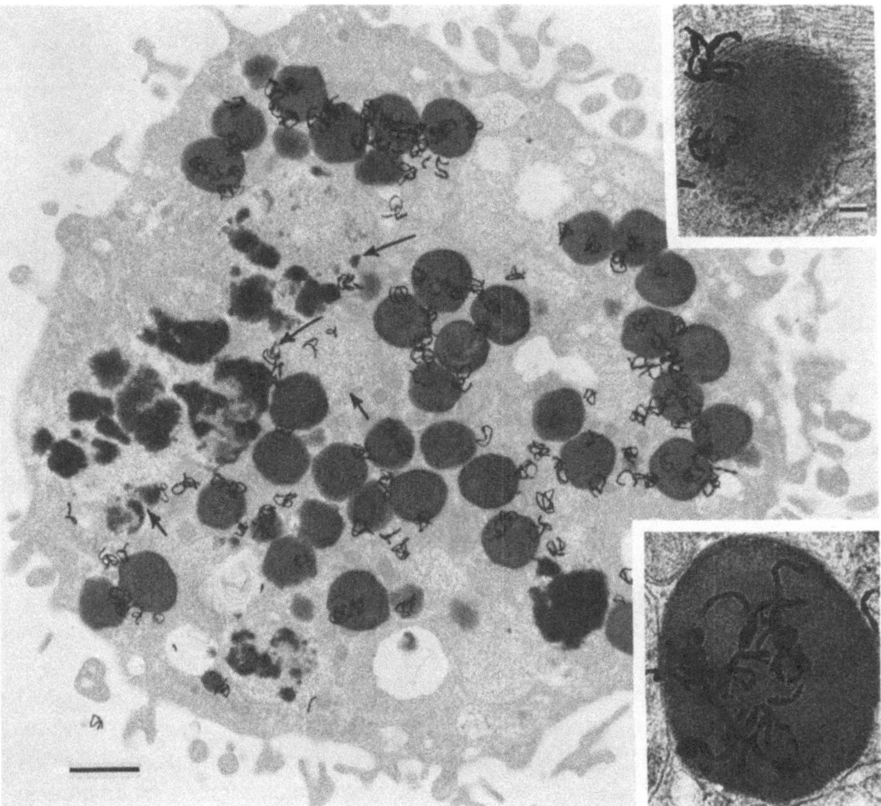

FIGURE 5. Electron microscopic autoradiograph of a macrophage purified from human lung and then incubated with [3H]AA *in vitro*. The cell contains numerous extensively labeled non-membrane-bound lipid bodies. Silver grains are also associated with carbon-filled phagosomes (arrows). The cell surface is not labeled. Scale bar = 1.0 μm. Insets show labeled lipid bodies at higher magnification. Scale bar = 0.1 μm. In the top inset, note the intimate relationship of the lipid body to intermediate filaments. (Reprinted from Dvorak *et al.*, 1983a, with permission.)

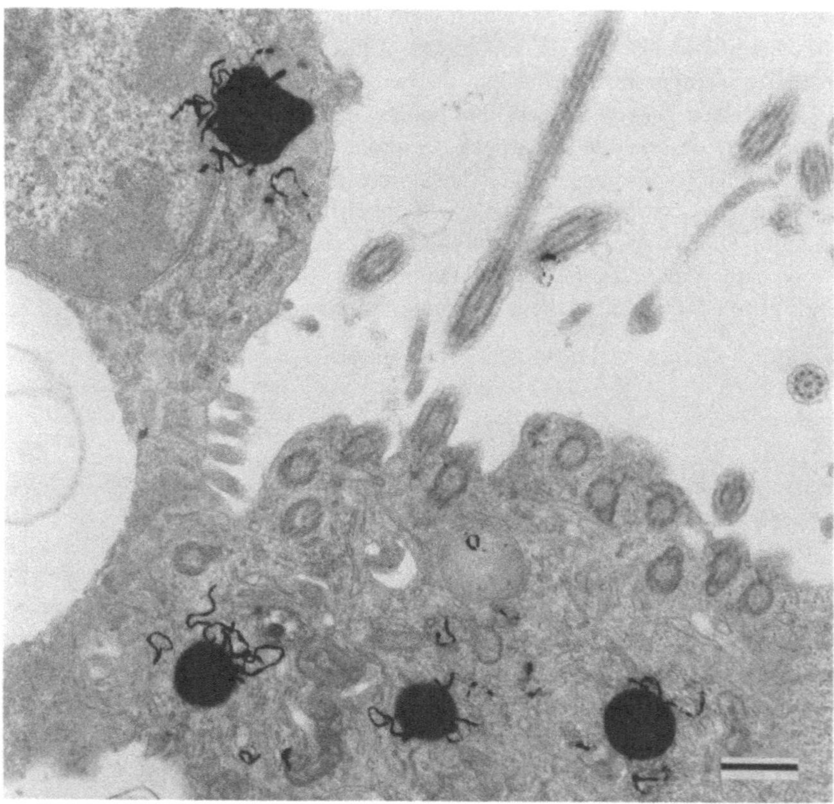

FIGURE 6. Electron microscopic autoradiograph of a ciliated respiratory epithelial cell contaminating a macrophage preparation purified from human lung and incubated with [³H]AA *in vitro.* Four lipid bodies are labeled with silver grains. Scale bar = 0.5 μm.

We also examined peritoneal macrophages isolated from normal mice, mice that had received i.p. thioglycollate, or guinea pigs that had received i.p. mineral oil. After 18-hr incubation with [³H]AA or [³H]palmitate *in vitro,* the cytoplasmic lipid bodies of such macrophages contained the bulk of incorporated ³H by autoradiography (Fig. 7). Like human lung macrophages, these cells also appeared to incorporate [³H]fatty acids into the plasma membrane. Nevertheless, the great majority of autoradiographic grains appeared over cytoplasmic lipid bodies, a few of which were observed releasing their contents into phagosomes that contained [³H]species by autoradiography. In addition to macrophages, the peritoneal cell preparations included occasional lymphoblasts. The lipid bodies of these cells also were intensely labeled with [³H]AA. As with human mast cells, exposure of peritoneal cells to [³H]AA or other precursors after fixation resulted in no significant cellular labeling, even when prolonged (>3 months) exposure times were used. In contrast to the pattern of labeling observed with [³H]lipid precursors, incubation of

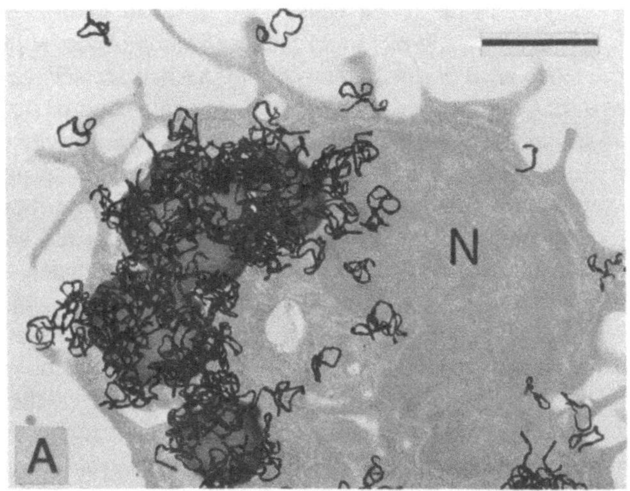

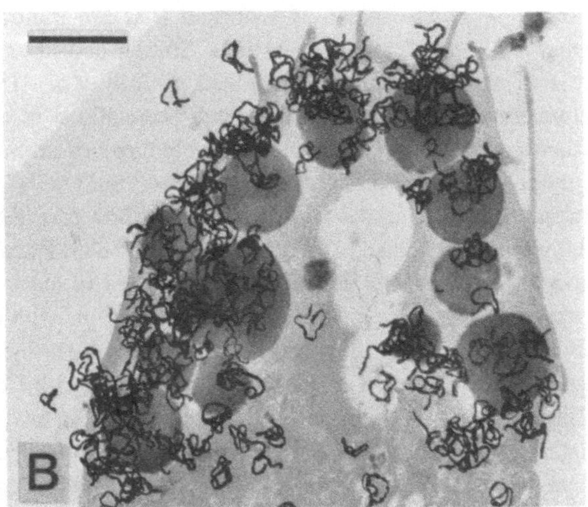

FIGURE 7. Electron microscopic autoradiographs of mouse resident peritoneal macrophages (A,B) incubated with [³H]AA *in vitro*. The cells contain numerous extensively labeled lipid bodies. Some label is also associated with the cell surface (A). N, nucleus. Scale bars = 1.0 μm. (Reprinted from Dvorak *et al.*, 1983a, with permission.)

guinea pig macrophages with [³H]fucose and [³H]glucosamine resulted in labeling of lysosomal granules and the cell surface but not of lipid bodies (Dvorak *et al.*, 1980).

Taken together, our autoradiographic findings indicate that a variety of cell types containing lipid bodies concentrate [³H]lipid precursors, including [³H]AA, predominantly into these structures. But what is the evidence that the distribution of silver grains observed by ultrastructural autoradiography accurately reflects the distribution of ^{3}H label in the living cells? Two different findings suggest that autoradiography provides a valid picture of the localization of the ^{3}H label *in vivo*. First, we determined to what extent amounts of [^{3}H]AA-labeled cellular lipids were lost or extracted during the chemical fixation and processing required for preparation of thin sections and autoradiography. By measuring the radioactivity of all fluids used to process glass coverslips with adherent [^{3}H]AA-labeled guinea pig peritoneal macrophages, we showed that approximately one-half of the incorporated radio-activity withstood fixation and processing and remained associated with the cells in Epon embedding plastic (Dvorak *et al.*, 1983a). Approximately half of the radioactivity lost in processing was removed during the osmium postfixation step, and only 25% of this ^{3}H was extractable by organic solvents. At least some and perhaps most of the remaining reduction in recovered radioactivity probably reflected losses of cellular material during the many washes required to process the specimens.

Second, we examined the distribution of ^{3}H label derived from [^{3}H]AA in mouse mast cell lines that contain few or no lipid bodies under usual conditions of culture. For these studies, we derived mast cell lines from the bone marrow of a normal C57BL/6 mouse or the spleen of a normal BALB/c mouse as previously described (Nabel *et al.*, 1981; Galli *et al.*, 1982). Similar cell lines have also been established in several other laboratories (reviewed in Galli *et al.*, 1984). Although the precise relationship of these cells to mast cell populations *in vivo* remains to be established, their properties most closely resemble those of immature mast cells and/or T-cell-regulated populations such as the "mucosal mast cells" that proliferate in the gut of murine rodents infected with certain intestinal parasites (reviewed in Galli *et al.*, 1984; also see Schrader, 1983; Sredni *et al.*, 1983; Razin *et al.*, 1984). Unlike the leukocyte and mast cell preparations examined in our earlier autoradi-ographic study, these mast cell lines were dividing populations with doubling times of approximately 4–6 days, and, thus, many of the cells were synthesizing new cellular membranes during the ~18-hr period of incubation with [^{3}H]lipid precur-sors. Both cell lines actively incorporated [^{3}H]AA or [^{3}H]oleic acid *in vitro* (Figs. 8, 9). Autoradiography localized ^{3}H label over the cells' cytoplasm, plasma mem-brane, and nuclear membrane; labeling of membrane-rich Golgi regions was par-ticularly intense (Fig. 9B). By contrast, the central regions of the nucleus and immature secretory granules exhibited little or no labeling. Because the procedures used for [^{3}H]AA labeling, cell fixation, and processing were the same as those employed in our studies of human lung mast cells and macrophages, the results demonstrate that these techniques can readily detect the ^{3}H labeling of cytoplasmic structures other than lipid bodies.

Although it appears likely that autoradiography accurately identifies the sub-cellular localization of label derived from [^{3}H]AA and other [^{3}H]lipid precursors, this approach obviously cannot reveal the species of [^{3}H]products present. However, preliminary information about the composition of lipid bodies has been obtained by combining autoradiography with biochemical analysis of ^{3}H-labeled cells. These studies suggest that the [^{3}H]species sequestered in cytoplasmic lipid bodies may vary according to cell type. For example, Bligh–Dyer extraction and thin-layer chromatography of human lung mast cells labeled with [^{3}H]AA (Peters *et al.*, 1984)

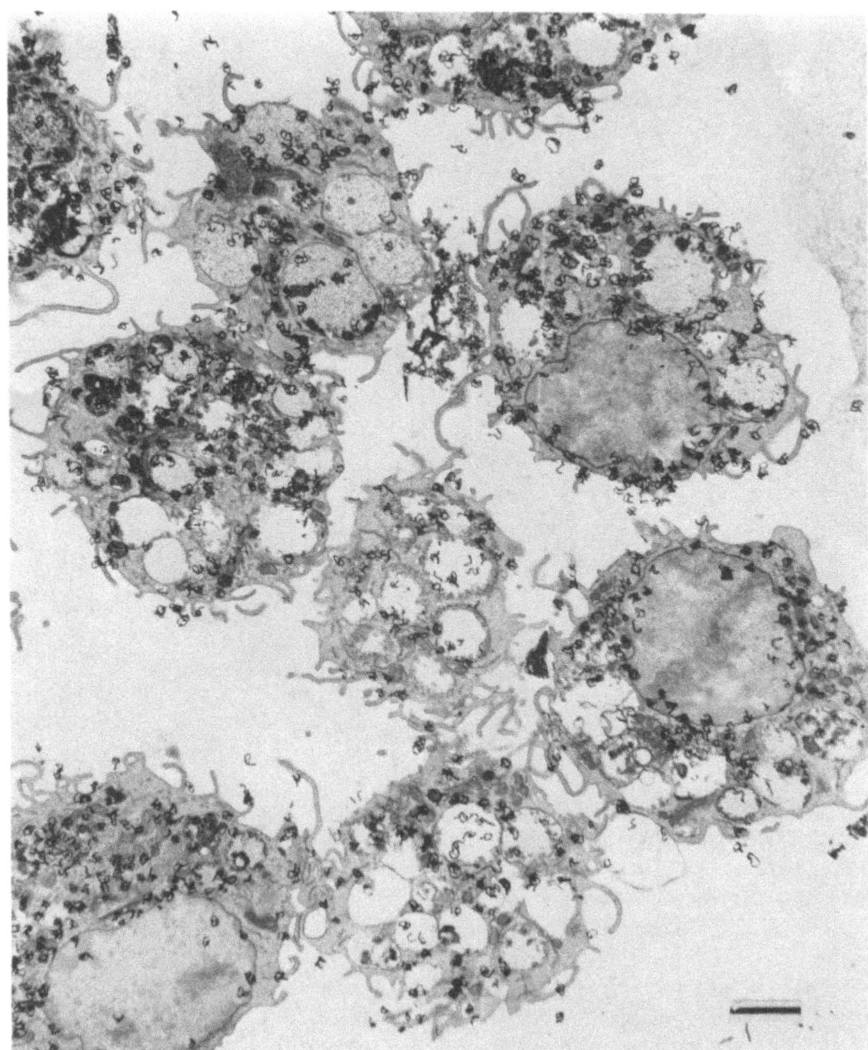

FIGURE 8. Electron microscopic autoradiographs of mast cells derived from normal BALB/c mouse spleen cells and maintained *in vitro* in supernatants derived from concanavalin-A-stimulated mouse splenocytes (Galli *et al.*, 1982). After incubation with [³H]AA for ~18 hr, large numbers of silver grains appear over the cytoplasm of the cells, which lack lipid bodies. Scale bar = 2.0 μm.

revealed that ³H was distributed predominantly in neutral lipids (68–80%) with smaller amounts in phospholipids (14–27%). By contrast, the [³H]lipids synthesized by 95% pure populations of mouse peritoneal macrophages included ~80% phospholipids and only 16.5% neutral lipids (Dvorak *et al.*, 1983a). Although these results reflect the [³H]lipid composition of the entire cells, the autoradiographic

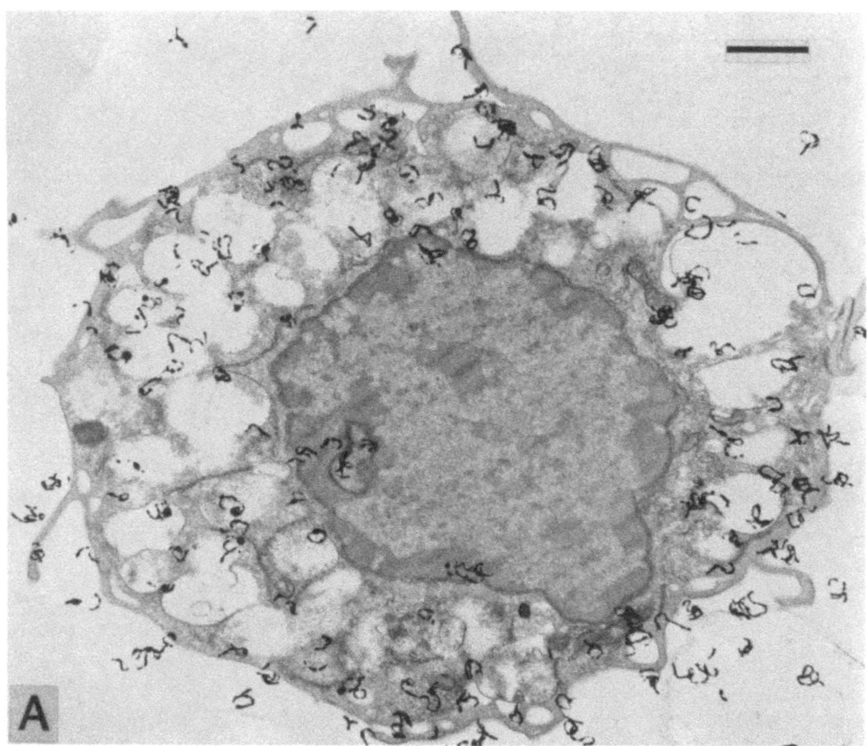

FIGURE 9. Electron microscopic autoradiograph of cultured mouse mast cells derived from normal C57BL/6 mouse bone marrow (A) or normal BALB/c spleen cells (B) and maintained as previously described (Galli *et al.*, 1982). After labeling with [³H]AA for ~18 hr *in vitro*, label appears over cytoplasmic structures, particularly the Golgi area (in B), as well as over plasma and nuclear membranes. Scale bars = 1.0 μm.

evidence presented above suggests that lipid bodies accounted for a major portion of the [³H]species analyzed. More definitive characterization of the biochemical composition of cytoplasmic lipid bodies awaits the purification of these inclusions from disrupted cells, work currently in progress.

3. PATHWAYS

In addition to defining the biochemical composition of lipid bodies, it will be important to characterize the subcellular pathways and biochemical transformations that govern the traffic of AA and other molecules into and out of these structures. For example, [³H]fatty acids or their products (Scott *et al.*, 1982) may enter lipid bodies directly. Alternatively, some traffic of [³H]fatty acids into lipid bodies may involve degradation of ³H-labeled cellular membranes. By

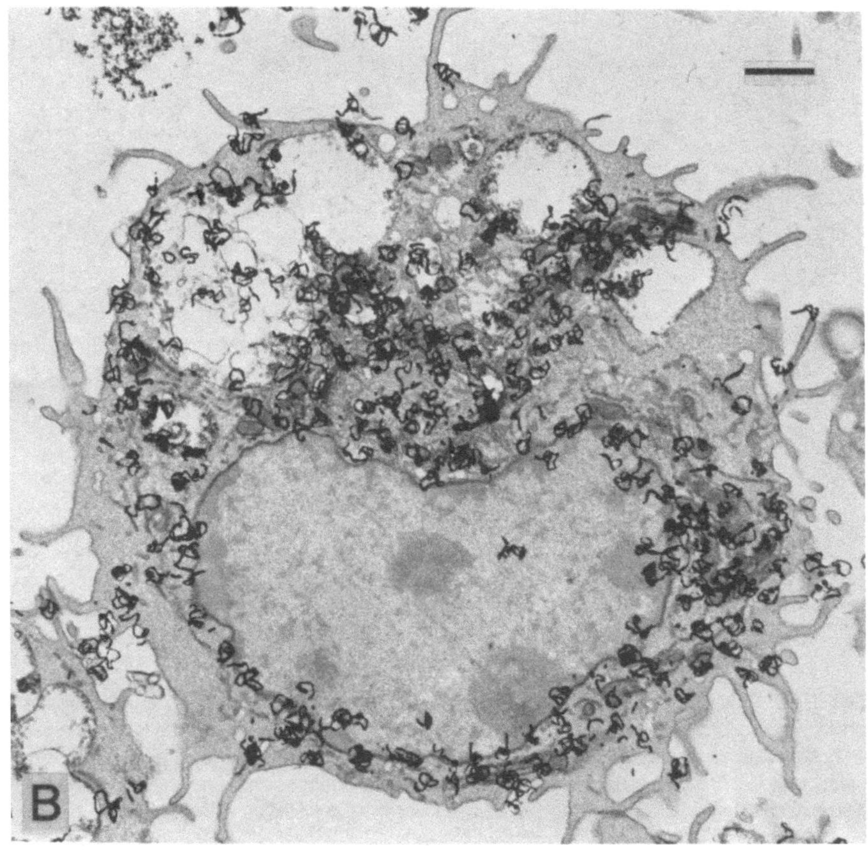

FIGURE 9. *(Cont.)*

whatever route they enter lipid bodies, our data indicate that [³H]species derived from exogenous [³H]AA can be mobilized from lipid bodies on appropriate cellular activation. First, phagocytosis of zymosan by mouse macrophages labeled with [³H]AA for ~18 hr was associated with a striking approximation of lipid bodies to phagolysosomes and the appearance of autoradiographic label over phagolysosomes (Fig. 10). Second, mouse macrophages labeled with [³H]AA released substantial amounts of [³H]products in association with phagocytosis (Table II). By contrast, zymosan induced little or no release of radioactivity from macrophages labeled with [³H]palmitic acid (Table II).

The finding that phagocytosis provokes the release of [³H]AA-derived species to a greater extent than metabolites of [³H]palmitic acid is in accord with previous work from other laboratories, which has also suggested that lipid metabolites released during macrophage phagocytosis are derived predominantly from AA (Bonney *et al.*, 1978; Scott *et al.*, 1980). Although the route of mobilization of [³H]species

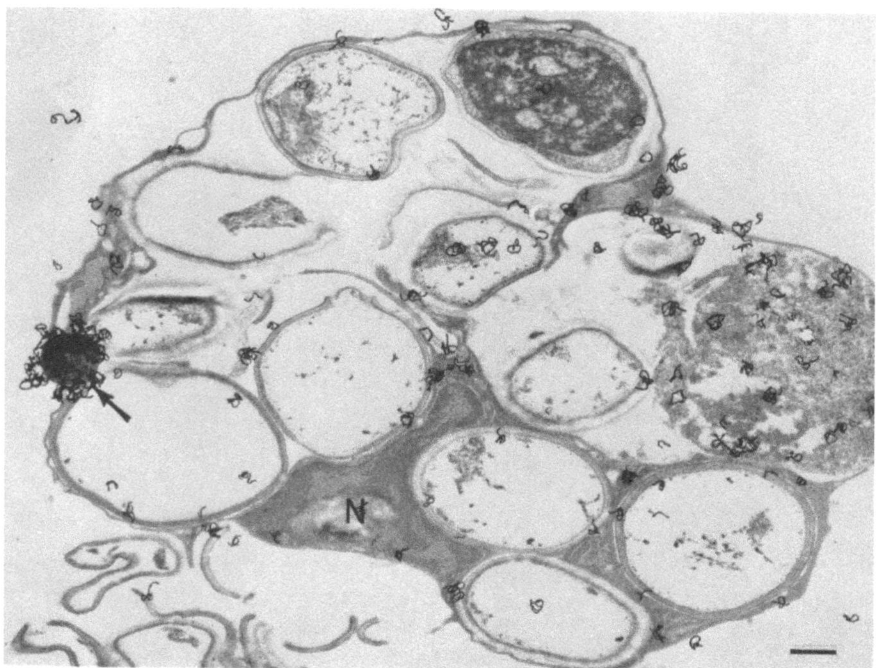

FIGURE 10. Resident mouse peritoneal macrophage exposed to a pulse of [³H]AA *in vitro* and then allowed to phagocytose zymosan particles. Cells such as these generally contain fewer lipid bodies than control cells (Fig. 7), although the lipid bodies present (e.g., arrow) are heavily labeled with ³H. Silver grains are also associated with phagosomes and the cell surface. N, nucleus. Scale bar = 1.0 μm. (Reprinted from Dvorak *et al.*, 1983a, with permission.)

from lipid bodies has not been defined, ultrastructural cytochemistry has shown that some lipid bodies contain nonspecific esterase (ANAE) and/or peroxidase activities in a number of different cell types (Cavallo, 1974; Monahan *et al.*, 1981; Dvorak *et al.*, 1983b; A. M. Dvorak unpublished data; Figs. 11, 12). The substrate specificities of these enzymes have not been fully defined, however, and their roles in lipid body metabolism are unknown.

We also evaluated the behavior of lipid bodies during human lung mast cell degranulation (Dvorak *et al.*, 1984). Autoradiograph grain count analysis showed that lipid bodies represented the major site of intracellular localization of [³H] AA-derived species in control mast cells incubated for ∼18 hr with [³H]AA. Indeed, other structures such as the specific cytoplasmic granules, the cytoplasm, and the nucleus were not significantly labeled. Stimulation of mast cell degranulation by anti-IgE provoked a characteristic sequence of ultrastructural changes that resulted in the fusion of cytoplasmic granule membranes with each other to form degranulation channels in communication with the extracellular space. Degranulation-related ultrastructural alterations affected 77% of the aggregate

TABLE II. Incorporation of [3H]Fatty Acids by Mouse Peritoneal Macrophages and Release of Radioactivity Associated with Zymosan Phagocytosis[a]

			Incorporation		Release	
Expt.	[3H]Fatty acid	Zymosan	DPM ($\times$ 10^{-4})	Percent	DPM ($\times$ 10^{-4})	Percent
1	[3H]AA					
	(0.13 μCi/ml)	−	14.4 ± 0.6	34.4 ± 0.3	0.7 ± 0.1	4.8 ± 0.4
		+	12.2 ± 0.4	33.5 ± 0.9	3.7 ± 0.2	30.4 ± 1.2
	[3H]PA	−	4.7 ± 0.6	14.6 ± 0.5	0.5 ± 0.1	10.2 ± 0.6
	(0.13 μCi/ml)	+	5.2 ± 0.7	15.8 ± 0.4	0.4 ± 0.04	8.8 ± 0.7
2	[3H]AA	−	1124 ± 150	53.0 ± 7.0	35.6 ± 3	3.2 ± 0.1
	(43.1 μCi/ml)	+	1039 ± 49	49.0 ± 2.2	130.5 ± 5	12.6 ± 0.4
	[3H]PA	−	409 ± 8	33.5 ± 0.5	22 ± 2	5.5 ± 0.5
	(24.4 μCi/ml)	+	229 ± 49	24.0 ± 4.0	23 ± 6	7.6 ± 0.9
	[3H]AA	−	0	0	0	0
	(43.1 μCi/ml after fixation)					

[a] Macrophages from the peritoneal cavities of mice injected 3 days previously with thioglycollate broth were labeled with [3H]arachidonic acid ([3H]AA) or [3H]palmitic acid ([3H]PA) for ~17 hr in medium containing 10% fetal calf serum. They were then washed three times in serum-free medium and then incubated for an additional 2 hr in serum-free medium with or without zymosan (25–40 μg/ml). To quantitate fatty acid incorporation or release, 200-μl aliquots of culture supernatants or solubilized adherent cells (0.05% Triton X-100 with scraping) were added to 4 ml of Aquasol for liquid scintillation counting. Reprinted from Dvorak *et al.* (1983a) with permission

volume of cytoplasmic granules within the first 20 min of stimulation. These changes were associated with a substantial (~72%) release of histamine, a mediator stored in the cytoplasmic granules.

In contrast to the dramatic changes affecting the specific granules of activated mast cells, the alterations involving lipid bodies were more subtle. Lipid bodies occurred in intimate proximity to degranulation channels in anti-IgE-stimulated cells, but ultrastructural evidence of actual discharge of lipid bodies into these structures was rare. Furthermore, stereological analysis showed that the aggregate volume of lipid bodies did not change significantly during the 20-min period of mast cell activation. Similarly, although a few degranulation channels appeared heavily labeled with products derived from [3H]AA, grain count analysis of the entire cellular population did not detect significant labeling of these structures.

Parallel biochemical analysis also suggested that mast cell degranulation may have been associated with relatively little net loss of lipid body content. Thus, stimulation with anti-IgE provoked the discharge of only a small fraction (mean of five experiments 3%, range 2–12%) of total cell-associated lipids (Peters *et al.*, 1984; Dvorak *et al.*, 1984). In addition, degranulation had relatively little effect on the pattern of mast cell total [3H]lipid content by TLC (Peters *et al.*, 1984; Dvorak *et al.*, 1984). Because only a small fraction of total cellular [3H]species was released on mast cell activation, the ultrastructural, autoradiographic, and biochemical evidence is consistent with either of two very different possibilities. On the one hand, [3H]species released on activation of human lung mast cells may

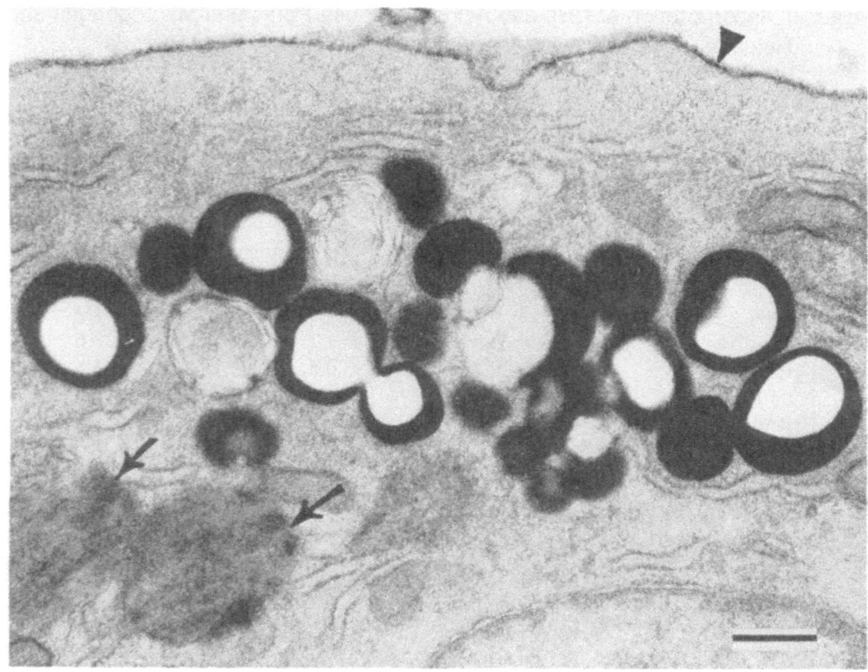

FIGURE 11. A cloned strain 2 guinea pig aortic endothelial cell from a population maintained under isologous conditions *in vitro* (Galli *et al.*, 1983 contains numerous lipid bodies, which exhibit intense nonspecific esterase activity (α-naphthyl acetate substrate). The plasma membrane (arrowhead) also exhibits reaction product, and two immature Weibel–Palade bodies (arrows) are focally positive. Scale bar = 0.5 μm.

have been largely derived from cytoplasmic lipid bodies, but the relatively small fraction of total cellular [^{3}H]lipids mobilized precluded the confirmation of this fact either by autoradiographic grain count analysis or by stereological analysis of lipid body volume. On the other hand, we showed that the processing of [^{3}H]AA-labeled cells for electron microscopy resulted in the loss of ~50% of cell-associated radioactivity (Dvorak *et al.*, 1983a).

Although we feel it is unlikely, we have not formally excluded the possibility that [^{3}H]species in the human mast cell plasma membrane are more susceptible to extraction than those sequestered in the cytoplasmic lipid bodies. If this is the case, then autoradiographic analysis would have revealed less activity in the plasma membrane than was actually present in the living cell. Under these circumstances, release of [^{3}H]lipids from the plasma membrane may not have been detected by the techniques employed even if this route accounted for all of the [^{3}H]products appearing in the supernatant of the stimulated cells.

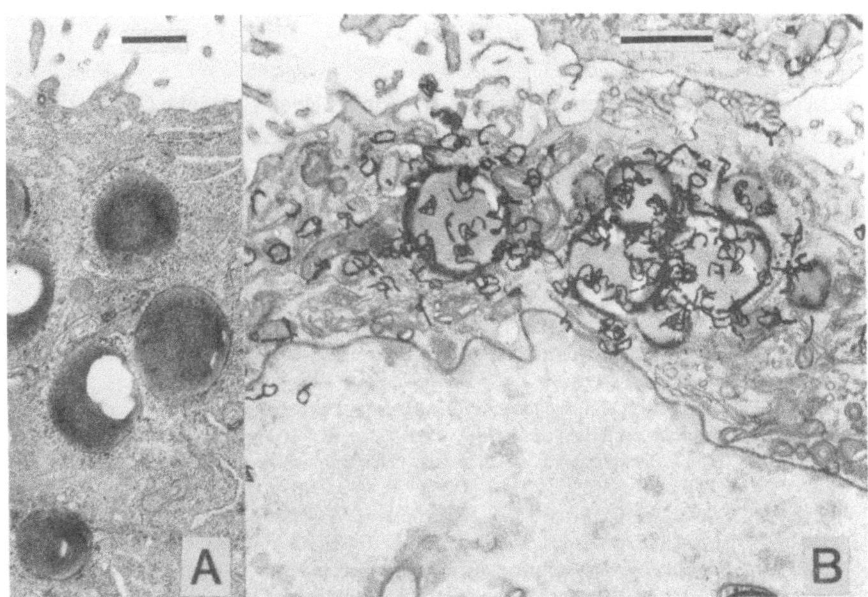

FIGURE 12. Electron micrographs of line 10 guinea pig hepatobiliary carcinoma cells. A: Lipid bodies exhibiting nonspecific esterase activity (α-napthyl acetate substrate). B: Autoradiograph of a cell labeled *in vitro* with [3H]AA showing 3H localized predominantly to the lipid bodies. Scale bars = 1.0 μm.

4. SUMMARY

In summary, lipid bodies can develop in the cytoplasm of a wide variety of cells *in vivo,* including mast cells, leukocytes, platelets, endothelial cells, tumor cells, and the macrophages and smooth muscle cells in the vascular lesions of atherosclerosis (Fowler *et al.,* 1978; Shio *et al.,* 1978). Although the development of lipid bodies may reflect the influence of multiple factors, our studies suggest that when these inclusions are present, they can represent the major repository of [3H]molecules in cells incubated with [3H]AA and other [3H]lipids. Thus, when labeled for ~18 hr *in vitro,* human mast cells, human, guinea pig, and mouse macrophages, a variety of other leukocytes, including neutrophils, eosinophils, and lymphocytes, and certain tumor cells (Fig. 12) incorporated 3H label derived from exogenous [3H]AA predominantly into cytoplasmic lipid bodies. The preferential incorporation of [3H]AA into lipid bodies also can occur during short incubation periods (1–4 hr). In addition to entering lipid bodies, current evidence suggests that [3H]lipids also may be mobilized from [3H]AA-labeled lipid bodies during the activation of certain cells, for example, in mouse macrophages undergoing phagocytosis. Efforts are now under way to characterize in more detail the lipids and enzymatic activities of lipid bodies and to determine the role of these structures in cellular AA metabolism.

ACKNOWLEDGMENTS. Supported by U.S. Public Health Grants CA 28834, AI 16925, AI 07290, HL 23586, and AI 20292. This is publication No. 587 from the O'Neill Laboratories at the Good Samaritan Hospital.

REFERENCES

Bonney, R. J., Wightman, P. D., Davies, P., Sadowski, S. J., Kuehl, F. A., and Humes, J. L., 1978, Regulation of prostaglandin synthesis and of the selective release of lysosomal hydrolases by mouse peritoneal macrophages, *Biochem. J.* **176:**433–442.

Cavallo, T., 1974, Fine structural localization of endogenous peroxidase activity in inner medullary interstitial cells of the rat kidney, *Lab. Invest.* **31:**458–464.

Crawford, M. A., 1983, Background to essential fatty acids and their prostanoid derivatives, *Br. Med. Bull.* **39:**210–213.

Dvorak, A. M., Dvorak, H. F., Peters, S. P., Schulman, E. S., MacGlashan, D. W., Jr., Pyne, K., Harvey, V. S., Galli, S. J., and Lichtenstein, L. M., 1983a, Lipid bodies: Cytoplasmic organelles important to arachidonate metabolism in macrophages and mast cells, *J. Immunol.* **131:**2965–2976.

Dvorak, A. M., Galli, S. J., Marcum, J. A., Nabel, G., Der Simonian, H., Goldin, J. M., Monahan, R. A., Pyne, K., Cantor, H., Rosenberg, R. D., and Dvorak, H. F., 1983b, Cloned mouse cells with natural killer function and cloned suppressor T cells express ultrastructural and biochemical features not shared by cloned inducer T cells, *J. Exp. Med.* **157:**843–861.

Dvorak, A. M., Hammond, M. E., Morgan, E. S., and Dvorak, H. F., 1980, Ultrastructural studies of macrophages: *In vitro* removal of cell coat with macrophage inhibition factor (MIF)-containing lymphocyte culture supernatants; chloroform extraction, phospholipase digestion, and autoradiographic studies, *J. Reticuloendothelial Soc.* **27:**119–142.

Dvorak, A. M., Hammel, I., Schulman, E. S., Peters, S. P., MacGlashan, D. J., Jr., Schleimer, R. P., Newball, H. H., Smith, J., Pyne, K., Dvorak, H. F., Lichtenstein, L. M., and Galli, S. J., 1984a, Differences in the behavior of cytoplasmic granules and lipid bodies during human lung mast cell degranulation, *J. Cell Biol.* **99:**1678–1687.

Dvorak, A. M., and Monahan, R. A., 1984b, Weight loss, fatigue, diarrhea, fever and lymphadenopathy in a forty-eight year old male industrial engineer: Whipple's Disease, *Norelco Reporter* **31:**10–21.

Fawcett, D. W., 1981, *The Cell*, ed. 2, W. B. Saunders, Philadelphia, pp. 655–667.

Fowler, S., Berberian, P. A., Haley, N. J., and Shio, H., 1978, Lysosomes, vascular intracellular lipid accumulation, and atherosclerosis, in: *International Conference of Atherosclerosis* (L. A. Carlson, R. Paoletti, C. E. Sirtori, and G. Weber, eds.), Raven Press, New York, pp. 19–27.

Galli, S. J., Dvorak, A. M., Marcum, J. A., Ishizaka, T., Nabel, G., Der Simonian, H., Pyne, K., Goldin, J. M., Rosenberg, R. D., Cantor, H., and Dvorak, H. F., 1982, Mast cell clones: A model for the analysis of cellular maturation, *J. Cell Biol.* **95:**435–444.

Galli, S. J., Isomura, T., Dvorak, A. M., and Dvorak, H. F., 1983, Lymph node cells (LNC) activated by antigen (Ag) nonspecifically damage cloned strain 2 guinea pig (S2GP) aortic endothelial cells (EC) *in vitro, Fed. Proc.* **42:**662.

Galli, S. J., Dvorak, A. M., and Dvorak, H. F., 1984, Basophils and mast cells: Morphologic insights into their biology, secretory patterns, and function, *Prog. Allergy* **34:**1–141.

Hawley, H. P., and Gordon, G. B., 1976, The effects of long chain free fatty acids on human neutrophil function and structure, *Lab. Invest.* **34:**216–222.

Monahan, R. A., Dvorak, H. F., and Dvorak, A. M., 1981, Ultrastructural localization of nonspecific esterase activity in guinea pig and human monocytes, macrophages, and lymphocytes, *Blood* **58:**1089–1099.

Moncada, S., and Vane, J. R., 1978, Pharmacology and endogenous role of prostaglandin endoperoxides, thromboxane A_2, and prostacyclin, *Pharmacol. Rev.* **30**:293–331.

Nabel, G., Galli, S. J., Dvorak, A. M., Dvorak, H. F., and Cantor, H., 1981, Inducer T lymphocytes synthesize a factor that stimulates proliferation of cloned mast cells, *Nature* **291**:332–334.

Peters, S. P., MacGlashan, D. W., Jr., Schulman, E. S., Schleimer, R. P., Hayes, E. C., Rokach, J., Adkinson, N. F., Jr., and Lichtenstein, L. M., 1983, Arachidonic acid metabolism in purified lung mast cells, *J. Immunol.* **132**:1972–1979.

Razin, E., Ihle, J. N., Seldin, D., Mencia-Huerta, J.-M., Katz, H. R., Leblanc, P. A., Hein, A., Caulfield, J. P., Austen, K. F., and Stevens, R. L., 1984, Interleukin 3: A differentiation and growth factor for the mouse mast cell that contains chondroitin sulfate E proteoglycan, *J. Immunol.* **132**:1479–1486.

Samuelsson, B., 1983, Leukotrienes: Mediators of immediate hypersensitivity reactions and inflammation, *Science* **220**:568–575.

Samuelsson, B., Granstrom, E., Green, K., Hamberg, M., and Hammarstrom, S., 1975, Prostaglandins, *Annu. Rev. Biochem.* **44**:669–695.

Schrader, J. W., 1983, Bone marrow differentiation *in vitro*, *CRC Crit. Rev. Immunol.* **4**:197–277.

Scott, W. A., Zrike, J. M., Hamill, A. L., Kempe, J., and Cohn, Z. A., 1980, Regulation of arachidonic acid metabolites in macrophages, *J. Exp. Med.* **152**:324–335.

Scott, W. A., Pawlowski, N. A., Andreach, M., and Cohn, Z., 1982, Resting macrophages produce distinct metabolites from exogenous arachidonic acid, *J. Exp. Med.* **155**:535–547.

Shio, H., Haley, N. J., and Fowler, S., 1978, Characterization of lipid-laden aortic cells from cholesterol-fed rabbits. II. Morphometric analysis of lipid-filled lysosomes and lipid droplets in aortic cell populations, *Lab. Invest.* **39**:390–397.

Sredni, B., Friedman, M. M., Bland, C. E., and Metcalfe, D. D., 1983, Ultrastructural, biochemical and functional characteristics of histamine-containing cells cloned from mouse bone marrow: Tentative identification as mucosal mast cells, *J. Immunol.* **131**:915–922.

Uchida, T., Kao, H., Quispe-Sjogren, M., and Peters, R. L., 1983, Alcoholic foamy degeneration— a pattern of acute alcoholic injury of the liver, *Gastroenterology* **84**:683–692.

Weller, P. F., and Dvorak, A. M., 1985, Arachidonic acid incorporation by cytoplasmic lipid bodies of human eosinophils, *Blood* in press.

Willis, A. L., Hassam, A. G., Crawford, M. A., Stevens, P., and Denton, J. P., 1981, Relationships between prostaglandins, prostacyclin, and EFA precursors in rabbits maintained on EFA-deficient diets, *Prog. Lipid Res.* **20**:161–168.

23

Differences between the Effects of f-Met-Leu-Phe and Leukotriene B$_4$ on Phosphoinositide Turnover and Their Relationship to Calcium Mobilization and Protein Kinase C Activation

RAMADAN I. SHA'AFI, MARIO VOLPI, and PAUL H. NACCACHE

1. INTRODUCTION

It is generally accepted that the mobilization and metabolism of arachidonic acid are involved in the activation and/or modulation of the neutrophil's responses and that the major biological activities of arachidonic acid are mediated by the ability of leukotriene B$_4$ to alter calcium homeostasis (for a review and original citations, see Bach, 1983). This conclusion is based on various experimental results. First, the biological activities of exogenously added arachidonic acid and some of its metabolites have been demonstrated. Second, the activation of the neutrophils by chemotactic factors can be modulated by certain lipase and lipoxygenase inhibitors. Third, chemotactic factors cause the release of arachidonic acid previously incorporated in phospholipids and stimulate its subsequent metabolism including

RAMADAN I. SHA'AFI and MARIO VOLPI • Department of Physiology, University of Connecticut Health Center, Farmington, Connecticut 06032.   PAUL H. NACCACHE • Department of Pathology, University of Connecticut Health Center, Farmington, Connecticut 06032.

the generation of leukotriene B_4. Fourth, the exogenous addition of leukotriene B_4 to neutrophils increases the intracellular concentration of free calcium, with characteristics similar to those of its biological activities. This increase occurs through the release of calcium from internal stores, by displacement of previously bound calcium from membranous and other sites, and from the extracellular medium through an increase in the plasma membrane permeability to calcium.

Though a great deal has been learned about the interrelationships among leukotriene B_4, calcium mobilization, and neutrophil activation during the past few years, several important problems remained to be studied. For example, what are the similarities and differences between the biochemical and biophysical events that are elicited by leukotriene B_4 and f-Met-Leu-Phe in the neutrophils, and do these two stimuli use the same signal to release calcium? Leukotriene B_4, like f-Met-Leu-Phe, mobilizes cell calcium and causes an increase in the association of actin with the cytoskeleton (Yassin *et al.*, 1984) and thus is a potent chemotactic agent. On the other hand, it is unlike f-Met-Leu-Phe in that the addition of leukotriene B_4 does not induce a significant metabolic burst (Palmblad *et al.*, 1981). In addition, as judged by the extent and the slope of the degranulation dose–response curves, leukotriene B_4 is a less potent secretagogue than f-Met-Leu-Phe. With regard to the signal responsible for calcium mobilization, the most commonly accepted hypothesis is that occupancy of receptors by soluble stimuli activates phospholipase C, which results in the generation, among other things, of inositol-1,4, 5-trisphosphate (IP_3) from the breakdown of phosphatidylinositol-4,5-bisphosphate (TPI), and that calcium is released as the result of both the breakdown of TPI and the action of the generated IP_3 (Michell and Kirk, 1982; Streb *et al.*, 1983).

The present studies were undertaken to investigate the similarities and differences in the biochemical events inititated by f-Met-Leu-Phe and leukotriene B_4 and to examine the mechanism by which leukotriene B_4 mobilizes calcium. The results to be reported demonstrate that, unlike f-Met-Leu-Phe, leukotriene B_4 mobilizes calcium without causing a breakdown of TPI or generating phosphatidic acid or 1,2-diacylglycerol. These results strongly suggest that the stimulation of at least some of the neutrophil responses elicited by leukotriene B_4 proceeds without the activation of the phospholipase C pathway.

2. MOBILIZATION OF CALCIUM BY f-MET-LEU-PHE AND LEUKOTRIENE B_4 AND ITS RELATIONSHIP TO PHOSPHOINOSITIDE METABOLISM

The addition of leukotriene B_4 and f-Met-Leu-Phe to neutrophils causes a rise in the intracellular concentration of free calcium as measured by the fluorescent probe quin-2 (Fig. 1). The calcium reponse is rapid, reaching maximum value within 20 sec, and dose dependent. The ED_{50} of the quin-2 responses to leukotriene

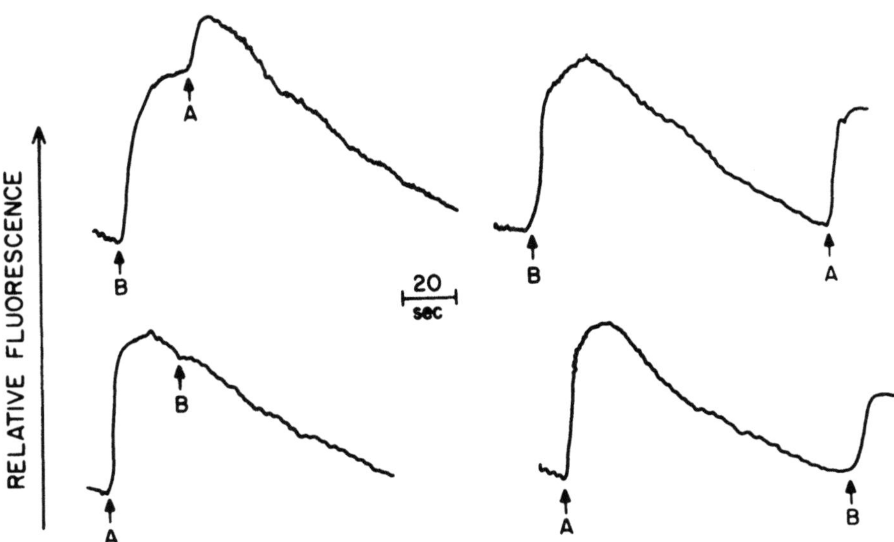

FIGURE 1. The effects of the addition of f-Met-Leu-Phe (A, 10^{-9} M) and leukotriene B_4 (B, 7.2 × 10^{-8} M) on the fluorescence of quin-2-loaded rabbit neutrophils. The cells were loaded with quin-2 as described previously (White *et al.*, 1983), and the fluorescence signal of the calcium-sensitive probe quin-2 was monitored using an SLM (Model 8000) fluorescence spectrophotometer (White *et al.*, 1983). The arrows indicate the time at which the stimulus was added. Note that the addition of either f-Met-Leu-Phe (A) or leukotriene B_4 (B) increases the signal, indicating an increase in the intracellular concentration of free calcium. The two stimuli appear to release the same pool(s) of calcium, and this pool is able to recover.

B_4 and f-Met-Leu-Phe are 1.2 × 10^{-10} M and 2 × 10^{-10} M, respectively. The two stimuli appear to release the same pool(s) of calcium.

It has been hypothesized that the initial lipolytic response to stimulation does not require a rise above basal intracellular free calcium concentration and that the breakdown of TPI, and therefore the generation of IP_3, is directly receptor linked and ultimately responsible in turn for the mobilization of calcium necessary for cell activation. In order to examine the relationship between calcium mobilization and phosphoinositide metabolism, we have measured the effect of f-Met-Leu-Phe and leukotriene B_4 on the levels of phosphatidylinositol-4,5-bisphosphate (TPI) and phosphatidylinositol 4-monophosphate (DPI). The results are summarized in Fig. 2. Note that whereas f-Met-Leu-Phe causes a rapid breakdown of TPI and DPI, leukotriene B_4 has no significant effect on the turnover of these two phosphoinositides. The same results were obtained in cells that had been labeled with [³H]arachidonic acid instead of ^{32}P (data not shown).

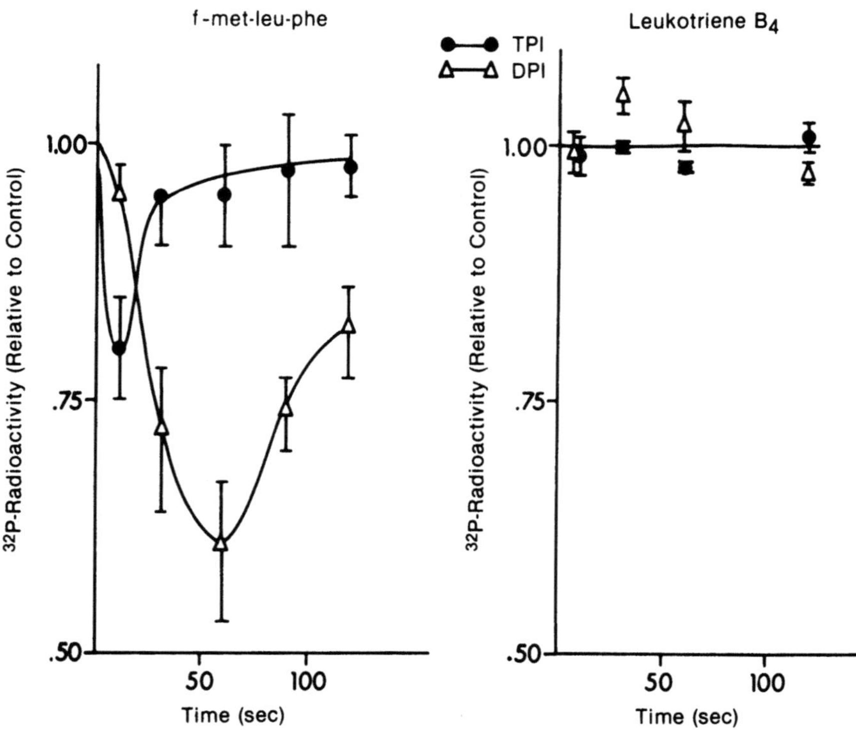

FIGURE 2. Time course of ^{32}P radioactivity changes in phosphatidylinositol-4,5-bisphosphate, TPI (●), and phosphatidylinositol-4-monophosphate, DPI (△), following the addition of f-Met-Leu-Phe (10^{-8} M) and leukotriene B$_4$ (7×10^{-8} M). The cells were labelled with ^{32}P radioactivity for 1 hr as described previously (Volpi *et al.*, 1983). At the end of the incubation period, the cells were pelleted by centrifugation, and the supernatant was removed by suction. The cells were then resuspended in Hanks' balanced salt solution (10^8 cells/ml). Samples of 1 ml each (10^8 cells) of ^{32}P-prelabeled neutrophils were reacted with the stimuli for a preset time, and the reaction was then stopped by the addition of 5 ml of hexane/isopropanol/concentrated HCl to give a final concentration of 0.1 M (300 : 200 : 4 v/v). The extracted lipids of each sample were dissolved in 100 μl of hexane/isopropanol (3 : 2), and 10 μl was used for spotting on precoated TLC plates, silica gel. The plates were developed in one dimension in chloroform/methanol/20% methylamine (60 : 36 : 10 v/v).

The ability of leukotriene B$_4$ to mobilize calcium in the absence of a breakdown of TPI, and thus of the generation of IP$_3$, suggests any one of three possibilities. First, leukotriene B$_4$ could mobilize calcium by acting as a calcium ionophore. This view is supported by the finding that it is able, albeit at high concentrations, to transport calcium in model systems (Serhan *et al.*, 1982). In spite of this evidence, it is unlikely that at physiological concentrations it acts as an ionophore. This conclusion is based on many experimental observations (for review and original citations, see Feinstein and Sha'afi, 1983; Sha'afi and Naccache, 1984). For example, leukotriene B$_4$ is extremely potent, and its effect is cell specific and is sensitive to relatively minor stereochemical modifications. In addition, neutrophils become deactivated after exposure to leukotriene B$_4$ or to f-Met-Leu-Phe, and specific binding sites for leukotriene B$_4$ have been demonstrated. Second, leukotriene B$_4$ could mobilize calcium by the same mechanism as IP$_3$ by utilizing common "receptors." This is unlikely since leukotriene B$_4$ is cell specific whereas IP$_3$ is not. Third, leukotriene B$_4$ could mobilize calcium by yet another unknown mechanism, possibly through the "G proteins" system. The "G proteins," previously described in the regulation of the catalytic unit of the adenylate cyclase (Gilman *et al.*, 1984), probably represent branch points for the transduction of information across the plasma membrane. A "G protein" system has been recently identified in the plasma membrane of human neutrophils (Lad *et al.*, 1984). In addition, leukotriene B$_4$ is known to increase rapidly the intracellular concentration of cAMP (Claesson, 1982), an effect potentially mediated through the activation of the "G proteins."

3. EFFECTS OF LEUKOTRIENE B$_4$ AND f-MET-LEU-PHE ON THE PRODUCTION OF 1,2-DIACYLGLYCEROL AND THE RELEASE OF ARACHIDONIC ACID

The breakdown of phosphoinositides by phospholipase C generates 1,2-diacylglycerol (1,2-DG) in addition to IP$_3$. Although IP$_3$ may be related to calcium release, 1,2-DG is closely involved in the activation of protein kinase C, which can also be activated by the addition of calcium and phosphatidylserine (Kishimoto *et al.*, 1983). Recently, it has been found that direct activation of protein kinase C by phorbol-12-myristate-13-acetate (PMA) causes significant neutrophil aggregation, induces slow lysosomal enzyme release, and increases the oxygen consumption and cytoskeletal actin, all without a detectable rise in the intracellular free calcium concentration (Sha'afi *et al.*, 1983). Based on these and other studies, it has been hypothesized that activation of the protein kinase C system is closely related to lysosomal enzyme release in neutrophils (Sha'afi *et al.*, 1983; White *et al.*, 1984). In order to examine this hypothesis further, we have investigated the effect of f-Met-Leu-Phe and leukotriene B$_4$ on the production of 1,2-diacylglycerol, and the results are summarized in Table I. Note that whereas the addition of f-Met-Leu-Phe produces an increase in the generation of 1,2-DG, that of leukotriene B$_4$, on the other hand, causes no significant change in 1,2-DG. In the case of leukotriene B$_4$, longer times of incubation and different concentrations of the stimulus were also used, and the results were the same (data not shown). The inability of leukotriene B$_4$ to generate 1,2-DG is consistent with its lack of effect on a TPI breakdown. The ability of leukotriene B$_4$ to increase the intracellular concentration of free calcium without generating significant amounts of 1,2-DG strongly suggests that it activates the protein kinase C system to a much smaller extent than f-Met-Leu-Phe. This may also explain some of the differences outlined earlier between the characteristics of the secretory activities of leukotriene B$_4$ and f-Met-Leu-Phe.

The effects of these two stimuli on the release of arachidonate from

TABLE I. Effects of the Addition of f-Met-Leu-Phe and Leukotriene B_4 on the Production of 1,2-Diacylglycerol and the Release of Arachidonic Acid

Stimulus	1,2-Diacylglycerol[a]	Arachidonic acid[b]
f-Met-Leu-Phe (2×10^{-8} M)	1.25 ± 0.05 (6)	1.6 ± 0.1 (4)
Leukotriene B_4 (7×10^{-8} M)	0.97 ± 0.02 (8)	1.10 ± 0.03 (4)

[a] The cells were labeled with [^{3}H]arachidonic acid, and the lipids were extracted, spotted, and developed as described in Fig. 2. A known amount of 1,2-diacylglycerol was spotted along with the radioactive lipids in each lane, and the positions of the various spots were identified by iodine vapor. The amount of 1,2-diacylglycerol produced was measured 30 sec after the addition of the stimulus, and the results are expressed relative to control. The values represent the mean $\pm$ standard error of the mean, and the number in parentheses refers to the number of different experiments. The change in 1,2-diacylglycerol produced by f-Met-Leu-Phe is statistically significant ($P < 0.02$ using paired sample t-test).

[b] The amount of arachidonate released was measured 60 sec after the addition of the stimulus. The results are expressed relative to control, and the values represent the mean $\pm$ standard error of the mean.

[^{3}H]arachidonic-acid-labeled rabbit neutrophils were also investigated, and the results are included in Table I. Whereas leukotriene B_4 releases a small but statistically significant amount of arachidonic acid, f-Met-Leu-Phe produces a much larger increase. The change in arachidonate release induced by f-Met-Leu-Phe but not that by leukotriene B_4 continues to increase with time. The leukotriene B_4-induced release of arachidonic acid is most likely mediated through the action of phospholipase A_2, which has been shown to be stimulated by leukotriene B_4 (Bormann *et al.*, 1984).

4. RELATIONSHIP BETWEEN CALCIUM MOBILIZATION AND THE GENERATION OF PHOSPHATIDIC ACID PRODUCED BY LEUKOTRIENE B_4 AND f-MET-LEU-PHE

It has been proposed that phosphatidic acid, formed by the phosphorylation of diacylglycerol, may assist in mobilizing calcium in neutrophils in view of its ionophoretic properties (Serhan *et al.*, 1982). The latter results have, however, been challenged by Holmes and Yoss (1983), and the controversy has yet to be resolved. In order to get a better understanding of this problem, we have investigated the effects of various concentrations of f-Met-Leu-Phe and leukotriene B_4 on the levels of both the intracellular free calcium and phosphatidic acid. In these experiments, the concentration of free calcium was measured using the calcium-sensitive fluorescent probe quin-2, and the level of phosphatidic acid was measured 20 sec after the addition of the stimulus. This time was chosen since it represents the time at which the quin-2 signal is maximal. The results of these studies are summarized in Figs. 3 and 4. The data in Fig. 3 clearly show that the addition of f-Met-Leu-Phe to rabbit neutrophils causes the generation of a significant amount of phosphatidic acid. The amount of phosphatidic acid production increases with the time of incubation of the cells with f-Met-Leu-Phe (Volpi *et al.*, 1983). In addition, a significant and time-dependent increase in lysophosphatidic acid was observed fol-

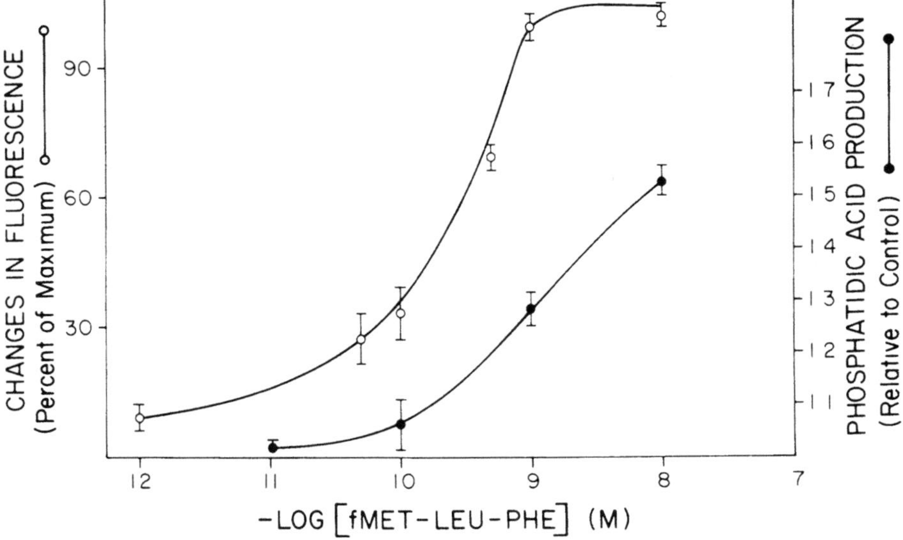

FIGURE 3. Dose–response curve of the effect of f-Met-Leu-Phe on phosphatidic acid production and calcium mobilization as measured by the fluorescence of quin-2. The experimental procedures are the same as those in Fig. 2 except that the reaction was stopped 20 sec following the addition of the stimuli.

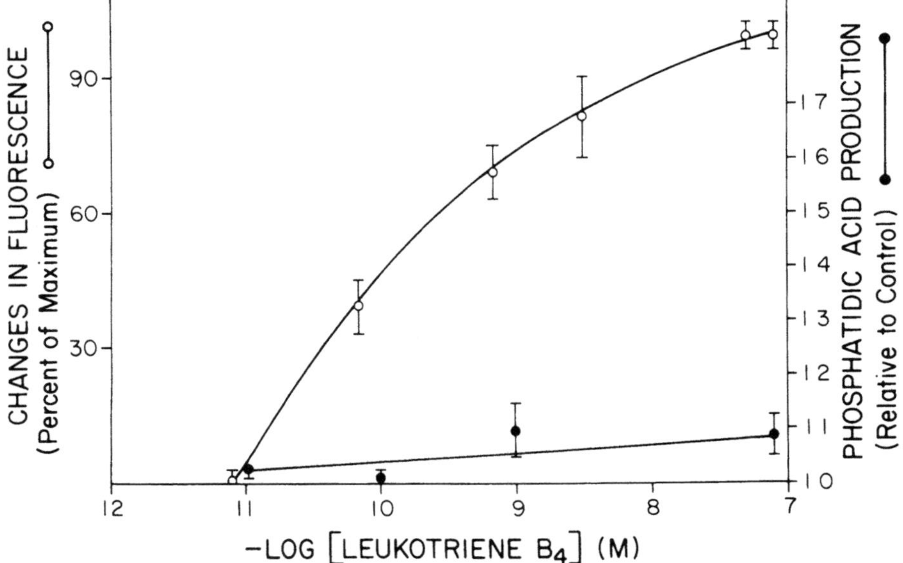

FIGURE 4. Dose–response curve of the effect of leukotriene B_4 on phosphatidic acid production and calcium mobilization as measured by the fluorescence of quin-2. Conditions are the same as in Fig. 4.

lowing stimulation by f-Met-Leu-Phe (data not shown). The means ($\pm$ standard error) of the ED_{50} values for calcium mobilization are $1.2 \pm 0.5 \times 10^{-10}$ M for leukotriene B_4 and $2.0 \pm 0.2 \times 10^{-10}$ M for f-Met-Leu-Phe. Note that, consistent with the early results, little if any phosphatidic acid can be recovered from the leukotriene B_4-stimulated cells. On the other hand, although the incubation of the cells with f-Met-Leu-Phe for 20 sec generates significant amounts of phosphatidic acid, the dose–response curve is shifted to the right relative to that for calcium mobilization. On the basis of these results, it is unlikely that calcium mobilization by leukotriene B_4 or even by f-Met-Leu-Phe is mediated through the action of phosphatidic acid.

5. COMPARISON BETWEEN THE EFFECTS OF LEUKOTRIENE B_4 AND f-MET-LEU-PHE

Despite a great deal of similarities, significant quantitative, as well as qualitative differences between the actions of f-Met-Leu-Phe and leukotriene B_4 do exist. These differences are very important since, in addition to explaining the action of leukotriene B_4, they also may provide some insight into the biochemical and biophysical events that are needed to elicit the various neutrophil responses. A summary of the main similarities and differences in neutrophil responses and biochemical changes produced by leukotriene B_4 and f-Met-Leu-Phe is given in Table II.

TABLE II. Comparison of Neutrophil Responses and Biochemical Changes Produced by f-Met-Leu-Phe and Leukotriene B_4

Response	f-Met-Leu-Phe	Leukotriene B_4
Chemotaxis	Yes	Yes
Aggregation	Yes	Yes[a]
Degranulation	Yes	Yes[a]
Oxygen consumption	Yes	No
Calcium mobilization	Yes	Yes
TPI and DPI breakdown	Yes	No
Generation of 1,2-diacylglycerol	Yes	No
Generation of phosphatidic acid	Yes	No
Release of arachidonate	Yes	Yes[b]
Activation of Na^+,K^+-ATPase	Yes	Yes
Activation of phospholipase A_2	Yes	Yes
Increased cAMP	Yes	Yes
Increased cytoskeletal actin	Yes	Yes
Phosphorylation by protein kinase C of a 50 K protein	Yes	?
Increased Na^+ influx[c]	Yes	Yes

[a] The action of leukotriene B_4 is significantly weaker than that of f-Met-Leu-Phe.
[b] The change is small but significant.
[c] This increase is inhibited by amiloride.

6. SUMMARY

Several important conclusions can be drawn from these studies. First, leukotriene B$_4$ causes the release of calcium without stimulating the breakdown of TPI or the generation of IP$_3$. The mechanism by which it does so is not known but may involve the action of "G proteins." Second, calcium mobilization by f-Met-Leu-Phe and leukotriene B$_4$ cannot be mediated by phosphatidic acid. Third, arachidonic acid release by leukotriene B$_4$ is secondary to the action of phospholipase A$_2$, whereas that released by f-Met-Leu-Phe is probably a result of the action of both phospholipase A$_2$ and phospholipase C. Fourth, because leukotriene B$_4$ mobilizes calcium but does not generate 1,2-diacylglycerol, it should be expected to activate the protein kinase C system to a much lesser extent than f-Met-Leu-Phe. Fifth, the findings that significant amounts of lysophosphatidic acid and arachidonic acid are produced following stimulation by f-Met-Leu-Phe suggest that part of the arachidonate released comes from phosphatidic acid, possibly through the action of a phosphatidic-acid-specific phospholipase A$_2$.

ACKNOWLEDGMENTS. This work was supported in part by NIH grants AI-13734 and AM-31000.

REFERENCES

Bach, M. K., 1983, *The Leukotrienes and Their Structure, Actions and Role in Disease*, Upjohn, Kalamazoo, MI.

Bormann, B. J., Huang, C.-K., Mackin, W. M., and Becker, E. L., 1984, Receptor-mediated activation of phospholipase A$_2$ in rabbit neutrophil plasma membrane, *Proc. Natl. Acad. Sci. U.S.A.* **81:**767–770.

Claesson, H.-F., 1982, Leukotrienes A$_4$ and B$_4$ stimulate the formation of cyclic AMP in human leukocytes, *FEBS Lett.* **139:**305–308.

Feinstein, M. B., and Sha'afi, R. I., 1983, Role of calcium in arachidonic acid metabolism and in the actions of arachidonic acid-derived metabolites, in *Calcium and Cell Function*, Vol. IV (W. Y. Cheung, ed.), Academic Press, New York, pp. 337–376.

Gilman, A. G., 1984, G proteins and dual control of adenylate cyclase, *Cell* **36:**577–579.

Holmes, R. P., and Yoss, N. L., 1983, Failure of phosphatidic acid to translocate Ca^{2+} across phosphatidylcholine membranes, *Nature* **305:**637–638.

Kishimoto, A., Kajikawa, N., Shiota, M., and Nishizuka, Y., 1983, Proteolytic activation of calcium-activated, phospholipid-dependent protein kinase by calcium-dependent neutral protease, *J. Biol. Chem.* **258:**1156–1164.

Lad, R. M., Glovsky, M. M., Smiley, P. A., Klempner, M., Reisinger, D. M., and Richard, J. H., 1984, The β-adrenergic receptor in the human neutrophil plasma membrane: Receptor–cyclase uncoupling is associated with amplified GTP activation, *J. Immunol.* **132:**1466–1471.

Michell, R. H., and Kirk, C. J., 1982, The unknown meaning of receptor-stimulated inositol lipid metabolism, *Trends. Pharmacol. Sci.* **3:**1490–1499.

Palmblad, J., Maimsten, C. L., Uden, A., Radmark, O., Engsted, E. L., and Samuelsson, B., 1981, Leukotriene B$_4$ is a potent and stereospecific stimulator of neutrophil chemotaxis and adherence, *Blood* **58:**658–661.

Serhan, C. N., Fridovich, J., Goetzl, E. J., Dunham, P. B., and Weissmann, G., 1982, Leukotriene

B_4 and phosphatidic acid are calcium ionophores: Studies employing arsenazo III in liposomes, *J. Biol. Chem.* **257**:4746–4752.

Sha'afi, R. I., and Naccache, P. H., 1985, Relationships between calcium, arachidonic acid metabolites and neutrophil activation, in : *Calcium in Biological Systems* (R. P. Rubin, G. Weiss, and J. W. Putney, Jr., eds.), Plenum Press, New York, pp. 137–145.

Sha'afi, R. L., White, J. R., Molski, T. F. P., Shefcyk, J., Volpi, M., Naccache, P. H., and Feinstein, M. B., 1983, Phorbol 12-myristate 13-acetate activates rabbit neutrophils without an apparent rise in the level of intracellular free calcium, *Biochem. Biophys. Res. Commun.* **114**:638–645.

Streb, H., Irvine, R. F., Berridge, M. J., and Shulz, I., 1983, Release of Ca^{2+} from a non-mitochondrial intracellular store in pancreatic acinar cells by inositol-1,4,5-triphosphate, *Nature* **306**:67–69.

Volpi, M., Yassin, R., Naccache, P. H., and Sha'afi, R. I., 1983, Chemotactic factor causes rapid decrease in phosphatidyl inositol 4.5-bis phosphate and phosphatidyl inositol 4-monophosphate in rabbit neutrophils, *Biochem. Biophys. Res. Commun.* **112**:957–964.

White, J. R., Naccache, P. H., Molski, T. F. P., Borgeat, P., and Sha'afi, R. I., 1983, Direct demonstration of increased intracellular concentration of free calcium in rabbit and human neutrophils following stimulation by chemotatic factors, *Biochem. Biophys. Res. Commun.* **113**:44–50.

White, J. R., Huang, C.-K., Hill, J. M., Naccache, P. H., Becker, E. L., and Sha'afi, R. I., 1984, Effect of phorbol 12-myristate 13-acetate and its analogue 4α-phorbol 12,13-didecanoate on protein phosphorylation and lysosomal enzyme release in rabbit neutrophils, *J. Biol. Chem.* **259**:8605–8611.

Yassin, R., Shefcyk, J., Tao, W., White, J. R., Molski, T. F. P., Naccache, P. H., and Sha'afi, R. I., 1984, Actin association with the cytoskeleton in human and rabbit neutrophils: Possible signal for mediating actin polymerization, *Fed. Proc.* **43**:1507.

24

Prostaglandylinositol Cyclic Phosphate

An Antagonist to Cyclic AMP

HEINRICH KARL WASNER

1. INTRODUCTION

About 10 years ago studies were started in Earl Sutherland's laboratory in search of a regulatory molecule with the ability to switch off the functions initiated by cAMP. I should like to dedicate this chapter to his memory. He would have been 70 years old in 1985.

I describe here an intracellular regulator, cyclic PIP, which I had tentatively named cAMP antagonist (Wasner, 1975b) since it fulfills these requirements. In recent years several putative second messengers for insulin have been reported (Larner, 1979; Jarett, 1979; Seals, 1980) with properties similar to cyclic PIP. They are considered to be peptides and thus are clearly distinct from cyclic PIP, which is not a peptide and appears to be a more potent regulator. The finding that prostaglandin E plays a role in cyclic PIP synthesis (Wasner, 1975a, 1981) was confirmed recently (Begum, 1983). This chapter describes the present concept of the structure of cyclic PIP.

2. ISOLATION AND ASSAY OF CYCLIC PIP

A small molecule has been reported (Wasner, 1975, 1981) that can be extracted from hepatocytes shortly after α-adrenergic stimulation. In the presence of epinephrine there is a rapid increase of cyclic PIP, reaching a maximum after 3–5 min

HEINRICH KARL WASNER ● Diabetes-Forschungsinstitut, Biochemische Abteilung, 4000 Düsseldorf 1, West Germany.

and declining thereafter to reach base levels after 15 min. In contrast, cyclic PIP is almost undetectable in the absence of hormone. Figure 1A shows the assay to measure cyclic PIP (Wasner, 1975b). The cAMP-dependent protein kinase is inhibited in a dose-dependent manner. This inhibition is noncompetitive with respect to cAMP and ATP, and the separated catalytic subunit is equally inhibited. One unit of cyclic PIP is defined as the amount that inhibits the protein kinase by 50% in an assay volume of 0.1 ml. Cyclic PIP also stimulates the phosphoprotein phosphatase, the enzyme counterregulating the protein kinase. Figure 1B shows that cyclic PIP stimulates this enzyme sevenfold in a similar dose-dependent manner.

3. PURIFICATION OF CYCLIC PIP

Figure 2A shows the initial purification step, gel filtration on Sephadex G-15. Peaks A, B, and C represent ATP, AMP, and cAMP, respectively. Cyclic PIP (peak I) is eluted close to ATP, suggesting a molecular weight in the range of 500–600. The shoulder at the ascending part of peak I can be attributed to a hydrolysis product of cyclic PIP. On rechromatography of cyclic PIP (peak I), this degradation material can be detected again, and its proportion increases with time after isolation of cyclic PIP.

As is seen in Fig. 2B, on anion-exchange chromatography cyclic PIP is eluted as a sharp peak following the elution of carboxylic acids, which peak in fraction 20–21. It elutes just ahead of cAMP, indicating a similar negative charge for cyclic PIP.

The elution profile from silicic acid adsorbtion chromatography is shown in Fig. 2C. On elution of the adsorbent with benzene/ethyl acetate and increasing

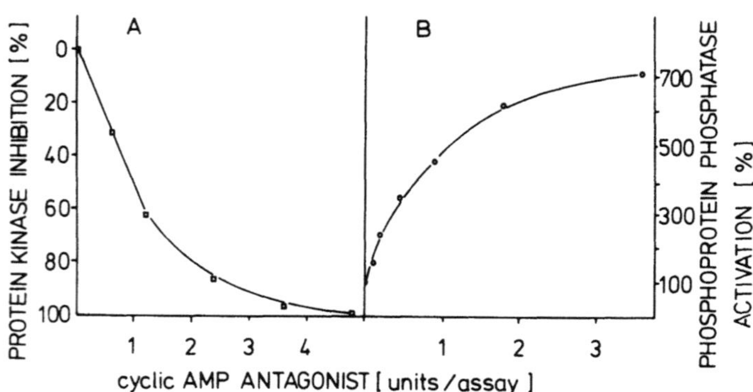

FIGURE 1. A: Concentration profile of the inhibition of cAMP-dependent protein kinase by cyclic PIP. Per 0.1-ml assay volume were added 5-, 10-, 15-, 20-, and 25-μl aliquots of a solution of purified cyclic PIP. B: Concentration profile of the activation of phosphoprotein phosphatase by cyclic PIP; 10^{-5} M phosphohistone was used as substrate in an assay volume of 0.1 ml.

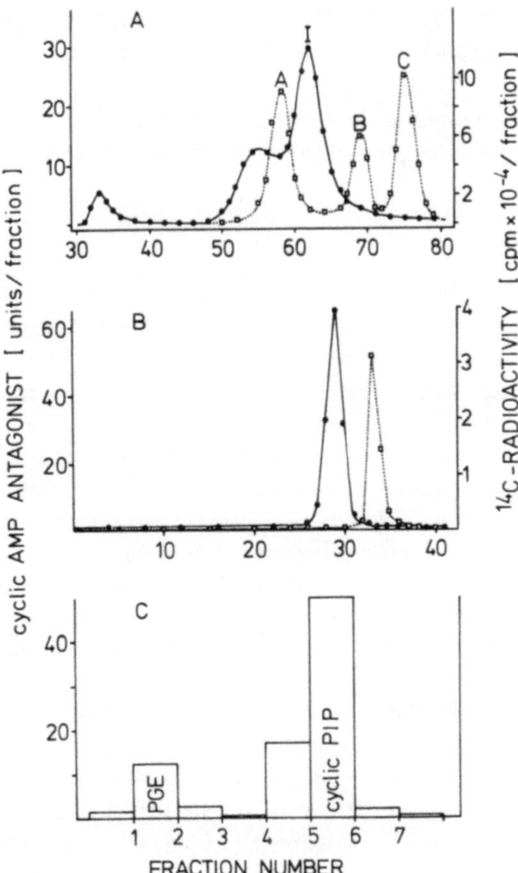

FIGURE 2. A: Chromatography on a Sephadex G-15 column (1.5 × 90 cm). When 2-ml fractions are collected, cyclic PIP peaks in fraction 62 and has a shoulder in fraction 54. The elution profiles of [^{14}C]ATP (A), [^{14}C]AMP (B), and [^{14}C]cAMP (C) are shown. B: Elution profile of cyclic PIP from a QAE-Sephadex anion-exchange column (0.9 × 30 cm) by a linear salt gradient. The dotted line shows the elution profile of cyclic AMP. C: Chromatography of cyclic PIP on a silicic acid column (0.9 × 1.5 cm). On increasing the methanol concentration stepwise from 2 to 16, 28, 44, 56, and 68% in the eluent benzene/ethyl acetate/methanol, PGE elutes at 9% and cyclic PIP at 56%.

methanol concentrations, prostaglandin E is eluted at a methanol concentration of 9%, and cyclic PIP at a concentration of 56%, which indicates that cyclic PIP is much more polar than prostaglandin E.

A further purification step is metal chelate affinity chromatography. Here organophosphates are adsorbed by forming less soluble metal phosphates, which can be eluted by a pH gradient.

4. THE PRESENT STATE OF KNOWLEDGE OF THE CHEMICAL STRUCTURE OF CYCLIC PIP

I shall briefly summarize the evidence suggesting that prostaglandin E, *myo*-inositol, and phosphate are the primary constituents of cyclic PIP.

4.1. Prostaglandin E

It can be demonstrated that synthesis of cyclic PIP requires the presence of PGE, since no cyclic PIP can be detected in PGE-depleted hepatocytes except after addition of exogenous PGE to the cells (Wasner, 1981). Further, PGE-depleted hepatocytes show an almost fivefold increase in cAMP synthesis on adrenergic stimulation when compared with untreated control cells (Wasner, 1976). With this approach the incorporation of tritium-labeled PGE into cyclic PIP can be demonstrated. In a control experiment using hormonally nonstimulated cells, only one peak, PGE, is detected on gel filtration. However, on hormonal stimulation of such hepatocytes, two additional peaks are observed, both with higher molecular weights: the peak with the highest molecular weight represents cyclic PIP, and the intermediate peak has been attributed to its degradation product, a dephosphorylated cyclic PIP. The specific activity—that is, the radioactivity per unit of cyclic PIP—remains constant throughout the purification.

On the basis of the specific radioactivity of exogenously added PGE, it can be estimated that 1 unit of cyclic PIP is equivalent to 5 pmol. Thus, full inhibition of protein kinase would be achieved at a concentration of 2.5×10^{-7} M. Under defined conditions, the amount of isolated cyclic PIP is in the range of 150 units per rat liver. This is equivalent to an extractable cyclic PIP concentration of 0.7×10^{-7} M. Thus, the amount of cyclic PIP extracted is in the range of action.

Cyclic PIP decomposes, resulting in loss of activity and leading to liberation of native PGE. This is shown as follows. When previously purified cyclic PIP is maintained at acid pH and then rechromatographed on a reversed-phase column, the remaining cyclic PIP elutes again unadsorbed with unchanged specific activity; when a gradient from water to ethanol is applied to the column, radioactive material is eluted in an amount equivalent to the loss of cyclic PIP. Mass spectrometry showed this product to be native PGE.

We have, further, been able to demonstrate that antibodies against PGE bind cyclic PIP, although with a lower affinity than PGE. This result has been obtained by a radioactive dilution assay and by antibodies immobilized on Sepharose, retaining cyclic PIP.

4.2. Phosphate

It became clear during the purification that cyclic PIP is a polar compound. Its polar as well as anionic nature persists after esterification of the carboxyl group of PGE, strongly suggesting the presence of an inorganic anionic group. Presence of phosphate can be directly shown by [^{32}P]phosphate incorporation into cyclic PIP and the demonstration of a constant ratio of radioactivity per unit of cyclic PIP.

4.3. *myo*-Inositol

The presence of *myo*-inositol in cyclic PIP was first indicated by analysis of its dephosphorylated product, the degradation product previously mentioned. Like cyclic PIP, it decomposes to yield prostaglandin and a still unidentified moiety. We therefore subjected this purified product to mass spectrometry after acid hydrolysis, trimethylsilylation, and gas chromatography. We found this moiety to be consistent with inositol. This was confirmed by studies with tritium-labeled *myo*-inositol, which was incorporated into cyclic PIP.

4.4. Structure

This leaves us with the question: How are PGE, inositol, and phosphate linked to give cyclic PIP? As indicated, cyclic PIP and its degradation product, dephosphorylated cyclic PIP, decompose to yield free PGE. Thus, apparently, either inositol phosphate or inositol is linked to PGE. Which functional group of PGE carries the inositol phosphate moiety? To elucidate this question the approach was the following:

1. It is possible to prepare a methyl ester of cyclic PIP in which the carboxyl group of PGE is esterified.
2. By KOH treatment of cyclic PIP, the C_{11} hydroxyl group of the PGE moiety is removed. This treatment, however, although not destroying the integrity of cyclic PIP, does reverse its regulatory properties.
3. It is possible to make a methoxim derivative.

These three experiments exclude binding of the PGE to the inositol phosphate via its carboxyl, carbonyl, or C_{11} hydroxyl group, therefore suggesting that the inositol phosphate is bound to the PGE via its C_{15} hydroxyl group. This is further supported by the finding that catalytic hydrogenation of the double bond of PGE_1 yields an acid-stable cyclic PIP (Fig. 3).

5. SUMMARY

Cyclic PIP is rapidly synthesized in response to catecholamines and in response to insulin stimulation. It is composed of PGE, *myo*-inositol, and phosphate and may be named prostaglandylinositol cyclic phosphate (cyclic PIP) (Fig. 3). It is rapidly degraded.

Its regulatory properties have been briefly described. The most important are inhibition of cAMP-dependent protein kinase in a variety of tissues and activation

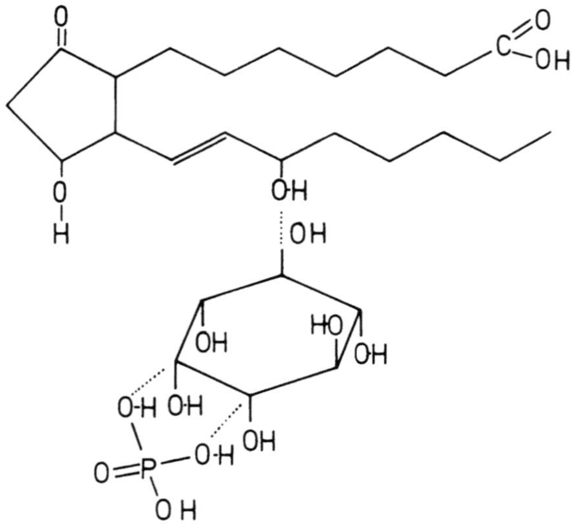

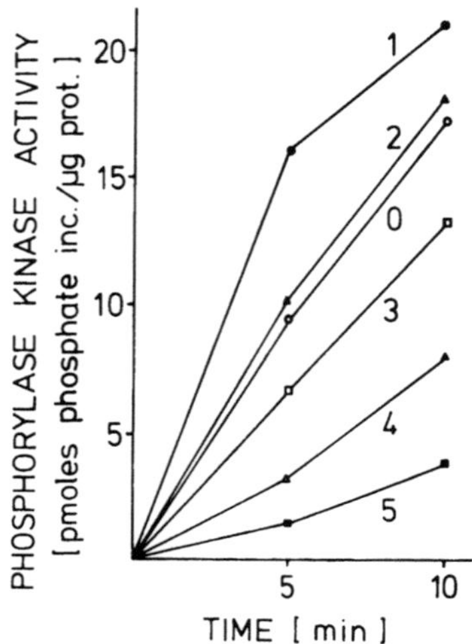

FIGURE 3. Possible structure of cyclic PIP. It is presently not clear which of the six isomer hydroxyl groups of *myo*-inositol binds the PGE and which bind the phosphate. It may be speculated that the phosphate is bound to the C_1 and C_2 hydroxyl groups since a 1,2-cyclic inositol phosphate is known to exist.

of phosphoprotein phosphatase. By modulation of the activity of these two regulatory enzymes, the following activities have been measured: a fivefold activation of pyruvate dehydrogenase in adipocyte mitochondria, complete inhibition of phosphorylase *b* of muscle (Fig. 4), complete inhibition of adenylate cyclase in hepatocytes (Wasner, 1981), and, finally, a complete blockade of the cAMP signal in *Dictyostelium discoideum* (H.K. Wasner, unpublished data).

Thus, the cAMP antagonist cyclic PIP is ubiquitous, and it clearly counter-regulates the regulatory actions of cAMP so far investigated. And, finally, the possibility exists that cyclic PIP may be the biochemical mediator of those intracellular regulations attributed to PGE such as the antilipolytic effect in adipocytes (Steinberg, 1964) and the inhibition of hepatic glucose output (Brass, 1984).

FIGURE 4. Phosphorylation of phosphorylase *b* by phosphorylase kinase in the presence of cAMP-dependent protein kinase (curve O) and cAMP (curve 1) and in additional presence of increasing concentrations of cPIP (curves 2–5).

REFERENCES

Begum, N., Tepperman, H., and Tepperman, J., 1983, Effect of indomethacin and prostaglandin E_2 on insulin–generation of pyruvate dehydrogenase activator by liver and adipocyte plasma membrane, *Diabetes* **32**:34a.

Brass, E. P., Maureen, J. G., and Robertson, R. P., 1984, Inhibition of glucagon-stimulated hepatic glycogenolysis by E-series prostaglandins, *FEBS Lett.* **169**:293–296.

Jarett, L., and Seals, J. R., 1979, Pyruvate dehydrogenase activation in adipocyte mitochondria by an insulin-generated mediator from muscle, *Science* **206**:1407.

Larner, J., Galasko, G., Cheng, K., De Paoli-Roach, A. , Huang, L., Daggy, P., and Kellogg, J., 1979, Generation by insulin of a chemical mediator that controls protein phosphorylation and dephosphorylation, *Science* **206**:1408–1410.

Seals, J. R., and Czech, M. P., 1980, Evidence that insulin activates an intrinsic plasma membrane protease in generating a secondary chemical mediator, *J. Biol. Chem.* **255**:6529–6531.

Steinberg, D., Vaughan, M., Nestel, P. J., Strand, O., and Bergstrom, S., 1964, Effects of the prostaglandins on hormone-induced mobilization of free fatty acids, *J. Clin. Invest.* **43**:1533.

Wasner, H. K., 1975a, On the biochemistry of a new hormone-messenger involving the alpha-receptor in rat liver, *Communications 10th FEBS Meeting, Paris*, Abstract 1367.

Wasner, H. K., 1975b, Regulation of protein kinase and phosphoprotein phosphatase by cyclic AMP and cyclic AMP antagonist, *FEBS Lett.* **57**:60–63.

Wasner, H. K., 1976, Preparation of a liver plasma membrane with an adrenalin responsive adenylyl cyclase after inhibition of prostaglandin synthesis by indomethacin, *FEBS Lett.* **72**:127–130.

Wasner, H. K., 1980, cyclic AMP antagonist—the hormone-messenger of insulin, *Akt. Endokrinol. Stoffwechsel.* **1**:207–208.

Wasner, H. K., 1981, Biosynthesis of cyclic AMP antagonist in hepatocytes from rats after adrenalin- or insulin-stimulation (isolation, purification and prostaglandin E-requirement for its synthesis), *FEBS Lett.* **133**:260–264.

III

Cardiovascular–Pulmonary Interactions of Prostaglandins, Prostacyclin, and Leukotrienes

25

Regulatory Mechanisms in Prostacyclin–Blood Vessel Interactions

BABETTE B. WEKSLER

1. INTRODUCTION

The mechanisms governing prostaglandin production and action in vascular tissue are steadily becoming clarified. The former concept of a direct balance between vascular prostacyclin (PGI_2) and platelet thromboxane (TxA_2) in mediating vascular tone or hemostasis now appears simplistic and has been replaced. Current knowledge suggests instead a network of interactions among eicosanoid products of vascular cells, blood elements, and plasma factors (Fig. 1).

Prostacyclin, the main cyclooxygenase product of large vessel endothelium, is released into the bloodstream at the luminal surface and modulates the functions of platelets and leukocytes. By raising platelet cAMP levels, PGI_2 inhibits platelet activation, disaggregates platelet clumps, and decreases platelet secretion. Conversely, cyclic endoperoxides released by platelets during activation can be utilized by nearby endothelial cells (Marcus *et al.*, 1978) or leukocytes (Defreyn *et al.*, 1982) to synthesize PGI_2. This effect is enhanced by thromboxane synthetase inhibitors (Fitzgerald *et al.*, 1983). Prostacyclin affects neutrophil leukocytes by preventing their aggregation, adhesion to plastic (Boxer *et al.*, 1980), and chemotaxis (Weksler *et al.*, 1977) and can inactivate free oxygen radicals produced by neutrophils (Weksler *et al.*, 1979). Leukotriene C_4, synthesized and released by neutrophils, can stimulate PGI_2 production by endothelium (Cramer *et al.*, 1983). A minor product of normal endothelium but a more important of product of atherosclerotic vessel (Rolland *et al.*, 1984) or microvascular endothelium (Gerritsen *et*

BABETTE B. WEKSLER ● Division of Hematology–Oncology, Department of Medicine, Cornell University Medical College, New York, New York 10021.

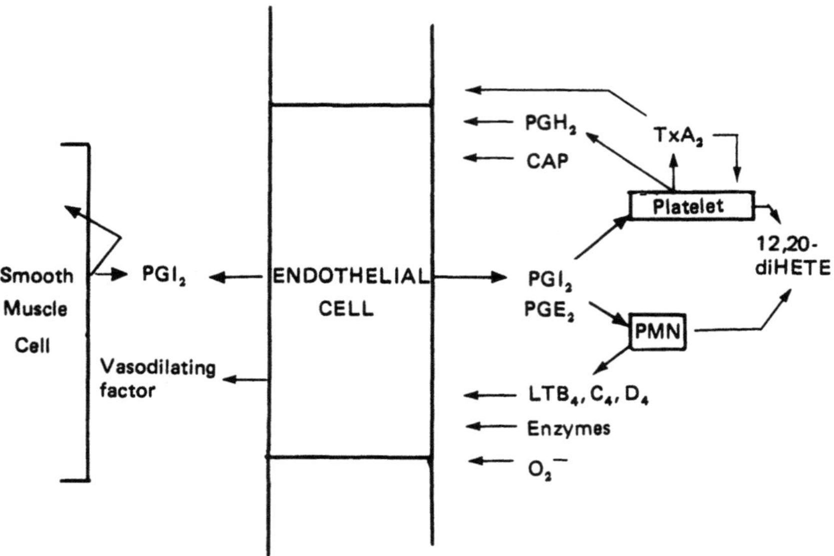

FIGURE 1. Eicosanoid interactions between vascular wall and blood cells.

al., 1983), PGE$_2$ can enhance platelet activation. Interaction between platelets and neutrophils releases an unusual 12,20-diHETE (Marcus *et al.*, 1984), which causes leukocyte aggregation. Since vessel wall can also produce 12-HETE (Greenwald *et al.*, 1979; Salzman *et al.*, 1980), a precursor for this 12,20-diHETE, it is possible that endothelial interaction with neutrophils might also result in 12,20-diHETE release.

2. CYCLOOXYGENASE REGULATION

Modulation of cyclooxygenase activity is a major factor in regulating prostaglandin production in vascular tissue. This enzyme is rapidly inactivated following initial activation by various stimuli, probably because of autooxidation by free-radical-containing intermediates. Thus, monolayers of human endothelial cells release markedly diminished amounts of PGI$_2$ on repeated stimulation by arachidonate following initial exposure to arachidonate, thrombin, or ionophore A23187 (Brotherton and Hoak, 1983). Repeated exposure to PGH$_2$, however, results in non-diminished production of PGI$_2$, indicating that no inactivation of PGI$_2$ synthase has occurred. Only if cyclooxygenase is protected by inhibition with the reversible cyclooxygenase inhibitor ibuprofen during the initial exposure to arachidonate does repeat stimulation with this fatty acid lead to a full PGI$_2$ response (Brotherton and Hoak, 1983). Similar rapid inactivation of cyclooxygenase has been observed in isolated perfused aortas (Kent *et al.*, 1983) and jugular vein (Hoak *et al.*, 1984).

In our studies of aspirin inhibition of vascular cyclooxygenase activity, we have obtained evidence that another governing factor may be a differential resynthesis of this enzyme in vascular smooth muscle and endothelium following its inactivation with aspirin. Exposure of vascular smooth muscle cells and endothelial cells to aspirin *in vitro* completely inhibits PGI_2 production from arachidonate 1 hr later, but smooth muscle cells recultured without aspirin show little recovery of PGI_2 synthetic capacity 48 hr later, whereas endothelial cells have fully recovered by 36 hr (Jaffe and Weksler, 1979). *In vivo* studies give similar results. Following daily administration of lowdose aspirin to surgical patients preoperatively, we observe a 73% inhibition of PGI_2 production in venous fragments obtained at surgery 3 hr after the last dose of aspirin. At 24 hr after the last dose, venous fragments show 52% inhibition of PGI_2 production, whereas the endothelial surface of the same venous segments shows no inhibition and produces the same amount of PGI_2 as venous segments from nonaspirinated control patients (Weksler *et al.*, 1985).

3. VASCULAR PROSTAGLANDINS AND CHOLESTERYL ESTER REGULATION

Since circulating levels of vascular PGI_2 are very low and probably do not have systemic effects on regulation of vascular tone or platelet function, it is of interest to consider other, more local, functions of PGI_2 within the vessel wall itself. Possible functions include local effects on vascular tone, influence on stored cholesterol, and modulation of smooth muscle cell proliferation. In uninjured vessels, it is not clear whether PGI_2 of smooth muscle origin reaches the bloodstream, since removal of endothelium results in a prompt drop in PGI_2 released at the luminal surface (Eldor *et al.*, 1981; Goldsmith *et al.*, 1981). After endothelial injury or removal, the migration of neointimal smooth muscle cells to the vessel surface soon furnishes PGI_2 at this site. In the presence of hypercholesterolemia, however, neointimal PGI_2 release is blunted (Eldor *et al.*, 1982).

Exploration of the interactions between PGI_2 and the accumulation of cholesterol by vascular smooth muscle has led to the observation that PGI_2 modulates the biological activity of cholesteryl-ester-metabolizing enzymes. Addition of nanomolar amounts of PGI_2 to cultures of vascular smooth muscle cells of rabbit or bovine origin stimulated the activity of both the lysosomal (acid) cholesteryl ester hydrolase and the cytoplasmic (neutral) cholesteryl ester hydrolase without having any effect on the microsomal acyl-CoA cholesterol acyl transferase (ACAT), which reesterifies free cholesterol (Hajjar *et al.*, 1982; Hajjar and Weksler, 1983). As shown in Table I, enzyme activities increased two- to threefold and were accompanied by a rise in intracellular cAMP levels. The effect of PGI_2 on cholesteryl ester hydrolysis was mediated by stimulation of adenylate cyclase and subsequent rise in cAMP, since treatment of cells with a specific inhibitor of adenylate cyclase activation, 2′,5′-dideoxyadenosine, prevented the rise in cholesteryl ester hydrolysis, whereas addition of dibutyryl cAMP to the cells in the absence of PGI_2

TABLE I. Effects of Prostacyclin on Cholesteryl Ester Hydrolytic Enzymes and cAMP Levels in Vascular Smooth Muscle Cells[a]

Additive	ACEH μU/mg DNA	NECH μU/mg DNA	cAMP pmol/10^5 cells
Buffer	3.4 ± 0.6	33 ± 5	3.0 ± 0.3
20 nM PGI$_2$	13.4 ± 0.9	N.D.	8.9 ± 0.4
250 nM PGI$_2$	N.D.	75 ± 5	14.7 ± 1.4

[a] Data adapted from Hajjar *et al.* (1982) and Hajjar and Weksler (1983).

mimicked the effect on cholesteryl ester hydrolysis. Moreover, enzyme activity could be stimulated by addition of sodium arachidonate to the smooth muscle cells; aspirin treatment of the cells prevented the stimulating effect of arachidonate but not of exogenous PGI$_2$. More recently, studies with metabolites of PGI$_2$ have shown that several metabolites including 2,3-dinor-6-keto-PGF$_{1\alpha}$ have similar effects (Etingen *et al.*, 1984). Other prostaglandins such as PGE$_1$ and PGE$_2$ do not stimulate cholesteryl ester hydrolysis in vascular smooth muscle cells, even at higher concentrations (Hajjar *et al.*, 1982; Hajjar and Weksler, 1983), although PGE$_2$ can depress ACAT activity (Hajjar and Weksler, 1983).

The implications of these studies for the modulation of atherosclerotic changes in the vascular wall through the manipulation of PGI$_2$ are intriguing. These studies clearly show that nanomolar amounts of exogenous, or amounts of endogenous PGI$_2$ produced through stimulation of the cells, can increase cholesteryl ester breakdown and stimulate egress of cholesterol from smooth muscle cells over longer periods of time (Hajjar *et al.*, 1982; Hajjar and Weksler, 1983; Etingen *et al.*, 1984). These effects are augmented by theophylline derivatives, but the changes in intracellular cAMP levels observed in the absence of phosphodiesterase inhibitors are sufficient to mediate the increased cholesteryl ester hydrolysis. Reports of the prevention of experimental atherosclerosis in animals fed high-cholesterol diets when the animals were treated with interferon (Kuo *et al.*, 1984) may have pertinence, since it has been reported that interferon α strongly and selectively stimulates endothelial cell PGI$_2$ synthesis (Eldor *et al.*, 1984), and it is likely that it has a similar action on the vascular smooth muscle cell. Thus, therapeutic strategies to preserve or augment PGI$_2$ synthesis in vascular smooth muscle could have a direct antiatherogenic affect. This hypothesis remains to be tested in appropriate models.

REFERENCES

Boxer, L. A., Allen J. M., Schmidt, M., Yoder, M., Baehner, R. L., 1980, Inhibition of polymorphonuclear leukocyte adherence by prostacyclin, *J. Lab. Clin. Med.* **95:**672–678.

Brotherton, A. F. A., and Hoak, J. C., 1983, Prostacyclin biosynthesis in cultured vascular endothelium is limited by deactivation of cyclooxygenase, *J. Clin. Invest.* **72:**1255–1262.

Cramer, E. B., Pologe, L., Pawlowski, N. A., Cohn, Z. A., and Scott, W. B., 1983, Leukotriene C promotes prostacyclin synthesis by human endothelial cells, *Proc. Natl. Acad. Sci. U.S.A.* **80:**4109–4113.

Defreyn, G., Deckmyn, H., and Vermylen, J., 1982, A thrombin synthetase inhibitor reorients endoperoxide metabolism in whole blood towards prostacyclin and prostaglandin E_2, *Throm. Res.* **26:**389–400.

Eldor, A., Falcone, D. J., Hajjar, D. P., Minick, C. R., and Weksler, B. B., 1981, Recovery of prostacyclin production by de-endothelialized rabbit aorta. Critical role of neointimal smooth muscle cells, *J. Clin. Invest.* **67:**735–741.

Eldor, A., Falcone, D. J., Hajjar, D. P., Minick, C. R., and Weksler, B. B., 1982, Diet-induced hypercholesterolemia inhibits the recovery of prostacyclin production by injured rabbit aorta, *Am. J. Pathol.* **107:**186–190.

Eldor, A., Fridman, R., Vlodavsky, I., Hyam, E., Fuks, Z., and Panet, A., 1984, Interferon enhances prostacyclin production by cultured vascular endothelial cells, *J. Clin. Invest.* **73:**251–257.

Etingin, O., Hajjar, D. P., and Weksler, B. B., 1984, Hydrolytic metabolites of prostacyclin alter cholesteryl ester metabolism in arterial cells, *Clin. Res.* **32:**473A.

Fitzgerald, G. A., Brash, A. R., Oates, J. A., and Pedersen, A. K., 1983, Endogenous prostacyclin biosynthesis and platelet function during selective inhibition of thromboxane synthase in man, *J. Clin. Invest.* **72:**1336–1343.

Gerritsen, M. E., and Cheli, C. D., 1983, Arachidonic acid and prostacyclin endoperoxide metabolism in isolated rabbit and coronary microvessels and isolated cultivated coronary microvessel endothelial cells, *J. Clin. Invest.* **72:**1658–1671.

Goldsmith, J. C., Jafvert, C. T., Lollar, P., Owen, W. G., and Hoak, J. C., 1981, Prostacyclin release from cultured and *ex vivo* bovine vascular endothelium, *Lab. Invest.* **45:**191–197.

Greenwald, J. E., Bianchine, J. R., and Wong, I. K., 1979, The production of the arachidonate metabolite HETE in vascular tissue, *Nature* **281:**588–590.

Hajjar, D. P., Weksler, B. B., Falcone, D. J., Hefton, J. M., Tack-Goldman, K., and Minick, C. R., 1982, Prostacyclin modulates cholesteryl ester hydrolytic activity by its effect on cyclic adenosine monophosphate in rabbit aortic smooth muscle cells, *J. Clin. Invest.* **70:**479–488.

Hajjar, D. P., and Weksler, B. B., 1983, Metabolic acitivity of cholesteryl esters in aortic smooth muscle cells is altered by prostaglandins I_2 and E_2, *J. Lipid. Res.* **24:**1176–1185.

Hoak, J. C., Fry, G. L., Czervionke, R. L., and Haycraft, D. L., 1984, Prostacyclin production in the isolated jugular vein segment, *Clin. Res.* **32:**551A.

Jaffe, E. A., and Weksler, B. B., 1979, Recovery of endothelial prostacyclin (PGI_2) production after inhibition by low dose aspirin, *J. Clin. Invest.* **63:**532–535.

Kent, R. S., Diedrich, S. L., and Whorton, A. R., 1983, Regulation of vascular prostaglandin synthesis by metabolites of arachidonic acid in perfused rabbit aorta, *J. Clin. Invest.* **72:**455–465.

Kuo, P. T., Wilson, A. C., Goldtsein, R. C., and Schaub, R. G., 1984, Suppression of experimental atherosclerosis in rabbits by interferon-inducing agents, *J. Am. Coll. Cardiol.* **3:**129–134.

Marcus, A. J., Safier, L. B., Ullman, H. L., Broekman, M. J., Islam, N., Oglilsby, T. D., and Gorman, R. R., 1984, 12S,20-Dihydroxyicosatetraenoic acid. A new icosanoid synthesized by neutrophils from 12-S hydroxyicosatetraenoic acid produced by thrombin- or collagen-stimulated platelets, *Proc. Natl. Acad. Sci. U.S.A.* **81:**903–907.

Marcus, A. J., Weksler, B. B., and Jaffe, E. A., 1978, Enzymatic conversion of prostaglandin endoperoxide H_2 (PGH_2) and arachidonic acid to prostacyclin by cultured human endothelial cells, *J. Biol. Chem.* **253:**7138–7141.

Rolland, P. H., Jouve, R., Pellegrin, E., Mercier, C., and Serradimigni, A., 1984, Alterations in prostacyclin and prostaglandin E_2 production. Correlation with changes in human aortic atherosclerotic disease, *Arteriosclerosis* **4:**70–78.

Salzman, P. M., Salmon, J. A., and Moncada, S., 1980, Prostacyclin and thromboxane A_2 synthesis by rabbit pulmonary artery, *J. Pharmacol. Exp. Ther.* **215:**240–247.

Weksler, B. B., Knapp, J. M., and Jaffe, E. A., 1977, Prostacyclin (PGI$_2$) synthesized by cultured endothelial cells modulates polymorphonuclear leukocyte function, *Blood* **50**:287.

Weksler, B. B., Knapp, J. M., and Tack-Goldman, K., 1979, Prostacyclin (PGI$_2$) is destroyed by superoxide produced by human polymorphonuclear leukocytes, *Clin. Res.* **27**:466A.

Weksler, B. B., Tack-Goldman, K., Subramanian, V. A., and Gray, W. A., Jr., 1985, Cumulative inhibitory effect of low-dose aspirin on vascular prostacyclin and platelet thromboxane production in patients with atherosclerosis, *Circulation* **71**:332–340.

26

A Comparison of the Contractile Responses of Rodent and Human Pulmonary Vascular Segments to Eicosanoids

YVONNE T. MADDOX, CYNTHIA M. CUNARD, RON SHAPIRO, AKIRA KAWAGUCHI, MITCHELL GOLDMAN, RICHARD R. LOWER, and PETER W. RAMWELL

1. INTRODUCTION

Eicosanoids such as thromboxanes (Txs) and leukotrienes (LTs) are powerful spasmogens. The pulmonary hemodynamic actions of leukotrienes are complex. For example, in the isolated rat lung preparation, the perfusion pressure is increased by LTC_4 in a dose-dependent manner, and this response is blocked by FPL-55712, a leukotriene receptor antagonist (Iacopino *et al.*, 1983). However, in the closed-chest rat with an intact circulation, the intravenous administration of LTC_4 produces a dose-dependent decrease in pulmonary artery pressure. We concluded that the direct action of LTC_4 on the pulmonary vasculature in this preparation (Iacopino *et al.*, 1984) is obscured by the concurrent decrease in cardiac output.

In order to further analyze the site of action of leukotrienes and thromboxanes, it is useful to determine their effects on pulmonary vascular segments. One anecdotal account suggests that leukotrienes have little or no action on spiral strip

YVONNE T. MADDOX, CYNTHIA M. CUNARD, and PETER W. RAMWELL • Department of Physiology and Biophysics, Georgetown University Medical Center, Washington, D. C. 20007.   RON SHAPIRO, AKIRA KAWAGUCHI, MITCHELL GOLDMAN, and RICHARD R. LOWER • Division of Cardiac Surgery, Medical College of Virginia, Richmond, Virginia 23298.

preparations of human pulmonary vascular tissue (Hanna *et al.*, 1981), but another study using human pulmonary vessels indicates that venous but not arterial strips are contracted semiisometrically by LTC_4 (Schellenberg and Foster, 1984). However, an important factor is the endothelium, and the endothelium has recently been recognized as a regulator of some of the effects of vascular agonists (Furchgott and Zawadzki, 1980; DeMey and Vanhoutte, 1982). Therefore, if the endothelium is to be considered, it is necessary to employ vascular ring preparations, since little endothelium remains after spiral vascular strips are prepared in the conventional manner (Furchgott and Zawadzki, 1980). The present study was undertaken with vascular rings and is designed to evaluate the response of rat pulmonary segments and to compare their response to the corresponding human vessels, which have limited availability.

2. MATERIALS AND METHODS

Segments of pulmonary arteries and veins were obtained from five subjects (male, 17–30 years old) and male Sprague–Dawley rats (Charles River, 250–400 g). Ring segments (length, 2–3 mm, rat; 3–4 mm, human) were prepared and carefully mounted, to avoid stripping the endothelium, in a tissue bath that contained Krebs–Ringer bicarbonate solution (KRB) with glucose, gassed with 5% CO_2 in 95% O_2, and kept at 37°C. After an initial preload of 1.5 g (human) or 1 g (rat), the tissue was equilibrated until basal tone stabilized. This usually took 1 to 2 hr. Tension was measured isometrically with a force displacement transducer connected to a preamplifier and a recorder (Harvard).

Cumulative concentration–response curves to LTC_4, U46619, and norepinephrine were generated. The responses are reported as tension generated in grams. The concentration producing 50% of the maximal contraction (EC_{50}) was calculated from regression lines.

The release of prostacyclin and thromboxane A_2 from segments of human pulmonary arteries and veins was measured by radioimmunoassay (RIA) of 6-keto-$PGF_{1\alpha}$ and TxB_2, respectively. Vessel segments (3 mm) were incubated in triplicate in 500 μl of KRB for 15 min at 37°C, and aliquots were analyzed in duplicate. After incubation, vessel segments were air dried and weighed. The RIA results are reported as picograms per milligram dry weight of tissue.

Leukotriene C_4 was kindly provided by Dr. J. Rokach of Merck Frosst Laboratories (Quebec, Canada); U46619, TxB_2, and 6-keto-$PGF_{1\alpha}$ were gifts of Dr. J. Pike of the Upjohn Co. (Kalamazoo, MI). FPL-55712 was a gift from Dr. P. Sheard of the Fisons Co. (Leicestershire, U.K.). The antibody to TxB_2 was a gift from Dr. L. Levine, Brandeis University (Waltham, MA). We developed the antibody to 6-keto-$PGF_{1\alpha}$ by immunizing rabbits with 6-keto-$PGF_{1\alpha}$ conjugated to bovine serum according to the method of Smith *et al.* (1978). The cross reactivity has been described (Pomerantz *et al.*, 1980).

Statistical analysis was performed by use of Student's unpaired *t*-test and analysis of variance.

3. RESULTS

Pulmonary arterial segments from rat and human contracted in a dose-dependent manner to the thromboxane mimic U46619 and to LTC_4 (Table I).

There was no significant difference in the maximum tension generated to U46619 between pulmonary arteries (PA) of the two species. In addition, the EC_{50} values (10^{-8} M) were not significantly different, i.e., 2.9 ± 1.8 and 2.5 ± 0.5 for human and rat, respectively.

The rat PA and human PA produced similar contractile responses to LTC_4, but the rat PA was significantly more sensitive (Table I). The EC_{50} values (10^{-8} M) for LTC_4 in the PA of the rat and human were 1.7 ± 0.4 and 3.6 ± 1.6, respectively.

The pulmonary vein (PV) of the rat did not respond to U46619, and the human PV generated less force than the human PA. In addition, the rat PV was unresponsive to LTC_4, and the human PV generated less tension with this agonist than the human PA (data not shown).

The force of contraction generated to norepinephrine (NE) by segments of rat PA was similar to that produced with U46619, but U46619 was less potent. The EC_{50} values (10^{-8} M) were 5.1 ± 1.4 for NE compared to 2.5 ± 0.5 for U46619 (Table I).

Indomethacin (5.5 μM) did not significantly alter the contractile response to either U46619 or LTC_4 in rat or human vessels (Fig. 1). The SRS-A receptor antagonist FPL-55712 (4 μM) inhibited the response to LTC_4 in the human pulmonary arterial segments (Fig. 2).

Segments of human pulmonary arteries and veins released 6-keto-$PGF_{1\alpha}$ and TxB_2 (Table II). There was no significant difference between arterial and venous segments in the basal release of either cyclooxygenase product. In addition, arachidonic acid (10^{-5} M) markedly enhanced the production of 6-keto-$PGF_{1\alpha}$ and TxB_2 in the arterial and venous preparations. Leukotriene C_4 significantly increased, by threefold, 6-keto-$PGF_{1\alpha}$ release from arterial segments; LTC_4 was ineffective in stimulating TxB_2 and had no effect on 6-keto-$PGF_{1\alpha}$ release from venous segments.

Acetylcholine (ACh) produced a dose-dependent relaxation of the human PA

TABLE I. EC_{50} Values and Maximum Developed Tension (T_{max}) for U46619, LTC_4, and Norepinephrine (NE) by Pulmonary Artery Rings from Human and Rat[a]

		U46619	LTC_4	NE
Human	EC_{50} (10^{-8} M)	2.9 ± 1.8	3.6 ± 1.6	—
	T_{max}(g)	1.2 ± 0.3	0.45 ± 0.05	—
		(4)	(4)	
Rat	EC_{50} (10^{-8} M)	2.5 ± 0.5	1.7 ± 0.4	5.1 ± 1.4
	T_{max}(g)	1.8 ± 0.2	0.75 ± 0.05	1.8 ± 0.2
		(8)	(9)	(8)

[a] Values expressed as mean $\pm$ S.E.M.; numbers in parentheses refer to the number of vessels used.

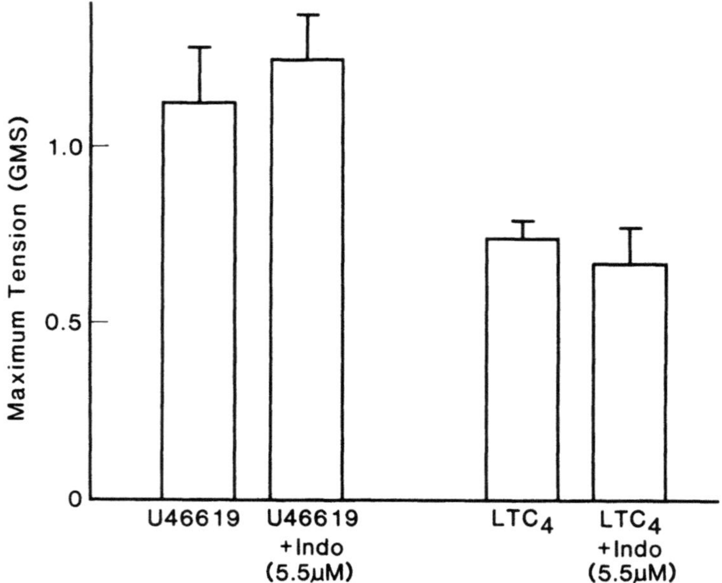

FIGURE 1. The effect of indomethacin on the maximum tension generated in rat pulmonary artery segments with U46619 and LTC$_4$. Data are presented as mean ±S.E.M. of five experiments.

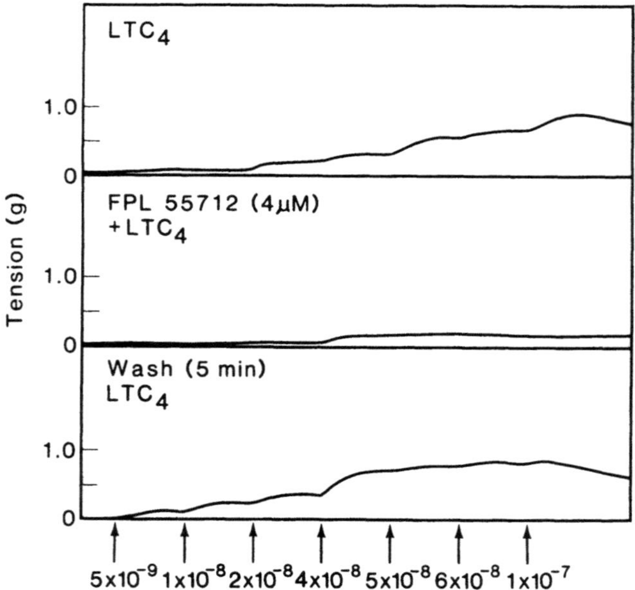

FIGURE 2. Dose–response tracings for LTC$_4$ in the human PA before, during, and after FPL-55712. FPL-55712 was added to the tissue bath 15 min before LTC$_4$ in the second tracing.

TABLE II. Comparison of the Release of Selected Cyclooxygenase Products by Human Pulmonary Artery and Vein Segments[a]

	6-Keto-PGF$_{1\alpha}$ (pg/mg dry tissue)		TxB$_2$ (pg/mg dry tissue)	
	PA	PV	PA	PV
Basal	48 ± 9	58 ± 9	11 ± 2	12 ± 1
Arachidonate (10^{-5} M)	2493 ± 267	2490 ± 108	250 ± 36	223 ± 21
LTC$_4$ (10^{-5} M)	157 ± 3	74 ± 6	21 ± 5	17 ± 11
Arachidonate plus LTC$_4$	3366 ± 54	2580 ± 124	312 ± 7	190 ± 6

[a] Vessel segments were incubated in triplicate at 37°C for 15 min. Data expressed as pg/mg dry tissue and represent mean ± S E M

precontracted with a submaximal dose of LTC$_4$ (Fig. 3A). Rubbing of the intimal surface of the PA abolished this relaxation and instead produced a dose-dependent contractile response. The human PV did not relax to ACh (Fig. 3B). When the venous segment was rubbed, ACh produced a contractile response similar to that seen in the PA.

4. DISCUSSION

Rat and human pulmonary artery preparations were found to contract to both LTC$_4$ and U46619, but the tension generated by the human pulmonary vein to both

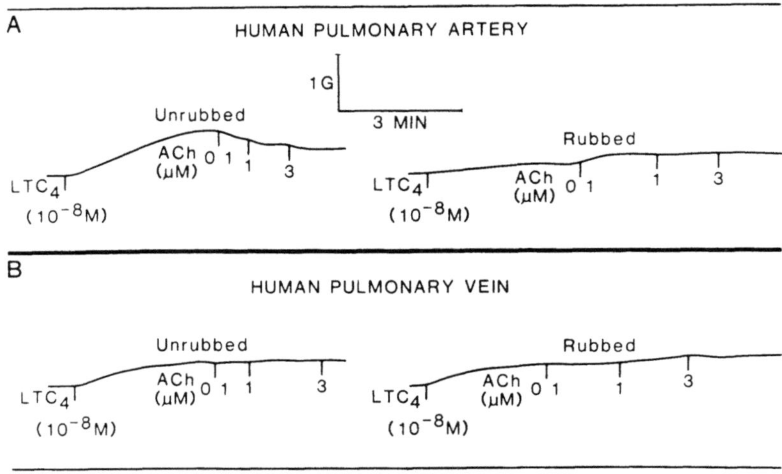

FIGURE 3. The effects of acetylcholine on LTC$_4$-precontracted human pulmonary artery and vein segments before and after intimal rubbing.

LTC$_4$ and U46619 was significantly less than that generated by the human pulmonary artery and, moreover, was not reproducible. Rat pulmonary veins were found to be unresponsive to both agonists.

In contrast to our studies reported here, Schellenberg *et al.* (1984) found spiral strips of human pulmonary vein to be significantly more responsive to LTC$_4$ than human pulmonary artery. This discrepancy may be related to the absence of the endothelium in the spiral preparation. In the present study, the acetylcholine-induced relaxation of human pulmonary arteries suggests that endothelial cells were present in our experiments. The endothelial cells appear to be involved in facilitating the contractile response to many agonists, i.e., norepinephrine and arachidonic acid (DeMey and Vanhoutte, 1982). Thus, the presence of endothelium could explain the increased reactivity observed in intact human pulmonary arteries in our studies in contrast to those of Schellenberg *et al.* (1984).

The lack of effect of indomethacin indicated that cyclooxygenase products did not mediate LTC$_4$- or U46619-induced contractility. This suggests that the contractile response to these two agonists in the pulmonary artery was a direct effect. The fact that FPL-55712 inhibited the LTC$_4$-induced response supports the idea that LTC$_4$ exerts its actions on the pulmonary artery through SRS-A-responsive receptors.

The new observation that human pulmonary arterial and venous segments produce 6-keto-PGF$_{1\alpha}$ and TXB$_2$ is in agreement with data obtained in sheep pulmonary artery slices (McDonald *et al.*, 1983). However, in the sheep study, the ratio of 6-keto-PGF$_{1\alpha}$ to TXB$_2$ was 9 : 1, which is different from the ratio of 3 : 1 observed in the present study. Howev..., the release of substantially more TXB$_2$ from human pulmonary vessels could account for this difference, since there was no significant difference in 6-keto-PGF$_{1\alpha}$ levels between the two species. Leukotriene C$_4$ significantly enhanced the release of prostacyclin but not thromboxane from segments of human pulmonary arteries. These results are in accord with a recent study that demonstrated that LTC$_4$ promotes prostacyclin synthesis by endothelial cells from human umbilical vein (Cramer *et al.*, 1983).

Acetylcholine releases a factor from endothelial cells that relaxes vascular smooth muscle (Furchgott and Zawadzki, 1980; Furchgott, 1983; Griffith *et al.*, 1984). The relaxation observed with acetylcholine in LTC$_4$-precontracted human pulmonary artery but not in the pulmonary vein implies that the artery has an endothelium-dependent relaxant factor (EDRF) whereas the vein does not. This is in accord with results obtained with canine pulmonary veins in which acetylcholine produced no relaxation but only contraction whether endothelial cells were present or absent (DeMey and Vanhoutte, 1982). These data are of interest in view of the suggestion that the reduction of pulmonary artery pressure by acetylcholine in pulmonary hypertension is mediated by a direct dilatory effect on the pulmonary arterial tree (Chand and Altura, 1981).

In conclusion, the present study indicates that there are similarities in the responsiveness of isolated pulmonary vascular segments obtained from rodents and humans. These results may be important given that the rat is frequently used as a model for studying experimental pulmonary hypertension.

ACKNOWLEDGMENT. Y. T. Maddox was supported by a fellowship from the American Heart Association, Nation's Capital Affiliate.

REFERENCES

Chand, N., and Altura, B. M., 1981, Endothelial cells and relaxation of vascular smooth muscle cells: Possible relevance to lipoxygenases and their significance in vascular diseases, *Microcirculation* 1:297–317.

Cramer, E. B., Pologe, L., Pawlowski, N. A., Cohn, A. A., and Scott, W. A., 1983, Leukotriene C promotes prostacyclin synthesis by human endothelial cells, *Proc. Natl. Acad. Sci. U.S.A.* 80:4109–4113.

De Mey, J. G., and Vanhoutte, P. M., 1982, Heterogeneous behavior of the canine arterial and venous wall: Importance of the endothelium, *Circ. Res.* 51:439–447.

Furchgott, R. F., 1983, Role of endothelium in responses of vascular smooth muscle, *Circ. Res.* 53:557–573.

Furchgott, R. F., and Zawadzki, J. V., 1980, The obligatory role of endothelial cells in the relaxation of arterial smooth muscle by acetylcholine, *Nature* 288:373–376.

Griffith, T. M., Edwards, D. H., Lewis, M. J., Newby, A. C., and Henderson, A. H., 1984, The nature of the endothelium-derived vascular relaxant factor, *Nature* 308:645–647.

Hanna, C. J., Bach, M. K., Pare, P. D., and Schellenberg, R. R., 1981, Slow-reacting substances (leukotrienes) contract human airways and pulmonary vascular smooth muscle *in vitro*, *Nature* 290:343–344.

Iacopino, V. J., Fitzpatrick, T. M., Ramwell, P. W., Rose, J. C., and Kot, P. A., 1983, Cardiovascular responses to leukotrienes in the rat, *J. Pharmacol. Exp. Ther.* 227:244–247.

Iacopino, V. J., Compton, S. K., Fitzpatrick, T. M., Ramwell, P. W., Rose, J. C., and Kot, P. A., 1984, Responses to leukotriene C_4 in perfused rat lung, *J. Pharmacol. Exp. Ther.* 229:654–657.

McDonald, J. W., Ali, M., Morgan, E., Towsend, E. R., and Cooper, J. D., 1983, Thromboxane synthesis by sources other than platelets in association with complement-induced pulmonary leukostasis and pulmonary hypertension in sheep, *Circ. Res.* 52:1–6.

Pomerantz, K., Maddox, Y., Maggi, F., Ramey, E., and Ramwell, P., 1980, Sex and hormonal modification of 6-keto-$PGF_{1\alpha}$ release by rat aorta, *Life Sci.* 27:1233–1236.

Schellenberg, R. R., and Foster, A., 1984, Differential activity of leukotrienes upon human pulmonary vein and artery, *Prostaglandins* 27:475–482.

Smith, B. J., Ogletree, M. L., Lefer, A. M., and Nicolaou, K. C., 1978, Antibodies which antagonise the effects of prostacyclin, *Nature* 274:64–65

27

Leukotriene C₄ Is Released from the Anaphylactic Heart: A Case for Its Direct Negative Inotropic Effect

ROBERTO LEVI, YUICHI HATTORI,
JAMES A. BURKE, ZHAO-GUI GUO,
UGHETTA HACHFELD del BALZO,
WILLIAM A. SCOTT, and CAROL A. ROUZER

1. INTRODUCTION

Data from our laboratory indicate that the heart reacts as a target organ in systemic hypersensitivity reactions (Capurro and Levi, 1975; Graver *et al.*, 1983). Cardiac dysfunction observed during anaphylaxis in the guinea pig (Capurro and Levi, 1975; Zavecz and Levi, 1977) resembles that reported in humans (Bernreiter, 1959; Both and Patterson, 1970; Criep and Woehler, 1971; Petsas and Kotler, 1973; Sullivan, 1982) and is caused by mediators released intracardially and reaching the heart from the lung (Zavecz and Levi, 1977). "Cardiac anaphylaxis" (Feigen and Prager, 1969) is characterized by tachycardia, arrhythmias, contractile failure, coronary constriction, and mediator release (Capurro and Levi, 1975; Levi and Allan, 1980; Levi

ROBERTO LEVI, YUICHI HATTORI, JAMES A. BURKE, ZHAO-GUI GUO, UGHETTA HACHFELD del BALZO, WILLIAM A. SCOTT, AND CAROL A. ROUZER ● Department of Pharmacology, Cornell University Medical College and the Rockefeller University, New York, New York 10021.   Y. H. is on leave from Department of Pharmacology. Hokkaido University School of Medicine, Sapporo, Japan.   Z.-G. G. is on leave from Hunan Medical College, Changsha, Hunan, People's Republic of China.   Present address of W. A. S.: Squibb Institute for Medical Research, Princeton, New Jersey 08540.   Present address of C. A. R.: Department of Physiological Chemistry II, Karolinska Institutet, S-104 01 Stockholm, Sweden.

et al., 1982). Tachycardia and arrhythmias are caused by the release of endogenous cardiac histamine, since they are reproduced by the intracardiac administration of histamine and abolished by antihistamines (Levi and Allan, 1980; Levi *et al.*, 1982). On the other hand, anaphylactic coronary constriction is markedly reduced by cyclooxygenase inhibitors such as indomethacin or aspirin or by thromboxane synthetase inhibitors such as 1-(2-isopropylphenyl)imidazole (Allan and Levi, 1981). Furthermore, the intracardiac administration of U 46619, a stable thromboxane analogue, causes coronary constriction (Allan and Levi, 1980a). Thus, prostanoate compounds, particularly thromboxane, contribute to the fall in coronary flow rate that characterizes cardiac anaphylaxis (Levi *et al.*, 1982). Other potent coronary-constricting agents, such as platelet-activating factor (PAF, AGEPC), are also likely to contribute to anaphylactic coronary constriction (Levi *et al.*, 1984).

2. CARDIAC ANAPHYLAXIS, CONTRACTILE FAILURE, AND ITS INHIBITION BY THE END-ORGAN SRS ANTAGONIST FPL 55712

In experiments designed to identify the mediator(s) responsible for the contractile failure of anaphylaxis, we first investigated the role of histamine. Since the inotropic response to histamine is the resultant of two components, positive (H_2-mediated) and negative (H_1-mediated) (Zavecz and Levi, 1978; Guo *et al.*, 1984), we assessed whether H_1 blockers antagonize the negative inotropic effect of anaphylaxis. However, in the presence of chlorpheniramine, anaphylaxis caused an even greater decrease in contractility (Fig. 1), an indication that histamine is not responsible for the negative inotropic symptom of cardiac anaphylaxis. Chlorpheniramine did not directly decrease myocardial contractility, nor has chlorpheniramine ever been shown to induce the release of negative inotropic mediators. Therefore, the further decline in contractility in the presence of chlorpheniramine is probably a consequence of the faster ventricular rate at which the anaphylactic heart beats when treated with an H_1 blocker. Indeed, chlorpheniramine antagonizes the histamine-induced, H_1-mediated atrioventricular conduction block, a characteristic feature of cardiac anaphylaxis, but not the concomitant H_2-mediated sinus tachycardia (Levi and Capurro, 1973). Thus, ventricular rate is much faster in the presence than in the absence of chlorpheniramine. This causes a greater shortening of diastolic time and, in turn, a more pronounced decline in contractility.

Various cyclooxygenase products have a negative inotropic effect on the guinea pig heart (Allan and Levi, 1980a). Among these, PGD_2 and thromboxane are released during cardiac anaphylaxis (Anhut *et al.*, 1978; Allan and Levi, 1981). We therefore investigated whether indomethacin would antagonize the anaphylactic contractile failure (Levi *et al.*, 1976; Allan and Levi, 1980b, 1981). As shown in Fig. 1, indomethacin not only failed to antagonize the negative inotropic effect of anaphylaxis but actually potentiated it and prolonged it. This indicates that cyclooxygenase products cannot be the cause of the decrease in ventricular contractility. Moreover, since indomethacin did not increase ventricular rate during ana-

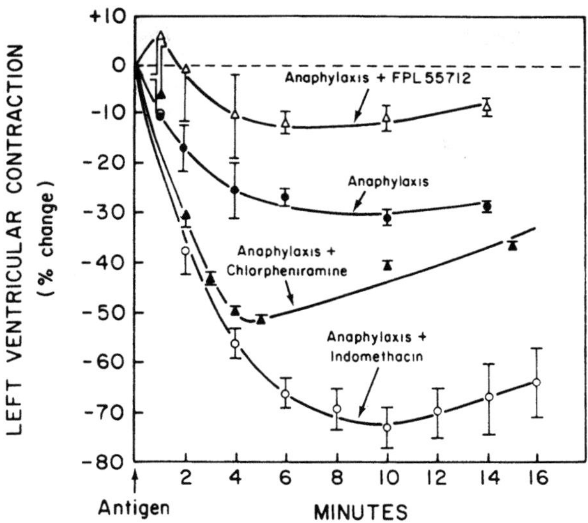

FIGURE 1. Anaphylaxis in the isolated guinea pig heart. Time course of changes in left ventricular contractile force following antigen challenge either in the absence (anaphylaxis) or in the presence of one of the drugs indicated. Concentrations of the drugs were: FPL 55712, 0.48 μM; chlorpheniramine, 1 μM; indomethacin, 2.8 μM. The hearts, obtained from guinea pigs passively sensitized with homologous anti-DNP-bovine γ-globulin, were challenged with DNP-bovine serum albumin (for method of sensitization see Levi *et al.*, 1984). Points are means (±S.E.M.; N = 4–6).

phylaxis, the further decline in contractility in the presence of indomethacin may be attributable to an enhanced release of negative inotropic mediators. One such mediator could be slow-reacting substance of anaphylaxis (SRS-A). Indeed, indomethacin has been reported to enhance the release of SRS-A from the anaphylactic heart (Liebig *et al.*, 1975).

To assess the contribution of SRS-A to the contractile failure of cardiac anaphylaxis, we used the specific anti-SRS compound FPL 55712 (Augstein *et al.*, 1973). As shown in Fig. 1, FPL 55712 markedly reduced and abbreviated the negative inotropic effect of anaphylaxis (Burke and Levi, 1980; Levi *et al.*, 1982; Ezeamuzie and Assem, 1983), an indication that SRS-A, known to be released in large amounts from the anaphylactic heart (Liebig *et al.*, 1975; Levi and Burke, 1980), may be a major cause of contractile failure.

3. THE NEGATIVE INOTROPIC AND CORONARY VASOCONSTRICTING EFFECTS OF PARTIALLY PURIFIED SRS-A AND HPLC-PURIFIED SRS

We sought confirmation of the putative role of SRS-A as the mediator of contractile failure by investigating the effects of SRS-A directly administered into the heart. As shown in Fig. 2, partially purified histamine-free SRS-A obtained from the guinea pig lung or HPLC-purified SRS from mouse peritoneal macrophages (Rouzer *et al.*, 1980) caused a dose-dependent negative inotropic efect that was antagonized by the SRS end-organ antagonist FPL 55712 (Burke and Levi, 1980; Levi *et al.*, 1982).

Thus, the facts that: (1) indomethacin worsens the anaphylactic contractile failure as it enhances SRS-A release, (2) the anti-SRS compound FPL 55712 greatly diminishes anaphylactic contractile failure, and (3) the intracardiac administration of SRS causes a dose-dependent negative inotropic effect that is antagonized by FPL 55712, indicate that SRS-A is an important mediator of anaphylactic cardiac failure.

With the discovery of the molecular structure of SRS-A and the identification of leukotrienes C_4, D_4, and E_4 as its components (Murphy *et al.*, 1979; Lewis *et al.*, 1980; Morris *et al.*, 1980; Samuelsson *et al.*, 1980), we next determined, in collaboration with E. J. Corey, that synthetic LTC_4, D_4, and E_4 have direct negative

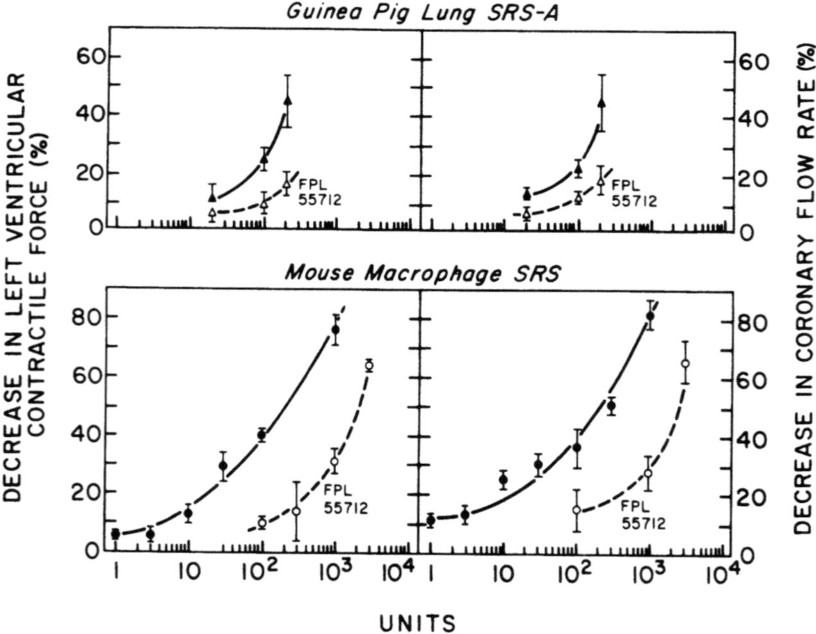

FIGURE 2. Dose–response curves for the negative inotropic and coronary vasoconstricting effects of partially purified guinea pig lung SRS-A (top panels) and of HPLC-purified mouse peritoneal macrophage SRS (lower panels) on isolated guinea pig hearts perfused at constant pressure (40 cm H_2O) in a Langendorff apparatus, either in the presence or in the absence of FPL 55712 (0.48 μM). The SRS-A was obtained from guinea pig lungs immunologically challenged in the presence of indomethacin (2.8 μM), followed by ethanol extraction, desalting, and histamine separation by Amberlite XAD-7 column chromatography (Levi *et al.*, 1982). The SRS was derived from macrophages challenged *in vitro* with unopsonized zymosan and purified by extraction, Sephadex G25 column chromatography, and HPLC in two different solvent systems (Rouzer *et al.*, 1980). Points are means (± S.E.M.; *N* = 3–4).

inotropic effects on guinea pig heart and human myocardium *in vitro* (Burke *et al.*, 1981, 1982). These findings explained and confirmed our previous results with SRS-A.

4. IDENTIFICATION OF LTC$_4$ AS THE PRIMARY LEUKOTRIENE RELEASED DURING CARDIAC ANAPHYLAXIS

We next turned our attention to the identification of leukotrienes released from the anaphylactic guinea pig heart. By HPLC analysis of the coronary effluents during cardiac anaphylaxis, we determined that LTC$_4$ is the primary SRS-A leukotriene released from the heart; LTD$_4$ and E$_4$ were not detected (Fig. 3). By bioassay, approximately 400 pmol of LTC$_4$ were found to be released per gram of heart. This value is somewhat higher than that measured by radioimmunoassay by Aehringhaus *et al.* (1983), probably reflecting a different degree of cardiac sensitization. We used a passive sensitization technique with a known concentration of guinea pig anti-DNP antibodies, whereas Aehringhaus and colleagues used active sensitization with ovalbumin. Recently, Aehringhaus *et al.* (1984) have reported that indomethacin enhances the release of LTC$_4$-like immunoreactivity from the anaphylactic heart, confirming their earlier finding with SRS-A (Aehringhaus *et al.*, 1983). Therefore, it is most probable that the indomethacin-induced potentiation of the negative inotropic effect of anaphylaxis results from the enhanced release of LTC$_4$. This in turn supports the view that LTC$_4$ is a major contributor to anaphylactic contractile failure.

5. THE NEGATIVE INOTROPIC AND CORONARY VASOCONSTRICTING EFFECTS OF SYNTHETIC LTC$_4$

Since the administration of either SRS (Fig. 2) or leukotrienes (Burke *et al.*, 1982) into the Langendorff heart perfused at constant pressure caused a decrease in both contractility and coronary flow, we evaluated the possibility that the decrease in contractility might be secondary to the decrease in coronary flow. For this, we studied the effects of LTC$_4$ in the isolated guinea pig heart perfused at constant flow. As shown in Fig. 4, LTC$_4$ decreases contractility independently of coronary flow changes, since the flow was maintained constant. On the other hand, the increase in perfusion pressure indicates that, as a function of dose, LTC$_4$ progressively constricts the coronary vessels (Fig. 4).

6. COMPARISON OF THE INOTROPIC EFFECTS OF LTC$_4$, PAF, PGD$_2$

We next took into consideration the remaining possibility, i.e., that LTC$_4$-induced coronary constriction may cause regional shunting and ischemia and that

  ROBERTO LEVI *et al.*

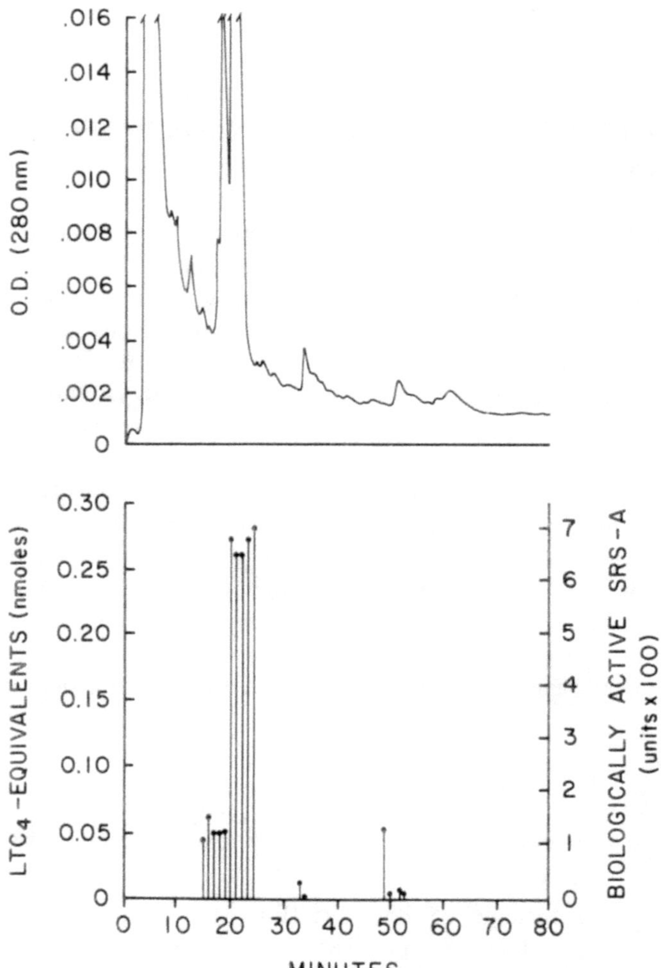

FIGURE 3. Elution profiles on HPLC of coronary effluents from anaphylactic guinea pig hearts assessed by UV absorption (upper panel) and bioassay (lower panel). Pooled coronary effluents from four isolated guinea pig hearts undergoing anaphylaxis *in vitro* were extracted with chloroform after the addition of ethanol and acidification. The extract was dried, partially purified by silicic acid chromatography, and subjected to RP-HPLC using C_{18} columns. Elution was done with a mixture of methanol/water/acetic acid (65/35/1) at pH 5.4 at a flow rate of 1 ml/min. The UV absorbance was monitored at 280 nm. Retention times of leukotriene standards were: LTC_4, 18–24 min; LTD_4, 34–35 min and LTE_4, 46–47 min. Eluates were dried and resuspended in Tyrode's solution prior to bioassay on the isolated guinea pig ileum in the presence of pyrilamine (1 μM) and atropine (1 μM). FPL 55712 (0.48 μM) completely abolished the response of the ileum to the HPLC eluates. Cardiac sensitization as in legend to Fig. 1.

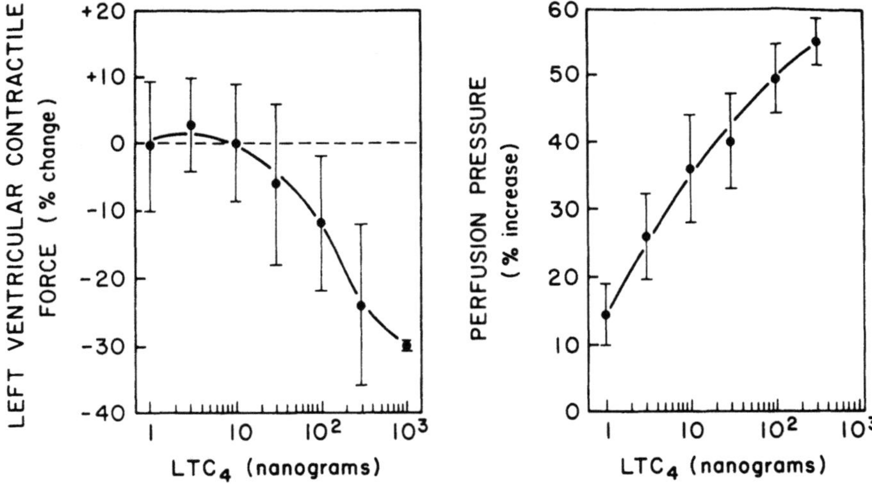

FIGURE 4. Dose–response curves for the negative inotropic and coronary vasoconstricting effects of LTC$_4$ on isolated guinea pig hearts perfused in a Langendorff apparatus at constant flow. Flow rate was adjusted to give a perfusion pressure of 40 mm Hg. The LTC$_4$ was administered by intracoronary bolus injections at the doses indicated. Points are means ($\pm$ S.E.M.; $N = 3$).

contractile failure may in part result from it. To test this, we used isolated non-coronary-perfused cardiac preparations from the guinea pig such as the electrically paced left atrium and right ventricular papillary muscle (Burke *et al.*, 1982). As shown in Fig. 5, LTC$_4$ caused a dose-dependent decrease in contractile force in both of these preparations, demonstrating beyond any doubt that LTC$_4$, like LTD$_4$ (Burke *et al.*, 1982; Hattori and Levi, 1984), has a primary negative inotropic effect. Further, this effect is antagonized by FPL 55712 but not by indomethacin (Burke *et al.*, 1982; Hattori and Levi, 1984).

As shown in Fig. 5, a linear correlation exists between the decrease in left ventricular contractile force and the decrease in coronary flow caused by LTC$_4$ in the isolated guinea pig heart. Further, it is apparent that the decrease in contractility in the left atrium and papillary muscle is about 20% smaller than that in the whole heart. Thus, although LTC$_4$ undoubtedly has a direct negative inotropic effect, in the entire heart this may be amplified by local ischemia resulting from the concomitant coronary-constricting effect of LTC$_4$. As seen in Fig. 5, this reasoning probably applies also to PAF, another mediator of cardiac anaphylaxis (Levi *et al.*, 1984). In the guinea pig, PAF has cardiac effects similar to those of leukotrienes but indepedent of them (Levi *et al.*, 1984). Indeed, PAF has potent negative inotropic effects in animals (Bessin *et al.*, 1983; Feuerstein *et al.*, 1984; Levi *et al.*, 1984; Vemulapalli *et al.*, 1984) and humans (Gateau *et al.*, 1984). Again, as for LTC$_4$, it is possible that local ischemia resulting from coronary vasoconstriction may magnify the primary negative inotropic effect of PAF. In view of the obser-

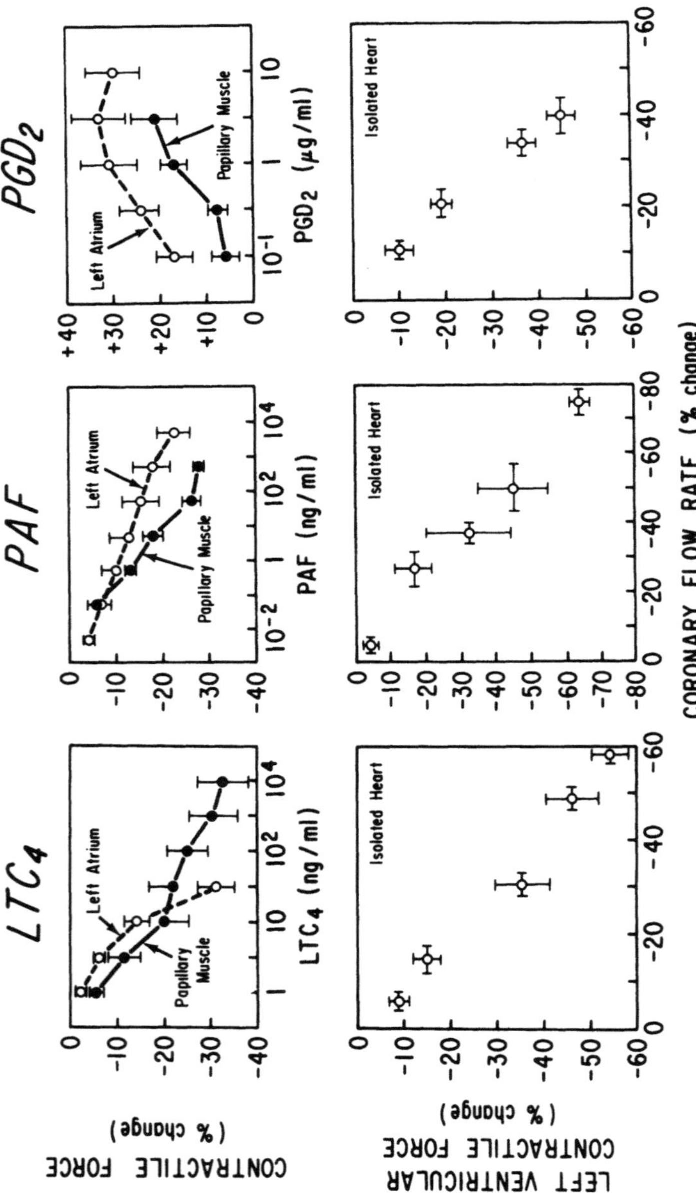

FIGURE 5. Upper panels: Dose–response curves for the inotropic effect of LTC₄, PAF, and PGD₂ on the electrically paced guinea pig left atrium and right ventricular papillary muscle. Lower panels: Relationships between the negative inotropic and coronary vasoconstricting effects of LTC₄, PAF, and PGD₂ on isolated guinea pig hearts perfused at constant pressure. In the lower panels, the three mediators were administered by bolus intracoronary injections at increasing doses of 0.3–30 ng (LTC₄), 5 pg–1.5 μg (PAF), and 0.01–10 μg (PGD₂). Maximum decreases in contractility (ordinates) are plotted against corresponding maximum decreases in coronary flow (abscissae). Points are means (±S.E.M.; N = 5–6) (Burke *et al.*, 1982; Levi *et al.*, 1984; Levi and Hattori, 1985).

vations with LTC$_4$ and PAF, one could speculate that other anaphylactic mediators, such as PGD$_2$, that have both negative inotropic and coronary-constricting effects in the whole heart (Anhut *et al.*, 1978; Allan and Levi, 1980a) may also have a direct negative inotropic effect in non-coronary-perfused myocardial preparations. As shown in Fig. 5, however, this is not the case: although PGD$_2$ decreases ventricular contractility *pari passu* with coronary flow in the whole heart, PGD$_2$ increases contractility in the non-coronary-perfused left atrium and papillary muscle. Furthermore, whereas indomethacin does not antagonize the negative inotropic effect of LTC$_4$ and PAF in the isolated heart (Burke *et al.*, 1982; Levi *et al.*, 1984), it antagonizes the coronary vasoconstricting effect of PGD$_2$ and transforms the decrease in ventricular contractility caused by PGD$_2$ into an increase (Levi and Hattori, 1985). This suggests that the negative inotropic effect of PGD$_2$ on the whole heart may be secondary to the coronary vasoconstriction caused by a cyclooxygenase product and that this coronary constriction conceals the primary positive inotropic action of PGD$_2$. On the contrary, as already discussed, the negative inotropic effects of LTC$_4$ and PAF persist after cyclooxygenase inhibition and are present in isolated non-coronary-perfused preparations (Burke *et al.*, 1982; Levi *et al.*, 1984). Thus, our data clearly indicate that LTC$_4$ has a primary, direct negative inotropic effect and that PAF is another mediator with similar, but independent, effects on the myocardium.

7. POSSIBLE MECHANISMS OF THE NEGATIVE INOTROPIC EFFECT OF LEUKOTRIENES

We have recently initiated research designed to understand the mechanism of the negative inotropic effect of leukotrienes. Since Ca^{2+} movements play a major role in excitation–contraction coupling, we have hypothesized that leukotrienes may interfere with such movements. Indeed, we have found that the negative inotropic effect of LTD$_4$ is attenuated at higher extracellular Ca^{2+} concentrations (Hattori and Levi, 1984). Therefore, we have assessed the effect of leukotrienes under conditions in which myocardial contractility is maintained exclusively by slow inward Ca^{2+} current (Hattori and Levi, 1984). One such condition (Houki, 1973; Shigenobu *et al.*, 1974) is shown in Fig. 6 (panel B), i.e., histamine-restored contractility in papillary muscles rendered inexcitable by a high extracellular potassium concentration. From comparison of panels A and B in Fig. 6, it is evident that the negative inotropic effect of LTC$_4$ is much greater when contractility depends exclusively on the slow inward Ca^{2+} current (panel B, maximum 80% decrease in contractility) than in control conditions (panel A, maximum 30% decrease). Notably, the effects of leukotrienes on Ca^{2+}-dependent contractility are reversed by increasing extracellular Ca^{2+} (Hattori and Levi, 1984) and are antagonized by compound FPL 55712 (Fig. 6). Thus, our results are compatible with the possibility that the cardiodepressant effect of leukotrienes may be attributable to an inhibition of transsarcolemmal Ca^{2+} influx. Specific LTD$_4$ binding sites in membranes pre-

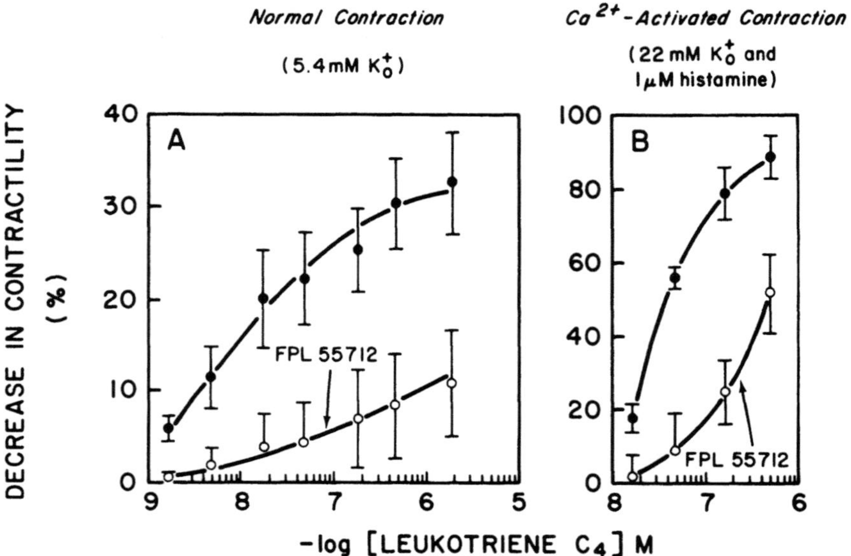

FIGURE 6. Concentration–response curves for the negative inotropic effect of LTC$_4$ in guinea pig papillary muscles electrically paced at 2 Hz in normal Tyrode's solution (normal contraction, A) and in potassium-depolarized papillary muscles electrically paced at 0.2 Hz whose contractility had been restored by histamine (i.e., Ca^{2+}-activated contraction, B). Points are means (±S.E.M.; N = 4–6) (Hattori and Levi, 1984).

pared from guinea pig ventricular myocardium have been identified by Hogaboom *et al.* (1983), who have also suggested that the binding occurs primarily in the cardiac sarcolemma. Thus, it is possible that leukotrienes may bind to specific myocardial receptors and reduce transsarcolemmal Ca^{2+} influx (Hattori and Levi, 1984).

8. SUMMARY AND CONCLUSIONS

In all of the animal species in which the cardiovascular effects of LTC$_4$ and D$_4$ have been studied, these eicosanoids have been shown to cause profound functional alterations. Significant decreases in coronary flow (Burke *et al.*, 1982; Panzenbeck and Kaley, 1983; Michelassi *et al.*, 1983; Woodman and Dusting, 1983; Ezra *et al.*, 1983), left ventricular contractility (Burke *et al.*, 1982; Panzenbeck and Kaley, 1983; Michelassi *et al.*, 1983), and cardiac output (Pfeffer *et al.*, 1983; Panzenbeck and Kaley, 1983; Smedegard *et al.*, 1982) as well as prolonged hypotension (Smedegard *et al.*, 1982) have been reported. Although the potent coronary vasoconstricting effect of LTC$_4$ and LTD$_4$ may contribute to the overall decrease in contractility, the decrease in left ventricular systolic shortening caused by LTC$_4$ and D$_4$ in the whole animal is far greater than that expected on the basis of the reduction in coronary flow (Michelassi *et al.*, 1983). Moreover, although in the isolated heart the decreases in coronary flow and contractility caused by LTC_4 are linearly correlated, a negative inotropic effect of leukotrienes can be clearly demonstrated in the non-coronary-perfused left atrium and right ventricular papillary muscle preparations. Thus, leukotrienes have direct negative inotropic effects (Burke *et al.*, 1982; Hattori and Levi, 1984). The fact that the decrease in contractility in the left atrium and papillary muscle is smaller than that in the whole heart (Burke *et al.*, 1982; Hattori and Levi, 1984) suggests that local ischemia caused by coronary constriction (Ezra *et al.*, 1983) may amplify the direct negative inotropic effect of leukotrienes. Furthermore, the cardiodepressant effect of leukotrienes is not mediated by the release of vasoactive cyclooxygenase products (Burke *et al.*, 1982; Michelassi *et al.*, 1982; Panzenbeck and Kaley, 1983; Hattori and Levi, 1984).

Possible mechanisms of the direct negative inotropic effect of leukotrienes may include an interference with the Ca^{2+} movements that underlie excitation–contraction coupling. Indeed, our recent evidence (Hattori and Levi, 1984) indicates that the negative inotropic effect of LTD_4 in the papillary muscle decreases with increasing extracellular Ca^{2+} concentrations. Furthermore, LTC_4 and LTD_4 inhibit the restoration of contractility induced by isoproterenol or histamine in guinea pig papillary muscle depolarized by high K^+ (Hattori and Levi, 1984). Since (1) this restoration results from the activation of a slow inward Ca^{2+} current (Thyrum, 1974; Inui and Imamura, 1976: Tritthart *et al.*, 1976), (2) LTC_4 antagonizes this Ca^{2+}-activated contraction much more than normal contraction, and (3) this effect of LTC_4 is reversed by increasing extracellular Ca^{2+} concentrations, it is plausible that the negative inotropic effect of leukotrienes is related to an inhibition of trans-sarcolemmal Ca^{2+} influx.

ACKNOWLEDGMENTS. The LTC_4 used in the experiments depicted in Fig. 4 and part of Fig. 5 was a gift of Professor E. J. Corey, Harvard University. The LTC_4 used in the other part of Fig. 5 and in Fig. 6 was a gift of Dr. J. Rokach, Merck-Frosst Laboratories. Dr. Arleen B. Rifkind critically reviewed this chapter, and Ms. Claudia B. Gross provided excellent secretarial assistance. The studies presented here were supported by N.I.H. grants GM 20091 and HL 18828. Dr. Guo was supported in part by the China Medical Board, Inc.

REFERENCES

Aehringhaus, U., Peskar, B. A., Wittenberg, H. R., and Wolbling, R. H., 1983, Effect of inhibition of synthesis and receptor antagonism of SRS-A in cardiac anaphylaxis, *Br. J. Pharmacol.* **80**:73–80.

Aehringhaus, U., Dembinska-Kiéc, A., and Peskar, B. A., 1984, Effects of exogenous prostaglandins on the release of leukotriene C₄-like immunoreactivity and on coronary flow in indomethacin-treated anaphylactic guinea-pig hearts, *Naunyn Schmiedebergs Arch. Pharmacol.* **326**:368–374.

Allan, G., and Levi, R., 1980a, The cardiac effects of prostaglandins and their modification by the prostaglandin antagonist N-0164, *J. Pharmacol. Exp. Ther.* **214**:45–49.

Allan, G., and Levi, R., 1980b, Pharmacological studies on the role of prostaglandins in cardiac hypersensitivity reactions, in: *Prostaglandins in Cardiovascular and Renal Function* (A. Scriabine, A. M. Lefer, and F. A. Keuhl, Jr., eds.), Spectrum, New York, pp. 223–237.

Allan, G., and Levi, R., 1981, Thromboxane and prostacyclin release during cardiac immediate hypersensitivity reactions *in vitro, J. Pharmacol. Exp. Ther.* **217:**157–161.

Anhut, H., Peskar, B. A., and Bernauer, W., 1978, Release of 15-keto-13,14-dihydro-thromboxane B_2 and prostaglandin D_2 during anaphylaxis as measured by radioimmunoassay, *Naunyn Schmiedebergs Arch. Pharmacol.* **305:**247–252.

Augstein, J., Farmer, J. B., Lee, T. B., Sheard, P., and Tattersall, M. L., 1973, Selective inhibitor of slow reacting substance of anaphylaxis, *Nature* **245:**216–217.

Bernreiter, M., 1959, Electrocardiogram of patient in anaphylactic shock, *J.A.M.A.* **170:**1628–1630.

Bessin, P., Bonnet, J., Apffel, D., Soulard, C., Desgroux, L., Pelas, I., and Benveniste, J., 1983, Acute circulatory collapse caused by platelet-activating factor (PAF-Aether) in dogs, *Eur. J. Pharmacol.* **86:**403–413.

Booth, B. H., and Patterson, R., 1970, Electrocardiographic changes during human anaphylaxis, *J.A.M.A.* **211:**627–631.

Burke, J.A., and Levi, R., 1980, Slow reacting substance of anaphylaxis (SRS-A): Direct and indirect cardiac effects, *Fed. Proc.* **39:**389.

Burke, J. A., Levi, R., and Corey, E. J., 1981, Cardiovascular effects of pure synthetic leukotrienes C and D, *Fed. Proc.* **40:**1015.

Burke, J. A., Levi, R., Guo, Z.-G., and Corey, E. J., 1981, Leukotrienes C_4, D_4 and E_4: Effects on human and guinea-pig cardiac preparations *in vitro. J. Pharmacol. Exp. Ther.* **221:**235–241.

Capurro, N., and Levi, R., 1975, The heart as a target organ in systemic allergic reactions, *Circ. Res.* **36:**520–528.

Criep, L. H., and Woehler, T. R., 1971, The heart in human anaphylaxis, *Ann. Allergy* **29:**399–409.

Ezeamuzie, I. C., and Assem, E. S. K., 1983, Effects of leukotrienes C_4 and D_4 on guinea-pig heart and the participation of SRS-A in the manifestations of guinea-pig cardiac anaphylaxis, *Agents Actions* **13:**182–187.

Ezra, D., Boyd, L. M., Feuerstein, G., and Goldstein, R. E., 1983, Coronary constriction by leukotriene C_4, D_4, and E_4 in the intact pig heart, *Am. J. Cardiol.* **51:**1451–1454.

Feigen, G. A., and Prager, D. J., 1969, Experimental cardiac anaphylaxis: physiologic, pharmacologic and biochemical aspects of immune reactions in the isolated heart, *Am. J. Cardiol.* **24:**474–491.

Feuerstein, G., Boyd, L. M., Ezra, D., and Goldstein, R. E., 1984, Effect of platelet-activating factor on coronary circulation of the domestic pig, *Am. J. Physiol.* **246:**H466–H471.

Gateau, O., Arnoux, B., Deriaz, H., Viars, P., and Benveniste, J., 1984, Acute effects of intratracheal administration of PAF-acether (platelet-activating factor) in humans, *Am. Rev. Respir. Dis.* **129:**A3.

Graver, L. M., Levi, R., Chenouda, A. A., Becker, C. G., and Gay, W. A., 1983, IgE-mediated hypersensitivity in human heart tissue, *Circulation* **68**(III):322.

Guo, Z.-G., Levi, R., Graver, L. M., Robertson, D. A., and Gay, Jr., W. A., 1984, Inotropic effects of histamine in human myocardium: Differentiation between positive and negative components, *J. Cardiovasc. Pharmacol.* **6:**1210–1215.

Hattori, Y., and Levi, R., 1984, Negative inotropic effect of leukotrienes: Leukotrienes C_4 and D_4 inhibit calcium-dependent contractile responses in potassium-depolarized guinea-pig myocardium, *J. Pharmacol. Exp. Ther.* **230:**646–651.

Hogaboom, G. K., Mong, S., Clark, M., and Crooke, S. T., 1983, Identification of specific leukotriene-D_4 binding sites in guinea-pig heart, *Pharmacologist* **25:**201.

Houki, S., 1973, Restoration effects of histamine on action potential in potassium-depolarized guinea-pig papillary muscle, *Arch. Int. Pharmacodyn. Ther.* **206:**113–120.

Inui, J., and Imamura, H., 1976, Restoration by histamine of the calcium-dependent electrical and mechanical response in the guinea-pig papillary muscle partially depolarized by potassium, *Naunyn Schmiedebergs Arch. Pharmacol.* **294:**261–269.

Levi, R., and Allan, G., 1980, Histamine-mediated cardiac effects, in: *Drug-Induced Heart Disease* (M. Bristow, ed.), Elsevier/North-Holland Biomedical Press, Amsterdam, pp. 377–395.

Levi, R., and Burke, J. A., 1980, Cardiac anaphylaxis: SRS-A potentiates and extends the effects of released histamine, *Eur. J. Phamacol.* **62**:41–49.

Levi, R., and Capurro, N., 1973, Histamine H_2-receptor antagonism and cardiac anaphylaxis, in: *Proceedings International Symposium on Histamine H_2-Receptor Antagonists* (C. J. Wood and M. A. Simkins, eds.), S.K.&F., London, pp. 175–184.

Levi, R., and Hattori, Y., 1985, Prostaglandin D_2 and cardiac contractility: a negative inotropism secondary to coronary vasoconstriction conceals a primary positive inotropic ation, *Fed. Proc.* **44**:6191.

Levi, R., Allan, G., and Zavecz, J. H., 1976, Prostaglandins and cardiac anaphylaxis, *Life Sci.* **18**:1255–1264.

Levi, R., Burke, J. A., and Corey, E. J., 1982, SRS-A, leukotrienes, and immediate hypersensitivity reactions of the heart, in: *Leukotrienes and Other Lipoxygenase Products, Advances in Prostaglandin, Thromboxane, and Leukotriene Research*, Vol. 9 (B. Samuelsson and R. Paoletti, eds.), Raven Press, New York, pp. 215–222.

Levi, R., Burke, J. A., Guo, Z.-G., Hattori, Y., Hoppens, C. M., McManus, L. M., Hanahan, D. J., and Pinckard, R.N., 1984, Acetyl glyceryl ether phosphorylcholine (AGEPC): A putative mediator of cardiac anaphylaxis in the guinea pig, *Circ. Res.* **54**:117–124.

Lewis, R. A., Austen, K. F., Drazen, J. M., Clark, D. A., Marfat, A., and Corey, E. J., 1980, Slow reacting substances of anaphylaxis: Identification of leukotrienes C-1 and D from human and rat sources, *Proc. Natl. Acad. Sci. U.S.A.* **77**:3710–3714.

Liebig, R., Bernauer, W., and Peskar, B. A., 1975, Prostaglandin, slow-reacting substance, and histamine release from anaphylactic guinea-pig hearts, and its pharmacological modification, *Naunyn Schmiedebergs Arch Pharmacol.* **289**:65–76.

Michelassi, F., Landa, L., Hill, R. D., Lowenstein, E., Watkins, W. D., Petkau, A. J., and Zapol, W. M., 1982, Leukotriene D_4: A potent coronary artery vasoconstrictor associated with impaired ventricular contraction, *Science* **217**:841–843.

Michelassi, F., Castorena, G., Hill, R. D., Lowenstein, E., Watkins, W. D., Petkau, A. J., and Zapol, W. M., 1983, Effects of leukotrienes B_4 and C_4 on coronary circulation and myocardial contractility, *Surgery* **94**:267–275.

Morris, H. R., Taylor, G. W., Piper, P. J., and Tippins, J. R., 1980, Structure of slow-reacting substance of anaphylaxis from guinea-pig lung, *Nature* **285**:104–106.

Murphy, R. C., Hammarström, S., and Samuelsson, B., 1979, Leukotriene C: A slow-reacting substance from murine mastocytoma cells, *Proc. Natl. Acad. Sci. U.S.A.* **76**:4275–4279.

Panzenbeck, M. J., and Kaley, G., 1983, Leukotriene D_4 reduces coronary blood flow in the anesthetized dog, *Prostaglandins* **25**:661–670.

Petsas, A. A., and Kotler, M. N., 1973, Electrocardiographic changes associated with penicillin anaphylaxis, *Chest* **64**:66–69.

Pfeffer, M. A., Pfeffer, J. M., Lewis, R. A., Braunwald, E., Corey, E. J., and Austen, K. F., 1983, Systemic hemodynamic effects of leukotrienes C_4 and D_4 in the rat, *Am. J. Physiol.* **244**:H628–H633.

Rouzer, C. A., Scott, W. A., Cohn, Z. A., Blackburn, P., and Manning, J. M., 1980, Mouse peritoneal macrophages release leukotriene C in response to a phagocytic stimulus, *Proc. Natl. Acad. Sci. U.S.A.* **77**:4928–4932.

Samuelsson, B., Borgeat, P., Hammarström, S., and Murphy, R. C., 1980, Leukotrienes: A new group of biologically active compounds, in: *Advances in Prostaglandin and Thromboxane Research*, Vol. 6 (B. Samuelsson, P. W. Ramwell, and R. Paoletti, eds.), Raven Press, New York, pp. 1–18.

Shigenobu, K., Schneider, J. A., and Sperelakis, N., 1974, Verapamil blockade of slow Na^- and Ca^{++} responses in myocardial cells, *J. Pharmacol. Exp. Ther.* **190**:280–288.

Smedegård, G., Hedqvist, P., Dahlén, S.-E., Revenäs, B., Hammarström, S., and Samuelsson, B., 1982, Leukotriene C_4 affects pulmonary and cardiovascular dynamics in monkey, *Nature* **295**:327–329.

Sullivan, T. J., 1982, Cardiac disorders in penicillin-induced anaphylaxis, association with intravenous epinephrine therapy, *J.A.M.A.* **248:**2161–2162.

Thyrum, P. T., 1974, Inotropic stimuli and systolic transmembrane calcium flow in depolarized guinea-pig atria, *J. Pharmacol. Exp. Ther.* **188:**166–179.

Tritthart, H., Volkmann, R., Weiss, R., and Eibach, H., 1976, The interrelationship of calcium-mediated action potentials and tension development in cat ventricular myocardium, *J. Mol. Cell. Cardiol.* **8:**249–261.

Vemulapalli, S., Chiu, P. J. S., and Barnett, A., 1984, Cardiovascular and renal action of platelet-activating factor in anesthetized dogs, *Hypertension* **6:**489–493.

Woodman, O. L., and Dusting, G. J., 1983, Coronary vasoconstriction induced by leukotrienes in the anaesthetized dog, *Eur. J. Pharmacol.* **86:**125–128.

Zavecz, J. H., and Levi, R., 1977, Separation of primary and secondary cardiovascular events in systemic anaphylaxis, *Circ. Res.* **40:**15–19.

Zavecz, J. H., and Levi, R., 1978, Histamine-induced negative inotropism: Mediation by H_1-receptors, *J. Pharmacol. Exp. Ther.* **206:**274–280.

28

Comparative Responses to Leukotriene D$_4$ in the Sheep and Cat

PHILIP J. KADOWITZ, DENNIS B. McNAMARA, and ALBERT L. HYMAN

1. SUMMARY

Pulmonary vascular responses to leukotriene (LT) D$_4$ were compared in the sheep and cat under conditions of controlled lobar blood flow. Intralobar injections of LTD$_4$ in the sheep caused dose-dependent increases in lobar arterial and small vein pressures without influencing left atrial or systemic arterial pressure. Leukotriene D$_4$ was very potent in increasing pulmonary vascular resistance in the sheep, with activity similar to U-46619, a thromboxane (Tx) A$_2$ mimic. Pulmonary vascular responses to LTD$_4$ in the sheep were similar when the lung was ventilated and when lobar ventilation was arrested and when the lobe was perfused with blood or with dextran. Pulmonary vasoconstrictor responses to LTD$_4$ but not to U-46619 in the sheep were reduced by inhibitors of cyclooxygenase and thromboxane synthesis. In contrast, LTD$_4$ had modest pressor activity in the pulmonary vascular bed of the cat, whereas U-46619 had marked activity in this species. Responses to LTD$_4$ in the cat were not altered by cyclooxygenase inhibitors.

It is concluded that LTD$_4$ has marked pulmonary vasoconstrictor activity in the sheep, increasing pulmonary vascular resistance by constricting intrapulmonary veins and upstream segments. In this species, responses to LTD$_4$ were independent

---

PHILIP J. KADOWITZ and DENNIS B. McNAMARA ● Department of Pharmacology, Tulane University School of Medicine, New Orleans, Louisiana 70112.   ALBERT L. HYMAN ● Departments of Pharmacology, Surgery, and Medicine, Tulane University School of Medicine, New Orleans, Louisiana 70112.

of changes in ventilation or the interaction with formed elements but were dependent on the formation of cyclooxygenase products including TxA_2. However, in the cat, LTD_4 had very weak pressor activity, and this activity was not dependent on the integrity of the cyclooxygenase system. These studies indicate that there is considerable species difference in responses to LTD_4, a major component of the slow reacting substance of anaphylaxis, in the pulmonary vascular bed.

2. INTRODUCTION

The leukotrienes are a family of biologically active substances formed from arachidonic acid via the 5-lipoxygenase pathway (Hammarström, 1983). In the lipoxygenase pathway, which is an alternative to the cyclooxygenase pathway, the substrate is converted to 5-hydroperoxyeicosatetraenoic acid, which is oxygenated to a labile epoxide intermediate named leukotriene (LT) A_4 (Borgeat *et al.*, 1976; Borgeat and Samuelsson, 1979a,b). This labile epoxide intermediate, which is analogous to the pivotal prostaglandin (PG) endoperoxide intermediate PGH_2 in the cyclooxygenase pathway, can be transformed enzymatically to LTB_4, which has potent chemotactic activity (Borgeat and Samuelsson, 1979a,b; Ford-Hutchinson *et al.*, 1980). Leukotriene A_4 can also be converted to LTC_4 by the addition of glutathione, and this leukotriene can be further metabolized to LTD_4 by a γ-glutamyl transpeptidase (Murphy *et al.*, 1979; Hammarström *et al.*, 1983; Orning *et al.*, 1980).

It has recently been reported that LTC_4 and LTD_4 are major components of the slow reacting substance of anaphylaxis (Murphy *et al.*, 1979; Morris *et al.*, 1980; Lewis *et la.*, 1980). Since SRS-A is a contractile substance that is released by immunologic challenge from the lung, it has been hypothesized that SRS-A is an important mediator of symptoms in asthma and other immediate-type hypersensitivity reactions (Kallaway and Trethewie, 1940; Brocklehurst, 1960; Dahlen *et al.*, 1983). The effects of the leukotrienes on the lung are of considerable interest because of their postulated role as a mediator in asthma (Dahlen *et al.*, 1983). Leukotrienes C_4 and D_4 have potent contractile activity on preparations of airway and vascular smooth muscle from the lung (Drazen *et al.*, 1980; Dahlen *et al.*, 1980; Hand *et al.*, 1981; Krell *et al.*, 1981; Jones *et al.*, 1982). These substances have significant bronchoconstrictor activity in a variety of species (Drazen *et al.*, 1980; Holroyde *et al.*, 1981; Smedegard *et al.*, 1982). However, little has been written about the effects of the leukotrienes on the pulmonary vascular bed. In the monkey, the predominant response to injection of LTC_4 is a fall in pulmonary arterial pressure, whereas aerosol administration of LTC_4 caused a marked rise in pulmonary arterial pressure (Smedegard *et al.*, 1982). In the rat, injections of LTC_4 caused a dose-related fall in pulmonary arterial pressure (Iacopino *et al.*, 1983). In contrast to studies with LTC_4 in the rat and monkey, LTD_4 caused a marked increase in pulmonary vascular resistance in the newborn lamb when injected into the pulmonary artery (Yokochi *et al.*, 1982). However, less is known about responses to

LTD$_4$ on the pulmonary vascular bed of the mature animal (Kadowitz and Hyman, 1984). The purpose of this chapter is to describe and compare responses to LTD$_4$ in the pulmonary vascular bed of cat and sheep under conditions of controlled blood flow using recently described methods (Hyman and Kadowitz, 1975, 1979; Kadowitz and Hyman, 1983).

3. RESULTS

3.1. Responses to LTD$_4$ in the Sheep

Lobar vascular responses to LTD$_4$ in the intact-chest sheep were studied in 11 animals, and these data are presented in Table I. Under constant-flow conditions, intralobar injections of LTD$_4$ in doses of 0.1–1 µg caused significant dose-related increases in lobar arterial and small vein pressures without changing left atrial pressure. In the doses employed in the present study in the sheep, LTD$_4$ had no significant effect on systemic arterial pressure. The increases in lobar arterial and small vein pressures were rapid in onset, and mean vascular pressures returned to base-line value over a 0.5 to 4-min period depending on the dose of the leukotriene. The lobar arterial-to-small-vein pressure gradient and the gradient from small vein to left atrium pressure increased significantly at all doses of LTD$_4$ studied (Table II).

3.2. Influence of Inhibitors in the Sheep

In order to determine if pulmonary vascular responses to LTD$_4$ in the sheep were dependent on formation of products in the cyclooxygenase pathway, the effects of sodium meclofenamate, a cyclooxygenase inhibitor, and of OKY1581, a thromboxane synthesis inhibitor, were investigated. After administration of sodium meclo-

TABLE I. Influence of Intralobar Injections of Leukotriene D$_4$ on Mean Vascular Pressures in the Sheep[a]

	Pressure (mm Hg)			
	Lobar artery	Small vein	Left atrium	Aorta
Control	15 ± 1	11 ± 1	5 ± 0	102 ± 4
LTD$_4$, 0.1 µg	26 ± 2*	15 ± 1*	5 ± 0	103 ± 4
Control	17 ± 1	12 ± 1	5 ± 1	100 ± 5
LTD$_4$, 0.3 µg	34 ± 2*	20 ± 2*	5 ± 1	100 ± 5
Control	15 ± 1	11 ± 1	5 ± 1	97 ± 5
LTD$_4$, 1 µg	39 ± 2*	22 ± 3*	5 ± 1	101 ± 4

[a] $n = 11$. *$P < 0.05$ when compared to corresponding control, paired comparison.

TABLE II. Influence of Intralobar Injections of Leukotriene D_4 on Mean Vascular Pressure Gradients in the Sheep Lung[a]

	Pressure gradient (mm Hg)		
	Lobar artery–left atrium	Lobar artery–small vein	Small vein–left atrium
Control	10 ± 1	4 ± 1	6 ± 3
LTD$_4$, 0.1 μg	21 ± 4*	11 ± 2*	10 ± 2*
Control	12 ± 1	5 ± 1	7 ± 2
LTD$_4$, 0.3 μg	29 ± 4*	14 ± 3*	15 ± 4*
Control	10 ± 2	4 ± 1	6 ± 3
LTD$_4$, 1 μg	34 ± 5*	17 ± 4*	17 ± 5*

[a] n = 10–11. *$P < 0.05$ when compared to corresponding control, paired comparison.

fenamate in a dose of 2.5 mg/kg i.v., the increases in lobar arterial pressure in response to LTD$_4$ were reduced markedly at each dose of the leukotriene studied. The thromboxane synthesis inhibitor OKY1581, in doses of 5–10 mg/kg i.v., also significantly reduced the increases in lobar arterial pressure in response to the three doses of LTD$_4$. However, the inhibitory effects of the cyclooxygenase inhibitor on responses to LTD$_4$ were greater than the inhibitory effects of the thromboxane synthesis inhibitor. Neither OKY1581 nor sodium meclofenamate had a significant effect on pulmonary vascular or systemic arterial pressure in the sheep. The effects of sodium meclofenamate and OKY1581 on pulmonary vascular responses to an agent whose actions mimic those of thromboxane A_2 were also investigated. U-46619, an agent whose actions are similar to those of thromboxane A_2 on smooth muscle, caused dose-dependent increases in lobar arterial and small vein pressures without affecting left atrial or systemic arterial pressure. The increases in lobar arterial pressure in response to U-46619 were not altered after administration of sodium meclofenamate, 2.5 mg/kg i.v., or OKY1581, 5–10 mg/kg, i.v.

In biochemical studies, the effects of OKY1581 on the metabolism of arachidonic acid and of the prostaglandin endoperoxide PGH$_2$ by microsomal fractions from sheep lung were investigated. The addition of [1-^{14}C]arachidonic acid (20 μM) to the microsomal fraction (200 μg protein) resulted in the formation of 6-keto-PGF$_{1\alpha}$, the stable breakdown product of PGI$_2$, 255 ± 21 pmol, and TxB$_2$, the stable breakdown product of TxA$_2$, 230 ± 19 pmol/hr in the absence of the inhibitor. Prostaglandins F$_{2\alpha}$, E$_2$, and D$_2$ were also formed. However, when OKY1581 was added to the incubation medium in concentrations of 10^{-9} M or greater, the formation of TxB$_2$ was reduced to 37% of control at 10^{-7} M and 31% of control at 10^{-6} M. Moreover, the synthesis of 6-keto-PGF$_{1\alpha}$ was not decreased at concentrations of OKY1581 up to 10^{-6} M. The formation of PGF$_{2\alpha}$, PGE$_2$, and PGD$_2$ was not decreased by OKY1581 in concentrations up to 10^{-6} M.

The influence of OKY1581 on endoperoxide metabolism by sheep lung microsomal fraction was also investigated. In the absence of inhibitor, 166 ± 15 pmol of 6-keto-PGF$_{1\alpha}$ and 161 ± 17 pmol of TxB$_2$ were formed per 2-min period when

10 μM PGH$_2$ was added to 200 μg microsomal protein. Prostaglandin F$_{2\alpha}$, PGE$_2$, and PGD$_2$ were also formed from PGH$_2$. However, addition of OKY1581 in concentrations of 10^{-9} M or higher reduced the formation of TxB$_2$. Formation of TxB$_2$ was reduced by more than 80% at the higher concentrations of the inhibitor. The formation of PGF$_{2\alpha}$, PGE$_2$, 6-Keto-PGF$_{1\alpha}$, or PGD$_2$ was not reduced by OKY1581.

The effects of the cyclooxygenase and thromboxane synthesis inhibitors on responses to arachidonic acid were also investigated in the sheep. Intralobar injections of arachidonic acid in doses of 30 to 100 μg caused a significant dose-dependent increase in lobar arterial pressure without affecting left atrial pressure. The increases in lobar arterial pressure in response to arachidonic acid were markedly decreased after administration of sodium meclofenamate, 2.5 mg/kg i.v. The increases in lobar arterial pressure in response to arachidonic acid were also decreased significantly after administration of OKY1581, 5–10 mg/kg i.v.

3.3. Influence of Ventilation on Responses to LTD$_4$ in the Sheep

The relationship between the effects of LTD$_4$ on ventilation and on the pulmonary vascular bed was studied in four sheep. In these experiments, responses to LTD$_4$ were obtained when the lung was ventilated and when lobar ventilation was arrested at end-expiration by inflating a balloon catheter in the left lower lobe bronchus. In these experiments, the left lower lobe was perfused with arterial blood to lessen the effects of hypoxia on the lung, and 1–3 ml of a 2% lidocaine viscous solution was instilled into the lobar bronchus to prevent coughing. The correlation between the increases in lobar arterial pressure in response to intralobar injections of LTD$_4$, 0.1–1 μg, when the lobe was ventilated and when lobar ventilation was arrested was very good. The correlation coefficient of the regression line was 0.90 ($P < 0.05$) with a slope of 0.83 that was not significantly different from the line of identity. These data indicate that responses to LTD$_4$ are similar when the lobe is ventilated and when ventilation is arrested.

3.4. Species Variation in the Response to LTD$_4$

In order to determine if responses to LTD$_4$ varied with species, the effects of LTD$_4$ on the pulmonary vascular bed were investigated in the intact-chest cat, and these data are summarized in Table III. Intralobar injections of LTD$_4$ in doses of 0.3, 1, and 4 μg caused small but significant dose-related increases in lobar arterial pressure without affecting left atrial pressure. Systemic arterial pressure was increased significantly in response to intralobar injections of the 1- and 3-μg doses of LTD$_4$. Although lobar vascular responses to LTD$_4$ were modest in the cat, U-46619 had marked vasoconstrictor activity (Table III). As shown earlier, both LTD$_4$ and U-46619 had marked vasoconstrictor activity in the sheep pulmonary vascular bed, and the dose–response curves for both substances in this species were superimposable. However, in the cat, U-46619 had far greater vasoconstrictor activity than did LTD$_4$.

TABLE III. Influence of Intralobar Injections of Leukotriene D_4 and U-46619 on Mean Vascular Pressures in the Cat[a]

	Pressure (mm Hg)		
	Lobar artery	Left atrium	Aorta
Control	14 ± 1	3 ± 1	110 ± 5
LTD_4, 0.3 μg	16 ± 1*	3 ± 1	113 ± 5
Control	13 ± 1	2 ± 1	110 ± 8
LTD_4, 1 μg	16 ± 1*	2 ± 1	119 ± 6*
Control	13 ± 1	3 ± 1	115 ± 10
LTD_4, 3 μg	21 ± 2*	3 ± 1	124 ± 11*
Control	14 ± 2	3 ± 1	130 ± 8
U-46619, 0.003 μg	20 ± 3*	2 ± 1	133 ± 7
Control	12 ± 1	3 ± 1	124 ± 9
U-46619, 0.01 μg	23 ± 3*	3 ± 1	130 ± 7
Control	11 ± 2	3 ± 1	118 ± 6
U-46619, 0.03μg	28 ± 4*	4 ± 2	124 ± 8

[a] n = 6–9. *$P < 0.05$ when compared to corresponding control, paired comparison.

In other experiments in the sheep or in the cat, responses to LTD_4 were similar when the lung was perfused with blood or with low-molecular-weight dextran. The role of the cyclooxygenase pathway in the mediation of pulmonary vascular responses to LTD_4 was also investigated in the cat. Administration of indomethacin or sodium meclofenamate, 2.5 mg/kg i.v., had no significant effect on pulmonary vasoconstrictor responses to U-46619 or LTD_4 in the cat. The increases in systemic arterial pressure in response to the 1- and 3-μg doses of LTD_4 were not altered by the cyclooxygenase inhibitors. However, the cyclooxygenase inhibitors in the doses employed significantly reduced the increases in lobar arterial pressure in response to intralobar injections of arachidonic acid. The cyclooxygenase inhibitors had no significant effect on pulmonary vascular or systemic arterial pressure in the cat.

4. DISCUSSION

Experiments in the intact-chest sheep demonstrate that intralobar injections of LTD_4 increase pulmonary lobar arterial pressure in a dose-related manner (Kadowitz and Hyman, 1984). Since pulmonary blood flow was maintained constant and left atrial pressure was unchanged, the increase in pressure gradient across the lung lobe suggests that pulmonary lobar vascular resistance was increased by LTD_4 (Kadowitz and Hyman, 1984). The increases in lobar arterial pressure in response to LTD_4 were associated with dose-related increases in small intrapulmonary vein pressure. In addition to increasing lobar arterial and venous pressures, LTD_4 increased the pressure gradient from lobar artery to small vein. These experiments in sheep suggest that LTD$_4$ increases pulmonary vascular resistance by constricting intrapulmonary veins and segments upstream to the small vein, which are believed to be small arteries (Kadowitz and Hyman, 1984). Results obtained in mature animals are consistent with results in the newborn lamb in which LTD$_4$ increased pulmonary and systemic vascular resistances and decreased cardiac output (Yokochi *et al.*, 1982; Kadowitz and Hyman, 1984). It has been reported that LTD$_4$ has potent coronary vasoconstrictor activity in the sheep that can be associated with left ventricular impairment (Michelassi *et al.*, 1982). However, in the sheep, LTD$_4$ had no significant effect on systemic arterial or left atrial pressures in the range of doses studied. The effects of LTD$_4$ on left atrial pressure in the newborn lamb were not measured, so the mechanism of the fall in cardiac output is uncertain (Yokochi *et al.*, 1982). The effects of LTD$_4$ on the systemic vascular resistance of the newborn lamb appear to be greater than those observed in the mature animal.

In terms of relative pressor activity in the pulmonary vascular bed of the sheep, LTD$_4$ was very potent, with activity paralleling that of U-46619, a stable prostaglandin analogue whose actions are thought to mimic those of thromboxane A$_2$ (Coleman *et al.*, 1981). Moreover, when compared to ther vasoactive hormones whose effects have been studied in the sheep, LTD$_4$ is far more active than other arachidonic acid metabolites, alveolar hypoxia, or histamine, which acts over a similar portion of the pulmonary vascular bed and is released along with the leukotrienes in immediate hypersensitivity reactions (Brocklehurst, 1960; Kadowitz *et al.*, 1974; Hyman and Kadowitz, 1975; Kadowitz and Hyman, 1983).

It has been reported that LTD$_4$ has potent contractile activity on isolated airway smooth muscle and lung parenchyma and that it increases bronchomotor tone (Dahlen *et al.*, 1980; Drazen *et al.*, 1980; Krell *et al.*, 1981; Holroyde *et al.*, 1981; Jones *et al.*, 1982). However, in the intact-chest sheep, the effects of LTD$_4$ on the pulmonary vascular bed appear to be independent of alterations in ventilation or those that occur as a consequence of changes in bronchomotor tone or lung volume, since similar responses were obtained when the lobe was ventilated or when lobar ventilation was arrested by obstruction of bronchial airflow. In previous studies, responses to a number of vasoactive substances including cyclooxygenase metabolites of arachidonic acid and histamine were similar when the lobe was ventilated and when lobar ventilation was arrested, suggesting that the actions of these vasoactive hormones on pulmonary vascular resistance appear to be independent of alterations in bronchomotor tone (Hyman *et al.*, 1978; Kadowitz and Hyman, 1983). In both the cat and in the sheep, pulmonary hypertensive responses to LTD$_4$ were similar when the lung was perfused with blood or low-molecular-weight dextran. Thus, responses to LTD$_4$ in both species are not dependent on the interaction with formed elements in blood.

In the sheep, pulmonary vasoconstrictor responses to LTD$_4$ were markedly attenuated after treatment with sodium meclofenamate, suggesting that responses to this lipoxygenase product are dependent on the formation of products in the cyclooxygenase pathway. In addition, vasoconstrictor responses to LTD$_4$ were decreased by OKY1581, a thromboxane synthesis inhibitor. These data suggest that a substantial portion of the pulmonary vasoconstrictor response to LTD_4 is mediated by the release of thromboxane A_2. The observation that meclofenamate had greater inhibitory effect on responses to LTD_4 than did OKY1581 suggests that pulmonary vasoconstrictor responses to this lipoxygenase metabolite are dependent on the formation of thromboxane A_2 and other cyclooxygenase products such as prostaglandins (PG) D_2 and $F_{2\alpha}$, which have substantial pressor activity in the pulmonary vascular bed (Kadowitz *et al.*, 1974; Kadowitz and Hyman, 1980). It has been shown that injections of SRS-A or synthetic LTC_4 and LTD_4 cause the release of prostaglandins and TxA_2 from isolated guinea pig lung (Engineer *et al.*, 1978; Piper and Samhoum, 1981). Cyclooxygenase inhibitors have been shown to reduce the contractile effects of LTD_4 on guinea pig lung parenchymal strips (Piper and Samhoum, 1981). The results of the present experiments in the sheep are consistent with data obtained with isolated guinea pig parenchyma and on bronchoconstrictor responses in the guinea pig indicating that responses to LTD_4 are dependent on the release of TxA_2 and prostaglandins (Piper and Samhoum, 1981; Weichman *et al.*, 1982).

A similar relationship between these inhibitors and responses to arachidonic acid was also observed in that there was a greater reduction in response to LTD_4 after treatment with meclofenamate than after OKY1581. These data confirm previous studies showing that pulmonary vasoconstrictor responses to arachidonic acid occur through formation of products in the cyclooxygenase pathway (Hyman *et al.*, 1978, 1980; Spannhake *et al.*, 1980) and extend these findings by showing that a portion of the response is attributable to TxA_2 formation.

Although responses to LTD_4 and arachidonic acid were markedly reduced by meclofenamate, this cyclooxygenase inhibitor had no significant effect on pulmonary vasoconstrictor responses to U-46619, an analogue whose actions are thought to mimic those of thromboxane A_2 (Coleman *et al.*, 1981). These data indicate that sodium meclofenamate inhibited cyclooxygenase activity in the pulmonary vascular bed and that the cyclooxygenase inhibitor did not influence vascular responses to the thromboxane mimic. In addition, vasoconstrictor responses to U-46619 were not altered by OKY1581 in doses that inhibited responses to LTD_4 and arachidonic acid. These results also suggest that the thromboxane synthesis inhibitor did not alter thromboxane receptor-mediated responses and that the effects of the inhibitor were caused by inhibition of the formation of thromboxane A_2.

The inhibition of thromboxane A_2 synthesis was also investigated in microsomal fractions from sheep lung (She *et al.*, 1981; Spannhake *et al.*, 1983). The results of these studies show that OKY1581 inhibited the formation of TxA_2 as measured by formation of its stable breakdown product TxB_2. Thromboxane B_2 formation was inhibited over a wide range of concentrations of OKY1581 when either arachidonic acid or the endoperoxide, PGH_2, was employed as substrate. Although TxB_2 formation was decreased by OKY1581, PGI_2 formation as measured by the production of 6-keto-$PGF_{1\alpha}$ was not inhibited even at very high concentrations of the thromboxane synthesis inhibitor. Prostaglandins E_2, $F_{2\alpha}$, and D_2 were formed when PGH_2 or arachidonic acid was added to the microsomal fractions. It is not known if this prostaglandin synthesis was enzymatic; however, the amount of these substances formed was not decreased by OKY1581 and, in the case of PGE_2, was enhanced by the inhibitor. Since the total amount of product formed from arachidonic acid (6-keto-$PGF_{1\alpha}$, TxB_2, $PGF_{2\alpha}$, PGE_2, and PGD_2) was not decreased although TxB_2 formation was reduced, it is unlikely that OKY1581 had a significant inhibitory effect on sheep lung cyclooxygenase activity. These experiments suggest that effects of OKY1581 on responses to LTD$_4$ and arachidonic acid are caused by inhibition of thromboxane synthetase activity and not by an effect on cyclooxygenase activity or on thromboxane receptor-mediated activity in the pulmonary vascular bed of the sheep. In other experiments in lung homogenates taken from sheep receiving OKY1581, 5–10 mg/kg i.v., TxB_2 formation was greatly reduced.

The results of studies in the sheep demonstrate that LTD$_4$ has very potent vasoconstrictor activity in the pulmonary vascular bed of this species and that this activity occurs for the most part through release of products in the cyclooxygenase pathway. However, the effects of LTD$_4$ in the pulmonary vascular bed of the sheep and the cat are different. In the cat, LTD$_4$ had only modest pressor activity equal to that of arachidonic acid and far less than that of $PGF_{2\alpha}$, PGD_2, or PGE_2 in that species (Kadowitz and Hyman, 1980). Furthermore, in this species, cyclooxygenase blockers did not modify responses to this lipoxygenase product.

Although the relative magnitude of responses to LTD$_4$ as well as the mechanism of action differ in the sheep and the cat, both species were extremely sensitive to the effects of U-46619. Thus, there appears to be true species variation in the pulmonary vascular response to this lipoxygenase metabolite. This variation was not observed with U-46619, which may operate via TxA_2 receptors in the pulmonary vascular bed. In addition to demonstrating marked species variation in the response to LTD$_4$, the present data may be interpreted to suggest that LTD$_4$ itself does not have potent vasoconstrictor activity in the lung when the cyclooxygenase system is blocked. Moreover, the remaining pressor activity of LTD$_4$ in the sheep after cyclooxygenase blockade and the pressor activity in the cat, which were very similar, suggest that the activity of this lipoxygenase metabolite is far less than that of products of the cyclooxygenase pathway such as TxA_2, $PGF_{2\alpha}$, PGD_2 (Kadowitz and Hyman, 1977, 1980). The data from the present study suggest that it would be difficult to formulate a unified hypothesis on the role of LTD$_4$, a major component of SRS-A, on the pulmonary circulation since species variation is so marked.

ACKNOWLEDGMENTS. The authors wish to thank Drs. Joshua Rokeach and Barry M. Weichman for the LTD$_4$ used in the study and Ms. Alice Landry for help with the biochemical experiments. We wish to thank Ms. Janice Ignarro for help in preparing the manuscript. This work was supported by National Heart, Lung and Blood Institute Grants HL11802, HL15580, HL18070, and HL29456.

REFERENCES

Borgeat, P., and Samuelsson, B., 1979a, Arachidonic acid metabolism in polymorphonuclear leuko-cytes: Unstable intermediates in formation of dihydroxy acids, *Proc. Natl. Acad. Sci. U.S.A.* **76**:3212–3217.

Borgeat, P., and Samuelsson, B., 1979b, Metabolism of arachidonic acid in polymorphonuclear leu-kocytes: Structural analysis of novel hydroxylated compounds, *J. Biol. Chem.* **254**:7865–7869.

Borgeat, P., Hamberg, M., and Samuelsson, B., 1976, Transformation of arachidonic acid and homo-γ-linolenic acid by rabbit polymorphonuclear leukocytes. Monohydroxy acids from novel lipox-ygenases, *J. Biol. Chem.* **251**:7816–7820.

Brocklehurst, W. E., 1960, The release of histamine and formation of a slow-reacting substance during anaphylactic shock, *J. Physiol. (Lond.)* **151**:416–435.

Coleman, R. A., Humphrey, P. P. A., Kennedy, I., Levy, G. P., and Lumley, P., 1981, Comparison of the actions of U-46619, a prostaglandin H_2-analogue, with those of prostaglandin H_2 and thromboxane A_2 on some isolated smooth muscle preparations, *Br. J. Pharmacol.* **73**:773–778.

Dahlen, S. E., Hedqvist, P., Hammarström, S., and Samuelsson, B., 1980, Leukotrienes are potent constrictors of human bronchi, *Nature* **288**:484–486.

Dahlen, S. E., Hansson, G., Hedqvist, P., Bjorck, T., Granstrom, E., and Dahlen, B., 1983, Allergic challenge of lung tissue from asthmatics elicits bronchial contraction that correlates with the release of leukotriene C_4, D_4 and E_4, *Proc. Natl. Acad. Sci. U.S.A.* **80**:1712–1716.

Drazen, J. M., Austen, K. F., Lewis, R. A., Clark, D. A., Goto, G., Marfat, A., and Corey, E. J., 1980, Comparative airway and vascular activities of leukotrienes C-1 and D *in vivo* and *in vitro*, *Proc. Natl. Acad. Sci. U.S.A.* **77**:4354–4358.

Engineer, D. M., Morris, H. R., Piper, P. J., and Sirois, P., 1978, The release of prostaglandins and thromboxanes from guinea-pig lung by slow-reacting substances of anaphylaxis and its inhibition, *Br. J. Pharmacol.* **64**:211–218.

Ford-Hutchinson, A. W., Bray, M. A., Doig, M. V., Shipley, M. E., and Smith, M. J. H., 1980, Leukotriene B: A potent chemokinetic and aggregating substance released from polymorphonuclear leukocytes, *Nature* **286**:264–265.

Hammarström, S., 1983, Leukotrienes, *Annu. Rev. Biochem.* **52**:355–377.

Hand, J.M., Will, J. A., and Buckner, C. K., 1981, Effects of leukotrienes on isolated guinea-pig pulmonary arteries, *Eur. J. Pharmacol.* **76**:439–443.

Holroyde, M. D., Altounyan, R. E. C., Cole, M., Dixon, M., and Elliot, E. V., 1981, Bronchocon-striction produced in man by leukotrienes C and D, *Lancet* **2**:17–18.

Hyman, A. L., and Kadowitz, P. J., 1975, Effect of alveolar and perfusion hypoxia and hypercapnia on pulmonary vascular resistance in the lamb, *Am. J. Physiol.* **228**:397–403.

Hyman, A. L., and Kadowitz, P.J., 1979, Pulmonary vasodilator activity of prostacyclin (PGI_2) in the cat, *Circ. Res.* **45**:404–409.

Hyman, A. L., Mathe, A. A., Leslie, C. A., Matthews, C. C., Bennett, J. T., Spannhake, E. W., and Kadowitz, P. J., 1978, Modification of pulmonary vascular responses to arachidonic acid by alterations in physiologic state, *J. Pharmacol. Exp. Ther.* **207**:388–401.

Hyman, A. L., Spannhake, E. W., and Kadowitz, P. J., 1980, Divergent responses to arachidonic acid in the feline pulmonary vascular bed, *Am. J. Physiol.* **239**:H40–H46.

Iacopino, V. J., Fitzpatrick, T. M., Ramwell, P. W., Rose, J. C., and Kot, P. A., 1983, Cardiovascular responses to leukotriene C_4 in the rat, *J. Pharmacol. Exp. Ther.* **227**:244–247.

Jones, T. R., Davis, C., and Daniel, E. E., 1982, Pharmacological study of the contractile activity of leukotrienes C_4 and D_4 on isolated human airway smooth muscle, *Can. J. Physiol. Pharmacol.* **60**:638–643.

Kadowitz, P. J., and Hyman, A. L., 1977, Influence of a prostaglandin endoperoxide analogue on the canine pulmonary vascular bed, *Circ. Res.* **40**:282–287.

Kadowitz, P. J., and Hyman, A. L., 1980, Comparative effects of thromboxane B_2 on the canine and feline pulmonary vascular bed, *J. Pharmacol. Exp. Ther.* **213**:300–305.

Kadowitz, P. J., and Hyman, A. L., 1983, Pulmonary vascular responses to histamine in sheep, *Am. J. Physiol.* **244**:H423–H428.

Kadowitz, P. J., and Hyman, A. L., 1984, Analysis of responses to leukotriene D$_4$ in the pulmonary vascular bed, *Circ. Res.* **55**:707–717.

Kadowitz, P. J., Joiner, P. D., and Hyman, A. L., 1974, Influence of prostaglandins E$_1$ and F$_{2\alpha}$ on pulmonary vascular resistance in the sheep, *Proc. Soc. Exp. Biol. Med.* **145**:1258–1261.

Kallaway, C. H., and Trethewie, E. F., 1940, The liberation of slow reacting smooth muscle-stimulating substances in anaphylaxis, *Q. J. Exp. Physiol.* **30**:121–145.

Krell, R. D., Osborn, R., Vickery, L., Falcone, K., O'Donnell, M., Gleason, J., Kinzig, C., and Bryan, D., 1981, Contraction of isolated airway smooth muscle by synthetic leukotrienes C$_4$ and D$_4$, *Prostaglandins* **22**:387–409.

Lewis, R. A., Austen, K. F., Drazen, J. M., Clark, D. A., Marfat, A., and Corey, E. J., 1980, Slow reacting substances of anaphylaxis: Identification of leukotrienes C and D from human and rat sources, *Proc. Natl. Acad. Sci. U.S.A.* **77**:3710–3714.

Michelassi, F., Landa, L., Hill, R. D., Lowenstein, E., Watkins, W. D., Petkau, A. J., and Zapol, W. M., 1982, Leukotriene D$_4$: A potent coronary artery vasoconstrictor associated with impaired ventricular contraction, *Science* **217**:841–843.

Morris, H. R., Taylor, G.W., Piper, P. J., and Tippins, J. R., 1980, Structure of slow-reacting substance of anaphylaxis from guinea-pig lung, *Nature* **285**:104–108.

Murphy, R. C., Hammarström, S., and Samuelsson, B., 1979, Leukotriene C: A slow reacting substance from murine mastocytoma cells, *Proc. Natl. Acad. Sci. U.S.A.* **76**:4275–4279.

Orning, L., Hammarström, S., and Samuelsson, B., 1980, Leukotriene D: A slow reacting substance from rat basophilic leukemic cells, *Proc. Natl. Acad. Sci. U.S.A.* **77**:2014–2017.

Piper, P. J., and Samhoum, M. N., 1981, The mechanism of action of leukotriene C$_4$ and D$_4$ in guinea-pig isolated perfused lung and parenchymal strips of guinea pig, *Prostaglandins* **21**:793–803.

She, H. S., McNamara, D. B., Spannhake, E. W., Hyman, A. L., and Kadowitz, P. J., 1981, Metabolism of prostaglandin endoperoxide by microsomes from cat lung, *Prostaglandins* **21**:531–541.

Smedegard, G., Hedqvist, P., Dahlen, S. E., Revenas, B., Hammarström, S., and Samuelsson, B., 1982, Leukotriene C$_4$ affects pulmonary and cardiovascular dynamics in monkey, *Nature* **295**:327–329.

Snedecor, C. W., and Cochran, W. G., 1967, *Statistical Methods*, 6th ed., Iowa State University Press, Ames, IA.

Spannhake, E. W., Hyman, A. L., and Kadowitz, P. J., 1980, Dependency of the airway and pulmonary vascular effects of arachidonic acid upon route and rate of administration, *J. Pharmacol. Exp. Ther.* **212**:584–590.

Spannhake, E. W., Colombo, J. L., Craigo, P. A., McNamara, D. B., Hyman, A. L., and Kadowitz, P. J., 1983, Evidence for modification of pulmonary cyclooxygenase activity by endotoxin in the dog, *J. Appl. Physiol.* **54**:191–198.

Weichman, B. M., Muccitelli, R. M., Osborn, R. R., Holden, D. A., Gleason, J. G., and Wasserman, M. A., 1982, *In vitro* and *in vivo* mechanisms of leukotriene-mediated bronchoconstriction in the guinea pig, *J. Pharmacol. Exp. Ther.* **222**:202–208.

Yokochi, K., Olley, P. M., Sideris, E., Hamilton, F., Huhtanen, D., and Coceani, F., 1982, Leukotriene D$_4$: A potent vasoconstrictor of the pulmonary and systemic circulations in the newborn lamb, in: *Leukotrienes and Other Lipoxygenase Products* (B. Samuelsson and R. Paoletti, eds.), Raven Press, New York, pp. 211–214.

29

Platelet-Activating Factor as a Modulator of Cardiac and Coronary Functions

GIORA FEUERSTEIN, DAVID EZRA,
PETER W. RAMWELL, GORDON LETTS, and
ROBERT E. GOLDSTEIN

1. INTRODUCTION

Platelet-activating factor (PAF) is a phospholipid that is produced by white blood cells and platelets (Demopoulos *et al.*, 1979; Benveniste *et al.*, 1982). It is released from these cellular elements during systemic anaphylaxis or stimulation by IgE or nonimmune stimuli (e.g., calcium ionophore) *in vitro*. Systemic anaphylaxis is accompanied by severe cardiovascular decompensation, which was previously attributed to histamine release. However, histamine antagonists cannot block the cardiovascular consequences of systemic anaphylaxis. Recently, it was shown that purified PAF (1-O-hexadecyl-2-acetyl-*sn*-glyceryl-3-phosphocholine) can produce anaphylacticlike hypotension in experimental animals (Feuerstein *et al.*, 1982; Benveniste *et al.*, 1983; Halonen *et al.*, 1980). Moreover, platelets were also associated with nonimmune pathophysiological processes: aggregation and thrombus formation on injured blood vessels. Thus, PAF may play a significant role in mediating the sequelae of platelet/blood vessel interactions, e.g., platelet-mediated coronary constriction. Since the heart shows profound pathological changes in

GIORA FEUERSTEIN, DAVID EZRA, and ROBERT E. GOLDSTEIN • Neurobiology Research Unit, Divisions of Cardiology and Clinical Pharmacology, Departments of Medicine and Pharmacology, Uniformed Services University of the Health Sciences, Bethesda, Maryland 20814.   PETER W. RAMWELL • Department of Physiology and Biophysics, Georgetown University Medical Center, Washington, D.C. 20007.   GORDON LETTS • Merck Frosst, Inc., Dorval, Québec H9R 4P8, Canada.

anaphylaxis (Zavecz and Levi, 1977; Cappuro and Levi, 1975), and platelet-related coronary spasm is suspected in patients suffering from ischemic heart disease, experiments (Addonazio *et al.*, 1982) were designed to study the direct effect of PAF on cardiac and coronary functions in the intact domestic pig heart.

2. MATERIALS AND METHODS

Domestic pigs (9–12 weeks old) weighing 25–35 kg were sedated with ketamine hydrochloride (20 mg/kg per hr i.v.) and anesthetized with pentobarbital sodium (2–4 mg/kg per hr i.v.). The pigs were ventilated with a Harvard respirator modified to deliver 3 cm H_2O positive end-expiratory pressure. Ventilatory rate and inspired oxygen concentration were adjusted to maintain arterial partial presssure of $O_2(Po_2)$ (80–120 mm Hg) and partial pressure of $CO_2(Pco_2$, 35–40 mm Hg; Blood Gas Analyzer model 213, Instrumentation Laboratories, Lexington, MA). Hematocrit was 25–32% (normal for pigs of this age). Catheters filled with heparinized saline were placed in the jugular vein and internal mammary artery. The left ventricular cavity was catheterized via the carotid artery. Rectal temperature was maintained at 37.5–38.0°C by an external heating pad. Mean systemic arterial blood pressure (MBP), left ventricular pressure, and lead II of the standard electrocardiogram (ECG) were monitored continuously. In addition, a surface ECG was recorded from the region supplied by the distal left anterior descending (LAD) coronary artery using a saline-soaked cotton wick electrode.

A left lateral thoracotomy was performed, and the heart was suspended in a pericardial cradle. A circumferential electromagnetic flow probe was placed around an exposed proximal portion of the LAD and attached to a Model 501D Carolina square-wave flowmeter (Carolina Medical Electronics, King, NC). A fine Tygon catheter used for the administration of PAF was introduced into the LAD several millimeters distal to the flow probe (Herd and Barger, 1964). Flow probe calibration was checked prior to study by comparison of flowmeter readings with the results of timed collection of blood flow. Pure synthetic 1-O-hexadecyl-2-acetyl-*sn*-glycero-3-phosphocholine (PAF) was kindly provided by Dr. F. Snyder (Oak Ridge Associated Universities, Oak Ridge, TN). The PAF was dissolved in sterile, pyrogen-free 0.9% NaCl (vehicle) and loaded into the LAD catheter in 0.1 ml, followed by a 0.5 ml vehicle flush (catheter volume 0.2 ml). The pH of all injection solutions was 7.15–7.20. After base-line recording of heart rate (HR), MBP, and mean LAD coronary blood flow (CBF), a 0.5-ml bolus of the vehicle was injected into the LAD. Twenty-five to 30 min later, individual bolus doses of PAF were injected over 20 sec in an ascending order: 0.03, 0.1, 0.3, 1, and 10 nmol. After each injection, all parameters were continuously monitored for 10–15 min or until the complete return of parameters to baseline.

In addition, experiments were conducted to study the effect of continuous infusion of PAF into the LAD: PAF, 1–6 nmol/min, was continuously infused for 5 to 8 min. In these experiments regional contractility was also assessed by the ultrasound technique measuring the shortening of the myocardial fibers in the area supplied by the LAD.

2.1. Assay of Plasma TXB_2 and 6-Keto-$PGF_{1\alpha}$

Blood was rapidly withdrawn from the catheter in the left ventricle or the coronary vein along the LAD. Plasma was separated by rapid centrifugation (Beckman microfuge B) and immediately frozen on dry ice. Thromboxane B_2 and 6-keto-$PGF_{1\alpha}$ were determined by radioimmunoassay as previously described (Granstrom and Kindahl, 1976).

2.2. Assay of Plasma Lactate

Plasma lactate was assayed by a routine biochemical method.

2.3. Assay of Plasma Leukotriene C_4 Immunoreactivity

Plasma was separated as previously described, and LTC_4-like immunoreactivity was determined as previously described (Hayes *et al.*, 1983).

3. RESULTS

Figure 1 shows the effect of bolus injection of 0.3 nmol PAF into the LAD. The CBF exhibited a biphasic response: first a brief increase followed by a more prolonged decrease. Both phases were observed at a time at which no significant systemic changes could be observed. Both changes in CBF were dose dependent: the first phase of increase in CBF reached a maximum of $+50\%$ at the highest dose (10 nmol, Fig. 2); the second phase showed a dose-dependent decrease in CBF (Fig. 3) that exceeded 90% at the highest dose (10 nmol). Although the dilatory phase of the LAD at all doses was not accompanied by systemic effects, the higher doses of PAF (1–10 nmol) also produced hypotension during the peak of reduction of CBF. During this latter phase, however, calculated coronary resistance showed a significant increase from 2.61 $\pm$ 0.26 to 18.4 $\pm$ 0.68 mm Hg/ml per min after the highest bolus dose of PAF.

Infusion of PAF at 1 nmol/min into the LAD (Fig. 4) had no significant effect on CBF, coronary resistance, or MAP; heart rate was increased (about 30%). End-diastolic myocardial segment length and extent of myocardial shortening in the territory of the LAD decreased throughout the infusion period, but complete recovery was observed about 10 min after the infusion. During infusion of a higher dose of PAF (3 nmol/min), local segment length and shortening were markedly diminished

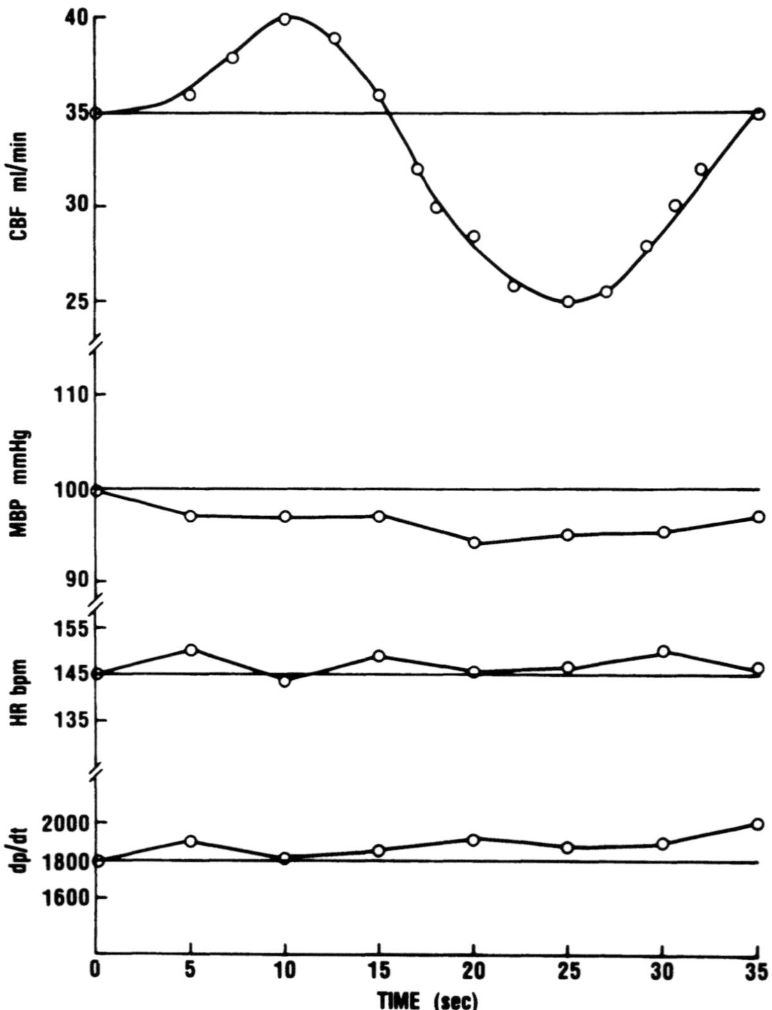

FIGURE 1. Effect of intracoronary PAF, 0.3 nmol, on coronary blood flow (CBF) and systemic hemodynamic variables in a domestic pig. MBP, mean blood pressure; HR, heart rate; dp/dt, peak rate of rise of left ventricular pressure; abscissa, time in seconds after PAF administration into the left anterior descending coronary artery.

along with reduction of left ventricular end-diastolic pressure (LVEDP). However, CBF and coronary resistance were not reduced (Fig. 5).

During PAF-induced reduction in CBF sufficient to produce cardiac ischemia, there was an increase in plasma lactate in the coronary vein, an increase in plasma thromboxane B_2 in both the arterial and coronary venous circulation, and no change in leukotriene C_4-like immunoreactivity (Fig. 6).

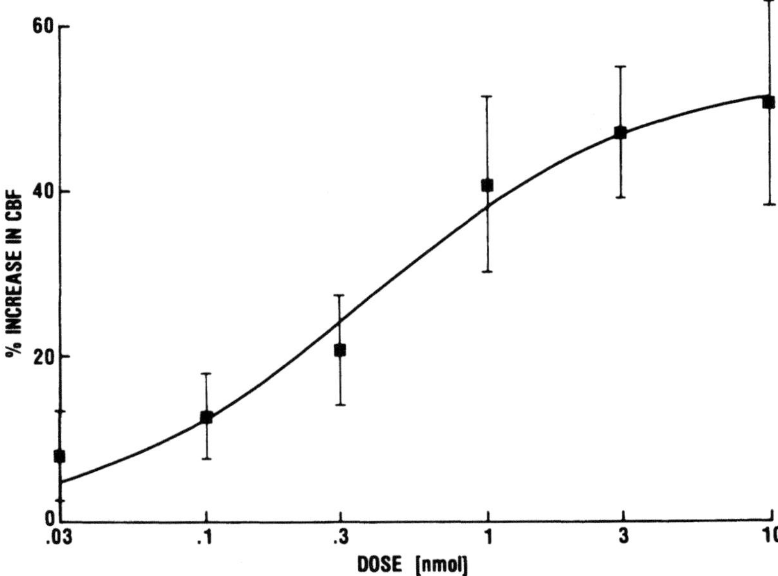

FIGURE 2. Initial increase in coronary blood flow (CBF) after intracoronary PAF administration (first phase in Fig. 1). Ordinate, percentage increase from basal level; abscissa, dose of PAF in nanomoles. Vertical bars represent standard errors. Curve is fitted by standard techniques. Data from five to seven pigs.

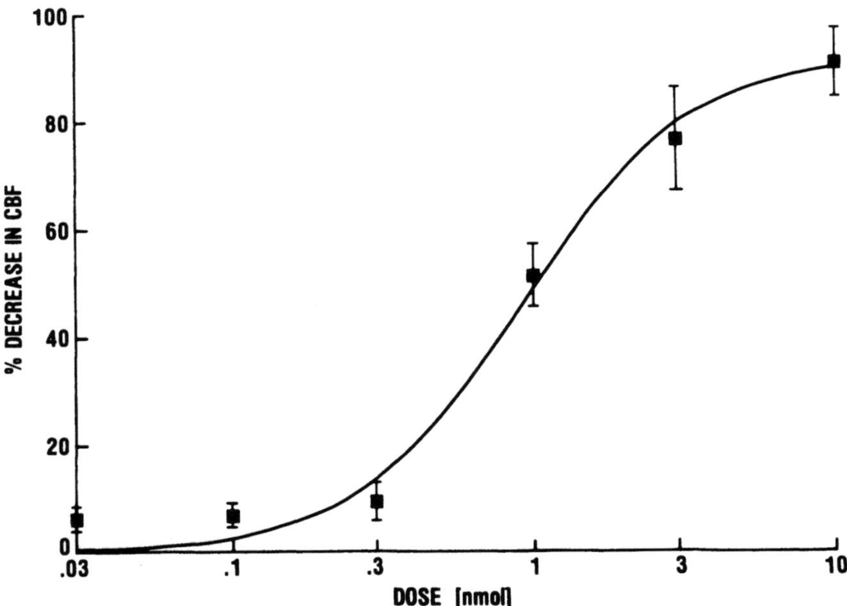

FIGURE 3. Subsequent decrease in coronary blood flow after intercoronary PAF administration (second phase in Fig. 1). Ordinate, percentage decrease from basal level; abscissa, dose of PAF in nanomoles. Curve is fitted as above. Data from five to seven pigs.

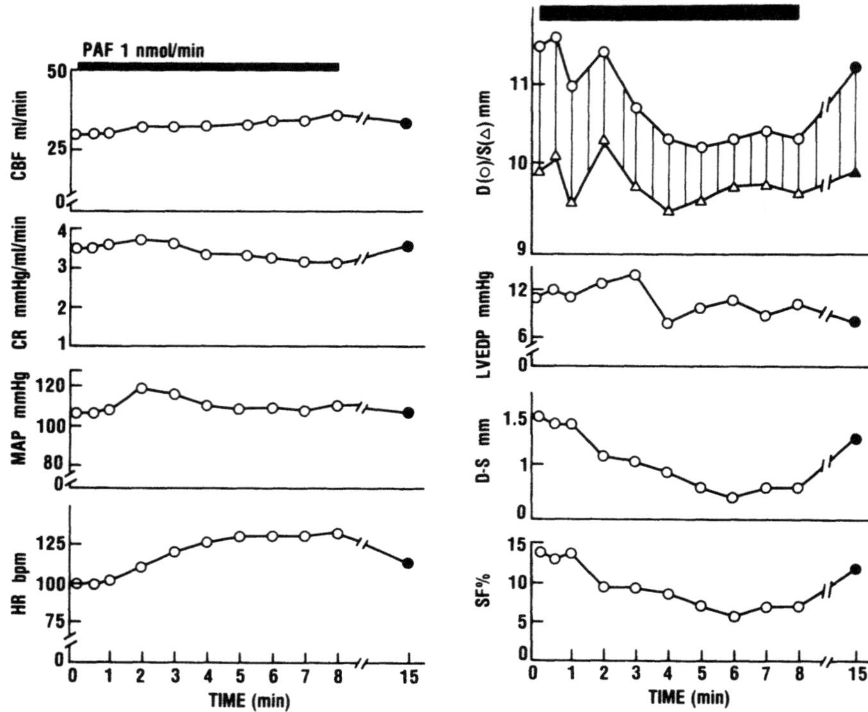

FIGURE 4. Effects of low-dose PAF infusion into the coronary artery of a domestic pig. LVEDP, left ventricular end-diastolic pressure; CBF, coronary blood flow; CR, coronary resistance; MAP, mean arterial presure; HR, heart rate (beats per minute); D/S, distance (mm) between the crystals in end-diastole/end-systole; D-S, shortening of the region between the crystals (mm); SF, shortening fraction, (D-S)/D, presented as a percentage. Black horizontal bar represents the length of the infusion of PAF (1 nmol/min). Closed symbols are data obtained after the infusion.

4. DISCUSSION

The data presented in this chapter suggest that PAF is a potent vasoactive substance with powerful effects on coronary and systemic hemodynamics. These data, obtained on an intact domestic pig heart, agree with recent studies conducted on isolated perfused guinea pig heart (Benveniste *et al.*, 1983; Levi *et al.*, 1984). However, unlike other vasoactive substances, bolus injections of PAF produced a complex effect on coronary tone. Initial brief coronary dilation was followed by an increase in coronary resistance and reduced coronary flow. The first phase, increase in coronary blood flow, seemed to be a direct effect of PAF since it was not blocked by indomethacin. The second phase, reduced coronary blood flow, might be partially attributed to systemic hypotension but primarily to increased

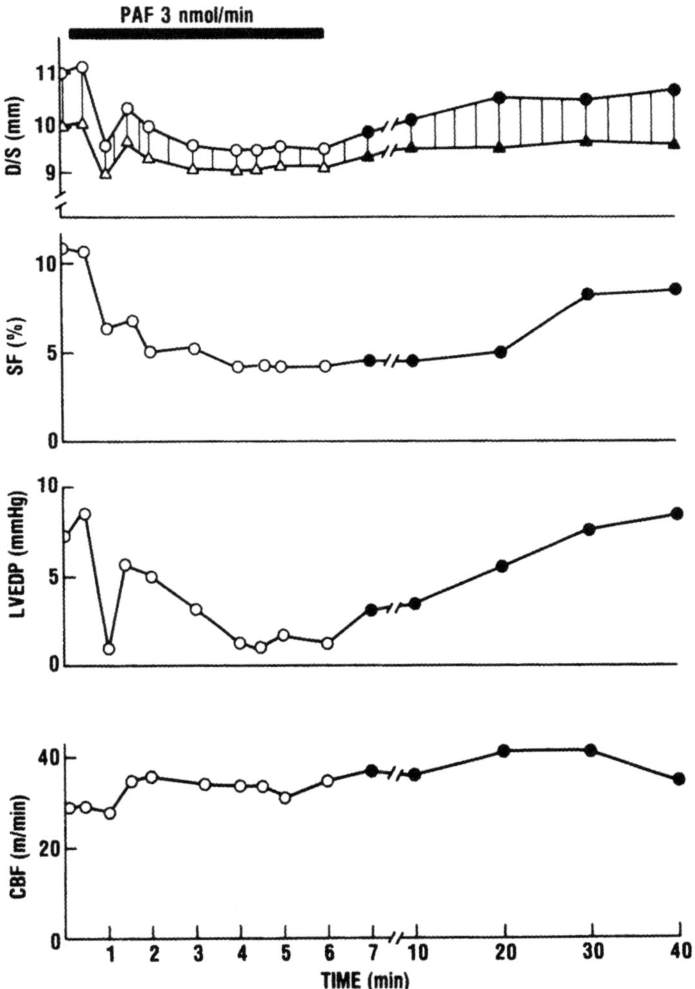

FIGURE 5. Effects of high-dose PAF infused into the coronary artery of a domestic pig heart. Symbols on ordinate are given in Fig. 4; open symbols are data obtained during infusion of PAF; closed symbols are data obtained after the infusion.

coronary resistance. The PAF-induced coronary constriction is probably mediated by a cyclooxygenase metabolite of arachidonate since it is effectively blocked by indomethacin (Feuerstein *et al.*, 1984a).

Continuous infusion of a low dose of PAF revealed additional aspects of PAF interference with cardiac function. First, the low dose of PAF infusion did not reduce CBF. On the contrary, some increase in CBF is observed as a result of reduced coronary resistance. Also, except for mild tachycardia, no changes in systemic variables were observed. However, contractile performance in the area

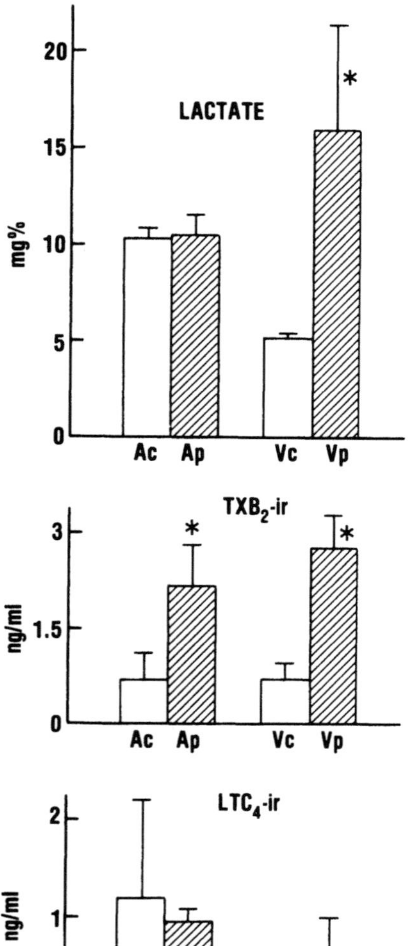

FIGURE 6. Plasma lactate, TXB$_2$, and LTC$_4$-like immunoreactivity (ir) during cardiac ischemia produced by PAF infusion into the coronary artery. A$_c$, arterial plasma levels at control period; A$_p$, arterial plasma levels during PAF injection; V$_c$, venous plasma levels during control period; V$_p$, venous plasma levels during PAF injection. The biochemical data were obtained from four pigs.

perfused with low-dose PAF was significantly affected. Local shortening fraction decreased along with a decrease in end-diastolic pressure and local segment length. The latter changes indicate a reduction in left ventricular diastolic filling, perhaps reflecting PAF-induced peripheral venous pooling. Although diminished ventricular filling can, in itself, decrease systolic shortening, it is possible that a direct, PAF-mediated negative inotropic action is also involved. Further studies are needed to resolve the origin of the diminished myocardial shortening. Nevertheless, our results do identify a PAF-related reduction in contractile performance that may figure significantly in hypotension associated with PAF administration. Infusion of a larger dose of PAF also reduced the local shortening fraction without reducing the CBF, but mild hypotension, tachycardia, and substantially reduced left ventricular end-diastolic pressure were observed.

These data taken together clearly indicate a substantial difference between leukotriene and PAF: the former substance generally cannot maintain a reduced coronary blood flow and contractility when infused in high doses to the coronary artery (Ezra *et al.*, 1984); PAF not only maintains its cardiac effects in most instances but also causes further hemodynamic deterioration, which ultimately leads to shock. Therefore, it is likely that PAF is the more important mediator of anaphylactic shock than the cysteinyl leukotrienes.

Infusion of PAF to the coronary artery also revealed that PAF is a potent stimulus to enhance the metabolism of arachidonate through the cyclooxygenase pathway. Plasma levels of thromboxane were markedly elevated in the coronary vein and systemic circulation. Therefore, it is pertinent to conclude the TXA_2 generation took place in extracardiac elements, probably platelets.

In summary, the experiments conducted on the intact domestic pig heart clearly indicate that PAF can produce all of the cardiac and systemic hemodynamic consequences of acute anaphylactic shock: hypotension, cardiac ischemia, arrhythmias, and contraction dysfunction. The interaction of PAF with the cardiac and vascular tissue is complex and involves both direct (e.g., vascular dilation) and indirect (coronary constriction) actions, the latter most probably thromboxane mediated.

Although PAF release has been related primarily to acute hypersensitivity reactions (e.g., anaphylaxis), it is important to note that PAF might play a role in many different pathological processes that are related to platelets (Dryfuss and Zahavi, 1973). Of particular interest in this regard are spastic events of the coronary arteries seen in variant angina syndromes. Hyperaggregability of platelets to injured blood vessels (i.e., minimal atherosclerosis, injured endothelium) might lead to local release of PAF. Since PAF is also a potent chemoattractant and activator of polymorphonuclear cells (Pinckard *et al.*, 1979), it is conceivable that many of these activated cellular elements (platelets, WBC) and their products (leukotrienes, PAF, thromboxane) act in concert to modulate coronary tone and cardiac function.

REFERENCES

Addonazio, V. P., Westein, L., Fisher, C. A., Feldman, P., Strauss, J. F., and Harber, A. H., 1982, Platelet mediated cardiac ischemia, *J. Surg. Res.* **33:**402–408.

Benveniste, J., Rubin, R., Chingard, M., Jouvin Marche, E., and LeCouedu, J. P., 1982, Release of platelet activating factor (PAF-acether) and 2-lyso PAF acether from three cell types, *Agents Actions* **12:**711–713.

Benveniste, J., Boullet, C., Brink, C., and Labat, C., 1983, The action of PAF-acether (platelet activating factor) on guinea pig isolated heart preparation, *Br. J. Pharmacol.* **80:**81–83.

Cappuro, N., and Levi, R., 1975, The heart as a target organ in systemic allergic reactions: Comparison of cardiac anaphylaxis *in vivo* and *in vitro*, *Circ. Res.* **36:**520–528.

Demopoulos, C. A., and Pinckard, R. N., and Hanahan, D. J., 1979, Platelet activating factor. Evidence for 1-O-alkyl-2-acetyl-*sn*-glyceryl-3-phosphocholine as the active component (a new class of lipid chemical mediators), *J. Biol. Chem.* **254**:9355–9358.

Dryfuss, F., and Zahavi, J., 1973, Adenosine diphosphate induced platelet aggregation in myocardial infarction and ischemic heart disease, *Atherosclerosis* **17**:107–111.

Ezra, D., Feuerstein, G., Erne, J., and Goldstein, R. E., 1984, Biphasic coronary flow response to sustained leukotriene D_4 administration, *J. Am. Coll. Cardiol.* **3**:502.

Feuerstein, G., Zukowska-Grojec, L., Krausz, M. M., Blank, M. L., Snyder, F., and Kopin, I. J., 1982, Cardiovascular and sympathetic effects of 1-O-hexadecyl-2-acetyl-*sn*-glycero-3-phosphocholine in conscious SHR and WKY rats, *Clin. Exp. Hypertens.* **A4**:1335–1350.

Feuerstein, G., Boyd, L. M., Ezra, D., and Goldstein, R. E., 1984a, Effect of platelet activating factor on coronary circulation of the domestic pig, *Am. J. Physiol.* **246**:H466–H471.

Feuerstein, G., Lux, W. E., Snyder, F., Ezra, D., and Faden, A. I., 1984b, Hypotension produced by platelet activating factor is reversed by thyrotropin releasing hormone, *Circ. Shock* **13**:255–260.

Granstrom E., and Kindahl, H., 1976, Radioimmunoassay for prostaglandin metabolites, *Adv. Prostaglandin Thromboxane Res.* **1**:81–92.

Halonen, M., Palmer, J. D., Lohman, I. C., McManus, L. M., and Pinckard, R. N., 1980, Respiratory and circulatory alterations induced by acetyl glyceryl ether phosphorylcholine (AGEPC), a mediator of IgE anaphylaxis in the rabbit, *Am. Rev. Respir. Dis.* **122**:915–924.

Hayes, E. C., Lombardo, D. L., Girard, Y., Maycock, A. L., Rokach, J., Rosenthal, A. S., Young, R. N., Egan, R. W., and Zweerink, H. J., 1983, Measuring leukotrienes of slow reacting substance of anaphylaxis: Development of a specific radioimmunoassay, *J. Immunol.* **131**:429–433.

Herd, J. A., and Barger, A. C., 1964, Simplified technique for chronic catheterization of blood vessels, *J. Appl. Physiol.* **19**:791–792.

Levi, R., Burke, J. A., Guo, L. G., Hattori, Y.,. Hoppens, C. M., McManus, L. M., Hanahan, D. J., and Pinckard, P. N., 1984, Acetyl glyceryl ether phosphorylcholine (AGEPC): A putative mediator of cardiac anaphylaxis in the guinea pig, *Circ. Res.* **54**:117–124.

Pinckard, R. N., Farr, R. S., and Hanahan, D. J., 1979, Physicochemical and functional identity of rabbit platelet activating factor (PAF-acether) released *in vivo* during IgE anaphylaxis with PAF released *in vitro* from IgE sensitized basophils, *J. Immunol.* **125**:1847–1857.

Zavecz, J. H., and Levi, B., 1977, Separation of primary and secondary cardiovascular events in systemic anaphylaxis, *Circ. Res.* **40**:15–19.

30

Distinct Sulfidopeptide Leukotriene Receptors

BARBARA J. BALLERMANN, TAK H. LEE, ROBERT A. LEWIS, and K. FRANK AUSTEN

1. INTRODUCTION

The oxidative metabolism of arachidonic acid by 5-lipoxygenase to form 5-hydroperoxy-6-*trans*-8,11,14-*cis*-eicosatetraenoic acid (5-HPETE) is followed by enzymatic conversion of 5-HPETE to 5,6-oxido-7,9-*trans*-11,14-*cis*-eicosatetraenoic acid (LTA$_4$). An epoxide hydrolase converts LTA$_4$ to 5S,12R-dihydroxy-6,14-*cis*-8,10-*trans*-eicosatetraenoic acid (LTB$_4$), whereas a glutathione-S-transferase adducts glutathione to yield 5S-hydroxy-6R-S-glutathionyl-7,9-*trans*-11,14-*cis*-eicosatetraenoic acid (LTC$_4$). Sequential cleavages by γ-glutamyltranspeptidase of glutamic acid and by a dipeptidase of glycine form 5S-hydroxy-6R-S-cysteinylglycyl-7,9-*trans*-11,14-*cis*-eicosatetraenoic acid (LTD$_4$) and 5S-hydroxy-6R-S-cysteinyl-7,9-*trans*-11,14-*cis*-eicosatetraenoic acid (LTE$_4$), respectively. Leukotrienes C$_4$, D$_4$, and E$_4$ are generically described as sulfidopeptide leukotrienes and constitute the activity previously termed slow reacting substance of anaphylaxis (Samuelsson, 1983; Lewis and Austen, 1984). The evidence indicating that the sulfidopeptide leukotrienes exert their physiological effects through interaction with several distinct receptors includes their differential functional activities on different tissues, the differential effects of pharmacological inhibitors on the agonist effects of the three sulfidopeptide leukotrienes, different receptor characteristics defined by radioligand

BARBARA J. BALLERMANN, TAK H. LEE, ROBERT A. LEWIS, and K. FRANK AUSTEN • Department of Medicine, Harvard Medical School, and Department of Rheumatology and Immunology, Brigham and Women's Hospital, Boston, Massachusetts 02115.

binding studies, and apparent differences in the subcellular distribution of sulfidopeptide leukotriene binding sites.

2. FUNCTIONAL EVIDENCE FOR DISTINCT SULFIDOPEPTIDE LEUKOTRIENE RECEPTORS

The molar concentrations of sulfidopeptide leukotrienes required to produce equivalent degrees of smooth muscle contraction *in vitro* are in a ratio for $LTC_4 : LTD_4 : LTE_4$ of $1 : 0.2 : 1.2$ in guinea pig ileum, $1 : 0.1 : 0.3$ in guinea pig lung parenchymal strips, and $1 : 1 : 0.1$ in guinea pig tracheal spirals (Lewis *et al.*, 1980; Drazen *et al.*, 1983; Lee *et al.*, 1984). The ratios of the molar concentrations of $LTC_4 : LTD_4 : LTE_4$ that reduce human and guinea pig myocardial contractility *in vitro* are $1 : 0.04 : 10$ and $1 : 0.1 : 200$, respectively (Burke *et al.*, 1982). The vasoconstrictor action of LTC_4 in the renal vasculature of the rat was more potent than that of LTD_4, whereas LTE_4 was almost without effect (Rosenthal and Pace-Asciak, 1983). After intradermal injection into humans, the potencies of the three sulfidopeptide leukotrienes were relatively equivalent in eliciting a wheal and flare (Soter *et al.*, 1983). Thus, the potency order of physiological leukotriene action in guinea pig ileum and heart and in human myocardium is $LTD_4 > LTC_4 > LTE_4$, in guinea pig lung parenchymal strips is $LTD_4 > LTE_4 > LTC_4$, in guinea pig trachea is $LTE_4 > LTD_4 = LTC_4$, and in the rat renal vasculature is $LTC_4 > LTD_4 \gg LTE_4$.

Further evidence for separate modes of action of LTC_4 and LTD_4 is provided by the distinct patterns of the time dependence of the elicited spasmogenic responses in smooth muscle. The contraction of guinea pig ileal smooth muscle strips produced by LTC_4 requires a 60-sec latent period after exposure to LTC_4, whereas the contraction of guinea pig ileum produced by LTD_4 is immediate. By use of radiolabeled LTC_4 and sequential RP-HPLC analysis of the molecular integrity of the agonist, the contractile response in guinea pig ileal strips was shown to be caused by a direct action of that compound and not by its conversion to LTD_4. Inhibition of the conversion of LTC_4 to LTD_4 by serine–borate complex did not extend the latency period required for the contractile response to LTC_4, further reinforcing the notion of different signaling mechanisms for LTC_4 and LTD_4 (Krilis *et al.*, 1983a).

Whereas LTC_4 and LTE_4 elicited monophasic contractions in guinea pig peripheral airway strips, LTD_4 evoked a biphasic response, that is, a partial dose-dependent contraction that plateaued at concentrations in the range of 10^{-11} M LTD_4, followed by a progressive response with concentrations of $LTD_4 > 10^{-10}$ M that reached its maximum at 10^{-8} M. Although FPL55712 inhibited the contractile response of guinea pig lung parenchymal strips to low concentrations of LTD_4, it was without effect on the spasmogenic response to high concentrations of LTD_4 or on any portion of the monophasic LTC_4 response. These initial physiological observations prompted the suggestion that LTC_4 and LTD_4 act through distinct receptors (Drazen *et al.*, 1980). More recent studies demonstrated that the LTC_4 response and the high-concentration LTD_4 response in guinea pig lung parenchymal strips are inhibited by the calcium channel blocker diltiazem in contrast to the inhibition of the response to low concentrations of LTD_4 only with FPL55712. Combination of the two inhibitors attenuated the contraction of lung parenchymal strips at all concentrations of LTD_4 (Drazen and Fanta, 1984), suggesting that the biphasic LTD_4 effects were mediated through two separate receptors and that the effect of high concentrations of LTD_4 might be mediated through the same receptor as that for LTC_4. The smooth muscle contractile action of LTE_4 is generally less potent than that of LTD_4 and LTC_4, which would be consistent with a low-affinity interaction with either or both of the receptors considered to mediate LTC_4 and LTD_4 action. Guinea pig tracheal smooth muscle, however, was shown to be more sensitive to the contractile influence of LTE_4 when compared to LTC_4 or LTD_4, indicating that a distinct LTE_4 subclass receptor might exist in this tissue (Drazen *et al.*, 1983).

In addition, exposure of guinea pig tracheal smooth muscle but not lung parenchymal strips to LTE_4 followed by washing and return of tension to base-line level results in hyperresponsiveness to subsequent stimulation with histamine, an effect that is not produced by LTD_4 or LTC_4 at concentrations resulting in equivalent initial contractile responses. This LTE_4-mediated effect was not inhibited by low-dose FPL55712, which blocked the contractile phase but was preempted by indomethacin, which had no effect on the LTE_4-induced contraction (Lee *et al.*, 1984). Thus, LTE_4-induced airway hyperresponsiveness to histamine was clearly separated both in terms of tissue localization and in terms of pharmacological inhibition from the general spasmogenic effects of the sulfidopeptide leukotrienes. It nonetheless may represent a different postreceptor response in the guinea pig trachea rather than yet another separate receptor.

The differences in the order of relative physiological potencies for the sulfidopeptide leukotrienes in various tissues, the differences in the time courses of spasmogenic responses to the three compounds in the same tissues, the difference in effectiveness of inhibitors, and the distinct effect of LTE_4 in producing hyperresponsiveness and in contracting tracheal tissue cannot be explained by interaction of these agonists with a single population of receptors even if different affinities for each leukotriene are postulated, nor can they be solely accounted for by differential rates of leukotriene metabolism (Krilis *et al.*, 1983a). Thus, early (Lewis *et al.*, 1980; Drazen *et al.*, 1980; Burke *et al.*, 1982) and continuing (Rosenthal and Pace-Asciak, 1983; Drazen and Fanta, 1984; Lee *et al.*, 1984) physiological and pharmacological studies of the sulfidopeptide leukotriene subclasses indicate at least three recognition mechanisms and, in view of the stereochemical requirements for agonist action (Lewis and Austen, 1984), suggest the involvement of two or more distinct receptors. The existence of more than one receptor does not rule out interaction of more than one sulfidopeptide leukotriene with each receptor, is compatible with different postreceptor response mechanisms, and does not exclude actions that bypass plasma membrane receptors.

3. RADIOLIGAND BINDING STUDIES WITH [^{3}H]LTC$_4$

Leukotriene C$_4$ binding sites have been characterized in intact cells, namely, a cultured smooth muscle cell line derived from the hamster vas deferens (DDT$_1$) (Krilis *et al.*, 1983b), in intact tissues such as guinea pig ileum smooth muscle segments (Krilis *et al.*, 1984) and whole isolated rat renal glomeruli (Ballermann *et al.*, 1985), and in crude particulate fractions of whole rat lung (Pong *et al.*, 1983), guinea pig lung (Bruns *et al.*, 1983; Hogaboom *et al.*, 1983), and guinea pig ileum (Krilis *et al.*, 1984; Nicosia *et al.*, 1984).

Specific [^{3}H]LTC$_4$ binding to intact DDT$_1$ cells in 200 mM tris-HCl buffer, pH 7.4, with 1 mM CaCl$_2$ in the presence of 1 mM serine–borate complex at 4°C was time dependent and rapidly reversible on addition of excess unlabeled LTC$_4$. The LTC$_4$ binding site in this cell line was saturable and displayed a high affinity for LTC$_4$ with an equilibrium dissociation constant (K_d) of 5 nM derived from equilibrium binding studies with increasing concentrations of [^{3}H]LTC$_4$. The binding site was highly specific for LTC$_4$ in that LTD$_4$, LTE$_4$, and LTB$_4$ competed only at concentrations 2.5–3.0 orders of magnitude above that of the homoligand. The putative sulfidopeptide leukotriene antagonist FPL55712 was also relatively ineffective in competing for this binding site, with 50% displacement of [^{3}H]LTC$_4$ occurring at 8 µM. Physiologically active analogues of LTC$_4$, 5R,6S-LTC$_4$, 11-*trans*-LTC$_4$, and the C1-monoamide derivative of LTC$_4$, were potent competitors for [^{3}H]LTC$_4$ binding compatible with this biological activity, whereas deamino-LTC$_4$, an analogue with minimal contractile activity, was ineffective in the binding competition assay. These studies with intact cells established the presence of a stereospecific subclass receptor for LTC$_4$ (Krilis *et al.*, 1983b) and were presumptive evidence for the existence of separate additional receptors.

Under assay conditions comparable to those used with DDT$_1$ cells, guinea pig ileal segments demonstrated similar [^{3}H]LTC$_4$ binding characteristics in terms of saturability, reversibility, and binding affinity (Krilis *et al.*, 1984): the K_d was 7.6 nM, and the density of binding sites was 0.23 pmol/mg protein. It was also observed that the density of LTC$_4$ binding sites in a subcellular fraction of guinea pig smooth muscle enriched for mitochrondria was 1.6 pmol/mg protein as compared to 0.5 pmol/mg protein in the plasma membrane fraction prepared from the same tissue (Krilis *et al.*, 1984).

Further, the competition studies for [^{3}H]LTC$_4$ with both subcellular fractions demonstrated a subclass specific receptor for LTC$_4$ in this tissue known to respond to the other natural sulfidopeptide leukotrienes, LTD$_4$ and LTE$_4$. Studies by others (Nicosia *et al.*, 1984) of a crude guinea pig ileum particulate fraction prepared by centrifugation of the homogenized tissue at 10,000 $\times$ g for 15 min suggested a nonhomogeneous population of LTC$_4$ binding sites on the basis of a nonlinear Scatchard plot and a biphasic [^{3}H]LTC$_4$ dissociation curve on the addition of excess unlabeled LTC$_4$; LTC$_4$ receptor affinities and receptor densities were not calculated. However, nonlinearity of the Scatchard plot in that study is most likely caused by failure to reach an equilibrium at low concentrations of radioligand, because incubation was only for 10 min. Biphasic dissociation curves are commonly observed when excess unlabeled ligand is used to displace the labeled probe (Pollet *et al.*, 1977). Such curves do not necessarily identify populations of receptors with different affinities but may reflect negative cooperativity or dynamic alterations in receptor affinity.

The $[^3H]LTC_4$ binding sites in crude particulate fractions of whole rat lung homogenates have been characterized with use of a 50 mM tris-HCl buffer, pH 7.4, at 4°C with 20 mM $CaCl_2$. Saturability, time dependence of binding, reversibility, and specificity for the homoligand relative to binding inhibition by LTD_4, LTE_4, and FPL55712 were similar to that described for intact DDT_1 cells. However, analysis of equilibrium binding data revealed an equilibrium dissociation constant of 41 nM and a receptor density of 31 pmol/mg protein (Pong *et al.*, 1983). Thus, under assay conditions similar to those used for the study of DDT_1 cells and guinea pig ileal segments, the LTC_4 receptor affinity in lung was lower by one order of magnitude as compared to 5 nM for DDT_1 cells and 7.6 nM for guinea pig ileum segments (Table I). Furthermore, the receptor density in lung particulate fractions was two orders of magnitude greater than that found in intact guinea pig ileum; this large LTC_4 receptor density in the rat lung particulate fraction may reflect subcellular receptors exposed during homogenization as well as some plasma membrane receptors.

Evaluation for LTC_4 receptors of a readily accessible portion of the renal vasculature, namely, whole rat renal glomeruli, was recently undertaken after physiological studies had demonstrated a direct vasoconstrictor effect of LTC_4 in the kidney (Rosenthal and Pace-Asciak, 1983; Badr *et al.*, 1984). Binding of $[^3H]LTC_4$ in 20 mM HEPES buffer at 4°C in the presence of 125 mM NaCl and 5 mM $CaCl_2$ was time dependent, reversible, and saturable. Equilibrium binding parameters for one homogeneous population of LTC_4 receptors in this tissue were an equilibrium dissociation constant of 30 nM and a receptor density of 8.5 pmol/mg glomerular protein (Ballermann *et al.*, 1985). The potency of competition for this binding site by LTD_4, LTE_4, and FPL55712 was three orders of magnitude less than that of the homoligand LTC_4, similar to findings reported in DDT_1 cells (Krilis *et al.*, 1983b), guinea pig ileum (Krilis *et al.*, 1984), and rat lung (Pong *et al.*, 1983).

Thus, LTC_4 receptors in nonvascular and vascular tissue exhibit a high degree of specificity for LTC_4 and do not bind LTD_4 with an affinity high enough to account for pharmacological LTD_4 action. The LTC_4 receptor affinity differs among tissues as evidenced by equilibrium dissociation constants of 5 and 7.6 nM in DDT_1 cells and guinea pig ileum smooth muscle, respectively, contrasted with 41 and 30 nM in crude rat lung homogenate particulate fraction and intact rat renal glomeruli, respectively, thereby indicating that LTC_4 receptors may display a degree of heterogeneity in different tissues or perhaps across species. The demonstration in guinea pig ileum of a large number of subcellular LTC_4 receptors raises the intriguing possibility that LTC_4 action may include mobilization of the LTC_4 receptors to the

TABLE I. Radioligand Binding Studies

Tissue	Radioactive probe	K_d (nM)	R_0	K_i (nM)				Effect of monovalent, divalent cations, and guanine nucleotides on radioligand binding	Reference
				LTC$_4$	LTD$_4$	LTE$_4$	FPL55712		
DDT$_1$ smooth muscle cell line	[^{3}H]LTC$_4$	5.0	0.4 pmol/10^6 cells	4.4	2,000	13,000	7,700	ND[a]	Krilis *et al.*, 1983b
GP[b] ileal smooth muscle (intact)	[^{3}H]LTC$_4$	7.6	0.23 pmol/ mg protein	ND	ND	ND	ND	ND	Krilis *et al.*, 1984
GP ileum crude particulate fraction	[^{3}H]LTC$_4$	ND	ND	100	60,000	inactive	10,000	Ca$^+$ > Mg^{2+} stimulatory	Nicosia *et al.*, 1984
Intact rat renal glomeruli	[^{3}H]LTC$_4$	30	8.5 pmol/mg protein	35	6,800	14,000	49,000		Ballermann *et al.*, 1985
Rat lung crude particulate fraction	[^{3}H]LTC$_4$	41	31 pmol/mg protein	40	4,000	>10,000	16,000	Ca^{2+}, Mg^{2+}, Mn^{2+} stimulatory	Pong *et al.*, 1983
GP lung crude particulate fraction	[^{3}H]LTC$_4$	ND	ND	8	12,000	40	14,000	Ca^{2+}, Mg^{2+} stimulatory; Na$^+$, GTP inhibitory	Bruns *et al.*, 1983
	[^{3}H]LTD$_4$	ND	ND	8	16,000	98	11,000		
GP lung crude particulate fraction	[^{3}H]LTC$_4$	ND	ND	75	>30,000	>30,000	>30,000		Hogaboom *et al.*, 1983
	[^{3}H]LTD$_4$	ND	ND	523	7.5	214	13,000	Gpp(NH)p inhibitory	
GP lung crude particulate fraction	[^{3}H]LTD$_4$	0.21	0.34 pmol/ mg protein	31	0.32	0.78	150	Ca^{2+}, Mg^{2+}, Mn^{2+} stimulatory; Na$^+$, GTP inhibitory	Pong and DeHaven, 1983

[a] ND, not done.
[b] GP, guinea pig.

plasma membrane or alternatively that LTC_4 acts at a subcellular site. Either of these possibilities may explain the characteristically slow onset of LTC_4-induced smooth muscle contraction.

4. RADIOLIGAND BINDING STUDIES WITH [³H]LTD₄

In addition to the demonstration that LTD_4 does not bind to LTC_4 receptors with high affinity, radioligand binding studies utilizing [³H]LTD₄ as a probe have shown distinct receptors for LTD_4 in crude particulate fractions of guinea pig lung (Pong and DeHaven, 1983; Hogaboom *et al.*, 1983) and on a plasma-membrane-enriched fraction of guinea pig ileum (Krilis *et al.*, 1985). The [³H]LTD₄ binding to guinea pig lung at 4°C and 20°C was saturable and reversible. The [³H]LTD₄ binding was enhanced by the divalent cations Mn^{2+}, Mg^{2+}, and Ca^{2+} and was inhibited by 200 μM GTP and in a concentration-dependent manner by Na^+ (0.5–50 mM). The inhibition of [³H]LTD₄ binding by GTP and Na^+ was attributed to a reduction in LTD_4 receptor affinity, since these agonists increased the dissociation rate of [³H]LTD₄ from the receptor. Binding parameters derived from [³H]LTD₄ saturation studies at equilibrium showed a very high affinity with a K_d at 20°C of 0.055 nM and at 0°C of 0.21 nM. The receptor density was 0.34 pmol/mg protein. Competition by LTC_4 for this site required concentrations two orders of magnitude higher than for the homoligand, whereas LTE_4 was nearly as effective as LTD_4 in competing with [³H]LTD₄. FPL55712 competed at concentrations in the range of 10^{-7} M. Although FPL55712 was 100-fold as effective in competing for this LTD_4 site as it was in competition for LTC_4 receptors in diverse tissues (Table I), it is still more than 3 logs less potent than the homoligand in binding assays of either receptor.

Radioligand binding studies by Hogaboom *et al.* (1983) in a crude particulate fraction of homogenized guinea pig lung with both [³H]LTC₄ and [³H]LTD₄ as radiolabeled probes demonstated separate binding sites for LTC_4 and LTD_4 in this tissue. The LTC_4 receptor characteristics were similar to those described in rat lung homogenate (Pong *et al.*, 1983), whereas the equilibrium dissociation constant of the LTD_4 receptor was estimated at 7.5 nM by binding competition. Furthermore, LTE_4 competed more effectively than LTC_4 for this site, and 100 μM of guanylyl imidodiphosphate [Gpp(NH)p] was found to inhibit [³H]LTD₄ binding, although there was no effect on [³H]LTC₄ binding (Table I).

Binding of [³H]LTD₄ and [³H]LTC₄ on crude guinea pig lung homogenate particulate fractions was also studied by Bruns *et al.* (1983), who observed that [³H]LTC₄ binding occurred mostly after conversion to [³H]LTD₄. Competition studies, which yielded a potency order for synthetic sulfidopeptide leukotrienes of $LTC_4 > LTD_4 > LTE_4$, were therefore similar for both radioactive probes. However, the apparent effectiveness of unlabeled LTC_4 in competing for the binding site makes it difficult to interpret competition data for this putative LTD_4 receptor, which differ markedly from those found by others (Pong and DeHaven, 1983;

Hogaboom *et al.*, 1983). Concentration-dependent inhibition of [^{3}H]LTD$_4$ binding by guanine nucleotides and by sodium ion was also found in this study (Bruns *et al.*, 1983).

The LTD$_4$ receptors identified in guinea pig plasma membrane fractions (Krilis *et al.*, 1985) have a K_d of 2.2 nM, which would be compatible with the fourfold greater potency of LTD$_4$ relative to LTC$_4$ in ileum tissue in the presence of sodium-containing buffers. The physiological data of a biphasic dose response to LTD$_4$ in guinea pig parenchymal strips and the very different radioligand binding affinities observed for lung and ileum suggest subclass heterogeneity with regard to the affinity of the LTD$_4$ receptor.

5. CONCLUDING COMMENTS

Experimental evidence from physiological and receptor binding studies, summarized in Tables II and III, indicates that separate receptors exist for sulfidopeptide leukotrienes C$_4$ and D$_4$, and possibly E$_4$. Separate receptors for LTC$_4$ and LTD$_4$ have been identified and characterized by physiological and radioligand binding criteria. The LTD$_4$ receptors appear to be plasma-membrane associated and display characteristics previously described for many hormone receptors such as inhibition of agonist binding by guanine nucleotides and Na$^+$, suggesting a postreceptor linkage to adenylate cyclase of an inhibitory nature. The LTC$_4$ receptors have a substantial subcellular distribution; agonist binding to the receptor is not influenced by monovalent ions or guanine nucleotides but is enhanced by divalent cations. Within these two major subclass receptors with specificity for LTC$_4$ and LTD$_4$, respectively, there is an additional heterogeneity in terms of affinities. The distinct binding sites each manifest a characteristic potency order for the displacement of the respective homoligand by other leukotrienes and by FPL55712 (Table I). The

TABLE II. Evidence for Subclass Specific Receptors for the Sulfidopeptide Leukotrienes: Physiological Studies

Tissue	LTC$_4$	LTD$_4$	LTE
Guinea pig ileum			
Equiactive molar ratios relative to LTC$_4$	1.0	0.2	1.2
Latency period for ileal response	60 sec	None	None
Guinea pig lung parenchyma			
Equiactive molar ratios relative to LTC$_4$	1.0	0.1	0.3
Nature of response	Monophasic	Biphasic	Monophasic
Pharmacologic inhibition	Diltiazem	FPL55712	FPL55712
Guinea pig trachea			
Equiactive molar ratios relative to LTC$_4$	1.0	1.0	0.1
Induction of hyperresponsiveness to histamine/GP trachea	None	None	Threefold

TABLE III. Evidence for Subclass Specific Receptors for the Sulfidopeptide Leukotrienes: Radioligand Binding Studies

Tissue	Radioligand	LTC_4		LTD_4		LTE_4	
		K_d(nM)	K_i(nM)	K_d(nM)	K_i(nM)	K_d(nM)	K_i(nM)
Hamster DDT_1 cell	[^{3}H]LTC_4	5.0	4.4	—	2,000	—	13,000
Guinea pig ileum segment	[^{3}H]LTC_4	7.6	—	—	—	—	—
12,000 × g fraction	[^{3}H]LTC_4	—	12.0	—	800	—	>10,000
Plasma membrane fraction	[^{3}H]LTC_4	8.5	—	—	—	—	—
Plasma membrane fraction	[^{3}H]LTD_4	—	—	2.2	—	—	—
Guinea pig lung homogenate 45,000 × g (without Na^+)	[^{3}H]LTD_4	—	31.0	0.22	0.32	—	0.78

effects of LTE_4 on guinea pig lung parenchymal strips may be mediated by LTD_4 receptors as evidenced by effective competition by LTE_4 for the LTD_4 receptor in crude lung particulate fractions (Pong and DeHaven, 1983).However, a third receptor or mechanism may mediate the distinct actions of LTE_4 on guinea pig tracheal spirals, where this agonist exhibits greater contractile potency than LTD_4 and LTC_4 and elicits hyperreactivity to histamine, which is not a nonspecific response to a prior contraction (Lee *et al.*, 1984).

REFERENCES

Badr, K. F., Baylis, C., Pfeffer, J. M., Pfeffer, M. A., Soberman, R. J., Lewis, R. A., Austen, K. F., Corey, E. J., and Brenner, B. M., 1984, Renal and systemic hemodynamic responses to intravenous infusion of leukotriene C_4 in the rat, *Circ. Res.* **54**:492–499.

Ballermann, B. J., Lewis, R. A., Corey, E. J., Austen, K. F., and Brenner, B. M., 1985, Identification and characterization of leukotriene C_4 receptors in isolated rat renal glomeruli, *Circ. Res.* **56**:324–330.

Bruns, R. F., Thomsen, W. J., and Pugsley, T. A., 1983, Binding of leukotrienes C_4 and D_4 to membranes from guinea pig lung: Regulation by ions and guanine nucleotides, *Life Sci.* **33**:645–653.

Burke, J. A., Levi, R., Guo, Z.-G., and Corey, E. J., 1982, Effects on human and guinea pig cardiac preparations *in vitro*, *J. Pharmacol. Exp. Ther.* **221**:235–241.

Drazen, J. M., and Fanta, C., 1984, Physiologic studies of receptor heterogeneity for the sulfidopeptide leukotrienes, *Clin. Res.* **32**:528A.

Drazen, J. M., Austen, K. F., Lewis, R. A., Clark, D. A., Goto, G., Marfat, A., and Corey, E. J., 1980, Comparative airway and vascular activities of leukotrienes C-1 and D *in vivo* and *in vitro*, *Proc. Natl. Acad. Sci. U.S.A.* **77**:4354–4358.

Drazen, J. M., Lewis, R. A., Austen, K. F., and Corey, E. J., 1983, Pulmonary pharmacology of the SRS-A leukotrienes, in: *Leukotrienes and Prostacyclin* (F. Berti, G. Folco, and G. P. Velo, eds.), Plenum Press, New York, pp. 125–134.

Hogaboom, G. K., Mong, S., Wu, H.-L., and Crooke, S., 1983, Peptidoleukotrienes: Distinct receptors for leukotrienes C_4 and D_4 in the guinea pig lung, *Biochem. Biophys. Res. Commun.* **116**:1136–1143.

Krilis, S., Lewis, R. A., Corey, E. J., and Austen, K. F., 1983a, Bioconversion of C-6 sulfidopeptide leukotrienes by the responding guinea pig ileum determines the time course of its contraction, *J. Clin. Invest.* **71**:909–915.

Krilis, S., Lewis, R. A., Corey, E. J., and Austen, K. F., 1983b, Specific binding of leukotriene C_4 on a smooth muscle cell line, *J. Clin. Invest.* **72**:1516–1519.

Krilis, S., Lewis, R. A., Corey, E. J., and Austen, K. F., 1984, Specific binding of leukotriene C_4 to plasma membrane and subcellular receptors of ileal smooth muscle cells, *Proc. Natl. Acad. Sci. U.S.A.* **81**:4529–4533.

Krilis, S., Lewis, R. A., Drazen, J. M., and Austen, K. F., 1985, Subclasses of receptors for the sulfidopeptide leukotrienes, in: *Prostaglandins and Membrane Ion Transport* (P. Braquet, R. P. Garay, J. C. Frölich, and S. Nicosia, eds.), Raven Press, New York, pp. 91–97.

Lee, T. H., Austen, K. F., Corey, E. J., and Drazen, J. M., 1984, LTE_4-induced airway hyper-responsiveness of guinea pig tracheal smooth muscle to histamine, and evidence for three separate sulfidopeptide leukotriene receptors, *Proc. Natl. Acad. Sci. U.S.A.* **81**:4922–4925.

Lewis, R. A., Drazen, J. M., Austen, K. F., Clark, D. A., and Corey, E. J., 1980, Identification of the C(6)-S-conjugate of leukotriene A with cysteine as a naturally occuring slow reacting substance of anaphylaxis (SRS-A). Importance of the 11-*cis* geometry, *Biochem. Biophys. Res. Commun.* **96**:271–277.

Lewis, R. A., and Austen, K. F., 1984, The biologically active leukotrienes: Biosynthesis, metabolism, receptors, functions and pharmacology, *J. Clin. Invest.* **73**:889–897.

Nicosia, S., Crowley, H. J., Oliva, D., and Welton, A. F., 1984, Binding sites for ^3H-LTC_4 in membranes from guinea pig ileal longitudinal muscle, *Prostaglandins* **27**:483–494.

Pollet, R. J., Standaert, M. L., and Haase, B. A., 1977, Insulin binding to the human lymphocyte receptor: Evaluation of the negative cooperativity model, *J. Biol. Chem.* **252**:5828–5834.

Pong, S. S., and DeHaven, R. N., 1983, Characterization of a leukotriene D_4 receptor in guinea pig lung, *Proc. Natl. Acad. Sci. U.S.A.* **80**:7415–7419.

Pong, S. S., DeHaven, R. N., Kuehl, R. A., Jr., and Egan, R. W., 1983, Leukotriene C_4 binding to rat lung membranes, *J. Biol. Chem.* **258**:9616–9619.

Rosenthal, A., and Pace-Asciak, C. R., 1983, Potent vasoconstriction of the isolated perfused rat kidney by leukotrienes C_4 and D_4, *Can. J. Physiol. Pharmacol.* **61**:325–328.

Samuelsson, B., 1983, Leukotrienes: Mediators of immediate hypersensitivity reactions and inflammation, *Science* **220**:568–575.

Soter, N. A., Lewis, R. A., Corey, E. J., and Austen, K. F., 1983, Local effects of synthetic leukotrienes, *J. Invest. Dermatol.* **80**:115–119.

31

Some Novel Aspects of the Function of Prostaglandin E_2 in the Coronary Circulation

THOMAS H. HINTZE and GABOR KALEY

1. INTRODUCTION

Recently, much work has centered on the effects and possible physiological and pathological importance of coronary prostacyclin synthesis (Fitzpatrick *et al.*, 1978; Hyman *et al.*, 1978; Needleman and Kaley, 1978; Dusting *et al.*, 1979; Lefer *et al.*, 1979; Aiken *et al.*, 1980; Jugdutt *et al.*, 1981). Prostacyclin, however, is not the only cyclooxygenase product that is produced in the coronary vasculature. In this context, recent work by Gerritsen and Printz (1981) has shown that PGE_2 is produced in large coronary arteries, coronary microvessels, and also in microvascular endothelial cells (Gerritsen and Cheli, 1984). Additionally, we have demonstrated that PGE_2 has potent effects on coronary blood flow (Hintze and Kaley, 1977) and under special conditions may activate cardiac ventricular receptors, leading to reflex effects on the circulation in general (Hintze and Kaley, 1984). Furthermore, Jugdutt *et al.* (1981) have shown that infusions of PGE_1 ameliorate some of the harmful consequences of ischemia on the myocardium. The purpose of this chapter is to review the effects of PGE_2 on the myocardium and coronary circulation and to describe our own, more recent observations on the effects of PGE_2 on blood flow distribution in the myocardium, on large coronary vessels in conscious dogs, and on the possible reflex regulation of circulatory function.

Prostaglandins, particularly prostacyclin, PGE_1, and PGE_2, have previously been shown to increase blood flow in the pump-perfused or autoperfused dog heart

THOMAS H. HINTZE and GABOR KALEY ● Department of Physiology, New York Medical College, Valhalla, New York 10595.

(Carlson and Örö, 1966; Maxwell, 1967; Nakano, 1968; Nutter and Crumly, 1972; Hutton *et al.*, 1973; Rowe and Afonso, 1974; Hintze and Kaley, 1977). Furthermore, the intracoronary injection of arachidonic acid leads to the synthesis of prostaglandins, including PGE_2, within the coronary circulation and to subsequent increases in coronary blood flow, which are eliminated by administration of prostaglandin synthesis inhibitors (Hintze and Kaley, 1977). Many autoregulatory functions of the coronary circulation, including reactive hyperemia, have been attributed to the stimulation of synthesis of prostaglandins (Afonso *et al.*, 1974; Alexander *et al.*, 1975). These results, however, have not been confirmed by other groups of investigators (Owen *et al.*, 1975; Rubio and Berne, 1975; Giles and Wilcken, 1977; Hintze and Kaley, 1977; Harlan *et al.*, 1978; Sunahara and Talesnik, 1979; Hintze and Vatner, 1984). The evidence concerning this question has been reviewed previously (Parratt and Marshall, 1978; Belloni, 1979). As a consequence of myocardial infarction, when coronary flow is interrupted, an increase in the myocardial production of prostaglandins, including PGE_2, has been observed (Berger *et al.*, 1976; Kraemer *et al.*, 1976; Coker *et al.*, 1981; Jugdutt *et al.*, 1981).

In studies that eventually led to the discovery of prostacyclin, it was noticed that whereas PGE_2 caused contraction of large coronary artery strips *in vitro*, application of arachidonic acid induced relaxation. The relaxation to arachidonic acid was prostaglandin dependent since it was eliminated by preincubation of the strips with indomethacin (Kulkarni *et al.*, 1976; Dusting *et al.*, 1977). This was clear evidence that large coronary vessels produce a prostaglandin other than PGE_2. The above authors also postulated that PGE_2 would constrict large coronary arteries *in vivo* and hence could be one of the etiologic factors responsible for coronary vasospasm. In support of this hypothesis, Uchida and Murao (1979) have shown that the cyclic reductions in coronary blood flow that occur following severe stenosis of a coronary artery in anesthetized dogs correlate well with the generation of PGE_2 in the coronary circulation and that both the reductions in flow and PGE_2 release can be eliminated by administration of a receptor blocker of prostaglandins.

Previous studies from our laboratory have described the Bezold–Jarisch-like reflex responses following the administration of prostacyclin in anesthetized (Hintze *et al.*, 1979, 1982, 1984; Kaley *et al.*, 1980, 1981; Hintze and Kaley, 1984) or conscious dogs (Hintze *et al.*, 1981). These observations have been confirmed by other investigators (Chapple *et al.*, 1978; Jentzer *et al.*, 1979; Chiba and Malik, 1980; Dusting and Vane, 1980; Chiavarelli *et al.*, 1982). In a study designed to locate the receptors responsible for the reflex bradycardia to prostacyclin and in order to discern the bradycardia more easily following intracoronary injection of these agents by reducing the interactions between cardiopulmonary receptors and systemic arterial baroreflexes, we observed that circumflex coronary artery injection of PGE_2 also leads to a vagal reflex bradycardia (Hintze and Kaley, 1984). Stimulation of myocardial prostaglandin synthesis following intracoronary arachidonic acid injection had similar effects on heart rate and reflexly reduced peripheral resistance as well (Hintze and Kaley, 1984). Coleridge and Coleridge (1980) have shown that vagal C-fiber afferents are activated following left atrial injections of PGE_2, resulting in a vagally mediated reflex bradycardia. An early study by Giles *et al.* (1969) also showed that injection of PGE_1 into anesthetized dogs resulted in a reflex bradycardia despite a reduction in systemic arterial pressure.

2. RESULTS

2.1. Effects of PGE_2 and Stimulation of Prostaglandin Synthesis by Arachidonic Acid on the Transmural Distribution of Myocardial Blood Flow

In these studies, radioactive microspheres were injected into the left atrium in order to measure the transmural distribution of blood flow following the intracoronary (left circumflex) artery injection of 200 μg of arachidonic acid or 2 μg of PGE_2 in anesthetized dogs. The methods used have been described previously (Hintze and Kaley 1977, 1984). The hearts were sectioned as shown in Fig. 1, and blood flow was quantitated using standard methods (Rudolph and Heymann, 1967; Domench *et al.*, 1969; Buckberg *et al.*, 1971; Becker *et al.*, 1973; Becker, 1976; Heymann *et al.*, 1977).

A record from a typical experiment is shown in Fig. 2. In these dogs ($N = 17$), PGE_2 increased coronary blood flow significantly as measured with an electromagnetic flowmeter, reduced mean arterial pressure, and caused a small baroreflex-mediated tachycardia (Table I). Injection of arachidonic acid increased coronary blood flow and reduced mean arterial and LV systolic pressure (Table I). Radioactive microspheres labeled with ^{125}I, ^{85}Sr, and ^{46}Sc were injected at the time of maximum increases in coronary blood flow to both arachidonic acid and PGE_2, as indicated in Fig. 2. Left circumflex coronary artery injection of arachidonic acid caused increases in blood flow in the area perfused by this artery in the epicardial, myocardial, and endocardial layers (Fig. 3). Prostaglandin E_2 also caused a uniform

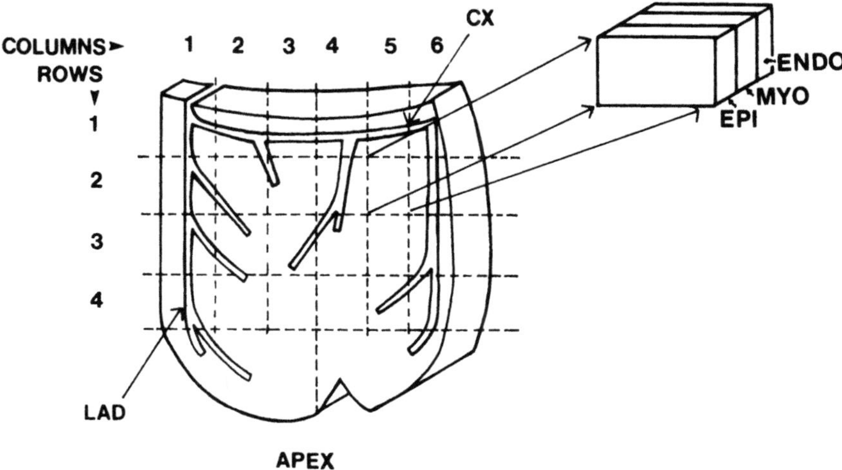

FIGURE 1. This figure illustrates the manner in which the left ventricular free wall was sectioned to study the distribution of radioactively labeled microspheres (9 μm, 3M Company) following the injection of PGE_2 and arachidonic acid. Prostaglandin E_2 and arachidonic acid were injected into the left circumflex coronary artery (CX) (row 1, column 2). A control injection of microspheres was given in each experiment, followed by injections of microspheres at the time of maximum flow changes following administration of PGE_2 and arachidonic acid. Blood flow in the area perfused by the left anterior descending coronary artery (LAD) (columns 1 and 2) also served

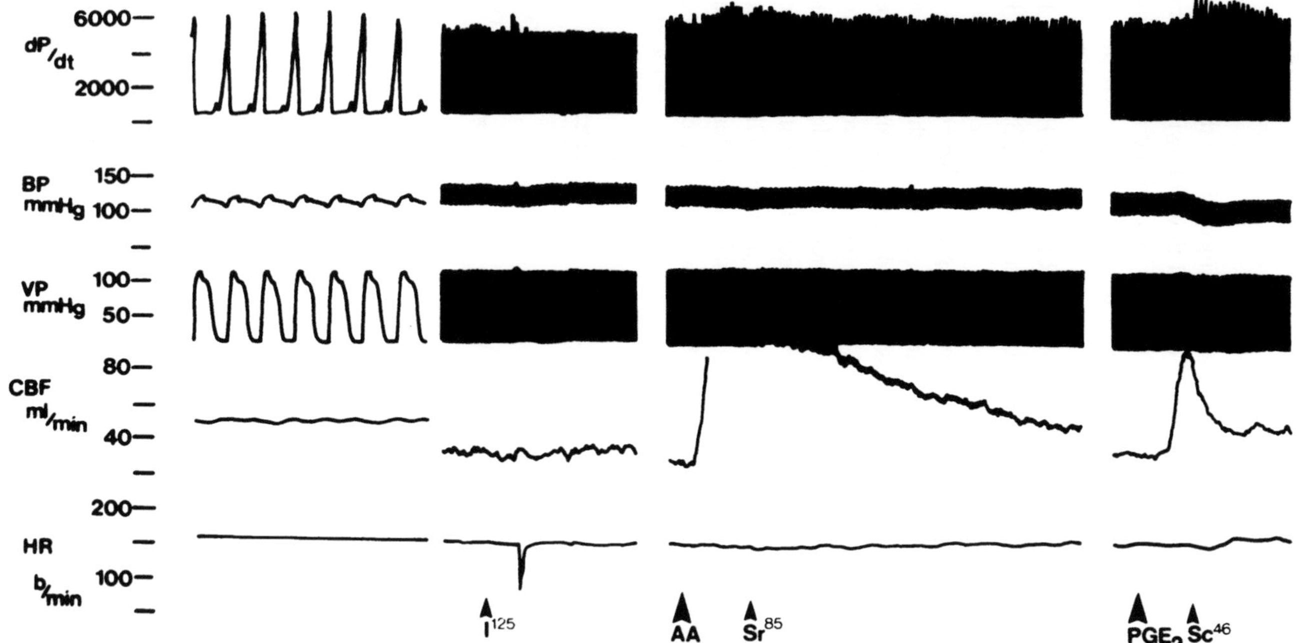

FIGURE 2. Radioactive microspheres labeled with ^{125}I, ^{85}Sr, and ^{46}Sc were injected at the arrows to measure the myocardial blood flow distribution during a control period and following left circumflex coronary artery injection of arachidonic acid (AA) and PGE_2. Microspheres were injected (as indicated by the arrows) at the time of maximum increases in circumflex coronary blood flow (CBF) during the continuous recording of left ventricular (LV) dP/dt, arterial pressure (BP), LV systolic pressure (VP), coronary blood flow, and heart rate (HR). The hemodynamic data for these experiments are summarized in Table II.

TABLE I. Effects of PGE$_2$ (2 μg) and Arachidonic Acid (200 μg) injected into the Left Circumflex Coronary Artery in Anesthetized Dogs[a]

	Control	PGE$_2$	Arachidonic acid
Mean arterial pressure (mm Hg)	121 ± 4.5	89 ± 5.0*	109 ± 6.1*
Heart rate (beats/min)	152 ± 5.9	158 ± 8.9	147 ± 6.6
LV systolic pressure (mm Hg)	149 ± 6.4	137 ± 12	141 ± 7.6*
LV dP/dt (mm Hg/sec)	5206 ± 232	5990 ± 611	5543 ± 348
Coronary blood flow (ml/min)	31.4 ± 2.1	93.8 ± 11.1*	126 ± 8.7*

[a] Absolute values are depicted in each case; *$P < 0.05$ from control.

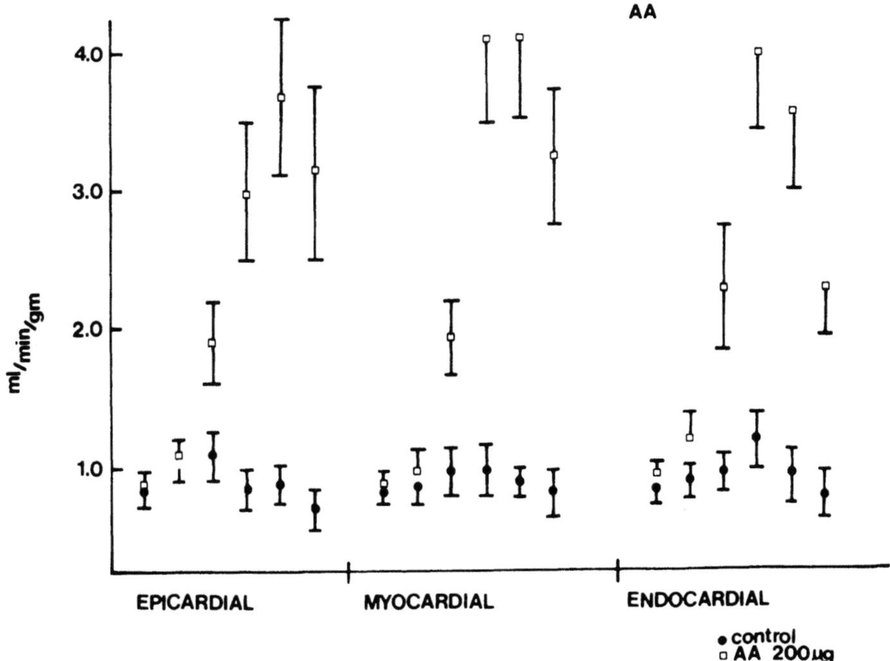

FIGURE 3. Blood flow in each layer of the heart, i.e., epicardium, myocardium, and endocardium, following the left circumflex coronary artery (CX) injection of arachidonic acid (AA). The closed circles represent blood flow in the control state, and the open squares represent blood flow following the injection of AA. Data are taken from row 3 as illustrated in Fig. 1. In each row the first two points, moving from left to right, correspond to the area perfused by the left anterior descending coronary artery (LAD), whereas the next four points correspond to the area perfused by the CX. Arachidonic acid caused a large increase in blood flow in the area perfused by the vessel in which it was injected and little or no increase in blood flow in the area perfused by the LAD. Blood flow increased uniformly across the myocardium. Values represent blood flow (ml/min per gram of tissue).

increase of blood flow in all layers of the left ventricular free wall (Fig. 4). Thus, it appears that generation of prostaglandins following the administration of arachidonic acid as well as injection of PGE_2 directly into the coronary arteries cause large, uniform, and approximately equivalent increases in blood flow to the epicardium and endocardium.

2.2. Effects of PGE_2 on Large Coronary Artery Dimensions in Conscious Dogs

In these experiments dogs were instrumented for the instantaneous and continuous measurement of large coronary artery diameters, coronary blood flow, and cardiac function and allowed to recover fully (Hintze *et al.*, 1985; Hintze and Vatner, 1984). Prostaglandin E_2 (5 μg/kg) injected intravenously caused a transient fall in mean arterial pressure that was closely followed by a reflex tachycardia and increases in LVdP/dt and large coronary artery diameter (Fig. 5). Injection of increasing doses of PGE_2 caused dose-related increases in large coronary vessel diameter. The large coronary artery dilation was unaffected by the previous administration of a prostaglandin synthesis inhibitor, indomethacin (5 mg/kg). At no time

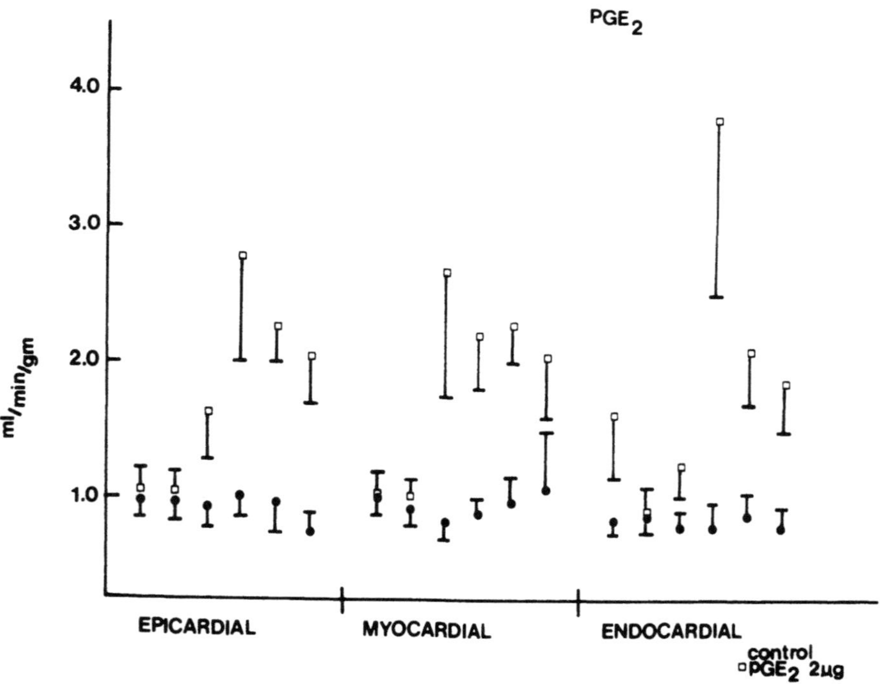

FIGURE 4. Injection of PGE_2 (open squares) caused an increase in blood flow in the area perfused by the artery in which it was injected, i.e., the area perfused by the left circumflex coronary artery (CX). The distribution of blood flow was uniform across the left ventricular free wall. Control values are shown by the closed circles. Each layer corresponds to row 3 as depicted in Fig. 1. Moving from left to right, the first two dots represent the area perfused by the left anterior descending coronary artery (LAD), whereas the next four dots represent area perfused by the CX. Values represent blood flow (ml/min per gram of tissue).

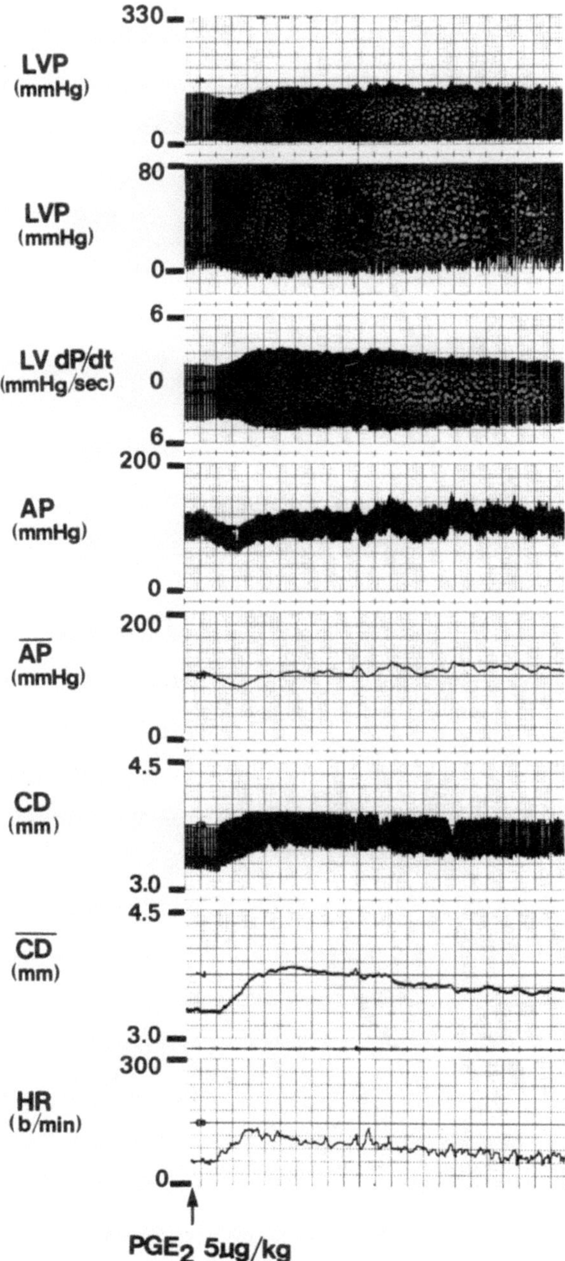

FIGURE 5. Intravenous injection of PGE$_2$ (5 μg/kg) into a conscious dog resulted in a transient reduction in phasic and mean arterial pressure (AP) and an initial reduction in left ventricular systolic pressure (LVP), accompanied by an increase in LV d*P*/d*t*, heart rate (HR), and a substantial increase in circumflex coronary artery diameter (CD). Large coronary artery diameter was measured instantaneously and continuously using a transient time ultrasonic dimension gauge and previously implanted 7 MHz piezoelectric transducers. Thus, although PGE$_2$ causes

TABLE II. Effects of Circumflex Coronary Artery (CX) and Left Anterior
Descending Coronary Artery (LAD) Injections of PGE_2 on Cardiovascular Function
in Anesthetized Dogs[a]

	Control	Change
Mean arterial pressure (mm Hg)		
LAD injection	139 ± 7.7	− 10 ± 2.9*
CX injection	149 ± 7.7	− 38 ± 6.0*
Heart rate (beats/min)		
LAD injection	176 ± 7.8	+ 1.0 ± 1.0
CX injection	176 ± 7.2	− 10 ± 2.1*
LV systolic pressure (mm Hg)		
LAD injection	173 ± 7.5	− 2.5 ± 4.4
CX injection	176 ± 7.5	− 30 ± 5.2*
LV dP/dt (mm Hg/sec)		
LAD injection	6100 ± 732	467 ± 111
CX injection	5800 ± 537	217 ± 480
Coronary blood flow		
LAD injection	33 ± 3.0	+ 46 ± 4.0*
CX injection	34 ± 2.7	+ 66 ± 6.6*

[a] Values reported for coronary blood flow represent flow in the LAD following LAD injection and flow in the CX
following CX injection; *$P < 0.05$ from control.

did PGE_2 cause a decrease in large coronary vessel diameter, as would be indicative
of large coronary vessel constriction. In contrast, using the same model in prior
studies, we have shown that large coronary vessel diameter is reduced by the potent
α-adrenergic agonist methoxamine. Recently, we have also found that U-46619, a
thromboxane-like substance, reduces large coronary vessel diameter and calculated
coronary cross-sectional area in conscious dogs (Hintze *et al.*, 1985).

2.3. Reflex Effects of Intracoronary PGE_2 in Dogs with Reduced Baroreflex Function

Left circumflex coronary artery injection of PGE_2 (10 μg) in dogs in which
the carotid arteries were ligated to eliminate carotid baroreceptor activation, in order
to observe the reflex response to cardiopulmonary receptor stimulation unopposed,
resulted in a reduction in heart rate and arterial pressure (Table II). Left anterior
descending coronary artery injection of PGE_2 had no effect on heart rate and caused
a significantly smaller hypotension. These data suggest that the ventricular receptors
that mediate the reflex responses to prostaglandins are located primarily in the
section of the myocardium perfused by the circumflex coronary artery (Hintze and
Kaley, 1984).

3. DISCUSSION

In this chapter we have reviewed briefly the effects of PGE_2 in the coronary
circulation, including the evidence that it is produced by coronary blood vessels

(Berger *et al.*, 1976; Uchida and Murao, 1979; Gerritsen and Printz, 1981), and pointed out some up to now unrecognized effects of PGE$_2$ in the heart.

Recent studies by Gerritsen and Cheli (1984) have established that PGE$_2$ is synthesized enzymatically by cultured coronary microvessel endothelial cells. Previous work by Gerritsen and Printz (1981) showed that coronary microvessels *in vitro* produce significant quantities of PGE$_2$, under some circumstances several fold more than PGI$_2$, and that large coronary blood vessels produce almost exclusively PGI$_2$. Other studies (Higgins *et al.*, 1971; Bloor *et al.*, 1973; Rowe and Afonso, 1974; Fitzpatrick *et al.*, 1978; Dusting *et al.*, 1979; Dusting and Vane, 1980), including our own (Hintze and Kaley, 1977), have demonstrated that injections of prostaglandins cause dose-related increases in coronary blood flow. Our recent observations indicate that PGE$_2$ induces a uniform increase in blood flow across the myocardium, distributing blood flow equally to the endocardium and epicardium. These effects of PGE$_2$ were similar to those that occur following the stimulation of prostaglandin synthesis *in vivo* by the intracoronary injection of arachidonic acid, indicating that prostaglandins are synthesized in all layers of the heart or, conversely, that prostaglandin synthesis is not restricted to any single layer of the myocardium. It is likely that injection of arachidonic acid does not lead to an exclusive synthesis of PGE$_2$ but rather to the synthesis of a variety of products.

The relaxation of large coronary artery strips following application of arachidonic acid was considered to be "paradoxical" because PGE$_2$, at one time thought to be the only important product of arachidonic acid metabolism by coronary vessels, caused contraction of large coronary artery strips *in vitro* (Strong and Bohr, 1967; Greenberg and Sparks, 1969). This observation has been repeated many times and eventually led to the identification of prostacyclin, a substance that induces relaxation of coronary artery strips (Kulkarni *et al.*, 1976; Dusting *et al.*, 1977). Unlike prostacyclin, PGE$_2$ increases resistance in isolated perfused cat coronary arteries (Ogletree *et al.*, 1978). However, in the normal healthy conscious dog chronically instrumented for the measurement of large coronary blood vessel diameter and fully recovered from surgery, PGE$_2$ causes dilation of large coronary arteries, not constriction. This response to PGE$_2$ may occur because of the presence of an uninterrupted layer of endothelial cells in coronary vessels *in vivo*, whereas contraction to PGE$_2$ may be the result of damage of the endothelium during preparation of isolated dog coronary vessels, an effect similar to the contraction observed by Furchgott and Zawadski (1980) following application of acetylcholine to a variety of blood vessel rings (or strips) whose endothelial cell layers are disrupted.

Another new aspect for the role of prostaglandins in circulatory control is their effect on reflex regulation of blood pressure and heart rate. It appears from our work (Hintze *et al.*, 1979) and the work of others (Coleridge *et al.*, 1976), that prostacyclin activates cardiac, "chemically" sensitive nerve endings (C-fiber afferents) to elicit reflex bradycardia and hypotension (Coleridge and Coleridge, 1980; Hintze and Kaley, 1984). The bradycardia to prostacyclin is dependent, to some degree, on the resting heart rate (Chiavarelli *et al.*, 1982). It involves primarily activation of the vagus nerves and the degree of bradycardia is limited by A–V nodal conduction. Our results indicate the PGE$_2$ is capable of eliciting similar reflex responses, although it may be less potent than PGI$_2$ or arachidonic acid. The bradycardia to PGE_2 is only evident when the competition between the systemic baroreflexes, which are activated by hypotension, and the cardiopulmonary receptor endings, which are activated by PGE_2, is eliminated. This bradycardia is not caused by an increase in coronary blood flow *per se,* since injection of nitroglycerin in doses that caused similar or greater increases in coronary blood flow caused a tachycardia (Hintze and Kaley, 1984). Previous studies by Leach *et al.* (1973) and Chen *et al.* (1979) have shown that in the rat, prostaglandins may cause reflex reductions in heart rate and peripheral resistance. Furthermore, the spontaneously hypertensive rat was particularly sensitive to the reflex actions of PGE_2 (Chen *et al.*, 1979). What is even more significant with respect to our own research is the major finding of the study by Gerritsen and Printz (1981) of a relatively active synthesis of PGE_2 by coronary microvessels, indicating that *in vivo* cardiac C-fiber endings would be exposed, in some instances perhaps primarily, to PGE_2.

REFERENCES

Afonso, S., Bandow, G. T., and Rowe, G. G., 1974, Indomethacin and the prostaglandin hypothesis of coronary flow regulation, *J. Physiol. (Lond.)* **241:**299–308.

Aiken, J. W., Shebuski, R. J., and Gorman, R. R., 1980, Blockage of partially obstructed coronary arteries with platelet thrombi: Comparison between its prevention with cyclooxygenase inhibitors versus prostacyclin, *Adv. Prostaglandin Thromboxane Res.* **7:**635–639.

Alexander, R. W. Kent, K. M., Pisano, J. J., Keiser, H. R., and Cooper, T., 1975, Regulation of postocclusion hyperemia by endogenously synthesized prostaglandins in the dog heart, *J. Clin. Invest.* **55:**1174–1181.

Becker, L., 1976, Effects of tachycardia on left ventricular blood flow distribution during coronary artery occlusion, *Am. J. Physiol.* **230:**1072–1077.

Becker, L. C., Ferreira, R., and Thomas, M., 1973, Mapping of left ventricular blood flow with radioactive microspheres in experimental coronary artery occlusion, *Cardiovasc. Res.* **7:**391–400.

Belloni, F. L., 1979, The local control of coronary blood flow, *Cardiovasc. Res.* **13:**63–85.

Berger, H. J., Zaret, B. L., Speroff, L., Cohen, L. S., and Wolfson, S., 1976, Regional and cardiac prostaglandin release during myocardial ischemia in anesthetized dog, *Circ. Res.* **38:**566–569.

Bloor, C. M., White, F. C., and Sobel, B. E., 1973, Coronary and systemic effects of prostaglandins in the unanesthetized dog, *Cardiovasc. Res.* **7:**156–166.

Buckberg, G. D., Luck, J. C., Payne, B. D., Hoffman, J. I. E., Archie, J. P., and Fixler, D. E., 1971, Some sources of error in measuring regional blood flow with radioactive microspheres, *J. Appl. Physiol.* **31:**598–604.

Carlson, L. A., and Örö, L., 1966, Effect of prostaglandin E_1 on blood pressure and heart rate in the dog, *Acta. Physiol. Scand.* **67:**89–99.

Chapple, D. J., Dusting, G. J., Hughes, R., and Vane, J. R., 1978, A vagal reflex contributes to the hypotensive effect of prostacyclin in anesthetized dogs, *Br. J. Pharmacol.* **68:**437–447.

Chen, D., Donald, D. E., and Romero, J. C., 1979, Role of vagal afferents in the vasodepressor effects of PGE_2 in the spontaneously hypertensive rat, *Am. J. Physiol.* **236:**H635–H639.

Chiavarelli, S., Moncada, S., and Mullane, K. M., 1982, Prostacyclin can either increase or decrease heart rate depending on the basal state, *Br. J. Pharmacol.* **75:**243–249.

Chiba, S., and Malik, K. U., 1980, Mechanism of the chronotropic effects of prostacyclin in the dog, *J. Pharmacol. Exp. Ther.* **213:**261–266.

Coker, S. J., Marshall, R. J., Parratt, J. R., and Zeitlin, I. J., 1981, Does the local myocardial release of prostaglandin E_2 or $F_{2\alpha}$ contribute to the early consequences of acute myocardial ischemia, *J. Mol. Cell Cardiol.* **13:**425–434.

Coleridge, H. M., and Coleridge, J. C. G., 1980, Cardiovascular afferents involved in regulation of peripheral vessels, *Annu. Rev. Physiol.* **42:**413–427.

Coleridge, H. M., Coleridge, J. C. G., Ginzel, K. H., Baker, D. G., Banzett, R. B., and Morrison, M. A., 1976, Stimulation of "irritant" receptors and afferent C-fibers in the lungs by prostaglandins, *Nature* **264**:451–453.

Domench, R. J., Hoffman, J. I. E., Noble, M. I., Saunders, K. B., Henson, J. R., and Subijanto, S., 1969, Total and regional coronary blood flow measured by radioactive microspheres in conscious and anesthetized dogs, *Circ. Res.* **25**:581–595.

Dusting, G. J., Moncada, S., and Vane, J. R., 1977, Prostacyclin (PGX) is the endogenous metabolite responsible for relaxation of coronary arteries induced by arachidonic acid, *Prostaglandins* **13**:3–15.

Dusting, G. J., Moncada, S., and Vane, J. R., 1979, Prostaglandins, their intermediates and precursors; cardiovascular actions and regulatory roles in normal and abnormal circulatory systems, *Prog. Cardiovasc. Dis.* **21**:405–430.

Dusting, G. J., and Vane, J. R., 1980, Some cardiovascular properties of prostacyclin (PGI$_2$) which are not shared with PGE$_2$, *Circ. Res.* **46**(Suppl. 1):183–187.

Fitzpatrick, T. M., Alter, I., Corey, E. J., Ramwell, P. W., Rose, J. C., and Kot, P. A., 1978, Cardiovascular responses to PGI$_2$ (prostacyclin) in the dog, *Circ. Res.* **42**:192–194.

Furchgott, R. L., and Zawadski, D., 1980, The obligatory role of endothelial cells in the relaxation of arterial smooth muscle by acetylcholine, *Nature* **288**:373–376.

Gerritsen, M. E., and Cheli, C. D., 1984, Arachidonic acid and endoperoxide metabolism in isolated rabbit coronary microvessels and isolated and cultivated coronary microvessel endothelial cells, *J. Clin. Invest.* **72**:1658–1671.

Gerritsen, M. E., and Printz, M. P., 1981, Sites of prostaglandin synthesis in bovine heart and isolated bovine coronary microvessels, *Circ. Res.* **49**:1152–1163.

Giles, R. W., and Wilcken, E. L., 1977, Reactive hyperaemia in the dog heart: Inter-relations between adenosine, ATP and aminophylline and the effect of indomethacin, *Cardiovasc. Res.* **11**:113–121.

Giles, T. D., Quiroz, A. C., and Burch, G. E., 1969, The effects of prostaglandin E$_1$ on the systemic and pulmonary circulations of intact dogs. The influence of urethane and pentobarbital anesthesia, *Experientia* **25**:1056–1069.

Greenberg, R. A., and Sparks, H. V., 1969, Prostaglandins and consecutive vascular segments of the canine hindlimb, *Am. J. Physiol.* **216**:567–571.

Harlan, D. M., Rooke, T. N., Belloni, F. L., and Sparks, H. V., 1978, Effects of indomethacin on coronary vascular response to increased myocardial oxygen consumption, *Am. J. Physiol.* **235**:4372–4378.

Heymann, M. A., Payne, B. D., Hoffman, J. I. E., and Rudolph, A. M., 1977, Blood flow measurements with radio-labelled particles, *Prog. Cardiovasc. Dis.* **20**:55–79.

Higgins, C. B., Vatner, S. F., Franklin, D., Patrick, T., and Braunwald, E., 1971, Effects of prostaglandin A$_2$ on systemic and coronary circulations in the dog, *Circ. Res.* **28**:638–647.

Hintze, T. H., and Kaley, G., 1977, Prostaglandins and the control of blood flow in the canine myocardium, *Circ. Res.* **40**:313–319.

Hintze, T. H., and Kaley, G., 1984, Ventricular receptors activated following myocardial prostaglandin synthesis initiate reflex hypotension, reduction in heart rate, and redistribution of cardiac output in the dog, *Circ. Res.* **54**:239–247.

Hintze, T. H., and Vatner, S. F., 1984, Reactive dilation of large coronary arteries in conscious dogs, *Circ. Res.* **54**:50–57.

Hintze, T. H., Martin, E. G., Messina, E. J., and Kaley, G., 1979, Prostacyclin (PGI$_2$) elicits reflex bradycardia in dogs: Evidence for vagal mediation, *Proc. Soc. Exp. Biol. Med.* **162**:96–100.

Hintze, T. H., Panzenbeck, M. J., Messina, E. J., and Kaley, G., 1981, Prostacyclin (PGI$_2$) lowers heart rate in the conscious dog, *Cardiovasc. Res.* **15**:538–546.

Hintze, T. H., Martin, E. G., Messina, E. J., and Kaley, G., 1982, Cardiovascular reflex actions of arachidonic acid metabolites: Activation of a vagal depressor reflex by prostacyclin (PGI$_2$) in the dog, in: *Cardiovascular Pharmacology of the Prostaglandins* (P. M. Vanhoutte, H. Denolin, and A. Goossens, eds.), Raven Press, New York, pp. 417–437.

Hintze, T. H., Kaley, G., and Panzenbeck, M. J., 1984, Mechanisms of reflex bradycardia and hypotension by metabolites of arachidonic acid in the cat, *Br. J. Pharmacol.* **82**:117–125.

Hintze, T. H., Kichuk, M. R., Stern, H., and Kaley, G., 1985, Dilation of large coronary arteries in conscious dogs by prostacyclin, *Adv. Prostaglandin Thromboxane and Leukotriene Res.* (G. V. R. Born *et al.*), Raven Press, New York, pp. 341–344.

Hutton, I., Parratt, J. R., and Lawrie, T. D. V., 1973, Cardiovascular effects of prostaglandin E_2 in experimental myocardial infarction, *Cardiovasc. Res.* **7**:149–155.

Hyman, A. L., Kadowitz, P. J., Lands, W. E. M., Crawford, G. G., Fried, J., and Barton, J., 1978, Coronary vasodilator activity of 13,14-dehydro prostacyclin methylester: Comparison with prostacyclin and other prostanoids, *Proc. Natl. Acad. Sci. U.S.A.* **75**:3522–3526.

Jentzer, J. H., Sonnenblick, E. H., and Kirk, E. S., 1979, Coronary and systemic vasomotor effects of prostacyclin: Implications for the ischemic myocardium, in: *Prostacyclin* (J. R. Vane and S. Bergström, eds.), Raven Press, New York, pp. 323–338.

Jugdutt, B. I., Hutchins, G. M., Bulkley, B. H., and Becker, L. C., 1981, Dissimilar effects of prostacyclin, prostaglandin E_1 and prostaglandin E_2 on myocardial infarct size after coronary occlusion in conscious dogs, *Circ. Res.* **49**:685–700.

Kaley, G., Hintze, T. H., and Messina, E. J., 1980, Role of prostaglandins in ventricular reflexes, *Adv. Prostaglandin Thromboxane Res.* **7**:601–608.

Kaley, G., Messina, E. J., and Hintze, T. H., 1981, Role of prostaglandins in the regulation of blood pressure and blood flow, *Adv. Physiol. Sci.* **7**:83–90.

Kraemer, R. J., Phernetton, J. M., and Folts, J. D., 1976, Prostaglandin-like substances in coronary venous blood following myocardial ischemia, *J. Pharmacol. Exp. Ther.* **199**:611–619.

Kulkarni, P. S., Roberts, R., and Needleman, P., 1976, Paradoxical endogenous synthesis of a coronary dilating substance from arachidonate, *Prostaglandins* **12**:337–352.

Leach, B. E., Armstrong, F. B., Germain, G. S., and Muirhead, E. E., 1973, Vasodepressor action of prostaglandin A_2 and E_2 in the spontaneously hypertensive rat (SHR): Evidence for an action mediated by the vagus, *J. Pharmacol. Exp. Ther.* **185**:479–485.

Lefer, A. M., Ogletree, M. L., Smith, J. B., Silver, M. J., Nicoloau, K. C., Barnette, W. E., and Gasic, G. C., 1979, Prostacyclin: A potentially valuable agent for preserving myocardial tissue in acute myocardial ischemia, *Science* **200**:52–54.

Maxwell, G., 1967, The effects of prostaglandin E_2 upon the general and coronary hemodynamics in the intact dog, *Br. J. Pharmacol. Chemother.* **31**:162–168.

Nakano, J., 1968, Effects of prostaglandin A_1, E_1 and $F_{2\alpha}$ on the coronary and peripheral circulations, *Proc. Soc. Exp. Biol. Med.* **127**:1160–1163.

Needleman, P., and Kaley, G., 1978, Cardiac and coronary prostaglandin synthesis and function, *N. Engl. J. Med.* **298**:1122–1128.

Nutter, P. O., and Crumly, H. J., 1972, Canine coronary vascular and cardiac responses to the prostaglandins, *Cardiovasc. Res.* **6**:217–225.

Ogletree, M. L., Smith, J. B., and Lefer, A. M., 1978, Actions of prostaglandins on isolated perfused cat coronary arteries, *Am. J. Physiol.* **235**:H400–H406.

Owen, T. L., Ehrhart, I. C., Weidner, W. J., Scott, J. B., and Haddy, F. J., 1975, Effects of indomethacin on local blood flow regulation in canine heart and kidney, *Proc. Soc. Exp. Biol. Med.* **149**:871–876.

Parratt, J. R., and Marshall, R. J., 1978, Are prostaglandins involved in the regulation of coronary blood flow? A review of the evidence, *Acta Biol. Med. Germ.* **37**:747–760.

Rowe, G. G., and Afonso, S., 1974, Systemic and coronary hemodynamic effects of intracoronary administration of prostaglandins E_1 and E_2, *Am. Heart J.* **88**:51–60.

Rubio, R., and Berne, R. M., 1975, Regulation of coronary blood flow, *Prog. Cardiovasc. Dis.* **18**:105–110.

Rudolph, A. M., and Heymann, M. A., 1967, The circulation of the fetus *in utero*, *Circ. Res.* **16**:163–184.

Strong, G. G., and Bohr, D. F., 1967, Effects of prostaglandins E_1, E_2, A_2 and $F_{2\alpha}$ on isolated vascular smooth muscle, *Am. J. Physiol.* **213**:725–733.

Sunahara, F. A., and Talesnik, J., 1979, Myocardial synthesis of prostaglandin-like substances and coronary reactions to cardiostimulation and to hypoxia, *Br. J. Pharmacol.* **65**:71–85.

Uchida, Y., and Murao, S., 1979, Effect of prostaglandin I_2 on cyclical reductions of coronary blood flow, *Jpn. Circ. J.* **43**:645–652.

32

Prostaglandin H_2 Causes Calcium Mobilization in Intact Human Platelets

L. D. BRACE, D. L. VENTON, and
G. C. Le BRETON

1. INTRODUCTION

It was originally suggested that thromboxane A_2 (TxA_2) production is a prerequisite for platelet activation induced by arachidonic acid (AA) addition. More recently, however, evidence has accumulated that suggests that, under the appropriate conditions, prostaglandin H_2 (PGH_2) can also play a role in the platelet activation process. In this regard, Needleman *et al.* (1976) have shown that when AA was added to washed platelets incubated with the thromboxane synthase inhibitor imidazole (Needleman *et al.*, 1976, 1977a,b; Moncada *et al.*, 1976, 1977), thromboxane synthesis was blocked, but platelet aggregation was not inhibited (Needleman *et al.*, 1977b). Based on these findings, they proposed that PGH_2 is itself active in causing platelet aggregation.

This notion was further supported by the subsequent studies of Grimm *et al.* (1981) using 7-(1-imidazolyl)heptanoic acid (7-IHA), a different thromboxane synthase inhibitor (Yoshimoto *et al.*, 1978; Kayama *et al.*, 1981). They demonstrated that a greater than 90% inhibition of TxA_2 formation by 7-IHA did not significantly inhibit platelet aggregation. In order to exclude the possibility that the observed aggregation was caused by the 2% residual thromboxane formation, they inhibited the platelet cyclooxygenase with indomethacin. It was found that although thromboxane formation was inhibited to a lesser extent than with 7-IHA, platelet aggre-

L. D. BRACE, and G. C. Le BRETON • Department of Pharmacology, The University of Illinois at Chicago, Health Sciences Center, Chicago, Illinois 60612.   D. L. VENTON • Department of Medicinal Chemistry, The University of Illinois at Chicago, Health Sciences Center, Chicago, Illinois 60612.

gation was nevertheless completely blocked. These results indicate that PGH_2 does indeed possess proaggregatory effects. Furthermore, a separate study by Rybicki and Le Breton (1983) showed that PGH_2 could directly lower cyclic AMP (cAMP) levels, which is consistent with the notion that PGH_2 promotes platelet activation.

On the other hand, certain inconsistencies have been associated with the use of thromboxane synthase inhibitors, which have prevented a clear demonstration of the effect: various inhibitors, e.g., azo analogue I (Gorman *et al.*, 1977; Fitzpatrick and Gorman, 1978), imidazole, 7-IHA, and UK 37,248 (Randall *et al.*, 1981), have all been shown to produce variable blockade of aggregation under certain circumstances. In this connection, a recent study by Smith (1982) investigated the mechanism of this inhibition by two classes of thromboxane synthase inhibitors. He found that the inhibitory properties of these compounds could be ascribed to increased production of PGD_2 and not to the absence of TxA_2 production.

In the present study, we measured the initial event in the process of platelet activation, i.e., the mobilization of internal calcium stores, to evaluate the relative ability of PGH_2 to stimulate platelet functional change. We have previously demonstrated that TxA_2 and/or PGH_2 causes calcium mobilization and shape change in intact human platelets (Brace *et al.*, 1982). This internal calcium redistribution could be blocked by indomethacin or the specific thromboxane receptor antagonist 13-azaprostanoic acid (13-APA). Similar results have recently been obtained by Kawahara *et al.* (1983) using the thromboxane A_2 analogue STA_2. However, none of these experiments has differentiated between the abilities of TxA_2 and PGH_2 to independently induce intraplatelet calcium redistribution.

In order to further address this question, we have measured intraplatelet calcium mobilization using the fluorescent calcium indicator chlortetracycline (CTC) in the presence of three separate classes of thromboxane synthase inhibitors. Our results are consistent with the concept that PGH_2 can directly stimulate platelet functional change.

2. MATERIALS AND METHODS

2.1. Platelet Preparation

Blood from healthy donors who denied having received medication for 10 days was collected into 0.38% citrate–phosphate–dextrose–adenine (CPDA 1) solution. Platelet-rich plasma (PRP) prepared from this blood was purchased from a commercial source. The PRP was centrifuged at $160 \times g$ for 10 min at room temperature to remove contaminating red and white blood cells.

2.2. Measurement of Intraplatelet Calcium Mobilization

Although CTC was originally used to monitor changes in mitochondrial membrane-bound calcium (Caswell and Hutchison, 1971), it has since been used to monitor changes in intracellular calcium binding in a wide variety of cell systems including platelets (Le Breton *et al.*, 1976; Feinstein, 1980; Owen and Le Breton, 1981). In the present study, PRP was incubated with 50 μM CTC (a concentration that we have shown does not interfere wth platelet activation) for 40 min at 25°C. The temperature of the PRP was then raised to 37°C for all subsequent experimental procedures. The CTC–platelet fluorescence was determined as previously described (Owen and Le Breton, 1981). Briefly, four 1-ml samples were withdrawn and layered over silicone oil (100 μl of a mixture of Dow 220 : Dow 550, 17 : 4) in the bottom of 1.5-ml plastic conical centrifuge tubes and then centrifuged at 7000 $\times$ g for 1 min. Immediately following centrifugation, the supernatant and silicone oil were aspirated, leaving the undisturbed platelet pellets in the tube tips. The tube tips were removed with a hot surgical blade and placed in acrylic holders for fluorescence determination using a photon-counting microspectrofluorometer. Control fluorescence values were normalized to 100,000 counts per second (cps). Additions of vehicles, agonists, and antagonists were made as described in the figure legends.

2.3. Measurement of Platelet Shape Change

Platelet shape change was measured according to the turbidometric method of Born (1962) with a Model 400 Lumi-aggregometer (Chrono-Log Corp., Havertown, PA.). With this method, shape change is represented by a decrease in light transmission.

Platelet shape change is generally regarded to be the most sensitive index of platelet activation. Therefore, all experiments were performed under conditions in which only the platelet shape change response was stimulated. Furthermore, the measurement of shape change alone precluded the possibility of nonspecific changes in CTC fluorescence that may result from platelet apposition (aggregation) or release of intraplatelet calcium-containing granules (secretion).

2.4. Measurement of Secretion

Secretion was measured simultaneously with platelet stimulation using a Lumi-aggregometer and the bioluminescent luciferin–luciferase assay (Feinman *et al.*, 1977). The PRP was preincubated with luciferin–luciferase for 2 min at 37°C before the addition of an agonist. ATP release from intracellular storage granules was used as an index of platelet secretion since it is cosecreted with ADP and calcium.

2.5. Measurement of TxB₂ Production

At the specified time intervals, 1-ml samples of PRP for thromboxane B_2 determination were withdrawn and added to 3 mM imidazole in a 1.5-ml conical plastic centrifuge tube, then immediately frozen in liquid nitrogen. The samples were thawed and diluted 1 : 5 in phosphate-buffered saline with gelatin (PBSG). Platelet-free plasma separated from unactivated platelets was also diluted 1 : 5 in

PBSG, and an equal volume was added to the tubes of the standard curve to correct for any nonspecific plasma protein binding and for circulating TxB_2. The TxB_2 in the diluted samples was measured by the radioimmunoassay of Fitzpatrick *et al.* (1977). Samples (0.1 ml) or TxB_2 standard solutions and $[^3H]TxB_2$ were added to TxB_2 antiserum (0.1 ml) at a dilution (1 : 1250) that bound 50% of the $[^3H]TxB_2$ in the absence of competition by unlabeled TxB_2. The final assay volume was 0.3 ml. The assay mixtures were mixed and incubated at 25°C for 1 hr and then at 4°C for 20–24 hr. Antibody-bound and free TxB_2 were separated by incubation with 1 ml of dextran-coated charcoal suspension containing 2.5 mg Norit A charcoal and 0.25 mg dextran T 70 in 1 ml PBSG buffer for 11 min at 0°C. Following centrifugation at $1500 \times g$ (0–5°C) for 12 min, the antibody-bound TxB_2 in the supernatant was measured by radioisotope counting in a Beckman Model LS6800 liquid scintilation spectrometer. The TxB_2 concentration in the experimental samples was determined by comparing the amount of $[^3H]TxB_2$ bound in the standard tubes with that in the sample tubes.

2.6. Materials

Human PRP was obtained from the Mid-American Red Cross (Chicago, IL) or Chicago Blood Services (Chicago, IL) within 3 hr of collection. Arachidonic acid (Sigma Chemical Co., St. Louis, MO) was dissolved in nitrogen-bubbled 75% ethanol : 25% normal saline to prepare a 50 mM stock solution. Chlortetracycline was dissolved in normal saline and the pH adjusted to 7.4 prior to addition to PRP. Luciferin–luciferase (Chrono-Log, Havertown, PA) was dissolved in sterile distilled water and used according to instructions. 13-Azaprostanoic acid (13-APA) was synthesized as previously described (Venton *et al.*, 1979) and dissolved in 50% ethanol : 50% normal saline to prepare a stock solution of 40 mM. A 40 mM stock solution of 2′,5′-dideoxyadenosine (DDA), an 8 mM stock solution of 7-(imidazoyl)heptanoic acid (7-IHA), a 0.3 M stock solution of imidazole, and a 0.2 M stock solution of sodium (E)-3-[4(3-pyridylmethyl)phenyl]-2-methylacrylate (OKY 1581) were prepared in normal saline. A 12 mM stock solution of 9,11-azoprosta-5,13-dienoic acid (azo analogue I) was prepared in 95% ethanol. Control and experimental samples received the same volume addition of ethanol, and the final ethanol concentration never exceeded 0.5%.

3. RESULTS

3.1. Effect of 7-IHA on Intraplatelet Calcium Mobilization and TxB₂ Formation

We have previously shown that a concentration of AA sufficient to cause platelet shape change (but not aggregation or secretion) in PRP also causes a significant mobilization of intraplatelet calcium (Brace *et al.*, 1982). This internal calcium redistribution upon AA addition could be blocked by inhibition of platelet cyclooxygenase with indomethacin, indicating that a cyclooxygenase metabolite of AA was responsible for the observed calcium flux. In addition, the specific TxA_2/PGH_2 receptor antagonist 13-APA also completely inhibited calcium mobilization, demonstrating that this calcium release by TxA_2 and/or PGH_2 occurred through a receptor-mediated event (Brace *et al.*, 1982).

In order to determine whether PGH_2 is itself capable of mobilizing intraplatelet calcium stores, TxA_2 synthesis was inhibited by incubating PRP with the thromboxane synthase inhibitor 7-IHA. Our results demonstrate that even though TxA_2 production was inhibited by greater than 95% (Fig. 1), only a partial inhibition of

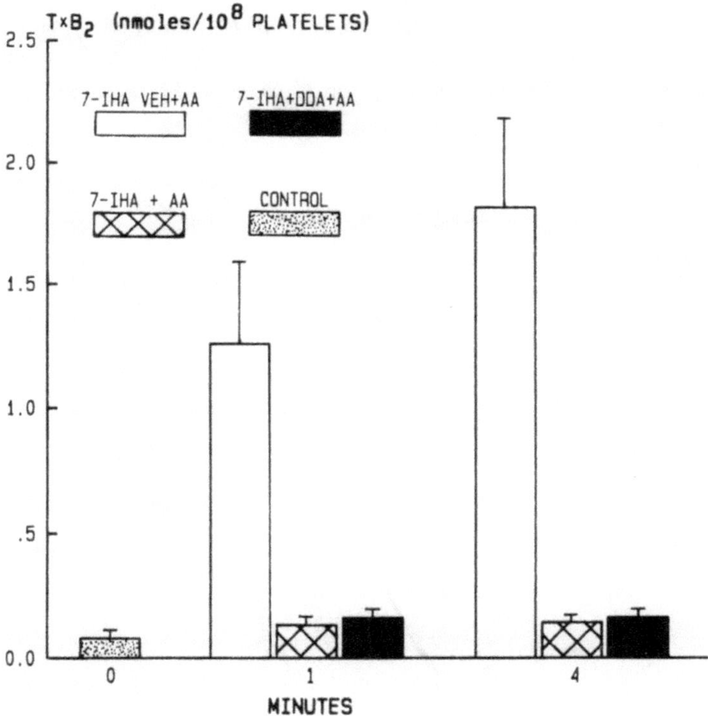

FIGURE 1. Inhibition of TxB_2 production by 7-IHA. Human PRP was incubated with CTC, and TxB_2 was measured as described in Section 2.5. Samples for TxB_2 determination were taken immediately prior to AA addition (0 min) and 1 and 4 min after addition. The CTC-treated PRP was incubated with 7-IHA vehicle and DDA vehicle for 2 min prior to AA addition (stippled bar). Inclusion of 100 μM DDA did not alter the TxB_2 level prior to AA addition (data not shown). Arachidonate (100 μM) was then added to CTC-treated PRP incubated with 20 μM 7-IHA (hatched bar) or 20 μM 7-IHA and 100 μM DDA (solid bar), and TxB_2 ws determined 1 and 4 min later. The TxB_2 production in the absence of 7-IHA (open bar) was not altered by deletion of 7-IHA vehicle and/or DDA (data not shown). The TxB_2 levels are expressed as nanomoles/10⁸ platelets. The data are expressed as mean $\pm$ S.E.M. of samples from at least five different blood donors. Treatment with 7-IHA suppressed TxB_2 production by more than 95% ($P < 0.001$).

calcium mobilization was observed compared to the absence of a thromboxane synthase inhibitor (Fig. 2). Specifically, calcium mobilization at 1 min was attenuated by 7-IHA, whereas calcium mobilization at 4 min was unaffected. These results indicate that inhibition of thromboxane synthase with 7-IHA inhibited the rate of AA-induced calcium release but not the extent of the calcium release.

Since PGH_2 is diverted to PGD_2 when thromboxane synthase is inhibited, and since PGD_2 stimulates adenylate cyclase activity and promotes cAMP production, we investigated whether cAMP production is associated with the partial inhibition of calcium mobilization. In these studies, platelets were incubated with the adenylate cyclase inhibitor $2',5'$-dideoxyadenosine (DDA) (Fain *et al.*, 1972) and the thromboxane synthase inhibitor 7-IHA prior to the addition of AA. Under these conditions, 7-IHA did not block calcium mobilization at either 1 min or 4 min following AA addition (Fig. 3) even though TxA_2 production was inhibited by greater than 95% (Fig. 1). These results are therefore consistent with the previous observation of Smith (1982) and indicate that diversion of PGH_2 to PGD_2 can lead to inhibition of PGH_2-induced platelet activation.

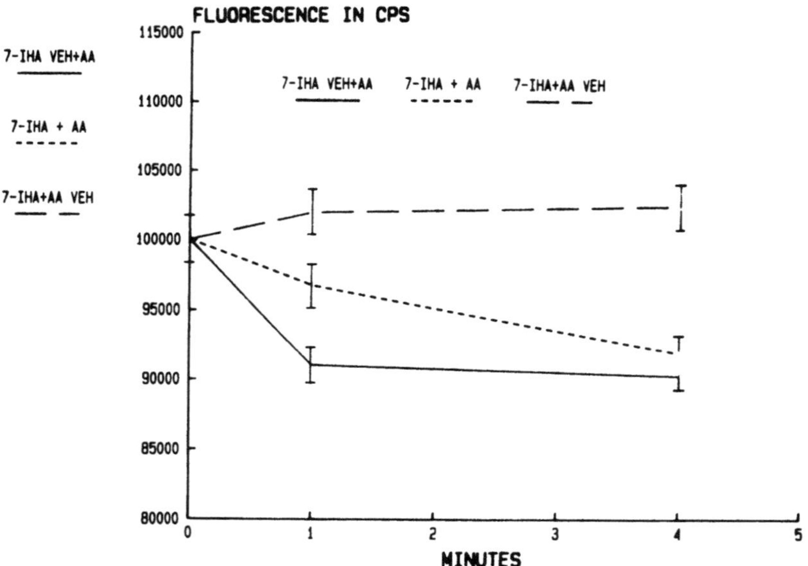

FIGURE 2. Effect of 7-IHA on AA-induced calcium mobilization. Fluorescence was measured in CTC-treated PRP platelets (see Section 2.2) immediately prior to AA or AA vehicle addition (0 min) and 1 and 4 min after additions. The PRP was incubated with 7-IHA vehicle (solid line) or 20 μM 7-IHA (dashed line) for 2 min prior to AA addition and 20 μM 7-IHA for 2 min prior to AA vehicle addition (broken line). A decrease in fluorescence counts from control (0 min) indicates mobilization of intraplatelet calcium. Fluorescence values are represented as counts per second (cps). The data are represented as mean ± S.E.M. of samples from at least five different blood donors. The 7-IHA caused inhibition of calcium mobilization 1 min after AA addition ($P < 0.001$).

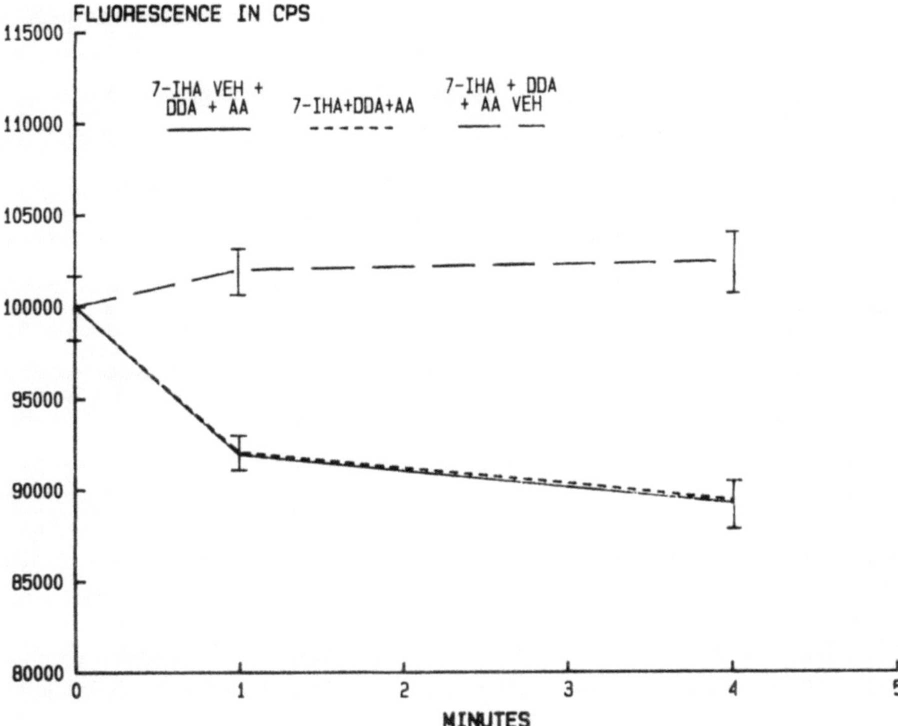

FIGURE 3. The effect of DDA on 7-IHA inhibition of AA-induced calcium mobilization. Experimental conditions are identical to Fig. 2 except that DDA was added. The PRP was incubated with 7-IHA vehicle plus 100 μM DDA (solid line) or 20 μM 7-IHA plus 100 μM DDA (dashed line) for 2 min prior to AA addition and 20 μM 7-IHA plus 100 μM DDA for 2 min prior to AA vehicle addition (broken line). Note that the solid and broken lines are nearly superimposed. Data are expressed as mean ± S.E.M. of samples from at least five different blood donors. Under these conditions, 7-IHA did not significantly inhibit AA-induced calcium mobilization.

3.2. Effect of Azo Analogue I and OKY 1581 on Calcium Mobilization and TxB₂ Production

To determine whether the above findings can be generalized to other thromboxane synthase inhibitors, we examined the effects of two additional classes of such inhibitors on calcium mobilization. In these studies, PRP was incubated with the endoperoxide structural analogue azo analogue I or the pyridine-based compound OKY 1581 (Miyamoto *et al.*, 1980; Tai *et al.*, 1980; Feuerstein and Ramwell, 1981). As can be seen in Table I, incubating PRP with either azo analogue I or OKY 1581 resulted in comparable results to those obtained using 7-IHA; i.e., there was partial inhibition of calcium mobilization 1 min after AA was added but not 4 min after AA addition. In both cases, TxA₂ production was inhibited by more than 95%.

TABLE I. Effects of Thromboxane Synthase Inhibitors on AA-Induced Calcium Mobilization and TxB$_2$ Production[a]

Time (min)		Vehicle	7-IHA	Azo analogue I	OKY 1581
0	CPS	100,000 ± 2,000	100,000 ± 1,006	100,000 ± 1,155	100,000 ± 1,659
	TxB	0.07 ± 0.02	0.06 ± 0.02	0.09 ± 0.03	0.07 ± 0.02
1	CPS	91,010 ± 1,162	96,738 ± 1,409	95,194 ± 740	93,861 ± 1,262
	TxB	1.25 ± 0.35	0.13 ± 0.03	0.15 ± 0.04	0.12 ± 0.03
4	CPS	90,271 ± 1,002	91,959 ± 1,240	89,921 ± 827	90,182 ± 955
	TxB	1.81 ± 0.39	0.11 ± 0.03	0.17 ± 0.04	0.11 ± 0.03

[a] Chlortetracycline-treated PRP was incubated with the appropriate thromboxane synthase inhibitor vehicle (vehicle), 20 μM 7-IHA, 30 μM azo analogue I, or 500 μM OKY 1581 for 2 min prior to AA addition. Samples for TxB$_2$ determination were withdrawn, and fluorescence measurements were made immediately prior to AA addition (0 min) and 1 and 4 min after AA addition. Inhibitor vehicles did not affect calcium mobilization or TxB$_2$ production. Data are expressed as mean ± S.E.M. of samples from at least five different blood donors. CPS, fluorescence counts per second; TxB, TxB$_2$ (nmol/10^8 platelets).

In order to evaluate the involvement of PGD$_2$ in the observed inhibition of calcium mobilization, platelets were again pretreated with DDA. The DDA-pretreated PRP was then incubated with either azo analogue I or OKY 1581 to inhibit thromboxane synthesis prior to the addition of AA. As in the studies using 7-IHA, it was found that in the presence of DDA, neither of the thromboxane synthase inhibitors blocked calcium mobilization; i.e., the results obtained 1 and 4 min after AA addition were comparable to those obtained with the addition of AA alone (Table II). Therefore, these findings using three separate classes of thromboxane synthase inhibitors provide evidence that both PGH$_2$ and TxA$_2$ are effective in promoting internal calcium redistribution.

3.3. Effect of 13-APA on PGH$_2$-Induced Calcium Mobilization

In order to determine whether the calcium mobilization induced by PGH$_2$ is a receptor-mediated event, we employed the specific TxA$_2$/PGH$_2$ receptor antagonist

TABLE II. Effect of DDA on the Inhibition of AA-Induced Calcium Mobilization by Thromboxane Synthase Inhibitors[a]

Time (min)		Vehicle	7-IHA	Azo analogue I	OKY 1581
0	CPS	100,000 ± 1,459	100,000 ± 934	100,000 ± 1,282	100,000 ± 1,284
	TxB	0.07 ± 0.01	0.09 ± 0.01	0.09 ± 0.01	0.08 ± 0.01
1	CPS	91,972 ± 720	91,905 ± 910	88,563 ± 1,050	88,840 ± 873
	TxB	0.94 ± 0.22	0.16 ± 0.02	0.13 ± 0.01	0.10 ± 0.01
4	CPS	89,262 ± 1,155	89,340 ± 1,084	86,741 ± 957	89,006 ± 827
	TxB	1.43 ± 0.31	0.15 ± 0.02	0.16 ± 0.02	0.11 ± 0.01

[a] Experimental conditions are identical to Table I except that 100 μM DDA was added to all incubation mixtures 2 min prior to AA addition. Alone or in combination with inhibitor vehicles, DDA did not affect calcium mobilization or thromboxane production induced by AA addition. Data are expressed as mean ± S.E.M. of samples from at least five different blood donors. CPS, fluorescence counts per second; TxB, TxB$_2$ (nmol/10^8 platelets).

13-APA (Le Breton *et al.*, 1979; Hung *et al.*, 1983). In these experiments, platelets were incubated with 13-APA to block the TxA_2/PGH_2 receptor, DDA to inhibit adenylate cyclase activity, and a thromboxane synthase inhibitor to block TxA_2 production. Table III illustrates that in the presence of 13-APA, PGH_2-induced calcium mobilization is completely inhibited. These results indicate that PGH_2 must, indeed, interact at a specific receptor site in order to mobilize internal calcium stores and cause platelet activation.

4. DISCUSSION

This study was undertaken to determine whether PGH_2 is capable of causing intraplatelet calcium mobilization. In this connection, we have shown that greater than 95% inhibition of thromboxane synthesis by 7-IHA, azo analogue I, or OKY 1581 decreased the rate of internal calcium release but not the extent of calcium release. Thus, when a thromboxane synthase inhibitor was present, calcium redistribution observed at 1 min was generally less than half of that observed when no thromboxane syntase inhibitor was prsent (Fig. 2, Table I). In contrast, the thromboxane synthase inhibitors caused no inhibition of calcium mobilization 4 min after the addition of AA.

It has been suggested that thromboxane synthase inhibition may promote the metabolism of PGH_2 to PGD_2. Prostaglandin D_2 is known to inhibit platelet aggregation (Smith *et al.*, 1974) and to stimulate adenylate cyclase activity, which in turn leads to an accumulation of intracellular cAMP. Furthermore, Smith *et al.*, (1976) provided evidence that sufficient PGD_2 to inhibit platelet aggregation may be formed from endoperoxides when thromboxane synthesis is blocked.

We have previously shown that another stimulator of adenylate cyclase, prostacyclin (PGI_2), causes an increase in intracellular calcium binding, which is consistent with the ability of PGI_2 to inhibit platelet functional change (Brace *et al.*, 1982; Owen and Le Breton, 1981). Thus, an increase in platelet cAMP leads to sequestration of intraplatelet calcium, reducing the availability of cytosolic calcium (Feinstein *et al.*, 1983) for platelet activation processes. On this basis, it would be

TABLE III. Effect of 13-APA on PGH_2-Induced Calcium Mobilization[a]

	0 min	1 min	4 min
Vehicles	100,000 ± 1,460	91,927 ± 1,000	89,262 ± 1,155
7-IHA	100,000 ± 1,350	101,182 ± 1,123	102,309 ± 1,592
Azo analogue I	100,000 ± 1,168	99,439 ± 1,513	102,284 ± 1,756
OKY 1581	100,000 ± 1,179	100,585 ± 1,201	105,446 ± 1,711

[a] Chlortetracycline-treated PRP was incubated with 100 μM DDA and the appropriate vehicles (vehicle) or 100 μM DDA plus 100 μM 13-APA and either 20 μm 7-IHA, 30 μM azo analogue I, or 500 μM OKY 1581 for 2 min prior to AA addition (0 min) and 1 and 4 min following AA addition. Fluorescence counts per second are expressed as mean ± S.E.M. of samples from four different blood donors.

expected that increased production of PGD_2 would also result in inhibition of calcium mobilization. We found that stimulation of cAMP production had indeed affected the results obtained in the present experiments; i.e., when platelets were pretreated with the cyclase inhibitor DDA, thromboxane synthase inhibition did not block calcium mobilization (Fig. 3, Table II). These results are in agreement with those of Smith (1982), who showed that the inhibition of platelet aggregation by thromboxane synthase inhibitors could be reversed by treating the platelets with an adenylate cyclase inhibitor. Hence, when cyclase activity is blocked, the stimulatory effects of PGH_2 are fully expressed. Taken together, these results indicate that PGH_2 can be as effective as TxA_2 in promoting calcium mobilization.

Studies in isolated calcium-accumulating platelet vesicles enriched in the dense tubular system (DTS) also support this contention. The DTS is thought to play an important role in intraplatelet calcium regulation and platelet activation in a manner anologous to the sarcoplasmic reticulum of muscle cells (White, 1972). Rybicki *et al.* (1983) showed that PGH_2 caused a significant release of calcium from these isolated vesicles only in the presence of an adenylate cyclase inhibitor. These results are therefore consistent with the present findings in intact platelets.

Finally, our results also indicate that PGH_2 produces calcium mobilization by interacting at the same or a stucturally similar site as TxA_2. Thus, the specific TxA_2/PGH_2 receptor antagonist 13-APA inhibits calcium mobilization in response to either TxA_2 or PGH_2 (Table III).

ACKNOWLEDGMENTS. This work was supported by grants from the American Heart Association (82-973) and the National Institutes of Health (HL 24530). G.C.L. is an Established Investigator of the American Heart Association.

REFERENCES

Born, G. V. R., 1962, Aggregation of blood platelets by adenosine diphosphate and its reversal, *Nature* **194**:927–929.

Brace, L. D., Venton, D. L., and Le Breton, G. C., 1982, Thromboxane A_2/prostaglandin H_2 mobilizes calcium in human blood platelets, *Circulation [Suppl.]* **64**:1198.

Caswell, A. H., and Hutchison, J. D., 1971, Visualization of membrane bound cations by a fluorescent technique, *Biochem. Biophys. Res. Commun.* **42**(1):43–49.

Fain, J. N., Pointer, R. H., and Ward, W. F., 1972, Effects of adenosine nucleosides on adenylate cyclase, phosphodiesterase, cyclic adenosine monophosphate accumulation and lipolysis in fat cells, *J. Biol. Chem.* **247**(21):6866–6872.

Feinman, R. D., Lubowsky, J., Charo, I., and Zabinski, M. P., 1977, The lumiaggregometer: A new instrument for simultaneous measurement of secretion and aggregation, *J. Lab. Clin. Med.* **90**(1):125–129.

Feinstein, M. B., 1980, Release of intracellular membrane-bound calcium precedes the onset of stimulus-induced exocytosis in platelets, *Biochem. Biophys. Res. Commun.* **93**(2):593–600.

Feinstein, M. B., Egan, J. J., Sha'afi, R. I., and White, J., 1983, The cytoplasmic concentration of free calcium in platelets is controlled by stimulators of cyclic AMP production (PGD_2, PGE_1, forskolin), *Biochem. Biophys. Res. Commun.* **113**(2):598–604.

Feuerstein, N., and Ramwell, P. W., 1981, OKY-1581, a potential selective thromboxane synthetase inhibitor, *Eur. J. Pharmacol.* **69**:533–534.

Fitzpatrick, F. A., and Gorman, R. R., 1978, A comparison of imidazole and 9,11-azoprosta-5,13-dienoic acid: Two selective thromboxane synthetase inhibitors, *Biochim. Biophys. Acta* **539**:162–172.

Fitzpatrick, F. A., Gorman, R. R., Mc Guire, J. C., Kelly, R. C., Wynalda, M. A., and Sun, F. F., 1977, A radioimmunoassay for thromboxane B₂, *Anal. Biochem.* **82**(1):1–7.

Gorman, R., Sun, F., Peterson, D., Sun, F., Miller, O., and Fitzpatrick, F., 1977, Inhibition of human platelet thromboxane synthetase by 9,11-azoprosta-5,13-dienoic acid, *Proc. Natl. Acad. Sci. U.S.A.* **74**(9):4007–4011.

Grimm, L. J., Knapp, D. R., Senator, D., and Halushka, P. V., 1981, Inhibition of platelet thromboxane synthesis by 7-(1-imidazolyl)heptanoic acid: Dissociation from inhibition of aggregation, *Thromb. Res.* **24**:307–317.

Hung, S. C., Ghali, N. I., Venton, D. L., and Le Breton, G. C., 1983, Specific binding of the thromboxane antagonist 13-azaprostanoic acid to human platelet membranes, *Biochim. Biophys. Acta* **728**:171–178.

Kawahara, Y., Yamanashi, J., Furuta, Y., Kaibuchi, K., Takai, Y., and Fukuzaki, H., 1983, Elevation of cytoplasmic free calcium concentration by stable thromboxane A₂ analogue in human platelets, *Biochem. Biophys. Res. Commun.* **117**(3):663–669.

Kayama, N., Sakaguchi, K., Kaneko, S., Kutoba, T., Fukuzawa, T., Kawamura, S., Yoshimoto, T., and Yamamoto, S., 1981, Inhibition of platelet aggregation by 1-alkylimidazole derivatives, thromboxane A synthetase inhibitors, *Prostaglandins* **21**(4):543–554.

Le Breton, G. C., Dinerstein, R. J., Roth, L. J., and Feinberg, H., 1976, Direct evidence for intracellular divalent cation distribution associated with platelet shape change, *Biochem. Biophys. Res. Commun.* **71**(1):362–370.

Le Breton, G. C., Venton, D. L., Enke, S. E., and Halushka, P. V., 1979, 13-Azaprostanoic acid: A specific antagonist of the human blood platelet thromboxane/endoperoxide receptor, *Proc. Natl. Acad. Sci. U.S.A.* **76**:4097–4101.

Miyamoto, T., Tanaiguchi, K., Tanouchi, T., and Hirata, F., 1980, Selective inhibitor of thromboxane synthetase: Pyridine and its derivatives, *Adv. Prostaglandin Thromboxane Res.* **6**:443–445.

Moncada, S., Needleman, P., Bunting, S., and Vane, J. R., 1976, Prostaglandin endoperoxide and thromboxane generating systems and their selective inhibition, *Prostaglandins* **12**(3):323–335.

Moncada, S., Bunting, S., Mullane, K., Thorogood, P., Vane, J. R., Raz, A., and Needleman, P., 1977, Imidazole: A selective potent antagonist of thromboxane synthetase, *Prostaglandins* **13**(4):611–618.

Needleman, P., Minkes, M., and Raz, A., 1976, Thromboxane: Selective biosynthesis and distinct biological properties, *Science* **193**:163–165.

Needleman, P., Bryan, B., Wyche, A., Bronson, S. D., Eakins, K., Ferrendelli, J. A., and Minkes, M., 1977a, Thromboxane synthetase inhibitors as pharmacological tools: Differential biochemical and biological effects on platelet suspensions, *Prostaglandins* **14**(5):897–907.

Needleman, P., Raz, A., Ferrendelli, J. A., and Minkes, M., 1977b, Application of imidazole as a selective inhibitor of thromboxane synthetase in human platelets, *Proc. Natl. Acad. Sci. U.S.A.* **74**(4):1716–1720.

Owen, N. E. and Le Breton, G. C., 1981, Ca²⁺ mobilization in blood platelets as visualized by chlortetracycline fluorescence, *Am. J. Physiol.* **241**:H613–H619.

Randall, M. J., Parry, M. J., Hawkeswood, E., Cross, P. E., and Dickenson, R. P., 1981, UK-37,248, a novel, selective thromboxane synthetase inhibitor with platelet antiaggregatory and anti-thrombotic activity, *Thromb. Res.* **23**:145–162.

Rybicki, J. P., and Le Breton, G. C., 1983, Prostaglandin H₂ directly lowers human platelet cAMP levels, *Thromb. Res.* **30**:407–414.

Rybicki, J. P., Venton, D. L., and Le Breton, G. C., 1983, The thromboxane antagonist, 13-azaprostanoic acid, inhibits arachidonic acid-induced Ca²⁺ release from isolated platelet membrane vesicles, *Biochim. Biophys. Acta* **751**:66–73.

Smith, J. B., 1982, Effect of thromboxane synthetase inhibitors on platelet function: Enhancement by inhibition of phosphodiesterase, *Thromb. Res.* **28**:477–485.

Smith, J. B., Ingerman, C. M., and Silver, M. J., 1974, Prostaglandin D_2 inhibits the aggregation of human platelets, *Thromb. Res.* **5**:291–299.

Smith, J. B., Ingerman, C. M., and Silver, M. J., 1976, Formation of prostaglandin D_2 during endoperoxide-induced platelet aggregation, *Thromb. Res.* **9**:413–418.

Tai, H., Lee, N., and Tai, C. L., 1980, Inhibition of thromboxane synthesis and platelet aggregation by pyridine and its derivatives, *Adv. Prostaglandin Thromboxane Res.* **6**:447–452.

Venton, D. L., Enke, S. E., and Le Breton, G. C., 1979, Azaprostanoic acid derivatives. Inhibitors of arachidonic acid induced platelet aggregation, *J. Med. Chem.* **22**(7):824–830.

White, J. G., 1972, Interaction of membrane systems in blood platelets, *Am. J. Pathol.* **66**(2):295–312.

Yoshimoto, T., Yamamoto, S., and Hayaishi, O., 1978, Selective Inhibition of prostaglandin endoperoxide thromboxane isomerase by 1-carboxylalkylimidazoles, *Prostaglandins* **16**(4):529–540.

33

Control of Ca²⁺ Mobilization and Polyphosphoinositide Metabolism in Platelets by Prostacyclin

GEORGE B. ZAVOICO, STEPHEN P. HALENDA, DAVID CHESTER, and MAURICE B. FEINSTEIN

1. INTRODUCTION

Cyclic AMP is perhaps the most important inhibitory regulator of platelet function. Dibutyryl cAMP, as well as agents that stimulate adenylate cyclase, can inhibit the characteristic responses of platelets to stimulation by agonists, i.e., shape change, aggregation and secretion, increased lipid metabolism, protein phosphorylation, and cytoskeleton assembly. Prostacyclin (PGI_2) is the most potent physiological activator of platelet adenylate cyclase, and its role as a biological regulator of platelet function appears to be directly related to this action (Weksler, 1982). It has also been suggested that the antithrombotic effectiveness of certain agents observed clinically may be attributable to their antiphosphodiesterase activity, which enhances the action of prostacyclin, released from blood vessel walls, on platelet cAMP levels (Weksler, 1982).

Our interest in the mechanism of action of PGI_2 has centered on its actions vis-à-vis calcium because of the important involvement of Ca^{2+} in many aspects of cellular function. In order to understand how PGI_2 interacts with Ca^{2+}, it is equally important to understand the precise roles of Ca^{2+} in platelet physiology, which at present remain only partly understood, and our knowledge is constantly evolving (Kaibuchi *et al.*, 1982; Rink *et al.*, 1981, 1982; Feinstein *et al.*, 1981).

GEORGE B. ZAVOICO, STEPHEN P. HALENDA, DAVID CHESTER, and MAURICE B. FEINSTEIN • Department of Pharmacology, University of Connecticut Health Center, Farmington, Connecticut 06032.

Platelet activation is associated with two Ca^{2+}-dependent pathways for protein phosphorylation (Fig. 1) that appear to be involved in secretion: phosphorylation of M_r 20,000 myosin light chains by Ca^{2+}/calmodulin-dependent myosin light chain kinase (Hathaway and Adelstein, 1979) and phosphorylation of a M_r 47,000 protein by Ca^{2+}/phosphatidylserine-dependent protein kinase C (Takai *et al.*, 1979; Kawahara *et al.*, 1980; Kikkawa *et al.*, 1983). In platelets permeabilized by digitonin or an intense electric field, the presence of adequate Ca^{2+} alone directly causes myosin phosphorylation (Daniel *et al.*, 1982) and secretion (Knight *et al.*, 1982). However, the responses of intact platelets to agonists such as thrombin are more complex and are characterized by the mobilization of both Ca^{2+} and 1,2-diacylglycerol. A synthetic diglyceride (oleoylacetyldiacylglycerol) can by itself partially activate platelets (Rink *et al.*, 1983; Kaibuchi *et al.*, 1982). 1,2-Diacylglycerol is

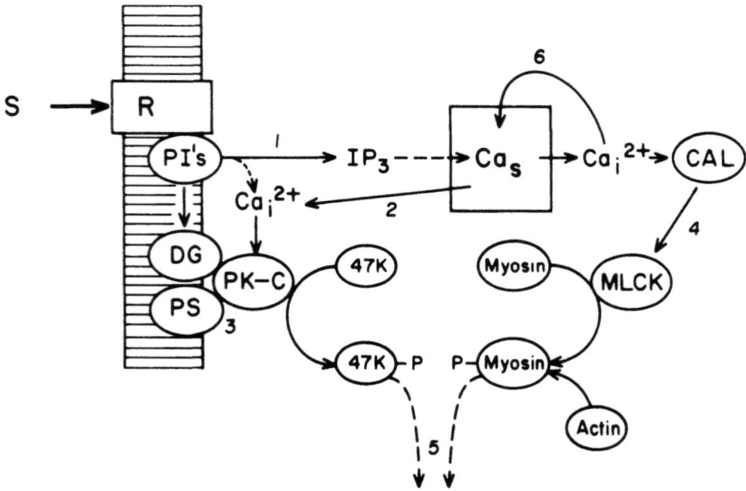

FIGURE 1. Proposed sequence of initial events in the activation of platelets. The binding of stimulatory agonists (S) such as thrombin to membrane receptors (R) initiates the hydrolysis and subsequent resynthesis of phosphoinositides (PIs) and the increase of $[Ca^{2+}]_i$. Phosphatidylinositol-4,5-diphosphate is probably the first phospholipid broken down by phosphodiesterase activity (phospholipase C) to produce 1,2-diacylglycerol (DG) and inositol 1,4,5-trisphosphate (IP_3). This reaction has been suggested to release lipid-bound Ca^{2+} and to regulate membrane Ca^{2+} permeability. (2) Calcium is mobilized from an internal source by mechanisms that are unknown at present; IP_3 has been suggested to be a second messenger that releases Ca^{2+} from a nonmitochondrial source such as a Ca^{2+}-accumulating isolated membrane fraction. (3) The increase of $[Ca^{2+}]_i$ acting in concert with 1,2-DG and phosphatidylserine (PS) activates protein kinase C, which phosphorylates a cytosolic M_r 47,000 protein. (4) Calcium also binds to calmodulin (CAL), which can then activate myosin light chain kinase (MLCK) to phosphorylate the M_r 20,000 light chains of myosin. The ATPase activity of phosphorylated myosin can then be activated by actin. (5) Actomyosin and M_r 47,000 protein may both participate in reactions that enable secretion to occur. (6) Calcium can be resequestered within the platelet. Speculative reactions are shown as dashed lines.

generated by phosphodiesteratic hydrolysis of phosphoinositides and functions co-operatively with Ca^{2+} (Kaibuchi *et al.*, 1982, 1983; Kikkawa *et al.*, 1983), in part by greatly increasing the affinity of protein kinase C for Ca^{2+} (Takai *et al.*, 1979).

Although the phosphodiesterase activity that hydrolyzes phosphoinositides to produce diacylglycerol is itself Ca^{2+} dependent, the enzymatic activity is highly sensitive to lipid conformation and can be expressed at low free Ca^{2+} concentrations (e.g., 0.1 μM) under favorable conditions (Irvine *et al.*, 1984). In intact platelets, the production of diacylglycerol requires receptor–agonist interaction and cannot be effectively duplicated solely by increasing the availability of Ca^{2+} as through the application of calcium ionophores (Kaibuchi *et al.*, 1982). As a result of receptor-mediated generation of diacylglycerol, the responses of intact platelets to agonists are elicited at lower concentrations of cytoplasmic calcium, $[Ca^{2+}]_i$, than are required in permeabilized or ionomycin-treated platelets (Rink *et al.*, 1981; Knight *et al.*, 1982; Kaibuchi *et al.*, 1982, 1983).

The role of phosphoinositides in platelet function is of additional interest since it has been proposed that the mobilization of Ca^{2+} in many cells [including platelets (Vickers *et al.*, 1981)] occurs through the receptor-linked activation of hydrolysis of polyphosphoinositides (Michell, 1975; Downes and Michell, 1982). Recently, it has been proposed that inositol-1,4,5-trisphosphate (IP$_3$) produced by the hydrolysis of phosphatidylinositol-4,5-diphosphate (PIP$_2$) may be a second messenger that is responsible for mediating receptor-linked mobilization of intracellular stores of Ca^{2+} (Berridge, 1983). Initial experimental support for this theory has been obtained in pancreatic acinar cells (Streb *et al.*, 1983) and rat hepatocytes (Joseph *et al.*, 1984). Inositol-1,4,5-trisphosphate has been detected in stimulated platelets (Agranoff *et al.*, 1983; Vickers and Mustard, 1984) and causes the release of Ca^{2+} from isolated platelet membranes (O'Rourke *et al.*, 1985). In this chapter we present evidence that PGI$_2$, by stimulating adenylate cyclase, suppresses the breakdown of PIP$_2$ and the mobilization of Ca^{2+} and diacylglycerol in thrombin-stimulated platelets.

2. RESULTS AND DISCUSSION

2.1. Inhibition of Calcium Mobilization by PGI$_2$ and Other Agents That Increase cAMP

One of the mechanisms by which cAMP could regulate platelet functions is by controlling the level of Ca^{2+} in the cytosol. Although this has long been assumed to be so, no direct evidence was available until recently. Owen and Le Breton (1981), using chlortetracycline as a probe for membrane-bound Ca^{2+}, obtained evidence that increased cAMP stabilized internal Ca^{2+} stores that were susceptible to release by epinephrine, A23187, and the PG endoperoxide analogue U46619. Direct measurements of dynamic changes in free cytoplasmic Ca^{2+} (i.e, $[Ca^{2+}]_i$) subsequently became feasible with the introduction of the fluorescent Ca^{2+} detector quin-2 (Tsien *et al.*, 1982).

From use of quin-2 to measure $[Ca^{2+}]_i$, it was found that adenylate cyclase stimulants had two types of effects on $[Ca^{2+}]_i$: (1) they inhibited the rise of Ca^{2+} that would normally occur in response to platelet stimulation by thrombin (Rink and Smith, 1983; Feinstein *et al.*, 1983; Yamanishi *et al.*, 1983) and (2) they stimulated resequestration of Ca^{2+} that had already been released into the cytosol by thrombin (Feinstein *et al.*, 1983; Yamanishi *et al.*, 1983). Pretreatment of platelets with PGI_2, PGE_1, (Rink and Smith, 1983; Feinstein *et al.*, 1983), PGD_2, or forskolin (Feinstein *et al.*, 1983) was found to inhibit the rate and extent of rise in $[Ca^{2+}]_i$ caused by thrombin. Prostaglandin D_2 and PGI_2 (and PGE_1) act, respectively, on different receptors linked to adenylate cyclase, whereas forskolin acts directly on the catalytic unit of the enzyme (Seamon and Daly, 1981) and/or on the stimulatory guanine nucleotide regulatory protein, N_s (Darfler *et al.*, 1982). The effects of low to moderate concentrations of prostaglandins on $[Ca^{2+}]_i$ were substantially potentiated by the cAMP phosphodiesterase inhibitor theophylline. Dibutyryl cAMP also produced a time- and concentration-dependent inhibition of calcium mobilization (Fig. 2), providing the most direct evidence that cAMP-dependent reactions can regulate $[Ca^{2+}]_i$ in intact platelets. Prostacyclin was by far the most potent inhibitor of Ca^{2+} mobilization and caused a faster rate of fall of $[Ca^{2+}]_i$ than PGD_2. This is in keeping with the relative potency of PGI_2 as a stimulator of adenylate cyclase (Gorman *et al.*, 1977). The I_{50} for inhibition of the

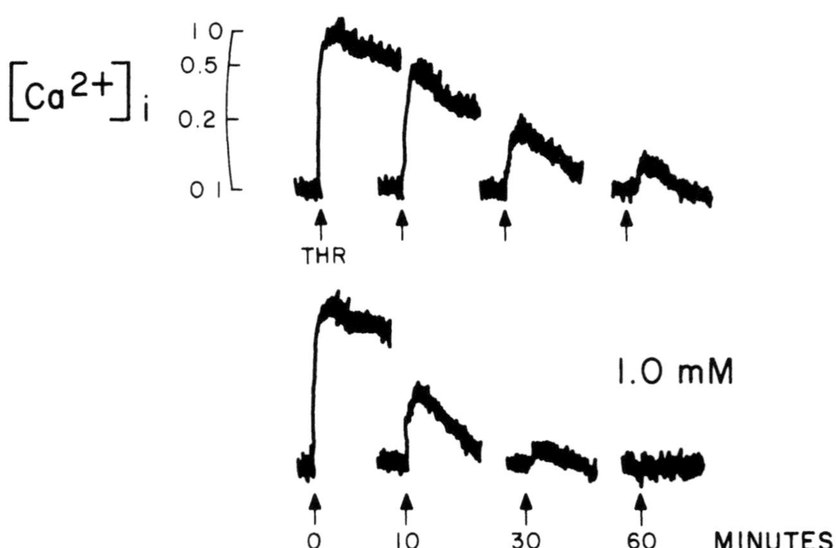

FIGURE 2. Inhibition of Ca^{2+} mobilization (μM) by dibutyryl cAMP (Db cAMP). Washed platelets incubated with either 0.25 or 1.0 mM (Db cAMP) for the times indicated were then stimulated with 1.0 U/ml thrombin at 23°C.

increase in Ca^{2+} caused by 1.0 U/ml thrombin was about 0.5–1.0 nM PGI_2, and 95% inhibition was attained at 10 nM PGI_2. Prostacyclin was even more effective at lower thrombin concentrations; e.g., 1.0 nM PGI_2 completely blocked the rise of $[Ca^{2+}]_i$ evoked by 0.1 U/ml thrombin. The dose–response curve to PGI_2 was not affected by the lack of extracellular Ca^{2+} ($+$EGTA), indicating that PGI_2 had no selective action on Ca^{2+} influx and was fully capable of preventing the mobilization of internal stores of Ca^{2+}.

When platelets were first stimulated with thrombin to allow normal mobilization of Ca^{2+}, the subsequent addition of PGI_2 (or PGD_2, PGE_1, or forskolin) caused the elevated $[Ca^{2+}]_i$ to rapidly fall back to the normal prestimulus level at either 37°C (Feinstein *et al.*, 1983) or 23°C (Zavoico and Feinstein, 1984). A similar effect with PGE_1 was reported by Yamanishi *et al.* (1983).

Stimulation of adenylate cyclase might prevent the thrombin-induced rise in $[Ca^{2+}]_i$ by inhibiting the initial receptor-linked reactions that mobilize Ca^{2+} stores and/or enhancing Ca^{2+} sequestration. Under certain conditions, cAMP-dependent protein kinase has been reported to stimulate calcium uptake by platelet microsomal membrane vesicles (Käser-Glanzmann *et al.*, 1977), although this is disputed by Le Peuch *et al.* (1983). A mitochondrial uncoupler (e.g., FCCP) did not inhibit the PGI_2-stimulated fall of $[Ca^{2+}]_i$ (Halenda *et al.*, 1985), and a Na^+/Ca^{2+} exchange mechanism does not seem to be involved either since the response to PGI_2 was not inhibited by lack of Na^+ in the medium. On the contrary, we found that PGI_2 was more effective in the absence of Na^+.

Another possible mechanism is that the production of a messenger necessary for the mobilization of Ca^{2+} is terminated by a cAMP-dependent reaction, thereby allowing the normal transport processes to act unimpeded to restore $[Ca^{2+}]_i$ to its basal level.

Cyclic AMP can inhibit phospholipase C activity (Rittenhouse-Simmons, 1979) and therefore can prevent the hydrolysis of the phosphoinositides that have been proposed to be involved in Ca^{2+} mobilization. Furthermore, if cAMP were to inhibit PIP_2 breakdown, then the formation of IP_3, a reputed second messenger for mobilization of Ca^{2+} (Berridge, 1983; Streb *et al.*, 1983; Joseph *et al.*, 1984), would be reduced.

Prostaglandin E_1 and dibutyryl cAMP have been reported to inhibit diacylglycerol production (Rittenhouse-Simmons, 1979; Imai *et al.*, 1983) and phosphoinositide metabolism (Rendu *et al.*, 1983). In contrast, the initial breakdown of PIP_2 caused by PAF-acether in horse platelets was reported to be insensitive to a high concentration of PGI_2 (Billah and Lapetina, 1983). We have measured Ca^{2+} mobilization and lipid metabolism concurrently in quin-2-loaded human platelets stimulated by thrombin. Prostacyclin, PGD_2, and dibutyryl cAMP all inhibited $[^{32}P]PIP_2$ breakdown, $[^3H]$diacylglycerol and $[^{32}P]$phosphatidic acid formation, and Ca^{2+} mobilization concurrently (Halenda *et al.*, 1985). Prostacyclin inhibited the initial fall of $[^{32}P]PIP_2$ as well as the subsequent increase above control levels that normally occurs. The inhibition of PIP_2 turnover would also suppress the formation of IP_3 which can release Ca^{2+} from an isolated Ca^{2+}-accumulating membrane fraction (O'Rourke *et al.*, 1985). The suppression of inositol-1,4,5-trisphosphate production might account for the depression of Ca^{2+} mobilization. Further work will be necessary to determine if a direct relationship exists. Prostacyclin inhibited Ca^{2+} mobilization and lipid metabolism induced by 1.0 U/ml thrombin over a range from 0.5 to 10 nM (Fig. 3).

The rise of $[Ca^{2+}]_i$ was prevented by PGI_2 somewhat more effectively than diacylglycerol production or breakdown of PIP_2. This suggests that cyclic AMP may control $[Ca^{2+}]_i$ by mechanisms in addition to inhibition of phosphoinositide breakdown. The ability of PGI_2 to simultaneously suppress increases in $[Ca^{2+}]_i$

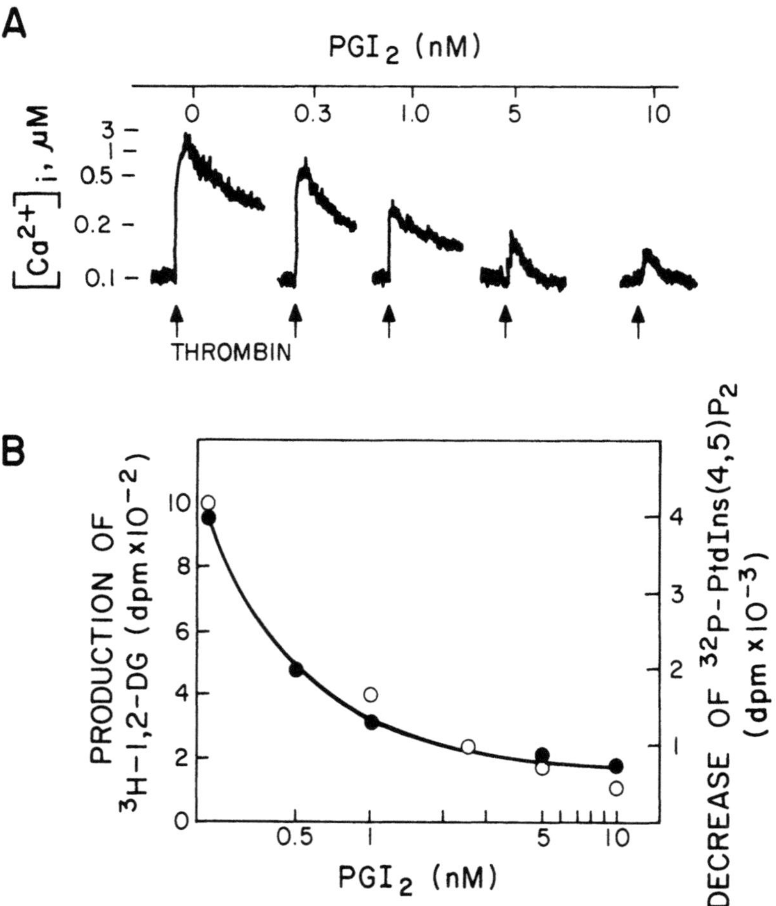

FIGURE 3. Inhibition of $[Ca^{2+}]_i$ mobilization, PIP_2 breakdown (●), and 1,2-diacylglycerol (○) formation by PGI_2. Platelets were loaded with quin-2 and either [^{3}H]arachidonic acid (to measure DG) or [^{32}P]phosphate (to measure PIP_2). Stimulation was by 1.0 U/ml thrombin.

and the production of diacylglycerol probably contributes substantially to its ability to strongly inhibit platelet responses to stimulation.

2.2. Antagonism of Adenylate Cyclase Stimulators by Epinephrine and ADP

In intact platelets epinephrine and ADP can reduce the concentration of cAMP that has previously been elevated by adenylate cyclase stimulants (Mills, 1975). They also inhibit basal and hormone-stimulated adenylate cyclase activity in platelet particulate fractions (Jakobs et $al.$, 1978; Cooper and Rodbell, 1979). Both antagonists act through specific receptors coupled to guanine nucleotide-binding proteins (N_i) that inhibit the enzyme (Smith and Limbird, 1982; Cooper and Rodbell, 1979). Therefore, in the simultaneous presence of prostaglandins and epinephrine (or ADP), the rate of cyclic AMP production should represent the net balance between the effects of the various receptors on the stimulatory (N_s) and inhibitory (N_i) (Murayama and Ui, 1983; Jakobs et $al.$, 1981; Rodbell, 1980) GTP-binding protein regulators of adenylate cyclase activity.

This dynamic balance between the opposing forces acting on the enzyme was clearly reflected in the changes in $[Ca^{2+}]_i$ levels that were produced in stimulated platelets. Epinephrine and ADP antagonized the effects of PGI_2, PGD_2, and forskolin on $[Ca^{2+}]_i$ in thrombin-stimulated platelets, but they were unable to reverse the effects of dibutyryl cAMP (Fig. 2), presumably because the cAMP analogue can bypass adenylate cyclase to directly activate cAMP-dependent kinases (Zavoico and Feinstein, 1984). The inhibition of Ca^{2+} mobilization by the adenylate cyclase stimulants was largely prevented if epinephrine (or ADP) was added prior to prostaglandins (or forskolin), and the platelets then were stimulated with thrombin (Fig. 4A). In addition, the ability of PGD_2 and PGI_2, added after stimulation by thrombin, to rapidly restore elevated $[Ca^{2+}]_i$ to its basal level was also counteracted by epinephrine and ADP (Fig. 4A). Epinephrine and ADP by themselves had little or no effect on $[Ca^{2+}]_i$ when added to washed platelets in the absence of thrombin or prostaglandins; sometimes a transient rise of $[Ca^{2+}]_i$ of <100 nM was observed.

Epinephrine and ADP can also reverse the effects of PGI_2, PGD_2, or forskolin (Zavoico and Feinstein, 1984). Platelets were first incubated with PGD_2 or PGI_2 and then stimulated with thrombin. After the abortive rise in $[Ca^{2+}]_i$ had subsided, the addition of epinephrine (or ADP) caused a further increase in $[Ca^{2+}]_i$ up to levels ranging from 300 nM to about 700 nM (Fig. 4B). Both the rate and extent of rise of $[Ca^{2+}]_i$ under these conditions were substantially less than the response of untreated platelets to thrombin. This was probably because of the onset of desensitization to thrombin by the time epinephrine (or ADP) was added. The partial rise in $[Ca^{2+}]_i$ was accompanied by secretion, but at a lower rate than normal for the amount of thrombin present (Fig. 4B). Preliminary experiments (unpublished) indicate that epinephrine also partially reverses the depression of PIP_2 breakdown caused by PGI_2.

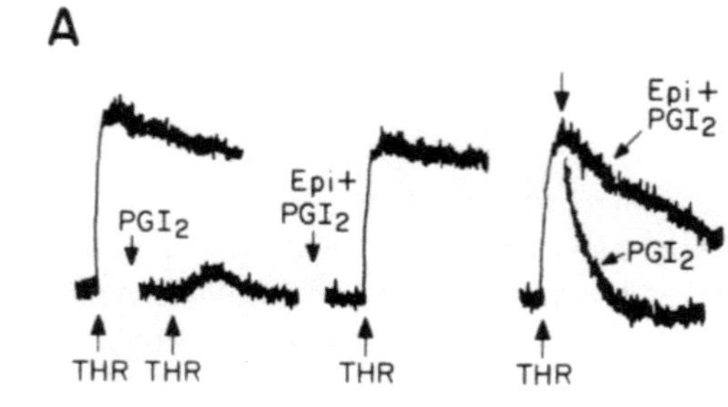

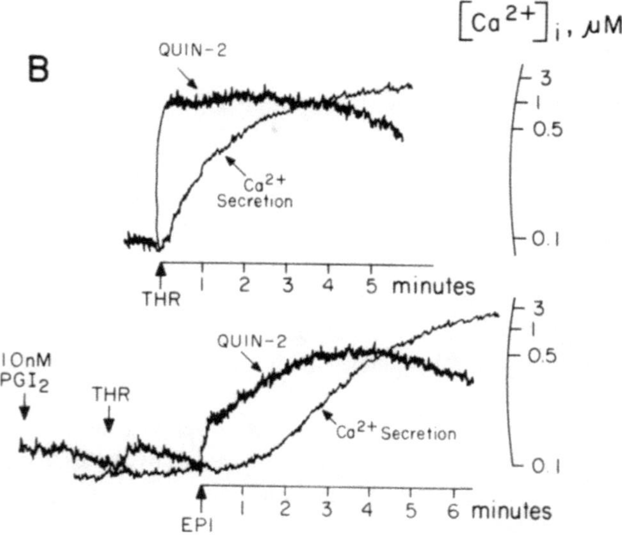

FIGURE 4. Antagonism of PGI_2 by epinephrine. (A) Measurement of $[Ca^{2+}]_i$ with quin-2. Control response to thrombin (1.0 U/ml), inhibition by 10 nM PGI_2, and reversal of PGI_2 effect by 1 μM epinephrine. Last panel shows rapid fall of $[Ca^{2+}]_i$ when PGI_2 is added at peak of response to thrombin (10–15 sec) and the inhibition of the PGI_2 effect by epinephrine. (B) Simultaneous measurement of $[Ca^{2+}]_i$ with quin-2 and dense granule secretion with an extracellular Ca^{2+} electrode. Upper panel is control response to 1.0 U/ml thrombin. Lower panel shows inhibition of $[Ca^{2+}]_i$ mobilization and secretion by PGI_2 and the reversal of both effects by epinephrine (1 μM). Note that the rise in $[Ca^{2+}]_i$ after epinephrine was slower and smaller than the control response. Furthermore, secretion occurred after a much longer lag period and was much slower than normal, although the same total amount of calcium was secreted. Temperature was 23°C; extracellular Ca^{2+} was 50 μM.

Epinephrine and ADP were able to increase $[Ca^{2+}]_i$ even in the absence of Ca^{2+} in the medium, but intact phosphodiesterase activity was necessary since the effect was inhibited by theophylline. Although ADP and epinephrine can reduce elevated cAMP back to basal levels, the presence of a phosphodiesterase inhibitor slows the fall, and the new steady-state level of cAMP attained remains well above normal (Mills, 1974). The response to epinephrine (and ADP) also depended on the occupancy of thrombin receptors, since no effects on $[Ca^{2+}]_i$ were elicited if thrombin was first removed by hirudin. These experiments imply that the mechanisms for mobilizing calcium, although inhibited by a cAMP-dependent reaction, remain potentially operant for some time as long as thrombin continues to be associated with its receptor.

All of the effects of epinephrine on $[Ca^{2+}]_i$ were blocked by yohimbine, an α_2-adrenoreceptor antagonist, but not by corynanthine, which is a highly selective α_1 antagonist (Weitzell et al., 1979). α-Adrenergic receptors mediate two types of physiological responses: (1) inhibition of adenylate cyclase (Fain and Garcia-Sainz, 1980) and (2) increasing $[Ca^{2+}]_i$ (Charest et al., 1983) and increased turnover of phosphoinositides (Fain and Garcia-Sainz, 1980). Different subsets of α receptors mediate these effects since α_1 receptors are linked to the Ca^{2+} and lipid effects and α_2 receptors have been suggested to act solely on adenylate cyclase (Fain and Garcia-Sainz, 1980). Platelet aggregation appears to be mediated by α_2 receptors since it is much more effectively inhibited by the α_2 antagonist yohimbine than by prazosin, an α_2 antagonist (Grant and Scrutton, 1979). This is consistent with our finding that epinephrine had little effect on Ca^{2+} by itself and strongly suggests that epinephrine may produce aggregation of platelets in plasma by affecting adenylate cyclase or through some action involving GTP-binding proteins that is as yet unknown.

REFERENCES

Agranoff, B. W., Murthy, P., and Seguin, E. B., 1983, Thrombin-induced phosphodiesteratic cleavage of phosphatidylinositol bisphosphate in human platelets, *J. Biol. Chem.* **258**:2076–2078.

Berridge, M. J., 1983, Rapid accumulation of inositol trisphosphate reveals that agonists hydrolyse polyphosphoinositides instead of phosphatidylinositol, *Biochem. J.* **212**:849–858.

Billah, M. M., and Lapetina, E. G., 1983, Platelet-activating factor stimulates metabolism of phosphoinositides in horse platelets: Possible relationship to Ca^{2+} mobilization during stimulation, *Proc. Natl. Acad. Sci. U.S.A.* **80**:965–968.

Charest, R., Blackmore, P. F., Berthon, B., and Exton, J. H., 1983, Changes in free cytosolic Ca^{2+} in hepatocytes following α_2-adrenergic stimulation. Studies on Quin-2 loaded hepatocytes, *J. Biol. Chem.* **258**:8769–8773.

Cooper, D. M. F., and Rodbell, M., 1979, ADP is a potent inhibitor of human platelet adenylate cyclase, *Nature* **282**:517–518.

Daniel, J. C., Purdon, A. D., and Molish, I. R., 1982, Platelet myosin phosphorylation as an indicator of cellular calcium concentration, *Fed. Proc.* **41**:1118.

Darfler, F. J., Mahan, L. C., Koachman, A. M., and Insel, P. A., 1982, Stimulation by forskolin of intact S49 lymphoma cells involves the nucleotide regulatory protein of adenylate cyclase, *J. Biol. Chem.* **257**:11901–11907.

Downes, C. P., and Michell, R. H., 1982, Phosphatidylinositol 4-phosphate and phosphatidylinositol 4,5-biphosphate: lipids in search of a function, *Cell Calcium* **3**:467–502.

Fain, J. N., and Garcia-Sainz, J. A., 1980, Role of phosphatidylinositol turnover in $alpha_1$ and of adenylate cyclase inhibition in $alpha_2$ effects of catecholamines, *Life Sci.* **26**:1183–1194.

Feinstein, M. B., Egan, J. J., Sha'afi, R. I., and White, J., 1983, The cytoplasmic concentration of free calcium in platelets is controlled by stimulators of cyclic AMP production (PGD_2, PGE_1, forskolin), *Biochem. Biophys. Res. Commun.* **113**:598–604.

Feinstein, M. B., Rodan, G. A., and Cutler, L. S., 1981, Cyclic AMP and calcium in platelet function, in: *Platelets in Biology and Pathology-2* (J. L. Gordon, ed.), Elsevier/North-Holland, Amsterdam, pp. 437–472.

Gorman, R. R., Bunting, S., and Miller, O. V., 1977, Modulation of human platelet adenylate cyclase by prostacyclin (PGX), *Prostaglandins* **13**:377–388.

Grant, J. A., and Scrutton, M. C., 1979, Novel α_2-adrenoreceptors primarily responsible for inducing human platelet aggregation, *Nature* **277**:659–661.

Halenda, S. P., Zavoico, G. B., Chester, D. W., and Feinstein, M. B., 1985, Interrelationships between Ca^{2+} mobilization, phosphoinositide metabolism and diacylglycerol formation in thrombin-stimulated platelets, (submitted for publication).

Hathaway, D. R., and Adelstein, R. S., 1979, Human platelet myosin light chain kinase requires the calcium binding protein calmodulin for activity, *Proc. Natl. Acad. Sci. U.S.A.* **76**:1653–1657.

Imai, A., Hattori, H., Takahashi, M., and Nozawa, Y., 1983, Evidence that cyclic AMP may regulate Ca^{2+}-mobilization and phospholipases in thrombin-stimulated human platelets, *Biochem. Biophys. Res. Comm.* **112**:693–700.

Irvine, R. F., Letcher, A. J., and Dawson, R. M. C., 1984, Phosphatidylinositol 4,5-bisphosphate phosphodiesterase and phosphomonoesterase activities of rat brain, *Biochem. J.* **218**:177–185.

Jakobs, K. H., Aktories, K., and Schultz, G., 1981, Inhibition of adenylate cyclase by hormones and neurotransmitters, *Adv. Cyclic Nucl. Res.* **14**:173–187.

Jakobs, K. H., Saur, W., and Schultz, G., 1978, Characterization of α- and β-adrenergic receptors linked to human platelet adenylate cyclase, *Naunyn-Schmiedeberg's Arch. Pharmacol.* **302**:285–291.

Joseph, S. K., Thomas, A. P., Williams, R. J., Irvine, R. F., and Williamson, J. R., 1984, Myo-inositol 1,4,5-trisphosphate. A second messenger for the hormonal mobilization of intracellular Ca^{2+} in liver, *J. Biol. Chem.* **259**:3077–3081.

Kaibuchi, K., Sano, K., Hoshijima, M., Takai, Y., and Nishizuka, Y., 1982, Phosphatidylinositol turnover in platelet activation; calcium mobilization and protein turnover. *Cell Calcium* **3**: 323–335.

Kaibuchi, K., Takai, Y., Sawamura, M., Hoshijima, M., Fujikura, T., and Nishizuka, Y., 1983, Synergistic functions of protein phosphorylation and calcium mobilization in platelet activation, *J. Biol. Chem.* **258**: 6701–6704.

Käser-Glanzmann, R., Jakábová, M., George, J. N., and Lüscher, E. F., 1977, Stimulation of calcium uptake in platelet membrane vesicles by adenosine 3',5'-cyclic monophosphate and protein kinase, *Biochim. Biophys. Acta.* **466**:429–440.

Kawahara, Y., Takai, Y., Minakuchi, R., Sano, K., and Nishizuka, Y., 1980, Phospholipid turnover as a possible transmembrane signal for protein phosphorylation during human platelet activation by thrombin, *Biochem. Biophys. Res. Commun.* **97**:309–317.

Kikkawa, U., Takai, Y., Tanaka, Y., Miyake, R., and Nishizuka, Y., 1983, Protein kinase C as a possible receptor protein of tumor-promoting phorbol esters, *J. Biol. Chem.* **258**:11442–11445.

Knight, D. E., Hallam, T. J., and Scrutton, M. C., 1982, Agonist selectivity and second messenger concentration in Ca^{2+}-mediated secretion, *Nature* **296**:256–257.

Le Peuch, C. J., Le Peuch, D. A. M., Katz, S., DeMaille, J. G., Hincke, M. T., Bredoux, R., Enouf, J., Levy-Toledano, S., and Caen, J., 1983, Regulation of calcium accumulation and efflux from platelet vesicles. Possible role for cyclic-AMP-dependent phosphorylation and calmodulin, *Biochim. Biophys. Acta.* **731**:456–464.

Michell, R. H., 1975, Inositol phospholipids and cell surface receptor function, *Biochim. Biophys. Acta.* **415**:81–147.

Mills, D. C. B., 1974, Factors influencing the adenylate cyclase system in human blood platelets, in: *Platelets and Thrombosis* (S. Sherry and A. Scriabine, eds.), University Park Press, Baltimore, pp. 45–67.

Mills, D. C. B., 1975, Initial biochemical responses of platelets to stimulation, in: *Biochemistry and Pharmacology of Platelets*, Ciba Foundation Symposium 35 (new series), Elsevier, New York, pp. 153–173.

Murayama, T., and Ui, M., 1983, Loss of the inhibitory function of the guanine nucleotide regulatory component of adenylate cyclase due to its ADP ribosylation by islet-activating protein, pertussis toxin, in adipocyte membranes, *J. Biol. Chem.* **258**:3319–3326.

O'Rourke, F. A., Halenda, S. P., Zavoico, G. B., and Feinstein, M. B., 1985, Inositol 1,4,5-trisphosphate releases Ca^{2+} from a Ca^{2+}-transporting membrane vesicle fraction derived from human platelets, *J. Biol. Chem.* **260**:956–962.

Owen, N. E., and Le Breton, G. C.,1981, The involvement of calcium in epinephrine or ADP potentiation of human platelet aggregation, *Amer. J. Physiol.* **241**:H613–H619.

Rendu, F., Marche, P., Maclouf, J., Girard, A., and Levy-Toledano, S., 1983, Triphosphoinositide breakdown and dense body release as the earliest events in thrombin-induced activation of human platelets, *Biochem. Biophys. Res. Commun.* **116**:513–519.

Rink T. J., and Smith, S. W., 1983, Inhibitory prostaglandins suppress Ca^{2+} influx, the release of intracellular Ca^{2+} and the responsiveness to cytoplasmic Ca^{2+} in human platelets, *J. Physiol.* **338**:66–67P.

Rink, T. J., Sanchez, A., and Hallam, T. J., 1983, Diacylglycerol and phorbol ester stimulate secretion without raising cytoplasmic free calcium in human platelets, *Nature* **305**:317–319.

Rink, T. J., Smith, S. W., and Tsien, R. Y., 1981, Intracellular free calcium in platelet shape change and aggregation, *J. Physiol.* **324**:53–54P.

Rink, T. J., Smith, S. W., and Tsien, R. Y., 1982, Cytoplasmic free Ca^{2+} in human platelets: Ca^{2+} thresholds and Ca-independent activation for shape-change and secretion, *FEBS Letts.* **148**:21–26.

Rittenhouse-Simmons, S., 1979, Production of diglyceride from phosphatidylinositol in activated platelets, *J. Clin. Invest.* **63**:580–587.

Rodbell, M., 1980, The role of hormone receptors and GTP-regulatory proteins in membrane transduction, *Nature* **284**:17–22.

Seamon, K. B., and Daly, J. W., 1981, Forskolin: A unique diterpene activator of cyclic AMP-generating systems, *J. Cycl. Nucl. Res.* **7**:201–224.

Smith, S. K., and Limbird, L. E., 1982, Evidence that human platelet α-adrenergic receptors coupled to inhibition of adenylate cyclase are not associated with the subunit of adenylate cyclase ADP-ribosylated by cholera toxin, *J. Biol. Chem.* **257**:10471–10478.

Streb, H., Irvine, R. F., Berridge, M. J., and Schulz, I., 1983, Release of Ca^{2+} from a nonmitochondrial intracellular store in pancreatic acinar cells by inositol-1,4,5-trisphosphate, *Nature* **306**:67–69.

Takai, Y., Kishimoto, A., Iwasa, Y., Kawahara, Y., Mori, T., and Nishizuka, Y., 1979, Calcium-dependent activation of a multifunctional protein kinase by membrane phospholipids, *J. Biol. Chem.* **254**:3692–3695.

Tsien, R. Y., Pozzan, T., and Rink, T. J., 1982, Calcium homeostasis in intact lymphocytes: Cytoplasmic free calcium monitored with a new, intracellularly trapped fluorescent indicator, *J. Cell. Biol.* **94**:325–334.

Vickers, J. D., and Mustard, J. F., 1984, Thrombin-induced increase in inositol phosphates during aggregation and release from washed rabbit platelets, *Fed. Proc.* **43**:979.

Vickers, J.D., Kinlough-Rathbone, R. L., and Mustard, J. F., 1982, Changes in phosphatidylinositol-4,5-bisphosphate 10 seconds after stimulation of washed rabbit platelets with ADP, *Blood* **60**:1247–1250.

Weitzell, R., Tanaka, T., and Starke, K., 1979, Pre- and postsynaptic effects of yohimbine stereoisomers on noradrenergic transmission in the pulmonary artery of the rabbit, *Nauyn-Schmiedeberg's Arch. Pharmacol.* **308**:127–136.

Weksler, B., 1982, Prostacyclin, in: *Progress in Hemostasis and Thrombosis, Vol. 6* (T. H. Spaet, ed.), Grune and Stratton, Inc., New York, pp. 113–138.

Yamanishi, J., Kawahara, Y., and Fukuzaki, H., 1983, Effect of cyclic AMP on cytoplasmic free calcium in human platelets stimulated by thrombin: Direct measurement with Quin2, *Thrombosis Res.* **32:**183–188.

Zavoico, G. B., and Feinstein, M. B., 1984, Cytoplasmic Ca^{2+} in platelets is controlled by cyclic AMP: Antagonism between stimulators and inhibitors of adenylate cyclase, *Biochem. Biophys. Res. Commun.* **120:**579–585.

34

Heme–Polyenoic Acid Interaction and Prostaglandin Synthesis

GUNDU H. R. RAO and JAMES G. WHITE

1. BACKGROUND

Yoshimoto and associates (1970) were first to demonstrate that heme is required for enhancing the activity of prostaglandin synthetase enzymes. Ogino *et al.* (1978) found very little iron in purified endoperoxide synthetase but found that the enzyme required metalloporphyrins for its activity. Ultrastructural studies from our laboratory using an iron-reactive drug, diaminobenzidine, reported that enzymes involved in endoperoxide generation are localized in the dense tubular system (DTS) of human platelets. The fact that heme iron acts as a catalyst for the isolated enzymes and that heme-complexing compounds react with these enzymes in the platelet DTS prompted us to evaluate the role of iron and heme in arachidonic acid oxidation. Using a cell-free system, we showed that ferrous iron and heme can induce oxidation of polyenoic acids and the extent of oxidation was related to the degree of unsaturation in the fatty acids. The iron-induced oxidation of fatty acids was not inhibitable by radical scavengers or by superoxide dismutase. Heme or iron chelators, as well as several of the known inhibitors of cyclooxygenase, blocked this reaction. Ferrous iron chelators and heme-complexing agents inhibited prostaglandin synthesis as well as platelet function. *In vivo* studies using a short-acting drug, ibuprofen, and an irreversible inhibitor of cyclooxygenase, aspirin, confirmed our findings that the majority of prostaglandin inhibitors exert their effect by interfering with heme–arachidonic acid interaction. Based on this evidence, we hypothesized that polyenoic acid released from platelet membranes following activation could compete for the heme site and inhibit the conversion of arachidonic acid.

Recent studies have shown that eicosapentaenoic acid (EPA, 20:5ω3) and

GUNDU H. R. RAO and JAMES G. WHITE ● Department of Laboratory Medicine and Pathology, Medical School, University of Minnesota, Minneapolis, Minnesota 55455.

docosahexaenoic acid (DHA, 22:6ω3) can be incorporated into platelet membrane phospholipids and released in response to stimulation by potent agonists. The released fatty acids could compete with arachidonic acid for cyclooxygenase. Effects of these fatty acids on thromboxane synthesis and platelet function have been evaluated in the present study. Both fatty acids blocked arachidonate-induced aggregation of platelets and prevented the second-wave response to adenosine diphosphate and epinephrine. Inhibition induced by these fatty acids could be overcome by a higher concentration of agonists. Docosahexaenoic acid, a poor substrate for cyclooxygenase even at a concentration of 150 μM, was as potent as aspirin in preventing the conversion of radiolabeled arachidonic acid to thromboxane. Unlike aspirin-treated platelets, DHA-treated cells, when washed and stirred with labeled arachidonic acid, generated as much thromboxane as control platelets. These results and our earlier observations on the role of heme in modulation of cyclooxygenase activity suggest that polyenoic acids, if released in sufficient quantities in the vicinity of cyclooxygenase enzymes, could effectively compete with the favored substrate, arachidonic acid, thereby preventing conversion to endoperoxides.

2. INTRODUCTION

Intense interest in platelet and endothelial prostaglandin biosynthesis has developed in recent years because alterations in the synthetic pathway of these metabolites may lead to thrombotic or hemorrhagic disorders (Allen, 1974; Marcus, 1978; Burch and Majerus, 1979; Gorman, 1979; Gryglewski, 1980; Mustard and Kinlough-Rathbone, 1980; Harlan and Harker, 1981; Buchanan and Hirsch, 1982; Granstrom et al., 1982). As a result, there has been a considerable effort to develop nontoxic drugs that may help prevent or control the contribution of platelets and endothelial cells to thrombotic and hemorrhagic diseases, atherosclerosis, and inflammatory conditions (Hamberg et al., 1974; Needleman et al., 1976; Bills et al., 1977; Moncada et al., 1977; Weksler et al., 1977; Forster, 1980; Packham and Mustard, 1980; Packham, 1983).

Stimulation of platelets or endothelial cells initiates activation of phospholipases resulting in the release of polyenoic acids (Bills et al., 1977; Weksler et al., 1977). Arachidonic acid, the major natural polyenoic acid released from these cells, is transformed by cyclooxygenase into cyclic endoperoxides (prostaglandin G_2 and prostaglandin H_2) (Hamberg et al., 1974). The endoperoxides are immediately converted to thromboxane A_2 in platelets and to prostacyclin (PGI_2) in endothelial cells (Needleman et al., 1976; Moncada et al., 1977). Although information about cellular production and drug-induced inhibition of eicosanoids and their role in pathophysiology is available, little is known about natural mechanisms modulating the synthesis of these biologically active compounds. Our interest in the role of iron in arachidonic acid metabolism began with localization of prostaglandin-synthesizing enzymes in the platelet dense tubular system (Gerrard et al., 1976). Since that observation, we have maintained an ongoing investigation to define the role of heme in polyenoic acid metabolism. These studies led to a hypothesis suggesting that heme–arachidonic acid interaction was the critical step in platelet fatty acid oxidation by cyclooxygenase (Peterson *et al.*, 1979). Here we review our findings as they relate to heme–arachidonic acid interaction and the mechanism of action of cyclooxygenase inhibitors.

3. LOCALIZATION OF PROSTAGLANDIN SYNTHETASE IN PLATELETS

The enzyme catalyzing the initial oxygenation of polyenoic acids is generally referred to as fatty acid cyclooxygenase. This step is followed by the cleavage of 15-hydroperoxide of PGG to generate PGH. Both of these reactions require heme and are catalyzed by a homogeneous preparation of an enzyme (prostaglandin endoperoxide synthetase) purified from bovine vesicular gland microsomes (Ohki *et al.*, 1979). Earlier work on platelet prostaglandin production included studies on intact cells as well as lysates and microsomal preparations. Since platelet microsomal fractions include a mixture of plasma membranes, organelle membranes, membranes of the open canalicular system, and membranes of the dense tubular system, Gerrard *et al.*, (1976) conducted studies to localize the site for prostaglandin synthesis. Janszen and Nugteren (1971) had used a histochemical procedure employing di-aminobenzidine to localize prostaglandin biosynthesis in columnar epithelial cells of the sheep vesicular glands. White (1972), using this technique in an earlier study, demonstrated peroxidase in platelets in the membranes of the dense tubular system. Ultrastructural studies by Gerrard *et al.* (1976) using diaminobenzidine reactions demonstrated that prostaglandin synthetase activity was confined to the peroxidase-positive areas of the dense tubular system in platelets. Aminotriazole, which specifically inhibited this reaction, blocked platelet arachidonic acid conversion as well as the aggregatory response of platelets to the action of arachidonate. Recent studies by Carey *et al.* (1982) have confirmed these findings. They used high-voltage free-flow electrophoresis to separate and isolate membranes of the dense tubular system from surface membranes. Their studies suggest that a sequence of activities leading to the biosynthesis of endoperoxides and thromboxanes from the substrate arachidonic acid is associated with the membranes of the dense tubular system.

4. POLYENOIC ACID OXIDATION IN A CELL-FREE SYSTEM

Demonstration of the requirement for hemoglobin or metalloporphyrin for the activation of prostaglandin synthesis (Yoshimoto *et al.*, 1970; Ogino *et al.*, 1978) and localization of the enzymes in specific internal membrane systems (Gerrard *et al.*, 1976) by using an iron-complexing agent prompted many workers to explore the role of free radicals in prostaglandin synthesis and platelet function (Panganamala *et al.*, 1974, 1976; Rao *et al.*, 1976; Egan *et al.*, 1976). In an attempt to demonstrate the role of oxygen radicals, Rao *et al.* (1978a) showed that a combination of nitroblue tetrazolium and vitamin E inhibited platelet function. However, they were not able to demonstrate any inhibitory effect on platelet function by radical scavengers or by enzymes such as superoxide dismutase and catalase. Studies using a cell-free system containing platelet lysates showed that arachidonic acid oxidation could be monitored by a combination of nitroblue tetrazolium (NBT) and vitamin E (Rao *et al.*, 1978a). Fatty-acid-induced NBT reduction was not inhibitable by superoxide dismutase, suggesting that superoxide anion was not involved in this reaction. For further elucidation of this phenomenon, ferrous iron was used to oxidize polyenoic acid (Rao *et al.*, 1978a). Fatty acid oxidation was measured in two different ways: (1) from formation of conjugated intermediates by the increase in UV absorbance at 232 nm and (2) by measuring thiobarbituric-acid-positive compounds. Ferrous-iron-induced oxidation of fatty acids required oxygen, and the degree of oxidation related to the number of double bonds. Ferrous iron was converted to ferric iron during the oxidation of fatty acids.

Utilizing nitroblue reduction as an endpoint to monitor ferrous-iron-induced oxidation of arachidonic acid, Peterson *et al.* (1978) showed that the molar ratio of Fe^{2+} to arachidonic acid in this rapid reaction was 1 : 1, suggesting an interaction of one fatty acid molecule with one molecule of iron. Since ferrous-iron-induced oxidation of fatty acids monitored by NBT reduction was not inhibited by superoxide dismutase or catalase, the radical initiating the reduction of NBT was not one of the reduced oxygen species. Therefore, they hypothesized that ferrous-iron-induced oxidation of fatty acid led to formation of an active species of arachidonic acid responsible for NBT reduction (Rao *et al.*, 1976).

Although no direct proof was provided for the generation of an arachidonic acid radical in these studies, recent investigations by Mason *et al.* (1980) demonstrated formation of a carbon-centered radical intermediate in the prostaglandin synthetase oxidation of arachidonic acid by ESR spectroscopy. Peterson and associates (1979), based on studies of several known cyclooxygenase inhibitors, proposed a molecular model to explain the mechanism by which indomethacin exerts its inhibitory effects on the ferrous-iron-induced oxidation of arachidonic acid (Fig. 1). They demonstrated that ibuprofen and tolmetin, two other prostaglandin synthetase inhibitors, also block the reaction of Fe^{2+} with arachidonic acid, suggesting a general mechanism of action for this class of drugs.

5. FERROUS IRON AND HEME-COMPLEXING AGENTS AS INHIBITORS OF PLATELET FUNCTION

Gerrard *et al.* (1976) demonstrated that aminotriazole, an inhibitor of peroxidase and an iron chelator, blocked arachidonic acid conversion to prostaglandins. Peterson and associates (1978), who used the NBT reduction assay to monitor ferrous-iron-induced oxidation of arachidonic acid, speculated that NBT traps the arachidonic acid radical and vitamin E enhances the availability of NBT for this

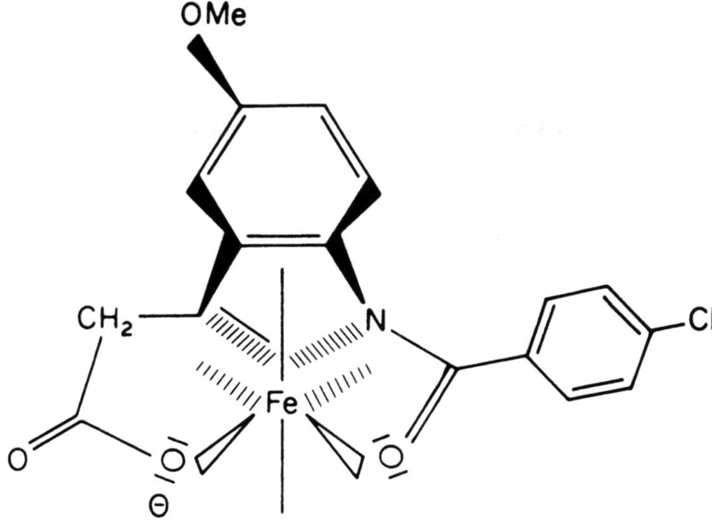

FIGURE 1. A postulated model for the interaction of ferrous iron and indomethacin, a cyclooxygenase inhibitor. The carboxylate and amide groups of indomethacin are shown interacting with the valences of iron to form this complex. (From Peterson *et al.*, 1979, with permission.)

process. Thus, the combination of these agents prevents arachidonic acid metabolism. Extending this hypothesis, Rao *et al.* (1979) demonstrated that NBT and vitamin E block conversion of labeled arachidonic acid to prostaglandins by sheep vesicular gland enzymes. To further elucidate the mechanism of action of cyclooxygenase inhibitors, a cell-free system was used in which ferrous-iron-induced oxidation of arachidonic acid was followed in the presence of NBT (Peterson *et al.*, 1978, 1979). Inhibitors such as indomethacin, ibuprofen, and tolmetin were used in these studies to demonstrate affinity of these drugs for Fe^{2+}. The results demonstrated that indomethacin formed a one-to-one complex with ferrous iron and inhibited arachidonic acid oxidation and NBT reduction. They further demonstrated that the Fe^{2+}–indomethacin complex dissociated in the presence of arachidonic acid. In these studies, ibuprofen and tolmetin also exerted inhibitory effects on the interaction of Fe^{2+} with arachidonic acid. In an earlier study, Vargaftig *et al.* (1978) demonstrated that *ortho*-phenanthroline, a ferrous iron chelator, could bind to the enzyme and protect it from the inhibitory effect of aspirin. Van der Ouderaa *et al.* (1977) found that during the purification of the enzyme the heme dissociates, and reconstitution of heme is critical for cyclooxygenase activity, thereby confirming the observations of several investigators regarding the heme requirement for maximal enzyme activity (Ohki *et al.*, 1979; Yoshimoto *et al.*, 1970; Ogino *et al.*, 1978).

Since heme iron is required for the activity of cyclooxygenase, Peterson *et al.* (1978, 1979) substituted heme for ferrous iron in their studies using the NBT reduction assay developed by Rao *et al.* (1978a) to evaluate heme interaction with arachidonic acid. Their results demonstrated that ferrous heme is essential for arachidonic acid oxidation. In isolated enzyme systems, ferric heme is commonly used for enhancing the activity of prostaglandin synthesizing enzymes. Studies by Rao and associates (1978a), however, demonstrated that ferrous iron or ferrous heme was essential for the oxidation of polyenoic acids. Therefore, a mechanism must be available to convert the ferric heme to ferrous heme. Peterson *et al.* (1980a,b,c) showed that fatty acid peroxides and physiologically available compounds such as epinephrine, tryptophan, and ascorbic acid could reduce ferric heme to the ferrous state, suggesting that reduction of heme is critical for the activation of prostaglandin synthetase as well as lipoxygenases. Results of these studies provided additional evidence for the earlier hypothesis on the role of ferrous heme in polyenoic acid oxidation and suggested a rationale for the functional regulation of cyclooxygenase activity.

Peterson and Gerrard (1979) built a model to show that, sterically, it was possible for arachidonic acid to interact with heme. According to this model, arachidonic acid will have to attach by its carboxylate group to Fe^{2+} ligand on one

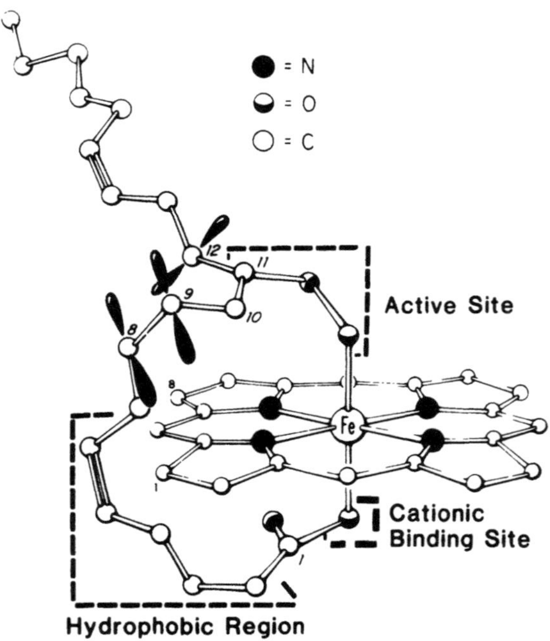

FIGURE 2. A proposed model for the interaction of heme, arachidonic acid, and oxygen. In the model, an oxygen from the carboxylic acid residue of the arachidonic acid binds to the Fe^{2+} ligand below the plane of the porphyrin ring of the heme. The fatty acid then wraps around the heme, and the C_{11} interacts with the oxygen bound to the Fe^{2+} ligand above the plane of the porphyrin ring. (From Peterson and Gerrard, 1979, with permission.)

side of the heme and wind around the protoporphyrin ring. The C_{11} double bond would then interact with the oxygen bound to the second available ligand on the other side of heme (Fig. 2).

Since heme iron has only two available ligands, one below and the other above the plane of the porphyrin ring, Rao *et al.* (1980) evaluated bidentate and unidentate compounds for their inhibitory effect on platelet arachidonic acid conversion. According to the proposed model of Peterson and Gerrard (1979) for heme–arachidonic acid interaction, both unidentate and bidentate chelators would be expected to inhibit the complex mentioned above. Results obtained with agents such as 2,2'-dipyridyl, 4,4'-dipyridyl, 1,17-phenanthroline, and 1,7-phenanthroline suggest that unidentate and bidentate chelators are equally effective in inhibiting this reaction, since an effective chelator would need only one functional position to interfere with the heme–arachidonic acid interaction.

6. COMPETITION BY DRUGS FOR THE ACTIVE SITE ON THE ENZYME

Vargaftig *et al.* (1975) demonstrated that metal-complexing agents inhibited the arachidonic-acid-induced platelet aggregation and concluded from their results that copper was involved in the bioconversion of arachidonic acid. In a separate study, they showed that phenanthrolines, as well as salicylate, prevented the inhibition of cyclooxygenase by aspirin, suggesting a competition for a single site as opposed to multiple-site interaction. McDonald and Ali (1978) also observed a protective effect of salicylate and sulfinpyrazone on aspirin-induced inhibition of cyclooxygenase. To evaluate if the salicylate was competing for the same site as acetylsalicylic acid, Peterson *et al.* (1981) tested these drugs for their effect on heme–arachidonic acid interaction. Their results showed that salicylic acid, a weak inhibitor of platelet cyclooxygenase, was as potent as acetylsalicylic acid in preventing the oxidation of arachidonic acid by ferrous iron.

Recent *in vivo* studies have shown that salicylate, sulfinpyrazone, indomethacin, and ibuprofen can effectively block the irreversible inhibition of cyclooxygenase by aspirin (Rajtar *et al.*, 1981; Cerletti *et al.*, 1982; Livio *et al.*, 1982; Rao *et al.*, 1983a). Cerletti *et al.* (1982) have suggested that there are two active sites on platelet cyclooxygenase for drug interaction. Studies done by several investigators suggest a common site on the cyclooxygenase for drug interaction (Rao *et al.*, 1980, 1983a,b; Vargaftig, 1978; McDonald and Ali, 1978; Peterson *et al.*, 1981; Rajtar *et al.*, 1981; Livio *et al.*, 1982). This active site seems to be metal sensitive, and available evidence suggests it to be a heme protein. Therefore, most inhibitors of prostaglandin synthesis appear to exert their inhibitory effect by interfering with the redox state of the heme essential for activity of the enzyme. Although the presence of additional groups, such as acetyl or fluoride, on the drug and their reaction with other sites on the enzyme may confer longlasting inhibitory effects (Rao *et al.*, 1983a,b).

7. COMPETITION BY SUBSTRATES FOR THE ACTIVE SITE ON THE ENZYME

Although eicosatetraenoic acid is the favored substrate of platelet and endothelial cell cyclooxygenase, this enzyme will convert other polyenoic acids to their metabolites if they are offered as a substrate (Struijk *et al.*, 1966; Needleman *et al.*, 1979; Whitaker *et al.*, 1979; Spector *et al.*, 1983; Fischer and Weber, 1983, 1984). Fish fats are rich in eicosapentaenoic acid, docosapentaenoic acid, and docosahexaenoic acid (Culp *et al.*, 1980). Several recent studies have suggested that subjects whose diet consists mainly of fish as the major source of nutrients have a low risk of cardiovascular disease (Dyerberg *et al.*, 1975, 1978; Hornstra, 1975; Bloch *et al.*, 1979; Dyerberg and Bang, 1979; Jakubowski and Ardlie, 1979; Lands *et al.*, 1980; Needleman *et al.*, 1980; Siess *et al.*, 1980; Brox *et al.*, 1981; Hornstra *et al.*, 1981; Saynor and Verel, 1981; Thorngren and Gustafson, 1981; Hay *et al.*, 1982; Morita *et al.*, 1983). It has been proposed that the beneficial effect of altered dietary fat arises from the conversion of EPA to eicosanoids such as prostacyclin (PGI_3) and thromboxane (TxA_3) (Needleman *et al.*, 1979, 1981; Whitaker *et al.*, 1979; Nidy and Johnson, 1978; Smith *et al.*, 1979; Hamberg, 1980). These products differ in their properties from naturally occurring prostacyclin (PGI_2) and thromboxane A_2 (TxA_2) derived from eicosatetraenoic acid (AA, 20:4ω3). Prostacyclin (PGI_3) seems to be as potent as PGI_2 in inhibiting platelet function, whereas thromboxane A_3 (TxA_3) has been shown to be less active than TxA_2 in causing platelet aggregation (Needleman *et al.*, 1979, 1980; Jakubowski and Ardlie, 1979). A recent study by Corey *et al.*, (1983) showed that DHA is a potent competitive inhibitor of arachidonic acid transformation by sheep vesicular gland cyclooxygenase. Aveldano and Sprecher (1983) have further demonstrated that DHA could be converted to their respective hydroxy acids by platelet lipoxygenase.

These studies suggest that polyenoic acids could alter prostaglandin metabolism in several ways. First, they can substitute for arachidonic acid and form prostaglandins of the 3 series. Second, they can compete for the enzyme and act as competitive inhibitors of the natural substrate arachidonic acid. Third, they can form hydroperoxy acids via lipoxygenase and exert inhibitory effects on platelet function through the actions of these metabolic products. However, considerable confusion exists as to which of these mechanisms the beneficial effect of a fish diet could be attributed to.

Rao *et al.* (1983b) in a recent study have offered yet another possibility to explain the mechanism of inhibition by polyenoic acid. If sufficient polyenoic acids are released in the vicinity of the enzyme, fatty acids may compete with each other for the active site of cyclooxygenase and act as competitive inhibitors. Results of their study show that both EPA and DHA can effectively inhibit arachidonic-acid-induced aggregation of platelets and the second-wave response of platelets to the action of epinephrine and adenosine diphosphate (Fig. 3). Docosahexaenoic acid was found to be as potent as aspirin in inhibiting the conversion of labeled arachidonic acid to thromboxane. However, unlike aspirin, DHA-treated platelets, on

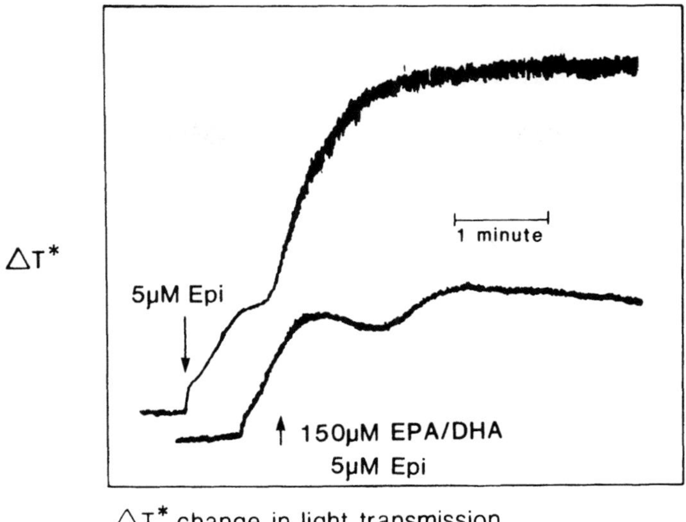

ΔT* change in light transmission

FIGURE 3. Eicosapentaenoic acid (EPA) as well as docosahexaenoic acid (DHA) at 150 μM concentration blocked the second-wave response of platelets in plasma to the action of 5 μM epinephrine.

washing, could convert normal levels of thromboxane from arachidonic acid, suggesting the transient nature of DHA-induced inhibition.

A similar reversible inhibitory effect has been demonstrated by a ferrous iron chelator, dipyridyl (Rao *et al.*, 1980). Since EPA and DHA are relatively poor substrates for cyclooxygenase, an alternative mechanism in which all three fatty acids are favored equally in a reaction is appealing (Corey *et al.*, 1983; Hamberg, 1980; Rao *et al.*, 1978a, 1983b). Since, in the course of ferrous-iron- or heme-mediated oxidation, EPA and DHA are oxidized to a greater extent than arachidonic acid, it is reasonable to speculate that all three fatty acids compete for the heme site on the enzyme whether or not they are converted to significant quantities of vasoactive metabolites (Rao *et al.*, 1978a, 1983b; Peterson *et al.*, 1980a,b,c). If sufficient quantities of these fatty acids (EPA, DPA, and DHA) are incorporated into platelet and endothelial membrane phospholipids and released in significant quantities in response to the normal physiological stimuli, then the fatty acids themselves or their metabolites generated by the actions of cyclooxygenase or lipoxygenase may exert inhibitory effects on platelet and endothelial cell function.

In conclusion, our studies suggest a critical role for ferrous heme in polyenoic acid oxidation and the formation of eicosanoids. Ferrous-iron- or heme-mediated oxidation of fatty acids is not inhibitable by superoxide dismutase or catalase but is effectively blocked by iron chelators and heme-complexing agents. Based on our results, we have presented a novel concept of heme–arachidonic acid interaction. Results of our studies suggest that the majority of the cyclooxygenase inhibitors block and prevent the formation of metabolites of polyenoic acids by interfering with this critical mechanism. We have extended these ideas further to provide an alternative explanation as to a possible mechanism by which polyenoic acids could compete for this heme-associated site on the enzyme cyclooxygenase and prevent the conversion of the favored substrate, arachidonic acid.

REFERENCES

Allen, J. E., 1974, Prostaglandins in hematology, *Arch. Intern. Med.* **133**:86–96.

Aveldano, M. I., and Sprecher, H., 1983, Synthesis of hydroxy fatty acids from 4,7,10,13,16,19-[1-^{14}C]docosahexaenoic acid by human platelets, *J. Biol. Chem.* **258**:9339–9343.

Bills, T. K., Smith, J. B., and Silver, M. J., 1977, Selective release of arachidonic acid from the phospholipids of human platelets in response to thrombin, *J. Clin. Invest.* **60**:1–6.

Bloch, K. L., Culp, B., Madison, D., Randall, O. S., and Lands, W. E. M., 1979, The protective effects of dietary fish oil on focal cerebral infraction, *Prostaglandins Med.* **5**:247–258.

Brox, J. H., Killie, J. E., Gunnes, S., and Nordoy, A., 1981, The effect of cod liver oil and corn oil on platelets and vessel wall in man, *Thromb. Haemostas.* **46**:604–611.

Buchanan, M. R., and Hirsch, J., 1982, Prostaglandins: Their role in hemostasis and thrombosis, in: *Acetylsalicylic Acid: New Uses for an Old Drug* (H. J. M. Barnett, J. Hirsch, and J. F. Mustard, eds.), Raven Press, New York, pp. 33–43.

Burch, J. W., and Majerus, P. W., 1979, Prostaglandins in platelet function, *Semin. Hematol.* **16**:196–207.

Carey, F., Menashi, S., and Crawford, N., 1982, Localization of cyclooxygenase and thromboxane synthetase in human platelet intracellular membranes, *Biochem. J.* **204**:847–851.

Cerletti, C., Livio, M., and DeGaetano, G., 1982, Non-steroidal anti-inflammatory drugs react with two sites on platelet cyclooxygenase. Evidence from *in vivo* drug interaction studies in rats, *Biochim. Biophys. Acta* **714**:122–128.

Corey, E. J., Shih, C., and Cashman, J. R., 1983, Docosahexaenoic acid is a strong inhibitor of prostaglandin but not leukotriene biosynthesis, *Proc. Natl. Acad. Sci. U.S.A.* **80**:3581–3584.

Culp, B. R., Lands, W. E. M., Lucches, B. R., Pitt, R., and Romson, J., 1980, The effect of dietary supplementation of fish on experimental myocardial infarction, *Prostaglandins* **20**:1021–1031.

Dyerberg, J., and Bang, H. O., 1979, Hemostatic function and platelet polyunsaturated fatty acids in Eskimos, *Lancet* **2**:433–435.

Dyerberg, J., Bang, H. O., and Hjorne, N., 1975, Fatty acid composition of the plasma lipids in Greenland Eskimos, *Am. J. Clin. Nutr.* **28**:958–966.

Dyerberg, J., Bang, H. O., Stofferson, S., Moncada, S., and Vane, J. R., 1978, Eicosapentaenoic acid and prevention of thrombosis and atherosclerosis, *Lancet* **2**:117–119.

Egan, R. W., Paxton, J., and Kuehl, F. A., Jr., 1976, Mechanism for reversible self-deactivation of prostaglandin synthetase, *J. Biol. Chem.* **251**:7329–7335.

Fischer, S., and Weber, P. C., 1983, Thromboxane A$_3$ (TxA$_3$) is formed in human platelets after dietary eicosapentaenoic acid (C20:5ω3), *Biochim. Biophys. Res. Commun.* **116**:1091–1099.

Fischer, S., and Weber, P. C., 1984, Prostaglandin I$_3$ is formed *in vivo* in man after dietary eicosapentaenoic acid, *Nature* **307**:165–168.

Forster, W., 1980, Effect of various agents on prostaglandin biosynthesis and anti-aggregatory effect, *Acta Med. Scand.* **642**:35–46.

Gerrard, J. M., White, J. G., Rao, G. H. R., and Townsend, D., 1976, Localization of platelet prostaglandin production in the platelet dense tubular system, *Am. J. Pathol.* **83**:283–299.

Gorman, R. R., 1979, Modulation of human platelet function by prostacyclin and thromboxane A$_2$, *Fed. Proc.* **38**:83–88.

Granstrom, E., Diczfalusy, U., Hamberg, M., Hansson, G., Malmsten, C., and Samuelsson, B., 1982, Thromboxane A_2: Biosynthesis and effect on platelets, in: *Prostaglandins and the Cardiovascular System* (J. A. Oates, ed.), Raven Press, New York, pp. 15–58.

Gryglewski, R. J., 1980, Prostaglandins, platelets and atherosclerosis, *CRC Crit. Rev. Biochem.* **7**:291–338.

Hamberg, M., 1980, Transformation of 5,8,11,14,17-eicosapentaenoic acid in human platelets, *Biochim. Biophys. Acta* **618**:389–398.

Hamberg, M., Svensson, J., Wakabayashi, T., and Samuelsson, B., 1974, Isolation and structure of two prostaglandin endoperoxides that cause platelet aggregation, *Proc. Natl. Acad. Sci. U.S.A.* **71**:345–349.

Harlan, J. M., and Harker, L. A., 1981, Hemostasis, thrombosis and thromboembolic disorders. The role of arachidonic acid metabolism in platelet–vessel wall interactions, *Med. Clin. North Am.* **68**:855–880.

Hay, C. R. M., Durber, A. D., and Saynor, R., 1982, Effect of fish oil on platelet kinetics in patients with ischemic heart disease, *Lancet* **2**:1269–1272.

Hornstra, G., 1975, Specific effects of types of dietary fats on arterial thrombosis, in: *The Role of Fats in Human Nutrition* (A. J. Vergroessen, ed.), Academic Press, New York, pp. 303–330.

Hornstra, E., Christ-Hazelhof, Haddeman, E., tenHoor, F., and Nugteren, D. H., 1981, Fish oil feeding lowers thromboxane and prostacyclin production by rat platelets and aorta and does not result in the formation of prostaglandin I_3, *Prostaglandins* **21**:727–738.

Jakubowski, J. A., and Ardlie, N. G., 1979, Evidence for the mechanism by which eicosapentaenoic acid inhibits human platelet aggregation and secretion implications for prevention of vascular disease, *Thromb. Res.* **16**:205–217.

Janszen, F. H. A., and Nugteren, D. H., 1971, Histochemical localization of prostaglandin synthetase, *Histochemie* **27**:159–164.

Lands, W. E. M., Pitt, B., and Culp, B. R., 1980, Recent concepts on platelet functions and dietary lipids in coronary thrombosis, vasospasm and angina, *Herz* **5**:34–42.

Livio, M., Delmaschio, D., Cerletti, C., and deGaetano, G., 1982, Indomethacin prevents the long lasting inhibitory effect of aspirin on human platelet cyclooxygenase activity, *Prostaglandins* **23**:787–796.

Marcus, A. J., 1978, The role of lipids in platelet function with particular reference to the arachidonic acid pathway, *J. Lipid Res.* **19**:793–826.

Mason, R. P., Kalyanaraman, B., Tainer, B. E., and Eling, T. E., 1980, A carboncentered free radical intermediate in the prostaglandin synthetase oxidation of arachidonic acid. Spin trapping and oxygen uptake studies, *J. Biol. Chem.* **255**:5019–5022.

McDonald, J. M., and Ali, M., 1978, Salicylate and sulfinpyrazone reduce inhibition by aspirin of platelet cyclooxygenase, *Clin. Res.* **26**:856A.

Moncada, S., Herman, A. G., Higgs, E. A., and Vane, J. R., 1977, Differential formation of prostacyclin (PGX or PGI_2) by layers of the arterial wall. An explanation for the anti-thrombotic properties of vascular endothelium, *Thromb. Res.* **11**:323–344.

Morita, I., Saito, Y., Chang, W. C., and Murota, S., 1983, Effects of purified eicosapentaenoic acid on arachidonic acid metabolism in cultured murine aortic smooth muscle cells, vessel walls and platelets, *Lipids* **18**:42–49.

Mustard, J. F., and Kinlough-Rathbone, R. L., 1980, Prostaglandins and platelets, *Annu. Rev. Med.* **31**:89–96.

Needleman, P., Minkes, M., and Raz, A., 1976, Thromboxanes: Selective biosynthesis and distinct biological properties, *Science* **193**:163–165.

Needleman, P., Ray, L., Minkes, M. S., Ferrendelli, J. A., and Sprecher, H., 1979, Triene prostaglandins: Prostacyclins and thromboxane biosynthesis and unique biological properties, *Proc. Natl. Acad. Sci. U.S.A.* **76**:944–948.

Needleman, P., Whitaker, M. O., Wyche, A., Watters, K., Sprecher, H., and Raz, A., 1980, Manipulation of platelet aggregation by prostaglandins and their fatty acid precursors: Pharmacological basis for a therapeutic approach, *Prostaglandins* **19**:165–181.

Needleman, P., Wyche, A., LeDuc, L., Sankarappa, S. K., Jackschik, B. A., and Sprecher, H., 1981, Fatty acids as sources of potential "magic bullets" for the modification of platelet and vascular function, *Prog. Lipid Res.* **20:**415–422.

Nidy, E. G., and Johnson, R. A., 1978, Synthesis of prostaglandin I₃ (PGI₃), *Tetrahedron Lett.* **27:**2375–2378.

Ogino, N., Ohki, S., Yamamato, S., and Hayaishi, O., 1978, Prostaglandin endoperoxide synthetase from bovine vesicular gland microsomes. Inactivation and activation by heme and other metalloporphyrins, *J. Biol. Chem.* **253:**5061–5068.

Ohki, S., Ogino, N., Yamamato, S., and Hayaishi, O., 1979, Prostaglandin hydroperoxidase, an integral part of prostaglandin endoperoxide synthetase from bovine vesicular gland microsomes, *J. Biol. Chem.* **254:**829–836.

Packham, M. A., 1983, Platelet function inhibitors, *Thromb. Haemostas.* **50:**610–619.

Packham, M. A., and Mustard, J. F., 1980, Pharmacology of platelet affecting drugs, *Circulation* **62:**26–41.

Panganamala, R. V., Brownlee, N. R., Sprecher, H., and Cornwell, D. G., 1974, Evaluation of superoxide anion and singlet oxygen in the biosynthesis of prostaglandins from eicosa-8,11,14-trienoic acid, *Prostaglandins* **7:**21–29.

Panganamala, R. V., Sharma, H. R., Heikkila, R. E., Geer, J. C., and Cornwell, D. G., 1976, Role of hydroxyl radical scavengers dimethylsulfoxide, alcohols and methional in the inhibition of prostaglandin biosynthesis, *Prostaglandins* **11:**599–608.

Peterson, D. A., and Gerrard, J. M., 1979, A hypothesis for the interaction of heme and arachidonic acid in the synthesis of prostaglandins, *Med. Hypotheses* **5:**683–694.

Peterson, D. A., Gerrard, J. M., Rao, G. H. R., Krick, T. P., and White, J. G., 1978, Ferrous iron mediated oxidation of arachidonic acid. Studies employing nitroblue tetrazolium (NBT), *Prostaglandins Med.* **1:**304–317.

Peterson, D. A., Gerrard, J. M., Rao, G. H. R., and White, J. G., 1979, Inhibition of ferrous iron induced oxidation of arachidonic acid by indomethacin, *Prostaglandins Med.* **2:**97–108.

Peterson, D. A., Gerrard, J. M., Rao, G. H. R., Mills, E. L., and White, J. G., 1980a, Interaction of arachidonic acid and heme iron in the synthesis of prostaglandins, *Adv. Prostaglandin Thromboxane Res.* **6:**157–161.

Peterson, D. A., Gerrard, J. M., Rao, G. H. R., and White J. G., 1980b, Epinephrine and other activators of prostaglandin endoperoxide synthetase can reduce Fe^{+++}-heme to Fe^{++}-heme, *Prostaglandins Med.* **5:**357–365.

Peterson, D. A., Gerrard, J. M., Rao, G. H. R., and White, J. G., 1980c, Reduction of ferric heme to ferrous by lipid peroxides: Possible relevance to the role of peroxide tone in the regulation of prostaglandin synthesis, *Prostaglandins Med.* **4:**73–79.

Peterson, D. A., Gerrard, J. M., Rao, G. H. R., and White, J. G., 1981, Salicylic acid inhibition of the irreversible effect of acetylsalicylic acid on prostaglandin synthetase may be due to competition for the enzyme cationic binding site, *Prostaglandins Med.* **6:**161–164.

Rajtar, G., Cerletti, C., Livio, M., and deGaetano, G., 1981, Sulphinpyrazone prevents *in vivo* the inhibitory effect of aspirin on rat platelet cyclooxygenase activity, *Biochem. Pharmacol.* **30:**2773–2776.

Rao, G. H. R., Gerrard, J. M., Eaton, J. W., and White, J. G., 1976, Detection of arachidonic acid (AA) free radical formation, an essential step in platelet prostaglandin (PG) synthesis, *Blood* **48:**977A.

Rao, G. H. R., Gerrard, J. M., Eaton, J. W., and White, J. G., 1978a, Arachidonic acid peroxidation, prostaglandin synthesis and platelet function, *Photochem. Photobiol.* **28:**845–850.

Rao, G. H. R., Gerrard, J. M., Eaton, J. W., and White, J. G., 1978b, The role of iron in prostaglandin synthesis: Ferrous iron mediated oxidation of arachidonic acid, *Prostaglandins Med.* **1:**55–70.

Rao, G. H. R., Burris, S. M., Gerrard, J. M., and White, J. G., 1979, Inhibition of prostaglandin (PG) synthesis in sheep vesicular gland microsomes (SVGM) by nitroblue tetrazolium (NBT) and vitamin E, *Prostaglandins Med.* **2:**111–121.

Rao, G. H. R., Cox, C. A., Gerrard, J. M., and White, J. G., 1980, Effect of 2,2'-dipyridyl and related compounds on platelet prostaglandin synthesis and platelet function, *Biochim. Biophys. Acta* **628**:468–479.

Rao, G. H. R., Johnson, G. J., Reddy, K. R., and White, J. G., 1983a, Ibuprofen protects platelet cyclooxygenase from irreversible inhibition by aspirin, *Arteriosclerosis* **3**:383–388.

Rao, G. H. R., Radha, E., and White, J. G., 1983b, Effect of docosahexaenoic acid (DHA) on arachidonic acid metabolism and platelet function, *Biochem. Biophys. Res. Commun.* **117**:549–556.

Saynor, R., and Verel, D., 1981, Effect of a fish oil on blood lipids and coagulation, *Thromb. Haemostas.* **46**:65A.

Siess, W. B., Scherer, B., Bohlig, B., Roth P., Kurzmann, I., and Weber, D. C., 1980, Platelet membrane fatty acids, platelet aggregation and thromboxane formation during a mackerel diet, *Lancet* **1**:441–444.

Smith, D. R., Weathery, B. C., Salmon, J. A., Ubatula, F. B., Gryglewski, R. J., and Moncada, S., 1979, Preparation and biochemical properties of PGH$_3$, *Prostaglandins* **18**:423–438.

Spector, A. A., Kaduce, T. L., Figard, P. H., Norton, K. C., Hoak, J. C., and Czervionke, R. L., 1983, Eicosapentaenoic acid and prostacyclin production by cultured endothelial cells, *J. Lipid Res.* **24**:1595–1604.

Struijk, C. B., Beerthuis, R. K., Pabon, H. J. J., and van Dorp, D. A., 1966, Specificity in the enzymatic conversion of polyunsaturated fatty acids into prostaglandins, *Rec. Trav. Chim.* **85**:1233–1250.

Thorngren, M., and Gustafson, A., 1981, Effects of 11 week increase in dietary eicosapentaenoic acid on bleeding time, lipids and platelet aggregation, *Lancet* **11**:1190–1193.

Van der Ouderaa, F. J., Buytenhek, M., Nugteren, D. H., and van Dorp, D. A., 1977, Purification and characterization of prostaglandin endoperoxide synthetase from sheep vesicular glands, *Biochim. Biophys. Acta* **487**:315–331.

Vargaftig, B. B., 1978, The inhibition of cyclooxygenase of rabbit platelets by aspirin is prevented by salicylic acid and by phenanthrolines, *Eur. J. Pharmacol.* **50**:231–241.

Vargaftig, B. B., Trainer, Y., and Chignard, M., 1975, Blockade by metal complexing agents and by catalase of the effects of arachidonic acid on platelets. Relevance to the study of anti-inflammatory mechanisms, *Eur. J. Pharmacol.* **33**:19–29.

Weksler, B. B., Marcus, A. J., and Jaffe, E. A., 1977, Synthesis of prostaglandin I$_2$ (prostacyclin) by cultured human and bovine endothelial cells, *Proc. Natl. Acad. Sci. U.S.A.* **74**:3922–3926.

Whitaker, M. O., Syche, A., Fitzpatrick, F., Sprecher, R. H., and Needleman, P., 1979, Triene prostaglandins: Prostaglandin D$_3$ and icosapentaenoic acid as potential antithrombotic substances, *Proc. Natl. Acad. Sci. U.S.A.* **76**:5919–5923.

White, J. G., 1972, Interaction of membrane systems in blood platelets, *Am. J. Pathol.* **66**:295–312.

Yoshimoto, A., Ito, H., and Tomita, K., 1970, Cofactor requirements of the enzyme synthesizing prostaglandin in bovine seminal vesicles, *J. Biochem.* **68**:487–499.

35

Cardiac Effects of Peptidoleukotrienes

DAVID M. ROTH, CARL E. HOCK,
DAVID J. LEFER, and ALLAN M. LEFER

1. INTRODUCTION

The leukotrienes are a group of newly identified eicosanoids of the lipoxygenase pathway of arachidonic acid metabolism (Samuelsson, 1981). The peptide-containing leukotrienes (LT) (i.e., LTC_4, LTD_4, and LTE_4) are known to possess potent biological properties including bronchoconstriction (Holroyde *et al.*, 1981), enhancement of vascular permeability (Peck *et al.*, 1981), and vasoconstriction (Yokochi *et al.*, 1982). In this regard, it is well documented that leukotrienes C_4 and D_4 are potent coronary constrictors in a number of animal species (Woodman and Dusting, 1983; Terashita *et al.*, 1981). Previous work in our laboratory has shown that the leukotrienes are potent constrictors of isolated perfused cat coronary arteries (Roth and Lefer, 1984). Leukotrienes have also been claimed to exert direct negative inotropic effects on the myocardium (Burke *et al.*, 1982). The purpose of this investigation was to examine the inotropic effects of the leukotrienes in relation to their potent coronary constricting activity and to clarify the relationship between coronary vasoactivity and inotropicity in cardiac preparations isolated from three commonly used mammalian species.

2. METHODS

Three species of animals were studied: adult male cats, male Hartley guinea pigs, and Sprague–Dawley rats. The experimental preparations employed were the

---

DAVID M. ROTH, CARL E. HOCK, DAVID J. LEFER, and ALLAN M. LEFER • Department of Physiology, Jefferson Medical College, Thomas Jefferson University, Philadelphia, Pennsylvania 19107.

isolated Langendorff perfused hearts and the isolated papillary muscle preparation. Leukotrienes C_4 and D_4 (Merck-Frosst Laboratories, Dorval, Canada) were stored in distilled water at $-71°C$ until use.

In the heart perfusion experiments, hearts were excised and placed in ice-cold Krebs–Henseleit (K–H) solution. The hearts were attached to the perfusion apparatus and perfused through the aorta. Experiments were carried out under both constant-pressure and constant-flow perfusion conditions. The hearts were attached to a Statham FT-03 strain gauge via a suture in the apex of the heart in order to monitor myocardial contractile force. The leukotrienes were infused into the aortic inflow tract of the heart, and changes in coronary perfusion pressure under constant-flow perfusion or in coronary flow under constant-pressure perfusion were monitored.

In the papillary muscle experiments, right papillary muscles of the cat or left papillary muscles of the guinea pig and rat were dissected and mounted in 10-ml chambers containing oxygenated K–H solution. The preparations were stimulated at 2 V above threshold with a duration of 17 msec and a frequency of 1/sec. Resting tension of each muscle was set at a point just below that which produced maximum contractile force. Changes in contractile force were monitored on the addition of the leukotrienes to the bath. Experiments were also carried out in which known negative inotropic agents (e.g., sodium pentobarbital) were added to the bath in order to test the responsiveness of the preparation.

3. RESULTS

The relationship between coronary perfusion pressure and contractile force in response to LTC_4 (10, 25, 50, and 100 ng/ml) is illustrated in Fig. 1. In constant-flow-perfused rat hearts, it is apparent that there is a close linkage between the increase in coronary perfusion pressure and the reduction in contractile force. Furthermore, the close relationship is dose dependent and linear at the LTC_4 concentrations studied, with a correlation coefficient of 0.80. Although complete dose–response relationships were not determined for cat and guinea pig hearts, coronary perfusion pressure/contractile force relationships were similar to those of the rat for these species.

Direct effects of both LTC_4 and LTD_4 on isolated, electrically paced papillary muscles from the three species studied are summarized in Table I. Leukotriene C_4, given at a concentration that produced near-maximal reduction in contractile force in Langendorff perfused hearts (Fig. 1), had no significant inotropic effect in any of three species studied. Similarly, LTD_4 also did not exhibit direct inotropic activity at 50 ng/ml. In contrast, sodium pentobarbital (100 μg/ml) exerted significant direct negative inotropic action in all three species, indicating that the papillary muscle preparations were responsive to negative inotropic agents.

Figure 2 represents a typical recording of the effect of LTC_4 on constant-flow-perfused rat hearts both before and after treatment with the leukotriene antagonist FPL-55712. Pretreatment with 10 μg/ml of the antagonist blocked both the coronary

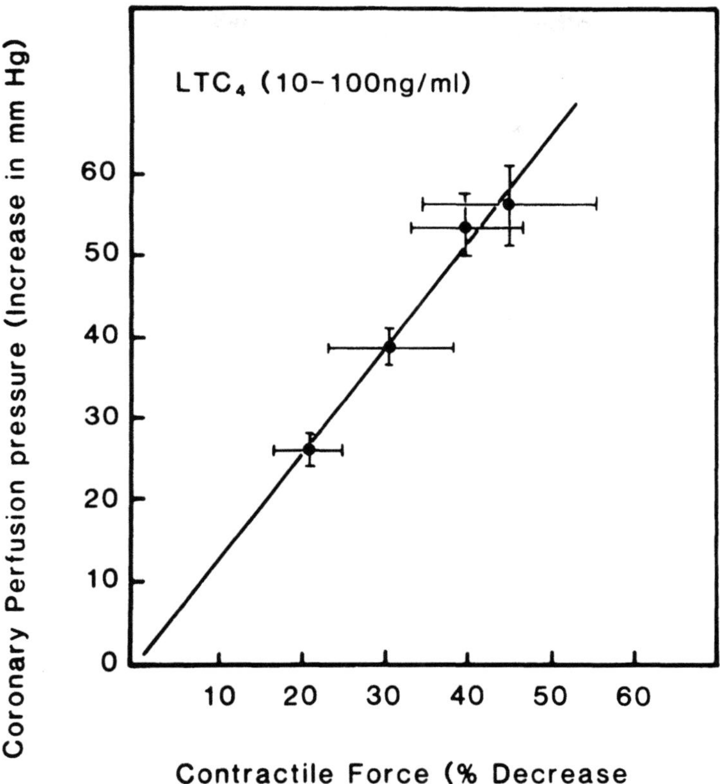

FIGURE 1. Relationship between coronary perfusion pressure and contractile force in constant-flow-perfused rat hearts. Data points represent changes in perfusion pressure and contractile force on the addition of LTC$_4$ (10, 25, 50, and 100 mg/ml). All values are means ± S.E.M. for six hearts in each group.

TABLE I. Inotropic Effects of Leukotrienes in Isolated Papillary Muscle[a]

Species	LTC$_4$ (50 ng/ml)	LTD$_4$ (50 ng/ml)	Pentobarbital (100 μg/ml)
Cat	−3 ± 2	+2 ± 2	−66 ± 7*
Guinea pig	+5 ± 4	−4 ± 3	−61 ± 5*
Rat	−2 ± 4	0 ± 3	−56 ± 9*

[a] All values are mean percentage changes ± S E M. for six to eight muscles: *$P < 0.01$ from control

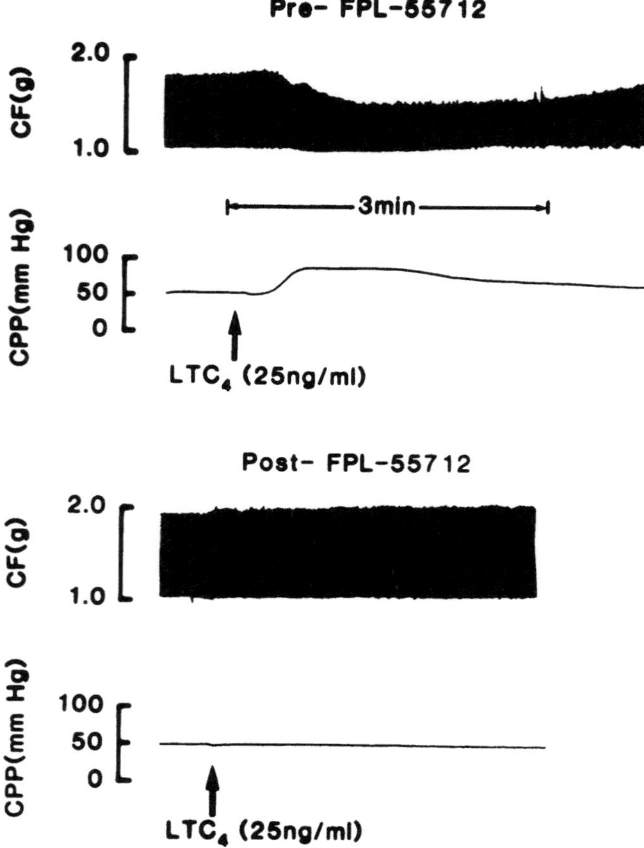

FIGURE 2. Representative recordings of changes in coronary perfusion pressure and contractile force in constant-flow-perfused rat hearts in response to LTC$_4$ (25 ng/ml) in the presence and absence of FPL-55712 (10 μg/ml).

constrictor and the cardiodepressant actions of LTC$_4$. Prior to FPL-55712 administration, 25 ng/ml of LTC$_4$ increased coronary perfusion pressure by 48% and reduced contractile force by 50%. However, after FPL-55712, coronary perfusion pressure increased by only 2%, and contractile force was reduced by only 1%, indicating almost complete blockade of the cardiac actions of LTC$_4$. The coronary constrictor and cardiodepressant effects of LTD$_4$ (25 ng/ml) were also completely antagonized by FPL-55712.

In the guinea pig there was a similar antagonism of the cardiac actions of LTC$_4$ by FPL-55712, as illustrated in Fig. 3. Infusion of LTC$_4$ (25 ng/ml) produced a 30 $\pm$ 2 mm Hg increase in coronary perfusion pressure and a 67 $\pm$ 4% reduction

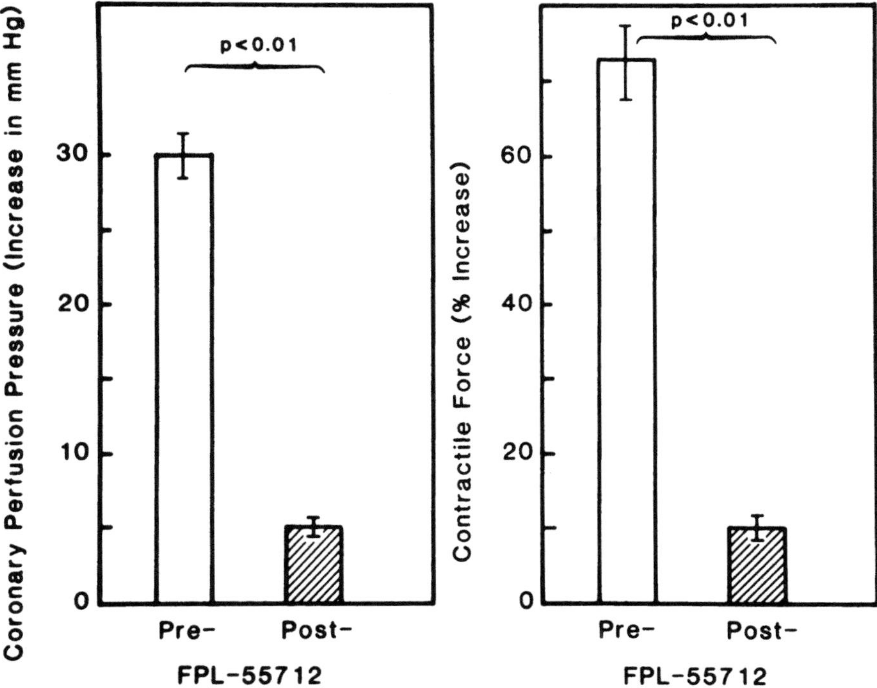

FIGURE 3. Results from experiments in which LTC_4 (25 ng/ml) was added to constant-flow-perfused hearts of guinea pigs before and after treatment of the hearts with FPL-55712 (10 μg/ml). The bars represent changes in coronary perfusion and changes in contractile force on the addition of LTC_4 before and after FPL-55712 administration.

in contractile force. Pretreatment with FPL-55712 (10 μg/ml) significantly reduced both the rise in coronary perfusion pressure and the reduction in contractile force (5 ± 1 mm Hg and 10 ± 2%, respectively; $P < 0.01$). These results together with those illustrated in Fig. 2 indicate that FPL-55712 almost totally abolishes the coronary vasoconstrictor and cardiodepressant actions of LTC_4 in both the rat and guinea pig heart.

Figure 4 illustrates the cardiac effects of another vasoconstrictor eicosanoid, 9,11-methanoepoxy-PGH_2 (U-46619), on both the isolated electrically driven cat papillary muscle and the isolated Langendorff perfused cat heart. This endoperoxide/thromboxane A_2 mimetic produced a marked coronary vasoconstriction and concomitant reduction in contractile force. The large reduction in contractile force is entirely a result of the coronary vasoconstriction since U-46619 has no direct action on isolated papillary muscles. These results illustrate that eicosanoid-induced profound reduction in contractile force can be explained solely on the basis of the coronary vasoactivity of the agents in question.

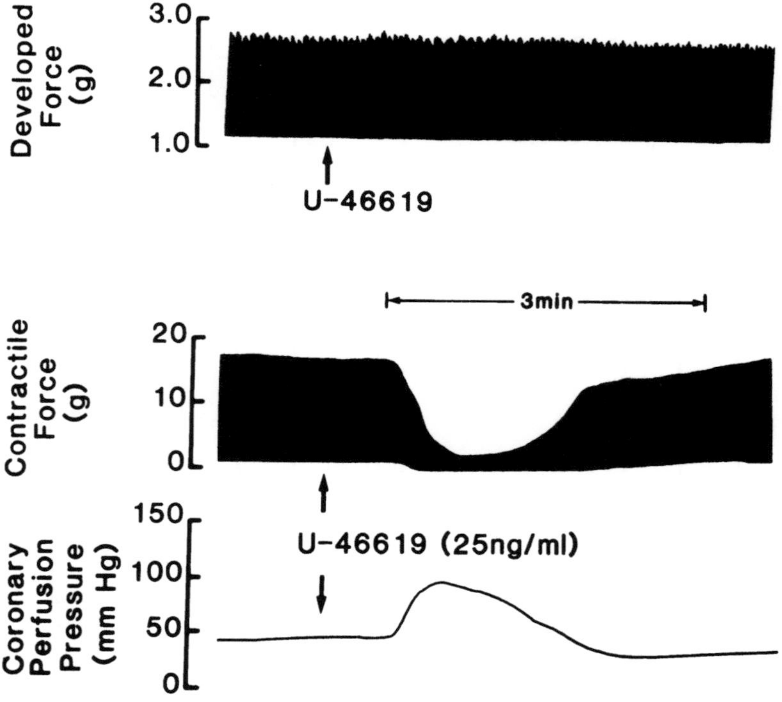

FIGURE 4. Representative recordings of developed force in isolated cat papillary muscles, contractile force, and coronary perfusion pressures in isolated constant-flow-perfused cat hearts on the addition of U-446619 (25 ng/ml) to the preparations.

4. DISCUSSION

Our results indicate that the peptide leukotrienes (i.e., LTC_4 and LTD_4) are potent coronary constricting agents that induce a marked concomitant reduction in contractile force in rat, cat, and guinea pig hearts. However, the reduction in cardiac contractile force is secondary to the coronary vasoconstriction, and these leukotrienes do not have a direct negative inotropic effect at concentrations that are normally employed. These findings differ from those of Burke *et al.* (1982) using LTC_4 and LTD_4 and Michaelassi *et al.* (1983) using LTC_4. These investigators suggested significant direct negative inotropic actions for the peptide leukotrienes in addition to their coronary vasoconstrictor effects and attempted to differentiate between distinct receptor populations mediating the two cardiac effects (Michaelassi *et al.*, 1983; Burke *et al.*, 1984). However, their conclusions are based on inconclusive data.

We have demonstrated that concentrations of LTC_4 and LTD_4 (50 ng/ml) that produce potent coronary constriction, both in isolated perfused coronary arteries

(Roth and Lefer, 1984) and in Langendorff perfused myocardial preparations, do not have significant direct negative inotropic effects. This lack of direct negative inotropicity on electrically paced papillary muscles was common to all three species studied (Table I). In the constant-flow-perfused heart preparations, a reduction in contractile force was observed only in conjunction with an increase in coronary perfusion pressure, and in the case of constant-pressure-perfused hearts, a decrease in contractile force occurred only with a decrease in coronary flow. Furthermore, the close relationship between coronary constriction and contractile force was maintained at all concentrations tested. When the LTC_4-induced coronary vasoconstriction is blocked by the leukotriene antagonist FPL-5712, the usual reduction in contractile force is abolished. The data on FPL-55712 antagonism in conjunction with the lack of direct negative inotropicity clearly indicate that the coronary constrictor effects of the leukotrienes are responsible for their negative inotropic effects.

In addition, other vasoconstrictor eicosanoid agents such as U-46619 produce cardiac effects very similar to those of LTC_4 and LTD_4 in the Langendorff perfused heart preparations. These results indicate that the relationship between direct coronary vasoconstriction and a secondary reduction in contractile force is not unique to the leukotrienes.

When taken together, our findings show that LTC_4 and LTD_4 decrease cardiac contractile force secondary to their potent coronary vasoactivity rather than through a direct inotropic mechanism. Our findings support those of Letts and Piper (1982) using electrically driven guinea pig ventricular strips.

ACKNOWLEDGMENT. This work was supported in part by a grant from the W. W. Smith Charitable Trust.

REFERENCES

Burke, J. A., Levi, R., Gui, Z., and Corey, E. J., 1982, Leukotrienes C_4, D_4 and E_4: Effects on human and guinea pig cardiac preparation *in vitro*, *J. Pharmacol. Exp. Ther.* **221**:235–241.

Burke, J. A, Levi, R., and Gleason, J. G., 1984, Antagonism of the cardiac effects of leukotriene C_4 by compound SKF 88046: Dissociation of effects on contractility and coronary flow, *J. Cardiovasc. Pharmacol.* **6**:122–125.

Holroyde, M. C., Altounyan, P. E. C., Cole, M., Dixon, M., and Elliott, E. V., 1981, Leukotrienes C and D induce bronchoconstriction in man, *Agents Actions* **11**:573–574.

Letts, L. G., and Piper, P. J., 1982, The actions of leukotrienes C_4 and D_4 on guinea pig isolated hearts, *Br. J. Pharmacol.* **76**:169–176.

Michaelassi, F., Costorena, G., Hill, R. D., Lowenstein, E., Watkins, D., Petkau, J., and Zapol, W. M., 1983, Effects of leukotrienes B_4 and C_4 on coronary circulation and myocardial contractility, *Surgery* **94**:267–275.

Peck, M. J., Piper, P. J., and Williams, T. J., 1981, The effect of leukotrienes C_4 and D_4 on the microvasculature of guinea pig skin, *Prostaglandins* **21**:315–321.

Roth, D. M., and Lefer, A. M., 1984, Studies on the mechanism of leukotriene induced coronary artery constriction, *Prostaglandins* **26**:573–581.

Samuelsson, B., 1981, Leukotrienes: Mediators of allergic reactions and inflammation, *Int. Arch. Allergy Appl. Immunol.* **66**(Suppl. 1):98–106.

Terashita, Z., Fukui, H., Hirata, M., Terao, S., Ohkawa, S., Nishikawa, K., and Kikuchi, S., 1981, Coronary vasoconstriction and PGI_2 release by leukotrienes in isolated guinea pig hearts, *Eur. J. Pharmacol.* **73**:357–361.

Woodman, O. L., and Dusting, G. J., 1983, Coronary vasoconstriction induced by leukotrienes in the anaesthetized dog, *Eur. J. Pharmacol.* **86**:125–128.

Yokuchi, K., Olley, P. M., Sideris, E., Hamilton, F., Hutanen, D., and Coceani, F., 1982, Leukotriene D_4: A potent vasoconstrictor of the pulmonary and systemic circulation in the newborn lamb, in: *Leukotrienes and Other Lipoxygenase Products* (B. Samuelsson and R. Paoletti, eds.), Raven Press, New York, pp. 211–214.

36

Effects of Selenium Deficiency on Arachidonic Acid Metabolism and Aggregation in Rat Platelets

N. W. SCHOENE, V. C. MORRIS, and
O. A. LEVANDER

1. INTRODUCTION

Evidence from investigations in the early 1970s indicated the possibility of an interaction between the metabolism of arachidonic acid and the activity of selenium (Se)-dependent glutathione peroxidase (GSHPx, E.C. 1.11.1.9). Lands *et al.* (1971) reported that the addition of GSHPx to preparations of prostaglandin (PG) synthase and soybean lipoxygenase inhibited the activity of these two dioxygenases. These initial experiments were followed by a series of studies (see review, Warso and Lands, 1983) that demonstrated the peroxide requirement for activation and maintenance of the catalytic activity of PG synthase *in vitro*. In 1973, it was discovered that GSHPx contained selenium as a necessary component for activity (Rotruck *et al.*, 1973). It is this requirement that has been exploited by investigators to study the relationship between GSHPx and arachidonic acid metabolism, since GSHPx activity can be manipulated by dietary selenium (see reviews, Sunde and Hoekstra, 1980; Schoene, 1985).

Recently, Bryant and co-workers (Bryant and Bailey, 1980; Bryant *et al.*, 1983) used the Se-deficient rat model to demonstrate that a decrease in GSHPx activity in the platelet resulted in an alteration in the metabolism of arachidonic acid by 12-lipoxygenase. Exogenously added arachidonic acid was converted to isomeric trihydroxyeicosatrienoic acids (THETEs) in the Se-deficient platelets, whereas

---

N. W. SCHOENE, V. C. MORRIS, and O. A. LEVANDER ● Beltsville Human Nutrition Research Center, Beltsville, Maryland 20705.

only trace amounts of these fatty acids were found in control platelets (Bryant and Bailey, 1980). They also demonstrated an increase in the accumulation of 12-hydroperoxy-5,8,11,14-eicosatetraenoic acid (12-HPETE) in the platelets from Se-deficient rats compared to platelets from Se-supplemented rats (Bryant *et al.*, 1983). These authors concluded that GSHPx plays a significant role in the reduction of 12-HPETE to 12-hydroxy-5,8,11,14-eicosatetraenoic acid (12-HETE) in platelets. When GSHPx activity is reduced, increased amounts of 12-HPETE can lead to the formation of THETEs, rearrangement products of the hydroperoxy fatty acid.

The presence or absence of GSHPx activity can not only affect the type of arachidonic acid metabolite produced but may also affect lipoxygenase activity in the cell. A 12-lipoxygenase that produces the leukotrienes has been partially purified from rat lung and has been shown to be sensitive to GSHPx in a manner similar to that seen for PG synthase and soybean lipoxygenase (Yokoyama *et al.*, 1983).

The production of arachidonic acid metabolites by PG synthase in platelets may also be affected by GSHPx activity. Platelets from Se-deficient rats exhibited increased aggregation to ADP, collagen, and arachidonic acid compared to platelets from Se-supplemented rats (Masukawa *et al.*, 1983). These workers also found that arachidonic-acid-induced respiratory distress in mice was enhanced in a selenium-deficient group compared to a control group. This type of respiratory distress is caused by platelet thrombi in the lung and is most likely caused by increased production of thromboxane A_2 (TxA_2). As reported earlier in preliminary form (Schoene *et al.*, 1984), we have investigated the production of TxB_2 in platelets from Se-deficient and -supplemented rats. Aggregation studies were also performed with platelets from the two groups of animals in order to relate TxB_2 production with platelet function.

2. MATERIALS AND METHODS

Weanling male Fischer 344 rats were obtained from Harlan/Sprague–Dawley, Indianapolis, IN, and were housed in pairs in stainless steel cages with free access to food and water. Diets were prepared as described previously (Levander *et al.*, 1983), with the Se-supplemented diet containing 0.5 μg selenium as sodium selenite per gram of diet. This level represents a nutritionally generous but nontoxic amount of Se. The Se-deficient diet contained no additional selenium.

After 12 weeks on diet, rats were anesthetized with diethyl ether, and blood was drawn by cardiac puncture into a plastic syringe containing 0.1 volume of 3.8% trisodium citrate in saline. Platelet-rich plasma was prepared by centrifuging at 150 g for 10 min at 25°C. All subsequent procedures with platelets were conducted at either 25°C or 37°C in plastic or siliconized glassware. Platelets were filtered through a Sepharose 2B column equilibrated with a modified Tyrodes buffer adjusted to pH 7.1 and containing no calcium (Lages *et al.*, 1975). Platelet aggregation studies were performed with a Chrono-Log aggregometer (Model 440, dual-channel unit, Havertown, PA). Sepharose 2B, ADP, TxB_2, and collagen were obtained from Sigma Chemical, St. Louis, MO. Hydroxy fatty acid standards were a generous gift of Dr. Robert Bryant. Separation and quantitation by high-performance liquid chromatography (HPLC) and radioimmunoassay (RIA), respectively, of arachidonic acid metabolites from platelet incubations required the use of pooled samples. Incubations were terminated after 3 min, the pH adjusted to 3.0 with formic acid, and arachidonic acid metabolites were extracted from the buffer using an octade-cylsilyl (ODS) column (Waters Associates, Milford, MA) according to the method described by Powell (1980).

Arachidonic acid metabolites were then separated by our own modification of a previously reported technique (Van Rollins *et al.*, 1980). A Hewlett-Packard 1084B with variable wavelength detector was equipped with a Spherisorb-ODS column (4.6 mm $\times$ 15 cm, 5-μm particle size, Robert E. Gourley, Laurel, MD). All solvents were HPLC grade. The TxB_2 peak was collected and quantitated by RIA. The TxB_2 antibody was purchased from Seragen, Boston, MA, and $[^3H]TxB_2$ from New England Nuclear, Boston, MA. The RIA was performed according to established procedures (Granstrom and Kindahl, 1976). Data were analyzed for significant differences with Student's *t*-test (two-tailed).

3. RESULTS AND DISCUSSION

After 12 weeks, specific activity of GSHPx in the livers of rats fed the Se-deficient diet was decreased to 17 $\pm$ 2 (mean $\pm$ S.E.M., mU/mg protein) compared to 893 $\pm$ 44 for the activity of the enzyme in livers from rats fed the Se-supplemented diet. Values for GSHPx activity in platelets for the two groups are not available since all platelet material was needed for aggregation and incubation experiments. From past studies with Se-deficient rats (Levander *et al.*, 1983), it can reasonably be assumed that GSHPx in platelets was reduced by the Se-deficient diet to a value 10–15% of the activity in control platelets.

Gel-filtered platelets were used to study the effects of selenium deficiency on platelet aggregation in order to obtain reproducible results in aggregation studies. Problems associated with citrate, extracellular calcium, and adjustment of platelet concentrations could be controlled with greater ease by removing plasma from the platelets by gel filtration. Platelets from rats are known to exhibit differences in responses to ADP and epinephrine (Dodds, 1978; Macmillan and Sim, 1970) when compared to platelets from humans and other laboratory animals. *In vivo* platelet function studies have revealed that rat platelets are quite sensitive to the concentration of extracellular calcium (Mallarkey and Smith, 1984). Also, rat platelets will not undergo secondary (irreversible) aggregation in response to ADP when extracellular calcium is low (Packham, 1983). For this reason, calcium was added to suspensions of gel-filtered platelets for all of our aggregation and incubation experiments.

The results from aggregation studies are shown in Table I. Differences between the responses of platelets from Se-supplemented and Se-deficient rats could be seen

TABLE I. Effect of Selenium Deficiency on
Aggregation of Gel-Filtered Rat Platelets Induced by
Collagen and ADP

Stimulant	+ Se diet	− Se diet
Collagen (20 μl/ml)[a]	15.6 ± 1.7[b]	35.8 ± 2.4
ADP (0.5 μM)	18.2 ± 1.3	39.2 ± 2.6

[a] Collagen suspension, 2 mg protein/ml.
[b] Mean $\pm$ S.E.M. for four to five experiments with platelets from individual rats. Platelet counts were adjusted to 3×10^8/ml and recalcified (3 mM) for all series of aggregations. Aggregation was measured as percentage change in transmittance to submaximal doses of stimulant. There was a significant difference between the two dietary groups ($P < 0.001$, two tailed t-test).

when submaximal doses of collagen and ADP were used. Aggregation studies were performed with samples from individual rats. Our results obtained with gel-filtered platelets agree with those obtained by Masukawa *et al.* (1983) with platelet-rich plasma. However, we did not observe the difference in lag time reported by these workers in response to collagen. Platelets from the Se-deficient animals in their study exhibited a decrease in lag time compared to platelets from the control animals.

The aggregatory response of platelets to submaximal doses of collagen is known to be dependent on the release and subsequent oxidation to TxB_2 of arachidonic acid (Ali and McDonald, 1977). It has been demonstrated that the response of platelets in citrated plasma or in media that contain no calcium differs considerably from that when the media contain a physiological concentration of ionized calcium (Packham, 1983). The presence or absence of citrate could explain some of the differences in the results obtained in our laboratory compared to that of Masukawa *et al.* (1983). Both laboratories demonstrated a difference in the response of platelets to ADP between the two dietary groups. ADP can induce platelet aggregation independent of arachidonic acid metabolism (Packham *et al.*, 1977). It is possible that alterations in membrane receptors and/or properties occur in Se-deficient platelets and could account for the increased responsiveness to ADP observed in these platelets compared to platelets from the Se-supplemented animals.

Incubation studies were conducted with pooled samples of platelets from each of the two groups. This was necessary in order to have enough sample for analysis by HPLC of the arachidonic acid metabolites released during incubation of the platelets. With this methodology, we were able to separate TxB_2 from the other oxygenated metabolites of arachidonic acid produced during platelet stimulation by collagen. These other products are 12-hydroxy-5,8,10-heptadecatrienoic acid (HHT), 12-HPETE/HETE, and possibly, from the Se-deficient platelets, THETEs.

The TxB_2 was collected from the HPLC effluent and quantitated by RIA. The values obtained are shown in Table II. Platelets from the Se-deficient rats produced significantly greater amounts of TxB_2 than did the platelets from the Se-supplemented rats. This increase in the production of TxB_2 may be an indication of

**TABLE II. Effect of Selenium
Deficiency on Collagen-Induced
Synthesis of TxB$_2$ in Gel-Filtered
Rat Platelets**

TxB$_2$ produced (pmol/ml)	
+ Se diet	-- Se diet
19 ± 2[a]	50 ± 3

[a] Mean ± S.E.M. for three experiments Gel-filtered platelets in 2 ml of Tyrodes (3 × 10^8/ml) were recalcified and incubated at 37° with 20 μl of collagen suspension (2 mg protein/ml) for 3 min and extracted as described, TxB$_2$ was determined by RIA after purification by HPLC. There was a significant difference between the two groups ($P < 0.01$).

stimulation of PG synthase by the higher "peroxide tone" in the Se-deficient (GSHPx-poor) platelets. In contrast to the results reported by Bryant and Bailey (1980), we observed no detectable THETE peaks with HPLC in extracts from collagen-stimulated, Se-deficient platelets. However, our HPLC methodology may not be sensitive enough to detect the production of THETEs from endogenous substrate. Different experimental conditions can account for the disparity in results between our laboratory and that of Bryant and co-workers. We used collagen to stimulate release and oxidation of arachidonic acid in gel-filtered platelets. They used exogenously added [1-^{14}C]arachidonic acid to follow oxidation of this acid in platelets washed by centrifugation. Aside from the many technical problems associated with the study of arachidonic acid metabolites, however, experimental evidence obtained from the Se-deficient rodent indicates a relationship between arachidonic acid metabolism and GSHPx activity.

4. SUMMARY

Platelets from rats fed a Se-deficient diet aggregated to a significantly greater extent to both collagen and ADP stimulation than did platelets from rats fed a Se-supplemented diet. Thromboxane B$_2$ formation was studied by incubating the gel-filtered platelets from the two dietary groups with collagen and calcium. Thromboxane B$_2$ was quantitated by RIA after extraction and purification by HPLC. Activity of the PG synthase was enhanced as indicated by the significant increase in the production of TxB$_2$ in the platelets from Se-deficient rats compared to Se-supplemented rats. This increased activity of PG synthase is possibly caused by higher levels of lipid peroxides in the Se-deficient rat platelets. It can be concluded that selenium deficiency, by lowering GSHPx activity in the platelet, could contribute to altered production of arachidonic acid metabolites with pathophysiological consequences on platelet function.

REFERENCES

Ali, M., and McDonald, J. W. D., 1977, Effects of sulfinpyrazone on platelet prostaglandin synthesis and platelet release of serotonin, *J. Lab. Clin. Med.* **89**:868–875.

Bryant, R. W., and Bailey, J. M., 1980, Altered lipoxygenase metabolism and decreased glutathione perioxidase activity in platelets from selenium-deficient rats, *Biochem. Biophys. Res. Commun.* **92**:268–276.

Bryant, R. W., Simon, T. C., and Bailey, J. M., 1983, Hydroperoxy fatty acid formation in selenium deficient rat platelets: Coupling of glutathione peroxidase to the lipoxygenase pathway, *Biochem. Biophys. Res. Commun.* **117**:183–189.

Dodds, W. J., 1978, Platelet function in animals: Species specificities, in: *Platelets: A Multidisciplinary Approach* (G. de Gaetano and S. Garattini, eds.), Raven Press, New York, pp. 45–59.

Granstrom, E., and Kindahl, H., 1976, Radioimmunoassay for prostaglandin metabolites, *Adv. Prostaglandin Thromboxane Res.* **1**:81–92.

Lages, B., Scrutton, M. C., and Holmsen, H., 1975, Studies on gel-filtered human platelets: Isolation and characterization in a medium containing no added Ca^{2+}, Mg^{2+}, or K^+, *J. Lab. Clin. Med.* **85**:811–825.

Lands, W. E. M., Lee, R. E., and Smith, W. L., 1971, Factors regulating the biosynthesis of various prostaglandins, *Ann. N.Y. Acad. Sci.* **180**:107–122.

Levander, O. A., DeLoach, D. P., Morris, V. C., and Moser, P. B., 1983, Platelet glutathione peroxidase activity as an index of selenium status in rats, *J. Nutr.* **113**:55–63.

Macmillan, D. C., and Sim, A. K., 1970, A comparative study of platelet aggregation in man and laboratory animals, *Thromb. Diathes. Haemorrh.* **24**:385–394.

Mallarkey, G., and Smith, G. M., 1984, *In vivo* platelet aggregation in the rat: Dependence on extracellular divalent cation and inhibition by nonsteroidal anti-inflammatory drugs, *Br. J. Pharmacol.* **81**:31–39.

Masukawa, T., Goto, J., and Iwata, H., 1983, Impaired metabolism of arachidonate in selenium deficient animals, *Experientia* **39**:405–406.

Packham, M. A., 1983, Platelet function inhibitors, *Thromb. Haemostas.* **50**:610–619.

Packham, M. A., Rinlough-Rathbone, R. L., Reimers, H.-J., Scott, S., and Mustard, J. F., 1977, Mechanisms of platelet aggregation independent of adenosine diphosphate, in: *Prostaglandins in Hematology* (M. J. Silver, J. B. Smith, and J. J. Kocsis, eds.), Spectrum Publication, New York, pp. 247–276.

Powell, W. S., 1980, Rapid extraction of oxygenated metabolites of arachidonic acid from biological samples using octadecylsilyl silica, *Prostaglandins* **20**:947–957.

Rotruck, J. T., Pope, A. L., Ganther, H. E., Swanson, A. B., Hafeman, D. G., and Hoekstra, W. G., 1973, Selenium: Biochemical role as a component of glutathione peroxidase, *Science* **179**:588–590.

Schoene, N. W., Morris, V. C., and Levander, O. A., 1984, Effects of selenium deficiency on aggregation and thromboxane formation in rat platelets, *Fed. Proc.* **43**:477.

Schoene, N. W., 1985, Selenium-dependent glutathione peroxidase and eicosanoid production, in: *Biochemistry of Arachidonic Acid Metabolism* (W. E. M. Lands, ed.), Martinus Nijhoff, Boston, pp. 151–160.

Sunde, R. A., and Hoekstra, W. G., 1980, Structure, synthesis and function of glutathione peroxidase, *Nutr. Rev.* **38**:265–273.

Van Rollins, M., Ho, S. H. K., Greenwald, J. E., Alexander, M., Dorman, N. J., Wong, L. K., and Horrocks, L. A., 1980, Complete separation by high-performance liquid chromatography of metabolites of arachidonic acid from incubation with human and rabbit platelets, *Prostaglandins* **20:**571–577.

Warso, M. A., and Lands, W. E. M., 1983, Lipid peroxidation in relation to prostacyclin and thromboxane physiology and pathophysiology, *Br. Med. Bull.* **39:**277–280.

Yokoyama, C., Mizuno, K., Mitachi, H., Yoshimoto, T., Yamanoto, S., and Pace-Asciak, C. R., 1983, Partial purification and characterization of arachidonate 12-lipoxygenase from rat lung, *Biochim. Biophys. Acta* **750:**237–243.

37

Different Functions of Phosphoinositide Turnover and Arachidonate Metabolism for Platelet Activation

WOLFGANG SIESS, PETER C. WEBER, and EDUARDO G. LAPETINA

1. INTRODUCTION

In a wide variety of cells including platelets, the hormone-induced breakdown of phosphoinositides by phospholipase C is closely associated with the mobilization of Ca^{2+} from intracellular stores (Michell, 1975). Recent studies gave new insights into the possible link between phosphoinositide breakdown and intracellular Ca^{2+} mobilization: myoinositol-1,4,5-trisphosphate, the phosphodiesteratic cleavage product of phosphatidylinositol-4,5-biphosphate, mobilizes Ca^{2+} from the endoplasmatic reticulum in pancreatic acinar cells and hepatocytes (Streb *et al.*, 1983; Joseph *et al.*, 1984). Besides myoinositol-1,4,5-trisphosphate, two other products of the phospholipase-C-induced degradation of the phosphoinositides have possible functions as intracellular activators: 1,2-diacylglycerol activates the Ca^{2+}, phospholipid-dependent protein kinase C, and phosphatidic acid has been implicated in Ca^{2+} fluxes and membrane fusion processes (Nishizuka, 1983; Sundler and Papahadjo-poulos, 1981; Fig. 1).

A different phospholipid response on cell stimulation is the release of arachidonic acid from phosphatidylethanolamine, -choline, and -inositol. Arachidonic

WOLFGANG SIESS and PETER C. WEBER ● Medizinische Klinik Innenstadt der Universität München, 8000 Munich 2, West Germany.   EDUARDO G. LAPETINA ● Department of Molecular Biology, The Wellcome Research Laboratories, Burroughs Wellcome, Research Triangle Park, North Carolina 27709.

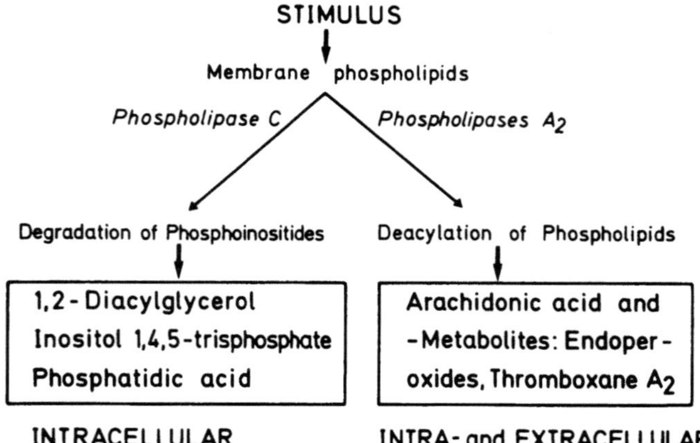

FIGURE 1. Stimulus-induced degradation of cellular membrane phospholipids leading to the formation of active metabolites.

acid is converted by cyclooxygenase to the biologically active prostaglandin endoperoxides PGG_2 and PGH_2 and by thromboxane synthetase to thromboxane A_2. Those substances can be released to the outside to activate other platelets. Endoperoxides and thromboxane A_2 can induce the whole cascade of platelet responses, i.e., platelet shape change, release reaction, and aggregation, most probably through a receptor-mediated mechanism (Parise *et al.*, 1984). Following specific binding of these agonists, an increase in Ca^{2+} in the platelet cytosol has been observed (Kawahara *et al.*, 1983). We have recently shown that arachidonic acid and endoperoxide analogues stimulate the formation of phospholipase-C-derived products in human platelets and that the degree of phospholipase C activation is closely related to protein kinase C activation and platelet functional changes (Siess *et al.*, 1983a,b; 1985).

2. PLATELET SHAPE CHANGE: DISSOCIATION OF PHOSPHOLIPASE C ACTIVATION FROM ARACHIDONATE METABOLISM

Platelet shape change is the first physiological platelet response, preceding other responses such as platelet aggregation and release reaction. Low concentrations of thrombin, platelet-activating factor, or endoperoxide analogues, which only induce platelet shape change, stimulate the rapid formation of 1,2-diacylglycerol and phosphatidic acid, indicating the activation of phospholipase C (Siess *et al.*, 1984,

1985). Activation of phospholipase C is paralleled by activation of protein kinase C and myosin light chain kinase during platelet shape change (Lapetina and Siegel, 1983; Siess *et al.*, 1985). Release of arachidonic acid is inconstantly observed; metabolism by cyclooxygenase or lipoxygenase does, however, not occur (Siess *et al.*, 1984, 1985). Inhibitors of the release of arachidonic acid (trifluoperazine) or platelet cyclooxygenase (aspirin, indomethacin) do not affect platelet shape change or phospholipase C activation induced by thrombin, platelet-activating factor, or endoperoxide analogues. Prostacyclin, however, prevents platelet shape change and phospholipase C activation induced by those agonists, probably by increasing cAMP levels in the platelet cytosol. Platelet shape change and phospholipase C activation are induced in the absence and independently of extracellular Ca^{2+} and Mg^{2+}. Epinephrine is a unique platelet stimulus, since it does not induce platelet shape change or phospholipase C activation (Siess *et al.*, 1984). The results indicate that platelet shape change is closely related to phospholipase C activation but independent of arachidonate metabolism.

The phospholipase-C-induced formation of intracellular activators such as 1,2-diacylglycerol, phosphatidic acid, and myoinositol-1,4,5-trisphosphate might be related to activation of protein kinases and Ca^{2+} mobilization occurring during platelet shape change (Fig. 2).

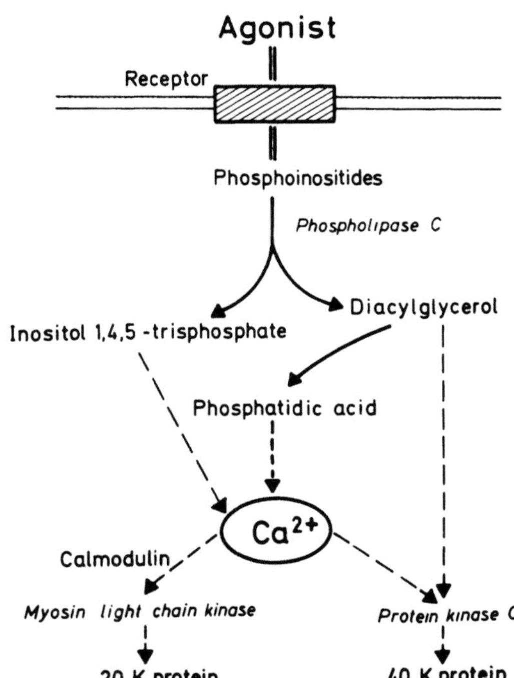

FIGURE 2. Phosphoinositide degradation as a signal translation mechanism for activation of protein kinases.

3. PLATELET RELEASE REACTION AND AGGREGATION: STIMULATION OF ARACHIDONATE METABOLISM AND FURTHER ACTIVATION OF PHOSPHOLIPASE C

With increased doses of an agonist such as thrombin, the platelet release reaction and aggregation occur subsequently to platelet shape change. Under conditions of a Ca^{2+}-free extracellular medium, platelet aggregation is coupled to the release reaction (Siess *et al.*, 1983a,b). With low concentrations of thrombin, two steps of platelet activation can be dissociated, and the activation of phospholipase C shows a biphasic pattern. During platelet shape change, activation of phospholipase C and some release of arachidonic acid are observed. When platelet aggregation starts, there is formation of arachidonate metabolites and a second sharp increase of 1,2-diacylglycerol and phosphatidic acid. Inhibition of cyclooxygenase reduces this second increase of 1,2-diacylglycerol and phosphatidic acid and also platelet aggregation (Siess *et al.*, 1984). Thus, cyclooxygenase products such as endoperoxides and thromboxane A_2 may act as feedback promoters and induce further stimulation of phospholipase C, leading to acceleration and amplification of the platelet response. In comparison with thrombin, the action of collagen seems to be much more dependent on such a cyclooxygenase-mediated amplification mechanism (Siess *et al.*, 1983a,b; Lapetina and Siess, 1983).

4. HOW MANY MECHANISMS FOR PLATELET ACTIVATION?

Phospholipase-C-induced breakdown of phosphoinositides seems to be an important mechanism for platelet activation. Phospholipase C activation is induced by a wide spectrum of platelet agonists and is closely related to each of the platelet responses: platelet shape change, release reaction, and aggregation. Degrees of phospholipase C activity, protein kinase C activity, and functional platelet changes correlate closely. Phospholipase C activation is independent of extracellular Ca^{2+} or Mg^{2+} (Fig. 3). It seems, then, that phosphoinositide degradation by phospholipase C may be an intrinsic mechanism for agonists to activate platelets independently of extracellular divalent cations. The degradation of phosphoinositides may be achieved mainly by exposure of the substrate to the cytosolic enzyme (Siess and Lapetina, 1983).

Epinephrine is a unique platelet stimulus that reveals a second mechanism of platelet activation. Epinephrine requires extracellular divalent cations for platelet activation; it is known to bind to α_2 receptors, to increase the Ca^{2+} uptake of human platelets, and to induce exposure of fibrinogen receptors to which fibrinogen binds in the presence of Ca^{2+} or Mg^{2+} leading to platelet aggregation. Epinephrine does not induce platelet shape change or phospholipase C activation (Siess *et al.*, 1984). It seems, then, that the increase of Ca^{2+} uptake, which is also observed for

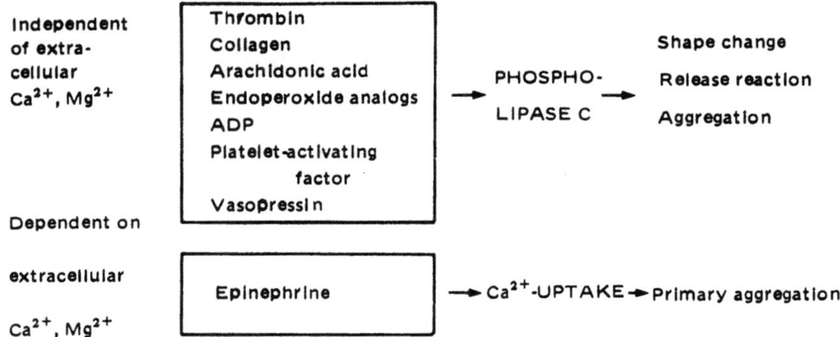

FIGURE 3. Stimulus-specific pathways for platelet aggregation.

other platelet agonists, may be a second, separate mechanism of platelet activation that is specifically related to platelet aggregation.

Besides those two "intrinsic" mechanisms of platelet activation, there exist three separate amplification mechanisms: the formation of cyclooxygenase products of arachidonic acid, the release of ADP, and the formation of platelet-activating factor. Those substances are released from stimulated platelets to the outside and activate, in turn, the "intrinsic" mechanisms, thereby leading to the whole cascade of platelet responses (Fig. 4).

5. SUMMARY

The molecular mechanism by which physiological agonists activate cells involves their binding to specific receptors followed by distinct changes in membrane phospholipids. In platelets, both the stimulation of the phosphoinositide breakdown

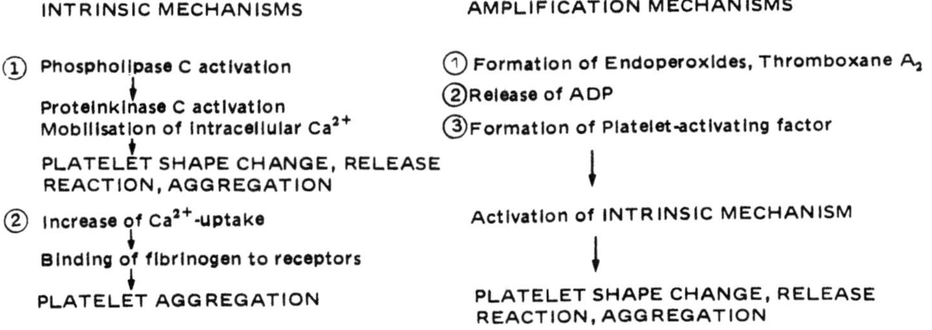

FIGURE 4. Possible mechanisms of platelet activation.

by phospholipase C and the release of arachidonic acid from phospholipids are observed following platelet stimulation with various agonists such as thrombin, collagen, ADP, platelet-activating factor, arachidonic acid, and endoperoxide analogues. The present chapter reviews studies from our laboratories concerning the respective roles of phosphoinositide breakdown and arachidonate metabolism for platelet activation. The results indicate that phosphoinositide breakdown induced by phospholipase C may serve as a signal-transducing mechanism for platelet activation, whereas arachidonate metabolites derived by cyclooxygenase may act as an amplification mechanism of platelet activation.

REFERENCES

Joseph, K. S., Thomas, A. P., Williams, R. J., Irvine, R. F., and Williamson, J. R., 1984, Myoinositol 1,4,5-trisphosphate: A second messenger for the hormonal mobilization of intracellular Ca^{++} in liver, *J. Biol. Chem.* **259**:3077–3081.

Kawahara, Y., Yamanishi, J., Furuta, Y., Kaibuchi, K., Takai, Y., and Fukuzaki, H., 1983, Elevation of cytoplasmic free calcium concentration by stable thromboxane A_2 analogue in human platelets, *Biochem. Biophys. Res. Commun.* **117**:663–669.

Lapetina, E. G., and Siegel, F. L., 1983, Shape change induced in human platelets by platelet-activating factor. Correlation with the formation of phosphatidic acid and phosphorylation of a 40,000 dalton protein, *J. Biol. Chem.* **258**:7241–7244.

Lapetina, E. G., and Siess, W., 1983, The role of phospholipase C in platelet responses, *Life Sci.* **33**:1011–1018.

Michell, R. H., 1975, Inositol phospholipids and cell surface receptor function, *Biochim. Biophys. Acta* **415**:81–147.

Nishizuka, Y., 1983, Phospholipid degradation and signal translation for protein phosphorylation, *Trends Biochem. Sci.* **8**:13–15.

Parise, L. K., Venton, D. L., and Le Breton, G. C., 1984, Arachidonic acid-induced platelet aggregation is mediated by a thromboxane A_2/prostaglandin H_2 receptor interaction, *J. Pharmacol. Exp. Ther.* **228**:240–244.

Siess, W., and Lapetina, E. G., 1983, Properties and distribution of phosphatidylinositol-specific phospholipase C in human and horse platelets, *Biochim. Biophys. Acta* **752**:329–338.

Siess, W., Cuatrecases, P., and Lapetina, E. G., 1983a, A role for cyclooxygenase products in the formation of phosphatidic acid in stimulated platelets. Differential mechanism of action of thrombin and collagen, *J. Biol. Chem.* **258**:4683–4686.

Siess, W., Siegel, F. L., and Lapetina, E. G., 1983b, Arachidonic acid stimulates the formation of 1,2-diacylglycerol and phosphatidic acid in human platelets. Degree of phospholipase C activation correlates with protein phosphorylation, platelet shape change, serotonin release and aggregation, *J. Biol. Chem.* **258**:11236–11242.

Siess, W., Weber, P. C., and Lapetina, E. G., 1984, Activation of phospholipase C is dissociated from arachidonate metabolism during platelet shape change induced by thrombin or platelet-activating factor. Epinephrine does not induce phospholipase C activation or platelet shape change, *J. Biol. Chem.* **259**:8286–8292.

Siess, W., Boehlig, B., Weber, P. C., and Lapetina, E. G., 1985, Prostaglandin endoperoxide analogs stimulate phospholipase C and protein phosphorylation during platelet shape change, *Blood* (in press).

Streb, H., Irvine, R. F., Berridge, M. J., and Schulz, I., 1983, Release of Ca^{++} from a nonmitochondrial intracellular store in pancreatic acinar cells by inositol-1,4,5-trisphosphate, *Nature* **306**:67–69.

Sundler, R., and Papahadjopoulos, D., 1981, Control of membrane fusion by phospholipid head groups. Phosphatidate/phosphatidylinositol specificity, *Biochim. Biophys. Acta* **649**:743–750.

38

Comparative Absorption and Lymphatic Transport of (ω-3) Eicosapentaenoic Acid, (ω-6) Arachidonic Acid, and (ω-9) Oleic Acid

GEORGE V. VAHOUNY, ISABEL S. CHEN, S. SATCHITHANANDAM, MARIE M. CASSIDY, and ALAN J. SHEPPARD

1. INTRODUCTION

There is a rapidly expanding literature on the dietary and metabolic effects of the ω-3 class of fatty acids, which include, among others, linolenic, eicosapentaenoic, and docosahexaenoic acids (Goodnight *et al.*, 1982; Holman, 1982; Willis, 1981). Interest in this area has been heightened by the finding that population groups consuming greater quantities of fish and other marine animals, which are richer in ω-3 fatty acids, have a low incidence of ischemic heart disease. These include Greenland Eskimos (Bang and Dyerberg, 1980), coastal-dwelling Turks (Yotakis, 1981), and Japanese residing in fishing villages (Hirai *et al.*, 1980). In general, these populations have lower levels of fasting very-low-density and low-density lipoprotein cholesterol, increased levels of high-density lipoprotein cholesterol (Bang

GEORGE V. VAHOUNY, ISABEL S. CHEN, and S. SATCHITHANANDAM • Department of Biochemistry, The George Washington University School of Medicine and Health Sciences, Washington, D.C. 20037.   MARIE M. CASSIDY • Department of Physiology, The George Washington University School of Medicine and Health Sciences, Washington, D.C. 20037   ALAN J. SHEP-PARD • Division of Nutrition, Food and Drug Administration, Washington, D.C. 20204.

and Dyerberg, 1972), and a prolonged bleeding time (Dyerberg and Bang, 1979; Goodnight *et al.*, 1981). These effects are largely attributed to the intake of marine fats, which contain, in addition to "common" fatty acids, higher levels of ω-3 fatty acids than are found in Western-type diets (Bang *et al.*, 1976).

The antithrombotic response to ω-3 fatty acid intake appears to be related in part to the eicosapentaenoic acid (EPA) content of the diet. EPA can substitute for more typical fatty acids, such as arachidonate, in plasma membrane phospholipids, including those of the platelet membranes (Goodnight *et al.*, 1981). This can modify platelet behavior (Jakobowski and Ardlie, 1980; Needleman and Sprecher, 1979; Needleman *et al.*, 1979, 1980), increase platelet survival, and decrease platelet counts (Goodnight *et al.*, 1981; Hay *et al.*, 1982). Platelets enriched in EPA exhibit reduced thromboxane A_2 (TxA_2) production and platelet aggregation. This is presumably because of decreased availability and conversion of arachidonate to TxA_2 (Culp *et al.*, 1979; Witaker *et al.*, 1979), and the possible conversion of EPA to an antiaggregatory product (Needleman and Sprecher, 1979; Witaker *et al.*, 1979).

The antithrombotic response may also be related to the presence of EPA in plasma membranes of vascular tissues. Some studies (Jakobowski and Ardlie, 1980) have suggested that release of EPA from membrane phospholipids of vascular tissues results in formation of PGI_3 and may also reduce the conversion of arachidonate to the normal prostacyclin, PGI_2 (Spector *et al.*, 1983). There is, however, some uncertainty regarding the extent of PGI_3 formation from EPA and the overall effect of EPA on PGI_2 formation from arachidonate (e.g., Spector *et al.*, 1983).

The mechanism(s) underlying the hypolipidemic response to EPA-containing diets is not yet known (Goodnight *et al.*, 1982). There is evidence that ingestion of fish oil does not result in typical fat tolerance curves in humans (Harris *et al.*, 1982). Furthermore, ingestion of marine fat does not result in rapid accumulation of ω-3 fatty acids in tissue lipids (Garton *et al.*, 1952; Goodnight *et al.*, 1982). These data suggest that, among other potential metabolic effects of EPA (e.g., Goodnight *et al.*, 1982), the bioavailability and absorption of marine oil fatty acids may be limited. Eicosapentaenoic acid is largely found in the 1 and 3 positions of the triglycerides of marine oils (Bottino *et al.*, 1967; Brockerhoff *et al.*, 1966). There is convincing evidence that the ω-3 fatty acids, and particularly EPA, in marine glycerides are resistant to lipolysis by pancreatic lipase (Bottino *et al.*, 1967; Brockerhoff, 1965) and to routine procedures of hydrolysis and derivatization during compositional analysis (Yurkowski and Brockerhoff, 1966). It has also been suggested that the absorbability and route of transport (e.g., lymphatic versus portal) may be unlike that of typical saturated and monounsaturated fatty acids (Goodnight *et al.*, 1982).

The inefficient digestion of EPA-containing glycerides or the limited absorbability of these polyunsaturated fatty acids should produce a persistent oil phase in the luminal contents of the intestine and modify the normal oil–micellar partitioning, which is required for efficient absorption of lipids, including cholesterol. This, in turn, could contribute to reduced absorption of lipids and to the overall hypolipidemic responses attributed to the Eskimo-type diet.

There is a sparsity of information on the absorption and lipoprotein transport of polyunsaturated fatty acids in general and EPA in particular. The present chapter addresses this specific question.

2. EXPERIMENTAL PROCEDURES

Bovine serum albumin fraction V (fatty acid poor), sodium taurocholate, oleic acid, arachidonic acid, and eicosapentaenoic acid were obtained from Sigma Chemical, St. Louis, MO. Isotopic fatty acids labeled in the carboxyl position and [1,2-^{3}H]cholesterol were obtained from New England Nuclear, Boston, MA. Other chemicals and solvents were of highest purity.

The test emulsions used for intragastric administration were prepared as described earlier (Vahouny *et al.*, 1980a) and included the following components per 1.5 ml physiological saline: 25 mg bovine serum albumin; 86 mg sodium taurocholate; 0.3 mmol of either oleic, arachidonic, or eicosapentaenoic acid. In addition, each emulsion contained 1 μCi of the respective [1-^{14}C]-labeled fatty acid and 20 μCi of [1,2-^{3}H]cholesterol. The emulsions were prepared immediately before use and rehomogenized prior to each administration.

Adult male albino rats of the Wistar strain (Charles River Laboratories) weighing approximately 200 g were allowed food and water *ad libitum* prior to use. Under sodium pentabarbital anesthesia, animals were subjected to cannulation of the left thoracic lymphatic duct cephalad to the cysterna chyli as described earlier (Vahouny *et al.*, 1980a). An indwelling catheter was placed in the pyloris of the stomach for continuous administration of saline–5% glucose and the test emulsions. After an overnight fast, the saline–glucose infusion (3 ml/hr) was interrupted, 1.5 ml of the appropriate test emulsion was infused via the stomach catheter, and the saline–glucose infusion was reestablished.

Lymph was collected on ice at 2-hr intervals for the first 8 hr and as a single 8 to 24-hr fraction. Aliquots (50–100 μl) were taken for direct liquid scintillation spectrometry, and the remainder was pooled as a single 24-hr sample. Approximately 8 ml of each lymph sample was subjected to ultracentrifugal separation of major lymph lipoproteins as described previously (Havel *et al.*, 1955). These included: chylomicrons ($d < 1.006$ g/ml; 3×10^6 g avg min); very-low-density lipoproteins (VLDL; $d < 1.006$ g/ml; 1×10^8 g avg min); low-density lipoproteins (LDL; $1.006 < d < 1.063$ g/ml; 1.3×10^8 g avg min); and high-density lipoproteins (HDL; $1.063 < d < 1.21$ g/ml; 1.7×10^8 g avg min).

Aliquots (50–100 μl) of each lipoprotein fraction were taken for liquid scintillation spectrometry, and additional samples (1 ml) were extracted in 20 volumes of chloroform–methanol (2 : 1 v/v) according to Folch *et al.* (1957). The chloroform extracts were evaporated under nitrogen, lipids were solubilized in hexane, and these were separated into major lipid classes by thin-layer silicic acid chromatography using hexane : diethyl ether : acetic acid, 80 : 16 : 2 (Gartner and Vahouny, 1972). Silicic acid areas corresponding to authentic phospholipid, mono- and diglycerides, triglycerides, unesterified fatty acids, and esterified cholesterol were individually scraped into liquid scintillation vials for analysis of isotope distribution among lipoprotein lipid classes.

Comparable extraction and lipid separation procedures were employed for analysis of lipid fatty acids of the test emulsions, lymph samples, and individual lipoproteins by gas–liquid chromatography of the derivatized fatty acids (Gartner and Vahouny, 1972). The appropriate separated or thin-layer fractions were transmethylated using BF_3–methanol, and the extracted fatty acid methylesters were subjected to gas–liquid chromatography as described earlier (Gartner and Vahouny, 1972).

3. RESULTS

Animal weights at the time of surgery were identical for all experimental groups. The lymph volumes for the 24-hr collection period varied significantly between animals in the same group and among groups. Although animals administered arachidonic acid had a significantly lower lymph output than did oleic acid fed rats, it has been previously demonstrated (Treadwell and Vahouny, 1968) that lymph flows exceeding 2 ml/hr have little effect on overall lipid absorption. This is also obvious in the present studies when the overall absorption of arachidonate is compared to that of oleate, which is identical despite a twofold difference in lymph volumes.

The net absorption of oleic acid over the 24-hr collection period and the absorption of tracer levels of cholesterol from an oleic acid-containing meal were comparable to the levels reported earlier using this approach (Vahouny *et al.*, 1980b, 1983). The absorption of either arachidonic acid or EPA was as efficient as that for oleate (Table I), and the absorption of endogenous cholesterol from either fatty acid medium was not significantly different (analysis of variance) from that with the oleate emulsion.

TABLE I. Overall Lymphatic Absorption of Fatty Acids and Cholesterol

Group[a]	Absorption (% of dose)	
	Fatty acid	Cholesterol
Oleic acid	76.2 ± 1.7[b]	59.1 ± 3.2
Arachidonic acid	81.6 ± 4.0	45.8 ± 3.8
Eicosapentaenoic acid	85.6 ± 3.0	58.1 ± 5.9

[a] Animals were surgically provided with a lymphatic drainage catheter in the left thoracic lymphatic channel and an infusion catheter in the stomach. These were restrained and fasted overnight while providing a continuous gastric infusion (3 ml/hr) of 5% glucose in saline and the same fluids *ad libitum*. At 9:00 a.m., each animal received, via the infusion catheter, 1.5 ml of a lipid emulsion consisting of 86 mg sodium taurocholate, 25 mg albumin 20 µCi [1,2-^{3}H]cholesterol and 0.3 mmol (1 µCi) of the respective [1-^{14}C]-labeled fatty acid. Lymph was collected and analyzed as described in Section 2.
[b] Values represent means $\pm$ S.E.M. for four to five animals in each group.

A summary of fatty acid and cholesterol recoveries in lymph during the initial 8 hr and final 16 hr of study is shown in Fig. 1. With all three fatty acids, the majority of absorption occurred during the initial 8 hr after administration of the fatty acid. There were, however, significant differences between the polyunsaturates. The absorption of EPA was significantly greater than that of arachidonate during the initial collection period, and this was reversed during the final collection period. Despite this, the recoveries of administered cholesterol in the initial and final lymph collection periods were statistically similar for all three groups.

The distribution of absorbed fatty acids among the major lipoprotein classes in lymph is summarized in Fig. 2. These were statistically comparable for all three fatty acids, with 93–95% of the absorbed acids associated with $d < 1.006$ g/ml lipoproteins (chylomicrons and very-low-density lipoproteins). Of this, 77–84% was in the chylomicron fraction, with a tendency for the greater values to be associated with the polyunsaturated acids. For these studies, the sums of recoveries of isotopic fatty acids among the individual lipoprotein fractions were statistically identical to the fatty acid recovery by analysis of whole lymph (difference for all groups $3.1 \pm 1.3\%$).

Analyses of the fatty acid distributions in the major lipid fractions of the two major transport particles, chylomicrons and very-low-density lipoproteins, are shown in Table II. These largely reflect the distributions in whole lymph, with the largest proportion (85–91%) of the fatty acids associated with the lipoprotein triglycerides. Again, with both polyunsaturated acids, there was a greater recovery of fatty acid label in the partial glycerides and phospholipids than with oleate. Recovery of

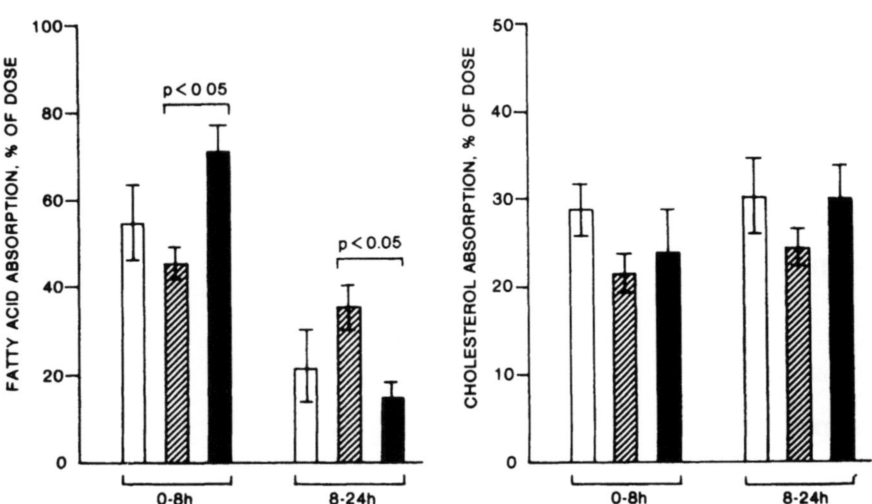

FIGURE 1. Recovery of administered oleic, arachidonic, and eicosapentaenoic acids (left panel) and of cholesterol in the presence of each fatty acid (right panel) during the initial 8 hr and final 16 hr of lymph collection. □ oleic acid; ◨ arachidonic acid; ■ eicosapentaenoic acid.

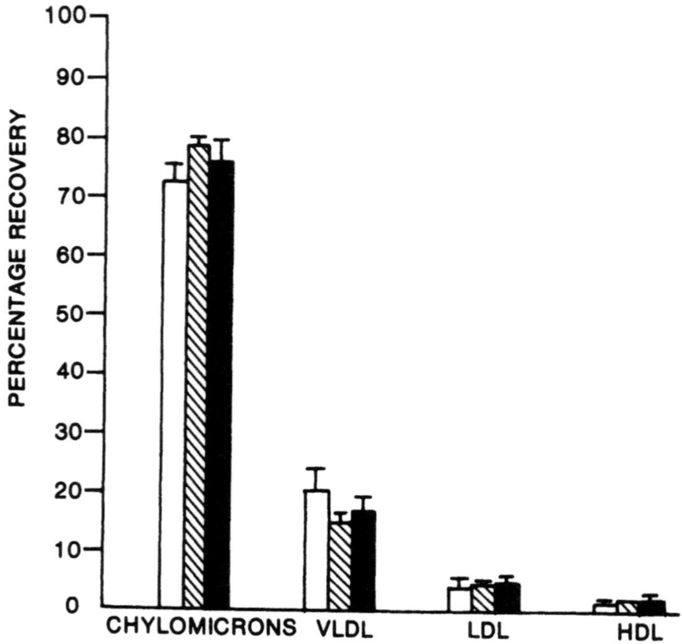

FIGURE 2. Recovery of administered oleic, arachidonic, and eicosapentaenoic acids in major lymph lipoprotein fractions. Details of lymph collection, lipoprotein separation, and analyses are given in Section 2. Figures represent means ± S.E.M. for four to five animals per group. □ oleic acid; ▨ arachidonic acid; ■ eicosapentaenoic acid.

TABLE II. Distribution of Fatty Acids among Individual Lipid Classes of Chylomicrons and VLDL

Measurement[a] (%)	Fatty acid administered		
	Oleic	Arachidonic	EPA
Chylomicron distribution			
TG	90.5 ± 1.1	88.9 ± 0.6	84.8 ± 1.1[b,c]
MG-DG	1.3 ± 0.1	3.3 ± 0.2[b]	5.0 ± 0.3[b,c]
UFA	2.8 ± 0.2	3.3 ± 0.3	2.1 ± 0.3[c]
CE	3.4 ± 0.8	0.8 ± 0.1[b]	4.2 ± 0.4[b]
PL	2.0 ± 0.2	3.6 ± 0.5[b]	3.9 ± 0.2[b]
VLDL distribution			
TG	89.2 ± 1.4	86.0 ± 0.9	88.0 ± 0.5
MG-DG	1.4 ± 0.1	3.4 ± 0.2[b]	3.4 ± 0.3[b]
UFA	1.6 ± 0.7	1.8 ± 0.1[b]	1.4 ± 0.1[b]
CE	5.1 ± 1.2	1.7 ± 0.1[b]	2.9 ± 0.5[b,c]
PL	2.6 ± 0.2	7.2 ± 1.0[b]	4.2 ± 0.2[b,c]

[a] Treatment of animals and collection and separation of lymph lipoproteins and lipids are described in Section 2. Values are mean ± S.E.M. TG, triglycerides; MG-DG, partial glycerides; UFA, unesterified fatty acids; CE, cholesterol esters; PL, phospholipids.
[b] $P < 0.05$ compared to oleic acid.
[c] $P < 0.05$ compared to arachidonic acid.

arachidonate as cholesterol esters was less than for the other fatty acids in both lipoprotein fractions, and this was reflected in analysis of whole lymph distribution. The recovery of EPA in cholesterol esters, in contrast, was lower only in the VLDL fraction of lymph and was not reflected in whole lymph distributions.

Distributions of absorbed cholesterol among the major lipoproteins of lymph are summarized in Fig. 3. As has been reported earlier with cholesterol and oleic acid administration (Vahouny *et al.*, 1980a), there is a smaller percentage of absorbed cholesterol associated with lymph chylomicrons in the rat and a greater recovery in the $d > 1.006$ g/ml lipoproteins. Similar results were obtained for cholesterol distributions when the infusion contained arachidonic acid but were not as pronounced with animals given EPA. The difference in the lipoprotein distributions of absorbed fatty acid and cholesterol is best reflected as the ratios of absorbed fatty acid and cholesterol. For the $d < 1.006$ g/ml lipoproteins, the ratio of fatty acid to cholesterol was 0.91 ± 0.03. For chylomicrons, this ratio was 1.20 ± 0.03, suggesting a greater incorporation of fatty acids, and for VLDL, the ratio was 0.78 ± 0.06, suggesting a relatively greater proportion of cholesterol. For the $d > 1.006$ g/ml lipoproteins, this ratio was 0.46 ± 0.09, demonstrating the predominance of absorbed cholesterol in these lipoproteins (LDL plus HDL). Again, the differences in cholesterol distribution among the lipoproteins for the

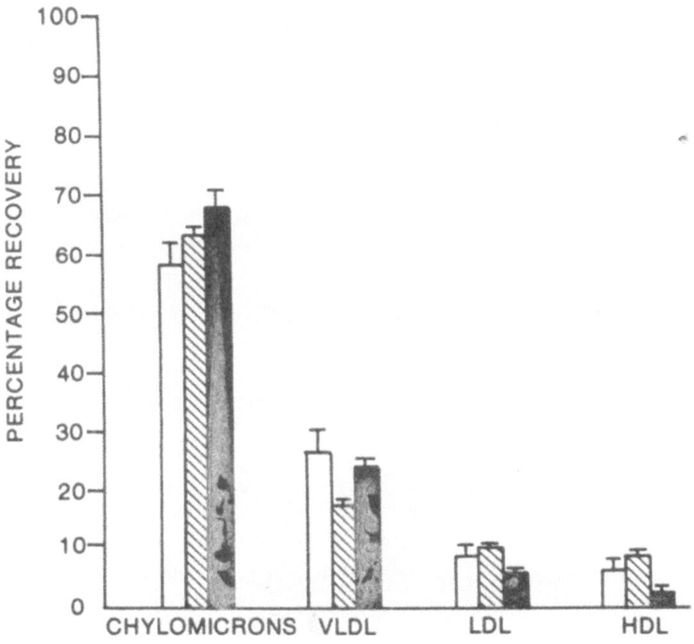

FIGURE 3. Recovery of administered cholesterol in major lymph lipoproteins. Details are given in Fig. 2 and the text. Cholesterol administered with □ oleic acid, ▨ arachidonic acid, ■ eicosapentaenoic acid.

TABLE III. Lymphatic Recovery of Absorbed Fatty Acid

Measurement (μmol)	Fatty acid administered		
	Oleic	Arachidonic	EPA
Fatty acid fed[a]	275.4	231.2	135.0
Fatty acid absorbed	209.0 ± 4.6	188.6 ± 9.2	115.6 ± 4.1
Lymph fatty acid	268.8 ± 11.7	254.4 ± 5.9	120.6 ± 4.9
Endogenous fatty acid	58.9 ± 14.2	56.8 ± 7.9	5.0 ± 2.0

[a] Derivatization of oleic acid was 100%, and purity by gas–liquid chromatography was 91.8%. For arachidonate, these values were 93.3% and 82.6%, respectively. For EPA, the values were 50% and 90%, respectively.

three fatty acid groups were not of sufficient magnitude to suggest major differences in absorption and transport characteristics for cholesterol.

Mass analyses of the administered fatty acids and of the fatty acids in lymph lipids are summarized in Table III. These studies were conducted by analyses of extent of derivatization of each fatty-acid-containing test emulsion (by isotope recoveries) and quantitative analysis of mass and purity by gas–liquid chromatography. As shown in Table III, 275.4 μmol of the weighed oleic acid (300 μmol) in the test emulsion was recovered as authentic oleic acid following derivatization and chromatography. Based on lymph isotope recoveries, 209.0 μmol of oleic acid was absorbed into lymph. Actual analysis demonstrated 268.8 μmol of oleic acid, indicating the presence of 58.9 μmol oleic acid from endogenous sources (e.g., intestinal secretion and filtered plasma lipoproteins). Similar calculations with arachidonic acid also suggested the presence of significant levels of endogenous fatty acid derived from sources other than the fatty acid administered.

With eicosapentaenoic acid, however, only half of the added level was recovered by derivatization, and of the mass derivatized, 90% was authentic EPA. Assuming equivalent properties of isotope and mass, it was calculated that 115.6 μmol of the absorbed EPA could be analyzed by the derivatization and chromatographic analysis. The actual recovery was 120.6 μmol, which was statistically the same as that calculated and demonstrated the absence of endogenous EPA in rat thoracic duct lymph.

4. DISCUSSION

There is a sparsity of information on the intestinal absorption and lymphatic transport of polyunsaturated fatty acids in general and of the ω-3 fatty acids in particular. In certain marine oils, the polyunsaturated fatty acids of the ω-3 class, such as eicosapentaenoic acid and docosahexaenoic acid, are largely found in the triglycerides. There is substantial evidence that ingestion of these oils does not result in typical fat tolerance curves in humans (Harris *et al.*, 1982), nor do the ω-3 fatty acids accumulate rapidly in tissue lipids (e.g., Garton *et al.*, 1952; Goodnight *et al.,* 1982). Data from *in vitro* studies demonstrated the inefficient lipolysis of these oils (Bottino *et al.,* 1967; Brockerhoff *et al.,* 1966) and the resistance to chemical hydrolysis and derivatization of the ω-3 acids (Bottino *et al.,* 1967; Yurkowski and Brockerhoff, 1966). It has also been suggested that the absorbability and route of transport (e.g., lymphatic or portal transport) of polyunsaturated acids may be dissimilar to that of typical fatty acids such as the long-chain saturated and monounsaturated acids (Goodnight *et al.,* 1982; Willis, 1981).

The current research has addressed the question of the efficiency of intestinal absorption of ω-3 fatty acids and the lipoprotein and lipid distributions of the absorbed fatty acids. These data on $\Delta^{5,8,11,14,17}$-eicosapentaenoic acid are compared to those of the common ω-9 octadecaenoic acid (oleic) and the ω-6 eicosatetraenoic acid ($\Delta^{5,8,11,14}$-arachidonic). Despite differences in the rates of appearance in lymph, it is apparent that both arachidonic and eicosapentaenoic acids are absorbed with equal efficiency, as is oleic acid. Furthermore, like oleic acid, these polyunsaturates are transported largely in the chylomicron fraction of lymph and primarily as the triglyceride component of this lipoprotein. Although there is evidence for a somewhat greater incorporation of the polyunsaturated acids into phospholipids and phospholipid precursors (mono- and diglycerides) in both chylomicron and VLDL fractions, these levels do not reflect the ultimate tissue lipid distributions of either arachidonic or eicosapentaenoic acids.

Additional evidence for comparability in the mechanisms of intestinal oil–water phase characteristics and micellar capacities between these acids is reflected in the effects of the individual fatty acids on the absorption and lymphatic transport of the tracer cholesterol. The efficiency of cholesterol absorption in this experimental model is dependent on micellar solubilization and on the presence of unesterified fatty acids (Treadwell and Vahouny, 1968). In general, the efficiency of cholesterol absorption is improved in the presence of mono- and diunsaturated acids compared to the saturated analogues (Treadwell and Vahouny, 1968), and the present study also suggests that cholesterol absorption is equally efficient from emulsions containing any of the polyunsaturated fatty acids studied.

An important aspect of the present study, and one that requires additional attention, was the quantitative analysis of the administered fatty acids and the fatty acids of lymph lipids. These analyses allowed for corrections of data for the recoveries by derivatization and the purities of the administered fatty acids. With oleic and arachidonic acids, 92% and 77%, respectively, of the purchased materials (reported as 99% purity) were recovered as the authentic acids in the administered emulsions. Analysis of lymph content of these acids in the respective groups allowed for calculations of the contributions of the absorbed and endogenous sources of each acid in thoracic duct lymph. Thus, approximately 22% of the lymph oleic acid and 23% of the lymph arachidonic acids during the 24-hr collection period following acute administration of these acids were derived from endogenous sources.

The boron trifluoride–methanol derivatization procedure, however, resulted in methylation of only half of the weighed quantity of EPA (reported as 90% purity). These studies provide no indication of the actual content of EPA in the weighed preparations. However, similar difficulties in deacylation and derivatization of esterified EPA have been alluded to in the extensive and carefully controlled studies of Brockerhoff and co-workers (Brockerhoff, 1965; Brockerhoff *et al.*, 1966; Yurkowski and Brockerhoff, 1966). Despite this, based on the actual analysis of the administered EPA and the EPA content of total lymph lipids, it was apparent that all of the lymph EPA was accounted for by absorption of the administered acid.

These studies have largely described the efficiency and route of absorption of eicosapentaenoic acid and the lipoprotein and lipoprotein lipid distributions of this polyunsaturated fatty acid. Questions regarding the digestibility and absorption of EPA-rich glycerides of marine oils and the metabolic disposition of EPA-rich chylomicrons have yet to be addressed. Based on previous reports (Yurkowski and Brockerhoff, 1966) and on studies currently under way, routine analytical procedures for fatty acids may not be appropriate for quantitative analysis of EPA, and this issue may preclude definitive studies on further metabolism of EPA.

ACKNOWLEDGMENT. This work was supported in part by United States Department of Agriculture Grant 82-CRCR1-1001 and resources of the Food and Drug Administration.

REFERENCES

Bang, H. O., and Dyerberg, J., 1972, Plasma lipids and lipoproteins in Greenland Eskimos, *Acta Med. Scand.* **192**:85–94.

Bang, H. O., and Dyerberg, J., 1980, Lipid metabolism and ischemic heart disease, in: *Advanced Nutrition Research*, Vol. 3 (H. H. Draper, ed.), Plenum Press, New York, pp. 1–22.

Bang, H. O., Dyerberg, J., and Hjorne, N., 1976, The composition of food consumed by Greenland Eskimos, *Acta Med. Scand.* **100**:69–73.

Bottino, N. R., Vandenberg, G. A., and Reiser, R., 1967, Resistance of certain long-chain polyunsaturated fatty acids of marine oils to pancreatic lipase hydrolysis, *Lipids* **2**:489–493.

Brockerhoff, H., 1965, Stereospecific analysis of triglycerides: An analysis of human depot fat, *Arch. Biochem. Biophys.* **110**:586–592.

Brockerhoff, H., Hoyle, R. J., and Huang, P. C., 1966, Positional distribution of fatty acids of a polar bear and a seal, *Can. J. Biochem.* **44**:1519–1525.

Culp, B. R., Titus, B. G., and Lands, W. E. M., 1979, Inhibition of prostaglandin synthesis by eicosapentaenoic acid, *Prostaglandins Med.* **3**:369–378.

Dyerberg, J., and Bang, H. O., 1979, Atherogenesis and haemostasis in Eskimos—the role of prostaglandin-3 family, *Haemostasis* **8**:227–233.

Dyerberg, J., Jorgensen, J. A., and Arnfred, T., 1981, Human umbilical cord converts all *cis* 5,8,11,14,17-eicosapentaenoic acid to prostaglandin I$_3$, *Prostaglandins* **22**:857–862.

Folch, J., Lees, M., and Sloane-Stanley, G. H., 1957, Simple method for the isolation and purification of total lipids from animal tissues, *J. Biol. Chem.* **226**:497–509.

Gartner, S., and Vahouny, G. V., 1972, Effects of epinephrine and cAMP on perfused rat hearts, *Am. J. Physiol.* **222**:1121–1124.

Garton, G. A., Hilyditch, T. P., and Meara, M. L., 1952, The composition of depot fat of a pig fed on a diet rich in whale oil, *Biochem. J.* **50**:517–524.

Goodnight, S. H., Jr., Harris, W. S., and Connor, W. E., 1981, The effect of dietary ω-3 fatty acids on platelet composition and function in man: A prospective, controlled study, *Blood* **58**:880–884.

Goodnight, S. H., Jr., Harris, W. S., Connor, W. E., and Illingworth, D. E., 1982, Polyunsaturated fatty acids, hyperlipidemia and thrombosis, *Arteriosclerosis* **2**:87–113.

Havel, R. J., Eder, H. A., and Bragdon, J. H., 1955, The distribution and chemical composition of ultracentrifugally separated lipoproteins in human serum, *J. Clin. Invest.* **34**:1345–1353.

Harris, W. S., Connor, W. E., and Goodnight, S. H., Jr., 1982, Dietary fish oils, plasma lipids and platelets in man, in: *Essential Fatty Acids and Prostaglandins* (R. T. Holman, ed.), Pergamon Press, Oxford, pp. 75–79.

Hay, C. R. M., Durber, A. P., and Saynor, R., 1982, Effect of fish oil on platelet kinetics in patients with ischaemic heart disease, *Lancet* **2**:1269–1272.

Hirai, A., Hamazaki, T., Terano, T., Nishikawa, T., Tamura, Y., Kumagai, A., and Sajiki, J., 1980, Eicosapentaenoic acid and platelet function in Japanese, *Lancet* **2**:1132–1133.

Holman, R. T. (ed.), 1982, *Essential Fatty Acids and Prostaglandins. Progress in Lipid Research,* Pergamon Press, Oxford.

Jakobowski, J. A., and Ardlie, N. G., 1980, Cyclic AMP-independent inhibition of human platelet function by eicosapentaenoic acid, *Thromb. Res.* **17**:751–752.

Needleman, P., and Sprecher, H., 1979, Mechanism underlying the inhibition of platelet aggregation by eicosapentaenoic acid and its metabolites, *Adv. Prostaglandin Thromboxane Res.* **6**:61–68.

Needleman, P., Raz, A., Minkes, M. S., Ferrendelli, J. A., and Sprecher, H., 1979, Triene prostaglandins: Prostacyclin and thromboxane biosynthesis and unique biological properties, *Proc. Natl. Acad. Sci. U.S.A.* **76**:944–948.

Needleman, P., Whitaker, M. C., and Sprecher, H., 1980, Manipulation of platelet aggregation by prostaglandins and their fatty acid precursors—pharmacological basis for a therapeutic approach, *Prostaglandins* **19**:165–181.

Spector, A. A., Kaduce, T. L., Figard, P. H., Norton, K. C., Hoak, J. C., and Czervionke, R. L., 1983, Eicosapentaenoic acid and prostacycl in production by cultured human endothelial cells, *J. Lipid Res.* **24**:1595–1604.

Treadwell, C. R., and Vahouny, G. V., 1968, Cholesterol absorption, in: *Handbook of Physiology—Alimentary Canal,* Sec. 6, Vol. II, American Physiological Society, Washington, D.C., pp. 1407–1438.

Vahouny, G. V., Blendermann, E. M., Gallo, L. L., and Treadwell, C. R., 1980a, Differential transport cholesterol and oleic acid in lymph, *J. Lipid Res.* **21**:415–424.

Vahouny, G. V., Roy, T., Gallo, L. L., Story, J. A., Kritchevsky, D., and Cassidy, M. M., 1980b, Dietary Fiber. III. Effects of chronic intake on cholesterol absorption and metabolism in the rat, *Am. J. Clin. Nutr.* **33**:2182–2191.

Vahouny, G. V., Satchithanandam, S., Cassidy, M. M., Lightfoot, F. G., and Furda, I., 1983, Comparative effects of chitosan and cholestyramine on lymphatic absorption of lipids in the rat, *Am. J. Clin. Nutr.* **38**:278–284.

Witaker, M. O., Wyche, A., Fitzpatrick, F., Sprecher, H., and Needleman, P., 1979, Triene prostaglandins: Prostaglandin D_3 and eicosapentaenoic acid as potential antithrombotic substances, *Proc. Natl. Acad. Sci. U.S.A.* **76**:5919–5923.

Willis, A. L., 1981, Nutritional and pharmacological factors in eicosanoid-biology, *Nutr. Rev.* **39**:289–301.

Yotakis, L. D. O., 1981, The preventive effects of polyunsaturated fats on thrombosis, *Thromb. Haemostas.* **46**:65–68.

Yurkowski, M., and Brockerhoff, H., 1966, Fatty acid distribution of triglycerides determined by deacylation with methyl magnesium bromide, *Biochim. Biophys. Acta* **125**:55–59.

IV

Recent Clinical Applications of Eicosanoids

39

Status of Clinical Trials Evaluating Eicosanoids in Atherosclerotic Vascular Disease and Thrombosis

LAURENCE A. HARKER

1. INTRODUCTION

The indications for the use of agents that modify eicosanoid metabolism in the management of patients with arterial vascular disease are limited despite extensive basic, experimental animal and clinical studies. Aspirin is clearly beneficial in men with transient ischemic attacks and patients with unstable angina. It is also antithrombotic when used in combination with oral anticoagulants in patients with artificial heart valves and in patients with arteriovenous cannulas undergoing hemodialysis (and manifest associated underlying platelet dysfunction). In general, the negative clinical trials have been uninformative because of uncertainty with respect to the role of eicosanoids in the pathogenesis of the outcome events, inadequately defined mechanisms of drug actions, and uncertainty regarding the effective doses for the drugs in the trials. In the absence of established mechanisms of drug action, dose regimens appear to require objective assays with thrombotic endpoints rather than the use of *in vitro* biochemical measurements of less certain interpretation.

LAURENCE A. HARKER ● Roon Research Center for Arteriosclerosis and Thrombosis, Scripps Clinic and Research Foundation, La Jolla, California 92037.

2. PATHOPHYSIOLOGY OF ARTERIAL INSUFFICIENCY SYNDROMES

2.1. Transient Ischemic Attacks

Evidence from clinical and pathological studies suggests that platelets may be involved in the pathogenesis of ischemic syndromes involving the cerebral circulation. For example, patients with transient ischemic attacks frequently have ulcerated atherosclerotic lesions in the extracranial portion of the basilar or internal carotid arteries, and these attacks may be the result of microembolization from thrombi forming on such lesions. More direct evidence is derived from observations of the passage of microemboli through the retinal arterioles during attacks of amaurosis fugax and the histological demonstration that some of these microemboli are platelet aggregates. Alternatively, vasoconstriction at stenotic sites of diseased vessels may also be responsible for transient ischemia. Indeed, it should be noted that these transient symptoms may arise from a variety of other causes, including hemodynamic changes, cardiac arrhythmias, prolapse of the mitral valve, atrial myxomas, nonarteriosclerotic vasculopathies, embolism of atheroma, and cervical spondylosis. Hence, any large clinical series may contain patients with cerebral ischemia of varying etiology. The relative frequencies of these different causes are unknown.

2.2. Acute Myocardial Infarction

Recent angiographic studies confirm that acute transmural myocardial infarction is generally associated with thrombotic occlusion of the subtending diseased coronary artery (DeWood *et al.*, 1980; Rentrop *et al.*, 1981). An occlusive thrombus can be demonstrated by coronary arteriography in 90% of patients with transmural infarction if the angiography is performed within the first 4 hr following onset of symptoms. Although occlusion of a stenotic coronary artery may theoretically occur in several ways, the frequency of reperfusion achieved by fibrinolytic therapy suggests that denudation of an atherosclerotic plaque with secondary thrombosis is the typical course. Occlusion by intimal hemorrhage or thrombosis induced by discharge of the contents of the plaque is probably rare.

Coronary artery spasm may be important in some forms of myocardial ischemia. In that respect, Maseri and co-workers showed that spontaneous coronary artery spasm precipitated angina and that repeated episodes could lead to myocardial infarction in some patients (Maseri *et al.*, 1978).

The interaction of circulating blood elements and some of their secreted products with vascular components may be important in the development of coronary vascular spasm (Del Zoppo and Harker, 1984). Thromboxane A_2 is a potent labile vasoconstrictor released during the activation of platelets, leukocytes, and atherosclerotic arteries. Serotonin increases coronary vascular tone, whereas locally released adenosine and adenosine nucleotides induce vasodilatation. ADP- and ATP-mediated smooth muscle relaxation are dependent on an intact and functional endothelium, whereas adenosine and AMP produce a direct effect on smooth muscle cells. Adenosine also effects relaxation of the vascular smooth muscle by inhibiting the release of norepinephrine from sympathetic nerve terminals. Of those substances, thromboxane A_2-induced vasoconstriction probably predominates during platelet deposition.

Also, leukocytes and a diseased vascular wall may directly or indirectly modulate coronary artery tone in association with myocardial ischemia, either alone or in concert with thrombus formation. Leukotrienes cause potent dose-dependent coronary artery vasoconstriction and decreased myocardial contractility in isolated animal preparations. Although their role in coronary artery spasm has not been defined *in vivo*, it is possible that they may contribute to sustained arterial spasm. Thus, vascular spasm may precede or follow acute thrombosis of atherosclerotic coronary arteries and thereby contribute to the extent of infarction.

2.3. Sudden Death

The immediate factor or factors that precipitate dysrhythmias in victims of out-of-hospital ventricular fibrillation remain largely undefined (Harlan and Harker, 1983). Traditional risk factors associated with coronary atherosclerosis (smoking, hypertension, and hyperlipidemia) also are associated with sudden death. Postmortem studies of people who died suddenly of cardiac causes do not show thrombosis of coronary arteries. Such patients do have advanced disease with significant stenoses in several vessels, however, and the presence of platelet aggregates has been documented after death.

Experimental studies indicate that activated platelets may provoke terminal dysrhythmias, probably by producing myocardial ischemia through thromboxane A_2-induced coronary artery spasm or occlusion of myocardial microcirculation by microemboli. In that context, inhibition of thromboxane A_2 synthesis reduces the frequency of arrhythmias or ventricular fibrillation in several animal models. In other experimental studies, platelet emboli in the microcirculation cause ventricular fibrillation and sudden death. Those results suggest that platelet microemboli from coronary mural thrombi may obstruct the coronary microcirculation and precipitate arrhythmias. The relative importance for a role of platelets versus vasospasm in sudden cardiac death is unclear.

2.4. Angina Pectoris

Evidence in favor of a role for thrombosis in angina pectoris is not convincing (Harlan and Harker, 1983). However, recent *in vivo* studies measuring platelet function during exercise or cardiac pacing show changes in coronary sinus levels of thromboxane B_2, the stable metabolite of thromboxane A_2, following pacing-induced ischemia. Elevations of thromboxane B_2 in the coronary sinus correlated with the time course of increased lactate production, which suggests a relationship of thromboxane A_2 to myocardial ischemia.

Patients with unstable angina and recent chest pain also have higher thromboxane B_2 levels in the coronary sinus than in the ascending aorta, which suggests that recent episodes of angina in such patients are associated with local release of thromboxane A_2. The capacity of aspirin to reduce the frequency of myocardial infarction in patients with unstable angina (Lewis *et al.*, 1983) may be related to its blockade of platelet thromboxane A_2 production and provides additional support for that concept. This same rationale may underlie the benefit of aspirin in transient ischemic attacks (Fields, 1977; Canadian Cooperative Study Group, 1978).

A role for platelets in variant angina (characterized by rest pain, ST-segment elevation, and angiographic evidence of coronary artery spasm) is of great interest. It has been suggested that platelet activation, aggregation, and release of thromboxane A_2 might trigger episodes of spasm. Elevations in thromboxane B_2 and increased levels of circulating platelet aggregates have been described during episodes of rest pain and ST-segment elevation in patients with variant angina. However, double-blind crossover studies using inhibitors of platelet thromboxane synthesis in those patients show no reduction in angina or electrocardiographic changes despite the blockade of thromboxane A_2 release from platelets (Robertson *et al.*, 1981).

3. CLINICAL TRIALS USING EICOSANOIDS

Aspirin has been shown in well-designed clinical trials to be beneficial in four clinical settings (Table I): (1) reduction of thromboembolic complications associated with artificial heart valves (Dale, 1976); (2) prevention of stroke and death in men with transient ischemic attacks (Fields, 1977; Canadian Cooperative Study Group, 1978); (3) decrease in thrombotic occlusion of arteriovenous silastic cannula in uremic patients receiving hemodialysis (Harter *et al.*, 1979); and (4) reduction in myocardial infarction in patients with unstable angina (Lewis *et al.*, 1983). The antithrombotic effects of aspirin in patients with old-style Starr–Edwards mitral valves (Dale, 1976) was evident in association with oral anticoagulant therapy. However, this combination is associated with an unacceptably high frequency of

TABLE I. Positive Aspirin Clinical Trials

Prosthetic heart valves (with anticoagulants)
    Dale (1976)
Transient ischemic attacks
    Aspirin in Transient Ischemic Attacks (1977)
    Canadian Cooperative Study Group (1978)
A–V cannula (with platelet dysfunction)
    Harter *et al.* (1979)
Unstable angina
    Lewis *et al.* (1983)

gastrointestinal bleeding. The capacity of aspirin in low doses to reduce the thrombotic complications of an arteriovenous cannula (Harter *et al.*, 1979) was shown in patients undergoing chronic hemodialysis, and thus associated with significant platelet dysfunction. It cannot be assumed that a similar dose would be antithrombotic in the absence of underlying platelet dysfunction.

3.1. Transient Ischemic Attacks

The Aspirin in Transient Ischemic Attacks (AITIA) Study (Fields, 1977) was designed to test, in a double-blind manner, the effectiveness of aspirin in the prophylaxis of cerebrovascular ischemic events. Only patients with carotid system transient ischemic attacks were admitted to the study. Subjects received either 650 mg aspirin twice a day or appropriate placebos.

Analysis of the first 6 months of follow-up showed a statistically significant difference in favor of aspirin when death, cerebral or retinal infarction, and the occurrence of transient ischemic attacks were grouped together as endpoints. When stroke or death was considered as an endpoint, the beneficial effects of aspirin were confined to men (47% versus 0% reduction). When the occurrence of transient ischemic attacks was included as an endpoint and results were classified as favorable or unfavorable (death, nonfatal cerebral or retinal infarction, or continued transient ischemic attacks), there was no sex difference in the response to aspirin.

The Canadian Cooperative Study Group (CCSG) was a double-blind, randomized, multicenter trial to assess the relative efficiency of aspirin and sulfinpyrazone, singly and in combination, in the reduction of continuing transient ischemic attacks, stroke, and death (Canadian Cooperative Study Group, 1978). Five hundred eighty-five patients (after 64 exclusions) with one or more cerebral or retinal ischemic attacks within 3 months before entry were followed in a randomized clinical trial for an average of 26 months. About 30 of the patients were women. The four treatment regimens were placebo, aspirin (325 mg four times daily), sulfinpyrazone (200 mg four times daily), and aspirin plus sulfinpyrazone at the same dosage. For the entire group, aspirin reduced the risk of continued transient ischemic attack, stroke, or death by 19%. If only stroke or death were considered, aspirin reduced the risk by 31%. There was no statistically significant reduction in these events attributable to sulfinpyrazone. A striking difference was found between male and female patients in the therapeutic response to aspirin in that there was no observed benefit among women in terms of stroke or death, but there was a risk reduction of 48% among men.

3.2. Secondary Prevention of Myocardial Infarction

The use of eicosanoids in the prevention of myocardial infarction is based on the premise that platelet mechanisms are important in acute coronary artery thrombosis. Three platelet-inhibiting agents have been evaluated in various prospective randomized double-blind clinical trials: aspirin, sulfinpyrazone, and dipyridamole.

Clinical trials designed to assess the effects of those agents on secondary prevention of myocardial infarction have used mortality, nonfatal recurrent myocardial infarction, and coronary incidence (coronary mortality plus nonfatal myocardial infarction) as outcome events.

Five studies have addressed the efficacy of aspirin alone in the treatment of patients following acute myocardial infarction (Elwood *et al.*, 1974; Coronary Drug Project Research Group, 1976; Breddin *et al.*, 1979; Elwood and Sweetnam, 1979; Aspirin Myocardial Infarction Study Research Group, 1980). Four studies reported between 1974 and 1979 compared several dose regimens of aspirin versus placebo, and all reported a trend in favor of aspirin therapy with respect to all endpoints. In contrast, the Aspirin Myocardial Study (AMIS), the largest and only definitive trial, demonstrated no effect of aspirin on any outcome events including total coronary incidence in 2267 patients. The study may have been compromised by late patient entry after infarction. Subsequent analysis of the combined results of all the aspirin trials has suggested that there may be a 10% reduction in coronary mortality in post-myocardial infarction patients receiving aspirin. However, that analysis is flawed by the dissimilarities in the studies, including the variable entry time (mean, 7 days to 7 years) and dose regimen (300 to 1500 mg aspirin per day). Thus, a benefit of aspirin therapy is not established.

There are several possible explanations for the lack of benefit observed with aspirin. Aspirin inhibits platelet aggregation *in vitro* and *in vivo* by selective irreversible acetylation of platelet cyclooxygenase, thereby blocking the synthesis of the proaggregatory substance thromboxane A_2. A similar action on vessel wall cyclooxygenase results in concurrently decreased production of prostacyclin, a potent vasodilator and inhibitor of platelet aggregation. Since cyclooxygenase is renewable in vascular tissues but not in platelets, low and infrequent doses of aspirin (1–2 mg/kg per day) should, in theory, inhibit platelet aggregation without substantially affecting the production of prostacyclin.

Although the clinical trials assessing aspirin in acute myocardial infarction employed comparatively large doses (300 mg to 1.5 g per day), which affect both platelet and endothelial systems, no dose–response effects were noted in the outcome events. In this regard, however, recent studies involving drugs that produce selective blockade of thromboxane synthetase are of interest. These drugs prevent thromboxane A_2 formation without impairing prostacyclin generation. Despite this theoretically attractive mechanism of action, such drugs fail to show antithrombotic effects in experimental models (Lewis and Taylor, 1983). The lack of antithrombotic efficacy has been attributed to the persistent production of the proaggregatory endoperoxides PGG_2 and PGH_2 in the absence of thromboxane A_2, but that interpretation has not been established. Moreover, aspirin has been shown to have antithrombotic effects independent of its inhibition of platelet cyclooxygenase, particularly when used in combination with dipyridamole or sulfinpyrazone. Thus, measurements of aspirin's effects on thromboxane A_2 production do not appear to be predictive of antithrombotic efficacy *in vivo*.

Two clinical trials testing sulfinpyrazone have been reported: the Anturane

Reinfarction Trial (Anturane Reinfarction Trial Research Group, 1980; ART) and the Anturane Reinfarction Italian Study (Anturane Reinfarction Italian Study, 1982; ARIS). The ART investigators reported a significant reduction in sudden death in the first 6 months of follow-up with no additional benefit thereafter. Patients were enrolled 25 to 35 days after the primary cardiac event. No effect on other measured endpoints was detected in that study. In a review of ART, the Food and Drug Administration noted that the exclusion of patients from statistical analysis and irregularities in assignment of the cause of death in some cases represented major sources of potential bias (Temple and Pledger, 1980). Consequently, sulfinpyrazone was not approved for the prevention of reinfarction. A separate analysis of the original ART data, however, supported the original report with regard to a decreased incidence of sudden death and cardiac mortality in the first 6 months following myocardial infarction (Anturane Reinfarction Trial Policy Committee, 1982). Enigmatically, the Italian study in 1982 (Anturane Reinfarction Italian Study, 1982) demonstrated no effect on sudden death, although a significant reduction in the occurrence of fatal recurrent myocardial infarction was reported. The second study followed fewer patients than the first (727 versus 1558), but the general study design was similar. The conflicting results leave many issues unresolved.

The combination of aspirin and dipyridamole was employed in the Persantine–Aspirin Reinfarction Study (PARIS). In that study (Persantine–Aspirin Reinfarction Study Research Group, 1980), the combination was more effective than aspirin alone in the reduction of coronary incidence during the first 2 years of therapy. A trend in favor of either aspirin or the combination for the reduction of overall total and coronary artery-related mortality was also reported. The late entry of patients following myocardial infarction may have compromised the study. Indeed, subsequent analysis indicated that a subgroup of patients who were entered less than 6 months following myocardial infarction had a significant reduction in 3-year coronary mortality. Because the effects of overall final mortality were not conclusive, a new prospective study enrolling patients between 3 weeks and 6 months following myocardial infarction is in progress. At this time, there is sufficient evidence for a general recommendation that dipyridamole and aspirin alone or in combination should be used for secondary prevention of myocardial infarction.

3.3. Saphenous Vein Bypass Grafts

The combination of aspirin and dipyridamole has recently been shown by Chesebro and co-workers to prevent both early and late postoperative occlusion of aortocoronary-artery bypass grafts (Chesebro *et al.*, 1982, 1984). The rate of postoperative aortocoronary saphenous vein bypass graft occlusion varies from 8% to 18% at 1 month to 16% to 23% at 2 to 6 months. Early graft occlusion generally results from acute thrombosis, whereas late graft occlusion is probably consequent to platelet-mediated intimal lesion formation. The aspirin and dipyridamole combination has been shown to synergistically normalize platelet survival in patients with coronary atherosclerosis and in animal models of arterial thrombosis (Harlan and Harker, 1983). The prospective randomized double-blind trial compared placebo with dipyridamole administered 2 days before operation plus aspirin (325 mg) begun 7 hr after operation and continued with dipyridamole at a lower dose three times daily thereafter. Graft occlusion rate in the first month (Chesebro *et al.*, 1982) was reduced from the typical 10% to 3% without an attendant increase in postoperative hemorrhage. Although previous investigators reported less or no benefit, none had initiated antiplatelet therapy before bypass graft surgery. In the follow-up study (Chesebro *et al.*, 1984), the combination also reduced graft occlusion during the subsequent year. Thus, the combined regimen of aspirin (325 mg) and dipyridamole (75 mg) three times daily appears to be effective in bypass graft surgery to prevent both early and late graft occlusion.

3.4. Progression of Atherosclerosis

Strategies designed to modify symptomatic stenotic coronary atherosclerotic lesions include saphenous vein bypass, percutaneous transluminal coronary angioplasty (PTCA), and the use of drugs. Aortocoronary saphenous vein bypass grafting is firmly established in the management of patients with angina pectoris. In that context, the recently reported randomized Controlled Artery Surgery Study (CASS) has demonstrated that there is no significant difference in annual mortality between subgroups of patients receiving saphenous vein bypass or medical management alone for class I or II angina (Canadian Cardiovascular Society).

Percutaneous transluminal coronary angioplasty mechanically disrupts atherosclerotic coronary stenosis, thereby reestablishing luminal diameter and coronary flow. The procedure is not without risk. Since PTCA produces endothelial denudation and disruption of the atheromatous plaque, subendothelial connective tissue is exposed to blood elements, with resultant thrombus formation. Anecdotal reports suggest that combined eicosanoid anticoagulant prophylaxis may not prevent post-PTCA occlusion. Rethrombosis may require emergency coronary artery bypass grafting. Unfortunately, PTCA does not modify the established atherosclerotic disease process in native coronary vessels and is feasible for selected lesions.

Restenosis following PTCA may be a platelet-dependent process. Pathological evidence suggests that early restenosis may result from endothelial denudation with intraluminal thrombosis, and late restenosis may involve a platelet-dependent proliferative response of smooth muscle cells. Several groups have proposed the use of eicosanoids in reducing post-PTCA restenosis; however, the results of controlled clinical trials have not been reported to date.

No controlled clinical trials have been carried out to assess whether eicosanoids are effective in reducing the process of coronary artery atherosclerosis. The negative results in the AMIS, ART, and PARIS trials suggest that the native disease processes are not altered by those drug regimens. However, the demonstrated effect of antiplatelet agents on late lesion formation in vein grafts may be relevant.

Thus, considerable experimental evidence suggests that the interaction of platelets and endothelium is important in the pathogenesis of coronary atherosclerosis and in acute myocardial infarction. Although some studies suggest that platelets are involved in sudden cardiac death and coronary artery spasm, none of the studies demonstrates conclusively that platelet involvement is causal rather than secondary.

A number of trials have been conducted to determine the efficacy of eicosanoids in patients with acute myocardial infarction. The results are not sufficiently conclusive to justify recommendations for therapy in primary or secondary prevention. Aspirin/dipyridamole therapy initiated before and continued after coronary artery bypass surgery appears to significantly reduce graft occlusion.

ACKNOWLEDGMENTS. This is manuscript No. 3527-BCR from the Research Institute of Scripps Clinic and Research Foundation, La Jolla, CA. This work was supported in part by research grant HL 31950 from the National Institutes of Health.

REFERENCES

Anturane Reinfarction Italian Study, 1982, Sulfinpyrazone in post-myocardial infarction, *Lancet* **1**:237–242.

Anturane Reinfarction Trial Policy Committee, 1982, The Anturane Reinfarction Trial: Reevaluation of outcome, *N. Engl. J. Med.* **306**:1005–1008.

Anturane Reinfarction Trial Research Group, 1980, Sulfinpyrazone in the prevention of sudden death after myocardial infarction, *N. Engl. J. Med.* **302**:250–256.

Aspirin Myocardial Infarction Study Research Group, 1980, A randomized, controlled trial of aspirin in persons recovered from myocardial infarction, *J.A.M.A.* **243**:661–669.

Breddin, K., Loew, D., Lechner, K., Uberla, K., and Walter, E., 1980, Secondary prevention of myocardial infarction. Comparison of acetyl-salicylic acid, phenprocoumon and placebo. *Haemostasis* **9**:325–344.

Canadian Cooperative Study Group. 1978, A randomized trial of aspirin and sulfinpyrazone in threatened stroke, *N. Engl. J. Med.* **299**:53–59.

Chesebro, J. H., Clements, I. P., Fuster, V., Elveback, L. R., Smith, H. C., Bordsley, W. T., Frey, R. L., Holmes, D. R., Vlietstra, R. E., Pluth, J. R., Wallace, R. B., Puga, F. J., Orszulak, T. A., Piehler, J. M., Schaff, H. V., and Danielson, G. K., 1982, A platelet-inhibitor drug trial in coronary artery bypass operations. Benefit of perioperative dipyridamole and aspirin therapy on early postoperative vein-graft patency, *N. Engl. J. Med.* **307**:73–78.

Chesebro, J. H., Fuster, V., Elveback, L. R., Clements, I. P., Smith, H. C., Holmes, D. R., Bardsley, W. T., Pluth, J. R., Wallace, R. B., Puga, F. J., Orszulak, T. A., Piehler, J. M., Danielson, G. K., Schaff, H. V., and Frye, R. L., 1984, Effect of dipyridamole and aspirin on late vein-graft patency after coronary bypass operations, *N. Engl. J. Med.* **310**:209–214.

Coronary Drug Project Research Group, 1976, Aspirin in coronary heart disease, *J. Chronic Dis.* **29**:625–642.

Dale, J., 1976, Arterial thromboembolic complications in patients with Staff–Edwards aortic ball valve prostheses, *Am. Heart J.* **91**:653–659.

Del Zoppo, G. J., and Harker, L. A., 1984, Blood/vessel interaction in coronary disease, *Hosp. Pract.* **19**:163–182.

DeWood, M. A., Spores, J., Notske, R., Mouser, L. T., Burroughs, R., Golden, M. S., and Lang, H. T., 1980, Prevalence of coronary occlusion during the early hours of transmural myocardial infarction, *N. Engl. J. Med.* **303**:897–902.

Elwood, P. C., Cochrane, A. L., Burr, M. L., Sweetnam, P. M., Williams, G., Welsby, E., Hughes, S. J., and Renton, R., 1974, A randomized controlled trial of acetyl salicylic acid in the secondary prevention of mortality from myocardial infarction, *Br. Med. J.* **1**:436–440.

Elwood, P. C., and Sweetnam, P. M., 1979, Aspirin and secondary mortality after myocardial infarction, *Lancet* **2**:1313–1315.

Fields, W. S., 1977, Controlled trial of aspirin in cerebral ischemia, *Stroke* **8**:308.

Harlan, J. M., and Harker, L. A., 1983, Thrombosis and coronary artery disease, in: *Current Concepts,* Upjohn, Kalamazoo, MI., pp. 1–56.

Harter, H. R., Burch, J. W., Majerus, P. W., Stanford, N., Delmez, J. A., Anderson, C. B., and Weerts, C. A., 1979, Prevention of thrombosis in patients on hemodialysis by low-dose aspirin, *N. Engl. J. Med.* **301**:577–579.

Lewis, H. D., Davis, J. W., Archibald, D. G., Steinke, W. E., Smitherman, T. C., Doherty, J. E., Schnaper, H. W., LeWinter, M. M., Linares, E., Pouget, J. M., Sabharwal, S. C., Chesler, E., and DeMots, H., 1983, Protective effects of aspirin against acute myocardial infarction and death in men with unstable angina, *N. Engl. J. Med.* **309**:396–403.

Lewis, P., and Taylor, H. M., 1983, Dazoxiben—clinical prospects for a thromboxane synthetase inhibitor, *Br. J. Clin. Pharmacol.* **15**:1S–140S.

Maseri, A., L'Abbate, A., Baroldi, G., Chierchia, S., Marzilli, M., Ballestra, A. M., Severi, S., Parodi, O., Biagini, A., Distante, A., and Pesola, A., 1978, Coronary vasospasm as a possible cause of myocardial infarction. A conclusion derived from the study of "preinfarction" angina, *N. Engl. J. Med.* **299**:1271–1277.

Persantine–Aspirin Reinfarction Study Research Group, 1980, Persantine and aspirin in coronary heart disease, *Circulation* **62**:449–461.

Rentrop, K. P., Blanke, H., Karsch, K. R., Kaiser, H., Kostering, H., and Leitz, K., 1981, Selective intracoronary thrombolysis in acute myocardial infarction and unstable angina pectoris, *Circulation* **63**:307–317.

Robertson, R. M., Robertson, D., Roberts, L. J., Maas, R. L., FitzGerald, G. A., Friesinger, G. C., and Oates, J. A., 1981, Thromboxane A_2 in vasotonic angina pectoris: Evidence from direct measurements and inhibitor trials, *N. Engl. J. Med.* **304**:998–1003.

Temple, R., and Pledger, G. W., 1982, The FDA's critique of the Anturane Reinfarction Trial, *N. Engl. J. Med.* **303**:1488–1492.

40

Eicosanoids and Allograft Rejection

M. L. FOEGH, M. R. ALIJANI, G. B. HELFRICH,
B. S. KHIRABADI, G. E. SCHREINER, and
PETER W. RAMWELL

1. INTRODUCTION

The relationship of fatty acids to immunity is of considerable interest (Meade and Mertin, 1978). Their role is now believed to be mediated by their oxidation products, specifically, the eicosanoids. The metabolites of arachidonic acid and drugs that affect their metabolism have far-reaching implications in organ transplantation. One of the major problems in evaluating the role of these eicosanoids is their remarkably diverse effects on many biological systems. The diversity of action appears to be receptor mediated, and receptor classification awaits the synthesis of specific eicosanoid antagonists. Diametrically opposite properties of individual metabolites are apparent in other fields and are well recognized, as is the case with the steroids. Here we are concerned with eicosanoids that are immunosuppressive and others that clearly promote organ rejection.

2. MECHANISMS

The antirejection or protective effects of prostacyclin (PGI_2), prostaglandin E_2 (PGE_2), and prostaglandin D_2 (PGD_2) can be envisaged as being exerted through

M. L. FOEGH, G. E. SCHREINER • Department of Medicine, Division of Nephrology, Georgetown University Medical Center, Washington, D.C. 20007.   M. R. ALIJANI, G. B. HELFRICH • Department of Surgery, Division of Transplantation, Georgetown University Medical Center, Washington, D.C. 20007.   B. S. KHIRABADI, PETER W. RAMWELL • Department of Physiology and Biophysics, Georgetown University Medical Center, Washington, D.C. 20007.

a common mechanism, namely, the elevation of intracellular cyclic adenosine monophosphate (cAMP), which stabilizes cells such as platelets, leukocytes, and mast cells. Further, both PGI_2 and PGE_2 inhibit lymphocyte proliferation in response to both mitogens and antigens (Kelly et al., 1979; Leung and Minich, 1980). The involvement of platelets in hyperacute rejection is well known, but platelets are also involved in both acute and chronic rejection, as determined by studies with $[^{111}In]$-labeled platelets (Smith et al., 1979; Leithner et al., 1983; Oluwole et al., 1981). Thus, both PGI_2 and PGE_2 might also exert some of their antirejection effect through increasing platelet cAMP and preventing platelet aggregation. The potent vasodilating properties of these three prostaglandins may play a role in prolonging graft survival by maintaining blood flow through the graft during the rejection. It should be noted that PGI_2 and PGE_2 are released by endothelium, whereas PGD_2 synthesis occurs in all organs.

In contrast, the lipoxygenase products, such as the HETEs and the peptidoleukotrienes, collectively may facilitate rejection. The 5-lipoxygenase product leukotriene B_4 (LTB_4) increases calcium flux into cells (Naccache et al., 1981). Leukotrienes C_4, D_4, and E_4, like TxA_2, are potent smooth muscle constrictors (Maddox et al., 1984, Maddox et al., this volume). In addition, LTB_4 has chemotactic and chemokinetic stimulatory activity (Ford-Hutchinson et al., 1980). Thus, these compounds possess properties likely to curtail graft survival. However, there are in vitro experiments showing that LTB_4 induces suppressor lymphocytes (Rola-Pleszczynski et al., 1983; Payan and Goetzl, 1983).

3. DISCUSSION

The involvement of arachidonic acid metabolites in allograft rejection was originally suggested by Quagliata et al. (1973), who found PGE_1 to prolong skin allograft survival. The finding that the rat cardiac allograft undergoing rejection was associated with increased PGE synthesis was shown by Jaffe et al. (1975). Several years later, increase in urine immunoreactive thromboxane B_2 (i-TxB_2) was found to be an indicator of kidney allograft rejection in patients (Foegh et al., 1981) and probably also in patients undergoing cardiac allograft rejection (Foegh et al., 1985). Thromboxane A_2 is a potent platelet-aggregating agent; TxA_2 has also been shown indirectly to enhance lymphocyte proliferation and lymphocyte cytotoxicity (Rola-Pleszczynski et al., 1983). This effect might be exerted through inhibition of cAMP accumulation (Gorman et al., 1979) or through a calcium ionophore effect (Kawahara et al., 1983). A protective role of the prostaglandins is strongly supported by experimental allograft models in which both PGE_2 (Quagliata, 1973) and its analogues (Kort et al., 1982; Strom and Carpenter, 1983) as well as PGI_2 (Shaw, 1983) have been shown to improve allograft survival. Prostacyclin has been used as a treatment in kidney allograft rejection (Sinzinger et al., 1981). Prostaglandin $F_{2\alpha}$ has not been employed because of its venoconstrictor effects.

A different approach is the use of cyclooxygenase inhibitors, which prevent the formation of both the deleterious compound (TxA_2) and the immunosuppressive prostaglandins. This indiscriminate inhibition of both classes of compounds is reflected in the failure to improve graft survival with indomethacin (Fig. 1). A potentially important exception among cyclooxygenase inhibitors is aspirin (Jamieson *et al.*, 1979; Shaw, 1984) Aspirin affects the prostaglandin endoperoxide production by acetylating cyclooxygenase. This phenomenon is not unique, since such acetylation occurs with any enzyme containing a serine group at the active site. Another indiscriminate approach is by using inhibitors of both cyclo- and lipoxygenase enzymes. For example, BW755 has little selectivity and inhibits both the cyclooxygenase and lipoxygenase pathways. It is not surprising that BW755 does not improve cardiac allograft survival in rats (Shaw *et al.*, 1984).

Corticosteroids, which are used in the daily treatment of patients receiving an organ allograft, inhibit phospholipase A_2 and arachidonate release. This leads to decreased release of both cyclooxygenase and lipoxygenase products. Corticosteroids are thought to exert this effect through the synthesis of a polypeptide termed lipomodulin (Hirata *et al.*, 1980) or macrocortin (Carnuccio *et al.*, 1981), which inhibits phospholipase A_2. We followed the daily serum levels of lipomodulin in a renal transplant patient following treatment with methylprednisolone (1 g i.v. daily for 3 days). An increase was observed in the serum lipomodulin as determined by Dr. Hirata by RIA (Hirata *et al.*, 1981) on a blind basis. An increase in serum lipomodulin level was found to occur more than 36 hr following initiation of treatment with methylprednisolone (Fig. 2).

A dietary means of manipulating the availability of free fatty acids has been described in both men and rats (Mertin and Hunt, 1976). A diet rich in polyunsaturated fatty acids will prolong allograft survival. However, in the clinical trial, the improved graft survival did not extend beyond 6 months.

The next natural step was the use of a more specific inhibitor of the deleterious cyclooxygenase product TxA_2. Experiments *in vitro* with human peritoneal macrophages show not only an inhibition of TxA_2 but also a shunting of arachidonate into PGI_2 and PGE_2 (Foegh *et al.*, 1983). The use of a specific TxA_2 synthase

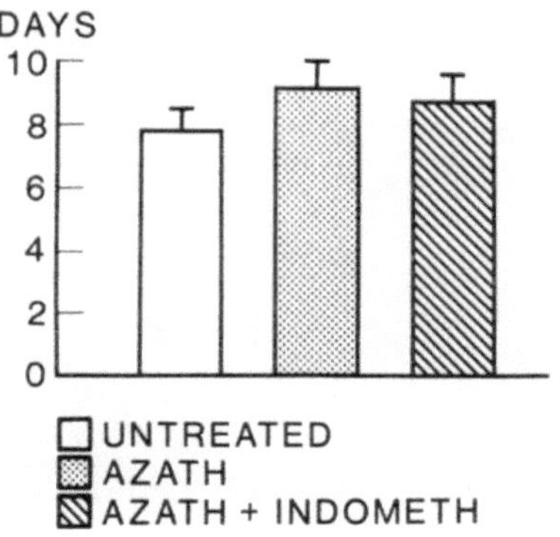

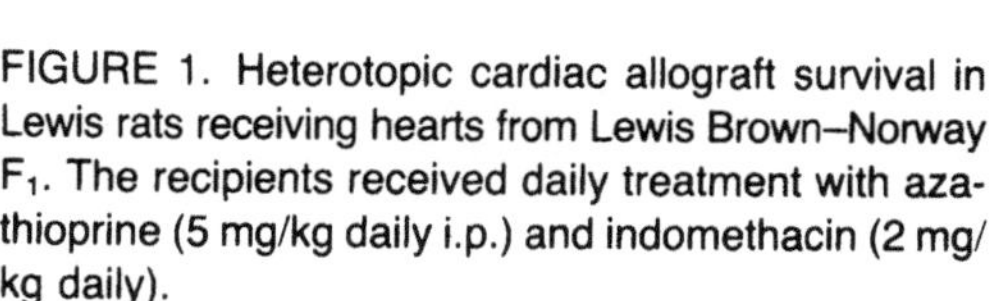
FIGURE 1. Heterotopic cardiac allograft survival in Lewis rats receiving hearts from Lewis Brown–Norway F_1. The recipients received daily treatment with azathioprine (5 mg/kg daily i.p.) and indomethacin (2 mg/kg daily).

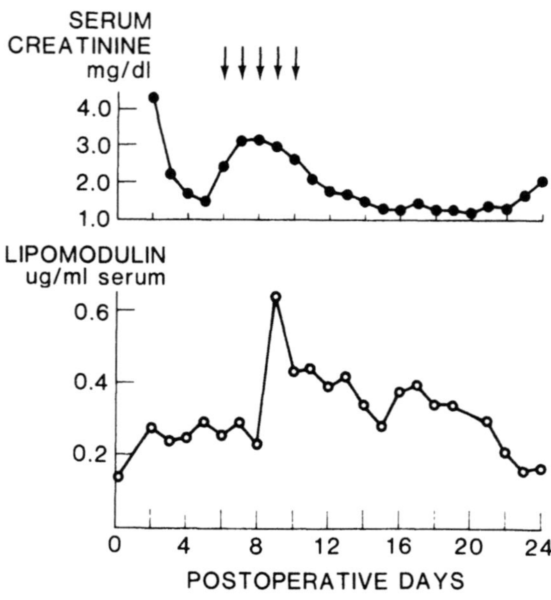

FIGURE 2. Daily serum creatinine levels and daily serum lipomodulin levels in a kidney transplant patient followed from the day of transplantation (day 0). Arrows indicate treatment with methylprednisolone, 1 g i.v.

inhibitor (OKY 1581, ONO Pharmaceutical, Japan) in our experimental rat cardiac allograft model significantly increased allograft survival. The use of L640,035 (Merck-Frosst, Canada), a thromboxane antagonist, improved allograft survival even further (Foegh *et al.*, 1985a,b; Table I). The effect of this latter drug may be partially exerted through its antagonist effect on spasmogenic cylooxygenase products other than TxA_2 (Carrier *et al.*, 1984) since it is not specific.

In summary, there is evidence that eicosanoids and, in particular, arachidonate

TABLE I. Heterotopic Cardiac Allograft Survival
in Lewis Rats Receiving Hearts
from Lewis Brown–Norway F_1 Rats[a]

Treatment	Allograft survival (days)
Untreated	8 ± 0.6
Azathioprine	9 ± 0.8
OKY − 1581 + azathioprine	12 ± 1.3*

[a] Recipients received daily treatment with azathioprine (5 mg/kg daily i.p.) and the thromboxane synthetase inhibitor OKY 1581 (6 mg/kg daily, continuous infusion) or the thromboxane antagonist L-640,035 (50 mg/kg daily orally). *$P < 0.05$.

metabolites are involved in the success or failure of organ transplantation. Those metabolites that elevate intracellular cAMP appear to prolong graft survival, whereas those metabolites that are vasoconstrictors and aggregate platelets reduce graft survival.

This is a simplistic scheme, but it possesses the advantage of promoting systematic experimentation along well-defined lines and further offers the prospect of new immunosuppressive agents.

ACKNOWLEDGMENTS. This work was supported by a grant from the American Heart Association # 84-1147 and a grant from NIH HL-32319.

REFERENCES

Carnuccio, R. M., DiRosa, M., Flower, R. J., and Pinto, A., 1981, The inhibition by hydrocortison of prostaglandin biosynthesis in rat peritoneal leucocytes is correlated with intracellular macrocortin level, *Br. J. Pharmacol.* **74:**322–324.

Carrier, R., Cragoe, E. J., Ethier, D., Ford-Hutchinson, A. W., Girard, W., Hall, R. A., Hamel, D., Rokach, J., Share, N. N., Stone, C. A., and Yusko, P., 1984, Studies on L-640,035: A novel antagonist of contractile prostanoids in the lung, *Br. J. Pharmacol.* **82:**389–395.

Feuerstein, N., and Ramwell, P. W., 1981, OKY-1581 a potent selective thromboxane synthetase inhibitor, *Eur. J. Pharmacol.* **69:**533–534.

Foegh, M. L., Winchester, J. F., Zmudka, M., Helfrich, G. H., Colley, C., Ramwell, P. W., and Schreiner, G. E., 1981, Urine i-TXB₂ in renal allograft rejection, *Lancet* **2:**431–434.

Foegh, M., Maddox, Y., Winchester, J., Rakowski, T., Schreiner, G., and Ramwell, P. W., 1983, Prostacyclin and thromboxane release from human peritoneal macrophages, *Adv. Prostaglandin Thromboxane Leukotriene Res.* **12:**45–49.

Foegh, M. L., Alijani, M. R., Helfrich, G. H., Khirabadi, B. S., Goldman, M. H., Lower, R. R., and Ramwell, P. W., 1985a, Thromboxane and leutrotrienes in clinical and experimental transplant rejection, *Adv. Prostaglandin Thromboxane Leukotriene Res.* **13:**209–217.

Foegh, M. L., Khirabadi, B., and Ramwell, P. W., 1985b, Prolongation of experimental cardiac allograft survival with thromboxane related drugs, *Transplantation* (in press).

Ford-Hutchinson, A., Bray, M., Doig, M., Shipley, M., and Smith, M., 1980, Leukotrine B; a potent chemokinetic and aggregating substance released from polymorphonuclear leukocytes, *Nature* **286:**264.

Gorman, R. R., Wienenga, W., and Miller, O. V., 1979, Independence of the cyclic AMP-lowering activity of thromboxane A₂ from platelet release reaction, *Biochim. Biophys. Acta* **572:**95–104.

Hirata, F., Schiffman, E., Venkatasubramanian, K., Salomon, D., and Axelrod, J., 1980, A phospholipase A₂ inhibitory protein in rabbit neutrophils induced by glucocorticoids, *Proc. Natl. Acad. Sci. U.S.A.* **77:**2533–2536.

Hirata, F., Carmine, R., Nelson, C. A., Axelrod, J., Shiffman, E., Warabi, A., Blas, A., Nirenberg, M., Manganiello, V., Vaughan, M., Kumagai, S., Green, I., Decker, J., and Steinberg, A., 1981, Presence of autoantibody for phospholipase inhibitory protein, lipomodulin, in patients with rheumatic disease, *Proc. Natl. Acad. Sci. U.S.A.* **78:**3190–3194.

Jaffe, B. M., Moore, T. C., and Vigran, T. S., 1975, Tissue levels of prostaglandin E following heterotopic rat heart allografting, *Surgery* **78:**481–484.

Jamieson, S. W., Burton, N., and Reitz, B., 1979, Platelets, sulfinpyrazone and organ graft rejection, in: *Cardiovascular Actions of Sulfinpyrazon: Basic and Clinical Research* (M. McCregor, J. Mustard, M. Oliver, and S. Sherry, eds.), Symposia Specialist, Miami, pp. 229–243.

Kawahara, Y., Yamarishi, J., Furuta, Y., Kaibuchi, K., Takai, Y., and Fukuzahi, H., 1983, Elevation of cytoplasmic free calcium concentration by stable thromboxane A_2 analogue in human platelets, *Biochem. Biophys. Res. Commun.* **117:**663–669.

Kelly, J. P., Johnson, M. C., and Parker, C. W., 1979, Effect of inhibition of arachidonic acid metabolism on mitogenesis in human lymphocytes: Possible role of thromboxanes and products of the lipoxygenase pathway, *J. Immunol.* **122:**1563–1571.

Kort, W. J., Bonta, I. L., Adolfs, M. J. P., and Westbroek, D. L., 1982, Synergism of (15 S-15-methylprostaglandin E_1 with either azathioprine or prednisolone on the survival of heart allografts in rats, *Prostaglandins Leukotrienes Med.* **8:**661–664.

Leithner, C., Sinzinger, H., and Schwarz, M., 1981, Treatment of chronic kidney transplant rejection with prostacyclin—reduction of platelet deposition in the transplant; prolongation of platelet survival and improvement of transplant function, *Prostaglandins* **22:**783–788.

Leithner, C. G., Sinzinger, H., Schwartz, M., Pohanka, E., and Syre, G. H., 1983, Increased deposition of indium-111 labelled platelets in chronically rejected kidney transplants, *Clin. Nephrol.* **18:**311–313.

Leung, K. H., and Minich, E., 1980, Prostaglandin modulation of development of cell-mediated immunity in culture, *Nature* **288:**597–600.

Maddox, Y., Shapiro, R., Goldman, M., Lower, R., and Ramwell, P., 1984, Contractile effects of leukotriene C_4 and the thromboxane mimic U46619 on human pulmonary arteries and veins *in vitro*, in: *Clinically Related Lipids*, Vol. 2 (T. J. Powles, R. S. Bochman, K. V. Honn, and P. Ramwell, eds.), Alan R. Liss, New York, pp. 47–65.

Maddox, Y., Cunard, C., Shapiro, R. Mitchell, G., Lower, R., and Ramwell, P., 1985, A comparison of the contractile responses of rodent and human pulmonary vascular segments to icosanoids, in this volume.

Mead, C. J., and Mertin, J., 1978, Fatty acids and immunity, *Adv. Lipid Res.* **16:**127–65.

Mertin, J., and Hunt, R., 1976, Influence of polyunsaturated fatty acids on survival of skin allografts and tumor incidence in mice, *Proc. Nat. Acad. Sci.* **3:**928–931.

Naccache, P. H., Sha'afi, R. I., Borgeat, P., Goetzl, E. J., 1981, Mono- and dihydroxy eicosatetraenoic acids alter calcium homeostasis in rabbit neutrophils, *J. Clin. Invest.* **67:**1584–1588.

Oluwole, S., Wang, T., Fawwaz, R., Sataki, K., Nowygrod, R., Reemtsme, K., and Hardy, M. A., 1981, Use of Indium-III-labelled cells in measurement, of cellular dynamics of experimental cardiac allograft rejection, *Transplant* **31:**51–55.

Payan, O. G., and Goetzl, E., 1983, Specific suppression of human T-lymphocyte function by leukotriene B_4, *J. Immunol.* **131:**551–553.

Quagliata, F., Lawrence, V. J. W., and Phillips-Quagliata, J. M., 1973, Prostaglandin E_1 as a regulator of lymphocyte function, *Cell Immunol.* **6:**457–65.

Rola-Pleszczynski, M., Borgeat, P., and Sirois, P., 1983, Leukotriene B_4 induces human suppressive lymphocytes, *Biochem. Biophys. Res. Commun.* **108:**1531–1537.

Shaw, J. F. L., 1983, Prolongation of rat cardiac allograft survival by treatment with prostacyclin or aspirin during acute rejection, *Transplant* **5:**526–529.

Shaw, J. F. L., 1984, Drugs affecting the prostaglandin synthetic pathway and rat heart allograft survival, in: *Platelets, Prostaglandins and Cardiovascular System*, Abstract, Florence, p. 130.

Sinzinger, H., Leithner, C., and Schwarz, M., 1981, Monitoring of human kidney transplants using quantification of autologous 111 indium-oxine platelet label deposition—beneficial effect of PGI_2-treatment in acute (AR) and chronic (CR) rejection, *Thomb. Haemostas.* **46:**263.

Smith, N., Chandler, S., Hawker, R. J., Hawke, L. M., and Barnes, A. D., 1979, Indium labelled autologous platelets as diagnostic aid after renal transplantation, *Lancet* **1:**1241–1242.

Strom, T. B., and Carpenter, C. B., 1983, Prostaglandin as an effective antirejection therapy in rat renal allograft recipients, *Transplant* **35:**279–281.

41

Prostaglandins in Human Seminal Fluid and Its Relation to Fertility

M. BYGDEMAN, E. BENDVOLD, C. GOTTLIEB, K. SVANBORG, and P. ENEROTH

1. INTRODUCTION

Human seminal fluid contains a number of prostaglandins (PGs) in large amounts. The first identified compounds were PGE_1, PGE_2, PGE_3, $PGF_{1\alpha}$, and $PGF_{2\alpha}$ (Samuelsson, 1963). More recently, 19-hydroxy-PGE_1, 19-hydroxy-PGE_2, 19-hy-droxy-$PGF_{1\alpha}$, and 19-hydroxy-$PGF_{2\alpha}$ have been identified. The 8β-isomers of these compounds are also present (Taylor, 1979; Taylor and Kelly, 1974, 1975). Although 50 years have now elapsed since von Euler (1936) first identified a new type of compound in human semen, which he named prostaglandin, the physio-logical function of the PGs in semen is still far from clarified. In this chapter, different possibilities that we have started to evaluate during the last years are discussed.

2. METHOD OF PROSTAGLANDIN MEASUREMENT

Different methods have been proposed to measure quantitatively the prostaglandins present in human seminal fluid. The one used in the present studies has been developed in our laboratory and allowed the measurement of the different groups of prostaglandins in individual semen samples. The method has been described in more detail previously (Svanborg *et al.*, 1982). Briefly, PGE and 19-hydroxy-PGE were converted to their respective prostaglandin B (PGB) compounds

M. BYGDEMAN, E. BENDVOLD, C. GOTTLIEB, K. SVANBORG, and P. ENEROTH ● Department of Obstetrics and Gynecology, Karolinska Hospital, S-104 01 Stockholm, Sweden.

and were quantified by their ultraviolet absorption after purification with two chromatography steps; PGF and 19-hydroxy-PGF were determined by gas chromatography, which also allowed measurement of the 8β-isomers. Labeled PGE_1 and $PGF_{1\alpha}$ were used for correction of losses during the procedure. Determination of samples in duplicate showed a precision of 5.6% (range 0.9–9.1%). The recovery of added PG was on average 99.2% (range 89.9–107.1%). Semen samples were obtained by masturbation from fertile men who had recently proven their fertility and from men in infertile marriages.

The PGEs and especially the 19-hydroxy-PGEs were rapidly degraded during storage. Within a few hours, a substantial part seemed to be converted to the corresponding PGB compounds. Since semen coagulation normally delays dividing the samples into fractions for 30 to 60 min, an analysis of PG concentration and sperm characteristics in the same ejaculate has been difficult. In a recent study (Bendvold *et al.*, 1984), semen coagulation was prevented by addition of phenylmethylsulfonylfluoride, and the semen samples were divided into fractions immediately following ejaculation. The results showed that neither addition of the protease inhibitor nor the spontaneous coagulation significantly changed the PG concentration. It was also demonstrated that with the method used for the analysis of the PG concentration, storage of the semen for up to a few hours did not influence the results; therefore, dividing the semen could await spontaneous liquefaction. The main reason was that in the method used all PGEs and 19-hydroxy-PGEs are converted to corresponding PGB compounds by alkali treatment, and the spontaneous conversion taking place during short-term storage (1 to 3 hr) was therefore of minor importance.

3. CONCENTRATION OF PROSTAGLANDINS IN HUMAN SEMINAL FLUID

The total prostaglandin concentration in semen from fertile men was very high in comparison with other body fluids. One ejaculate contained approximately 1.0 mg of the different prostaglandins together. The concentration varied considerably between individuals but was approximately 65 µg/ml for the PGEs, 2.8 µg/ml for the PGFs, 320 µg/ml for the 19-hydroxy-PGEs, and 15 µg/ml for the 19-hydroxy-PGFs (Svanborg *et al.*, 1982). Similar figures have been reported by Templeton *et al.* (1978). Semen samples from men in infertile marriages contained, in general, lower concentrations of PGEs and 19-hydroxy-PGEs but higher concentrations of 19-hydroxy-PGFs (Table I; Svanborg *et al.*, 1982).

The concentrations of different prostaglandins varied considerably between different individuals. The variation with time in the same individual was, however, less marked, as illustrated in a recent study in which nine healthy men delivered repeated semen samples with different time intervals for up to almost 1 year (Bendvold *et al.*, 1984a). This finding is of importance because it makes possible studies of the influence of different treatments on PG concentration in semen. A

TABLE I. Prostaglandin Concentration in Human Seminal Plasma[a]

Groups of men	Mean value (μg/ml) and range			
	PGE	PGF	19-OH-PGE	19-OH-PGF
Infertile, unselected	43	2.6	260	17.0
	(7–117)	(1.4–4.8)	(113–427)	(9.4–34.1)
	$n = 17$	$n = 11$	$n = 19$	$n = 10$
Fertile	62	2.8	326	14.9
	(15–144)	(1.0–5.3)	(155–638)	(7.0–20.6)
	$n = 10$	$n = 8$	$n = 10$	$n = 8$

[a] From Svanborg *et al.* (1982).

prerequisite is, however, that sperm characteristics remain unchanged, because a relationship between PG concentration and sperm density and sperm motility can be demonstrated. Spontaneous or induced variations in sperm characteristics may therefore result in variations in PG concentration and be misjudged as being a direct result of treatment.

4. ORIGIN OF PROSTAGLANDINS IN SEMEN

Some investigators have reported at least minor concentrations of PGE and PGF in the prostate gland and the testis. Incubating eicosatrienoic acid with homogenates of human seminal vesicles, prostate gland, and testis, however, Hamberg (1976) demonstrated that only the seminal vesicles were capable of converting the acid into PGE_1 in significant amounts.

Eliasson (1959), using the split ejaculate technique, found that the highest prostaglandin concentration measured by bioassay was found in the same fraction as fructose and therefore proposed that they originated from the seminal vesicles. Information regarding the production site of especially 19-hydroxy-PGE and 19-hydroxy-PGF is, however, scarce. To evaluate the origin of seminal prostaglandins, genital tract fluids were analyzed for content of prostaglandins in two clinical situations (Bendvold *et al.*, 1985). Six patients delivered semen samples before and after vasectomy. In these patients, the prostaglandin concentration remained essentially unchanged although sperm density decreased to zero, indicating a successful surgery. Similar results were recently reported by Gerozissis *et al.* (1982) using radioimmunoassay for the prostaglandin measurement. These authors found comparable prostaglandin concentrations in semen obtained 1 year after vasectomy with that in ejaculates from a group of fertile men.

In another patient a silicone prosthesis was introduced into the straight part of the ductus deferens, and secretory products from the testis and epididymis and the ejaculate representing mainly the secretion from the seminal vesicles and the prostate gland could be separately collected. The secretion from the testis and epididymis contained no detectable amounts of prostaglandins, whereas in the ejaculate from the same patient the concentration was within normal limits. The results of these studies show that the testis and the epididymis do not contribute significantly to the content of prostaglandins in human seminal fluid and support the assumption that the seminal vesicles are the main source of all prostaglandins.

5. REGULATION OF PROSTAGLANDIN PRODUCTION

Very little is known about the factors regulating the concentration of prostaglandins in human seminal plasma. Sturde (1971) has reported that androgen treatment for 6 to 12 weeks in men resulted in a large increase in prostaglandin concentration measured by bioassay. A rapid increase in 19-hydroxy-PGE production was reported by Skakkebaeck *et al.* (1976) in castrated or hypogonadal men after testosterone replacement therapy. The effect on 19-hydroxy-PGE concentration was more marked than that on PGE concentration.

Previous studies have shown that treatment with aspirin, a weak prostaglandin biosynthesis inhibitor, will result in a reduction of PGE and PGF concentrations in human semen measured by bioassay or physicochemical methods (Collier and Flower, 1971; Horton *et al.*, 1973). We have recently evaluated the effect of the strong biosynthesis inhibitor naproxen on the concentration of seminal prostaglandins as well as on traditional sperm parameters. Six patients delivered semen samples before, during, and after a 14-day period of medication with naproxen at a dosage of 250 mg three times a day. The results showed a significant decrease in the concentrations of all four main groups of prostaglandins through the period of medication. After 2 days of medication, the level of PGE was reduced to 22%, 19-hydroxy-PGE to 39%, PGF to 14%, and 19-hydroxy-PGF to 58% of the initial concentration. Corresponding levels of reduction after 7 days of medication were between 12 and 31% of the initial concentration. Measurement of prostaglandin concentration 1 week after cessation of medication revealed values close to initial concentrations. No changes in sperm density, motility, etc. were observed. Whether the treatment has any effect on fertility remains to be evaluated (Fig. 1).

6. PROSTAGLANDINS AND MALE FERTILITY

The high concentration of prostaglandins in the seminal fluid of men compared with that in most tissues and body fluids has led to the logical assumption that these compounds are intimately involved in reproductive processes. Early results based on bioassays have shown a positive relationship between prostaglandin concentration and the ability of the couple to affect conception. These results were later confirmed by quantitative chemical methods. Human males who were infertile for no apparent reason possessed significantly lower concentrations of seminal prostaglandins, especially PGEs and 19-hydroxy-PGEs, than men with normal fertility (Svanborg *et*

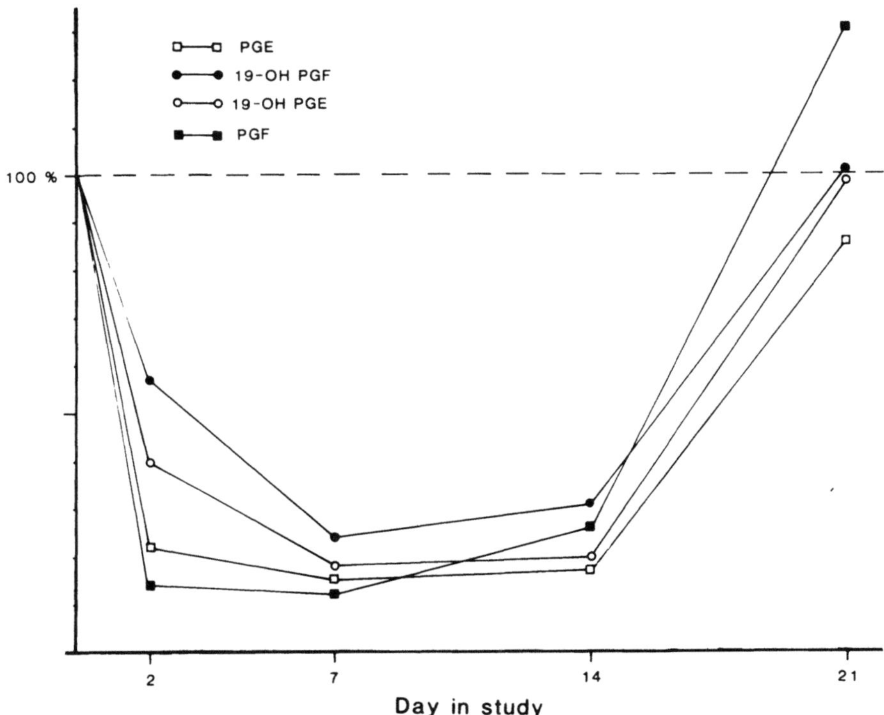

FIGURE 1. The concentration of the four main groups of protaglandins before, during, and after treatment with naproxen. Each volunteer received 250 mg three times daily for 14 days. The treatment started on day 0 after the first semen sample was collected. Mean values from six patients. the initial value is defined as 100%.

al., 1982). Bygdeman *et al.* (1970) showed that seminal fluid from fertile men contained about 55 μg total PGE/ml, an amount that was significantly higher than the corresponding value of 18 μg/ml in semen of men from infertile marriages where no other abnormalities could be detected.

Evidence for a correlation between fertility and seminal prostaglandin concentration has also been found in sheep. Following artificial insemination, the fertility of rams increased by more than 15% if PGE_2 and $PGF_{2\alpha}$ were added to the diluted ram semen in amounts comparable with the total amount of prostaglandins in one ejaculate (Dimov and Georgiew, 1977).

The precise mechanism by which seminal prostaglandins influence fertility is unknown. In fact, the correlation need not be one of cause and effect; the infertility may be caused by some other factors that also produce low prostaglandin levels in semen. If the high prostaglandin concentration in the human seminal fluid were of importance for normal fertility, at least three modes of action might be suggested. The prostaglandins may act on the spermatozoa themselves, on the male reproductive tract at or shortly before ejaculation, or on the female reproductive tract after ejaculation.

6.1. Relation between PG Concentration and Sperm Characteristics

In a number of ejaculates, both PG concentration and sperm qualities were analyzed. A correlation was found between the density of sperm and the PG concentration. When the number of spermatozoa was >250 million/ml, the mean content of PGE was low, 17 μg/ml, whereas if the sperm density was normal (20–250 million/ml), the mean PGE concentration was 69 μg/ml. The corresponding values for 19-hydroxy-PGE were 153 and 262 μg/ml, respectively (Bendvold *et al.*, 1984). Both differences were statistically significant. Similar results have also been reported by Kelly *et al.* (1979). The opposite relation may also be present. In some patients, oligozoospermia was associated with a high PG concentration. It is interesting to speculate that a low sperm density is compensated for by a high PG concentration, which may, for instance, facilitate sperm penetration through the cervical canal.

A relationship between sperm motility and PG concentration could also be found (Table II). This was especially true for 19-hydroxy-PGE and 19-hydroxy-PGF. The concentration of 19-hydroxy-PGE tended to be higher and that of 19-hydroxy-PGF to be lower in ejaculates with a normal sperm motility. The ratio between the two groups of PGs was significantly higher if sperm motility was normal. Addition of 19-hydroxy-PGF to semen samples with normal sperm motility also resulted in a decreased activity of the sperm and reduced their capacity to penetrate cervical mucus *in vitro*. This was not the case for PGE_1 or $PGF_{2\alpha}$ (Table III). The results indicate that asthenozoospermia could be the result of spontaneous or induced changes in the semen PG concentration. It is also tempting to speculate that the concentration of the 19-hydroxy compounds in semen is of importance for sperm penetration under *in vivo* conditions.

6.2. Regulation of Ejaculation

Von Euler postulated in 1936 that seminal PGs, through their smooth muscle stimulatory and vasodilating actions, could be of importance for the emptying of

TABLE II. Relation between Prostaglandin Concentration and Sperm Motility[a]

No. of patients	Sperm motility	PGE (μg/ml)	19-OH-PGE (μg/ml)	19-OH-PGF (μg/ml)	$\dfrac{\text{19-OH-PGE}}{\text{19-OH-PGF}}$
15	<2	71 ± 37	228 ± 81	19.7 ± 7	12 ± 3
13	3	70 ± 35	$289 \pm 76*$	$10.9 \pm 3**$	$27 \pm 6***$

[a] Mean values $\pm$ standard deviation; statistical significance from *t*-test: $*P < 0.05$; $**P < 0.01$; $***P < 0.001$. From Bendvold *et al.* (1984).

TABLE III. Outcome of *in Vitro* Penetration Tests
(Kremer Test) Before and After Addition of Different PGs to the Semen Sample
(Mean Values of Three Experiments)

Compound	Amount added (μg)	Outcome of penetration tests[a]	
		Before	After addition
PGE$_1$	104	7.7	7.6
PGF$_{1\alpha}$	2	7	7
19-OH-PGF	13	7	5
19-OH-PGF	50–70	7	1.3

[a] The outcome of the test was graded in points between 0 and 9.

the genital glands and in that way regulate the ejaculatory process. This possibility
has not been studied further since then. Preliminary results from an ongoing study
supported this assumption. If male volunteers were treated with oral PGE$_2$ (1 mg),
sperm density decreased to approximately 50% (mean value) of the initial level 12
to 24 hr after treatment. The number of spermatozoa returned rapidly to normal.
In control experiments without treatment but with equally frequent semen sampling,
a reduction in sperm count was not observed.

6.3. Passive Sperm Transport

It is possible that changes in uterine contractility during intercourse may facilitate sperm transport. During sexual stimulation there is a marked increase in
uterine activity that changes to inhibition following female orgasm. A pressure
gradient between the vagina and the uterus may be the result, which would favor
a passive sperm transport mechanism. These changes in uterine and vaginal contractility may be caused by seminal prostaglandins, but experimental evidence to
support this assumption is still lacking in the human. Evidence that PGE$_1$, PGE$_2$,
and PGF$_{2\alpha}$ enhance sperm migration and hence fertilization has, however, been
provided in experiments in the rabbit (Spilman *et al.*, 1973).

7. SUMMARY AND CONCLUSIONS

Human seminal fluid contains a number of prostaglandins: PGE$_1$, PGE$_2$, PGE$_3$,
PGF$_{1\alpha}$, PGF$_{2\alpha}$, 19-hydroxy-PGE$_1$, 19-hydroxy-PGE$_2$, 19-hydroxy-PGF$_{1\alpha}$, 19-hy-
droxy-PGF$_{2\alpha}$, and their 8β-isomers in large amounts. The total PG content in one
ejaculate was approximately 1 mg. The concentrations of these compounds varied
considerably among individuals. The variation in concentration in the same individual was, however, less marked provided that sperm characteristics remained
constant. All of the PGs most likely originate from the seminal vesicles. Any
addition from the testis or epididymis was minimal. Treatment with a prostaglandin

biosynthesis inhibitor significantly reduced the semen PG concentration. The concentration of the PGEs, 19-hydroxy-PGEs, and 19-hydroxy-PGFs seemed related to fertility. A correlation between PG concentration and sperm density, sperm motility, and sperm penetration capacity in cervical mucus *in vitro* could be demonstrated. It is tempting to suggest that functional requisites of sperm depend on an optimal seminal PG concentration. It is also possible that the PGs are of importance for the ejaculatory process.

ACKNOWLEDGMENTS. These studies were supported by the Swedish Research Council project B84-17X-05696-05. We are grateful to Astrid Häggblad for skillful typing of the manuscript.

REFERENCES

Bendvold, E., Svanborg, K., Eneroth, P., Gottlieb, C., and Bygdeman, M., 1984, The natural variations in prostaglandin concentration in human seminal fluid and its relation to sperm quality, *Fertil. Steril.* **41**:743–748.

Bendvold, E., Svanborg, K., Bygdeman, M., and Norén, S., 1985, On the origin of prostaglandins in human seminal fluid, *Int. J. Androl.* **8**:37–43.

Bygdeman, M., Fredricsson, B., Svanborg, K., and Samuelsson, B., 1970, The relation between fertility and prostaglandin content of seminal fluid in man, *Fertil. Steril.* **26**:622–629.

Collier, J. G., and Flower, R. J., 1971, Effect of aspirin on human seminal prostaglandins, *Lancet* **2**:852.

Dimov, V., and Georgiew, G., 1977, Ram semen concentration and its effect on fertility, *J. Anim. Sci.* **44**:1050–1054.

Eliasson, R., 1959, Studies on prostaglandins, *Acta Physiol. Scand. [Suppl.]* **158**:1–73.

Gerozissis, K., Jouannet, P., Soufir, J. C., and Dray, F., 1982, Origin of prostaglandins in human semen, *J. Reprod. Fertil.* **65**:401–406.

Hamberg, M., 1976, Biosynthesis of prostaglandin E_1 by human seminal vesicles, *Lipids* **11**:249–250.

Horton, E. W., Jones, R. L., and Marr, C. G., 1973, Effects of aspirin on prostaglandin and fructose levels in human semen, *J. Reprod. Fertil.* **33**:385–393.

Kelly, R. W., Cooper, I., and Templeton, A. A., 1979, Reduced prostaglandin levels in the semen of men with very high sperm concentration, *J. Reprod. Fertil.* **56**:195–199.

Samuelsson, B., 1963, Isolation and identification of prostaglandins from human seminal plasma, *J. Biol. Chem.* **238**:3229–3234.

Skakkebaeck, N. E., Kelly, R. W., and Cocker, C. S., 1976, Prostaglandin concentrations in the semen of hypogonadal men during treatment with testosterone, *J. Reprod. Fertil.* **47**:119–121.

Spilman, C. H., Finn, A. E., and Norland, J. F., 1973, Effect of prostaglandins on sperm transport and fertilization in the rabbit, *Prostaglandins* **4**:57–64.

Sturde, H. C., 1971, Behaviour of sperm prostaglandins under therapy with androgen, *Arzneim. Forsch.* **21**:1302–1307.

Svanborg, K., Bygdeman, M., Eneroth, P., and Bendvold, E., 1982, Quantification of prostaglandins in human seminal fluid, *Prostaglandins* **24**:363–375.

Taylor, P. L., 1979, The 8-iso prostaglandins: Evidence for eight components in human semen, *Prostaglandins* **17**:259–267.

Taylor, P. L., and Kelly, R. W., 1974, 19-Hydroxylated E prostaglandins as the major prostaglandins in human semen, *Nature* **250**:665–667.

Taylor, P. L., and Kelly, R. W., 1975, The occurrence of 19-hydroxy-F prostaglandins in human semen, *FEBS Lett.* **57**:22–25.

Templeton, A. A., Cooper, I., and Kelly, R. W., 1978, Prostaglandin concentration in the semen of fertile men, *J. Reprod. Fertil.* **52**:147–150.

von Euler, U. S., 1936, On the specific vasodilating and plain muscle stimulating substances from accessory genital glands in man and certain animals (prostaglandin and vesiglandin), *J. Physiol. (Lond.)* **88**:213–234.

42

Arachidonate Lipoxygenase Products and Psoriasis

RICHARD D. CAMP, MALCOLM W. GREAVES, and ROBERT M. BARR

1. INTRODUCTION

Psoriasis is a common skin disease that, in many cases, appears to be genetically determined. Lesional skin is characterized by epidermal proliferation and inflammatory changes, of which intraepidermal neutrophil infiltration is a consistent finding and one of the earliest abnormalities seen in developing lesions (Chowaniec *et al.*, 1981). Psoriasis may manifest clinically in various forms, of which the chronic, stable, scaly plaque type is the most common. However, the inflammatory changes predominate in some patients, and in generalized pustular psoriasis, an uncommon form of the disease, intraepidermal neutrophil microabscesses are the major pathological feature (Baker and Wilkinson, 1979). These findings suggest that the production of neutrophil chemoattractants by the epidermis may be important in the pathogenesis of psoriasis.

2. NEUTROPHIL CHEMOATTRACTANTS IN PSORIASIS

A chemotactic substance with some of the properties of the complement fragment C5a has been found in psoriatic scale (Tagami and Ofuji, 1976, 1977; Tagami *et al.*, 1982), although others have suggested that this material might be epidermal-cell-derived thymocyte-activating factor (ETAF), which is believed to be identical

RICHARD D. CAMP, MALCOLM W. GREAVES, and ROBERT M. BARR ● Wellcome Laboratories for Skin Pharmacology, Institute of Dermatology, Homerton Grove, London E9 6BX, United Kingdom.

to interleukin 1 and which has neutrophil chemoattractant properties (Luger *et al.*, 1983).

The first demonstration of abnormal arachidonic acid metabolism in psoriasis was by Hammarström *et al.* (1975), who used gas chromatography–mass spectrometry (GC–MS) to measure arachidonic acid, 12-L-hydroxy-5,8,10,14-eicosatetraenoic acid (12-HETE), and prostaglandin levels in purified extracts of keratome slices from lesional and uninvolved psoriatic skin. Greatly elevated levels of unesterified arachidonic acid and 12-HETE were found in lesional as compared to uninvolved skin. Subsequently, the same group showed that topical glucocorticoid application was capable of lowering the levels of free arachidonic acid and 12-HETE in lesional skin before clinical improvement had taken place, and they proposed that the reduction of chemoattractant 12-HETE might partly explain the therapeutic effect of topical glucocorticoids in psoriasis (Hammarström *et al.*, 1977).

In recent pharmacological studies of psoriasis, a skin chamber sampling method has been developed. This method is less invasive than keratome slicing and excludes superficial psoriatic scale from the sample, the possible advantage of which is discussed below. Briefly, skin is abraded with a scalpel blade until a smooth surface with minimal punctate bleeding is obtained. A cylindrical plastic chamber is fixed to the abraded skin with cyanoacrylate adhesive, and phosphate-buffered saline is added. Acidic lipid extracts of the buffered saline, removed after different intervals, were assayed by an agarose microdroplet leukocyte chemokinesis assay (Smith and Walker, 1980). Samples from lesional psoriatic skin contained significantly greater leukocyte chemokinetic activity than those from the uninvolved skin of psoriatics or the skin of normal volunteers (Brain *et al.*, 1982a). Purification by reversed- and straight-phase high-performance liquid chromatography (HPLC) and assay of the evaporated fractions by the agarose microdroplet method indicated that the chemokinetically active material in acidic lipid extracts of lesional chamber fluid was a mixture of leukotriene B_4 (LTB_4) and monoHETE-like compounds. Leukocyte counting prior to extraction showed that there were few or no leukocytes in the chamber fluid samples. When they were present, there was no correlation between leukocyte numbers and chemokinetic activity. This indicated that the results obtained were not caused by eicosanoid synthesis by cells in the chamber fluid (Brain *et al.*, 1983, 1984a).

Acidic lipid extracts of the surface scale obtained by abrasion of psoriatic lesions were also shown to contain biologically active amounts of LTB_4 and monoHETE-like material when analyzed by a combination of HPLC and the assay for chemokinetic activity (Camp *et al.*, 1983a; Brain *et al.*, 1984b). Lipid extracts of scale and chamber fluid from psoriatic lesions were purified by straight-phase HPLC, and appropriate fractions were derivatized and analyzed by GC–MS. Conclusive identification of 8-HETE, 9-HETE, 11-HETE, 12-HETE, and 15-HETE in purified scale extracts was achieved by obtaining the full mass spectrum of each compound. Semiquantitative analysis of purified scale and chamber fluid extracts by selected ion-monitoring GC–MS without internal standards indicated that 12-HETE was the most abundant arachidonic acid metabolite and that its levels were higher in chamber fluid from lesional as compared to uninvolved skin (Camp *et al.*, 1983a). This lipoxygenase metabolite probably accounts for the major part of the monoHETE-like neutrophil chemokinetic activity found in the chamber fluid and scale extracts. The other monoHETE compounds appeared to be present in amounts too low to contribute to the monoHETE-like neutrophil chemokinetic activity seen. Of these compounds, 5-HETE was found in the lowest amounts, although it proved the most difficult to assay.

The most abundant monohydroxy fatty acid identified in psoriatic lesions was 13-hydroxy-9,10-octadecadienoic acid (13-HODD), a lipoxygenase metabolite of linoleic acid, but it was found to have no chemokinetic activity when tested in the agarose microdroplet assay (Camp *et al.*, 1983a). In view of reports that 15-HETE reduces 5-lipoxygenase activity *in vitro* (Vanderhoek *et al.*, 1980) and the fact that 13-HODD has the same ω configuration as 15-HETE, it was considered that 13-HODD might have 5-lipoxygenase-inhibiting properties. Subsequently, it was shown that 13-HODD inhibited the release of LTB$_4$ from calcium-ionophore-stimulated human leukocyte suspensions, although it was less potent that 15-HETE. It has been proposed that 13-HODD is an endogenous modulator of 5-lipoxygenase activity in human skin (Fincham and Camp, 1984).

Quantitative assay of arachidonic acid and 12-HETE in chamber fluid from lesional and uninvolved psoriatic skin has been done by GC–MS with the use of deuterated internal standards (Barr *et al.*, 1984). There were approximately three- to fourfold elevations of arachidonic acid and 12-HETE levels in lesional chamber fluid as compared to that from uninvolved skin. These are considerably lower elevations than those reported by Hammarström *et al.* (1975), who found approximately 25-fold and 80-fold increases in arachidonic acid and 12-HETE, respectively, in lesional compared with uninvolved skin. The sampling methods used are the likely explanation for these differences. In the chamber fluid method, the surface scale is removed, and the abraded surface is washed prior to collection of the final chamber fluid samples. It is likely that the levels of compounds measured in such chamber fluid reflect their immediate release from the deeper, viable layers of the skin. When keratome slices are taken from psoriatic lesions, as in the report of Hammarström *et al.* (1975), the samples contain large amounts of surface scale, whereas samples from clinically normal skin contain much less scale. It has been shown that psoriatic scale contains biologically active amounts of arachidonate lipoxygenase products (Camp *et al.*, 1983a), and it is possible that this lipid-rich surface layer acts as a "sump" for fatty acid metabolites released from deeper layers. Therefore, sampling by keratome slicing may be expected to show greater differences between eicosanoid levels in lesional and uninvolved psoriatic skin than the chamber fluid method, but the latter method may reflect more rapidly changes in lipoxygenase product levels following, for example, drug therapy.

A new quantitative GC–MS method for the assay of monohydroxy fatty acids apart from 5-HETE has now been developed and is being applied to clinical samples. The method involves group purification of hydroxy fatty acid derivatives by HPLC, followed by selected ion-monitoring capillary GC–MS with hydrogen as mobile phase, and postcolumn, in-line reduction of separated derivatives. This method enables simultaneous profiling and quantification of monohydroxy metabolites of a parent unsaturated fatty acid and is being used to define more precisely the activity and specifity of the lipoxygenase enzymes in psoriatic skin (Woollard and Mallet, 1984). It has as yet not been possible to quantify the LTB_4-like material in chamber fluid and small scale samples from psoriatics by GC–MS, largely because of the small amounts of material present. Semiquantitative assays for LTB_4 are carried out on HPLC-purified samples by the agarose microdroplet technique, which usually has an absolute lower limit of detection for LTB_4 of 3 pg.

The release of the sulfidopeptide leukotrienes in psoriatic skin lesions has also been investigated by radioimmunoassay. The antiserum, which was raised against LTC_4, reacted with both LTC_4 and LTD_4 but was otherwise highly selective. Chamber fluid from abraded, washed lesional skin contained significantly higher levels of LTC_4-immunoreactive material than chamber fluid from uninvolved skin. When pooled lesional chamber fluid samples were purified on C_{18} "Sep-Pak" cartridges and by reversed-phase HPLC, 70% of the immunoreactive material cochromatographed with LTC_4, and 20% with LTD_4 (Brain *et al.*, 1984c). Similar results were obtained when psoriatic scale homogenates were purified by "Sep-Pak" and HPLC (Brain *et al.*, 1985).

3. PROINFLAMMATORY PROPERTIES OF ARACHIDONATE LIPOXYGENASE PRODUCTS IN HUMAN SKIN

Intradermal injection of 50–500 ng LTB_4 in normal volunteers produced ill-defined indurated areas, the diameters of which were not dose related. Histological examination of biopsies taken after 4 hr showed perivascular neutrophil and mononuclear cell infiltrates with a few eosinophils but no epidermal abnormality (Camp *et al.*, 1983b). It was considered that the lipid nature of LTB_4 might allow its percutaneous absorption following topical application and that the consequent high epidermal concentration of LTB_4 might result in intraepidermal neutrophil infiltrates, mimicking the inflammatory events in psoriatic skin lesions. Leukotriene B_4 was applied to forearm skin of normal volunteers in ethanol, which was evaporated, and the residue was occluded for 6 hr. Single doses of LTB_4 from 5 to 500 ng produced erythema and swelling, the diameter of which was dose related. The reactions appeared between 12 and 24 hr, were maximal at 20 to 48 hr, and persisted for up to 7 days, leaving variable brownish pigmentation and slight scaling. Histological examination of biopsies taken 24 hr after application of 100 ng LTB_4 showed intraepidermal vacuoles of varying size, filled with neutrophils, below the stratum corneum. These changes are similar to those found in pustular psoriasis except that in the LTB_4 reaction, unlike psoriasis, there were no intraepidermal lymphocytes and no significant epidermal edema around the microabscesses. After 24 hr, neutrophils were no longer a feature, perivascular mononuclear cell infiltrates being the main change. There was no significant epidermal hyperplasia as is found in psoriasis, nor did any volunteer develop clinical evidence of a psoriatic lesion (Camp *et al.*, 1984).

Further work has determined the effects of single and multiple applications of LTB_4 to the skin of normal volunteers and to the uninvolved skin of psoriatics. In both groups, multiple LTB_4 applications at 24-hr intervals for 9 days to the same skin site were associated unexpectedly with a reduction of the visible inflammatory reaction after three or four applications. Histologically, there were significantly fewer intraepidermal neutrophils in skin sites to which 50 ng LTB_4 had been applied daily for 9 days than in fresh sites to which a single 50-ng dose had been applied at the end of the multiple-dose experiments. This ability to elicit a florid reaction to a single dose in a previously untreated site at a time when a state of tolerance has been induced in an adjacent multiple-dose site indicates that the tolerance is caused by a local mechanism in the repeatedly treated skin and not by desensitization of circulating neutrophils. The nature of this local "desensitizing" mechanism remains to be determined. No subject, either psoriatic or normal, developed clinical evidence of a psoriatic lesion at sites of single or multiple applications of LTB_4. These findings have led to the conclusion that, in the subjects studied, factors other than the release of LTB_4 alone are needed in the pathogenesis of psoriatic lesions (Wong *et al.*, 1985).

Intradermal infusion of microgram amounts of 12-HETE in normal human volunteers produced significant accumulations of neutrophils and mononuclear cells in the dermis but no significant epidermal changes (Dowd *et al.*, 1983). Topical application of 200 ng to 50 μg 12-HETE to localized areas of the forearms of normal volunteers followed by occlusion for 6 hr produced erythema, the diameter of which was dose related. The erythema was visible at 6 hr but had usually disappeared by 24 hr, being of much shorter duration, with less swelling, than the topical LTB_4 reactions. Histological examination of biopsies 6 hr after application of 20 μg 12-HETE showed mixed mononuclear and neutrophil infiltrates in the dermis, with accumulations of neutrophils in the epidermis in four out of ten volunteers. At 24 hr, all subjects biopsied showed intraepidermal neutrophils (Dowd *et al.*, 1984). To our knowledge, these are the first demonstrations of inflammatory responses to 12-HETE in human subjects. Work on the effects of 12-HETE in psoriatic subjects and on the effects of multiple applications is in progress.

The proinflammatory effects of LTC_4 and LTD_4 have also been determined in human skin by intradermal injection (Camp *et al.*, 1983b). These two leukotrienes caused dose-related erythema and whealing with amounts of 0.012–0.38 nmol, but further dose increases up to 7.5 nmol caused no further increases in the magnitude of the inflammatory reactions. No significant clinical evidence of vasoconstriction was observed, in contrast to the vasoconstrictor effects of these leukotrienes in guinea pig skin (Peck *et al.*, 1981). Soter *et al.* (1983) report the presence of pallor in the center of wheals caused by injection of LTC_4 and LTD_4, but, as discussed by Camp *et al.* (1983b), it is possible that this pallor is simply a function of the degree of edema and is created by extrinsic compression of the vasculature by edema fluid.

4. THE EFFECTS OF DRUGS ON ARACHIDONATE LIPOXYGENASE PRODUCTS IN PSORIASIS

The effects of treatment of patients with chronic plaque psoriasis with the aromatic retinoid etretinate have been determined (Wong *et al.*, 1983). Chamber fluid samples before and at intervals after oral treatment with etretinate were purified by HPLC and assayed for arachidonic acid and 12-HETE by quantitative GC–MS and for LTB_4-like material by the agarose microdroplet chemokinesis method. Over a period of 6 weeks' treatment, there was moderate clinical improvement, evident mainly as a reduction in scaling, but not clearance of the patients' psoriatic lesions. Chamber fluid samples taken from lesions during this 6-week period showed a decline in arachidonic acid and 12-HETE levels, whereas levels of these compounds in uninvolved skin were not affected. It was not possible to determine whether the reductions in arachidonic acid and 12-HETE were caused by a direct effect of etretinate or its metabolites on arachidonic acid release and oxidation or were secondary to the partial clearance of the psoriatic lesions. However, it was of interest that the levels of LTB_4-like material in the lesional chamber fluid did not decline during the period of treatment in spite of *in vitro* evidence that the carboxy metabolite of etretinate is a 5-lipoxygenase inhibitor (Bray, 1984).

It is proposed to determine the effects of specific lipoxygenase inhibitors in psoriasis if they become available. A study of the effects of topically applied 15-HETE is under way.

5. CONCLUSIONS

These results show that arachidonic acid, LTB_4-like material, and 12-HETE are present in increased amounts in psoriatic lesions relative to uninvolved skin. The cell of origin of LTB_4 is not certain. Evidence for the release of LTB_4-like material by calcium-ionophore-stimulated cultured human keratinocytes has been presented (Brain *et al.*, 1982b), and this finding appears to be supported by a recent report of LTB_4 release by freshly suspended trypsinized human keratinocytes (Grabbe *et al.*, 1984). Thus, it is possible that the LTB_4-like material identified in psoriasis is released by epidermal cells, but it is also possible that this material is produced by neutrophils after their arrival in the skin.

There is also evidence for the production of 12-HETE by epidermal cells (Hammarström *et al.*, 1979). In addition, chamber fluid from abraded, uninvolved psoriatic skin contained 12-HETE levels measurable by GC–MS (Camp *et al.*, 1983a; Barr *et al.*, 1984), showing that skin is able to produce this compound *in vivo* in the absence of inflammatory cells, partly as a result of the trauma of the sampling method. This contrasts with the findings on measurement of LTB_4, which was undetectable in the majority of chamber fluid samples from abraded, uninvolved psoriatic skin (Brain *et al.*, 1984a).

The major source of the elevated levels of arachidonic acid in psoriatic lesions has not been determined but is likely to be the epidermis. In psoriatic skin, arachidonic acid appears to be metabolized more actively by the 12-lipoxygenase pathway than by the cyclooxygenase pathway. Where prostaglandin and 12-HETE measurements have been made in the same psoriatic samples, the elevation of 12-HETE in lesional compared to uninvolved skin was greater than that of the prostaglandin. For example, Hammarström *et al.* (1975) reported approximately 80-fold increases in 12-HETE levels in lesional compared to uninvolved skin, whereas prostaglandin E_2 was elevated only 1.5-fold. Barr *et al.* (1984) reported three- to fourfold elevations of 12-HETE in chamber fluid from lesional as compared to uninvolved psoriatic skin, whereas prostaglandin E_2 levels were not elevated.

The inflammatory changes found on topical application of LTB_4 to human skin appeared to provide further important evidence for the role of LTB_4 in the pathogenesis of the inflammatory changes in psoriasis. However, the tolerance that developed after multiple applications, both in psoriatic and normal volunteers, suggests that LTB_4 is not capable of maintaining tissue neutrophil infiltrates, at least in the skin. Its inability to elicit psoriatic lesions after single or multiple topical applications to uninvolved psoriatic skin indicates, at least in the patients studied, that the release of LTB_4 alone, in the epidermis, is not an initiating event in the development of psoriatic lesions. Nevertheless, the release of LTB_4 and possibly 12-HETE might be two of several processes that, together, are needed for the initiation of a lesion, and the inhibition of their syntheses by a drug may therefore have therapeutic value. The therapeutic efficacy of benoxaprofen in psoriasis (Allen and Littlewood, 1982; Kragballe and Herlin, 1983) is an important precedent, but whether this effect was a result of lipoxygenase-inhibiting properties is now open to question in view of doubts about the potency of benoxaprofen in inhibiting 5-lipoxygenase (Masters and McMillan, 1984).

The presence of LTC_4- and LTD_4-immunoreactive material in psoriatic lesions and the evidence that they produce vasodilation in human skin (Camp *et al.*, 1983b; Bisgaard *et al.*, 1982) suggest that these compounds may play a role in mediating the erythema and increased blood flow (Klemp and Staberg, 1983) that characterize psoriatic lesions. The source of the sulfidopeptide leukotrienes in psoriatic lesions has not been established.

ACKNOWLEDGMENTS. The work carried out at the Institute of Dermatology was supported by grants from the Medical Research Council and the Wellcome Trust.

REFERENCES

Allen, B. R., and Littlewood, S. M., 1982, Benoxaprofen: Effect on cutaneous lesions in psoriasis, *Br. Med. J.* **285**:1241.

Baker, H., and Wilkinson, D. S., 1979, Psoriasis, in: *Textbook of Dermatology*, Vol. II (A. Rook, D. S. Wilkinson, and F. J. G. Ebling, eds.), Blackwell, Oxford, pp. 1315–1359.

Barr, R. M., Wong, E., Mallet, A. I., Olins, L. A., and Greaves, M. W., 1984, The analysis of arachidonic acid metabolites in normal, uninvolved and lesional psoriatic skin, *Prostaglandins* **28:** 57–66.

Bisgaard, H., Kristensen, J., and Sondergaard, J., 1982, The effect of leukotriene C_4 and D_4 on cutaneous blood flow in humans, *Prostaglandins* **23:**797–801.

Brain, S. D., Camp, R. D. R., Dowd, P. M., Kobza Black, A., Woollard, P. M., Mallet, A. I., and Greaves, M. W., 1982a, Psoriasis and leukotriene B_4, *Lancet* **2:**762–763.

Brain, S. D., Camp, R. D. R., Leigh, I. M., and Ford-Hutchinson, A. W., 1982b, The synthesis of leukotriene B_4-like material by cultured human keratinocytes, *J. Invest. Dermatol.* **78:**328.

Brain, S. D., Camp, R. D. R., Dowd, P. M., Kobza Black, A., Woollard, P. M., Mallet, A. I., and Greaves, M. W., 1983, Release of leukotriene B_4 (LTB_4) and monohydroxyeicosatetraenoic acids (HETEs) from the involved skin of patients with psoriasis, in: *Leukotrienes and Other Lipoxygenase Products* (P. J. Piper, ed.), John Wiley & Sons, Chichester, pp. 248–254.

Brain, S., Camp, R., Dowd, P., Kobza Black, A., and Greaves, M., 1984a, The release of leukotriene B_4 like material in biologically active amounts from the lesional skin of patients with psoriasis, *J. Invest. Dermatol.* **83:**70–73.

Brain, S. D., Camp, R. D. R., Cunningham, F. M., Dowd, P. M., Greaves, M. W., and Kobza Black, A., 1984b, Leukotriene B_4-like material in scale of psoriatic lesions, *Br. J. Pharmacol.* **83:**313–317.

Brain, S. D., Camp, R. D. R., Charleson, S., Dowd, P. M., Ford-Hutchinson, A. W., Greaves, M. W., and Kobza Black, A., 1984c, The release of LTC_4-like material from the involved lesional skin in psoriasis, *Br. J. Clin. Pharmacol.* **17:**650P.

Brain, S. D., Camp, R. D. R., Kobza Black, A., Dowd, P. M., Greaves, M. W., Ford-Hutchinson, A. W., and Charleson, S., 1985, Leukotrienes C_4 and D_4 in psoriatic lesions, *Prostaglandins* (in press).

Bray, M. A., 1984, Retinoids are potent inhibitors of the generation of rat leukocyte leukotriene B_4-like activity *in vitro*, *Eur. J. Pharmacol.* **98:**61–67.

Camp, R. D. R., Mallet, A. I., Woollard, P. M., Brain, S. D., Kobza Black, A., and Greaves, M. W., 1983a, The identification of hydroxy fatty acids in psoriatic skin, *Prostaglandins* **26:**431–448.

Camp, R. D. R., Coutts, A. A., Greaves, M. W., Kay, A. B., and Walport, M. J., 1983b, Responses of human skin to intradermal injection of leukotrienes C_4, D_4, and B_4, *Br. J. Pharmacol.* **80:**497–502.

Camp, R. D. R., Russell Jones, R., Brain, S., Woollard, P., and Greaves, M., 1984, Production of intraepidermal microabscesses by topical application of leukotriene B_4, *J. Invest. Dermatol.* **82:**202–204.

Chowaniec, O., Jablonska, S., Beutner, E. H., Proniewska, M., Jarzabek-Chorzelska, M., and Rzesa, G., 1981, Earliest clinical and histological changes in psoriasis, *Dermatologica* **163:**42–51.

Dowd, P. M., Woollard, P. M., Kobza Black, A., Camp, R. D. R., and Greaves, M. W., 1983, The effect of intradermal infusions of 12-hydroxyeicosatetraenoic acid (12-HETE) in normal human skin, *Br. J. Dermatol.* **109:**693–694.

Dowd, P. M., Kobza Black, A., Woollard, P., Camp, R. D., and Greaves, M. W., 1984, The *in vivo* properties of 12-hydroxyeicosatetraenoic acid (12-HETE) in normal skin, *J. Invest. Dermatol.* **82:**413–414.

Fincham, N., and Camp, R., 1984, Monohydroxy fatty acids in skin may modulate 5-lipoxygenase activity, *J. Invest. Dermatol.* **82:**401.

Grabbe, J., Czarnetzki, B. M., and Mardin, M., 1984, Release of lipoxygenase products of arachidonic acid from freshly isolated human keratinocytes, *Arch. Dermatol. Res.* **276:**128–130.

Hammarström, S., Hamberg, M., Samuelsson, B., Duell, E. A., Stawiski, M., and Voorhees, J. J., 1975, Increased concentrations of nonesterified arachidonic acid, 12L-hydroxy-5,8,10,14-eicosatetraenoic acid, prostaglandin E_2 and prostaglandin F_2 in epidermis of psoriasis, *Proc. Natl. Acad. Sci. U.S.A.* **72:**5130–5134.

Hammarström, S., Hamberg, M., Duell, E. A., Stawiski, M. A., Anderson, T. F., and Voorhees, J. J., 1977, Glucocorticoid in inflammatory proliferative skin disease reduces arachidonic and hydroxyeicosatetraenoic acids, *Science* **197:**994–996.

Hammarström, S., Lindgren, J. A., Marcelo, C., Duell, E. A., Anderson, T. F., and Voorhees, J. J., 1979, Arachidonic acid transformations in normal and psoriatic skin, *J. Invest. Dermatol.* **73:**180–183.

Klemp, P., and Staberg, B., 1983, Cutaneous blood flow in psoriasis, *J. Invest. Dermatol.* **81**:503–506.

Kragballe, K., and Herlin, T., 1983, Benoxaprofen improves psoriasis, *Arch. Dermatol.* **119**:548–552.

Luger, T. A., Charon, J. A., Colot, M., Micksche, M., and Oppenheim, J. J., 1983, Chemotactic properties of partially purified human epidermal cell derived thymocyte-activating factor (ETAF) for polymorphonuclear and mononuclear cells, *J. Immunol.* **131**:816–820.

Masters, D. J., and McMillan, R. M., 1984, 5-Lipoxygenase from human leukocytes, *Br. J. Pharmacol.* **81**:70P.

Peck, M. J., Piper, P. J., and Williams, T. J., 1981, The effect of leukotrienes C_4 and D_4 on the microvasculature of guinea pig skin, *Prostaglandins* **21**:315–321.

Smith, M. J. H., and Walker, J. R., 1980, The effects of some antirheumatic drugs on an *in vitro* model of human polymorphonuclear leucocyte chemotaxis, *Br. J. Pharmacol.* **69**:473–478.

Soter, N. A., Lewis, R. A., Corey, E. J., and Austen, K. F., 1983, Local effects of synthetic leukotrienes (LTC_4, LTD_4, LTE_4, and LTB_4) in human skin, *J. Invest. Dermatol.* **80**:115–119.

Tagami, H., and Ofuji, S., 1976, Leukotactic properties of soluble substances in psoriasis scale, *Br. J. Dermatol.* **95**:1–8.

Tagami, H., and Ofuji, S., 1977, Characterisation of a leukotactic factor derived from psoriatic scale, *Br. J. Dermatol.* **97**:509–518.

Tagami, H., Kitano, Y., Suehisa, S., Oku, T., and Yamada, M., 1982, Psoriatic leukotactic factor. Further physicochemical characterisation and effect on the epidermal cells, *Arch. Dermatol. Res.* **272**:201–213.

Vanderhoek, J. Y., Bryant, R. W., and Bailey, J. M., 1980, Inhibition of leukotriene biosynthesis by the leukocyte product 15-hydroxy-5,8,11,13-eicosatetraenoic acid, *J. Biol. Chem.* **255**:10064–10066.

Wong, E., Barr, R. M., Brain, S. D., Olins, L. A., and Greaves, M. W., 1983, The effect of etretinate on cyclo-oxygenase and lipoxygenase products of arachidonic acid in psoriatic skin, *Br. J. Dermatol.* **109**:703.

Wong, E., Camp, R., and Greaves, M. W., 1985, Topical application of leukotriene B_4 in normal and psoriatic subjects, *J. Invest. Dermatol.* (in press).

Woollard, P. M., and Mallet, A. I., 1984, A novel gas chromatographic-mass spectrometric assay for monohydroxy fatty acids, *J. Chromatogr.* **306**:1–21.

43

Coronary Leukotriene Levels and Myocardial Ischemia

ROBERT E. GOLDSTEIN, GIORA FEUERSTEIN, GORDON LETTS, PETER W. RAMWELL, and DAVID EZRA

1. INTRODUCTION

1.1. Potential for Involvement of Cysteinyl Leukotrienes in Myocardial Infarction

Several lines of evidence suggest that leukotrienes might be involved in the pathogenesis of myocardial infarction. Cysteinyl leukotrienes released by inflammatory cells have a potent capacity to cause vasoconstriction and edema formation. Administration of cysteinyl leukotrienes to blood-perfused, *in situ* hearts can produce profound coronary constriction accompanied by electrocardiographic evidence of myocardial ischemia, depression of myocardial contractility, and ventricular arrhythmias (Michelassi *et al.*, 1982; Woodman and Dusting, 1983; Ezra *et al.*, 1983). Figure 1 illustrates the influence of leukotriene C_4 as observed in our laboratory. In this example, a 1-μg bolus of leukotriene C_4, given directly into the left anterior descending coronary artery of a pig, caused transient arrest of regional blood flow and ST segment elevation characteristic of myocardial ischemia on a local surface electrocardiogram. Left ventricular end-diatolic pressure is not altered in this example. In other experiments, however, regional wall motion, recorded by

ROBERT E. GOLDSTEIN, GIORA FEUERSTEIN, and DAVID EZRA • Neurobiology Research Unit, Divisions of Cardiology and Clinical Pharmacology, Departments of Medicine and Pharmacology, Uniformed Services University of the Health Sciences, Bethesda, Maryland 20814.   GORDON LETTS • Merck Frosst, Inc., Dorval, Québec H9R 4P8, Canada.   PETER W. RAMWELL • Department of Physiology and Biophysics, Georgetown University Medical Center, Washington, D.C. 20007.

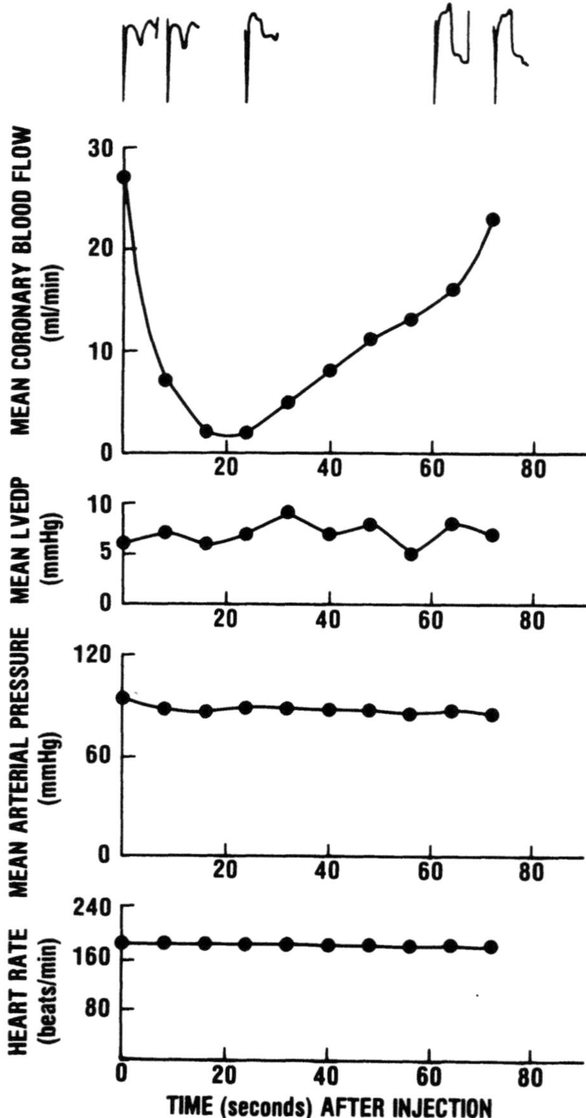

FIGURE 1. Leukotriene C$_4$, 1 μg, was administered as a bolus into the left anterior descending (LAD) coronary artery of an open-chest anesthetized pig. Within 20 sec, mean LAD blood flow (top graph) was virtually halted. A surface electrocardiogram recorded from territory perfused by the LAD (top, each complex positioned according to the time of recording) showed progressive ischemic abnormality. The LAD flow decrement soon reversed, returning to base-line within 2–4 min. The surface electrocardiogram also reverted to its base-line configuration within this time period. In this example, there was no simultaneous change in left ventricular end-diastolic pressure (LVEDP, second graph from top), mean systemic arterial pressure (third graph from top), or heart rate (bottom graph). Thus, the change in LAD flow reflected transient, severe leukotriene-induced coronary constriction. Intracoronary bolus administration of an equal volume of vehicle (0.1 ml saline, pH 7.20) was without effect.

epicardial sonomicrometers in myocardium supplied by the left anterior descending coronary artery, demonstrated transient, severe reduction in myocardial shortening in conjunction with marked leukotriene-induced coronary flow decrement (Michelassi *et al.*, 1982; Ezra *et al.*, 1984). Thus, release of cysteinyl leukotrienes within the heart might have substantial deleterious consequences for coronary and myocardial function.

We hypothesized that local leukotriene release may be stimulated by coronary occlusion. Such release might compound an ischemic insult by inducing inappropriate and detrimental constriction in the fine vessels of affected areas. This may lead to exaggeration of local flow deprivation. Arrhythmogenic and negative inotropic consequences of cysteinyl leukotriene exposure might also complicate incipient myocardial infarction.

Several authors have suggested that inflammatory cells and 5-lipoxygenase products figure significantly in the pathogenesis of myocardial infarction. Inhibition of leukocyte accumulation with ibuprofen (Romson *et al.*, 1982), drug-induced leukopenia (Mullane *et al.*, 1984), and 5-lipoxygenase inhibition with nafazatrom (Fiedler, 1983a, b; Mullane and Bednar, 1984) have each been associated with reduction in size of the infarct resulting from coronary occlusion.

Although multiple factors may be involved in mediating each of these infarct-sparing interventions, it is possible that reduction in leukocyte-produced cysteinyl leukotrienes played an important role in reducing the damage resulting from coronary underperfusion. Following coronary occlusion, coronary blood flow is often restored by development of collateral channels, lysis of thrombi, or relaxation of coronary spasm. In advanced stages of ischemic damage, microvessel resistance may be elevated. Termed the "no-reflow phenomenon," this failure of local reperfusion implies permanent loss of myocardial function (Kloner *et al.*, 1974). Cysteinyl leukotrienes might participate in this phenomenon by favoring microvessel constriction and edema in damaged regions. Moreover, decrease in local perfusion, contractile function, and electrical stability might occur with action of cysteinyl leukotrienes on myocardium with lesser degrees of injury. If correct, these concepts have potentially significant therapeutic implications. In addition to providing a possible rationale for the use of 5-lipoxygenase blockers such as nafazatrom or BW755C in treatment of myocardial infarction, this line of reasoning suggests that the infarct-sparing effects of drugs blocking leukotriene synthesis might be supplemented by agents (e.g., FPL-55712) interfering with leukotriene action.

1.2. Considerations in Evaluating the Role of Cysteinyl Leukotrienes

Although current evidence indicates that cysteinyl leukotrienes might complicate myocardial ischemia, a number of concerns must be addressed to substantiate such speculation. For example, the capacity of cysteinyl leukotrienes to produce sustained coronary constriction has not been evaluated previously. *In vivo* demonstrations of adverse effects of these substances have all utilized transient responses to bolus administration. Very recent work in our laboratory has shown that coronary constriction, electrocardiographic evidence of ischemia, and decreased contractility each reverse spontaneously within 5 min of initiating steady intracoronary infusion of leukotriene D_4 (LTD_4), even when initial abnormalities are severe (Ezra *et al.*, 1984). Such rapid tachyphylaxis is potentially significant. Pathophysiological circumstances involving leukotriene release would probably cause leukotriene exposure that persists for a prolonged period.

When postulating involvement of cysteinyl leukotrienes in ischemic myocardial damage, it is important to establish that ischemic insult leads to release of these materials in sufficient quantity and in an appropriate time frame to impact on the evolution of myocardial necrosis. Irreversible biochemical and morphological changes characterizing tissue necrosis begin in the innermost (endocardial) layers of the left ventricular wall 20–40 min after coronary occlusion (Reimer *et al.*, 1977; Fujiwara *et al.*, 1982). When occlusion persists, the necrotic region expands toward the outer (epicardial) surface, reaching its full extent after 2–5 hr. Thus, if cysteinyl leukotrienes participate in events leading to myocardial necrosis primarily through their action on vessels, they must reach levels sufficient to alter coronary vascular performance very soon after development of coronary occlusion. Previous studies have demonstrated enhanced arachidonate metabolism by ischemic myocardium. Experimental data suggest increased myocardial release of both prostacyclin and thromboxane A_2 during ischemia (Coker *et al.*, 1981; Sakai *et al.*, 1982; Tanabe *et al.*, 1982). The fact that arachidonic acid undergoes more rapid oxidative transformation via cyclooxygenase pathways favors the possibility that ischemic myocardium will also exhibit more rapid oxidative transformation via 5-lipoxygenase pathways. Questions remain, however, whether increased amounts of arachidonic acid are available for 5-lipoxygenase metabolism in ischemic myocardium and whether such enzyme systems are activated within ischemic tissue.

1.3. Arachidonate Metabolites in Coronary Venous Blood

To evaluate the possible role of cysteinyl leukotrienes in the evolution of myocardial necrosis, we measured leukotriene C_4 immunoreactivity (LTC_4-ir) in coronary venous blood draining from regions of myocardium rendered ischemic for up to 1 hr. The radioimmunoassay that we employed is highly sensitive to both LTD_4 and LTC_4 and moderately sensitive to LTE_4. Thus, our assay should detect a significant rise in any of the cysteinyl leukotrienes (Hayes *et al.*, 1983). Additional radioimmunoassays for 6-keto-$PGF_{1\alpha}$, the stable metabolite of prostacyclin, and thromboxane B_2 (TxB_2), the stable metabolite of thromboxane A_2, were performed to monitor simultaneous oxidative metabolism of arachidonic acid via cyclooxygenase pathways. We used an open-chest pig model previously shown to have high coronary sensitivity to the constrictive effects of the cysteinyl leukotrienes. Coronary flow was reduced by 67–77% to produce signs of severe ischemia. Persistent regional flow allowed repeated coronary venous sampling during ischemia and also accommodated a marked intolerance of the pig heart for total arrest of coronary flow. A

1-hr period of ischemia is generally sufficient to cause ultrastructural evidence of irreversible deterioration in pig hearts (Fujiwara *et al.*, 1982).

To evaluate the capacity of our radioimmunoassay system to detect leukotrienes, we measured LTC_4-ir values in coronary venous blood draining myocardium selectively perfused with pure, synthetic LTD_4 at a rate sufficient to produce an intermediate (approximately half-maximal) degree of coronary constriction. These values were compared with LTC_4-ir values measured in coronary venous effluent from myocardium subjected to ischemia.

2. METHODS

2.1. Preparation

Nine domestic pigs of either sex, 9–12 weeks in age and 25–30 kg in weight, were sedated with ketamine, 20 mg/kg i.m., and anesthetized with sodium pentobarbital 2–4 mg/kg per hr. Catheters were introduced into a carotid artery and jugular vein for pressure monitoring, blood sampling, and infusion. Animals were intubated and ventilated with a Harvard respirator modified to deliver 3 cm end-expiratory pressure. Rectal temperature, mean arterial pressure, and lead II of the electrocardiogram were recorded continuously. Rectal temperature was maintained by an external heating pad. A left thoracotomy was performed, and the heart suspended in a pericardial cradle. An electromagnetic flow probe was placed around a proximal portion of the left anterior descending (LAD) coronary artery, and an inflatable occlusion cuff was positioned 1–2 cm distal to the probe. A fine (non-occlusive) polyethylene catheter was introduced into a portion of the great cardiac vein that accompanies the LAD and drains myocardium perfused by the LAD segment distal to the occlusion cuff. In five pigs, pairs of hemispherical 2-mm ultrasound crystals (Triton Technology, San Diego, CA) were implanted approximately 4 mm deep and 10 mm apart within myocardium perfused by the distal LAD to record regional myocardial segment length. Throughout the remainder of the study, intercrystal distance was recorded continuously along with mean LAD coronary blood flow (CBF) and other parameters mentioned above.

Additional details relating to this preparation have been published previously (Ezra *et al.*, 1983).

2.2. Protocol

Once hemodynamic values stabilized, base-line recordings were performed, and base-line blood samples were collected simultaneously from coronary vein and carotid artery. Each 1.5-ml blood sample was centrifuged immediately for 1 min in a Beckman Microfuge. The resultant plasma was frozen, stored, and, when appropriate, shipped in dry ice. In five pigs additional 0.5-ml aliquots were placed immediately in 1 ml iced perchloric acid for subsequent lactate determination. On

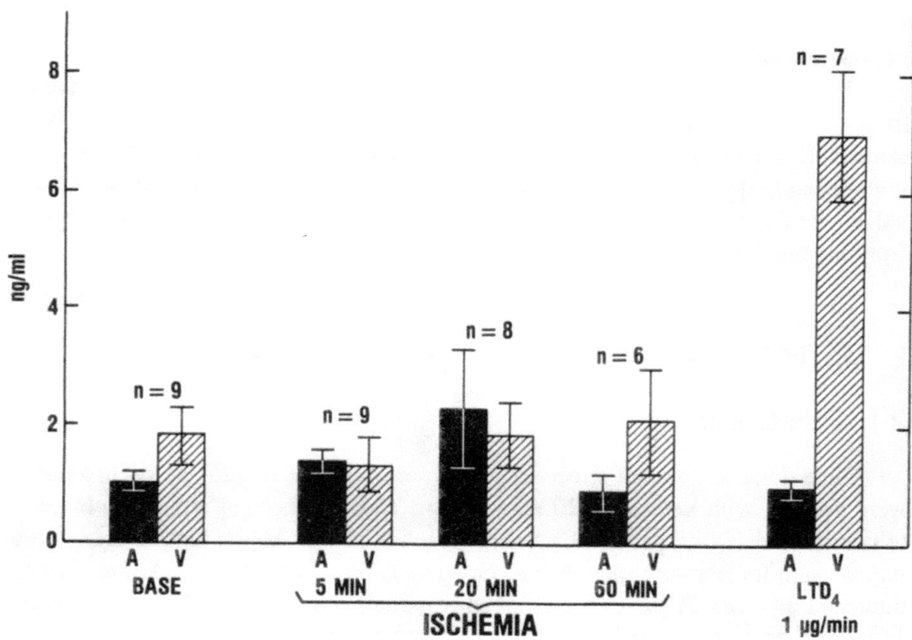

FIGURE 2. Leukotriene C$_4$ immunoreactivity (vertical axis) (ng/ml) in systemic artery (A) and in the vein (V) draining the territory perfused by the left anterior descending (LAD) coronary artery. Paired arterial and coronary venous mean values ($\pm$ S.E.) were obtained just before LAD constriction (BASE) and 5, 20, and 60 min after initiation of constriction sufficient to lower LAD flow by 67–77% (ISCHEMIA). For comparison, LTC$_4$ immunoreactivity is depicted (at extreme right) when pure, synthetic leukotriene D$_4$ (LTD$_4$) was infused directly into the LAD at 1 µg/min, sufficient to produce maximal flow decrement of approximately 50%. Although coronary venous LTC$_4$ immunoreactivity was significantly ($P < 0.01$) increased above base-line values during LTD$_4$ infusion, there was no difference between base-line values and measurements made during ischemia. Number (n) of contributing studies is depicted above each pair of bars. The seven pairs receiving LTD$_4$ infusion are different from the nine pigs subjected to ischemia.

completion of sampling, coronary constriction was produced by cuff inflation to reduce and maintain LAD flow at values 67–77% below preocclusion base-lines. Repeat measurements of hemodynamic parameters and paired, simultaneous blood sampling were performed 5, 20, and 60 min after initiation of coronary flow reduction. After the 60-min sampling was completed, unobstructed CBF was restored by deflation of the occlusion cuff. Final blood sampling and measurement of hemodynamic values were performed 1 min after CBF restoration.

Effects of steady intracoronary LTD$_4$ infusion were assessed in a separate series of seven pigs. Preparations were the same as those just described except that the occlusion cuff was omitted and a fine catheter inserted in the LAD 10–15 mm distal to the flow probe. Measurements and blood samples were obtained before and during control infusions of saline solution. Animals then received graded doses of LTD$_4$, 0.3, 1.0, and 3.0 µg per min, for 6 min interrupted by recovery periods of 20–40 min. Paired blood sampling was repeated at the peak of LTD_4 effect, after approximately 1–2 min of the infusion. A complete report of physiological and other data appears elsewhere (Ezra *et al.*, 1984). The only results described here are those obtained with radioimmunoassay for LTC_4-ir in pigs receiving 1 μg per min. These are presented for comparison with LTC_4-ir measured in pigs with ischemia.

2.3. Radioimmunoassays and Lactate Determination

Measurements of LTC_4-ir were performed by one of us (G.L.) according to the methods of Hayes *et al.* (1983). This assay can detect changes in plasma leukotriene C_4 (LTC_4) level as small as 0.5 ng/ml. Although maximally sensitive to LTC_4, the assay also cross reacts with other cysteinyl leukotrienes and their sulfone derivatives. Relative to LTC_4, the reactivities of these substances are: LTD_4, 50%; LTE_4, 7%; LTC_4-sulfone, 1%; LTD_4-sulfone, 15%; and LTE_4-sulfone, 3% (derived from data of Hayes *et al.*, 1983). The assay is insensitive to many other lipoxygenase products, including LTB_4 and its isomers.

Radioimmunoassays for 6-keto-$PGF_{1\alpha}$ and thromboxane B_2 were performed using standard techniques (Granstrom and Kindahl, 1976). Lactate determinations were made with a commercial kit (Sigma Chemical Co., St. Louis, MO).

2.4. Statistics

Comparisons between mean values were accomplished with analysis of variance and the Newman–Keuls test for statistical significance.

3. RESULTS

3.1. Hemodynamic Data and Electrocardiographic Findings

Three pigs died of ventricular fibrillation during the observation. Since their data did not differ from those of surviving animals, results at each point in time represent pooled means including all animals alive at the time of measurement. As shown in Table I, mean coronary blood flow in the LAD declined from 30 ml/min

TABLE I. Hemodynamic Data[a]

		CBF	MAP	HR	n[b]
Base line		30 ± 2	83 ± 4	108 ± 3	9
Ischemia	5 min	10 ± 2*	81 ± 3	110 ± 4	9
	20 min	8 ± 1*	85 ± 3	113 ± 4	7
	60 min	7 ± 2*	87 ± 4	114 ± 4	6
Flow restoration		74 ± 6*	74 ± 2*	123 ± 4*	6

[a] Means ± S.E.; *$P < 0.05$ versus base line.
[b] Three pigs died in ventricular fibrillation after 5 or 20 min postocclusion sampling.

TABLE II. Evidence for Myocardial Ischemia

Ischemic electrocardiographic changes (9/9)
Regional contractile depression (5/5)
Local lactate production (2/2)
Lethal arrhythmia (3/9)

at preocclusion base-line to values between 7 and 10 ml/min after partial coronary occlusion. The LAD flow rose above base-line at the time of blood sampling 1 min after cuff release. Mean arterial pressure and heart rate did not change significantly after occlusion. Arterial pressure decreased and heart rate increased with restoration of coronary flow.

Table II summarizes evidence of ischemia in myocardium perfused by the LAD. Horizontal or downsloping ST segment depression appeared within 2 min of coronary flow reduction on the electrocardiogram of each pig. At the same time, systolic shortening in the region perfused by the distal LAD diminished severely or ceased altogether, and local lactate consumption changed to marked lactate production after partial LAD occlusion. The development of lactate production in this and similar studies also demonstrated the capacity of our venous sampling technique to identify metabolic changes in the ischemic LAD region.

3.2. Coronary Levels of Leukotriene C_4 Immunoreactivity

Preocclusion arterial and coronary venous values of LTC_4-ir were very low (Fig. 2), somewhat below LTC_4-ir levels measured in human plasma examined using the same radioimmunoassay (Hayes *et al.*, 1983). These postthoracotomy values were not altered from prethoracotomy values. There was no consistent arteriovenous difference in LTC_4-ir.

The LTC_4-ir levels remained unchanged in plasma from systemic arterial and local coronary venous blood after 5, 20, and 60 min of myocardial ischemia (Fig. 2). Thus, there was no evidence that the early phases of myocardial ischemia are associated with local intracoronary release of cysteinyl leukotrienes or their sulfone

TABLE III. Coronary Eicosanoid Levels with Restoration of LAD Flow[a]

	LTC_4-ir ($n = 6$)		6-keto-$PGF_{1\alpha}$ ($n = 3$)		TxB_2 ($n = 6$)	
	A	V	A	V	A	V
Base line	1.0 ± 0.2	1.9 ± 0.5	0.3 ± 0.1	0.7 ± 0.2	0.4 ± 0.1	0.6 ± 0.2
Flow restoration	1.0 ± 0.2	1.4 ± 0.4	0.3 ± 0.1	2.4 ± 0.1	0.4 ± 0.1	0.3 ± 0.1

[a] Mean $\pm$ S.E. Number of animals with data after flow restoration (n) is shown in parentheses for each measurement. No statistically significant differences were observed for LTC_4-ir and TxB_2. Statistical analysis was not performed for 6-keto-$PGF_{1\alpha}$ because of the small number of data points. Abbreviations: LAD, left anterior descending coronary artery; LTC_4-ir, leukotriene C_4 immunoreactivity; n, number of data points; 6-keto-$PGF_{1\alpha}$, 6-keto-prostaglandin $F_{1\alpha}$; TxB_2, thromboxane B_2.

derivatives. Coronary flow restoration was similarly unassociated with any change in LTC_4-ir levels (Table III).

In contrast, coronary venous LTC_4-ir levels were substantiallay increased from 0.9 ± 0.1 to 7.0 ± 1.1 ng/ml by intracoronary LTD_4, 1 μg/min. These data demonstrate the capacity of the radioimmunoassay to detect cysteinyl leukotrienes in coronary venous blood when these are present in sufficient quantity to produce a moderate change (50% reduction) in coronary flow. Lack of alteration of arterial LTC_4-ir during LTD_4 infusion very likely represents redistribution, sequestration, or pulmonary degradation of the infused LTD_4.

3.3. Coronary Levels of 6-Keto-PGF$_{1\alpha}$ and Thromboxane B$_2$

Preocclusion data showed that 6-keto-PGF$_{1\alpha}$ was present in greater quantity in coronary venous blood than in systemic arterial blood (Fig. 3). Within 5 min of coronary flow reduction, coronary venous levels of 6-keto-PGF$_{1\alpha}$ rose and remained elevated to a similar degree for the remainder of the 60-min ischemic period. On

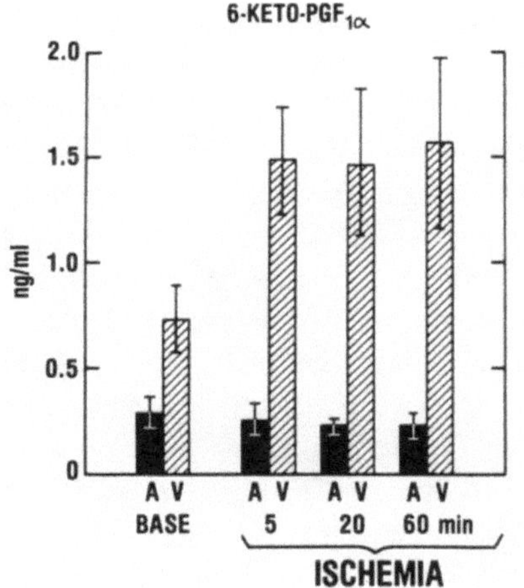
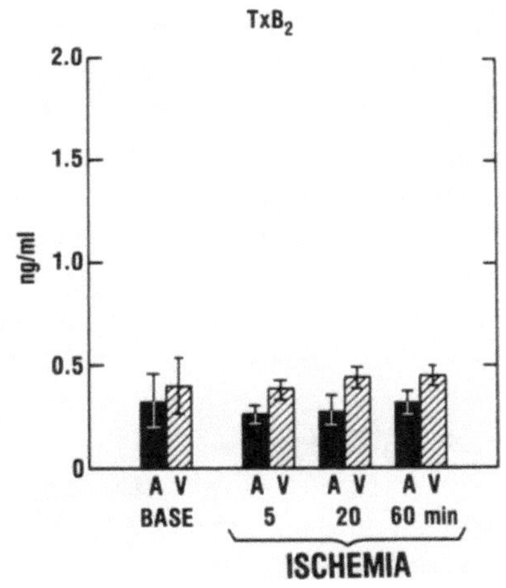

FIGURE 3. Concentrations of 6-keto-PGF$_{1\alpha}$ (left panel), the stable metabolite of prostacyclin, and thromboxane B$_2$ (TxB$_2$, right panel), the stable metabolite of thromboxane A$_2$ (ng/ml, vertical axis) for systemic artery (A) and coronary vein (V). Determinations were performed using aliquots of the same samples employed for leukotriene C$_4$ immunoreactivity. Samples were drawn before (BASE) and 5, 20, and 60 min after initiating constriction sufficient to lower coronary artery flow by 67–77% (ISCHEMIA). Values are means ($\pm$ S.E.) of six to eight studies (TxB$_2$), four studies (6-keto-PGF$_{1\alpha}$ base and 5 min ischemia), or three studies (6-keto-PGF$_{1\alpha}$, 20 and 60 min ischemia). The TxB$_2$ levels were unchanged during ischemia. Coronary venous levels for 6-keto-PGF$_{1\alpha}$ showed a strong and consistent tendency to rise; the small number of samples precluded statistical analysis.

restoration of unhindered coronary perfusion, coronary venous concentration of 6-keto-PGF$_{1\alpha}$ rose even further despite near doubling of the flow rate relative to the preocclusion period. Arterial concentrations of 6-keto-PGF$_{1\alpha}$ remained stable throughout the study.

The TxB$_2$ concentration was similar in arterial and coronary venous blood and did not change appreciably throughout the period of observation (Fig. 3).

4. DISCUSSION

4.1. Cysteinyl Leukotriene Levels during Myocardial Ischemia

Our data indicate that the local coronary venous concentration of cysteinyl leukotrienes is not increased during or immediately after a 1-hr period of myocardial ischemia. In contrast, the concentration of 6-keto-PGF$_{1\alpha}$ in the same blood samples rose during ischemia and increased even further with restoration of flow. These latter results, in agreement with the findings of other laboratories (Coker *et al.*, 1981; Sakai *et al.*, 1982), suggest that arachidonate release and oxidative transformation to prostacyclin was enhanced in the ischemic myocardium of our pigs. The apparent lack of concomitant arachidonate transformation to cysteinyl leukotrienes may be explained by lack of access of substrate (e.g., arachidonic acid, glutathione, or oxygen) to appropriate enzyme systems, unfavorable conditions for metabolism via 5-lipoxygenase pathways, or a simple lack of 5-lipoxygenase activity. Our studies do not discriminate among these possibilities. Lack of enzyme activity, however, is a particularly attractive hypothesis. Although sensitized cardiac tissues can release significant amounts of leukotrienelike material in response to antigenic challenge, perhaps because of the activity of mast cells (Capurro and Levi, 1975; Zaveca and Levi, 1977), the 5-lipoxygenase system of cells present in newly underperfused myocardium may not respond similarly to ischemic challenge. Leukotriene production is most characteristic of white blood cells, a cell type not present in abundance in the earliest phases of myocardial ischemia (Mullane *et al.*, 1984).

Our inability to detect a significant rise in cysteinyl leukotriene levels with ischemia does not seem related to insensitivity of assay methods. The large rise in local coronary venous LTC$_4$-ir during intracoronary LTD$_4$ infusion shows the capacity of our radioimmunoassay techniques to detect cysteinyl leukotrienes when present in quantity sufficient to produce important coronary constriction. These same results also tend to exclude rapid degradation or sequestration of leukotrienes within coronary vessels as an explanation for lack of rise in LTC$_4$-ir with ischemia.

4.2. Leukotrienes and White Blood Cells in Myocardial Infarction

White blood cells may play an important role in the pathogenesis of myocardial infarction. Early adherance of white blood cells to coronary endothelium damaged by ischemia may interfere with microcirculatory function by mechanically plugging vessels (Engler *et al.*, 1984) and by releasing edema-promoting substances, vasoconstrictors such as 12-hydroxyeicosatetraenoic acid (Trachte *et al.*, 1979; Mullane *et al.*, 1984), or platelet-activating factor (Feuerstein *et al.*, 1984). White cell production of free radical superoxide anion, hydrogen peroxide, and other cytotoxic substances may augment myocyte damage and augment progression to necrosis (Weissman *et al.*, 1980; Del Maestro *et al.*, 1980; Fantone and Ward, 1982). Early white cell release of leukotriene B_4 may be important in recruiting additional white cells, thereby increasing local superoxide anion generation. Nafazatrom, BW755C, and other inhibitors of 5-lipoxygenase may interfere with this process by blocking synthesis of leukotriene B_4 and, in the case of some of these agents, by direct inactivation of superoxide anion and other free radicals.

Although cysteinyl leukotrienes might theoretically participate in the pathogenesis of ischemic myocardial necrosis from sustained coronary occlusion, our results do not substantiate this hypothesis. Hence, blockade of synthesis or action of cysteinyl leukotrienes may have little impact on coronary or myocardial function within 1 hr of a coronary occlusion. By the time that white blood cells invade ischemic myocardium in sufficient quantity to manufacture significant amounts of cysteinyl leukotrienes, the myocardium may well have undergone irreversible changes of necrosis.

Cysteinyl leukotrienes may still have an important pathogenic role by stimulating spasm in large coronary arteries, thus initiating or sustaining coronary occlusion. This might occur when inflammatory processes affect previously stable atherosclerotic lesions. Such a spasm might represent the first step in the occurrence of myocardial infarction. In our opinion, present findings favor an initiating role rather than an exacerbating role for the cysteinyl leukotrienes in the pathogenesis of myocardial infarction.

ACKNOWLEDGMENTS. This work was supported by the Uniformed Services University of the Health Sciences Protocol RO 8346. The opinions expressed here are those of the authors. They do not reflect the views of the University or the Department of Defense. Experiments described here were conducted according to the principles set forth in *Guide for the Care and Use of Laboratory Animals*, Institute of Laboratory Animal Resources, National Research Council DHEW Pub. No. (NIH) 74-23. Dr. Ezra is a recipient of an International Fellowship in Clinical Pharmacology for the Merck Foundation. The authors are very grateful to Mr. John Czaja for his skilled technical assistance and to Ms. Janet Thomson for her help in preparing this manuscript.

REFERENCES

Capurro, N., and Levi, R., 1975, The heart as target organ in systemic allergic reactions: Comparison of cardiac anaphylaxis *in vivo* and *in vitro*, *Circ. Res.* **36**:520–528.

Coker, S. J., Parratt, J. R., McAledingham, I., and Zeitlin, I. J., 1981, Early release of thromboxane and 6-keto-$PGF_{1\alpha}$ from the ischaemic canine myocardium; relation to early post-infarction arrhythmias, *Nature* **291**:323–324.

Del Maestro, R., Thaw, H., Bjork, J., Planker, M., and Arfors, K., 1980, Free radicals as mediators of tissue injury, *Acta Physiol. Scand. [Suppl.]* **492**:43–47.

Engler, R. L., Dahlgren, M., Schmidt-Schoenbein, G., Witztum, K., Dobbs, A., and Peterson, M., 1984, Early accumulation of granulocytes in ischemic myocardium, *Clin. Res.* **32**:162A.

Ezra, D., Boyd, L. M., Feuerstein, G., and Goldstein, R. E., 1983, Coronary constriction by leukotriene C_4, D_4, and E_4 in the intact pig heart, *Am. J. Cardiol.* **51**:1451–1454.

Ezra, D., Feuerstein, G., Erne, J., and Goldstein, R. E., 1984, Biphasic coronary flow response to sustained leukotriene D_4 administration, *J. Am. Coll. Cardiol.* **3**:502.

Fantone, J. C., and Ward, P. A., 1982, Role of oxygen-derived free radicals and metabolites in leukocyte-dependent inflammatory reactions, *Am. J. Pathol.* **107**:397–418.

Feuerstein, G., Boyd, L. M., Ezra, D., and Goldstein, R. E., 1984, Effect of platelet-activating factor on coronary circulation of the domestic pig, *Am. J. Physiol.* **246**(*Heart Circ. Physiol.* **15**):H466–H471.

Fiedler, V. B., 1983a, Reduction of myocardial infarction and dysrhythmic activity by nafazatrom in the conscious rat, *Eur. J. Pharmacol.* **88**:263–267.

Fiedler, V. B., 1983b, The effects of oral nafaztrom ($=$Bay g6575) on canine artery thrombosis and myocardial ischemia, *Basic Res. Cardiol.* **78**:266–280.

Fujiwara, H., Ashraf, M., Sato, S., and Millard, R. W., 1982, Transmural cellular damage and blood flow distribution in early ischemia in pig hearts, *Circ. Res.* **51**:683–693.

Granstrom, E., and Kindahl, H., 1976, Radioimmunoassay for prostaglandin metabolites, *Adv. Prostaglandin Thromboxane Res.* **1**:81–92.

Hayes, E. C., Lombardo, D. L., Girard, Y., Maycock, A. L., Rokach, J., Rosenthal, A. S., Young, R. N., Egan, R. W., and Zweerink, H. J., 1983, Measuring leukotrienes of slow reacting subtances of anaphylaxis: Development of a specific radioimmunoassay, *J. Immunol.* **131**:429–433.

Kloner, R. A., Ganote, C. E., and Jennings, R. B., 1974, The "no-flow" phenomenon after temporary coronary occlusion in the dog, *J. Clin. Invest.* **54**:1496–1508.

Michelassi, F., Landra, L., Hill, R. D., Lowenstein, E., Watkins, W. D., Petkau, A. J., and Zapol, W. M., 1982, Leukotriene D_4: A potent coronary artery constrictor associated with impaired ventricular contraction, *Science* **217**:841–843.

Mullane, K., and Bednar, M., 1984, Contribution of leukocytes to ischemia-induced myocardial damage, *Clin. Res.* **32**:475A.

Mullane, K. M., Read, N., Salmon, J. A., and Moncada, S., 1984, Role of leukocytes in acute myocardial infarction in anesthetized dogs: Relationship to myocardial salvage by anti-inflammatory drugs, *J. Pharmacol. Exp. Ther.* **228**:510–522.

Reimer, K. A., Lowe, J. E., Rasmussen, M. M., and Jenning, R. B., 1977, The wavefront phenomenon of cell death. I. Myocardial infarct size vs. duration of coronary occlusion in dogs, *Circulation* **56**:786–794.

Romson, J. L., Hook, B. G., Rigor, V. H., Schork, M. A., Swanson, D. P., and Lucchesi, B. R., 1982, The effect of ibuprofen on accumulation of indium-111-labeled platelets and leukocytes in experimental myocardial infarction, *Circulation* **66**:1002–1011.

Sakai, K., Ito, T., and Ogawa, K., 1982, Roles of endogenous prostacyclin and thromboxane A_2 in the ischemic canine heart, *J. Cardiovasc. Pharmacol.* **4**:129–135.

Tanabe, M., Terashita, Z. I., Fujiwara, S., Shimamoto, N., Goto, N., Nishikawa, K., and Hirata, M., 1982, Coronary circulatory failure and thromboxane A_2 release during coronary occlusion and reperfusion in anesthetized dogs, *Cardiovasc. Res.* **16**:99–106.

Trachte, G. J., Lefer, A. M., Aharony, D., and Smith, J. B., 1979, Potent constriction of cat coronary arteries by hydroperoxides of arachidonic acid and its blockade by anti-inflammatory agents, *Prostaglandins* **18**:909–914.

Weissman, G., Smolen, J., and Korchak, H., 1980, Release of inflammatory mediators from stimulated neutrophils, *N. Engl. J. Med.* **303**:27–31.

Woodman, O. L., and Dusting, G. J., 1983, Coronary vasoconstriction induced by leukotrienes in the anesthetized dog, *Eur. J. Pharmacol.* **86:**125–128.
Zaveca, J. H., and Levi, R., 1977, Separation of primary and secondary cardiovascular events in systemic anaphylaxis, *Circ. Res.* **40:**15–19.

44

Decreased Concentrations of Prostaglandin $F_{2\alpha}$ in Delayed Labor

A Model for Studies on the Control of Prostaglandin Production by the Uterus

K. REDDI, A. KAMBARAN, S. M. JOUBERT,
R. H. PHILPOTT and R. J. NORMAN

1. INTRODUCTION

The mechanisms responsible for the onset of spontaneous human labor at term are poorly understood. Furthermore, millions of neonates die every year as a result of the complications of pre- and postterm delivery. Although there are numerous theories, little progress has been made in determining the precise factors that permit the delivery of a mature and healthy baby.

However, all investigators are agreed that prostaglandins are a key element in the final common pathway leading to myometrial contractions. The possible role of these compounds was first recognized by Karim and his colleagues (1966, 1967) who identified prostaglandinlike material in human amniotic fluid in labor. Subsequently, concentrations of prostaglandin E_2 (PGE) and prostaglandin $F_{2\alpha}$ (PGF_α) were shown to increase in maternal blood and amniotic fluid as labor progressed (Keirse and Turnbull, 1973; Keirse *et al.*, 1974; Mitchell *et al.*, 1978). Administration of prostaglandins and their analogues precipitated labor at all stages of gestation, whereas antiprostaglandin agents had the reverse effect (Zuckerman *et al.*, 1974).

K. REDDI, A. KAMBARAN, S. M. JOUBERT, R. H. PHILPOTT, and R. J. NORMAN • Medical Research Council Unit for Preclinical Diagnostic Research, University of Natal, Congella, South Africa.

Despite the evidence for the role of this group of compounds in causing uterine contractions, little is known about the control of prostaglandin synthesis and release by human uterine tissues. In the sheep, the fetus can activate prostaglandin synthesis by the effect of fetal cortisol on placental 17 α-hydroxylase activity and a subsequent increase in estrogen production by the redirected steroid pathways (Liggins *et al.*, 1973). No such mechanism exists in man. Fetal adrenal estrogen precursors (Turnbull *et al.*, 1977) or oxytocin from the fetal neurohypophysis (Chard *et al.*, 1971; Fuchs, 1983) have been suggested as possible regulators of prostaglandin biosynthesis and release, but evidence for these hypotheses has been difficult to investigate *in vivo*. As a further complication, Mitchell and his colleagues (1977) have suggested that local control of prostaglandins in the fetal membranes may play as important a role as hormones from mother or fetus. They showed that amniotomy, stretching of the cervical canal, or vaginal examination led to a rapid and sustained increase in prostaglandins in amniotic fluid and maternal blood independent of any observed changes in maternal circulating hormones. Thus, regulation of prostaglandin synthesis prior to labor may be under both hormonal and local control.

Prostaglandins are produced by all tissues in the body. Their actions are local, and they are rapidly metabolized to inactive compounds. Consequently, the measurement of primary prostaglandins in the blood is of little value in investigating production by the intrauterine tissues. The amnion, chorion, and decidua are major sites of prostaglandin production (Mitchell *et al.*, 1978; Keirse, 1983), and the amniotic fluid contains no prostaglandin-metabolizing enzymes (Keirse, 1978). This fluid is potentially a better source of information on the amount of PGE and PGF synthesis before and during labor. We have examined the concentrations of various prostanoids during dysfunctional labor, a condition in which poor uterine action leads to delay in delivery of the fetus. No biochemical abnormality has hitherto been identified in this condition, but we now demonstrate that it is associated with reduced production of prostaglandins.

2. PATIENTS AND METHODS

Patients selected for this study were in spontaneous labor at term with a cervical dilatation of 3–4 cm at entry. All patients had vaginal delivery of healthy, term neonates after 38–42 weeks of gestation; subjects with cephalopelvic disproportion or who required operative delivery were excluded. Details of labor were recorded on a partogram (Philpott, 1972), and patients were divided into those who progressed with a cervical dilatation rate of greater than 1 cm per hour (group I, $n = 10$) and those who progressed at less than 1 cm per hour (group II, $n = 12$). All patients in group II were given intravenous oxytocin to enhance uterine contractions. The two groups were equivalent with respect to maternal age, parity, mean duration of gestation, and birth weight of child. The duration of observed labor in group I was 3.2 hr and in group II 5.8 hr.

Amniotic fluid samples and intrauterine pressure readings were obtained from two Portex catheters placed via the cervix into the uterine cavity (Cowan *et al.*, 1982). The fetus was monitored by scalp electrodes and intermittent pH measurements where required. Prostaglandins E, F$_{2\alpha}$, 13,14-dihydro-15-keto-F$_{2\alpha}$ (PGFM), 6-keto-F$_{2\alpha}$ (PGIM), and thromboxane B$_2$ (TxB$_2$) were measured by specific radioimmunoassays as previously described (Norman *et al.*, 1981; Norman and Joubert, 1982; Reddi *et al.*, 1984).

3. RESULTS

The concentrations of PGE, PGF, and PGFM in amniotic fluid increased in patients with normal progress in labor (group I), but only PGE showed this trend in patients with delayed progress (group II) (Fig. 1). The administration of oxytocin, although increasing intrauterine pressure and hastening delivery, had no apparent effect on prostaglandin levels. The concentrations of PGIM and TxB$_2$ did not differ between the two groups and showed no rise with active labor. Concentrations of PGE, PGF, and PGFM in group I subjects in early labor were higher than in patients who were not in labor at the same gestational age (Norman and Joubert, 1982).

4. INTRAUTERINE PRODUCTION OF PROSTANOIDS

Prostanoids in the amniotic fluid are derived mainly from the fetal membranes rather than from fetal urine (Norman and Joubert, 1982; Robinson and Mitchell,

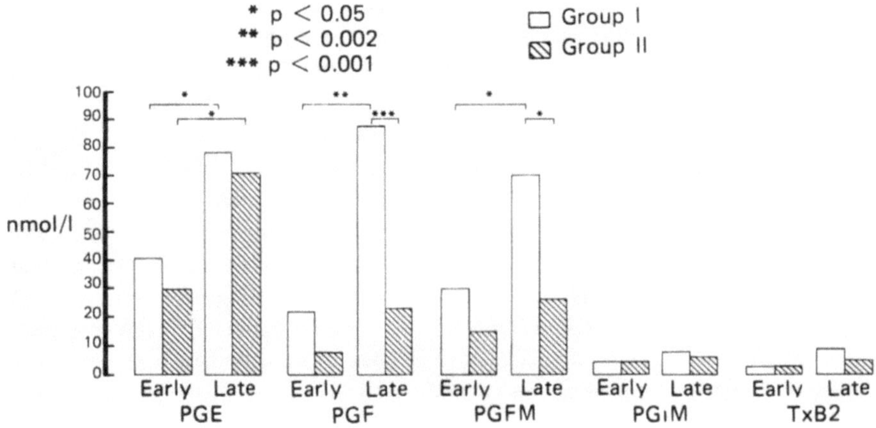

FIGURE 1. Amniotic fluid concentrations of various protanoids in early and late labor (nmol/liter). PGE, protaglandin E; PGF, prostaglandin F$_{2\alpha}$; PGFM, 13,14-dihydro-15-keto-prostaglandin F$_{2\alpha}$; PGIM, 16-keto-prostaglandin F$_{1\alpha}$; TxB$_2$, thromboxane B$_2$. Statistical differences were assessed by the Mann–Whitney sum test, and median values are shown in the figure.

1983). Prostaglandin and PGF are the most abundant prostanoids in amniotic fluid (Mitchell, 1981). Free arachidonic acid is the only precursor of these compounds and other prostanoids of the 2 series, and the concentration of arachidonic acid rises in amniotic fluid during spontaneous labor (Keirse *et al.*, 1977). Mobilization of free arachidonic acid from phosphatidylethanolamine and phosphatidylinositol in cell membranes is a crucial step in prostaglandin production. Bleasdale *et al.* (1983) have shown that in amnion, the amount of arachidonic acid in diacylphosphatidylethanolamine and phosphatidylinositol decreased by 42% and 35%, respectively, from early to late labor. Human chorioamnion contains phospholipase A_2 with a specific substrate preference for phosphatidylethanolamine containing arachidonic acid; this is calcium dependent (Okazaki *et al.*, 1978). Release of arachidonic acid from phosphatidylinositol requires at least three enzymes including (1) phospholipase C (specific to phosphaditylinositol); (2) diacylglycerol lipase; and (3) monoacylglycerol lipase (Bleasdale *et al.*, 1983) (Fig. 2). These enzymes show an increase in specific activity during gestation (with the exception of diacylglycerol lipase) and may be activated by any mechanism that releases calcium. Once arachidonic acid is liberated, production of prostaglandins and thromboxanes proceeds via cyclic endoperoxides. It seems likely that prostaglandin synthesis is not entirely dependent on availability of free arachidonic acid but rather depends on the activities of the enzymes in the prostaglandin synthetic pathyways (Keirse, 1983) (Fig. 3). The control of the relative proportions of PGE, PGF, prostaglandin, and TxB_2 from their common precursors is poorly understood.

In dysfunctional labor, the normal levels of PGE, PGIM, and TxB_2 suggest that mobilization of arachidonic acid is not impaired. However, preliminary work has suggested that there may be a difference between groups I and II with regard to prostaglandin synthesis-inhibiting substances that are found in the amniotic fluid (K. Reddi, R. J. Norman, W. M. Deppe, and S. M. Joubert, unpublished data). The defect may result from abnormal conversion of endoperoxides to PGF, a process that does not require a specific enzyme and is affected by numerous reducing agents (Nugteren and Hazelhof, 1973). The normal levels of PGIM and TxB_2 do not suggest a shunting of the pathway to prostacyclin or thromboxane A_2.

A further example of the low concentrations of PGF and PGFM relates to the different sites of PGE and PGF synthesis within the pregnant uterus. The major site of PGE synthesis is in the lower segment of the uterus (Norman *et al.*, 1981, 1983). It is possible that before the onset of labor, PGE production is predominant, and uterine contractions result from the switching on of other uterine tissues that make PGF. The change in the observed PGE : PGF ratio would result from activation of dormant tissues. In dysfunctional labor, this switch would not occur, leading to a reduced concentration of PGF in the upper part of the uterus where the major muscle fibers of the uterus are located.

The lack of effect of oxytocin on anmiotic fluid concentrations of PGF and PGFM is of interest in light of the evidence that this hormone increases prostaglandin production from decidua and membranes *in vitro* (Fuchs, 1983). The present study shows that improvement in intrauterine contractions and rapid delivery occurred in

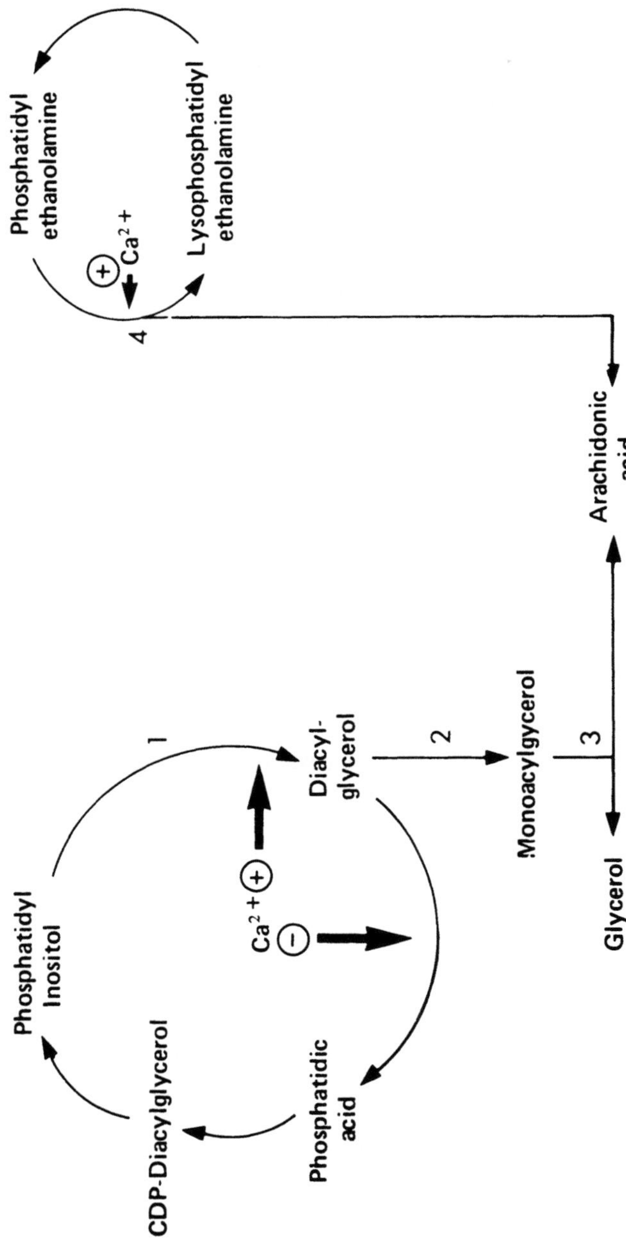

FIGURE 2. Proposed regulation of arachidonic acid mobilization in human fetal membranes (from Bleasdale et al., 1983). (1) Phospholipase C; (2) diacylglycerol lipase; (3) monocylglycerol lipase; (4) phospholipase A$_2$.

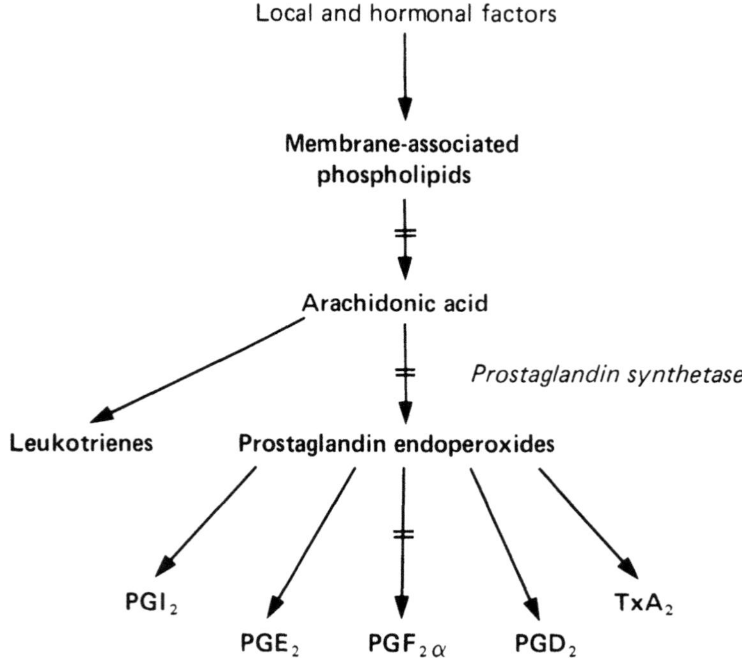

FIGURE 3. Possible sites of regulation of PGF$_{2\alpha}$ synthesis.

group II patients after administration of oxytocin to the mother despite continuing low levels of amniotic fluid PGF. This suggests that the higher levels of PGF in group I patients were not secondary to their more effective uterine contractions and that restoration of uterine pressure with oxytocin in group II patients did not influence PGF synthesis by the membranes. Oxytocin must act mainly through its direct action on the myometrium.

5. CONCLUSION

This study shows that an abnormality in the production of PGF is associated with dysfunctional hypotonic labor. This clinical condition may provide a model for studies of those factors thought to be involved in the regulation of prostaglandin production by intrauterine tissues in the human.

ACKNOWLEDGMENTS. Financial support was received from the South African Medical Research Council.

REFERENCES

Bleasdale, J. E., Okazaki, T., Sagawa, N., DiRenzo, G., Okita, J. R., MacDonald, P. C., and Johnston, J. M., 1983, The mobilization of arachidonic acid for prostaglandin production during parturition, in: *Initiation of Parturition: Prevention of Prematurity, Report of the Fourth Ross Conference on Obstetric Research* (P. C. MacDonald and J. Porter, eds.), Ross Laboratories, Columbus, OH, pp. 129–136.

Chard, T., Hudson, C. N., Edwards, C. R. W., and Boyd, N. R. H., 1971, Release of oxytocin and vasopressin by the human foetus during labour, *Nature* **234**:352.

Cowan, D. B., van Middlekop, A., and Philpott, R. H., 1982, Intrauterine pressure studies in African nulliparae. Normal labour progress, *Br. J. Obstet. Gynaecol.* **81**:257–263.

Fuchs, F., 1983, Endocrinology of parturition, in: *Endocrinology of Pregnancy* (F. Fuchs and A. Klopper, eds.), Harper & Row, Philadelphia, pp. 247–270.

Karim, S. M. M., 1966, Identification of prostaglandins in amniotic fluid, *Br. J. Obstet. Gynaecol.* **73**:903–908.

Karim, S. M. M., 1967, Prostaglandin content of amniotic fluid during pregnancy and labour, *Br. J. Obstet. Gynaecol.* **74**:230–234.

Keirse, M. N. J. C., 1978, Biosynthesis and metabolism of prostaglandins in the pregnant human uterus, *Adv. Prostaglandin Thromboxane Res.* **4**:87–102.

Keirse, M. N. J. C., 1983, Prostaglandins during human parturition, in: *Initiation of Parturition: Prevention of Prematurity, Report of the Fourth Ross Conference on Obstetric Research* (P. C. MacDonald and J. Porter, eds.), Ross Laboratories, Columbus, OH, pp. 137–144.

Keirse, M. N. J. C., and Turnbull, A. C., 1973, E prostaglandins in amniotic fluid during late pregnancy and labor, *Br. J. Obstet. Gynecol.* **80**:970–973.

Keirse, M. N. J. C., Ilint, A. P. F., and Turnbull, A. C., 1974, F prostaglandins in amniotic fluid during pregnancy and labor, *Br. J. Obstet. Gynecol.* **81**:131–135.

Keirse, M. N. J. C., Hicks, B. R., Mitchell, M. D., and Turnbull, A. C., 1977, Increase of the prostaglandin precursor, arachidonic acid, in amniotic fluid during spontaneous labour, *Br. J. Obstet. Gynaecol.* **84**:937–939.

Liggins, G. C., Fairclough, R. J., Grieves, S. A., Kendall, J. Z., and Knox, B. S., 1973, The mechanism of initiation of parturition in the ewe, *Recent Prog. Horm. Res.* **219**:111–149.

Mitchell, M. D., 1981, Prostaglandins during pregnancy and the perinatal period, *J. Reprod. Fertil.* **62**:305–315.

Mitchell, M. D., Keirse, M. N. J. C., Anderson, A. B. M., and Turnbull, A. C., 1977, Evidence for a local control of prostaglandins within the pregnant human uterus, *Br. J. Obstet. Gynaecol.* **84**:35–38.

Mitchell, M. D., Bibby, J., Hicks, B. R., and Turnbull, A. C., 1978, Specific production of prostaglandin E by human amnion *in vitro*, *Prostaglandins* **15**:377–382.

Norman, R. J., and Joubert, S. M., 1982, Fetal urinary prostaglandins, *Prostaglandins Leukotrienes Med.* **8**:171–172.

Norman, R. J., Bredenkamp, B. L. F., Joubert, S. M., and Beetar, C., 1981, Fetal prostaglandin levels in twin pregnancies, *Prostaglandins Med.* **6**:309–316.

Norman, R. J., Deppe, W. M., Coutts, P. C., and Joubert, S. M., 1983, Twin pregnancy as a model for studies in fetal cortisol concentrations in labour: Relation to prostaglandins, prolactin and ACTH, *Br. J. Obstet. Gynaecol.* **90**:1033–1039.

Nugteren, D. H., and Hazelhof, E., 1973, Isolation and properties of intermediates in prostaglandin biosynethsis, *Biochim. Biophys. Acta*, **326**:448.

Okazaki, T., Okita, J. R., MacDonald, P. C., and Johnston, J. M., 1978, Initiation of human parturition: X. Substrate specificity of phospholipase A$_2$ in human fetal membranes, *Am. J. Obstet. Gynecol.* **130**:432–438.

Philpott, R. H., 1972, Graphic records in labor, *Br. Med. J.* **4**:163–165.

Reddi, K., Kambaran, S. R., Norman, R. J., Joubert, S. M., and Philpott, R. H., 1972, Abnormal concentrations of prostaglandins in amniotic fluid during delayed labour in multigravid patients, *Br. J. Obstet. Gynaecol.* **91:**781–787.

Robinson, J. S., and Mitchell, M. D., 1983, Prostaglandins and parturition, in: *Initiation of Parturition: Prevention of Prematurity, Report of the Fourth Ross Conference on Obstetric Research* (P. C. MacDonald and J. Porter, eds.), Ross Laboratories, Columbus, OH, pp. 71–77.

Turnbull, A. C., Anderson, A. B. M., Flint, A. P. F., Jeremy, J. Y., Keirse, M. N. J. C., and Keirse, M. D., 1977, Human parturition, *Ciba Found. Symp.* **47:**427–451.

Zuckerman, H., Reiss, U., and Rubenstein, I., 1974, Inhibition of human premature labor by indomethacin, *Obstet. Gynecol.* **44:**787–792.

45

The Differentiated Contractile Effects of Prostaglandins on the Various Segments of the Pregnant Human Uterus

B. LINDBLOM, M. WIKLAND, N. WIQVIST, I. BRYMAN, and A. NORSTRÖM

1. INTRODUCTION

It is by now well established that prostanoids play a central role in the regulation of myometrial activity and thereby in the physiological control of human parturition (Liggins, 1979). Evidence for this concept is that exogenous PGs are effective for induction of labor (Karim *et al.*, 1969; Thiery and Amy, 1975) and that inhibitors of PG biosynthesis are able both to delay spontaneous labor (Lewis and Schulman, 1973) and to inhibit preterm labor (Zuckerman *et al.*, 1974; Wiqvist, 1979). To date, though, no evidence has emerged showing an increased formation of PG compounds prior to the spontaneous onset of parturition.

Normal labor is a complex kinetic process requiring adequate coordination of the uterine forces. The uterine wall of the upper segment becomes shorter as labor proceeds. Concomitant with this, there is a circumferential dilatation of the lower uterine segment and a gradual thinning of the uterine wall (Danforth and Ivy, 1949). Furthermore, the uterine cervix undergoes fundamental structural changes prior to term and during the course of parturition (Uldbjerg *et al.*, 1983).

It may therefore be hypothesized that the various uterine segments may react differently to endogenous myoactive products, e.g., prostanoids, and that the re-

B. LINDBLOM, M. WIKLAND, N. WIQVIST, I. BRYMAN, and A. NORSTRÖM ● Department of Obstetrics and Gynecology, University of Göteborg, Göteborg, Sweden.

sponse of a certain segment to a certain PG compound may change in temporal relation to the initiation of labor.

The aim of the present investigation was to study the effect of PGE_2, $PGF_{2\alpha}$, and PGI_2 on the contractility of the upper and lower uterine segments as well as the cervix before and during spontaneous labor. The effects were compared with the responses produced by oxytocin and norepinephrine.

2. MATERIAL AND METHODS

Tissue specimens were obtained from women who delivered in the 39th through 40th week of gestation. The women underwent elective or acute cesarean section. Clinical criteria for active labor were (1) regular spontaneous contractions (frequency $\geqslant 3$ per 10 min, duration >60 sec) and (2) cervical dilatation > 4 cm. None of the patients received oxytocin or other stimulatory or tocolytic drugs during the course of the delivery. The determination of gestational age was based on ultrasound examination in the 16th to 18th week of pregnancy.

Within a few minutes following delivery of the child and placenta, small specimens were excised from the fundal region of the uterine body ("upper segment") as well as from the lower edge of the transverse incision in the lower uterine segment. After identification of the internal os, cervical tissue specimens were isolated by the use of a biopsy needle with an inner diameter of 1.5 mm (Tru Cut, Travenol, Deersfield, IL, USA). All preparations were immediately immersed in ice-chilled oxygenated Krebs–Ringer bicarbonate (KRB) buffer and transferred to the laboratory. Under a stereomicroscope equipped with transluminant light, tissue strips with a cross-sectional area of approx 0.5 mm^2 and a length of 4 mm were prepared from the upper and lower segments. Cervical preparations measured 4–5 mm in length and had a cross-sectional area of 1–1.5 mm^2.

The strips were mounted in mantled 3-ml tissue chambers perfused by aerated (96% O_2, 4% CO_2) KRB buffer fortified with 10 mM d-glucose. Four specimens could be studied simultaneously, and the superfusion was maintained at a flow rate of 1–2 ml/min at a constant temperature of 37°C with a pH of 7.4.

The free end of each specimen was connected to a force transducer (Grass FT-03), from which the signals were amplified and registered on a polygraph (Grass Model 7D). For further methodological details see Wikland *et al.* (1982) and Bryman *et al.* (1984).

The strips were exposed to a drug during a 5- to 10-min period by injection of a minute amount directly into the tissue chamber. The data given in Section 3 represent the effect obtained with one or two strips from each uterine segment of an individual patient.

The following drugs were used: prostaglandin E_2 sodium salt, prostaglandin $F_{2\alpha}$ (Prostin®), both purchased from The Upjohn Co., Kalamazoo, MI, USA; PGI_2 sodium salt (kindly supplied by the Pharmaceutical Division, ICI, Ltd., Manchester, U.K.); crystalline 5,8,11,14-eicosatetraynoic acid (ETYA) (kindly supplied by F.

Hoffmann-LaRoche Ltd., Basel, Switzerland); indomethacin (kindly supplied by Merck Sharp & Dohme, Rahway, NJ, USA); norepinephrine (ACO Ltd., Sweden); and oxytocin (Syntocinon®, Sandoz AG, Basel, Switzerland).

3. RESULTS

Spontaneous contractile activity appeared in 70–90% of strips from the upper and lower segments and in 69% of cervical specimens. Oxytocin (OX) induced a marked excitatory effect on all segments at concentrations ranging from 0.01 to 100 mU/ml. Norepinephrine (NE) likewise had an excitatory effect at concentrations of 0.01–100 mM. There was no variation in the response to oxytocin or NE that could be related to the presence or absence of active labor.

Before labor, PGE_2 induced a stimulatory response in both segments when low concentrations were used. At higher concentrations ($\geq$10 ng/ml), a biphasic response (stimulation followed by inhibition) appeared. During labor, PGE_2 stim-

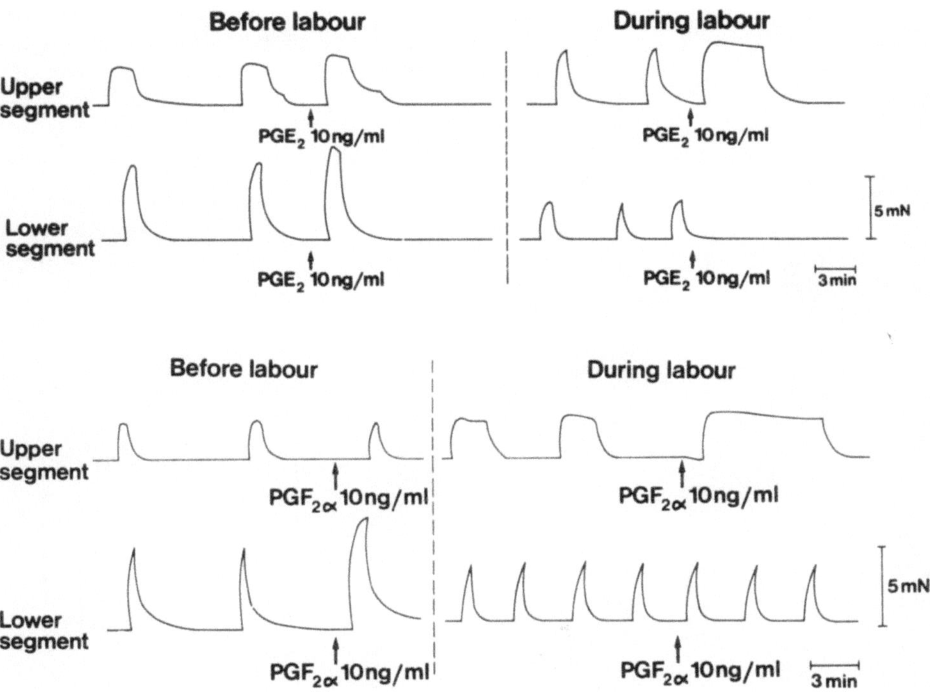

FIGURE 1. Responses of the upper and lower uterine segments to PGE_2 before and during labor. Note the powerful stimulatory response in the upper segment during labor and the relaxation of the lower segment. B: Responses of the upper and lower uterine segments to $PGF_{2\alpha}$ before and during labor. Note the absence of any excitatory effect in the upper segment before labor.

ulated the upper segment at all concentrations tested, whereas the lower segment showed an inhibitory or inconsistent response (Figs. 1A, 2).

Surprisingly, $PGF_{2\alpha}$ was without effect on the upper segment before labor but produced an excitatory response during labor. On the lower segment, $PGF_{2\alpha}$ was excitatory before labor, whereas during labor there was generally no effect (Figs. 1B, 3).

Prostacyclin induced a biphasic response (stimulation followed by inhibition) in preparations from the upper and lower segments but only at comparatively high concentrations. This effect occurred in 65% of cases before as well as during labor. In the remaining subjects, PGI_2 was without any effect. In a few cases the initial stimulation failed to occur in the lower segment, and thus there was an initial inhibition (Fig. 4).

The response of the upper and lower segments to PGI_2 is summarized in Fig. 5.

Cervical specimens responded to PGE_2 by inhibition before as well as during labor. These effects occurred at extremely low concentrations (0.0001–1.0 ng/ml). Prostaglandin $F_{2\alpha}$ caused an inhibitory response before as well as during labor. However, this effect occurred only at comparatively high concentrations (≥ 100 ng/ml). Prostacyclin always had an inhibitory action on the cervix, occurring at concentrations ranging from 0.1 to 100 ng/ml (Fig. 6).

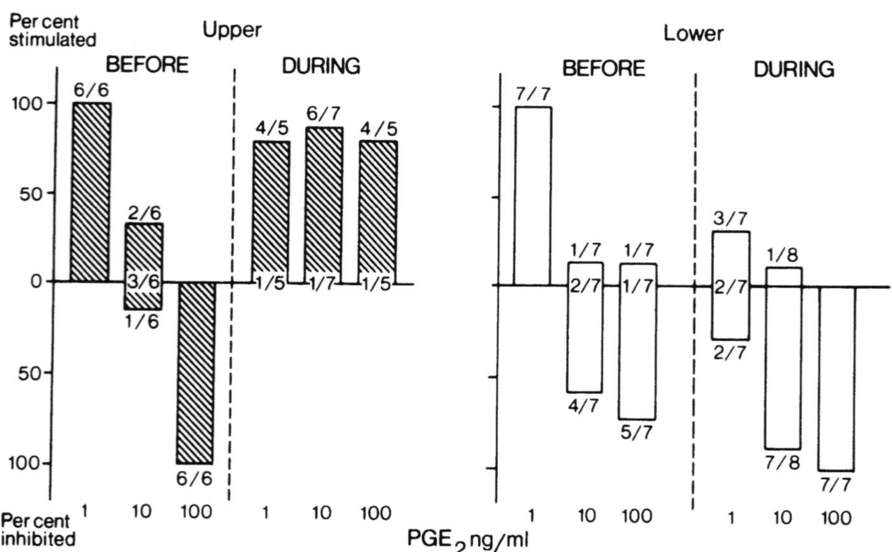

FIGURE 2. Diagrammatic illustration of the effects of PGE_2 at different concentrations. The figures at the bars indicate numbers of subjects. Figures at the zero line show the proportion of nonresponders.

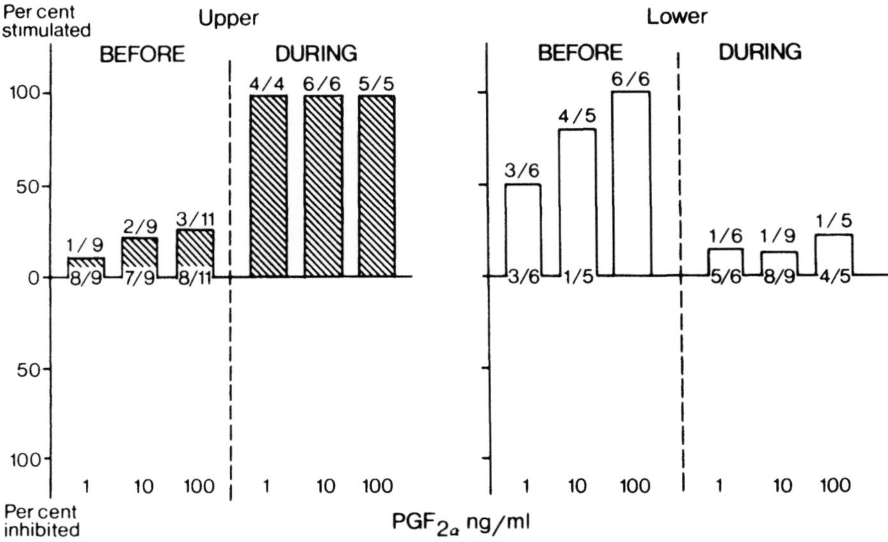

FIGURE 3. Diagrammatic illustration of the responses induced by various concentrations of PGF$_{2\alpha}$. Inhibitory responses were never observed. For explanation, see Fig. 2.

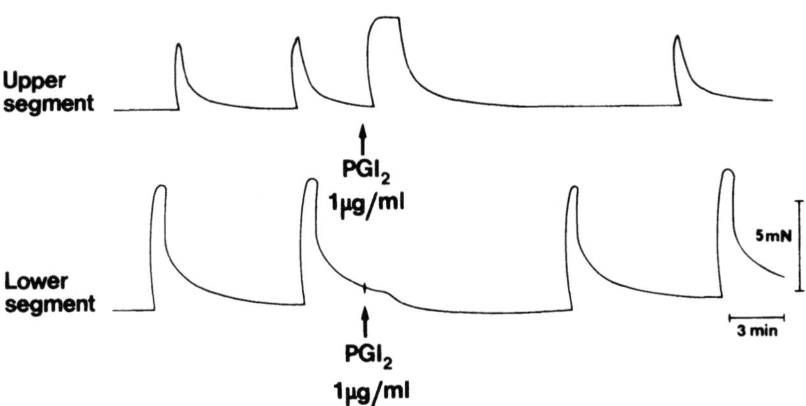

FIGURE 4. An example of the effect of PGI$_2$. In this particular case there was no stimulatory response in the lower segment, whereas in the upper segment there was a biphasic effect (stimulation followed by inhibition).

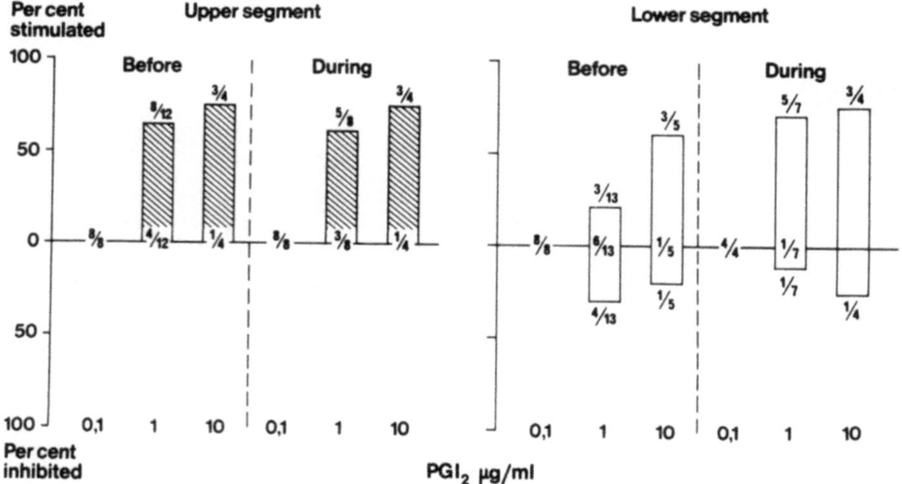

FIGURE 5. A summary of the effects of PGI$_2$ (note concentrations are in µg/ml). Only the initial effect is shown. In the upper segment the typical response is an initial stimulation. The lower segment reacts by either stimulation, inhibition, or no response before as well as during labor. For explanation, see Fig. 2.

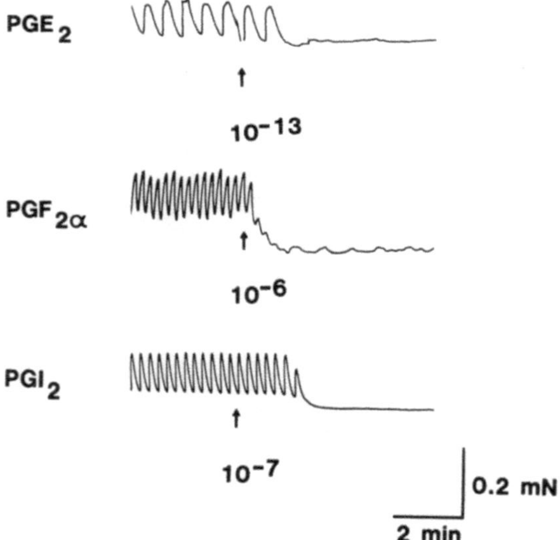

FIGURE 6. The effects of various PGs on cervical muscle strips: PGE$_2$ has an inhibitory effect at very low concentrations (1 pg/ml); also PGF$_{2\alpha}$ and PGI$_2$ act inhibitory, but only at high concentrations.

TABLE I. Response of Uterus and Cervix to Various Agents before Labor[a]

	OX	NE	PGE$_2$	PGF$_{2\alpha}$	PGI$_2$
Upper segment	+	+	+ / −	0	+ / −
Lower segment	+	+	+ / −	+	+ / −
Cervix	+	+	−	−	−

[a] Symbols. +, stimulation. −, inhibition; + / −, stimulation followed by inhibition, i.e., a biphasic response; 0, no response.

The compiled data are summarized in Tables I and II. During inhibition of spontaneous activity induced by indomethacin or ETYA (10–30 μg/ml), phasic contractile activity could be restored in the upper and lower segment by the addition of PGF$_{2\alpha}$ (10–100 ng/ml). However, PGF$_{2\alpha}$ was not able to reestablish contractile activity in cervical tissue during ETYA treatment.

4. DISCUSSION

Prostanoids are now considered to represent a link among various mechanisms regulating the timing and course of parturition. Theoretically, both the pattern of production of various PG compounds within a tissue and the myogenic response of that tissue may vary with the physiological condition. Furthermore, the action of the substances may be different when they act within the cell where they are synthesized and when they reach neighboring tissues by diffusion. The present data provide evidence that prostanoids, in contrast to oxytocin and NE, may induce opposite responses on the different segments of the term pregnant uterus. This differentiation of effects may be explained by an uneven distribution of PG receptors in different parts of the uterus, as demonstrated in the nonpregnant state (Crankshaw et al., 1979). Furthermore, the described shift in myometrial response to PGs at the time of spontaneous labor could depend on a change in the affinity for certain PG compounds (Bauknecht et al., 1981). Although PGE$_2$ and PGF$_{2\alpha}$ may induce different responses on the same specimen, PGI$_2$ generally induces the same qualitative response as PGE$_2$ (see Tables I, II), supporting the concept that the latter substances may act on the same receptor (Tougui et al., 1980).

Taken together, the present results favor the hypothesis that labor in the human female is associated with a critical change in the myometrial reaction to prostanoids.

TABLE II. Response of Uterus and Cervix to Various Agents during Labor[a]

	OX	NE	PGE$_2$	PGF$_{2\alpha}$	PGI$_2$
Upper segment	+	+	+	+	+ / −
Lower segment	+	+	−	0	+ / −
Cervix	+	+	−	−	−

[a] Symbols: +, stimulation; −, inhibition; + / −, stimulation followed by inhibition, i.e., a biphasic response; 0, no response

These alterations may involve suppression of the expulsive forces and tightening of the lower uterine segments during pregnancy. Prior to or accompanying the start of labor, this reactivity pattern is changed to a marked excitatory response towards PGE_2 and $PGF_{2\alpha}$ in the upper segment whereas the lower segment reacts in a manner that favors dilatation. The smooth muscle of the cervix changes its reaction to PGs quantitatively and qualitatively during pregnancy (Bryman *et al.*, 1984), although there is no evidence for a changed response prior to initiation of labor.

Finally, it is realized that information concerning the influence of more recently discovered arachidonate products must be gathered and taken into consideration in evaluating the role of prostanoids in human parturition.

ACKNOWLEDGMENTS. Thanks are due to Mrs. Ann Andersson for valuable technical assistance and to Mrs. Ann-Louise Dahl for expert secretarial help.

REFERENCES

Bauknecht, T., Krahe, B., Rechenbach, U., Zahradnik, H. P., and Breckwoldt, M., 1981, Distribution of prostaglandin E_2 and $PGF_{2\alpha}$ receptors in human myometrium, *Acta Endocrinol. (Kbh)* **98:**446–450.

Bryman, I., Sahni, S., Norström, A., and Lindblom, B., 1984, Influence of prostaglandins on contractility of the isolated human cervical muscle, *Obstet. Gynecol.* **63:**280–284.

Crankshaw, D. J., Crankshaw, J., Branda, L. A., and Daniel, E. E., 1979, Receptors for E type prostaglandins in plasma membrane of nonpregnant human myometrium, *Arch. Biochem. Biophys.* **198:**70–77.

Danforth, D. N., and Ivy, A. C., 1949, The lower uterine segment, *Am. J. Obstet. Gynecol.* **57:**831–841.

Karim, S. M. M., Trussel, R. R., and Hillier, K., 1969, Induction of labour with prostaglandin $F_{2\alpha}$, *J. Obstet. Gynaecol. Br. Commonw.* **76:**769–782.

Lewis, R. B., and Schulman, J. D., 1973, Influence of acetylsalicylic acid, an inhibitor of prostaglandin synthesis, on the duration of human gestation and labour, *Lancet* **2:**1159–1161.

Liggins, G. C., 1979, Initiation of parturition, *Br. Med. Bull.* **35:**145–150.

Thiery, M., and Amy, J.-J., 1975, Induction of labour with prostaglandins, in: *Advances in Prostaglandin Research, Prostaglandins and Reproduction* (S. M. M. Karim, ed.), MTP Press, Lancaster, pp. 149–228.

Tougui, Z., Do Kahc, L., and Harbon, S., 1980, Modulation of cyclic AMP content of the rat myometrium: Densensitization to isoproterenol, PGE_2 and prostacyclin, *Mol. Cell. Endocrinol.* **20:**17–22.

Uldbjerg, N., Ulmsten, U., and Ekman, G., 1983, The ripening of the human uterine cervix in terms of connective tissue biochemistry, *Clin. Obstet. Gynecol.* **26:**14–26.

Wikland, M., Lindblom, B., Wilhelmsson, L., and Wiqvist, N., 1982, Oxytocin, prostaglandins, and contractility of the human uterus at term pregnancy, *Acta Obstet. Gynecol. Scand.* **61:**467–472.

Wiqvist, N., 1979, The use of inhibitors of prostaglandin synthesis in obstetrics, in: *Human Parturition, Boerhaave Series,* Vol. 15 M. J. N. C. Keirse, A. M. B. Anderson, and J. Bennebroek Gravenhorst, eds.), Leiden University Press, Leiden, pp. 189–200.

Zuckerman, H., Reiss, U., and Rubinstein, I., 1974, Inhibition of human premature labor by indomethacin, *Obstet. Gynecol.* **44:**787–792.

V

Immune Reactions, Neoplasia, and Inflammatory Processes

46

Membrane Events and Guanylate Cyclase Activation in Mitogen-Stimulated Lymphocytes

JOHN W. HADDEN, ELBA M. HADDEN, and RONALD G. COFFEY

1. INTRODUCTION

The lymphocyte has provided an important model for the study of the early biochemical events associated with the initiation of cellular proliferation. Resting lymphocytes (G_0 or restricted G_1 phase), as obtained from human peripheral blood, can be induced in large numbers to undergo clonal proliferation using a variety of mitogens. The two most commonly employed plant lectin mitogens are phytohemagglutinin (PHA) and concanavalin A (Con A). The initiation process is characterized by rapid, saturable binding of the lectin to cellular receptors (<30 min), followed by a series of cellular events involving probably three cell types.

The thymus-dependent (T) cell triggered to proliferate is activated to enter the G_1 phase of the cell cycle, to synthesize RNA and protein, and to display on its surface new receptors including those for interleukin II (IL-2, formerly called T-cell growth factor). The adherent monocyte triggered by the lectin produces a soluble mediator termed interleukin I (IL-1, formerly called lymphocyte-activating factor), which in turn acts on a T helper cell and/or a large granular lymphocyte and induces it to synthesize and secrete interleukin II. The IL-2 then acts on the T cell originally primed to enter G_1 phase, and this cell enters the S phase of the cell cycle and replicates. The requirement for monocytes in this process can be obviated by their removal and replacement by exogenously added IL-1 or by using the mitogen

JOHN W. HADDEN, ELBA M. HADDEN, and RONALD G. COFFEY ● Program of Immunopharmacology, University of South Florida Medical College, Tampa, Florida 33612.

phorbol myristate acetate (PMA). The requirements for both the monocyte and the IL-2 producer cell can be obviated by their removal and the exogenous addition of preformed IL-2. The priming of accessory cells and the initiation of T-cell replication in this system offer a unique system for the characterization of the cellular interactions and early biochemical events in the proliferative process.

One unique feature of this system is that the lectin mitogen does not need to enter the cell in order to initiate the process. Phytohemagglutinin and Con A bound to sepharose beads will initiate it. Another feature is that the lectin need be present for a restricted time period, <3 hr for PHA and an initial 3 hr as a first signal and a second 3-hr signal (at 18–21 hr) for Con A. In both cases, the initial 3-hr signal is a calcium-dependent one.

Much of the work to date on the biochemical events of lymphocyte activation has focused on the first 1 hr. A variety of biochemical processes have been studied (see Table I for a partial listing). Unfortunately, the experiments have not been performed with isolated cell populations, so it is not clear which events are occurring in lectin-activated accessory cells and which occur in the T cells that ultimately proliferate. One objective of this type of study is to characterize the initial biochemical events so that cause-and-effect relationships can be ascribed and an obligate series of sequential reactions can be detected that are common to lectin mitogens

TABLE I. Early Events of Lymphocyte Activation by Lectin Mitogens

1 min	Lectin receptor binding
	Calcium uptake
	Arachidonate release from phosphatidyl inositol
2–5 min	Potassium uptake, (Na/K) ATPase increases
	Phosphatidylethanolamine methylation
	Arachidonate release from phosphatidylcholine
	Cyclooxygenase and lipoxygenase products produced
	Guanylate cyclase activation
5–15 min	Cyclic GMP increases
	Glucose and nucleoside transport increase
	Arachidonate and phosphate uptake and incorporation in membrane lipids
	Calcium-dependent protein kinase activation
	Membrane and nuclear protein phosphorylation
20–30 min	DNA-dependent RNA polymerase changes (I and III ↑, II ↓)
	Histone phosphorylation and acetylation
	Phosphoribosyl pyrophosphate synthetase activation
	Acyl-CoA transferase activity increases
Later changes	mRNA and protein synthesis
	Cyclic AMP increases
	Cyclic nucleotide protein kinases increase
	Ornithine decarboxylase increases with polyamine synthesis
	Cell receptors appear for insulin, transferrin, IL-2, etc.
	Increases in glylcolysis, glycogen accumulation
	DNA synthesis

and perhaps to other mitogens as well. As a result of the recent emergence of IL-2 as a second signal required to complete the $G_1 \rightarrow S \rightarrow G_2 \rightarrow$ mitosis sequence, the initiation events studied to date may refer either to accessory cell activation and/or to the induction of entry of the T cell into the G_1 phase of the cell cycle. More complete accounts of the biochemical events have been detailed elsewhere, and the reader is referred to these sources for complete referencing and for an appreciation of other views of this controversial area (Hadden, 1977; Hadden and Coffey, 1982; Hume and Weidemann, 1980; Resch, 1976; Strom *et al.*, 1977; Wedner and Parker, 1976).

2. CYCLIC NUCLEOTIDES IN THE EARLY EVENTS OF LYMPHOCYTE ACTIVATION

This brief review focuses on the possible roles of cyclic nucleotides, cyclic 3′,5′-adenosine and -guanosine monophosphate (cAMP and cGMP, respectively) and calcium in the initiation process. The majority of studies have employed PHA or Con A as the lymphocyte activator and resting human peripheral blood lymphocytes as the target cells, and no discussion is provided concerning the nuances of employing thymocytes, splenocytes, or lymph node lymphocytes. Early studies provided evidence that lectin-stimulated lymphocytes show early changes in cellular levels of cAMP or cGMP (Smith *et al.*, 1971; Hadden *et al.*, 1972), and two schools of thought developed independently, each dwelling on one or the other cyclic nucleotide as the initiator. Numerous publications ensued in support of each school. Without discussing the conceptual and technical aspects of this controversy, we here present our view, which we interpret to be supported by the preponderance of the evidence.

Phytohemagglutinin and Con A both induce early increases (5–20 min) in cellular levels of cGMP (Hadden *et al.*, 1972). In our experience, they increase cAMP levels only if impure agglutinating preparations or high concentrations (supraoptimal) are used. Perhaps the clearest demonstration of the distinction between the two responses was observed in studies comparing the action of succinylated Con A (S-Con A) to the native, tetrameric Con A (Hadden *et al.*, 1976). Native Con A, like PHA, shows a bell-shaped dose–response curve on lymphocyte proliferation with Con A doses from 50 μg/ml up to 250 μg/ml progressively inhibiting the proliferative response. Succinylated Con A, which does not induce patch and cap formation characteristic of the native Con A, shows a plateau type of dose–response curve with optimal proliferation occurring from 25 μg/ml up to 250 μg/ml. Both forms of Con A produced early increases in cellular levels of cGMP, and only the native form induced increases in cAMP at supraoptimal concentrations (progressively from 50 to 250 μg/ml). We interpret these data to indicate that increases in cellular levels of cGMP are a consistent finding in lectin-induced lymphocyte proliferation, whereas the cAMP increases and the concomitant events of patch and cap formation and an associated microtubular-based freezing of the membrane are epiphenomena of mitogen action related to the "turn off" of supraoptimal concentrations of Con A. In support of this intepretation are a number of observations indicating that agents that increase cAMP block mitogen action whereas those that increase cGMP promote it. Since these early observations, more than 20 laboratories (see Hadden and Coffey, 1982, for review) have confirmed that cGMP changes occur with these and other mitogens and have associated the changes with guanylate cyclase activation and cGMP-dependent protein kinase activation. In addition, the defective lymphoproliferative responses observed in cancer, aging, malnutrition, and immunotoxicant exposure have been correlated with impaired early cGMP responses.

It was clear from the onset of this work that cGMP increases alone were not sufficient to induce lymphocyte proliferation since hormonal influences that increase cGMP would promote but not initiate the process. Several lines of evidence indicated that calcium might be involved in the signal mechanism.

3. CALCIUM IN THE EARLY EVENTS OF LYMPHOCYTE ACTIVATION

Previous work had shown that mitogens induce calcium influx in lymphocytes and that lymphocyte proliferation, particularly the early activation process, is dependent on the presence of calcium in the extracellular medium (Alford, 1970; Whitney and Sutherland, 1973). Based on prior associations of calcium with the guanylate cyclase/cGMP system, we were led to consider this a central feature. We confirmed the observations that the calcium ionophore A23187 was mitogenic for human lymphocytes and showed that its action to induce early increases of cellular levels of cGMP, as with proliferation, were dependent on extracellular calcium (Coffey *et al.*, 1977). We showed that deletion of extracellular calcium markedly reduced the cellular cGMP increases that occurred with PHA and Con A. We further noted that following lectin priming in the absence of calcium, the subsequent addition of calcium led to a brisk increase in cGMP levels not evident in the controls. We found that if calcium was deleted and EGTA added to chelate cell-associated calcium, no cGMP increases occurred despite adequate cellular binding of the lectin. These observations led us to conclude that the membrane events induced by lectins that lead to cGMP generation were dependent on calcium.

The work at this point took two directions. In order to insure that the cGMP/calcium-related events were germane to T-cell activation, we probed the nucleus for evidence that these intracellular metabolic signals were involved. In the process, we confirmed that the early cGMP changes were not dependent on adherent accessory cells, indicating but not proving that the relationship of cGMP and calcium to early nuclear events would involve the G_1 activation events of the T cells. The second direction involved a probing of the components of a transmembrane signal that would give rise to guanylate cyclase activation. This approach proved to be more complex, since guanylate cyclase, unlike adenylate cyclase, is not activated by direct contact with a coupling subunit having receptor binding properties.

4. CYCLIC GMP AND CALCIUM IN LYMPHOCYTE NUCLEAR ACTIVATION

A number of workers (Pogo *et al.*, 1966; Johnson *et al.*, 1974) had shown that within 1 hr of stimulation by lectin mitogens a number of nuclear processes were initiated, including stimulation of RNA-dependent DNA polymerases I and III (with inhibition of polymerase II), acetylation of histones, phosphorylation of nonhistone acidic nuclear proteins, and RNA synthesis. To test the premise that cGMP and calcium were involved in these processes, we (Johnson and Hadden, 1975a; Ananthakrishnan *et al.*, 1981) incubated intact lymphocytes with PHA in tne presence and absence of calcium or EGTA and isolated the nuclei at 1 hr and measured RNA synthesis using the incorporation of labeled uridine triphosphate (UTP) into RNase-sensitive nuclear constituents. In the presence of PHA, RNA synthesis increased linearly over the 10-min incubation of the isolated nuclei. In the absence of extracellular calcium, the response was markedly reduced. With the further addition of EGTA, as with the cGMP response, no nuclear RNA synthesis occurred. These observations supported the hypothesis that the cGMP and calcium signal originating at the cell surface were intimately linked to nuclear activation.

We (Johnson and Hadden, 1975a; Ananthakrishnan *et al.*, 1981) further showed that under both isotonic conditions and high-salt conditions (to restrict reinitiation of RNA synthesis), addition of cGMP and calcium (at concentrations consistent with calculated intracellular levels following mitogen stimulation, i.e., 10^{-9} M and 10^{-4} M, respectively) to nuclei isolated from resting lymphocytes induced RNA synthesis. The degree of stimulation of RNA synthesis was comparable to that observed in the nuclei of lymphocytes primed by PHA.

Subsequent studies showed that cGMP binds to partially purified DNA-dependent RNA polymerases I, II, and III and increases or decreases their RNA synthesizing capabilities in a manner parallel to the changes that occur in nuclei of PHA-activated lymphocytes (i.e., I and III increase and II decreases within 1 hr) (Johnson and Hadden, 1977). Cyclic GMP also binds to a number of nonhistone acidic nuclear proteins, which are phosphorylated when lymphocytes are stimulated by lectins or other cGMP-raising agents (Johnson and Hadden, 1975b; Johnson *et al.*, 1975). No such changes occurred with addition of cAMP. Although much remains to be learned about the interrelationship of calcium with these events, our data indicate that the effects of calcium may relate to the increase in initiation sites that occurs with PHA activation and the effects of cGMP relate to the rate of transcription of these sites. The possible role of cGMP-dependent protein kinase in the protein phosphorylation associated with these events remains to be elucidated.

These observations, although only scratching the surface, are sufficient to support the hypothesis that cGMP and calcium are at least two components of lectin mitogen-induced membrane to nuclear signal.

5. GUANYLATE CYCLASE ACTIVATION IN MITOGEN ACTION

Initial studies indicated that PHA-induced early increases in lymphocyte cGMP levels were variable in magnitude and transient in timing. Since the catabolism of cGMP was shown not to be impaired following PHA, studies were initiated to analyze the effects of PHA on guanylate cyclase activation. To date we have performed more than 60 such experiments. Highly reproducible effects of PHA to stimulate membrane and to a lesser extent soluble guanylate cyclase have been observed, which persist up to 4 hr following the disruption of the cells and the isolation of the guanylate cyclase fractions by differential and isopycnic centrifugation. These observations indicate that the activation of guanylate cyclase by PHA is a relatively stable process and, thus, that the more variable changes in cellular levels of cGMP reflect only a small portion of a rapid turnover process with associated extensive intracellular destruction and binding, which probably obscure detection of free cGMP.

Experiments parallel to those preformed with cGMP levels and nuclear RNA synthesis in which calcium was excluded or chelated with EGTA (Coffey *et al.*, 1981) were performed with guanylate cyclase activity of broken cells, and it was shown that extracellular calcium (therefore, calcium influx) was largely responsible for PHA-induced guanylate cyclase activation. In the presence of EGTA, as with increases in cGMP levels and RNA synthesis, no activation of guanylate cyclase occurred.

Throughout our experiments with guanylate cyclase, three cationic environments were employed (i.e., magnesium, manganese, and calcium), since the appropriate intracellular cationic requirements of guanylate cyclase are thought to be calcium or magnesium, and manganese produces maximal activity. The levels of basal activity and stimulated activity in lymphocytes, as in other tissues, were greatest with manganese, less with magnesium, and least with calcium. Phytohemagglutinin-activated guanylate cyclase was measured with each cation. Although calcium stimulates the enzyme, the degree of stimulation, even with the concomitant presence of magnesium, compared to the increases that occur following PHA activation, is not of such a magnitude as to allow an interpretation that calcium itself is the primary stimulant. This observation led us to probe what other calcium-dependent metabolic processes might be involved.

Our initial observations characterizing the increase in guanylate cyclase activity induced by PHA demonstrated several distinguishing features (Coffey *et al.*, 1981). In the lymphocyte, approximately one-half of the guanylate cyclase is membrane bound, and the greatest (fourfold) and the earliest (2 min) increases following PHA occurred in a membrane fraction that was confirmed to be at least in part plasma membrane based on its density in sucrose gradients and its association with other known lymphocyte plasma membrane markers. The PHA-induced increases in the soluble enzyme were delayed and differed somewhat in cation dependence (principally manganese dependent). These observations have suggested that the soluble cytoplasmic enzyme is different in its mechanism of stimulation or represents an altered, translocated membrane enzyme.

Use of the reducing agent dithioerythritol showed that PHA activation of guanylate cyclase involved probable oxidation steps perhaps of the enzyme itself (Coffey *et al.*, 1981). Work of Goldberg and co-workers at that time (Haddox *et al.*, 1978; Graff *et al.*, 1978) gave us an important lead. Working with ascorbate and hydroperoxyeicosatetraenoic acids in spleen cells, which include lymphocytes, these investigators presented evidence that a variety of oxidation reactions could activate guanylate cyclase. These observations suggested to us that the release of oxidizing eicosanoids in mitogen-activated lymphocytes might well provide a mechanism of guanylate cyclase activation. At that time, the idea that arachidonic acid (AA), the precursor of these eicosanoids, was important was being treated in the context that AA incorporation into membrane phospholipids was critical (see Resch, 1976, for review). Its release had not been seriously considered, although in the activation of other cell types (e.g., platelets and mast cells) it seemed important.

A single experiment performed in February 1978 convinced us that it was relevant to actively pursue this avenue. We added the products of phospholipase, arachidonic acid, and lysolecithin to lymphocyte homogenates and observed that AA activated guanylate cyclase. Indomethacin, an inhibitor of cyclooxygenase and therefore of thromboxane and prostaglandin synthesis, had no effect on the activation of guanylate cyclase. 5,8,11,14-Eicosatetraenoic acid (ETYA), an inhibitor of lipoxygenase (as well as cyclooxygenase) and therefore of the synthesis of lipoxygenase-derived eicosanoids, completely blocked the activation of guanylate cyclase. These were the first observations directly implicating the lipoxygenase pathway in mitogen-induced lymphocyte activation. From this point, events moved rapidly in many laboratories, and lack of space makes a complete accounting impossible.

The strategy in detailing these events has involved three approaches. The first is that the lectin, particularly PHA, elicits certain changes in membrane enzyme activation and releases certain products; the second is that inhibitors of these processes block lymphocyte proliferation and, for our purposes, block guanylate cyclase activation; and the third is that readdition of the required product following inhibition restores the proliferative process and the activation of guanylate cyclase. Not all approaches have been sufficiently tested at this time to allow final conclusions, and the lack of specificity in the use of available inhibitors entails pitfalls. The story accumulated to date is of interest in detailing, for the first time, a tentative metabolic sequence for a transmembrane mitogen signal leading to the production of cGMP.

Perhaps the first event following lectin mitogen binding to the lymphocyte cell surface involves the methylation of membrane phospholipids. Hirata *et al.* (1980)

have presented evidence to show that the inhibition of the methylation of phosphatidylethanolamine (PE) blocks PHA- or Con A-induced arachidonic acid release and prevents lymphocyte proliferation. Correspondingly, lectins induce methylation of PE and increase synthesis of phosphatidylcholine (Fisher and Meuller, 1971; Hirata *et al.*, 1980; Parker *et al.*, 1979a,b). Hirata and co-workers (1980) have suggested that the methylation of phospholipids by lectin mitogens forms the basis for the calcium ionophoretic effect. Interestingly, nonlectin mitogens, the calcium ionophore A23187, and phorbol myristate acetate (PMA) do not induce phospholipid methylation but directly activate one or more phospholipases. A number of laboratories (Hirata *et al.*, 1980; Parker *et al.*, 1979a,b; Goetzl, 1981) have demonstrated that the lectin mitogens and PMA and the calcium ionophore all induce the release of arachidonic acid. In the case of the lectin mitogens, the sources of the arachidonic acid appear to be both phosphatidylinositol and phosphatidylcholine (Parker *et al.*, 1979a,b).

Based on the apparent sources of arachidonate, several mechanisms are implied: a phospholipase A_2 cleaving AA from the 2 position of phosphatidylcholine, a phospholipase C cleaving inositol phosphate from the 3 position of phosophatidylcholine, or inositol and a diglyceride lipase releasing AA from either the 1 or 2 position of the diglyceride. Of these three enzymes, only phospholipase C has been demonstrated in lymphocytes (Allan and Mitchell, 1974). A fourth possibility involves the suggested reversal of acyl-CoA transferase, which preferentially incorporates AA into the 2 position of phospholipids (Trotter *et al.*, 1982; K. Resch, personal communication). Functionally, this enzyme would perform as a phospholipase A_2 operating in reverse and coupled with an enzyme that cleaves arachidonoyl coenzyme A. The inhibitor sensitivity of the enzyme releasing AA from membrane phospholipids is that of phospholipase A_2. Cyclic AMP, quinacrine, hydrocortisone, dimethyl-DL-2,3-distearoyloxypropyl-2-hydroxyethylammonium acetate, and *p*-bromophenacylbromide inhibit phospholipases and are thought or have been shown to inhibit arachidonate release in lymphocytes. The acyl-CoA transferase is also sensitive to these inhibitors (K. Resch, personal communication). Although the specific enzyme has not been identified, it is clear that inhibitors of methylation and AA release block lymphocyte activation.

Mitogen-induced AA release results in the formation by the cyclooxygenase pathway of thromboxane A_2 and B_2 but little in the way of prostaglandins (Parker *et al.*, 1979a,b; Goetzl, 1981). Inhibition of this pathway with indomethacin has essentially no effect on, or increases, lymphocyte activation by lectin mitogens, as is the case for guanylate cyclase activation. More importantly, in mitogen-stimulated lymphocytes, AA gives rise to 5-, 11-, 12-, and 15-hydroxyeicosatetraenoic acids (HETEs) and leukotriene B_4 via the lipoxygenase pathway. Presumably, but difficult to measure, the precursor 5-, 11-, 12-, and 15-hydroperoxides are also formed. Inhibition of the lipoxygenase pathway inhibits the generation of these eicosanoids and inhibits lymphocyte proliferation induced by lectin. Further support for the relevance of these metabolites is provided by the observation that the addition of

AA, in preference to other fatty acids, to lectin-stimulated lymphocytes promotes lymphocyte proliferation (Kelly and Parker, 1979). This action is blocked by ETYA but not indomethacin. Arachidonic acid or one product of this pathway, 5-HETE, does not itself induce lymphocyte proliferation, and to date the precise mechanism of the calcium ionophoretic action thought necessary for proliferation remains unclear. In addition to monomethylethanolamine, phosphatidic acid (a calcium ionophore in other tissues) remains a possibility.

Inhibitors of phospholipid methylation and phospholipase inhibit guanylate cyclase activation in addition to lymphocyte proliferation (Coffey *et al.*, 1981). In addition, inhibitors of lipoxygenase (Coffey *et al.*, 1981) inhibit both guanylate cyclase activation and lymphocyte proliferation. The question remains as to what products of lipoxygenase are involved. The principal metabolite in lymphocytes is 5-HETE and presumably 5-HPETE, its precursor; the 5-HPETE is rapidly and spontaneously converted to 5-HETE, and for lack of availability of 5-HPETE we have been unable to test it. Dr. J. Martyn Bailey and Dr. Edward Goetzl have kindly provided us with 5-HETE, 12-HETE, and 11-HETE. In intact lymphocytes all three activate guanylate cyclase, and 5-HETE is the most active. In guanylate cyclase prepared from broken cell preparations, the addition of these HETEs to lymphocyte homogenates or to the membrane and soluble enzyme produces little or no effect (see R. C. Coffey and J. W. Hadden, this volume). Since the HETEs themselves do not have the oxidative potential thought important to activate guanylate cyclase directly, the HPETEs seem to be the most relevant to test, and plans are underway to do so. Leukotriene B$_4$ remains to be tested as well.

Reported by R. C. Coffey and J. W. Hadden in this volume is the effect of 15-HETE to block the effect of PHA to activate guanylate cyclase in lymphocytes. Dr. Bailey (Bailey *et al.*, 1982) previously showed that 15-HETE blocks lymphocyte proliferation induced by two mitogens, PHA and PMA. We demonstrated that both 15-HETE and ETYA block the effect of 5-HETE added extracellularly to activate guanylate cyclase. 15-Hydroxyeicosatetraenoic acid has been suggested to be a negative feedback regulator in the activation of this pathway by acting to inhibit the lipoxygenase responsible for the formation of 5- and 12-HETEs. The inhibition of the action of 5-HETE to stimulate guanylate cyclase is a new observation. The fact that 5-HETE presumably acts on intact cells implies that its role is extracellular and involves receptors. The observation that the effect is blocked by ETYA and 15-HETE implies that it promotes the conversion of AA to active eicosanoids or that it enhances release of AA, which in turn results in the further production of eicosanoids. In either case, it would appear to operate as a positive feedback regulator.

An important question is whether the action of 5-HETE is primarily directed to the same cell or to other cells. McCarty and Goetzl (1979) have shown that exogenous 5-HETE is chemokinetic forT lymphocytes. It is possible that the release of 5-HETE by lymphocytes acts in a lymphokinelike manner to promote the migration and aggregation of cells that are known to occur following mitogen stimulation. These possibilities of extracellular as well as intracellular signals now make it mandatory to employ isolated cell populations to probe the role of these metabolites in cell–cell interactions previously thought to be interleukin mediated.

It has been suggested at this meeting by Goldyne that T lymphocytes do not make lipoxygenase products and that monocytes do. This suggestion was made on the basis of experiments using double rosette-purified T cells (a small subpopulation of the total T-cell population) with or without monocytes. This suggestion is contradictory to many of the observations and interpretations presented in this chapter and deserves comment. The observations concerning release of arachidonate and the formation of eicosanoids (Goetzl, 1981; Parker *et al.*, 1979a,b) were performed following the removal of adherent cells (monocytes), and, as stated, guanylate cyclase activation, increases in cGMP, nuclear activation, and the triggering of T cells into the G_1 phase of the cell cycle all have been observed to occur in the absence of adherent cells. It may be that intracellular communication occurs and is necessary for the production of eicosanoids in T cell populations; however, rather than the monocyte, nonadherent accessory cells would appear to participate. The population of large granular lymphocytes as well as T helper cells would seem likely candidates. We plan to test these possibilities in the near future.

6. SUMMARY AND CONCLUSION

Lectin mitogens, PHA and Con A, activate T lymphocytes in a complex manner dependent on cell–cell interaction and intercellular molecular communication involving interleukins I and II and perhaps also AA, 5-HETE, cGMP, and other active substances known to be liberated by activated leukocytes. It is clear that lectin mitogens can activate T cells for RNA and protein synthesis in the absence of accessory cells and under similar conditions will induce calcium influx, cGMP increases, and activation of guanylate cyclase. These observations in conjunction with evidence indicating that cGMP and calcium are linked to nuclear activation processes known to be essential for RNA and protein synthesis allow the formulation of a hypothesis that cGMP and calcium are part of the initial triggering process ($G_0 \rightarrow G_1$) for T-lymphocyte activation. Second signals involving actions of IL-1 and IL-2 to provide a second signal ($G_1 \rightarrow M$) remain to be clarified mechanistically.

Based on the collected data, the calcium-dependent mechanisms by which PHA activates membrane guanylate cyclase involve phospholipid methylation, AA release by a yet to be defined phospholipase, and AA conversion to HPETEs, HETEs, and leukotriene B_4. Based on indirect evidence, HPETEs are good candidates for the proximal stimulants of guanylate cyclase. Evidence has been presented that 5-HETE is a positive feedback regulator acting extracellularly on the same or different cells to further promote AA conversion to eicosanoids. Evidence supports a role for 15-HETE as a negative feedback regulator to inhibit eicosanoid formation and guanylate cyclase activation. A tentative transmembrane signal sequence leading to cGMP formation can now be formulated.

REFERENCES

Alford, R. R., 1970, Metal cation requirements for phytohemagglutinin-induced transformation of human peripheral blood lymphocytes, *J. Immunol.* **104**:698.

Allan, D., and Mitchell, R. H., 1974, Phosphatidylinositol cleavage catalysed by the soluble fraction from lymphocytes, *J. Biochem* **142**:591.

Ananthakrishnan, R., Coffey, R. G., and Hadden, J. W., 1981, Cyclic GMP and calcium in lymphocyte activation by phytohemagglutinin, *Lymphocyte Diff.* **1**:183.

Bailey, J. M., Bryant, R. W., Low, C. E., Pupillo, M. B., and Vanderhoek, J. Y., 1982, Role of lipoxygenases in regulation of PHA and phorbol ester-induced mitogenesis, in: *Leukotrienes and Other Lipoxygenase Products* (B. Sammuelsson and R. Paoletti, eds.), Raven Press, New York, p. 341.

Coffey, R. G., Hadden, E. M., and Hadden, J. W., 1977, Evidence for cyclic GMP and calcium mediation of lymphocyte activation by mitogens, *J. Immunol.* **119**:1387.

Coffey, R. G., Hadden, E. M., and Hadden, J. W., 1981, Phytohemagglutinin stimulation of guanylate cyclase in human lymphocytes, *J. Biol. Chem.* **256**:4418.

Fisher, D. B., and Mueller, G. C., 1971, Studies on the mechanism by which phytohemagglutinin rapidly stimulates phospholipid metabolism of human lymphocytes, *Biochim. Biophys. Acta* **248**:434.

Goetzl, E. J., 1981, Selective feed-back inhibition of the 5-lipoxygenation of arachidonic acid in human T-lymphocytes, *Biochem. Biophys. Res. Commun.* **101**:344.

Graff, G., Stephenson, J. H., Glass, D. B., Haddox, M. K., and Goldberg, N. D., 1978, Activation of soluble splenic cell guanylate cyclase by prostaglandin endoperoxides and fatty acid hydroperoxides, *J. Biol. Chem.* **253**:7662.

Hadden, J. W., 1977, Cyclic nucleotides in lymphocyte proliferation and differentiation, in: *Immunopharmacology* (J. W. Hadden, R. G. Coffey, and F. Spreafico, eds.), Plenum Press, New York, p. 1.

Hadden, J. W., and Coffey, R. G., 1982, Cyclic nucleotides in mitogen-induced lymphocyte proliferation, *Immunol. Today* **3**:299.

Hadden, J. W., Hadden, E. M., Haddox, M. K., and Goldberg, N. D., 1972, Guanosine 3′:5′-cyclic monophosphate: A possible intracellular mediator of mitogenic influences in lymphocytes, *Proc. Natl. Acad. Sci. U.S.A.* **69**:3024.

Hadden, J. W., Hadden, E. M., Sadlik, J. R., and Coffey, R. G., 1976, Effects of concanavalin A and a succinylated derivative on lymphocyte proliferation and cyclic nucleotide levels, *Proc. Natl. Acad. Sci. U.S.A.* **73**:1717.

Haddox, M. K., Stephenson, J. H., Moser, M. E., and Goldberg, N. D., 1978, Oxidative–reductive modulation of guinea pig splenic cell guanylate cyclase activity, *J. Biol. Chem.* **253**:3143.

Hirata, F., Toyoshima, S., Axelrod, J., and Waxdal, M. J., 1980, Phospholipid methylation: A biochemical signal modulating lymphocyte mitogenesis, *Proc. Natl. Acad. Sci. U.S.A.* **77**:862.

Hume, D. A., and Weidemann, M. J., 1980, Mitogenic lymphocyte transformation, in: *Research Monographs in Immunology*, Vol. 2, Elsevier, Amsterdam, p. 183.

Johnson, L. D., and Hadden, J. W., 1975a, Cyclic GMP and lymphocyte proliferation: Effects on DNA-dependent RNA polymerase I and II activation, *Biochem. Biophys. Res. Commun.* **66**:1498.

Johnson, E. M., and Hadden, J. W., 1975b, Phosphorylation of lymphocyte nuclear acidic proteins: Regulation by cyclic nucleotides, *Science* **187**:1198.

Johnson, L. D., and Hadden, J. W., 1977, Modification of human DNA-dependent RNA polymerase activity by cyclic GMP, *Nucleic Acids Res.* **4**:4007.

Johnson, E. M., Karn, I., and Allfrey, V. G., 1974, Early nuclear events in the induction of lymphocyte proliferation by mitogens, *J. Biol. Chem.* **249**:4990.

Johnson, E. M., Inoue, A., Crouse, L., Allfrey, V. G., and Hadden, J. W., 1975, Cyclic GMP modifies DNA-binding by a calf thymus nuclear protein fraction, *Biochem. Biophys. Res. Commun.* **65**:714.

Kelly, J. P., and Parker, C. W., 1979, Effects of arachidonic acid and other unsaturated fatty acids on mitogenesis in human lymphocytes, *J. Immunol.* **122**:1556.

McCarty, J., and Goetzl, E. J., 1979, Stimulation of human T lymphocyte chemokinesis by arachidonic acid, *Cell. Immunol.* **43:**103.

Parker, C. W., Kelly, J. P., Falkenheim, S. F., and Huber, M. G., 1979a, Release of arachidonic acid from human lymphocytes in response to mitogenic lectins, *J. Exp. Med.* **149:**1487.

Parker, C. W., Stenson, W. F., Huber, M. G., and Kelly, J. P., 1979b, Formation of thromboxane B$_2$ and hydroxyarachidonic acids in purified human lymphocytes in the presence and absence of PHA, *J. Immunol.* **122:**1572.

Pogo, B. G. T., Allfrey, V. G., and Mirsky, A. E., 1966, RNA synthesis and histone acetylation during the course of gene activation in lymphocytes, *Proc. Natl. Acad. Sci. U.S.A.* **55:**805.

Resch, K., 1976, Membrane associated events in lymphocyte activation, in: *Receptors and Recognition,* Series A (P. Cuatrecasas and M. F. Greaves, eds.), Chapman and Hall, London, p. 59.

Smith, J. W., Steiner, A. L., Newberry, W. M., Jr., and Parker, C. W., 1971, Cyclic adenosine 3',5'-monophosphate in human lymphocytes. Alterations after phytohemagglutinin stimulation, *J. Clin. Invest.* **50:**432.

Strom, T. B., Lundin, A. P., III, and Carpenter, C. B., 1977, The role of cyclic nucleotides in lymphocyte activation and function, *Prog. Clin. Immunol.* **3:**115.

Trotter, J., Flesch, I., Schmidt, B., and Ferber, E., 1982, Acyltransferase-catalyzed cleavage of arachidonic acid from phospholipids and transfer to lysophosphatides in lymphocytes and macrophages, *J. Biol. Chem.* **257:**1816.

Wedner, H. J., and Parker, C. W., 1976, Lymphocyte activation, *Prog. Allergy* **20:**195.

Whitney, R. B., and Sutherland, R. M., 1973, Characteristics of calcium accumulation by lymphocytes and alterations in the process induced by phytohemagglutinin, *J. Cell. Physiol.* **82:**9.

47

Biochemical Cascade Involved in Mast Cell Activation for Mediator Release

TERUKO ISHIZAKA

1. INTRODUCTION

Mast cells and basophil granulocytes bear specific receptors for IgE, which bind IgE molecules with high affinity, and the reactions of cell-bound IgE antibodies with multivalent antigen initiate the release of a variety of preformed and newly generated mediators such as histamine and leukotrienes (Ishizaka and Ishizaka, 1975). Mast cells also secrete these inflammatory mediators in response to various nonspecific stimuli. These mediators cause allergic disorders in both experimental animals and men. Among the various mast-cell-derived mediators, oxidative products of arachidonic acid (AA) have drawn much attention in recent years. It has been shown that rat peritoneal mast cells and human lung mast cells, but not bone-marrow-derived mouse mast cells, generate prostaglandin D_2 (PGD_2) through the cyclooxygenase pathway. Conversely, mouse bone-marrow-derived mast cells, as well as human lung mast cells generate primarily leukotriene C_4 (LTC_4) through the lipoxygenase pathway, but peritoneal mast cells do not (Lewis et $al.$, 1981; Peters et $al.$, 1984; Razin et $al.$, 1983). Leukotrienes cause contraction of airway smooth muscles and are considered to be the most potent bronchoconstrictors in human asthma.

Over the past several years, we have analyzed IgE-mediated triggering mechanisms of mediator release in purified rat peritoneal mast cells. After human lung mast cells and cultured human basophils became available recently (Ishizaka et $al.$,

TERUKO ISHIZAKA ● Subdepartment of Immunology, The Johns Hopkins University School of Medicine, Good Samaritan Hospital, Baltimore, Maryland 21239.

1983; Ogawa *et al.*, 1983), we extended our studies to human mast cells and the basophil system. This chapter reviews the biochemical cascade involved in triggering mast cells for mediator release.

2. TRIGGERING SIGNALS OF IgE-DEPENDENT MEDIATOR RELEASE

In order to analyze the immunologic mechanisms involved in the initial triggering vents in mast cells for IgE-mediated histamine release, we prepared rabbit antibodies against IgE receptors on rat mast cells (Ishizaka *et al.*, 1977). Detailed analysis of membrane components bound to the antibody by SDS gel electrophoresis demonstrated that major antibodies in the preparation were antireceptor antibodies that were directed toward IgE binding sites in the receptor molecules (Conrad *et al.*, 1978). Using this antireceptor antibody preparation, we demonstrated that bridging of IgE receptors on rat mast cells by divalent antireceptor antibody or its $F(ab')_2$ fragments induced a marked increase in ^{45}Ca influx that was accompanied by histamine release, whereas the binding of Fab' monomer fragments with the receptor was not sufficient for triggering these reactions. However, when receptor-bound Fab' monomer was bridged by divalent antirabbit γ-globulin antibody, the mast cells were triggered for histamine release. This situation is similar to that of anti-IgE- or antigen-induced histamine release, in which receptor-bound IgE molecules were bridged by anti-IgE or antigen. From these results, we concluded that bridging of receptor molecules is responsible for triggering mast cells for mediator release (Ishizaka and Ishizaka, 1978).

In the IgE-mediated reactions, the IgE receptor serves as an anchor for a specific IgE antibody molecule. Once IgE antibody binds to the receptor with high affinity, cell-bound IgE molecules permit antigen to bridge adjacent receptor molecules, which in turn initiates the process of histamine release. Since no immunoglobulin other than IgE will bind to IgE receptors with high affinity, the triggering of histamine release through IgE receptors will be mediated only by IgE antibodies.

3. STIMULATION OF METHYLTRANSFERASES AND ADENYLATE CYCLASE BY BRIDGING OF IgE RECEPTORS

Since the bridging of IgE receptors induces a marked increase in ^{45}Ca influx (Ishizaka *et al.*, 1979), and Ca^{2+} influx is a prerequisite for mediator release, the possible participation of membrane-associated enzymes in Ca^{2+} influx was explored. Our results revealed that the bridging of IgE receptors on mast cells results in the activation of both methyltransferases and adenylate cyclase (Ishizaka *et al.*, 1981). When purified normal rat mast cells were challenged with either $F(ab')_2$ or Fab' fragments of antireceptor antibodies, $F(ab')_2$ fragments of anti-RBL induced a marked increase in incorporation of [^{3}H]methyl groups into phospholipids and a monophasic rise in intracellular cAMP. As shown in Fig. 1, both responses reached maximum 15 sec after the challenge. The phospholipid methylation is followed by an increase in ^{45}Ca uptake and release of both histamine and arachidonates. The ^{45}Ca uptake reached a plateau at 2 min, and the maximum release of histamine and arachidonates was observed at 3 min. In contrast, the Fab' monomer of anti-RBL did not induce any of these reactions.

It was also found that preincubation of purified mast cells with inhibitors of phospholipid methylation, such as 3-deazaadenosine together with L-homocysteine thiolactone, resulted not only in inhibition of phospholipid methylation but also in inhibition of all of biochemical events induced by receptor bridging. When the cells were preincubated with varying concentrations of 3-deazaadenosine plus 100 μM L-homocysteine thiolactone, phospholipid methylation, cAMP rise, ^{45}Ca influx, and release of both histamine and arachidonates were inhibited in a similar dose–response fashion. These results suggest that phospholipid methylation induced by receptor bridging is involved in Ca^{2+} influx and mediator release.

Subsequently, strong supporting evidence for the role of methyltransferases in IgE-mediated histamine release was provided by McGivney *et al.* (1981) using variants of rat basophilic leukemia (RBL) cells. They examined some histamine-

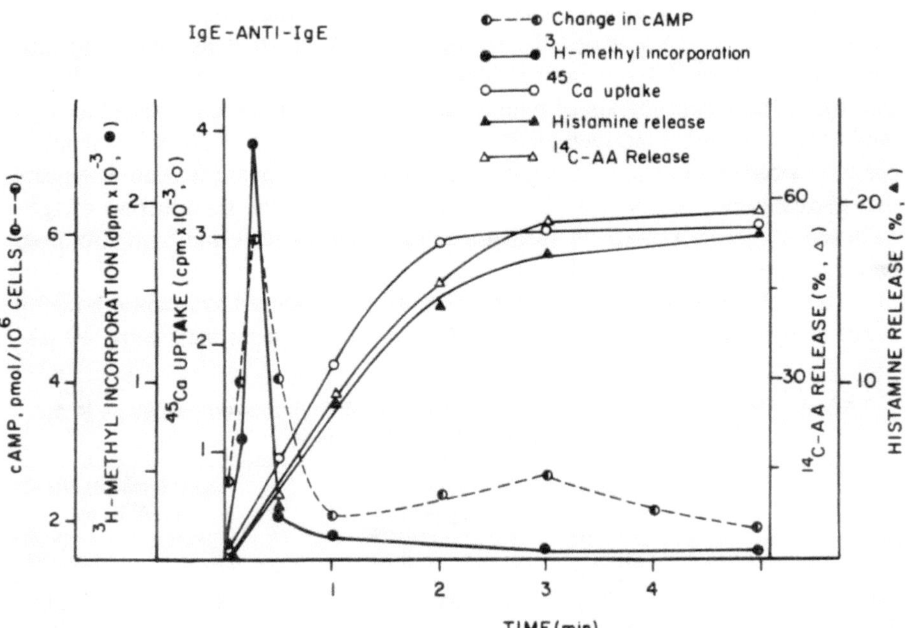

FIGURE 1. Kinetics of [^{3}H]methyl incorporation into phospholipids (●), ^{45}Ca uptake (○), change in cAMP level (◐), and the release of both histamine (▲) and [^{14}C]AA (△) in rat mast cells induced by bivalent anti-IgE. The same purified rat mast cells sensitized with IgE were used for all measurements.

releasing sublines of RBL cells for the capacity to increase phospholipid methylation and Ca^{2+} influx and to release AA and histamine and then established a number of variants that are defective at different stages in the histamine-releasing process. As shown in Table I, two cloned sublines of RBL cells, i.e., B1 and B6, are deficient in one of methyltransferases, and they did not give IgE-mediated histamine release. When these two sublines were hybridized, however, some of the hybrids such as A_1 and B_1 restored two methyltransferases. These hybrid cells sensitized with IgE released histamine on challenge with anti-IgE. The results support the concept that methyltransferases are involved in the transduction of IgE-mediated triggering signals to the process of mediator release.

Our recent studies have also shown that bridging of IgE receptors on human lung mast cells and cultured human basophils induced the activation of methyltransferases and adenylate cyclase (Ishizaka *et al.*, 1983). As shown in Fig. 2, challenge of purified human lung mast cells with $F(ab')_2$ fragments of anti-IgE resulted in a marked enhancement in phospholipid methylation and a monophasic rise in intracellular cAMP, which were followed by ^{45}Ca influx and histamine release. Phospholipid methylation reached maximum 30 sec after the challenge, whereas the cAMP level reached maximum at 1 min. The ^{45}Ca influx reached a plateau at 3 min, and the maximum histamine release was obtained within 5 to 8 min. As expected, Fab monomer fragments of anti-IgE failed to induce these responses. As demonstrated in rat mast cells, purified human lung mast cells preincubated with $[^{14}C]AA$ released the radiolabeled AA on challenge with anti-IgE. The kinetics of the AA release was identical to that of histamine release. It was also found that preincubation of human lung mast cells or cultured basophils with inhibitors of phospholipid methylation resulted in inhibition of all of biochemical events induced by receptor bridging in an identical dose–response fashion. It appears that phospholipid methylation plays an essential role in the transduction of IgE-mediated triggering signals for mediator release in mast cells and basophils from both experimental animals and man.

The biological significance of initial rise in cAMP induced by receptor bridging is not known. As demonstrated by Lewis *et al.* (1979) in rat mast cells, 10 μM

TABLE I. Histamine Release and Phospholipid Methyltransferase Levels in the Parental and Hybrid Cell Lines

Cell line	Phospholipid methyltransferase (pmol $[^3H]$methionine/mg protein)		Percent histamine release mediated by	
	I	II	IgE	Ionophore
2H3	1.71 ± 0.22	21.7 ± 1.9	65	72
B1	0.13 ± 0.04	30.2 ± 3.2	0	54
B6	1.04 ± 0.13	4.3 ± 1.3	0	46
Hybrid A1	0.82 ± 0.07	20.5 ± 4.7	43	52
Hybrid B1	0.70 ± 0.04	17.7 ± 5.0	24	30

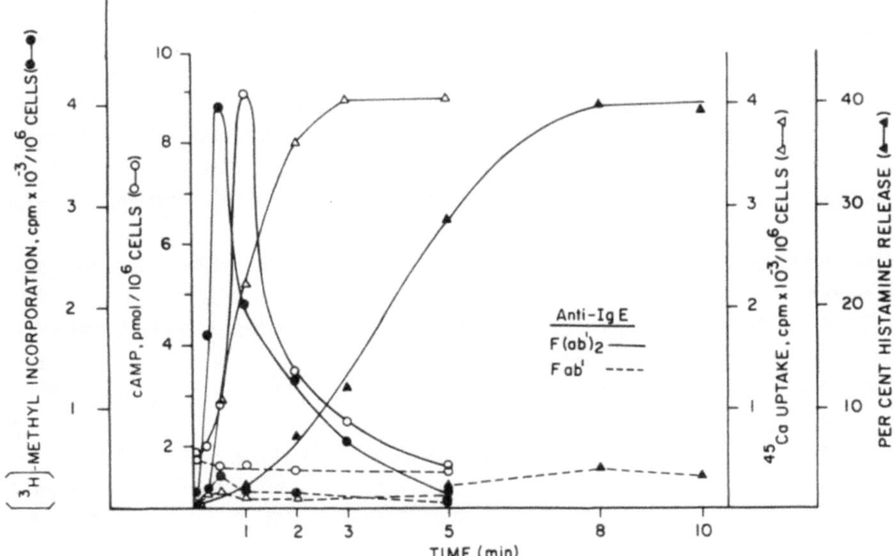

FIGURE 2. Kinetics of [³H]methyl incorporation (●), cAMP rise (o), ⁴⁵Ca uptake (△), and histamine release (▲) in human mast cells induced by either (F(ab')₂ fragments or Fab' fragments of anti-IgE. Purified human lung mast cells were sensitized with IgE overnight and challenged with the fragments of anti-IgE.

indomethacin, which completely inhibits prostaglandin synthesis, failed to inhibit the initial rise in cAMP induced by receptor bridging in both rat and human mast cells. The results indicate that the initial rise in cAMP is not caused by prostaglandin synthesis. Holgate *et al.* (1980) demonstrated that bridging of IgE molecules on rat mast cells resulted in an initial rise in cAMP, activation of cAMP-dependent protein kinase, and mediator release and speculated that an initial rise in cAMP may be an essential step in the histamine-releasing processes.

4. POSSIBLE PARTICIPATION OF MEMBRANE-ASSOCIATED PROTEOLYTIC ENZYMES IN INITIAL MAST CELL ACTIVATION

A question to be asked is whether methyltransferases are the first enzymes to be activated by receptor bridging or whether some other enzymes may be activated prior ot methyltransferases. Earlier studies by Austen and Brocklehurst (1960), Becker and Austen (1966), and others strongly suggested the possible involvement of a membrane-associated proteolytic enzyme in the early stage of antigen-induced histamine release from mast cells. In view of these findings, we explored the possible participation of a proteolytic enzyme in the activation of mast cells, and the results are summarized in Table II. When mast cells were challenged with either anti-IgE or antireceptor antibodies in the presence of an appropriate inhibitor of proteolytic enzymes, both phospholipid methylation and cAMP rise were inhibited. Diisopropylfluorophosphate (DFP), a potent inhibitor of serine esterases, inhibited both reactions, whereas diisopropylmethylphosphate (DMP), a nonphosphorylating analogue of DFP, had much less inhibitor effect. The effect of DFP was observed only when the inhibitor was present during the challenge. If the cells were first incubated with DFP, washed, and then challenged with the antibody in the absence of DFP, neither phospholipid methylation nor cAMP rise was inhibited. Furthermore, the presence of substrates or inhibitors of chymotrypsin or trypsin during the challenge inhibited both phospholipid methylation and the initial rise in cAMP. These results were confirmed using isolated plasma membranes of rat mast cells (Ishizaka *et al.*, 1981). It appears that a membrane-associated proteolytic enzyme is activated by receptor bridging, and a cleavage product formed by this enzyme may be involved in the activation of methyltransferases and adenylate cyclase. The putative proteolytic enzyme may be a key enzyme in the induction of initial triggering signals for mediator release.

TABLE II. Effect of Protease Inhibitors on Phospholipid Methylation and cAMP Rise

	Concentration (mM) for 50% inhibition of	
Protease substrates or inhibitors	Phospholipid methylation	Initial rise in cAMP
Diisopropylfluorophosphate (DFP)	0.1	0.1
p-Nitrophenylethylpentylphosphonate	0.3	0.5
Chymotrypsin substrate or inhibitor		
N-acetyl-DL-phenylalanine-β-naphthyl ester	0.2	0.3
Indole	0.6	0.7
L-Tosylamide-2-phenylethylchloromethylketone (TPCK)	0.8	0.6
Trypsin substrate or inhibitor		
p-Aminobenzamidine	0.7	0.5
p-Nitrophenyl-P′-guanidinobenzoate (NPGB)	0.4	0.5
Lima bean trypsin inhibitor	0.2	0.1
Chymase substates		
HO-Suc-Phe-Pro-Phe-NA	0.5	0.5
Suc-Ala-Ala-Pro-Phe-SBzL	0.4	0.3
Carboxypeptidase inhibitors		
N-CBZ-glycyl-L-phenylalanine	>2.0	>2.0
N-CBZ-glycyl-L-tryptophan	>2.0	>2.0

5. POSSIBLE ROLE OF PHOSPHOLIPID METHYLATION IN THE ACTIVATION OF A PHOSPHOLIPID– DIACYLGLYCEROL CYCLE

Bridging of IgE receptors on mast cells also induces alterations in phospholipid and diacylglycerol metabolism. Kennerly *et al.* (1979) observed that anti-IgE induced a selective ^{32}P incorporation into phosphatidic acid (PA), phosphatidylinositol (PI), and phosphatidylcholine (PC) in early stages of mast cell activation. Further studies by Sullivan (1981) demonstrated that the activation of rat mast cells resulted in accumulation of 1,2-diacylglycerol (DAG) followed by the formation of mono-acylglycerol (MAG) and free fatty acids. Since these products are surface active and promote membrane fusion, they suggested that DAG and MAG may induce degranulation and mediator release.

We wondered if the activation of diacylglycerol cycle may be biochemically connected to the activation of methyltransferases. In order to analyze possible interrelationships between the two pathways, we studied the effect of inhibitors of phospholipid methylation on ^{32}P incorporation into PA, PI, and PC. Thus, purified rat mast cells sensitized with IgE were incubated with various concentrations of 3-deazaadenosine (ADO) together with 100 μM L-homocysteine thiolactone and then challenged with anti-IgE. Measurements of ^{32}P incorporation into PA, PI, and PC and of phospholipid methylation in the same cells revealed that preincubation of rat mast cells with 3-deaza-ADO resulted in inhibition of both the incorporation of [^{3}H]methyl groups into phospholipids and the selective incorporation of ^{32}P into PA, PI, and PC in an identical dose–response fashion (Fig. 3). Another inhibitor of phospholipid methylation, 3-deaza-SIBA, gave identical results.

In order to exclude the possibility that the inhibitors of phospholipid methylation might have affected other enzymes involved in PI turnover, we examined whether the inhibition of ^{32}P incorporation by 3-deaza-ADO could be reversed by the addition of the methyl donor, S-adenosyl-L-methionine (SAM). Thus, rat mast cells sensitized with anti-DNP IgE antibody were preincubated with 3-deaza-ADO together with L-homocysteine thiolactone and then challenged with DNP-HSA in the presence or absence of 5 mM SAM. It was found that the addition of 5 mM SAM to the system reversed the inhibition of ^{32}P incorporation by 3-deaza-ADO, suggesting that accumulation of S-adenosyl-L-homocysteine or its analogues was responsible for the inhibition of ^{32}P incorporation. More recent experiments carried out in collaboration with Dr. T. Sullvan have shown that inhibitors of phospholipid methylation inhibited not only ^{32}P incorporation but also the formation of DAG. It was also found that inhibition of DAG formation by 3-deazaadenosine or 3-deaza-SIBA was reversed by the addition of S-adenosyl-L-methionine.

Hirata and Axelrod (1978) demonstrated that the first methyltransferase methylates phosphatidylethanolamine (PE) to form phosphatidyl-N-monomethylethanolamine (PME) and the second methyltransferase adds two methyl groups succes-

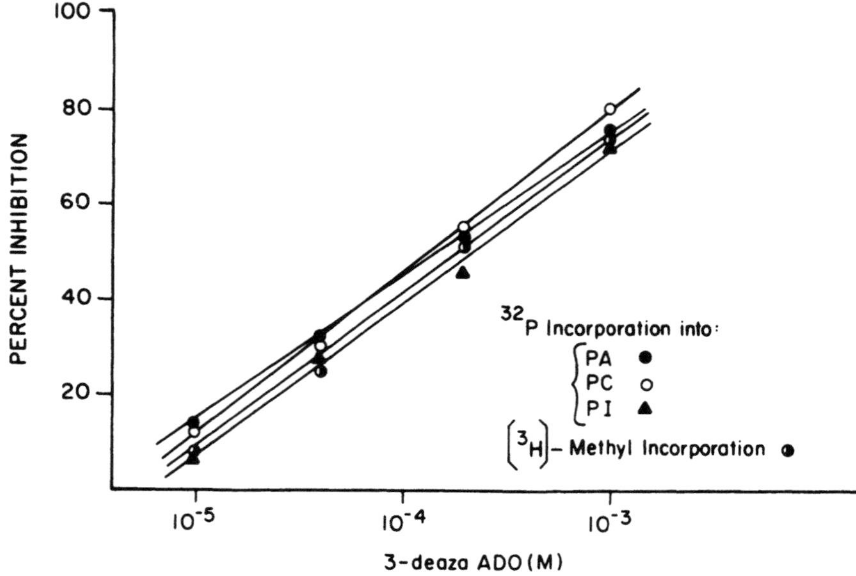

FIGURE 3. Inhibition of anti-IgE-induced [³H]methyl incorporation into phospholipids and ³²P incorporation by 3-deazaadenosine (ADO). Purified rat mast cells sensitized with IgE were preincubated with various concentrations of 3-deaza-ADO togther with 100 μM L-homocysteine thiolactone at 37°C for 1 hr and then challenged with anti-IgE. The ³²P incorporation into PA, PI, and PC was analyzed after 10 min of incubation, whereas [³H]methyl incorporation into phospholipids was determined 15 sec after the challenge.

sively to form phosphatidylcholine (PC). Thus, we asked whether the activation of both methyltransferases was required for stimulation of PI turnover. Since the first enzyme is Mg^{2+} dependent, purified rat mast cells were challenged in the presence of Mg^{2+} and EGTA or in the presence of Ca^{2+} and Mg^{2+}, and incorporation of both [³H]methyl groups and ³²P into phospholipids was analyzed by thin-layer chromatography.

The results are shown in Fig. 4. In the presence of both Ca^{2+} and Mg^{2+}, the two methyltransferases were activated to form PME, phosphatidyl-N,N-dimethyl-lethanolamine (PDE), and PC. In the presence of Mg^{2+} and EGTA, formation of PME was predominant, indicating that only the first methyltransferase was fully activated under this condition. However, ³²P incorporation into PA, PI, and PC was comparable in the two conditions. The results suggest that the activation of the first methyltransferase is sufficient for the activation of enzymes involved in PI turnover. Since accumulation of PME in the plasma membrane is known to reduce membrane viscosity and affect many membrane events, one may speculate that such changes in the plasma membrane may be responsible for the stimulation of a phospho-lipid–diacylglycerol cycle. Further studies indicated that inhibitors and substrates

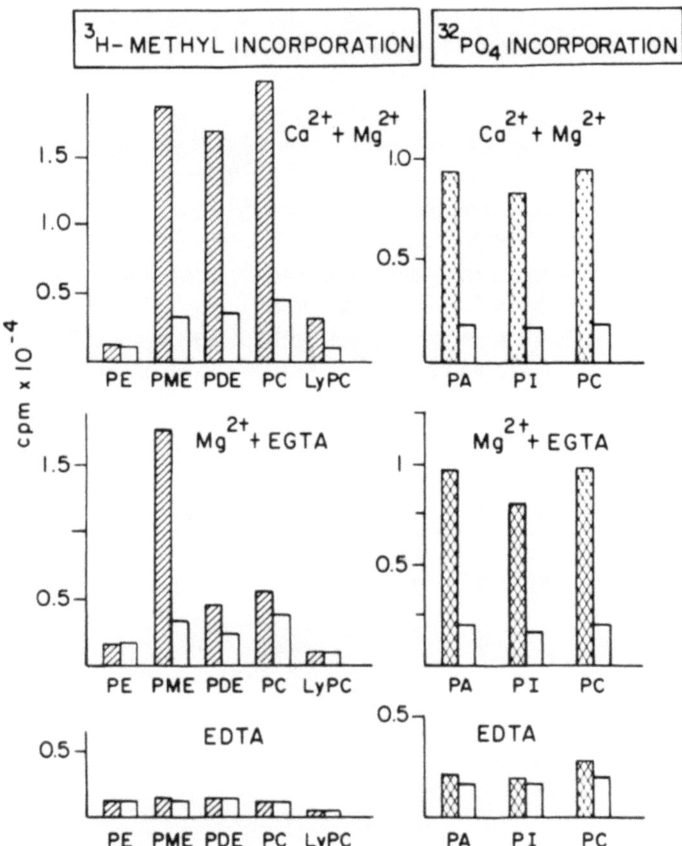

FIGURE 4. Possible participation of methyltransferase I but not of methyltransferase II in stimulation of PI turnover. Purified rat mast cells sensitized with anti-DNP IgE antibody were challenged with DNA-HSA in the presence of 1.6 mM Ca^{2+} and 1 mM Mg^{2+} (top); 2 mM Mg^{2+} plus 100 μM EGTA (middle), or 10 mM EDTA (bottom). Both [^{3}H]methyl incorporation (left) and ^{32}P incorporation (right) were analyzed by thin-layer chromatography. [^{3}H]Methyl incorporation ▨ was analyzed 15 sec after the challenge, and ^{32}P incorporation ▨ was measured at 10 min. □: Either [^{3}H]methyl incorporation or ^{32}P incorporation in control cells that were not challenged with DNP-HSA.

of proteolytic enzymes, which prevented phospholipid methylation, also inhibited ^{32}P incorporation into PA, PC, and PI. The concentrations of the inhibitors required for 50% inhibition of the ^{32}P incorporation were comparable to that required for 50% inhibition of phospholipid methylation. These results suggest that the activation of a proteolytic enzyme and methyltransferase I are involved in the activation of phospholipid diacylglycerol cycle.

Biochemical cascades involved in IgE-mediated activation of rat mast cells for mediator release are summarized in Fig. 5. Bridging of IgE receptors activates a membrane-associated putative proteolytic enzyme followed by activation of both methyltransferases and adenylate cyclase. Accumulation of monomethylated phospholipids in the plasma membrane by activation of methyltransferase I appears to be sufficient for stimulation of PI turnover. Once mast cells are activated through the biochemical pathways, lysophospholipid would be generated by the action of phospholipase A_2, and both DAG and MAG will be generated by action of phospholipase C and diacylglyceride lipase. Since these substances are powerful membrane fusogens, they may facilitate granule membrane fusion and mediator release. At the same time, free AA released through the two pathways will be metabolized to generate many biologically active lipid mediators such as prostaglandins and leukotrienes.

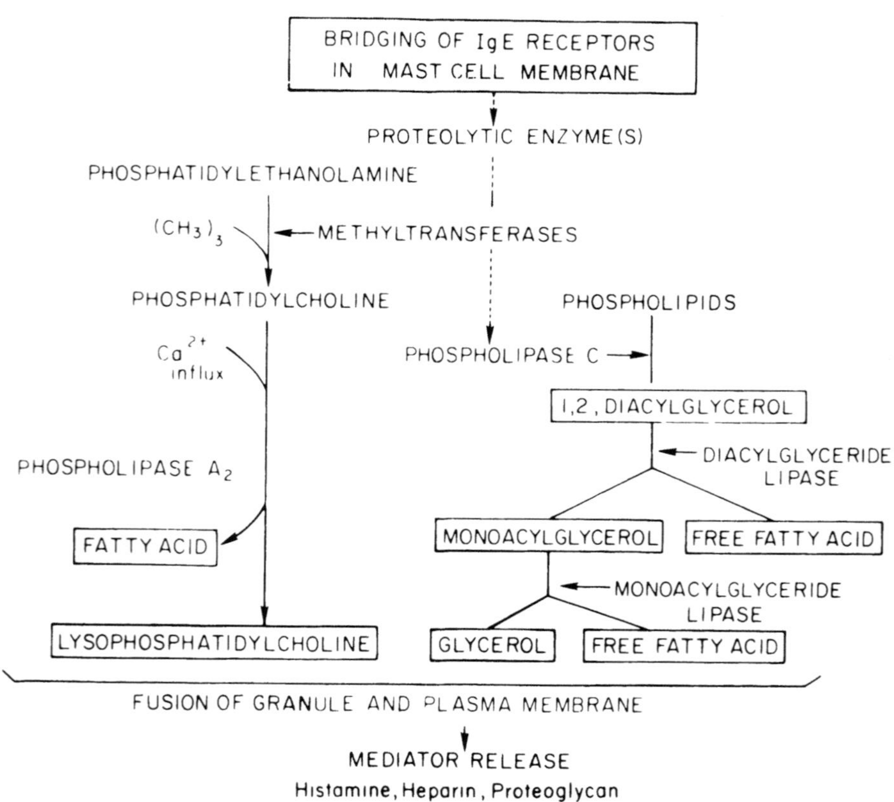

FIGURE 5. Biochemical cascade for IgE-dependent mediator release from mast cells.

5. BIOCHEMICAL EVENTS INVOLVED IN MEDIATOR RELEASE FROM MAST CELLS BY NONSPECIFIC STIMULI

Mediator release from mast cells is induced not only by specific ligand–receptor interaction but also by various nonspecific stimuli. Among them, important ligands from the biological viewpoint are complement-derived peptides C5a and C3a, which are generated during inflammation. Biochemical cascades activated by complement-derived peptides C5a and C3a are similar to IgE-mediated activation in rat mast cells. As shown in Fig. 6, challenge of rat mast cells with anaphylatoxin C5a resulted in stimulation of phospholipid methylation and an increase in ^{45}Ca uptake, which are followed by the release of histamine and arachidonate. As demonstrated in IgE-mediated activation, inhibitors of phospholipid methylation inhibited these biochemical events induced by the peptides, suggesting that phospholipid methylation is involved in transduction of C5a- and C3a-induced triggering signals for mediator release. An interesting finding was that the C5a- and C3a-induced cAMP rise was much slower than that induced by bridging of IgE receptors. Furthermore, the cAMP rise by C5a and C3a was completely inhibited by the presence of 10 μM indomethacin, indicating that the increase is the result of prostaglandin synthesis. Since indomethacin does not inhibit C5a-induced histamine release from rat mast cells, it appears that the increase in cAMP is not involved in C5a-induced histamine release.

Compound 48/80 and Ca^{2+} ionophore A23187 also induce histamine release from rat mast cells. Unlike IgE-mediated or anaphylatoxin-mediated histamine

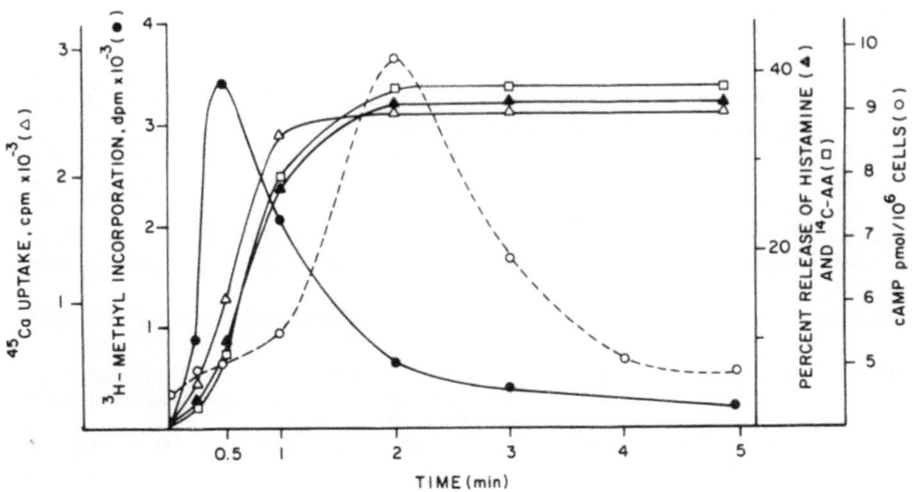

FIGURE 6. Kinetics of C5a-induced [^{3}H]methyl incorporation (●), ^{45}Ca uptake (△), release of histamine (▲), and [^{14}C]AA (□), and changes in cAMP (○ – – – ○).

release, however, both compound 48/80 and Ca^{2+} ionophore failed to induce phospholipid methylation. These compounds directly induce mobilization of intracellular Ca^{2+} and Ca^{2+} influx, which are followed by mediator release.

Our recent studies using quin-2 fluorescence demonstrated that phorbol 12-myristate 13-acetate induces histamine release from rat mast cells without increasing intracellular Ca^{2+} (White *et al.*, 1984). These results reflect the complexity of the activation processes for mediator release from mast cells.

ACKNOWLEDGMENTS. I would like to express my gratitude for the collaboration with the following investigators: Drs. K. Ishizaka, D. H. Conrad, and E. S. Schulman, Mr. A. R. Sterk, and Mrs. C. G. L. Ko, The Johns Hopkins University, Baltimore, MD; Drs. F. Hirata and J. A. Axelrod, NIMH, NIH, Bethesda, MD; Drs. J. R. White and R. I. Sha'afi, University of Connecticut Health Center, Farmington, CT; and Dr. T. J. Sullivan, University of Texas Health Science Center at Dallas, TX. Their excellent collaboration and assistance made it possible for me to carry out this series of studies. My special appreciation goes to Dr. T. E. Hugli, Scripps Clinic, La Jolla, CA and Dr. P. M. Henson, National Jewish Hospital, Denver, CO, for preparations of purified C5a and C3a, and Dr. J. C. Powers, Georgia Institute of Technology, Atlanta, GA for synthesized substates and inhibitors of chymase and tryptase. These materials were invaluable in my studies. This work was supported by research grant AI-10060 from USPHS and a grant from the Lillia Babbit Hyde Foundation. This is publication No. 560 from the O'Neill Laboratories at the Good Samaritan Hospital, Baltimore, MD.

REFERENCES

Austen, K. F., and Brocklehurst, W. W., 1960, Anaphylaxis in chopped guinea pig lung. I. Effect of peptidase substrate and inhibitors, *J. Exp. Med.* **113**:521–539.

Becker, E. L., and Austen, K. F., 1966, Mechanisms of immunologic injury of rat peritoneal mast cells. I. The effect of phosphonate inhibitors on homocytotropic antibody-mediated histamine release and the first component of rat complement, *J. Exp. Med.* **124**:379–395.

Conrad, D. H., Froese, A., Ishizaka, T., and Ishizaka, K., 1978, Evidence for antibody activity against the receptor for IgE in a rabbit antiserum prepared against IgE-receptor complexes, *J. Immunol.* **120**:507–512.

Hirata, F., and Axelrod, J. A., 1978, Enzymatic synthesis and rapid translocation of phosphatidylcholine by two methyltransferases in erythrocyte membranes, *Proc. Natl. Acad. Sci. U.S.A.* **75**:2348–2352.

Holgate, S., Lewis, R., and Austen, K. F., 1980, Role of adenylate cyclase in immunologic release of mediators from rat mast cells. Agonist and antagonist effect of purine and ribose-modified adenosine analogs, *Proc. Natl. Acad. Sci. U.S.A.* **77**:6800–6804.

Ishizaka, T., and Ishizaka, K., 1975, Biology of immunoglobulin E, *Prog. Allergy* **19**:60–121.

Ishizaka, T., and Ishizaka, K., 1978, Triggering of histamine release from rat mast cells by divalent antibodies against IgE-receptor, *J. Immunol.* **120**:800–805.

Ishizaka, T., Chang, T. H., Taggart, M., and Ishizaka, K., 1977, Histamine release from mast cells by antibodies against rat basophilic leukemia cell membrane, *J. Immunol.* **119**:1589–1596.

Ishizaka, T., Hirata, F., Ishizaka, K., and Axelrod, J. A., 1979, Induction of calcium influx across the rat mast cell membrane by bridging IgE receptors, *Proc. Natl. Acad. Sci. U.S.A.* **76**:5858–5862.

Ishizaka, T., Hirata, F., Sterk, A. R., Ishizaka, K., and Axelrod, J. A., 1981, Bridging of IgE receptors activates phospholipid methylation and adenylate cyclase in mast cell plasma membrane, *Proc. Natl. Acad. Sci. U.S.A.* **78**:6812–6816.

Ishizaka, T., Conrad, D. H., Schulman, D. S., Sterk, A. R., and Ishizaka, K., 1983, Biochemical analysis of initial triggering events of IgE-mediated histamine release from human lung mast cells, *J. Immunol.* **130**:2357–2362.

Kennerly, D. A., Sullivan, T. J., and Parker, C. W., 1979, Activation of phospholipid metabolism during mediator release from stimulated rat mast cells, *J. Immunol.* **122**:152–159.

Lewis, R. A., Holgate, S. T., Roberts, L. J., II, Maguire, J. F., Oates, J. A., and Austen, K. F., 1979, Effect of indomethacin on cyclic nucleotide levels and histamine release from rat serosal mast cells, *J. Immunol.* **123**:1633–1638.

Lewis, R. A., Holgate, S. T., Roberts, L. J., II, Oates, J. A., and Austen, K. F., 1981, Preferential generation of prostaglandin D_2 by rat and human mast cells, in: *Biochemistry of the Acute Allergic Reactions* (E. L. Becker, A. S. Simon, and K. F. Austen, eds.), Alan R. Liss, New York, pp. 239–254.

McGivney, A., Crew, F. T., Hirata, F., Axelrod, J. A., and Siraganian, R. P., 1981, Rat basophilic leukemia cell lines defective in phospholipid methyltransferase enzymes, Ca^{2+} influx and histamine release, Reconstitution by hybridization, *Proc. Natl. Acad. Sci. U.S.A.* **78**:6176–6180.

Ogawa, M., Nakahata, T., Leary, A. G., Sterk, A. R., Ishizaka, K., and Ishizaka, T., 1983, Suspension culture of human mast cells/basophils from umbilical cord blood mononuclear cells, *Proc. Natl. Acad. Sci. U.S.A.* **80**:4494–4498.

Peters, S. P., MacGlashan, D. W., Jr., Schulman, E. S., Schleimer, R. P., Hayes, E. C., Rokach, J., Adkinson, N. F., Jr., and Lichtenstein, L. M., 1984, Arachidonic acid metabolism in purified human lung mast cells, *J. Immunol.* **132**:1972–1979.

Razin, E., Mencia-Huerta, J. M., Stevens, R. L., Lewis, R. A., Liu, F. T., Corey, E. J., and Austen, K. F., 1983, IgE-mediated release of leukotriene C_4, chrondroitin sulfate E proteoglycan, β-hexosaminidase and histamine from cultured bone marrow-derived mouse mast cells, *J. Exp. Med.* **157**:189–201.

Sullivan, T. J., 1981, Diacylglycerol metabolism and the release of mediators from mast cells, in: *Biochemistry of the Acute Allergic Reactions* (E. L. Becker, A. S. Simon, and K. F. Austen, eds.), Alan R. Liss, New York, pp. 229–238.

White, J. R., Ishizaka, T., Ishizaka, K., and Sha'afi, R., 1984, Direct demonstration of increased intracellular concentration of free calcium as measured by quin-2 in rat peritoneal mast cell, *Proc. Natl. Acad. Sci. U.S.A.* **81**:3978–3982.

48

Stimulation of Lymphocyte Guanylate Cyclase by Arachidonic Acid and HETEs

RONALD G. COFFEY and JOHN W. HADDEN

1. INTRODUCTION

Mitogen activation of human peripheral blood lymphocytes (HPBL) occurs as a consequence of several early membrane events. One of the earliest is the Ca^{2+}-dependent stimulation of guanylate cyclase (Coffey *et al.*, 1981). Experiments with the T-cell mitogens phytohemagglutinin (PHA), concanavalin A (Con A), and phorbol myristate acetate (PMA) suggested that phospholipase and lipoxygenase activation are important in the sequence of early membrane events leading to guanylate cyclase stimulation, cGMP accumulation, and DNA synthesis (Coffey and Hadden, 1981, 1983a,b). Cyclooxygenase (Coffey *et al.*, 1981) and thromboxane synthetase activities (Gordon *et al.*, 1981) were not found to be essential for mitogenesis.

The present experiments concern the effects of phospholipase A_2, arachidonate, and several forms of hydroxyeicosatetraenoic acid (HETE) on the stimulation of guanylate cyclase in HPBL. Experiments involving addition of these agonists directly to cell homogenates, to separated membranes and soluble fractions, and to intact cells are described. An overview integrating the signals represented by cGMP, Ca^{2+}, and lipid metabolites in lymphocyte mitogenesis is presented by J. W. Hadden, E. M. Hadden, and R. G. Coffey (this volume) and by Hadden *et al.* (1979).

RONALD G. COFFEY and JOHN W. HADDEN • Program of Immunopharmacology, University of South Florida Medical College, Tampa, Florida 33612.

2. EFFECTS OF PHOSPHOLIPASE A$_2$

Our first experiment with phospholipase A$_2$ and the products of its action, lysolecithin and arachidonate, was suggested by a literature well summarized by Haddox and Goldberg (1977). This literature showed that both phospholipase A$_2$ and archidonate could, when added to cells or broken cell systems, stimulate guanylate cyclase. Our results indicated that the same is true for HPBL (Coffey and Hadden, 1981), in which cell homogenate cyclases dependent on either Mg^{2+} or Mn^{2+} were notably stimulated by both substances. The maximum stimulatory concentration of arachidonate was 10^{-5} M; higher concentrations inhibited guanylate cyclase. The stimulation by both substances was markedly inhibited by 5,8,11,14-eicosatetraynoic acid (ETYA), which blocks both cyclooxygenase and lipoxygenase pathways of arachidonate metabolism. On the other hand, it was not affected by indomethacin, which inhibits only the cyclooxygenase pathway. No significant stimulatory effects were produced by prostaglandins E$_1$ or F$_{2\alpha}$ (1–10 μM), or by phosphatidate, a product of phosphatidylinositol catabolism mediated by phospholipase C and diglyceride kinase activities.

Phospholipase A$_2$ from several sources (hog kidney, snake venoms) produced equivalent results when added to intact lymphocytes. The *Vipera russelli* enzyme (Sigma Chemical Co., St. Louis, MO) at 2.5 μg/ml caused maximal activation of 4.5- and threefold of the membrane and soluble Mg^{2+}-dependent guanylate cyclases, respectively. Phospholipase inhibitors such as dibutyryl cAMP (0.5 mM) and dimethyl-DL-2,3-distearoyloxypropyl-2-hydroxyethylammonium acetate (DDPH) (0.1 mM) greatly reduced or prevented guanylate cyclase activation in both membrane and soluble fractions. Sulfhydryl oxidation is important in guanylate cyclase activation by mitogens (Coffey *et al.*, 1981) and by most other agonists (Goldberg *et al.*, 1978; Brandwein *et al.*, 1981) and also by the events initiated with phospholipase A$_2$: complete inhibition of activation was obtained with 2.5 mM dithioerythritol. 5,8,11,14-Eicosatetraynoic acid (10 μM) inhibited, as it did with homogenates, whereas indomethacin (10 μM) and a thromboxane synthetase inhibitor had no effect on phospholipase A$_2$ activation of guanylate cyclase. These observations implicated lipoxygenase products of arachidonate in the activation of lymphocyte guanylate cyclase and, further, suggested that the arachidonate derived from phospholipase A$_2$ activation.

3. EFFECTS OF ARACHIDONATE

Having established that arachidonate represents the probable phospholipase A$_2$ product that is involved in guanylate cyclase stimulation, we further explored its effects when added directly to subcellular fractions or when preincubated with intact HPBL followed by homogenization (Fig. 1). When added directly to membrane or soluble fractions of HPBL, arachidonate stimulated only the soluble form of the enzyme. This appeared reasonable since lipoxyygenases, already implicated in mediating arachidonate stimulation in the homogenate experiments, are considered

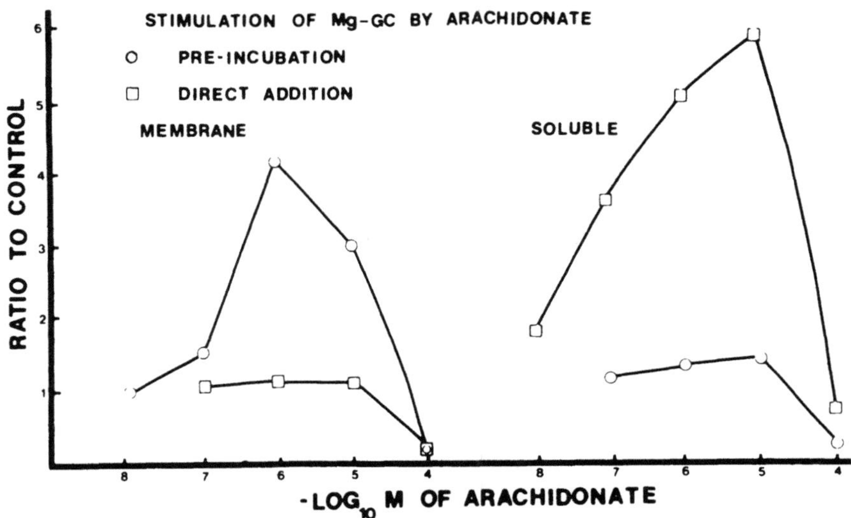

FIGURE 1. For the preincubation experiments, HPBL (10×10^6/ml) were incubated at 37°C in Hanks balanced salt solution for 30 min. Arachidonate (Sigma Chemical Co., St. Louis, MO) was dissolved in dichloromethane and added to glass tubes and dried with a stream of nitrogen. Lymphocytes were transferred to these tubes, incubation was continued for another 10 min, and cells were centrifuged at 2000 rpm, 2 min, 4°C. They were resuspended in 50 mM HEPES-Na^+, pH 7.5, 0.1 mM dithiothreitol, 0.1 mM EGTA (buffer A), homogenized, and the membrane and soluble fractions separated and assayed for Mg^{2+} (5 mM) -dependent guanylate cyclase activity as described earlier (Coffey *et al.*, 1981). For the direct addition experiments, fresh arachidonate solutions were pipetted to glass tubes, dried under N_2, and previously separated HPBL membrane or soluble ($48,000 \times g$, 60 min) fractions were added. After a 10-min incubation period, Mg^{2+}-GTP and other guanylate cyclase assay components were added, and the assay was conducted as described (Coffey *et al.*, 1981). The preincubation experiment has been performed four times, and the direct addition experiment twice. A representative experiment involving both protocols with the same cells is shown. All points are means of duplicates, which differed by less than 10%.

to be soluble enzymes. When added to intact cells, arachidonate stimulated only the membrane form of guanylate cyclase, with peak effects at 1 μM. In broken cells, arachidonate effects were maximal at 10 μM, suggesting a different mode of action. This suggestion was confirmed by inhibition studies described below.

4. EFFECTS OF HETES AND HPETES

Lipoxygenase activities vary according to cell type. Human leukocytes have several forms of lipoxygenase, producing 5-, 11-, 12-, and 15-hydroperoxyeicosatetraenoic acid (HPETE) (Goetzl, 1981; Parker *et al.*, 1979). Such hydroperoxy derivatives of arachidonate were shown by Goldberg and colleagues (1978) to activate spleen cell guanylate cyclase by an oxidative mechanism. Monohydroxy compounds, lacking oxidative potential, were ineffective. Based on this information and on own own data (Coffey and Hadden, 1981; Coffey *et al.*, 1981) showing inhibition by lipoxygenase inhibitors (ETYA and nordihydroguaiaretic acid, NDGA) of mitogen activation of guanylate cyclase, we postulated that HPETEs rather than HETEs were the proximal stimulants of guanylate cyclase. In apparent confirmation, we found only weak (<60%) stimulation of guanylate cyclases in a series of four experiments involving direct addition of the 5-, 11-, 12-, and 15-HETEs (0.1–10 μM) to either membrane or soluble fractions of HPBL.

Interestingly, in four experiments with intact cells, two- to sevenfold increases in membrane guanylate cyclase activity were observed following a 10-min incubation with 0.01 to 1.0 μM 5-HETE. No activation of soluble guanylate cyclase was seen. These responses resemble those of arachidonate, as seen in Table I. The effects of the other HETEs were similar but less marked: up to twofold stimulation was observed with 0.1–1 μM 12-HETE, and 2.5-fold with 10–20 μM 15-HETE in four experiments.

Further similarities in the effects of 5-HETE and arachidonate to stimulate membrane guanylate cyclase in intact cells are evident from inhibition studies (Table II). Two phospholipase A_2 inhibitors, P-bromophenacylbromide (PBPB) and quinacrine, effectively prevented activation by both agonists. Lipoxygenase inhibitors ETYA and NDGA were also potent inhibitors. Indomethacin inhibits cyclooxygenase completely at 1 μM, but effects of indomethacin on guanylate cyclase activation were observed only at 10 μM, a concentration capable of affecting phospholipase. We tentatively conclude that 5-HETE and archidonate, like PHA, act on lymphocyte surface receptors to activate phospholipase, arachidonate release, and arachidonate metabolism intracellularly by lipoxygenase to from a HPETE, which then functions by an oxidative mechanism to stimulate membrane guanylate cyclase. This type of transmembrane signal may mediate the effects of cholinergic and other agonists that activate guanylate cyclase by oxygen- and Ca^{2+}-dependent pathways (Craven and DeRubertis, 1982; White *et al.*, 1982; Vesin *et al.*, 1984).

Although 15-HETE can, at high levels, stimulate guanylate cyclase, it is more significantly a selective inhibitor of 5- and 12-lipoxygenases (Bailey *et al.*, 1982; Vanderhoek *et al.*, 1982). It does not inhibit 11- or 15-lipoxygenases or cyclooxygenase. It has been successfully employed to inhibit mitogen activation of mouse spleen cells (Bailey *et al.*, 1982) and is considered an immunosuppressant (Aldiger

TABLE I. Pretreatment of Lymphocytes[a]

	Guanylate cyclase (ratio to control)		Cellular cGMP
	Membrane	Soluble	
5-HETE, 0.01 μM	2.1	1.1	1.3
0.1 μM	4.6	0.8	1.5
1.0 μM	7.0	0.8	2.1
Arachidonate, 1.0 μM	5.0	0.8	1.5

[a] Guanylate cyclases were assayed in membrane and soluble fractions following incubation of intact cells with agonists for 10 min at 37°C, as described (Coffey *et al.*, 1981). Cyclic GMP levels were determined by radioimmunoassay following the three-column purification procedure (Coffey and Hadden, 1983b).

TABLE II. Inhibition of 5-HETE or Arachidonate Stimulation of Guanylate Cyclase[a]

Inhibitor	Percent inhibition	
	5-HETE	Arachidonate
PBPB, 10 μM	90 ± 9 (3)	84 (1)
QUIN, 10 μM	79 ± 20 (2)	68 (1)
INDO, 10 μM	63 ± 31 (3)	55 (1)
ETYA, 1 μM	100 (1)	90 (1)
NDGA, 10 μM	81 ± 15 (6)	76 (1)

[a] The HPBL were pretreated 10 min with inhibitor, then 10 min with 0.1 μM HETE or 1 μM arachidonate. Membrane Mg^{2+}-guanylate cyclase was assayed as described for Fig. 1 Effective concentrations of inhibitors are shown. In the case of ETYA, less inhibition was calculated at higher concentrations, where this substance is capable of stimulating activity alone Data represent means ± S.D. for the number of experiments in parentheses. QUIN, quinacrine (Mepacrine[8]); INDO, indomethacin

et al., 1984). We therefore tested the ability of 1 μM 15-HETE, incubated with lymphocytes for 10 min, to prevent guanylate cyclase activation by a range of concentrations of 5-HETE. The results of two experiments (Fig. 2) show complete inhibition by 5-HETE at 10^{-8} M and lesser inhibition at 10^{-7} to 10^{-6} M 5-HETE. The cation used to support guanylate cyclase activity in the assay did not affect the results. These data support the suggestion that 5- or 12-lipoxygenase activation by 5-HETE mediates the stimulation of guanylate cyclase.

15-Hydroxyeicosatetraenoic acid was also found to greatly reduce the stimulation of membrane guanylate cyclase when intact HPBL were incubated with mitogenic concentrations of PHA and PMA (Fig. 3). The inhibition was dose dependent and approached 100% for the Mg^{2+}-dependent guanylate cyclase when 10 μM 15-HETE was added. Lesser effects were seen with the Mn^{2+}-dependent enzyme.

Phytohemagglutinin stimulates primarily membrane guanylate cyclase, whereas PMA activates predominantly the soluble form of the enzyme (Coffey and Hadden, 1983a,b). Neither mitogen has significant effects if added directly to membrane or soluble fractions. Both mitogens stimulate guanylate cyclase and DNA synthesis in a Ca^{2+}-dependent fashion that is prevented by the phospholipase inhibitor quinacrine and the lipoxygenase inhibitor NDGA. However, PMA stimulation of guanylate cyclase is not inhibited by PBPB or ETYA, whereas PHA's effects are prevented by these well-recognized inhibitors of phospholipase and lipoxygenase, respectively. These differences and the reduced effectiveness of 15-HETE to block the activation of soluble guanylate cyclase by PMA indicate that the two mitogens employ similar but not identical pathways of lipid metabolism to accomplish the increase in lymphocyte cGMP levels. Novogrodsky *et al.* (1982) provided evidence linking hydroxyl radical formation to lymphocyte guanylate cyclase stimulation by PMA. The stimulation of soluble guanylate cyclase by either direct addition of arachidonate or by preincubation with PMA is significantly inhibited by 15-HETE but not by ETYA. The possibility therefore exists that arachidonate can be released from more than one lipid source by different mitogens and can be oxidized by several lipoxygenases with differing sensitivities to inhibitors. One or more oxidized

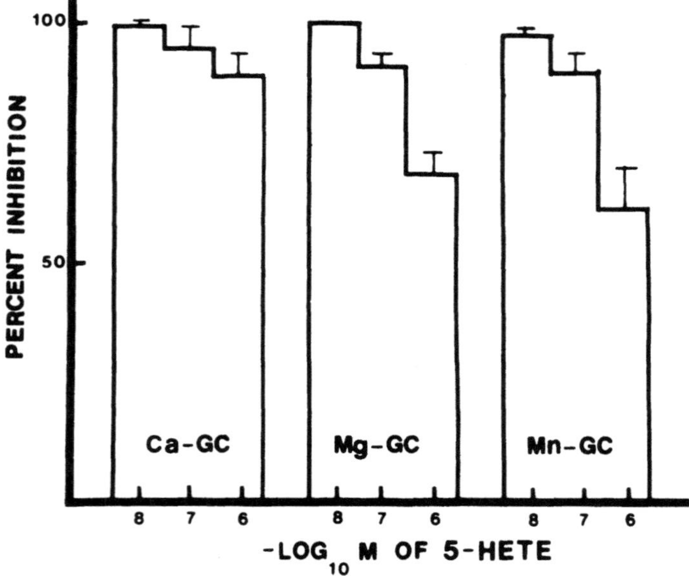

FIGURE 2. The HPBL were preincubated as described for Fig. 1 for 10 min with 15-HETE (1 μM) and then transferred to tubes containing the designated amounts of 5-HETE. Following a further 10-min, 37°C incubation, cells were centrifuged, and membrane guanylate cyclase was prepared and assayed as described for Fig. 1. The assay was conducted in the presence of either 5 mM Ca^{2+}, 5 mM Mg^{2+}, or 2 mM Mn^{2+}. The means of two experiments, $\pm$ S.D., are shown. Relative specific activities of HPBL guanylate cyclase are greater, with $Mn^{2+} > Mg^{2+} > Ca^{2+}$, yet similar inhibition of activation suggests a common enzyme activity expressed to varying degrees according to the divalent metal.

species of arachidonate may then interact with sulfhydryl or other oxidizable groups on guanylate cyclase, resulting in enhanced levels of the cGMP signal.

5. CONCLUSIONS

The experiments with human lymphocytes relate the initial effects of exogenously added phospholipase A_2 or arachidonate to those of mitogens in terms of guanylate cyclase activation. In all cases, the activation is dependent on a sequence of lipid metabolic steps requiring Ca^{2+} and oxygen and involving phospholipase activation, arachidonate release, and its metabolism via one or more lipoxygenase enzymes. The HPETEs so formed may serve as the proximal stimulants of guanylate cyclase.

When added to intact lymphocytes, arachidonate stimulates only the membrane-bound guanylate cyclase. The HETEs, formed by reduction of HPETEs, also stimulate membrane guanylate cyclase in a way that requires intact cells and both phospholipase and lipoxygenase activities.

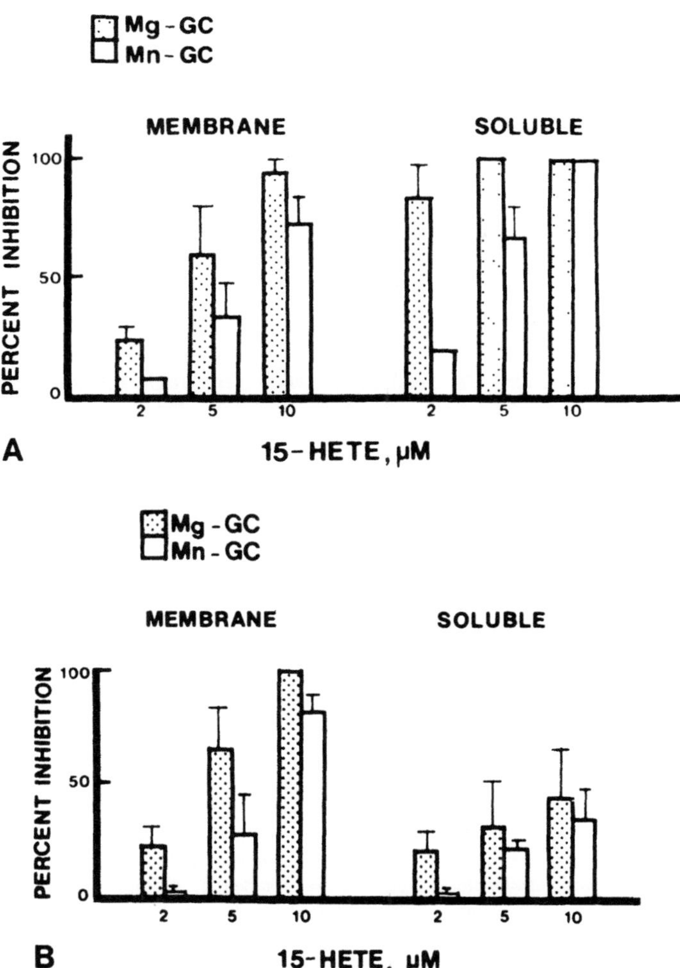

FIGURE 3. Effects of 15-HETE preincubation to inhibit the stimulation of guanylate cyclase by the mitogens PHA (A) and PMA (B) are shown as a function of 15-HETE concentration. Since some stimulation of guanylate cyclase by 15-HETE occurs, as discussed in the text, the percentage inhibition was calculated as $100\{1 - [(AI - I)/(A - C)]\}$ where A is agonist, I is inhibitor, C is control, and AI is agonist plus inhibitor. Three experiments are represented in each of A and B, and error terms represent S.D. Absence of error terms indicates only one experiment. The Ca^{2+}-dependent guanylate cyclase, not shown, was inhibited in a manner resembling that of the Mn^{2+}-dependent enzyme.

The effects of mitogens on broken cell systems are negligible. In contrast, HETEs and archidonate can, at high concentrations, modestly stimulate the soluble form of guanylate cyclase. This action is not inhibited by ETYA but is prevented by 15-HETE, a unique inhibitor specific for 5- and 12-lipoxygenases.

The effects of PMA, a tumor promoter and a mitogen for a subset of human T lymphocytes, differ from those of other T cell mitogens in several respects including the pattern of inhibition of guanylate cyclase stimulation. Although complex, the results with PMA are consistent with those of other mitogens in that Ca^{2+}-dependent phospholipases and lipoxygenases mediate both the activation of guanylate cyclase and DNA synthesis. The results with PMA differ somewhat from those with PHA in the subcellular form of guanylate cyclase being stimulated and in the nature of both the phospholipase and the lipoxygenase activities catalyzing the lipid changes that precede guanylate cyclase activation. Further studies are required to learn whether these different activities occur in different cell types or whether they represent different receptor responses within the same cell.

ACKNOWLEDGMENTS. We should like to acknowledge the generous gifts of HETEs by Dr. Edward Goetzl and Dr. J. Martyn Bailey. We also gratefully acknowledge the technical expertise of Christina S. Coffey and Dr. Elba M. Hadden.

REFERENCES

Aldiger, J. C., Gualde, N., Mexmain, C. R., Ratinaud, M. H., and Rigaud, M., 1984, Immunosuppression induced *in vivo* by 15-hydroxyeicosatetraeoic acid, *Prostaglandins Leukotrienes Med.* **13:**99–108.

Bailey, J. M., Bryant, R. W., Low, C. E., Pupillo, M. B., and Vanderhoek, J. Y., 1982, Regulation of T-lymphocyte mitogenesis by the leukocyte product 15-hydroxy-eicosatetraenoic acid (15-HETE), *Cell. Immunol.* **67:**112–120.

Brandwein, H. J., Lewicki, J. A., and Murad, F., 1981, Reversible inactivation of guanylate cyclase by mixed disulfide formation, *J. Biol. Chem.* **256:**2958–2962.

Coffey, R. G., and Hadden, J. W., 1981, Arachidonate and metabolites in mitogen activation of lymphocyte guanylate cyclase, in: *Advances in Immunopharmacology* (J. Hadden, L. Chedid, P. Mullen, and F. Spreafico, eds.), Pergamon Press, Oxford, pp. 365–373.

Coffey, R. G.,and Hadden, J. W., 1983a, Calcium and guanylate cyclase in lymphocyte activation, in: *Advances in Immunopharmacology,* Vol. 2 (J. W. Hadden, L. Chedid, P. Dukor, F. Spreafico, and D. Willoughby, eds.), Pergamon Press, Oxford, pp. 87–94.

Coffey, R. G., and Hadden, J. W., 1983b, Phorbol myristate acetate stimulation of lymphocyte guanylate cyclase and cyclic GMP phosphodiesterase and reduction of adenylate cyclase, *Cancer Res.* **43:**150–158.

Coffey, R. G., Hadden, E. M., and Hadden, J. W., 1981, Phytohemagglutinin stimulation of guanylate cyclase in human lymphocytes, *J. Biol. Chem.* **256:**4418–4424.

Craven, P. A., and DeRubertis, F. R., 1982, Relationship between calcium stimulation and cyclic GMP and lipid peroxidation in the rat kidney: Evidence for involvement of calmodulin and separate pathways of peroxidation in cortex versus inner medulla, *Metabolism* **31:**103–116.

Goetzl, E. J., 1981, Selective feed-back inhibition of the 5-lipoxygenation of arachidonic acid in human T-lymphocytes, *Biochem. Biophys. Res. Commun.* **101:**344–350.

Goldberg, N. D., Graff, G., Haddox, M. K., Stephenson, J. H., Glass, D. B., and Moser, M. E., 1978, *Adv. Cyclic Nucleotide Res.* **9:**101–130.

Gordon, D., Nouri, A. M. E., and Thomas, R. U., 1981, Selective inhibition of thromboxane biosynthesis in human blood mononuclear cells and the effects on mitogen-stimulated lymphocyte proliferation, *Br. J. Pharmacol.* **74:**469–475.

Hadden, J. W., Coffey, R. G., Ananthakrishnan, R., and Hadden, E. M., 1979, Cyclic nucleotides and calcium in lymphocyte regulation and activation, *Ann. N.Y. Acad. Sci.* **332:**241–254.

Haddox, M. K., and Goldberg, N. D., 1977, Cyclic GMP metabolism and involvement in biological regulation, *Annu. Rev. Biochem.* **46:**823–896.

Novogrodsky, A., Ravid, A., Rubin, A. L., and Stenzel, K. H., 1982, Hydroxyl radical scavengers inhibit lymphocyte mitogenesis, *Proc. Natl. Acad. Sci. U.S.A.* **79:**1171–1174.

Parker, C. W., Stenson, W. F., Huber, M. G., and Kelly, J. P., 1979, Formation of thromboxane B_2 and hydroxyarachidonic acids in purified human lymphocytes in the presence and absence of PHA, *J. Immunol.* **122:**1572–1577.

Vanderhoek, J. Y., Bryant, R. W., and Bailey, J. M., 1982, Regulation of leukocyte and platelet lipoxygenases by hydroxyeicosanoids, *Biochem. Pharmacol.* **31:**3463–3467.

Vesin, M. F., Bourgoin, S., Leiber, D., and Harbon, S., 1984, Lipoxygenase and cyclooxygenase products of arachidonic acid in uterus. Selective interaction with the cGMP and cAMP systems, *Prostaglandins Leukotrienes Med.* **13:**75–78.

White, A. A., Karr, D. B., and Patt, C. S., 1982, Role of lipoxygenase in the O_2-dependent activation of soluble guanylate cyclase from rat lung, *Biochem. J.* **204:**382–392.

49

Release of Leukotrienes B_4 and D_4 by the Anaphylactic Guinea Pig Lung *in Vitro*

HASSAN SALARI, PIERRE BORGEAT, PIERRE BRAQUET, YVON BROUSSEAU, and PIERRE SIROIS

1. INTRODUCTION

The lung possesses several enzymic systems for the metabolism of arachidonic acid (Samuelsson *et al.*, 1975; Hammarström, 1983). Among these, the 5-lipoxygenase has recently received considerable attention since some of its products, the leukotrienes (LTs), may play critical roles in inflammation (Samuelsson, 1982; Lewis and Austen, 1981) and immediate hypersensitivity reactions (Holroyde *et al.*, 1981; Dahlén *et al.*, 1980). Indeed, leukotriene B_4 (LTB$_4$) was shown to be a potent chemotactic substance towards polymorphonuclear leukocytes (PMNL) (Ford-Hutchison *et al.*, 1980), and leukotrienes C_4 (LTC$_4$), D_4 (LTD$_4$), and E_4 (LTE$_4$) were found to have strong myotropic activity on smooth muscles of the respiratory tract (Drazen *et al.*, 1980; Hedqvist *et al.*, 1980; Lewis *et al.*, 1980).

The formation of a slow-reacting substance during the anaphylactic reaction in perfused guinea pig lung was first reported in 1940 and later in 1960 (Kellaway and Trethewie, 1940; Brocklehurst, 1960). The chemical structure of slow-reacting substances as peptidolipids derived from arachidonic acid, i.e., LTC$_4$ and LTD$_4$,

HASSAN SALARI and PIERRE BORGEAT • Groupe de Recherche sur les Leucotriènes, Laboratoire d'Endocrinologie Moléculaire, Centre Hospitalier de l'Université Laval, Québec G1V 4G2, Canada. PIERRE BRAQUET • I. H. B. Research Laboratories, F-92350 Le Plessis-Robinson, France. YVON BROUSSEAU and PIERRE SIROIS • Département de Pédiatrie, Centre Hospitalier Universitaire, Université de Sherbrooke, Sherbrooke J1H 5N4, Canada.

was first reported by Murphy *et al.* (1979) and by Morris *et al.* (1980b). Since then, LTs have been detected in many experimental models of immediate-type hypersensitivity (Weichman *et al.*, 1982; Hansson *et al.*, 1983; Morris *et al.*, 1980a; Fleisch *et al.*, 1982). Most of these studies were carried out using bioassays or radioimmunoassays of LTs; in the present study, using reversed-phase high-performance liquid chromatography (RP-HPLC), we provide quantitative data on the 5-lipoxygenase products released by the anaphylactic guinea pig lung *in vitro*.

2. METHODS

2.1. Sensitization of Guinea Pigs and Antigen Challenge *in Vitro*

Guinea pigs of both sexes, weighing approximately 300–400 g, were sensitized to ovalbumin (grade II, Sigma Chemical Company, St. Louis, MO) by injection of a solution of the protein (100 mg s.c. and 100 mg i.p.). One week later, the animals were given another dose of ovalbumin (10 mg i.p.). After 2 weeks, the animals were killed by cervical dislocation, and the lungs were excised and washed with Krebs–bicarbonate through the pulmonary artery (10 ml/min, 10 min). The lungs were first perfused with Krebs–bicarbonate solution for 30 min and then with the same solution containing 100 μg/ml of ovalbumin for an additional 30 min at the flow rate of 2 ml/min (Engineer *et al.*, 1978). The perfusates were collected on dry ice and analyzed by RP-HPLC within the next 10 days.

2.2. Reverse-Phase High-Performance Liquid Chromatography

Samples of 60 ml, corresponding to 30 min of perfusion, were thawed, immediately mixed with 12 ml of isopropanol, and 300 ng of prostaglandin B$_2$ (PGB$_2$) was added as internal standard. The pH of the samples was lowered to 5.5 using dilute H$_3$PO$_4$, and the samples were heated to 90–100°C over 10 min. The samples were centrifuged to eliminate the precipitate, the supernatants were evaporated to dryness under reduced pressure, and the dry residue was recovered by rinsing the flask with 1 ml of water, 2 ml of methanol, and 1 ml of water successively. The samples were centrifuged again to remove any insoluble material, acidified to pH 3 with dilute H$_3$PO$_4$, and analyzed by RP-HPLC (details of these procedures will be reported separately). The RP-HPLC was performed using a Radial-Pak cartridge (C18, 100 × 8 mm i.d., 10-μm particle size) from Waters Associates. Analysis was performed using the three-solvent HPLC system already described (Borgeat *et al.*, 1984). The metabolites of 5-lipoxygenase were detected by UV photometry at 280 nm, and their quantities were determined by comparison of the peak areas with that of the internal standard (PGB$_2$) and correction for differences in molar extinction coefficients.

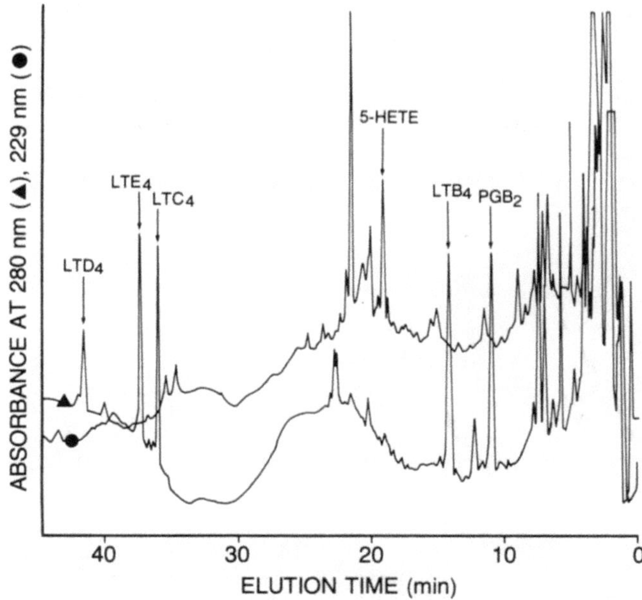

FIGURE 1. Reverse-phase HPLC chromatogram of synthetic leukotrienes (LTs, 100 ng each; PGB$_2$, 150 ng). The compounds were added to a biological sample free of detectable amount of LTs (a 30-min perfusate collected from an unsensitized unchallenged guinea pig lung) that was concentrated to a volume of 4 ml for analysis by RP-HPLC as described in Section 2.2. The traces show the UV absorbances at 280 nm and 229 nm. Attenuation settings were 0.05 and 0.02 absorbance units full scale at 229 and 280 nm, respectively.

3. RESULTS

In this study, we used RP-HPLC for the profiling of 5-lipoxygenase products in guinea pig lung perfusates collected before and during anaphylactic reactions. As shown in Fig. 1, the method of analysis used permitted the separation of the peptido-LTs, LTC$_4$, D$_4$, and E$_4$, as well as LTB$_4$ and PGB$_2$. Figure 2 shows the analysis of a 30-min perfusate of a sensitized guinea pig lung prior to antigen challenge (control). None of the above-mentioned products (except PGB$_2$, internal standard) were detectable in this sample. However, when the lungs were challenged with ovalbumin, two compounds were detected in the perfusates (Fig. 3). On the basis of their chromatographic properties and UV absorption at 280 nm, they were identified as LTB$_4$ and LTD$_4$ (the absorption traces at 229 and 280 nm demonstrate the specificity of the detection of LTs at 280 nm). In addition, the identity of LTB$_4$ was confirmed by GC–MS analysis, and the presence of LTD$_4$-like myotropic activity in the perfusate was previously demonstrated by bioassay on the guinea pig ileum (data not shown). Although the amounts of leukotrienes detected in

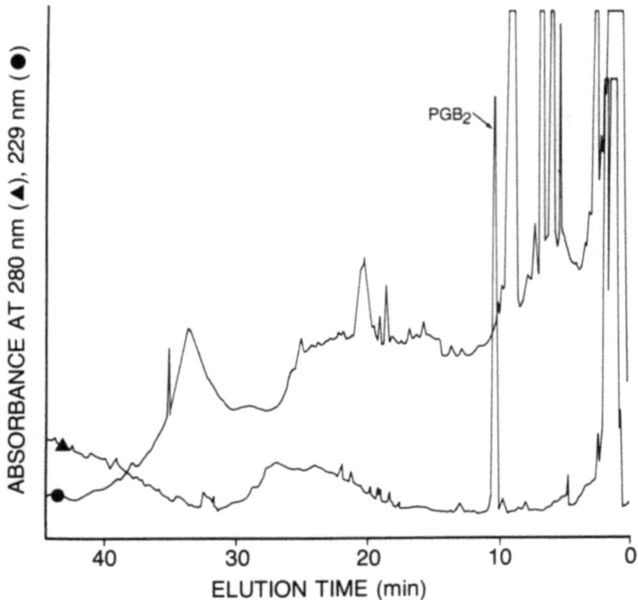

FIGURE 2. Reverse-phase HPLC chromatogram of a 30-min perfusate (no antigen challenge) of an ovalbumin-sensitized guinea pig lung. Prostaglandin B₂ was 300 ng. See Fig. 1 legend.

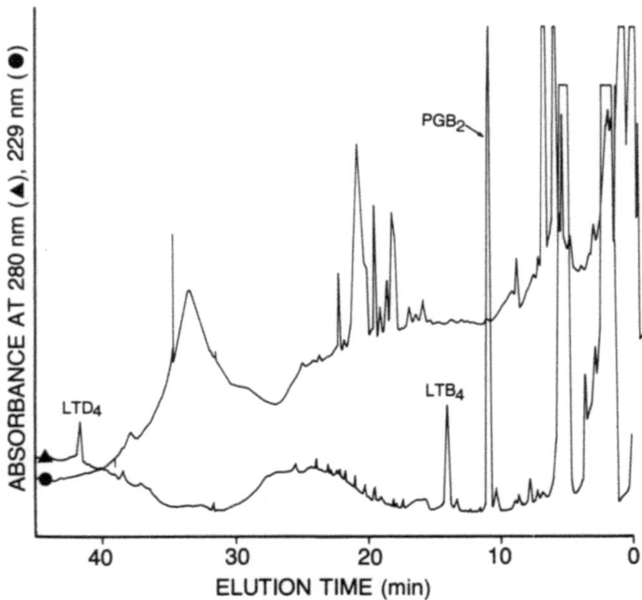

FIGURE 3. Reverse-phase HPLC chromatogram of a 30-min perfusate of an ovalbumin-sensitized guinea pig lung collected during antigen challenge. Prostaglandin B₂ was 300 ng. See Fig. 1 legend.

different anaphylactic lung perfusates were quite variable, the average release of LTB_4 and LTD_4 was 50–150 pmol/lung and 15–50 pmol/lung, respectively, during a 30-min challenge. Time-course studies (data not shown) indicated that under the experimental conditions used, the release of LTD_4 and B_4 was virtually completed 20 min after the beginning of antigen challenge.

The presence of 5-HETE could not be ascertained in any experiment, although small absorption peaks (229 nm) were observed at the proper retention time. Leukotriene C_4 and LTE_4 were occasionally detected in perfusates of anaphylatic lungs, but LTD_4 was always the major peptido-LT observed. The products of the ω-oxidation of LTB_4 (ω-hydroxy-LTB_4 and ω-carboxy-LTB_4) were not present in detectable amounts in any of the perfusates analyzed.

4. DISCUSSION

The originality of the present study lies in the methodology used for the analysis of 5-lipoxygenase products in guinea pig lung perfusates. Indeed, we used a physicochemical method of analysis as opposed to the biological or radioimmunologic assays most frequently used so far in this type of investigation. This RP-HPLC method is remarkably simple in regard to sample preparation, which requires only heat denaturation, centrifugation, and concentration of the sample before injection (4 ml) on the column. This methodology provided a reliable analysis and also brought about the possibility of profiling the various components of the sample.

The analysis of 5-lipoxygenase products in unfractionated and concentrated lung perfusates was made possible by the relatively low level of contaminants in the perfusates and by the high degree of specificity for the detection of LTs provided by dual-wavelength UV photometry (229 and 280 nm) and the solvent system used (Borgeat et al., 1984), which resolves the peptido-LTs from all other known derivatives of arachidonic acid.

In fact, the only purification step applied to the samples was a 10-min heating to destroy enzymatic activities and eliminate part of the proteins (which otherwise precipitate and clog the HPLC column). This procedure (heating) did not cause detectable loss of HETEs or LTB_4, C_4, D_4, and E_4 (to be discussed in detail elsewhere).

In this preliminary investigation, we have shown that LTB_4 is the major LT released by the guinea pig lung during antigen challenge *in vitro* and that LTD_4 was released in two to three times smaller amounts than LTB_4. These data are in agreement with those of Morris et al. (1979), who reported the presence of LTD_4 and LTB_4 in guinea pig lung perfusates under similar experimental conditions using mass spectrometry. Our data bring further support to the role of LTs in the bronchoconstriction and pulmonary PMNL accumulation associated with hypersensitivity reactions. Further studies involving the same methodology are in progress on the release of 5-lipoxygenase products by guinea pig and human lungs.

5. SUMMARY

We studied the synthesis and release of leukotrienes by guinea pig lungs on antigen challenge. Lungs of ovalbumin-sensitized animals were washed free of blood and perfused with a solution of the protein to induce an anaphylactic reaction. Analysis of lung perfusates by reverse-phase high-performance liquid chromatography showed the presence of 5S,12R-dihydroxy-6,8,10,14-(Z,E,E,Z)-eicosatetraenoic acid (LTB$_4$) and 5S,6R-cysteinylglycine-7,9,11,14-(E,E,Z,Z)-eicosatetraenoic acid (LTD$_4$). Anaphylactic lungs synthesized two to three times more LTB$_4$ than LTD$_4$ as measured by comparison of absorption peak areas on chromatograms. The amounts of leukotrienes synthesized by sensitized lungs on antigen challenge were 15–50 pmol of LTD$_4$ and 50–150 pmol of LTB$_4$; LTC$_4$ and LTE$_4$ were occasionally observed as trace components only. Leukotrienes were not detectable in the perfusate of sensitized lungs prior to antigen challenge. The liquid chromatographic method of analysis used did not require prior extraction, fractionation, or derivatization of the sample and provided a direct measurement of the relative amounts of the various LTs released by anaphylactic guinea pig lungs *in vitro*.

ACKNOWLEDGMENT. This study was supported by grants from the Quebec Lung Association and the Medical Research Council of Canada to P. Borgeat and P. Sirois.

REFERENCES

Borgeat, P., Fruteau de Laclos, B., Rabinovitch, H., Picard, S., Braquet, P., Hebert, J., and Laviolette, M., 1984, Eosinophil-rich human polymorphonuclear leukocyte preparations characteristically release leukotriene C$_4$ upon ionophore A23187 challenge, *J. Allergy Clin. Immunol.* **74**:310–315.

Brocklehurst, W. E., 1960, The release of histamine and formation of a slow-reacting substance (SRS-A) during anaphylactic shock, *J. Physiol. (Lond.)* **151**:416–435.

Dahlén, S. E., Hedqvist, P., Hammarström, S., and Samuelsson, B., 1980, Leukotrienes are potent constrictors of human bronchi, *Nature* **288**:484–486.

Drazen, J. M., Austen, K. F., Lewis, R. A., Clark, D. A., Goto, G., Marfat, A., and Corey, E. J., 1980, Comparative airway and vascular activities of leukotrienes C-1 and D *in vivo* and *in vitro*, *Proc. Natl. Acad. Sci. U.S.A.* **77**:4354–4358.

Engineer, D. M., Niederhauser, U., Piper, P. J., and Sirois, P., 1978, Release of mediators of anaphylaxis: Inhibition of prostaglandin synthesis and the modification of release of slow-reacting substances of anaphylaxis and histamine, *Br. J. Pharmacol.* **62**:61–66.

Fleisch, J. H., Haisch, K. D., and Spaethe, S. M., 1982, Slow reacting substance of anaphylaxis (SRS-A) release from guinea-pig lung parenchyma during antigen or ionophore-induced contraction, *J. Pharmacol. Exp. Ther.* **221**:146–151.

Ford-Hutchison, A. W., Bray, M. A., Doig, M. V., Shipley, M. E., and Smith, M. J. H., 1980, Leukotriene B, a potent chemokinetic and aggregating substance released from polymorphonuclear leukocytes, *Nature* **286**:264–265.

Hammarström, S., 1983, Leukotrienes, *Annu. Rev. Biochem.* **52**:355–377.

Hansson, G., Björck, T., Dahlén, S.-E., Hedqvist, P., Granström, E., and Dahlén, B., 1983, Specific allergen induces contraction of bronchi and formation of leukotrienes C$_4$, D$_4$ and E$_4$ in human asthmatic lung, *Adv. Prostaglandin Thromboxane Leukotriene Res.* **12**:153–159.

Hedqvist, P., Dahlén, S.-E., Gustafsson, L., Hammarström, S., and Samuelsson, B., 1980, Biological profile of leukotrienes C_4 and D_4, *Acta Physiol. Scand.* **110**:331–333.

Holroyde, M. C., Altounyan, R. E. C., Cole, M., Dixon, M., and Elliott, E. V., 1981, Bronchoconstriction produced in man by leukotrienes C and D, *Lancet* **2**:17–18.

Kellaway, C. H., and Trethewie, E. R., 1940, The liberation of a slow-reacting smooth-muscle stimulating substance in anaphylaxis, *Q. J. Exp. Physiol.* **30**:121–145.

Lewis, R. A., and Austen, K. F., 1981, Mediation of local homeostasis and inflammation by leukotrienes and other mast cell-dependent compounds, *Nature* **293**:103–108.

Lewis, R. A., Austen, K. F., Drazen, J. M., Clark, D. A., Marfat, A., and Corey, E. J., 1980, Slow reacting substances of anaphylaxis: Identification of leukotrienes C-1 and D from human and rat sources, *Proc. Natl. Acad. Sci. U.S.A.* **77**:3710–3714.

Morris, H. R., Taylor, G. W., Piper, P. J., and Tippins, J. R., 1979, Slow reacting substance of anaphylaxis. Studies on purification and characterization. *Agents Actions* **5**(Suppl. 6):27–36.

Morris, H. R., Taylor, G. W., Piper, P. J., and Tippins, J. R., 1980a, Structure of slow-reacting substance of anaphylaxis from guinea-pig lung, *Nature* **285**:104–106.

Morris, H. R., Taylor, G. W., Piper, P. J., Samhoun, M. N., and Tippins, J. R., 1980b, Slow-reacting substances (SRSs): The structure identification of SRSs from rat basophilic leukemia (RBL-1) cells, *Prostaglandins* **19**:185–201.

Murphy, R. C., Hammarström, S., and Samuelsson, B., 1979, Leukotriene C, a slow reacting substance (SRS) from mouse mastocytoma cells, *Proc. Natl. Acad. Sci. U.S.A.* **79**:4275–4279.

Samuelsson, B., 1982, Leukotrienes: A new group of biologically active compounds, in: *Advances in Pharmacology and Therapeutics II,* Vol. 4, *Biochemical and Immunological Pharmacology* (H. Yoshida, Y. Hagihara, and S. Ebashi, eds.), Permagon Press, Oxford, New York, pp. 55–75.

Samuelsson, B., Granström, E., Green, K., Hamberg, M., and Hammarström, S., 1975, Prostaglandins, *Rev. Biochem.* **44**:669–695.

Weichman, B. M., Hostelley, L. S., Bostick, S. P. Muccitelli, R. M., Krell, R. D., and Gleason, J. G., 1982, Regulation of the synthesis and release of slow-reacting substance of anaphylaxis from sensitized monkey lung, *J. Pharmacol. Exp. Ther.* **221**:295–302.

50

Mechanisms Involved in the Mediation of Contractions of Airway Smooth Muscle to Arachidonate Metabolites and Other Spasmogens

T. R. JONES and D. DENIS

1. INTRODUCTION

Airway smooth muscle (ASM) contractility, like that of other smooth muscle types, is dependent on the level of free intracellular Ca^{2+}, which is precisely regulated by the supply of ionized Ca^{2+} made available from intra- and extracellular pools (Filo *et al.*, 1965; Solmyo and Solmyo, 1968; Bolton, 1979). Membrane-potential-dependent and -independent mechanisms are involved in Ca^{2+} regulation in response to receptor stimulation by a variety of neural, humoral, and myogenic stimuli (Kirkpatrick *et al.*, 1975; Coburn and Yamaguchi, 1977; Farley and Miles, 1977; Bolton, 1979). These collectively serve to control smooth muscle contractility and also may influence the underlying state of tissue reactivity. Of particular importance in lung physiology and pathophysiology are the potent bronchoactive substances, namely, prostaglandins, endoperoxides, thromboxanes (Samuelsson *et al.*, 1978), and, more recently, leukotrienes (Samuelsson *et al.*, 1981). The exquisite sensitivity of respiratory tissue to eicosanoids and in particular the sulfidoleukotrienes is well recognized (Hedquist *et al.*, 1980; Drazen *et al.*, 1980; Piper *et al.*, 1981; Jones *et al.*, 1982a). However, little information is available on their precise mechanism

T. R. JONES and D. DENIS ● Department of Pharmacology, Merck Frosst Canada Inc., Pointe Claire-Dorval, Québec H9R 4P8, Canada.

of action, their receptors, or the specific Ca^{2+} pools required for full expression of their biological activity in the lung.

In the present study these contractile mechanisms were systematically investigated in guinea pig tracheal smooth muscle. Specifically, we determined the effects of Ca^{2+}-entry blockers and/or extracellular Ca^{2+} depletion on contractions induced by exogenously applied prostaglandins and leukotrienes and by endogenously released (antigen- and Ca^{2+}-ionophore-stimulated) arachidonate metabolites. For comparative evaluation of mechanisms, the effects of the above maneuvers were also determined against a number of biogenic mediators (histamine, acetylcholine) and a pharmacological stimulus (elevated K^{+}).

2. METHODS AND MATERIALS

Normal (unsensitized) and sensitized* adult male albino Hartley strain guinea pigs (350–450 g) were used as a tissue source in all experiments. Tracheal chains were prepared for isometric recording as described previously (Jones *et al.*, 1982b). Tissues were placed under a 1-g load, allowed to equilibrate 30–60 min, and then were primed two or three times with a maximal concentration of histamine (9×10^{-5} M). Most experiments were carried out in the presence of 1.4×10^{-6} M indomethacin in order to remove intrinsic tracheal tone. After this procedure, complete cumulative concentration–response curves to LTD_4, LTC_4, $PGF_{2\alpha}$, a prostaglandin endoperoxide analogue (U-44069); acetylcholine (ACh), and histamine (HIST) were obtained before and/or after pretreatment of the tissue with 10 μg/ml nifedipine, verapamil, or D-600. The Ca^{2+}-entry blockers were also examined against peak contractile responses to K^{+} and Ca^{2+} ionophore A23187 (3 μg/ml) on trachea from normal guinea pigs and specific antigen (0.1 μg/ml egg albumin) on trachea from sensitized guinea pigs.

The effects of removal of extracellular Ca^{2+} by placing the tissue in Ca^{2+}-free Krebs' solution with or without 0.125 mM EGTA for varying times (15–45 min) were also examined on the responses to the above-mentioned stimuli. Before exposure to Ca^{2+}-free Krebs' solution, all tissues were allowed to recover and/or equilibrate in normal Krebs' solution (2.5 mM Ca^{2+}).

3. RESULTS

The Ca^{2+}-entry blockers (nifedipine > D-600) produced a rightward shift of the concentration response curves to all agonists and depressed the maximum response (Table I). Similar results were obtained in eight experiments with 3×10^{-5} M verapamil (data not included in Table I). The rank order of susceptibility to

---

* 0.5 milliliters of a solution containing egg albumin (10 μg/ml) and Al(OH)$_3$ (100 μg/ml) dissolved in saline was given as an intraperitoneal injection no less than 14 days before sacrifice.

TABLE I. Effect of Ca^{2+}-Entry Blockers on Contraction of Guinea Pig Trachea to Different Spasmogens

		Treatment[d]			
		Nifedipine $(3 \times 10^{-5}$ M)		D-600 $(2 \times 10^{-5}$ M)	
Agonist	$pD_2{}^a$	D.R.[b]	% Max.[c]	D.R.	% Max.
LTD$_4$	8.6 ± 0.1 ($n = 10$)	9.0	71 ± 5	4.0	83 ± 4
LTC$_4$	8.5 ± 0.1 ($n = 10$)	80	73 ± 3	5.1	81 ± 7
U-44069	7.9 ± 0.2 ($n = 8$)	18.9	66 ± 8	6.0	69 ± 8
PGF$_{2\alpha}$	6.5 ± 0.1 ($n = 7$)	11.0	64 ± 5	4.0	84 ± 6
HIST	5.3 ± 0.1 ($n = 7$)	3.1	76 ± 3	4.7	80 ± 7
K$^+$	(80 mM) ($n = 6$)	—	0–5	—	0 ± 10

[a] pD_2 is negative log of the concentration producing 50% of the maximum response (EC_{50}) to each agonist.
[b] D.R. is dose ratio from EC_{50} in presence of antagonist to EC_{50} of control.
[c] % Max is percentage of maximum relative to control maximum of 100%.
[d] Indomethacin (1.4×10^{-6} M) and atropine (1.0×10^{-6} M) were present in all experiments.

Ca^{2+}-entry blockage was $K^+ \gg PGF_{2\alpha} > U\text{-}44069 > LTD_4 \gg LTC_4 > HIST > $ acetylcholine. Nifedipine and D-600 produced only a small two- to three-fold shift to the right of the concentration–response curve to acetylcholine and a 10–20% depression of the maximum. The Ca^{2+}-entry blockers inhibited a small (10–30%) component of the response to 0.1 μg/ml antigen and 3 μg/ml A-23187 in normal (2.5 mM Ca^{2+}) Krebs'.

Exposure of the tissues to Ca^{2+}-free Krebs' solution with 0.125 mM EGTA for 15 min dramatically reduced the contractile activity to leukotriene $E_4 = F_4 > C_4 > D_4 \gg U\text{-}44069 \gg PGF_{2\alpha} \gg HIST \gg Ach$ (Figs. 1, 2). Increasing the duration in Ca^{2+}-free Krebs' solution to 30 or 45 min virtually abolished the contractile activity to all agonists. Similar results were obtained with depolarizing concentrations of K^+ (80 mM) (Fig. 3) in the presence and absence of indomethacin. In the absence of EGTA, a significant contractile response could still be elicited with $ACh > HIST \gg K^+ > LTD_4 > LTC_4$ even after a 45-min exposure to Ca^{2+}-free Krebs' solution. Under these conditions, residual contractions to all agonists were reversed by 0.1–1 μg/ml nifedipine.

After a 45-min exposure to Ca^{2+}-free Krebs' (no EGTA), the response to A23187 was virtually abolished ($<5\%$), whereas a significant contraction ($24 \pm 7\%$) to specific antigen could still be elicited. Following a shorter exposure time to Ca^{2+}-free Krebs' (15 min) with or without EGTA, a significantly larger contractile response (61%) could still be elicited with antigen but not with the Ca^{2+} ionophore.

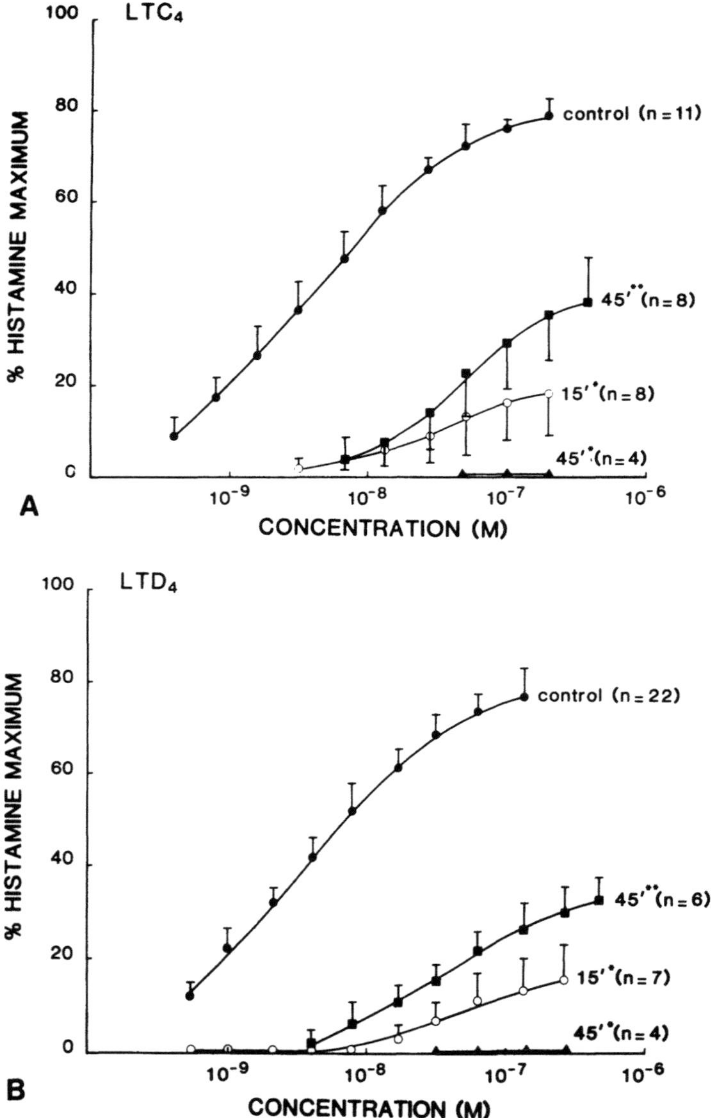

FIGURE 1. Concentration–response curves to LTC_4 (A), LTD_4 (B), LTE_4 (C), and LTF_4 (D) obtained on guinea pig trachea in normal (2.5 mM Ca^{2+}) Krebs' solution (●), in Ca^{2+}-free Krebs' solution (45 min) without EGTA (■), and in Ca^{2+}-free Krebs' solution with 0.1 mM EGTA for 15 (○), 30 (×), and 45 (▲) min. All studies were carried out in the presence of 1.4×10^{-6} M indomethacin.

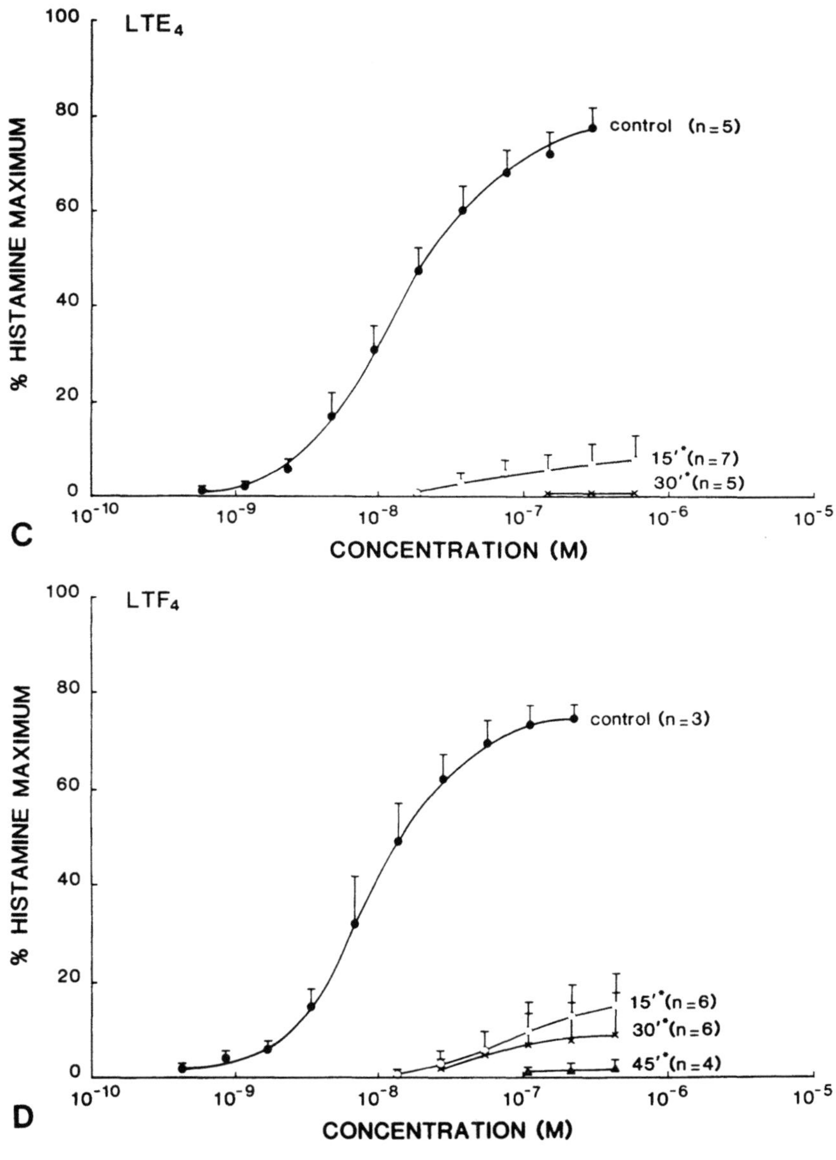

FIGURE 1. (*Cont.*)

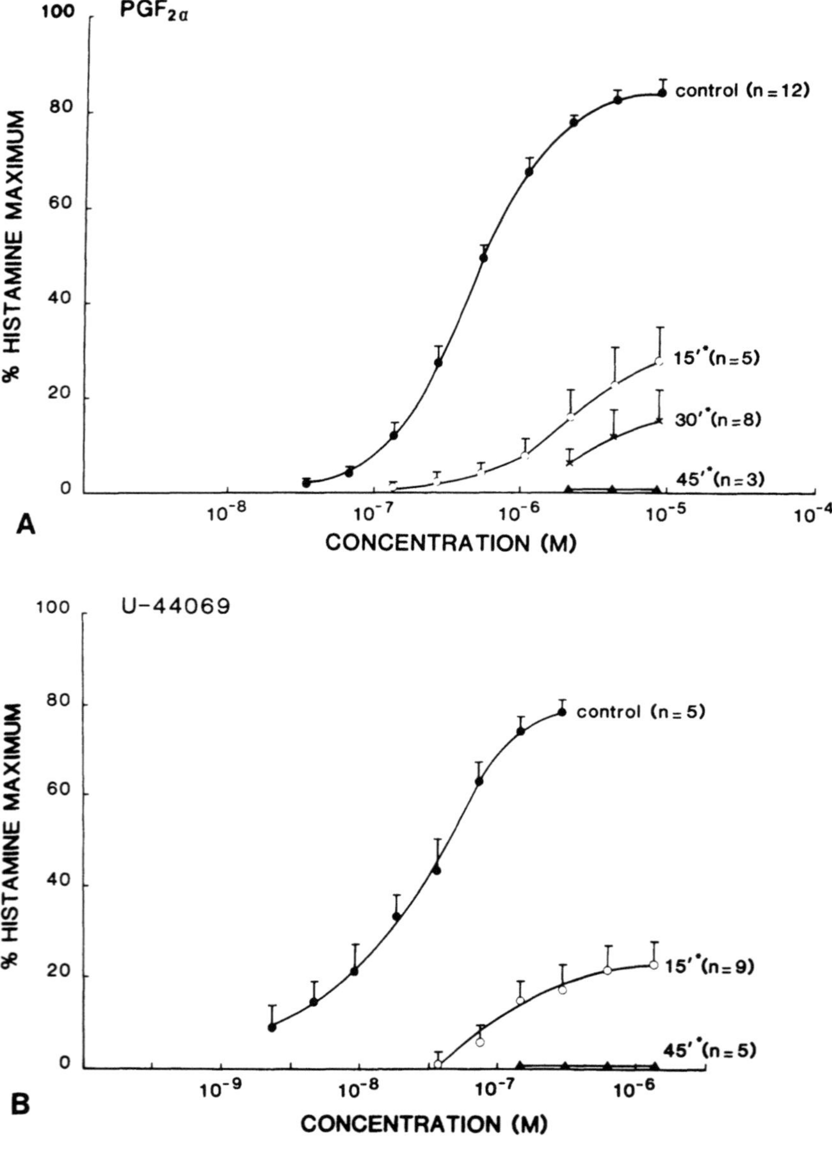

FIGURE 2. Concentration–response curves to $PGF_{2\alpha}$ (A), U-44069 (B), histamine (C), and ACh (D) obtained on guinea pig trachea in normal (2.5 mM Ca^{2+}) Krebs' solution (●), in Ca^{2+}-free Krebs' solution (45 min) without EGTA (■), in Ca^{2+}-free Krebs' solution with 0.1 mM EGTA for 15 (○), 30 (×), and 45 (▲) min. All studies were carried out in the presence of 1.4×10^{-6} M indomethacin.

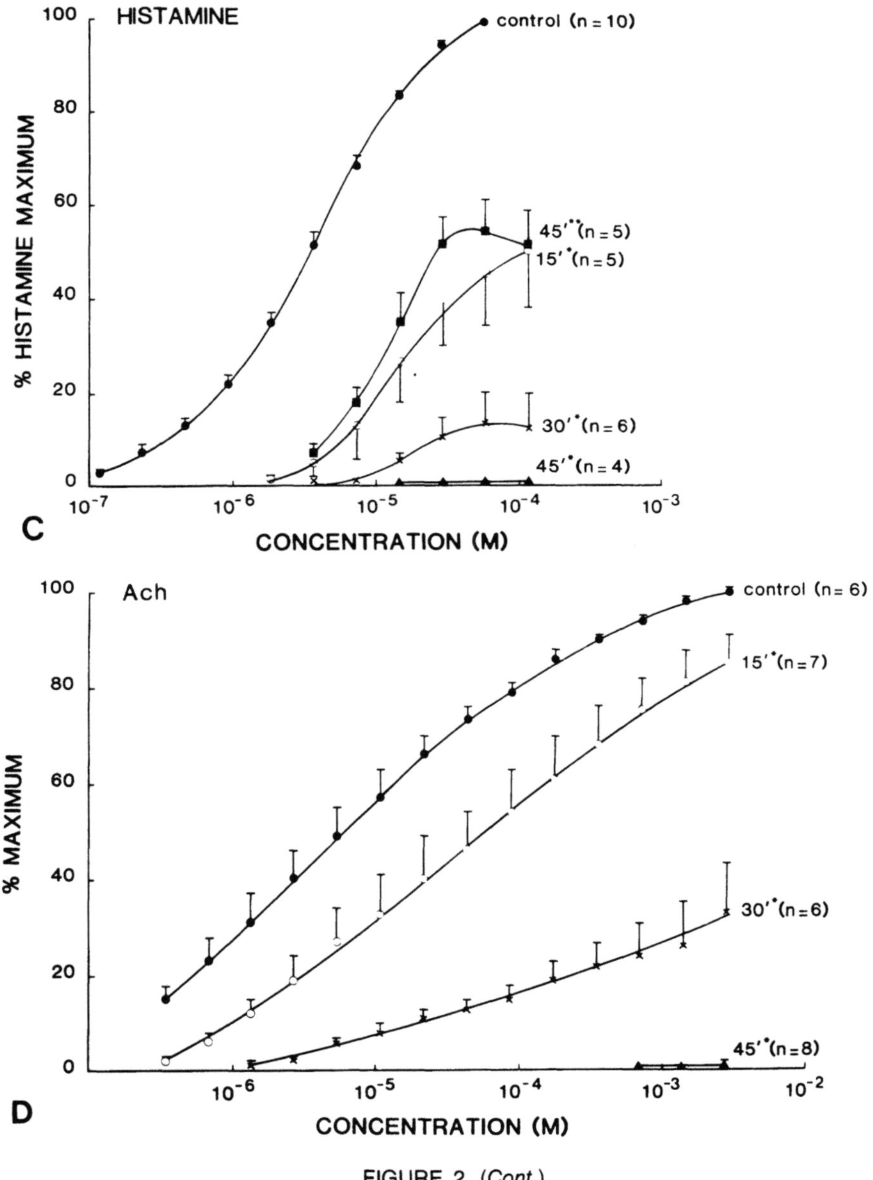

FIGURE 2. *(Cont.)*

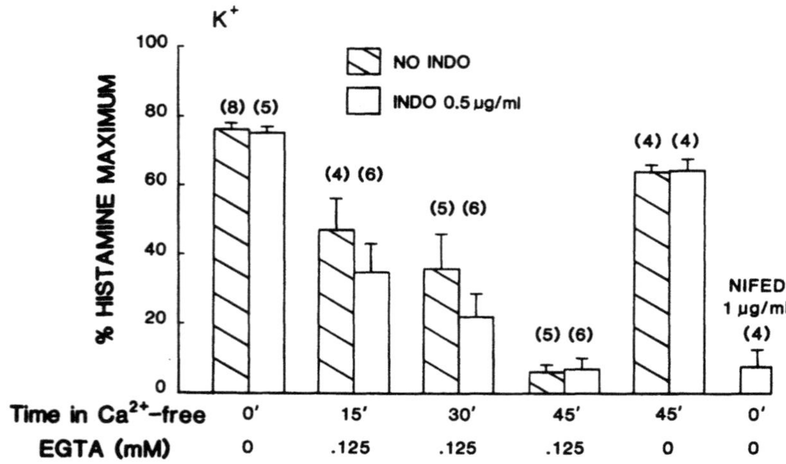

FIGURE 3. Contraction of guinea pig trachea to 80 mM K^+ in normal (2.5 mM Ca^{2+}) Krebs' solution (control) and after exposure of the tissues to Ca^{2+}-free Krebs' solution with or without EGTA for 15, 30, or 45 min. The results with nifedipine (1.0 µg/ml) were obtained in normal Krebs' solution.

Additional studies were carried out with leukotriene D_4 (a single response to 100 ng/ml) and antigen in which the effects of FPL-55712 or nifedipine were examined on the residual contraction to these stimuli after 15 min in Ca^{2+}-free Krebs' with EGTA. Results obtained are summarized in Table II. Under these conditions, the residual response to LTD_4 in Ca^{2+}-free Krebs' was somewhat larger than that observed when the leukotriene was added by cumulative addition (22% versus 37%). The residual responses to both LTD_4 and antigen were blocked by nifedipine but were only slightly decreased by pretreatment with FPL-55712.

TABLE II. Studies in Ca^{2+}-Free Krebs' Solution (0.125 mM EGTA)[a]

	Percentage histamine maximum ± S.E.M.			
Agonist	Control	Ca^{2+}-free[b]	Ca^{2+}-free + 0.1 µg/ml + nifedipine	Ca^{2+}-free 1.0 µg/ml FPL-55712
LTD_4	74 ± 2	37 ± 7[c]	12 ± 2	30 ± 2
(100 ng/ml)	(n = 12)	(n = 4)	(n = 3)	(n = 3)
Antigen	80 ± 6	61 ± 7	8 ± 2	39 ± 18[d]
(0.1 µg/ml)	(n = 10)	(n = 3)	(n = 3)	(n = 2)

[a] Indomethacin (1.4 × 10⁻⁶ M) was present in all experiments.
[b] All tissues exposed to Ca^{2+}-free Krebs' with 0.125 mM EGTA for 15 min before challenge.
[c] Residual response to a single high dose of LTD_4 (100 ng/ml) under these conditions was significantly larger than that observed following cumulative addition of LTD_4 (See Fig. 1A).
[d] Response after mepyramine (1 µg/ml) and FPL-55712 (1 µg/ml) in combination was 34 ± 10%.

4. DISCUSSION

The present study has demonstrated that a major component of the contractile response to the leukotrienes, $PGF_{2\alpha}$, and U-44069 was dependent on a store of Ca^{2+} that was readily depleted by exposing the tissue to Ca^{2+}-free EGTA medium. Although different receptors are involved in mediating contraction of airway smooth muscle to leukotrienes versus prostaglandins, the similarity in requirement for extracellular Ca^{2+} suggests some similarity in mechanism of action. In contrast, histamine and in particular acetylcholine required longer exposure times in Ca^{2+}-free Krebs' in order to eliminate their contractile activity. These findings are consistent with previous reports (Farley and Miles, 1977; Creese and Denborough, 1981). The commonly held interpretation of such findings is that acetylcholine and to a lesser extent histamine can use intracellular Ca^{2+} for initiation of their contractile response. Consistent with these observations is the relative inactivity of the Ca^{2+}-entry blockers against contractions induced by these agonists.

Despite the apparent dependence of leukotrienes and prostaglandins on extracellular Ca^{2+} for their contractile activity, only a component of the overall response in normal Krebs' (2.5 mM Ca^{2+}) was blocked by the Ca^{2+}-entry blockers. The degree of antagonism with these entry blockers is less than would be predicted from results obtained in Ca^{2+}-free Krebs'. These findings suggest that multiple modes of entry exist for extracellular Ca^{2+} in this tissue. One mechanism would appear to involve Ca^{2+} entry through channels that may be similar to those activated by K^+ and that are blocked by low doses of the Ca^{2+} entry blockers (Coburn and Yamaguchi, 1977; Bolton, 1979; Golenhofen, 1981). The other mechanism could involve entry of Ca^{2+} through channels (possibly specific receptor-operated channels—see Bolton, 1979) that, at least in Krebs' solution containing 2.5 mM Ca^{2+}, are less susceptible to blockade by the calcium antagonists used in the present study.

A major component of the response to leukotrienes and prostanoids would appear to be dependent on superficial extracellular Ca^{2+} stores since the response can be reduced by simple omission of Ca^{2+} from the Krebs' solution. Superficial extracellular Ca^{2+} does not appear to play as large a role in the mediation of response to histamine and acetylcholine. Differences between acetylcholine and histamine on one hand and leukotrienes on the other could be related to differences in receptor-mediated release of internal Ca^{2+} (Farley and Miles, 1977; Creese and Denborough, 1981) or, alternatively, could reflect a greater capability of the former agonists to mobilize a more tightly bound membrane store of Ca^{2+} that can be depleted only in the presence of EGTA (Jones *et al.*, 1984). Support for the latter hypothesis comes from the present observations that contractions to all spasmogens including histamine and acetylcholine were almost completely abolished if the various agonists were tested in Ca^{2+}-free medium containing EGTA. However, longer exposure times were required in order to eliminate contractile responses to histamine and, in particular, acetylcholine, so it is possible that such treatment depleted an intracellular store (Keatinge, 1972; Weichman *et al.*, 1983).

Also consistent with the hypothesis that some extracellular Ca^{2+} is very tightly bound are the observations that the contractile activity of K^+, which promotes entry of extracellular Ca^{2+} through channels that are blocked by nifedipine and D-600, still persists after simple omission of Ca^{2+}. A significant component of the response to K^+ also persisted after a 15- to 30-min exposure to Ca^{2+}-free Krebs' with EGTA. Furthermore, following simple omission of Ca^{2+} from the physiological buffer, residual contractions to the leukotrienes, histamine, and part of the contractions to acetylcholine were antagonized by low concentrations of nifedipine, indicating a requirement for a store of Ca^{2+} that ultimately entered smooth muscle cells through specific channels. Why this entry mechanism (i.e., blocked by Ca^{2+}-entry blockers) contributes little to these receptor-mediated contractile responses in normal (2.5 mM Ca^{2+}) Krebs' solution is an important question to be answered. Other possible explanations for the present findings are that the effects of EGTA may not be limited to chelation of extracellular Ca^{2+} only (Keatinge, 1972) and/or that some Ca^{2+}-entry blockers have actions unrelated to blockade of calcium channels (Janis and Triggle, 1983).

Similar to the leukotrienes, contractions induced by specific antigens were not completely inhibited in Ca^{2+}-free Krebs' (45 min). In contrast, responses to the Ca^{2+} ionophore A23187 were completely abolished in Ca^{2+}-free Krebs' solution with or without EGTA. These findings are consistent with previous reports (Burka and Paterson, 1981) that suggested similarities in the nature of the products (eicosanoids and leukotrienes) released by these stimuli but also demonstrated different requirements for extracellular Ca^{2+}. After exposure to Ca^{2+}-free Krebs' with EGTA (15 min), a significant response to antigen could be elicited that was reduced but not abolished in the presence of mepyramine and FPL-55712. This response was virtually abolished by low-dose nifedipine. These findings suggest that, as with the leukotrienes, antigen-induced contraction of guinea pig trachealis is dependent on a superficial and a tightly bound store of Ca^{2+}. The actual release process may still be directly mediated by intracellular Ca^{2+} stores within mast cells (Burka, 1984). However, rapid loss of the contractile response to antigen following a previous exposure to histamine or LTD_4 in Ca^{2+}-free Krebs' (T. R. Jones and D. Denis, unpublished observations) suggests that similar Ca^{2+} stores and/or mechanisms are involved in mediation of the response to the products released by antigen. Moreover, there has been some suggestion that multiple receptor subtypes mediate contraction of guinea pig trachea to leukotrienes (Drazen *et al.*, 1980; Krell *et al.*, 1981; Jones *et al.*, 1983). The present findings are consistent with these reports and indicate that the residual responses to LTD_4 and also to antigen in low Ca^{2+} are mediated through an FPL-55712-insensitive mechanism.

In summary, our findings suggest that extracellular stores of Ca^{2+} (superficial and membrane bound) are essential for initiating contraction to arachidonate metabolites and that at least in normal Krebs' (2.5 mM Ca^{2+}) Ca^{2+}-entry blockers are relatively weak antagonists of most spasmogens except K^+. Electrophysiological studies and studies with labeled $^{45}Ca^{2+}$ are required in order to define more precisely the roles for intra- and extracellular Ca^{2+} in this tissue.

REFERENCES

Bolton, T. B., 1979, Mechanism of action of transmitters and other substances on smooth muscle, *Physiol. Rev.* **59**:606–718.

Burka, J. F., 1984, Effects of calcium blockers and a calmodulin antagonist on contractions of guinea pig airways, *Eur. J. Pharmacol.* **99**:257–268.

Burka, J. F., and Paterson, N. A. M., 1981, A comparison of antigen-induced calcium ionophore A-23187-induced contraction of isolated guinea pig trachea, *Can. J. Physiol. Pharmacol.* **59**:1031–1042.

Coburn, R. F., and Yamaguchi, T., 1977, Membrane potential dependent and independent tension in the canine tracheal muscle, *J. Pharmacol. Exp. Ther.* **201**:276–284.

Creese, B. R., and Denborough, M. A., 1981, Sources of calcium for contraction of guinea pig isolated tracheal smooth muscle, *Clin. Exp. Pharmacol. Physiol.* **8**:175–182.

Drazen, J. M., Austen, K. F., Lewis, R. A., Clark, D. A., Goto, G., Marfat, A., and Corey, E. J., 1980, Comparative airway and vascular activities of leukotrienes C-1 and D *in vivo* and *in vitro*, *Proc. Natl. Acad. Sci. U.S.A.* **77**:4354–4358.

Farley, J. M., and Miles, P. R., 1977, Role of depolarization in acetylcholine-induced contractions of dog trachealis muscle, *J. Pharmacol. Exp. Ther.* **201**:199–205.

Filo, R. S., Bohr, D. F., and Rüegg, J. C., 1965, Glycerinated skeletal and smooth muscle: Calcium and magnesium dependence, *Science* **147**:1581–1583.

Golenhofen, K., 1981, Differentiation of calcium activation process in smooth muscle using selective antagonists, in: *Smooth Muscle* (E. Bülbring, ed.), University of Texas Press, Austin, pp. 157–170.

Hedqvist, P., Dahlén, D. W., Gustafsson, L., Hammarström, S., and Samuelsson, B., 1980, Biological profile of leukotriene C_4 and D_4, *Acta Physiol. Scand.* **110**:331–333.

Janis, R. A., and Triggle, D. J., 1983, New developments in Ca^{2+} channel antagonists, *J. Med. Chem.* **26**:775–785.

Jones, T. R., Davis, C., and Daniel, E. E., 1982a, Pharmacological study of the contractile activity of leukotriene C_4 and D_4 on isolated human airway smooth muscle, *Can. J. Physiol. Pharmacol.* **60**:638–643.

Jones, T. R., Masson, P., Hamel, R., Brunet, G., Holme, G., Girard, Y., Larue, M., and Rokach, J., 1982b, Biological activity of leukotriene sulfones on respiratory tissues, *Prostaglandins* **24**:279–291.

Jones, T., Denis, D., Hall, R., and Ethier, D., 1983, Pharmacological study of the effects of leukotrienes C_4, D_4, E_4 and F_4 on guinea pig trachealis: Interaction with FPL-55712, *Prostaglandins* **26**:833–843.

Jones, T. R., Denis, D., and Comptois, P., 1984, Study of mechanisms mediating contraction of leukotriene C_4, D_4 and other bronchoconstrictors on guinea pig trachea, *Prostaglandins* **27**:939–959.

Keatinge, W. R., 1972, Ca^{2+} concentration and flux in Ca^{2+}-deprived arteries, *J. Physiol. (Lond.)* **224**:35–59.

Kirkpatrick, C. T., Jenkinson, H. A., and Cameron, A. R., 1975, Interaction between drugs and potassium-rich solutions in producing contractions in bovine tracheal smooth muscle: Studies in normal and calcium-depleted tissues, *Clin. Exp. Pharmacol. Physiol.* **2**:559–570.

Krell, R. D., Osborne, R., Vichey, L., Falcone, K., O'Donnel, M., Gleason, J., Kinzig, C., and Bryan, D., 1981, Contraction of isolated airway smooth muscle by synthetic leukotriene C_4 and D_4, *Prostaglandins* **22**:387–409.

Piper, P. J., Samhoun, M. N., Tippins, J. R., Williams, T. J., Palmer, M. A., and Peck, M. J., 1981, Pharmacological studies on pure SRS-A, SRS and synthetic leukotriene C_4 and D_4, in: *SRS-A and Leukotrienes* (P. J. Piper, ed.), Wiley, Chichester, pp. 81–99.

Samuelsson, B., Goldyne, M., Granström, E., Hamberg, M., Hammarström, S., and Malmsten, C., 1978, Prostaglandins and thromboxanes, *Annu. Rev. Biochem.* **47**:997–1029.

Samuelsson, B., Borgeat, P., Hammarström, S., and Murphy, R. C., 1981, Leukotrienes: A new group of biologically active compounds, *Adv. Prostaglandin Thromboxane Res.* **6**:1–18.

Solmyo, A. P., and Solmyo, A. V., 1968, Vascular smooth muscle. I. Normal, structure, pathology, biochemistry and biophysics, *Pharmacol. Rev.* **20**:197–272.

Weichman, B. M., Muccitelli, R. M., Tucker, S. S., and Wasserman, M. A., 1983, Effect of calcium antagonists on leukotriene D_4-induced contractions of the guinea pig trachea and lung parenchyma, *J. Pharmacol. Exp. Ther.* **225**:310–315.

51

Macrophage Fcγ2b Receptor Expression and Receptor-Mediated Phospholipase Activity

Feedback Regulation by Metabolites of Arachidonic Acid

JOHN RHODES, JOHN SALMON, and JOHN WOOD

1. INTRODUCTION

Plasma membrane receptors specific for the Fc portion of immunoglobulins exist on myeloid and lymphoid cells and in a number of nonlymphoid tissues, including neonatal intestinal epithelial cells, placental trophoblasts, and virus-infected fibroblasts. It is likely that the diversity of such receptors in terms of distribution and specificity reflects a commensurate diversity of function.

Perhaps the best-characterized Fc receptor is the receptor for IgE on basophils and mast cells. This integral membrane glycoprotein has a molecular weight of 45–50,000. Its high affinity for the ligand has permitted purification to homogeneity in milligram quantities. The receptor mediates the immediate hypersensitivity (type I or anaphylactic) reaction that occurs when monomeric IgE already bound to cellular receptors is cross linked by specific antigen.

It is the cross linking of receptors that initiates a series of events leading to the release of histamine and other vasoactive mediators (Ishizaka and Ishizaka,

JOHN RHODES and JOHN WOOD ● Department of Experimental Immunobiology, Wellcome Research Laboratories, Beckenham, Kent BR3 3BS, United Kingdom.   JOHN SALMON ● Prostaglandin Research, Wellcome Research Laboratories, Beckenham, Kent BR3 3BS, United Kingdom.

1978). Receptors specific for IgG (FcγR) are present on B lymphocytes, whereas subsets of T lymphocytes possess receptors for IgG or IgM (FcμR). The function of these receptors is uncertain, but it seems likely that they are involved in immunoregulation as, for example, in the generation of suppressor T cells by anti-idiotypic antibody. The FcγR are thought to be associated with suppressor T cells, whereas FcμR are thought to define a subset of helper T cells (Moretta *et al.*, 1977). However, T cells can switch from expression of FcγR to FcμR on exposure to IgG immune complexes (Pichler *et al.*, 1978). Receptors specific for IgA have been described on monocytes and granulocytes (Fanger *et al.*, 1980).

Receptors for IgG are present on cells of myeloid origin (monocytes, macrophages, granulocytes) and enable these phagocytic cells to recognize microorganisms and other forms of antigen that have provoked an antibody response. One of the principal functions of these receptors is to facilitate the clearance of antigen during the effector phase of the immune response. All subclasses of IgG appear to mediate such endocytosis (Ralph *et al.*, 1980; Diamond and Yelton, 1981). In addition to internalization of the bound ligand (Silverstein *et al.*, 1977), Fcγ receptors also initiate a number of important cellular reactions, which include antibody-mediated cytolysis of target cells (Cerrotini and Brunner, 1974), release of hydrolytic enzymes (Gordon *et al.*, 1974), production of reactive intermediates of oxygen (Ezekowitz *et al.*, 1983), and synthesis and release of prostaglandins and leukotrienes (Bonney *et al.*, 1980; Scott *et al.*, 1980; Feuerstein *et al.*, 1981; Rouzer *et al.*, 1980).

Murine macrophages express at least three distinct receptors for IgG: a trypsin-sensitive receptor for IgG_{2a}, a trypsin-resistant receptor for IgG_1 and IgG_{2B}, and a receptor for aggregated IgG_3 (Unkeless *et al.*, 1981). Although all three classes of receptor can mediate endocytosis of particulate immune complexes, increasingly well-defined functional differences do exist among them. The receptor for IgG_{2a} is a more effective trigger of the respiratory burst in activated macrophages (Ezekowitz *et al.*, 1983), whereas only the receptor for IgG_1/IgG_{2b} appears to trigger the synthesis and release of prostaglandins and leukotrienes (Nitta and Suzuki, 1982).

The study of Fcγ receptor structure and function is most advanced in the case of the receptor for IgG_1/IgG_{2b}. This work has been facilitated by the production of a monoclonal antibody 2.4G2 specific for the receptor (Unkeless, 1979). The receptor is an integral membrane glycoprotein with an apparent molecular weight of 40–60,000. It has been purified and incorporated into lipid bilayers, where it exhibits the characteristics of a ligand-dependent, cation-selective ion channel (Ding-E Young *et al.*, 1983a,b). A molecule with IgG_{2b} binding activity has also been isolated from B-cell-derived chronic lymphocytic leukemia (CLL) cells and a macrophage cell line (Suzuki *et al.*, 1980a,b, 1982). This putative FcR exhibits phosphatidylcholine binding activity and typical phospholipase A_2 activity (i.e., it cleaves the esterified fatty acid from the C_2 position of the glycerol moiety of phosphatidylcholine). This phospholipase activity is enhanced by IgG_{2b}. The relationship between this molecule and the ligand-dependent, cation-selective ion channel defined by 2.4G2 awaits further clarification.

In the present chapter we examine the release and metabolism of arachidonic acid in resident murine peritoneal macorphages. We show that endogenous metabolites of arachidonic acid regulate changes in the expression of the IgG_1/IgG_{2b} receptor *in vitro* and that such changes produce commensurate effects on receptor-mediated arachidonic acid release. The data demonstrate a potential feedback regulation of receptor-triggered arachidonic acid metabolism by eicosanoids acting at the level of Fcγ2b receptor expression.

2. MATERIALS AND METHODS

2.1. Macrophage Cultures

Untreated female balb/c mice 8–20 weeks of age were sacrificed by CO_2 inhalation. The peritoneal cavity was washed with RPMI 1640 containing antibiotics and heat-inactivated fetal calf serum (FCS) at a concentration of 10%. Pooled suspensions of peritoneal cells were adjusted to 2×10^6 cells/ml, and 0.5-ml aliquots were placed directly into Lab-tek tissue culture chamber slides (Miles Ltd., Slough, England) or Linbro 24-well plates (Flow Ltd., Ayshire, Scotland) and allowed to adhere for 2 hr at 37°C. Nonadherent cells were then washed off, and the adherent cells cultured in 5% CO_2 in humidified air at 37°C.

2.2. Assay for IgG_1/IgG_{2b} Receptor Expression

Washed sheep red blood cells (SRBC) in phosphate-buffered saline (PBS) were specifically sensitized with mouse monoclonal IgG_{2b} anti-SRBC antibody (clone Sp2, SeraLab Ltd., Sussex, England) for 30 min at room temperature, after which the cells were washed twice and resuspended in HEPES-buffered RPMI. Washed macrophage monolayers were overlain with IgG_{2b}-coated SRBC at a concentration of 1% v/v for 1 hr at room temperature, after which monolayers were washed three times, fixed with 0.5% glutaraldehyde, and stained with citrate-buffered Giemsa. A dose-response curve was established by titrating the amount of monoclonal IgG_{2b} on the SRBC surface. Rosette formation was determined and expressed as a function of the dose of monoclonal antibody. In addition to the binding of immune complexes, receptor expression was also visualized by incubating macrophages with antibody alone for 1 hr at room temperature, after which the monolayers were washed and exposed to untreated SRBC for a further hour. Rosette formation in both systems was inhibited by pretreatment of the macrophages with the monoclonal antibody 2.4G2 (a gift of Dr. J. Unkeless, Rockefeller University). The cells were exposed to 2.4G2 at 100 μg/ml for 1 hr and then washed and subjected to the rosette assay.

2.3. Radioimmunoassay for Eicosanoids

Macrophages were cultured for 24 hr in medium containing FCS or for 5 hr in the presence of IgG_{2b}-sensitized SRBC in serum-free medium. Supernatants were then microfuged and frozen. The concentration of eicosanoids was determined by specific radioimmunoassay without prior extraction or chromatography. The specificity of these assays for the various eicosanoids has been established in previously published studies (Salmon, 1978; Salmon *et al.*, 1982).

2.4. [^{14}C]Arachidonic Acid Release Assay

Macrophage monolayers in Linbro 24-well plates were labeled with [^{14}C] arachidonic acid (Amersham International, England) at a concentration of 30 nCi per ml in RPMI 1640 containing 5% FCS for 2 hr at 37°C. The monolayers were then washed four times and exposed to EAIgG$_{2b}$ or other stimuli for 5 hr. Cell-free supernatants were harvested and subjected to liquid scintillation spectrometry.

2.5. Thin-Layer Chromatography of [^{14}C]Arachidonic Acid Metabolites

Macrophage monolayers in Linbro 12-well plates were labeled with [^{14}C]Arachidonic acid (Amersham International) at a concentration of 200 nCi/ml in RPMI 1640 containing 5% FCS for 4 hr at 37°C. Monolayers were then washed four times and exposed to test stimuli for 15 hr. Tissue culture supernantants were precipitated with acetone and then chloroform extracted before drying under nitrogen (Salmon and Flower, 1982). Samples, resuspended in chloroform/methanol (1 : 1), were chromatographed in silica-coated plates (Whatman, LKD5) in ethyl acetate–2,2,4-trimethylpentane–acetic acid–water (110 : 50 : 20 : 100 v/v/v/v) and then autoradiographed using Kodak X-Omat film at −70°C for 2–10 days.

2.6. Other Reagents

Cycloheximide (Calbiochem, La Jolla, CA) and tunicamycin (Sigma, Pool, Dorset) were added to macrophage monolayers after the initial 2-hr adherence and were present for the subsequent 24 hr of culture. Hydrocortisone (Sigma), indomethacin (Sigma), and BW 755C (Wellcome) were added to macrophage monolayers after the initial 2-hr adherence and were present for the subsequent 24 hr of

---

FIGURE 1. (A) Fcγ2b receptor expression progressively increases *in vitro*. Macrophages were cultured in medium containing 10% FCS. Rosette formation was determined as a function of the dose of IgG$_{2b}$ antibody at the times indicated. (B) The increase in Fcγ2b receptor expression depends on protein synthesis and glycosylation. EAIgG$_{2b}$ dose–response at 2 hr (○), at 24 hr (□), at 24 hr with cycloheximide (500 ng/ml) (●), and at 24 hr with tunicamycin (100 ng/ml) (■). (C) EAIgG$_{2b}$ rosette formation is inhibited by 2.4G2 monoclonal antibody at 2 hr and 24 hr. EAIgG$_{2b}$ dose–response at 2 hr (○), at 24 hr (□), at 2 hr after 2.4G2 treatment (1 hr at room temperature followed by washing) (●), and at 24 hr after 2.4G2 treatment (■). (D) Passive sensitization of macrophage with IgG$_{2b}$ reflects the increase in Fcγ2b receptor expresion occurring over 24 hr and is inhibited by 2.4G2. Macrophages were cultured for 2 hr (○) or 24 hr (□) and then sensitized with IgG$_{2b}$ for 1 hr at room temperature. The macrophages were then washed and exposed to untreated SRBC. Macrophages pretreated with 2.4G2 (●).

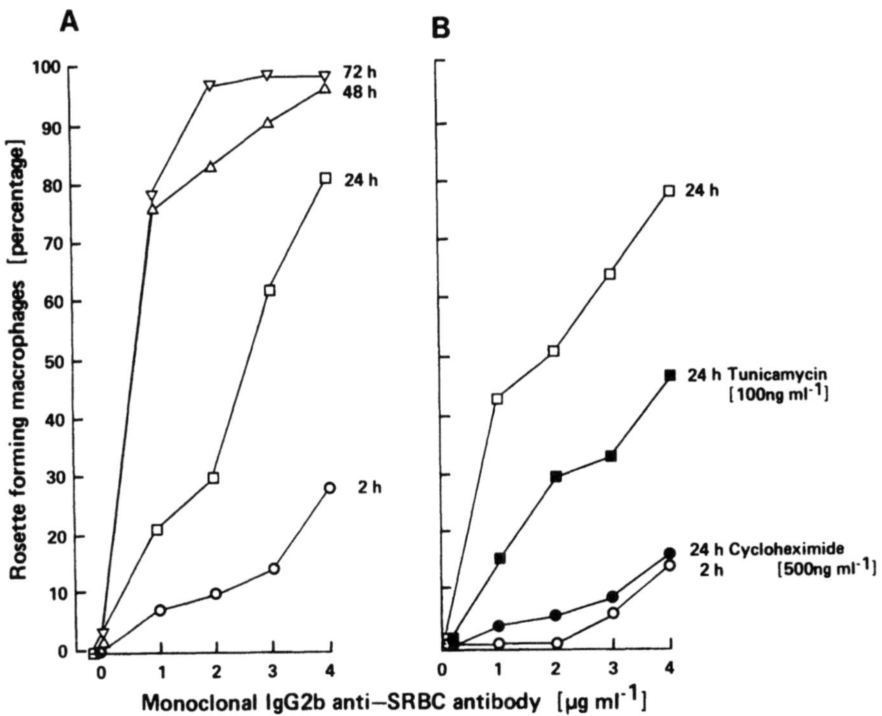
A
Rosette forming macrophages [percentage]
100
90
80
70
60
50
40
30
20
10
0
72 h
48 h
24 h
2 h
0 1 2 3 4
B
24 h
24 h Tunicamycin
[100ng ml-1]
24 h Cycloheximide
2 h
[500ng ml-1]
0 1 2 3 4
Monoclonal IgG2b anti-SRBC antibody [µg ml-1]

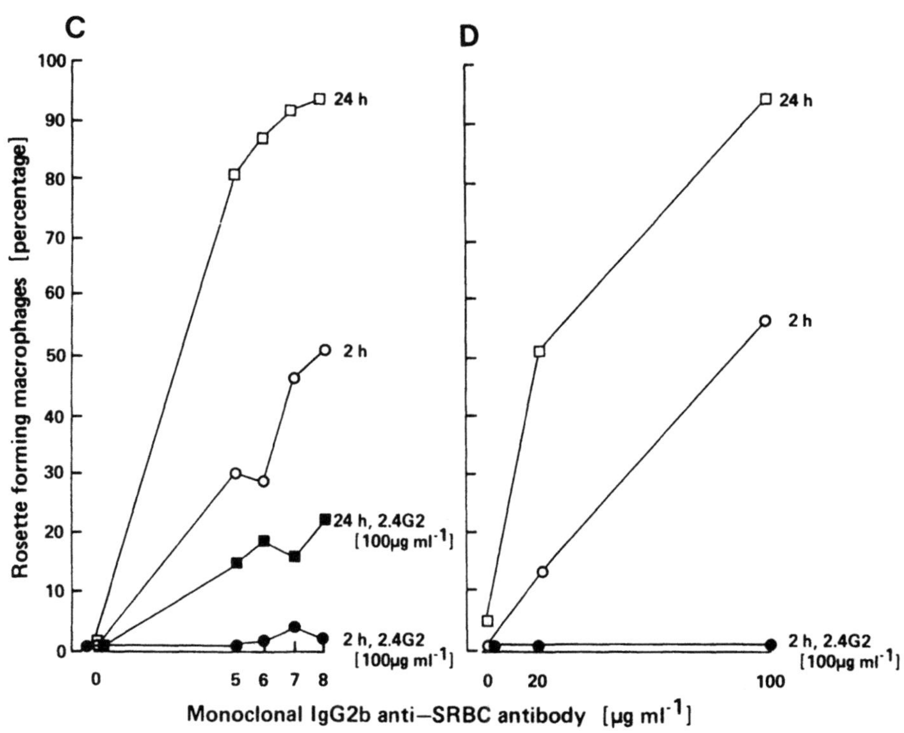
C
Rosette forming macrophages [percentage]
100
90
80
70
60
50
40
30
20
10
24 h
2 h
24 h, 2.4G2
[100µg ml-1]
2 h, 2.4G2
[100µg ml-1]
0 5 6 7 8
D
24 h
2 h
2 h, 2.4G2
[100µg ml-1]
0 20 100
Monoclonal IgG2b anti-SRBC antibody [µg ml-1]

culture. Prostaglandin E_2 (Upjohn, Kalamazoo, MI) and leukotrienes B_4, C_4, and D_4 (Miles Labs. Ltd., Slough, England) were added in the same manner at the concentrations specified in the following section.

3. RESULTS

3.1. Regulation of Macrophage Fcγ2b Receptor Expression *in Vitro*

Resident peritoneal macrophages in tissue culture in the presence of 10% FCS exhibit a progressive increase in the expression of receptors for IgG_{2b} over a 72-hr period (Fig. 1A). The increase in receptor expression over a 24-hr period was completely inhibited by cycloheximide and partially inhibited by tunicamycin, indicating a dependence on protein synthesis and glycosylation, respectively (Fig. 1B). Receptor expression at 2 hr and 24 hr was inhibited by pretreatment of the cells with the monoclonal antibody 2.4G2, which is specific for the macrophage receptor for IgG_1/IgG_{2b} (Fig. 1C). Passive sensitization of the macrophages with

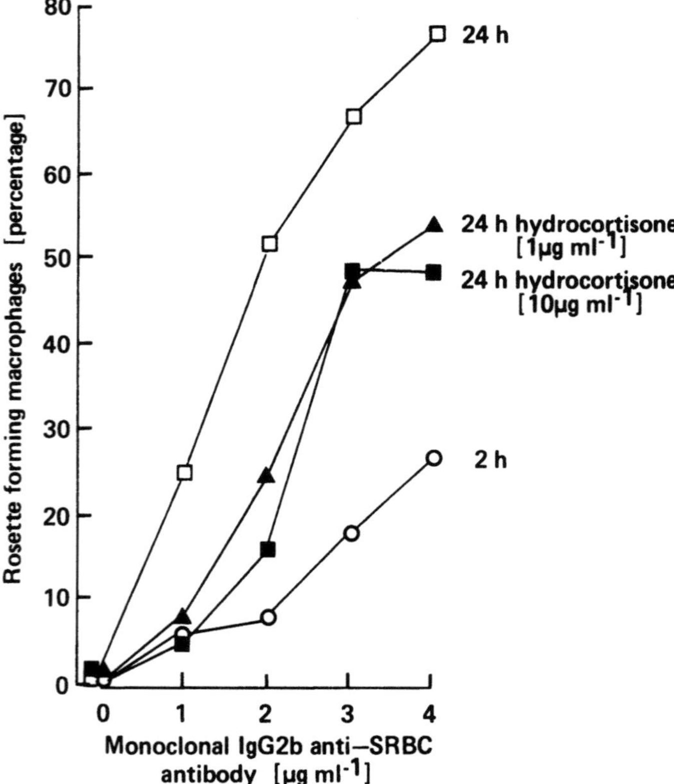

FIGURE 2. Inhibition of Fcγ2b receptor expression by hydrocortisone.

IgG$_{2b}$ antibody alone, subsequently detected by the addition of washed SRBC, was also inhibited by 2.4G2 (Fig. 1D).

The increase in Fcγ2b receptor expression occurring *in vitro* was found to be inhibited by hydrocortisone at 1 μg/ml. The degree of inhibition (around 50%) is comparable to the degree of inhibition produced by hydrocortisone on the synthesis of eicosanoids, and as in the latter system, the degree of inhibition is not affected by a tenfold increase in the dose of steroid (Fig. 2). Selective inhibition of the cyclooxygenase pathway by indomethacin resulted in enhancement of the increase in receptor expression (Fig. 3). This enhancement was partially reversed by PGE$_2$ at a concentration of 100 ng/ml. In contrast, inhibition of both lipoxygenase and cyclooxygenase pathways by BW 755C abolished the increase in Fcγ2b receptor expression. This effect was unaffected by LTB$_4$ (Fig. 4) but was completely reversed

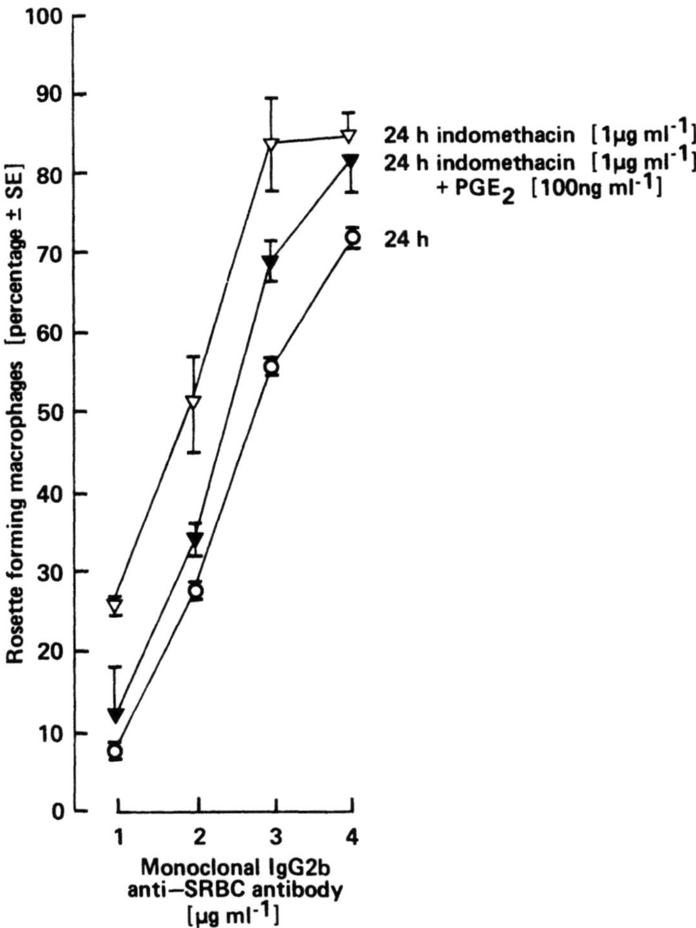

FIGURE 3. Enhancement of Fcγ2b receptor expression by indomethacin.

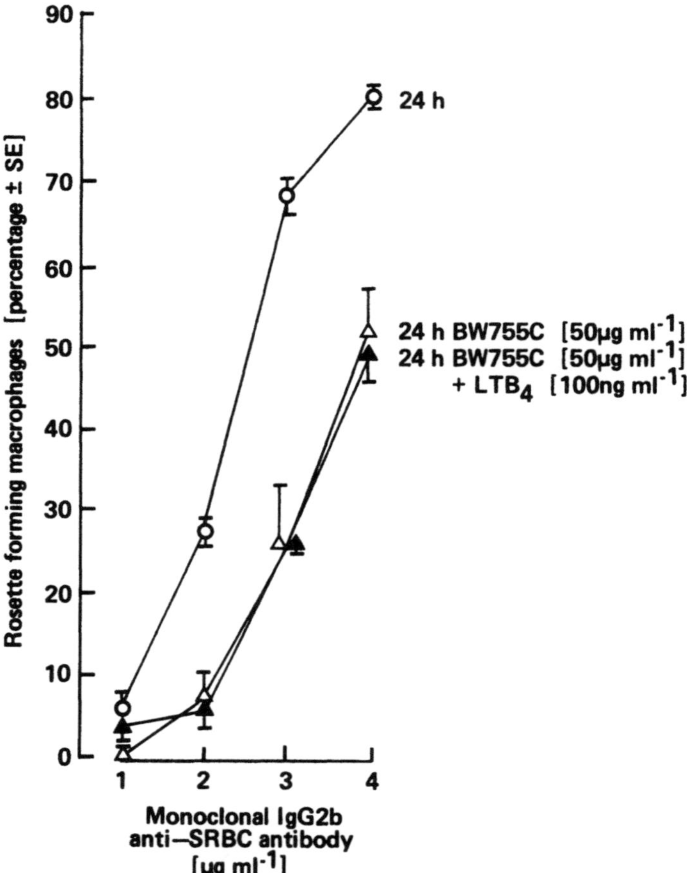

FIGURE 4. Influence of the lipoxygenase/cyclooxygenase inhibitor BW 755C on Fcγ2b receptor expression. Failure of LTB_4 to reverse BW 755C inhibition.

in the presence of LTD_4 at 100 ng/ml (Fig. 5). Leukotriene C_4 produced only a modest effect. Taken together, these data indicate that a lipoxygenase product produced by macrophages is responsible for the increase in Fcγ2b receptor expression and that this signal overrides the inhibitory influence of a cyclooxygenase product that down-regulates receptor expression.

3.2. Receptor-Mediated Phospholipase Activity

Phospholipase activity resulting from Fcγ2b receptor ligand interaction was determined by measuring the release of $[^{14}C]$arachidonic acid from prelabeled macrophages in response to IgG_{2b}-coated SRBC over a 5-hr period in serum-free medium. Under these conditions, the release of arachidonic acid and its metabolites

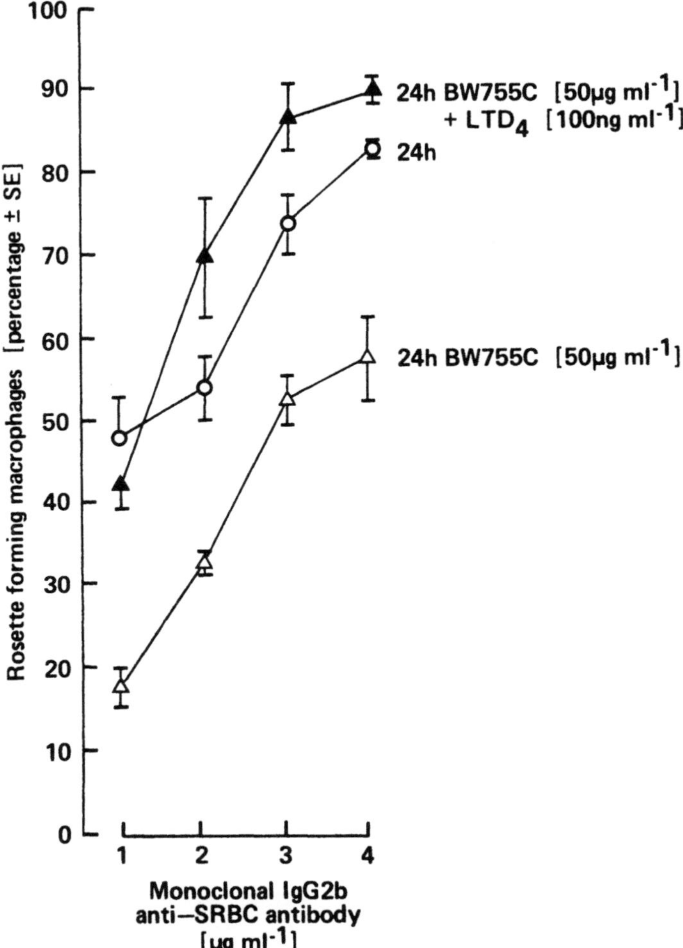

FIGURE 5. Inhibition of Fcγ2b receptor expression by BW755C. Reversal of inhibition by LTD₄.

was found to be a function of the dose of IgG_{2b} antibody (Fig. 6). Macrophages already cultured for 24 hr were used in these assays. The identity of the arachidonic acid metabolites produced in response to IgG_{2b} immune complexes was determined by radioimmunoassay and by thin-layer chromatography. Macrophages that had been in culture for 24 hr were washed and exposed to SRBC sensitized with an optimal concentration of IgG_{2b} antibody (100 μg/ml in serum-free HEPES-buffered medium. The reaction was allowed to proceed for 5 hr at 37°C in air, after which cell-free supernatants were obtained. Inhibitors of arachidonic acid metabolism were present where indicated for the entire 5 hr. The results obtained by radioimmunoassay are presented in Table I. The principal product was PGE_2, although sub-

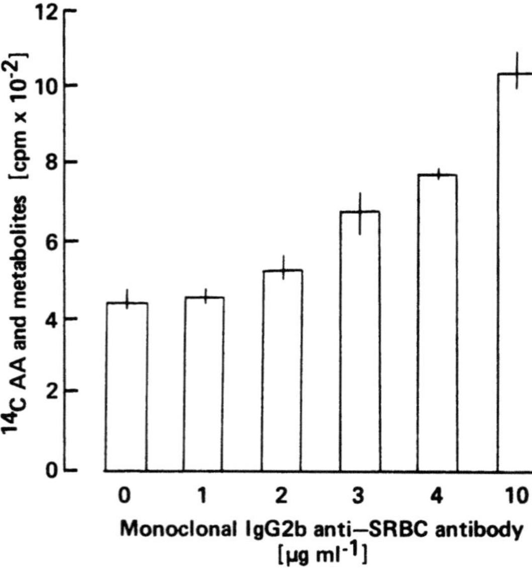

FIGURE 6. Release of [^{14}C]PGE$_2$ and [^{14}C]arachidonic acid from prelabeled macrophages in response to IgG$_{2b}$-coated sheep red blood cells.

stantial amounts of 6-keto-PGF$_{1\alpha}$, the stable metabolite of prostacyclin, were also produced. Small amounts of thromboxane B$_2$ were also detected, but the assay did not detect significant amounts of LTB$_4$ or LTC$_4$. Prostanoid production in response to IgG$_{2b}$ particulate immune complexes was completely inhibited by indomethacin and BW 755C. Results obtained by thin-layer chromatography are presented in Fig. 7. Prelabeled macrophages cultured for 5 hr in the presence of SRBC sensitized with an optimal dose of IgG$_{2b}$ released prostaglandins of the E series as expected, although much of the labeled material at this time was still in the form of unmetabolized arachidonic acid. Prostanoid production was significantly reduced by hydrocortisone. Zymosan stimulation also induced prostanoid production, whereas the calcium ionophore A23187 preferentially induced lipoxygenase products in the form of hydroxyeicosatetraenoic acids. Unsensitized SRBC had no effect.

3.3. Regulation of Receptor-Mediated Phospholipase Activity

Macrophages cultured for 48 hr with or without BW 755C were prelabeled with [^{14}C]arachidonic acid and then exposed to SRBC sensitized with suboptimal amounts of IgG$_{2b}$ for 1 hr at room temperature. After this time, the monolayers were washed and incubated for a further 5 hr in serum-free medium. Assays were then performed on cell-free supernatants. Results are shown in Fig. 8. In cultures that contained BW 755C for the initial 48 hr, the increase in receptor expression did not take place, and phospholipase activity in response to suboptimally sensitized EAIgG$_{2b}$ was found to be commensurately reduced in comparison with that of untreated macrophages.

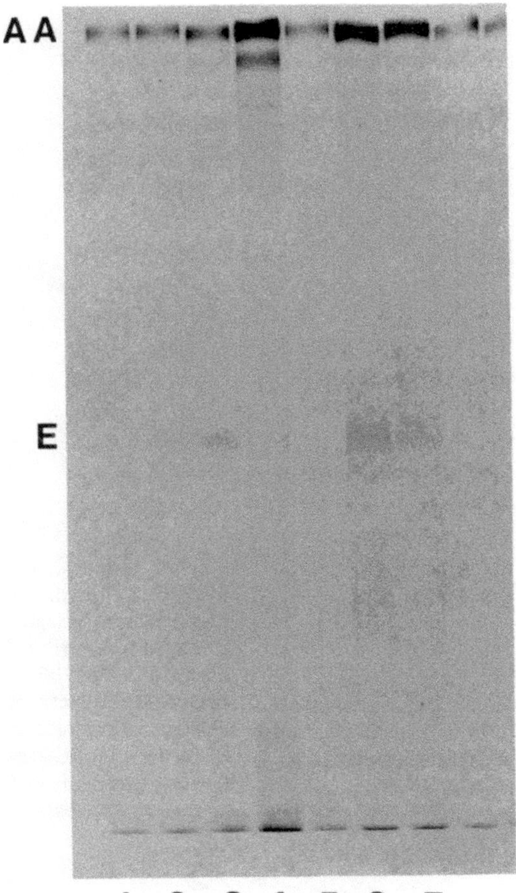

FIGURE 7. Thin layer chromatography of [^{14}C]arachidonic acid metabolism. 1, unstimulated macrophages; 2, 1% FCS; 3, zymogen; 4, calcium ionophore A23187; 5, untreated SRBC(E); 6, EAIgG2b; 7, EAIgG2b plus cortisone.

TABLE I. Eicosanoid Production by Macrophages Stimulated with IgG$_{2b}$ Immune Complexes (EAIgG$_{2b}$): Effects of Inhibitors of Arachidonic Acid Metabolism

	Eicosanoid release (ng/ml)			
Treatment	PGE$_2$	6-Keto-PGF$_{1\alpha}$	LTB$_4$	LTC$_4$
None	1.5 ± 0.2	<1	<0.2	<1
EAIgG$_{2b}$	22.0 ± 2.0	7.6 ± 1.4	<0.2	<1
EAIgG$_{2b}$ + BW755c, 50 μg/ml	1.8 ± 1.5	<1	<0.2	<1
EAIgG$_{2b}$ + indomethacin, 5 μg/ml	1.5 ± 0.35	<1	<0.2	<1

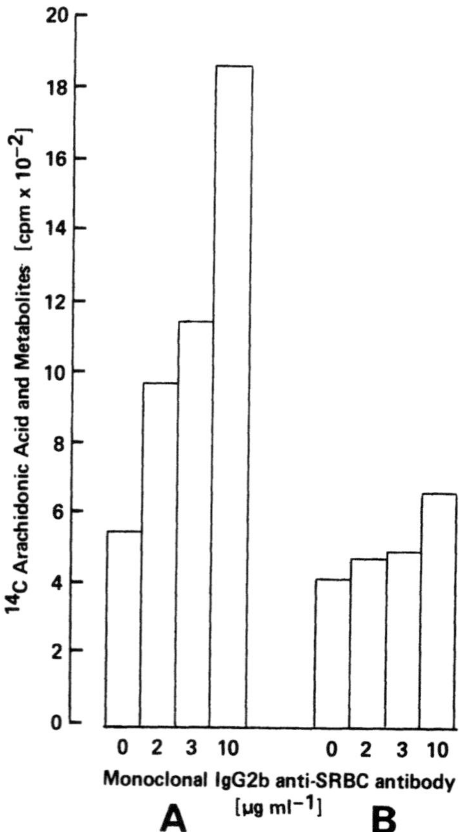

FIGURE 8. Inhibition of receptor-mediated phospholipase activity by BW 755C. Macrophages cultured in the absence (A) or presence (B) of BW 755C for 48 hr were prelabeled with [^{14}C]arachidonate and exposed to varying concentrations of IgG$_{2b}$-sensitized SRBC for 1 hr. Release of [^{14}C]AA and metabolites was measured over a 5-hr time period. For further details see text.

TABLE II. Eicosanoid Production by Macrophages Cultured in Medium Containing FCS: Effects of Inhibitors of Arachidonic Acid Metabolism

Treatment	Eicosanoid release (ng/ml)			
	PGE$_2$	6-Keto-PGF$_{1\alpha}$	LTB$_4$	LTC$_4$
None	97 ± 17	144 ± 20	<0.2	<0.3
BW 755C, 10 µg/ml	<1	<5	<0.2	<0.3
Indomethacin, 5 µg/ml	<1	<5	0.2	<0.3
Hydrocortisone, 1 µg/ml	26.5 ± 3	55 ± 12	<0.2	<0.3

3.4. Eicosanoid Production by Macrophages Cultured in the Presence of FCS

Macrophages cultured for 24 hr in the presence of 10% FCS produced substantial amounts of prostanoids (Table II). Unlike the $EAIgG_{2b}$ stimulus, where PGE_2 was the principal product, these conditions induced a greater production of PGI_2 detected as the stable metbolite 6-keto-$PGF_{1\alpha}$. Only small amounts of leukotrienes were detected in these assays. Prostanoid production was completely inhibited by indomethacin and BW 755C and substantially inhibited ($\sim70\%$) by hydrocortisone.

4. DISCUSSION

Macrophages synthesize eicosanoids from endogenous sources of arachidonic acid and release them into their environment. In chronic inflammatory reactions, the macrophage is likely to be a major source of these inflammatory mediators. Macrophage membrane phospholipids contain an unusually high proportion of esterified arachidonic acid (Scott *et al.*, 1980), and these cells also incorporate arachidonic acid into cytoplasmic organelles termed lipid bodies (Dvorak *et al.*, 1983). The fatty acid is released in response to a variety of stimuli through the activation of phospholipases (Wightman *et al.*, 1981). Effective experimental stimuli include zymozan, phorbol myristate acetate, the calcium inonophore A23187, and particulate immune complexes. Released arachidonic acid is metabolized via cyclooxygenase or lipoxygenase pathways to produce a mixture of prostanoids, hydroxyeicosatetraenoic acids (HETEs), and leukotrienes (Bonney *et al.*, 1980; Scott *et al.*, 1980; Pawlowski *et al.*, 1982; Rouzer *et al.*, 1980). The ratio of these products appears to depend on the nature of the stimulus.

Evidence exists to show that stimulation of the murine macrophage receptor for IgG_1/IgG_{2b} by particulate immune complexes is an effective trigger for phospholipase activation and the subsequent release of arachidonic acid metabolites (Nitta and Suzuki, 1982). The present data confirm these findings by means of [^{14}C]arachidonic acid release, thin layer chromatography of [^{14}C]arachidonic acid metabolites, and radioimmunoassay of eicosanoids. The expression of Fc receptors has long been known to be a dynamic property of the macrophage membrane (Arend and Mannick, 1973; Rhodes, 1975), and recent work has shown that murine macrophage activation *in vivo* is accompanied by an increase in the receptor for IgG_{2a} (Ezekowitz *et al.*, 1983). Increases in the receptor for IgG_1/IgG_{2b} also occur *in vivo*, however, in response to *C. parvum* (Glass *et al.*, 1983). The present study looks at changes in the expression of the receptor for IgG_1/IgG_{2b} *in vitro*. Receptor expression, characterized by $EAIgG_{2b}$ rosette formation inhibitable by the monoclonal antibody 2.4G2, markedly increased during 72 hr of culture in the presence of FCS. This increase was dependent on protein synthesis and glycoslylation. The increase in receptor expression was partially inhibited by hydrocortisone and completely inhibited by BW 755C (an inhibitor of lipoxygenase and cyclooxygenase pathways). In contrast, selective inhibition of the cyclooxygenase pathway by indomethacin enhanced the increase in receptor expression. The effects of indomethacin were partially reversible by PGE_2 whereas the effects of BW 755C were completely reversed by LTD_4. Leukotriene C had only a modest effect. Inhibition of receptor expression by BW 755C produced commensurate inhibition of arachidonic acid release in response to particulate IgG_{2b} immune complexes.

We also examined the production of eicosanoids by macrophages cultured in the presence of FCS in the absence of other stimuli. Our intention was to show that under the conditions and over the time course in which increases in receptor expression occur, macrophages produce significant amounts of eicosanoids and that this production is susceptible to inhibition by drugs that modulate receptor expression. Substantial amounts of PGE_2 and prostacyclin were produced. However, only very small amounts of leukotrienes were detected. Hydrocortisone and inhibitors of the lipoxygenase and cyclooxygenase pathways inhibited prostanoid production.

Taken together, our data indicate that increases in IgG_1/IgG_{2b} receptor expression *in vitro* are the result of an up-regulation of receptor expression by a lipoxygenase product(s). This signal overrides the inhibitory effects of a cyclooxygenase product. Exogenous LTD_4 has this property. However, we did not find substantial amounts of endogenous leukotrienes in macrophage cultures with or without IgG_{2b} immune complex stimulation. We are not currently able to assay LTD_4. Others have shown that murine macrophages produce substantial amounts of LTC_4 in response to IgE and IgG complexes (Rouzer *et al.*, 1982) and zymosan (Rouzer *et al.*, 1980). Our data, obtained *in vitro*, indicate a potential feedback regulation of receptor-triggered arachidonic acid metabolism by eicosanoids acting at the level of IgG_1/IgG_{2b} receptor expression.

ACKNOWLEDGMENTS. We are grateful to Magdy Fahmy, Susan Wishart, and Barbara Pearce for technical assistance.

REFERENCES

Arend, W. P., and Mannik, M., 1973, The macrophage receptor for IgG: Number and affinity of binding sites, *J. Immunol.* **110**:1455–1463.

Bonney, R. J., Davies, P., Kuehl, F. A., Jr., and Humes, J. L., 1980, Arachidonic acid oxygenation products produced by macrophages responding to inflammatory stimuli, *J. Reticubendothel. Soc.* **28**:113s–115s.

Cerrottini, J. D., and Brunner, K. T., 1974, Cell-mediated cytotoxicity, allograft rejection and tumor immunity, *Adv. Immunol.* **18**:67–132.

Diamond, B., and Yelton, D. E., 1981, A new Fc receptor on mouse macrophages binding IgG_3, *J. Exp. Med.* **153**:515–519.

Ding-E Young, J., Unkeless, J. C., Kaback, H. R., and Cohn, Z. A., 1983a, Macrophage membrane potential changes associated with γ2b/γ1 Fc receptor–ligand binding, *Proc. Natl. Acad. Sci. U.S.A.* **80**:1357–1361.

Ding-E Young, J., Unkeless, J. C., Young, T. M., Mauro, A., and Cohn, Z. A., 1983b, Role for mouse macrophage IgG Fc receptor as ligand-dependent ion channel, *Nature* **306**:186–189.

Dvorak, A. M., Dvorak, H. F., Peters, S. P., Shulman, E. S., Mc Glashan, Jr., O. W., Pyne, K., Harvey, V. S., Galli, S. J., and Lichtenstein, L. M., 1983, Lipid bodies: Cytoplasmic organelles important to arachidonate metabolism in macrophages and mast cells, *J. Immunol.* **131**:2965–2979.

Ezekowitz, R. A. B., Bampton, M., and Gordon, S., 1983, Macrophage activation selectively enhances expression of Fc receptors for IgG$_{2a}$, *J. Exp. Med.* **157**:807–812.

Fanger, M. W., Shen, L., Pugh, L., and Bernier, G. M., 1980, Subpopulations of human peripheral granulocytes and monocytes express receptors for IgA, *Proc. Natl. Acad. Sci. U.S.A.* **77**:3640–3644.

Feuerstein, N., Foegh, M., and Ramwell, P. W., 1981, Leukotrienes C$_4$ and D$_4$ induce prostaglandin and thromboxane release from rat peritoneal macrophages, *Br. J.. Pharmacol.* **72**:389–391.

Glass, E. J., Stewart, J., and Weir, D. M., 1983, Membrane changes in murine macrophages after *in vivo* stimulation and activation, *Immunology* **50**:165–173.

Gordon, S., Unkeless, J. C., and Cohn, Z. A., 1974, Induction of macrophage plasminogen activator by endotoxin stimulation and phagocytosis: Evidence for a two stage process, *J. Exp. Med.* **140**:995–1010.

Ishizaka, K., Ishizaka, T., 1978, Mechanisms of reaginic hypersensitivity and IgE antibody response, *Immunol. Rev.* **41**:109–148.

Moretta, L., Webb, J. R., Grossi, C. E., Lydyard, P. M., and Cooper, M. D., 1977, Functional analysis of two human T cell subpopulations: Help and suppression of B cell responses by T cells bearing receptors for IgM or IgG, *J. Exp. Med.* **146**:184–200.

Nitta, T., and Suzuki, T., 1982, Fcγ2b receptor-mediated prostaglandin synthesis by a murine macrophage cell line (P388 D1), *J. Immunol.* **128**:2527–2532.

Pawlowski, N. A., Scott, W. A., Andreach, M., and Cohn, Z. A., 1982, Uptake and metabolism of monohydroxyeicosatetraenoic acids by macrophages, *J. Exp. Med.* **155**:1653–1664.

Pawlowski, N. A., Kaplan, G., Hamill, A. L., Cohn, Z. A., and Scott, W. A., 1983, Arachidonic acid metabolism by human monocytes: Studies with platelet-depleted cultures, *J. Exp. Med.* **158**:393–412.

Pinchler, W. J., Lum, L., and Broder, S., 1978, Fc receptors on human T lymphocytes. I. Transition of Tγ to Tμ cells, *J. Immunol.* **121**:1540–1548.

Ralph, P., Nakoinz, I., Diamond, B., and Yelton, D., 1980, All classes of murine IgG antibody mediate macrophage phagocytosis and lysis of erythrocytes, *J. Immunol.* **125**:1885–1888.

Rhodes, J., 1975, Macrophage heterogeneity in receptor activity: The activation of macrophage Fc receptor function *in vivo* and *in vitro*, *J. Immunol.* **114**:976–981.

Rouzer, C. A., Scott, W. A., Hamill, A. L., and Cohn, Z. A., 1980, Dynamics of leukotriene C production by macrophages, *J. Exp. Med.* **152**:1236–1247.

Rouzer, C. A., Scott, W. A., Hamill, A. L., Liu, F.-T., Katz, D. H., and Cohn, Z. A., 1982, Secretion of leukotriene C and other arachidonic acid metabolites by macrophages challenged with IgE immune complexes, *J. Exp. Med.* **156**:1077–1086.

Salmon, J., 1978, A radioimmunoassay for 6 keto PGF$_{1\alpha}$, *Prostaglandins* **15**:383–393.

Salmon, J. A., and Flower, R. J., 1982, Extraction and thin-layer chromatography of arachidonic acid metabolites, *Methods Enzymol.* **86**:477–493.

Salmon, J., Simmons, P. M., and Palmer, R. M. J., 1982, A radioimmunoassay for LTB$_4$, *Prostaglandins* **24**:225–235.

Scott, W. A., Zrike, J. M., Hamill, A. L., Kempe, J., and Cohn, Z. A., 1980, Regulation of arachidonic acid metabolites in macrophages, *J. Exp. Med.* **152**:324–335.

Silverstein, S. C., Steinman, R. M., and Cohn, Z. A., 1977, Endocytosis, *Annu. Rev. Biochem.* **46**:669–772.

Suzuki, T., Sadasivan, R., Wood, G., and Bayer, W. L., 1980a, Isolation and characterization of biologically active Fc receptors of human B lymphocytes, *Mol. Immunol.* **17**:491–503.

Suzuki, T., Sadasivan, R., Saito-Taki, T., Steckschulte, D. J., Balentine, L., and Helmkamp, G. M., 1980b, Studies of Fcγ receptors of human B lymphocytes: Phospholipase A$_2$ activity of Fc receptors, *Biochemistry* **19**:6037–6044.

Suzuki, T., Saito-Taki, T., Sadasivan, R., and Nitta, T., 1982, Biochemical signal transmitted by Fcγ receptors: Phospholipase A_2 activity of Fcγ2b receptor of murine macrophage cell line P388 D1, *Proc. Natl. Acad. Sci. U.S.A.* **79**:591–595.

Unkeless, J. C., 1979, Characterization of a monoclonal antibody directed against mouse macrophages and lymphocyte Fc receptors, *J. Exp. Med.* **150**:580–596.

Unkeless, J. C., Fleit, H., and Mellman, I. S., 1981, Structural aspects and heterogeneity of immunoglobulin Fc receptors. *Adv. Immunol.* **31**:247–270.

Wightman, P. D., Humes, J. L., Davies, P., Bonney, R. J., 1981, Identification and characterization of two phospholipase A_2 activities in resident mouse peritoneal macrophages, *Biochem J.* **195**:427–433.

52

Differential Effects of Leukotriene B_4 and Its Analogues on Suppressor and Helper Cell Function

MAREK ROLA-PLESZCZYNSKI, LYNE GAGNON, VICTOR RADOUX, CLAUDINE POULIOT, and E. J. COREY

1. INTRODUCTION

Recent studies have demonstrated that leukotrienes (LTs), in addition to their myotropic properties, also strongly affect several leukocyte functions. For instance, LTB_4 was shown to stimulate neutrophil aggregation (Bass *et al.*, 1981), chemokinesis (Ford-Hutchinson *et al.*, 1980; Palmer *et al.*, 1980), chemotaxis (Palmer *et al.*, 1980), degranulation (Showell *et al.*, 1982a), hexose transport (Bass *et al.*, 1981), and cation fluxes (Molski *et al.*, 1981).

More recently, LTB_4 was also shown to affect several lymphocyte functions. We have demonstrated that LTB_4 can induce human T cells to exert a suppressor cell activity on lymphocyte proliferative responses to mitogens (Rola-Pleszczynski *et al.*, 1982; Rola-Pleszczynski and Sirois, 1983). Leukotriene B_4 was also shown to suppress LIF production (Payan and Goetzl, 1983). In addition, LTB_4 can strongly augment natural cytotoxicity of human peripheral blood lymphocytes against herpes simplex virus-infected target cells (Rola-Pleszczynski *et al.*, 1983) as well as un-

MAREK ROLA-PLESZCZYNSKI, LYNE GAGNON, CLAUDINE POULIOT, and VICTOR RADOUX • Immunology Division, Faculty of Medicine, University of Sherbrooke, Sherbrooke, Québec J1H 5N4, Canada.   E. J. COREY • Department of Chemistry, Harvard University, Cambridge, Massachusetts 02138.

infected tumor target cells (Rola-Pleszczynski *et al.*, 1984a), a process that appears to depend on the lipoxygenase pathway (Rola-Pleszczynski *et al.*, 1983, 1984a; Seaman, 1983). The cytotoxicity-enhancing properties of LTB_4 were shared to a significant extent by the parent molecule LTA_4 but only weakly by LTD_4, whereas HETEs and stereoisomers of LTB_4 were inactive (Rola-Pleszczynski *et al.*, 1984b).

With the aim of further elucidating the mechanisms through which LTB_4 regulates immune functions, we undertook to study the differential effects of LTB_4 and analogues on the generation of suppressor or helper activity.

2. MATERIALS AND METHODS

2.1. Leukotrienes

Leukotriene B_4 was obtained as a gracious gift from Drs. A. Ford-Hutchinson and J. Rokach (Merck-Frosst Laboratories, Dorval, Quebec). Leukotriene B_4 dimethylamide (LTB_4-dma) and the monocyclic analogues of LTB_4, SP-II-130 and DH-II-210, were synthesized by one of us (E.J.C.) as will be described elsewhere. They were appropriately diluted immediately before use in RPMI-1640 medium.

2.2. Leukocytes

Human peripheral blood mononuclear leukocytes (PBMLs) were obtained by density centrifugation of heparinized venous blood on a Ficoll–Hypaque gradient. They represented 75–85% lymphocytes and 15–25% monocytes. In selected experiments, they were passed through a nylon wool column to yield 99% lymphocytes, 95% of which were E-rosetting T cells.

2.3. Proliferation Assay

Mitogen-induced lymphocyte proliferation was assayed by measuring the uptake of [^{3}H]thymidine by PBML during the last 6 hr of 72-hr culture in the presence of either medium alone (supplemented with 10% human heat-inactivated AB serum) or phytohemagglutinin (PHA, 0.1% Burroughs-Wellcome, Research Triangle Park, NC). Cells were harvested at 72 hr, and total cell-associated radioactivity was measured in dpm.

2.4. Suppressor Cell Assay

Peripheral blood mononuclear leukocytes or nylon-wool-purified T cells were preincubated for 24 hr with varying concentrations of LTB_4 or analogues followed by three washes and addition of the preincubated cells to fresh mitogen-stimulated PBML cultures. Results are expressed as percentage suppression using the following formula:

$$\text{Percent suppression} = \left[1 - \frac{\text{(dpm after LT preincubation)}}{\text{(dpm of controls)}} \right] \times 100$$

3. RESULTS AND DISCUSSION

When used in a coculture system, LTB_4 was again observed, as reported previously, to induce suppressor cell function (Fig. 1). The dose–response curve was U-shaped, with maximal effects at 10^{-10} to 10^{-12} M concentrations of LTB_4. Leukotriene dimethylamide, which has been reported to behave as an antagonist of LTB_4 in neutrophil degranulation (Showell *et al.*, 1982b) while having weak agonist activity at concentrations about 10^{-6} M, behaved as an active analogue of LTB_4 in our system. Equimolar concentrations of LTB_4 and LTB_4-dma did not antagonize each other but were not synergistic either, suggesting that they may act on the same receptor.

Using synthetic monocyclic analogues of LTB_4, we attempted to further analyze the cellular interactions involved in LTB_4-induced suppressor activity. We had previously shown that the prime target for LTB_4 was a T cell that, in turn, after activation, induced suppression in a responder PBML culture (Rola-Pleszczynski *et al.*, 1982). When the responder PBML population was depleted of monocytes, the remaining nonadherent lymphocytes would no longer be suppressed by LTB_4-preactivated T cells (Fig. 2, top panel). Furthermore, when the responder PBML were treated with 2 μM indomethacin to inhibit cyclooxygenase, not only was

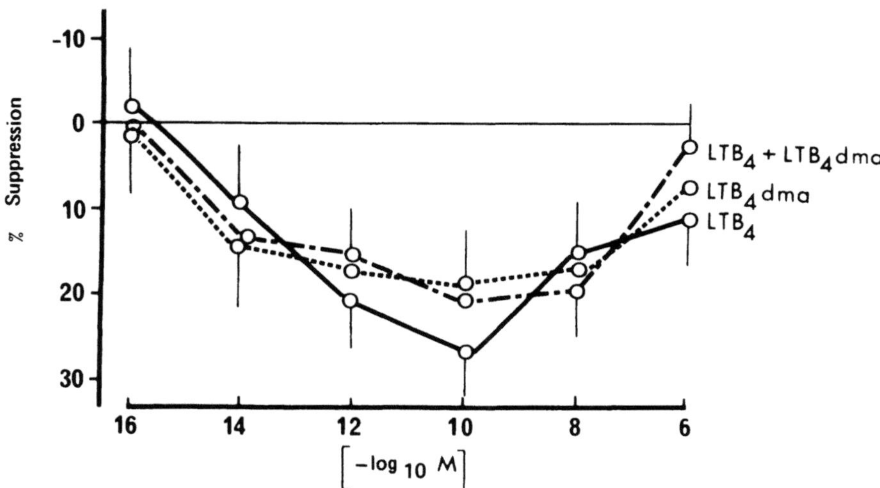

FIGURE 1. Dose–response effects of LTB_4, LTB_4-dma, and both substances together on induction of suppressor cell activity after preincubation for 24 hr. Preincubated cells were then added to a 72-hr PBML culture in the presence of PHA.

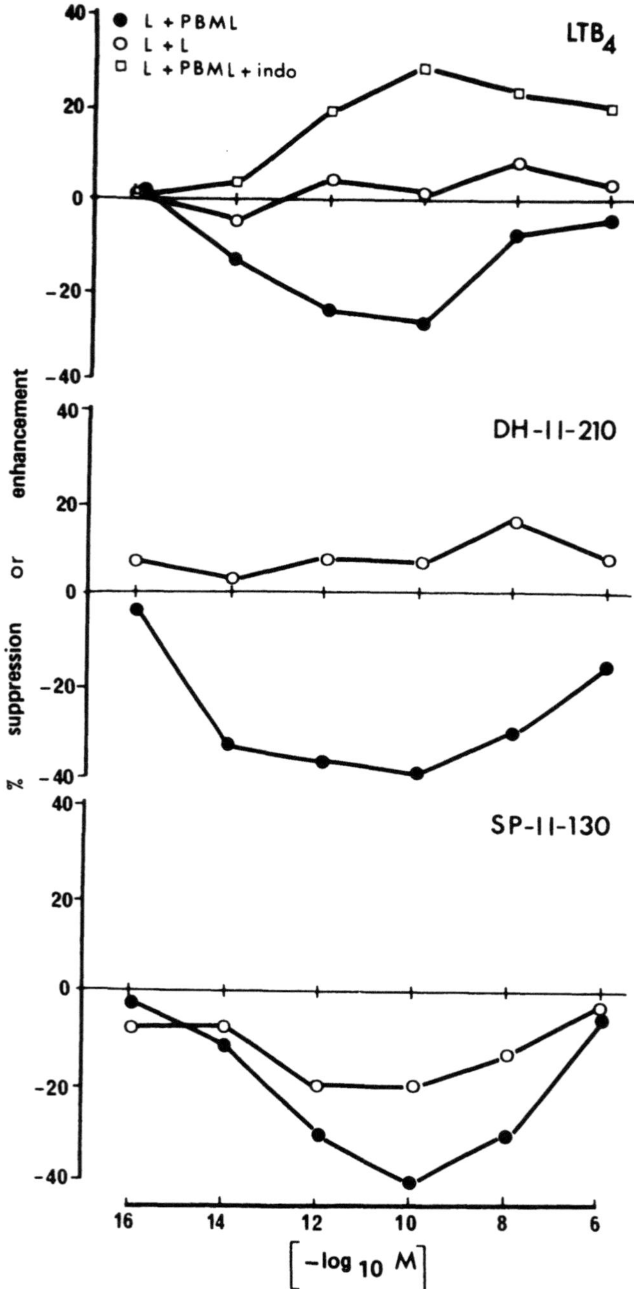

FIGURE 2. Dose–response effects of LTB$_4$ and the monocyclic analogues DH-ll-210 and SP-ll-130 on induction of suppressor T cell activity after 24-hr preincubation. The responder population consisted, as indicated, of unfractionated PBML (with or without 2μM indomethacin) or monocyte-depleted lymphocytes (L) and was stimulated for 72 hr with PHA.

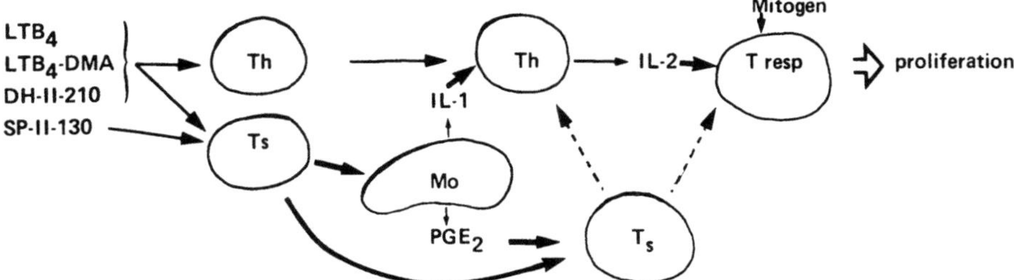

FIGURE 3. Hypothetical pathways of activation by LTB₄ and its analogues of suppressor and helper cell activities, depending on culture conditions.

suppression lost, but actual enhancement of PBML responses to PHA was observed, with maximal effects at 10^{-8} to 10^{-12} M concentrations of LTB₄. These findings suggested that LTB₄-activated suppressor cells required monocytes and cyclooxygenase metabolites of arachidonic acid to exert their effect. They also suggested that LTB₄-activated T cells could exert a helper function when suppressor function was prevented. We therefore compared the requirement for monocytes for the expression of suppression using two monocyclic analogues of LTB₄, namely, DH-II-210 and SP-II-130. Figure 2 shows that DH-II-210 and SP-II-130 were both quite active in inducing suppressor cell activity when tested in coculture with unfractionated PBML.

In marked contrast, however, DH-II-210, but not SP-II-130, could only exert its suppressive effect in the presence of monocytes. SP-II-130, on the other hand, induced suppressor cells that could further exert their suppressive effect even in the absence of monocytes (Fig. 2, lower panel). This suggested that SP-II-130 acted on a different subset of T cells, which would, in turn, during the subsequent coculture, exert their suppressive effect directly on mitogen-responsive lymphocytes. Alternatively, it could directly inhibit IL-2 production by helper cells.

4. CONCLUSION AND HYPOTHESIS

In studying the immunoregulatory activities of LTB₄, we have found that lymphocytes preincubated with LTB₄ may be induced to exert either positive (helper) or negative (suppressor) effects on the proliferative responses of PBML to mitogens depending on culture conditions. Suppressor function is at least partially dependent on monocytes and cyclooxygenase products. Helper function can be observed when monocytes are removed and especially when cyclooxygenase is inhibited. Although some synthetic analogues of LTB₄ mimic its activities, at least one monocyclic analogue, SP-II-130, may directly affect suppressor cells without the requirement for monocytes.

Figure 3 illustrates the hypothetical activation pathways induced by LTB₄ and its analogues as suggested by our studies.

5. SUMMARY

Leukotriene B_4 (LTB_4) can induce human T lymphocytes to exert suppressor cell activity at concentrations ranging from 1×10^{-8} to 1×10^{-12} M. The expression of suppressor cell activity required the cooperation of adherent phagocytic cells and cyclooxygenase metabolites. When the secondary, responder cultures were depleted of monocytes or treated with indomethacin, suppression could no longer be observed with the addition of LTB_4-pretreated T cells. In contrast, actual enhancement of lymphocyte proliferation to mitogens was seen, with peak effects at 1×10^{-10} M concentration of LTB_4. Leukotriene B_4 dimethylamide and DH-II-210, a synthetic analogue of LTB_4, mimicked the action of LTB_4 in inducing suppressor cell function and in requiring monocytes for its expression. In contrast, a different LTB_4 analogue, SP-II-130, also induced suppressor T cells but without the requirement for monocytes. Preincubated T cells would only exert suppressive activity whether the responding culture consisted of unfractionated peripheral blood mononuclear leukocytes or monocyte-depleted ($<1\%$) lymphocytes. Our data suggest that both suppressor and helper cells can be stimulated by LTB_4 and some of its analogues and that other analogues may interact more specifically with a more restricted lymphocyte population.

ACKNOWLEDGMENTS. This work was supported by grants from the Medical Research Council and the National Cancer Institute of Canada (M.R.-P.) and by the U. S. National Science Foundation (E.J.C.). M.R.-P. is a Research Scholar of the Fonds de Recherche en Santé du Québec; L.G. and V.R. are recipients of a studentship and a fellowship, respectively, from the IRSST and the Arthritis Society. The authors wish to express their gratitude to Miss Carole Jacques for preparing the manuscript.

REFERENCES

Bass, D. A., Thomas, M. J., Goetzl, E. J., DeChatelet, E. R., and McCall, C. E., 1981, Lipoxygenase-derived products of arachidonic acid mediate stimulation of hexose uptake in human polymorphonuclear leukocytes, *Biochem. Biophys. Res. Commun.* **100**:1–7.

Ford-Hutchinson, A. W., Bray, M. A., Doig, M. V., Shipley, M. E. and Smith, M. J., 1980, Leukotriene B, a potent chemokinetic and aggregating substance released from polymorphonuclear leukocytes, *Nature* **286**:264–265.

Molski, T. F. P., Naccache, P. H., Borgeat, P., and Sha'afi, R. I., 1981, Similarities in the mechanisms by which formyl-methionyl-leucyl-phenylalanine, arachidonic acid and leukotriene B_4 increase calcium and sodium influxes in rabbit neutrophils, *Biochem. Biophys. Res. Commun.* **103**:227–232.

Palmer, R. M. J., Stepney, R. J., Higgs, G. A., and Eakins, K. E., 1980, Chemokinetic activity of arachidonic acid lipoxygenase products on leukocytes of different species, *Prostaglandins* **20**:411–418.

Payan, D. G., and Goetzl, E. J., 1983, Specific suppression of human T lymphocyte function by leukotriene B_4, *J. Immunol.* **131**:551–553.

Rola-Pleszczynski, M., and Sirois, P., 1983, Leukotrienes as immunoregulators, in: *Leukotrienes and other Lipoxygenase Products* (P. J. Piper, ed.), Wiley and Sons, London, pp. 234–240.

Rola-Pleszczynski, M., Borgeat, P., and Sirois, P., 1982, Leukotriene B₄ induces human suppressor lymphocytes, *Biochem. Biophys. Res. Commun.* **108**:1531–1537.

Rola-Pleszczynski, M., Gagnon, L., and Sirois, P., 1983, Leukotriene B₄ augments human natural cytotoxic cell activity, *Biochem. Biophys. Res. Commun.* **113**:531–537.

Rola-Pleszczynski, M., Gagnon, L., and Sirois, P., 1984a, Natural cytotoxic cell activity enhanced by leukotriene B₄: Modulation by cyclooxygenase and lipoxygenase inhibitors, in: *Icosanoids and Cancer* (A. Crastes de Paulet, ed.), Raven Press, New York, pp. 235–240.

Rola-Pleszczynski, M., Gagnon, L., Rudzinska, M., Borgeat, P., and Sirois, P., 1984b, Human natural cytotoxic cell activity: Enhancement by leukotrienes A₄, B₄ and D₄ but not by steteoisomers of LTB₄ or HETEs, *Prostaglandins Leukotrienes Med.* **13**:113–117.

Seaman, W. E., 1983, Human natural killer cell activity is reversibly inhibited by antagonists of lipoxygenation, *J. Immunol.* **131**:2953–2957.

Showell, H. J., Naccache, P. H., Borgeat, P., Picard, S., Valerand, P., Beckker, E. L., and Sha'afi, R. I., 1982a, Characterization of the secretory activity of LTB₄ toward rabbit neutrophils, *J. Immunol.* **128**:811–816.

Showell, H. J., Otterness, I. G., Marfat, A., and Corey, E. J., 1982b, Inhibition of leukotriene B₄-induced neutrophil degranulation by leukotriene B₄-dimethylamide, *Biochem. Biophys. Res. Commun.* **106**:741–747.

53

Eicosanoid Precursor Fatty Acids in Plasma Phospholipids and Arachidonate Metabolism in Polymorphonuclear Leukocytes in Asthma

KARI PUNNONEN, RITVA TAMMIVAARA, and PEKKA UOTILA

1. INTRODUCTION

The cysteine-containing leukotrienes C_4, D_4, and E_4 as well as leukotriene B_4, a dihydroxy acid, are formed from arachidonic acid via the Ca^{2+}-dependent 5-lipoxygenase pathway (Borgeat *et al.*, 1983). They have been considered to be mediators of immediate hypersensitivity reactions. Sensitized human lung is capable of releasing leukotrienes when exposed to a specific antigen (Dahlén *et al.*, 1983), and LTC_4 and LTD_4 have been shown to be potent bronchoconstrictors both *in vivo* (Holroyde *et al.*, 1981) and *in vitro* (Dahlén *et al.*, 1980). Leukotriene B_4 is a potent chemokinetic and chemotactic substance (Ford-Hutchinson *et al.*, 1980). Also, stimulated human polymorphonuclear leukocytes (PMNL) are capable of producing substantial amounts of 5-lipoxygenase products, as well as other lipoxygenase and cyclooxygenase metabolites (Borgeat and Samuelsson, 1979; Samuelsson, 1983).

Because leukotrienes and other metabolites of arachidonic acid may be of importance in asthma, we have measured the relative amounts of arachidonic acid

KARI PUNNONEN and PEKKA UOTILA ● Department of Physiology, University of Turku, SF-20520 Turku 52, Finland.   RITVA TAMMIVAARA ● Department of Diseases of the Chest, University of Turku, SF-20520 Turku 52, Finland.

and other fatty acids in plasma phospholipids and the amounts of some arachidonate metabolites in the plasma and serum of asthmatic and healthy subjects. The metabolism of exogenous arachidonic acid was also studied in nonstimulated PMNL of asthmatic and healthy subjects.

2. MATERIALS AND METHODS

[^{14}C]Arachidonic acid was purchased from Radiochemical Centre (Amersham, England) and converted to the sodium salt. Unlabeled reference compounds, including 6-keto-PGF$_{1\alpha}$, PGF$_{2\alpha}$, PGE$_2$, TxB$_2$, 13,14-dihydro-15-keto-PGF$_{2\alpha}$, PGD$_2$, 15-keto-PGE$_2$, 13,14-dihydro-15-keto-PGE$_2$, PGA$_2$, and 5-HETE were either gifts from Dr. J. Pike (Upjohn, Kalamazoo, MI, USA) or were purchased from Sigma (St. Louis, MO, USA) or Cayman Chemical (Denver, CO, USA). Leukotrienes B$_4$, C$_4$, and D$_4$ were generous gifts from Dr. J. Rokach (Merck Frosst Canada Inc., Dorval, Quebec, Canada). [^{3}H]Thromboxane B$_2$ and [^{3}H]6-keto-PGF$_{1\alpha}$ were from Amersham (Buckinghamshire, England). The antiserum for TxB$_2$ was from Upjohn, and that for 6-keto-PGF$_{1\alpha}$ from Seragen (Boston, MA, USA). 6,9-Deepoxy-6,9-(phenylimino)-$\Delta^{6,8}$-prostaglandin I$_1$ (U-60,257), an inhibitor of leukotriene synthesis, was a generous gift from Dr. M. Bach (Upjohn, Kalamazoo, MI, USA) (Bach *et al.*, 1982).

Twenty-nine asthmatic patients with the intrinsic type of asthma from mild to moderate severity (aged 18 to 59 years; ten female and 19 male) and 17 healthy subjects (aged 21 to 45 years; one female and 16 male) were involved in this study. The blood samples were taken during long-term treatments for asthma and when the subjects had not taken aspirinlike drugs during the preceding 2 weeks. About 7 ml of venous blood was collected for plasma measurements into heparinized plastic tubes. In order to measure TxB$_2$ serum levels, about 3 ml of blood was collected into a glass tube and allowed to clot for 1 hr at 37°C. After this, indomethacin was added, and serum was separated by centrifugation. The serum samples were stored at -20°C until analyzed.

The plasma levels of 6-keto-PGF$_{1\alpha}$ and TxB$_2$ as well as the serum levels of TxB$_2$ were measured by radioimmunoassay (Puustinen and Uotila, 1983). Plasma samples were acidified and extracted with ethyl acetate, and serum samples were diluted with ethanol before the radioimmunoassay.

For PMNL experiments, about 10 ml of venous blood was collected into a plastic tube containing 2 ml of 2.94% sodium citrate. To separate PMNL, the blood samples were at first mixed with 2–3 ml of 6% dextran T 500 (Pharmacia, Uppsala, Sweden) in Dulbecco's phosphate-buffered saline (PBS, pH 7.4) (Dulbecco and Vogt, 1954), and erythrocytes were allowed to settle (Boyum, 1976). Leukocytes were further purified by a discontinuous Percoll® (Pharmacia) gradient. The cells were layered on a top of a discontinuous Percoll® gradient (densities 1.09, 1.07, and 1.06 g/ml), and the tubes were centrifuged for 30 min at 400 × *g*. After centrifugation, PMNL were between the gradients of 1.09 and 1.07 g/ml. The

PMNL were washed and resuspended in PBS (pH 7.4) containing 1 mM $CaCl_2$ and 5 mM glucose. The platelet : leukocyte ratio was always less than 1 : 5.

The PMNL suspension containing about 7×10^6 PMNL in PBS and 1 mM of reduced glutathione was preincubated for 10 min at 37°C in the presence or absence of 10 μM U-60,257. After the preincubation, [^{14}C]AA (8 μM, about 500,000 cpm) was added, and the incubation was continued for 2 min in a final volume of 0.5 ml. The incubation was terminated by the addition of 4.5 ml of ice-cold PBS, and the mixture was extracted with ethyl acetate (2 + 2 ml) first at neutral pH and then at pH 4.5. The pH of the remaining water phase was neutralized, and the water phase was stored at $-20°C$ until analyzed. Formed metabolites of [^{14}C]AA were analyzed as follows. The metabolites extracted with ethyl acetate at pH 7.4 were analyzed by one-dimensional thin-layer chromatography using petroleum ether : diethyl ether : acetic acid (50 : 50 : 1) as the mobile phase (Fig. 1) (Uotila et al., 1983). Synthetic LTB$_4$ and 5-HETE standards were used to identify these metabolites on the TLC plates. The 12-HETE, derived from the platelets, migrated in the vicinity of the mono-HETE group. Also, 15-HETE is known to migrate in the same area in the TLC system used (König et al., 1982). The cyclooxygenase metabolites extracted with ethyl acetate at pH 4.5 were analyzed by a two-dimensional TLC system (Fig. 1) (Uotila et al., 1983). First, the TLC plates were developed in the organic phase of ethyl acetate : isooctane : acetic acid : water

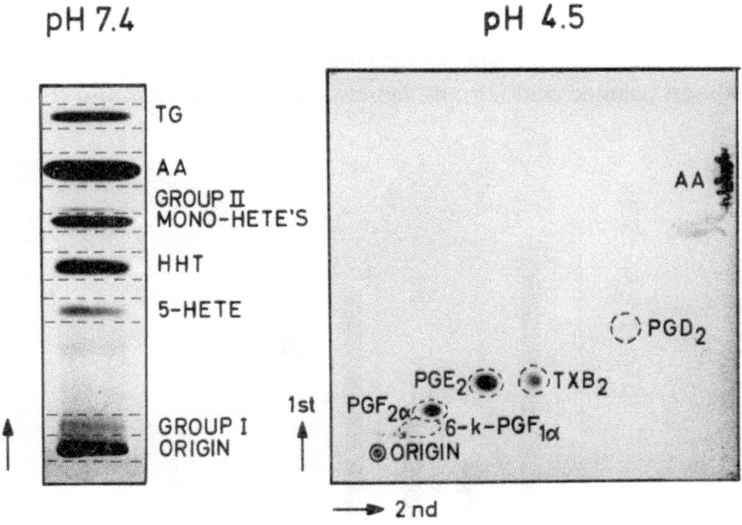

FIGURE 1. Autoradiograms of the TLC plates containing arachidonate metabolites formed in polymorphonuclear leukocytes (PMNL). The metabolites were extracted from the incubation mixture with ethyl acetate first at pH 7.4 and then at pH 4.5. The chromatographing of unlabeled standards is shown. These areas as well as those of unidentified metabolite groups were cut along the dashed lines for liquid scintillation counting.

(110 : 50 : 20 : 100) and then twice in the second direction in ethyl acetate : acetic acid (99 : 1). The metabolites were identified using unlabeled standards. Autoradiograms of the plates were developed, and radioactive areas were cut out for liquid scintillation counting. The water phase was analyzed using bioassay on superfused guinea pig ileum. The contractions were measured by the Harvard smooth muscle transducer. The dose–response curves were based on contractions elicited by synthetic LTC_4 and LTD_4.

The fatty acid composition of plasma phospholipids was analyzed as follows (Norred and Wade, 1972). Plasma samples (1 ml) were mixed with 17 volumes of chloroform : methanol (2 : 1), and the mixture was filtered through glass wool. The filtrate was washed with 0.2 volumes of 0.75% KCl and then evaporated to dryness. Acetone saturated with magnesium chloride was added, and the supernatant containing neutral lipids and free fatty acids was removed. To release the fatty acids from the precipitated phospholipids, 2 ml of 0.5 N NaOH in methanol was added, and the tubes were kept for 5 min in a boiling water bath. Then, to methylate the fatty acids, 1 ml of 1 N HCl and 2 ml of 10% BF_3, both in methanol, were added, and heating was continued for 2 min. After the addition of 2.5 ml of saturated NaCl and extraction with *n*-hexane the fatty acid methyl esters were separated and measured using a Packard 419 Becker gas chromatograph with a flame ionization detector (Hietanen *et al.*, 1982).

3. RESULTS

The two major fatty acids in plasma phospholipids of both asthmatic and healthy subjects were palmitic acid (16 : 0) and *cis*-linoleic acid (18 : 2ω6) (Fig. 2). The

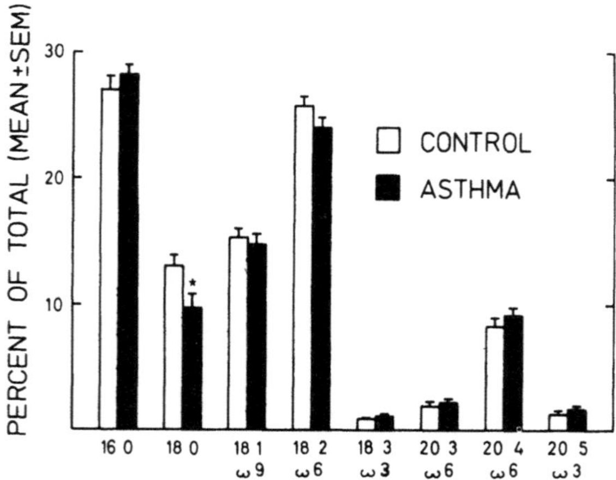

FIGURE 2. The fatty acid composition of plasma phospholipids in asthmatic and healthy subjects as percentage of total fatty acids in plasma phospholipids. Asthmatic patients ($n = 10$) were compared to control subjects ($n = 7$) with Student's *t*-test: *$2P < 0.05$.

amounts of the eicosanoid precursor fatty acids, arachidonic acid (20 : 4ω6), di-homogammalinolenic acid (20 : 3ω6), and eicosapentaenoic acid (20 : 5ω3) were rather similar in asthmatic patients and in healthy subjects. In plasma phospholipids of asthmatic persons, the percentage distribution was $9.2 \pm 0.5\%$ arachidonic acid, $2.3 \pm 0.3\%$ dihomogammalinolenic acid, and $1.6 \pm 0.3\%$ eicosapentaenoic acid. The corresponding values for healthy subjects were 8.3 ± 0.6, 1.9 ± 0.3, and $1.3 \pm 1.3\%$, respectively.

Plasma levels of TxB_2 and 6-keto-$PGF_{1\alpha}$ were not significantly different between healthy and asthmatic subjects. In asthmatics, the plasma level of TxB_2 was

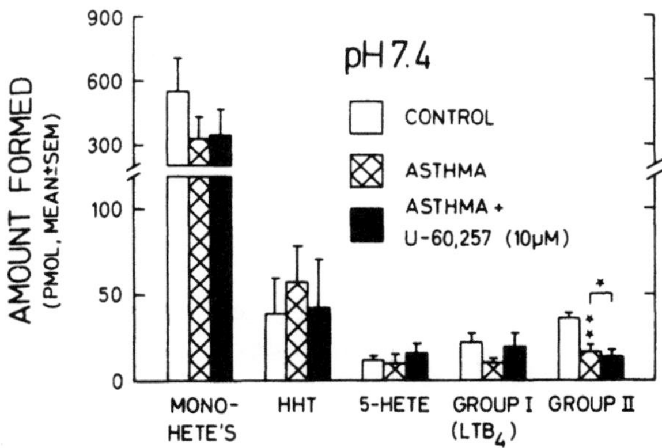

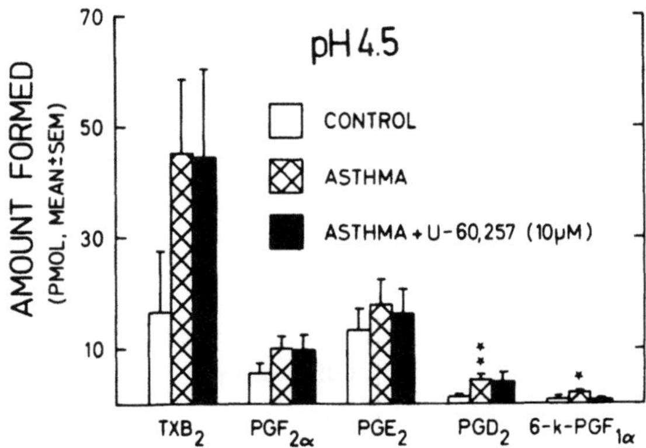

FIGURE 3. The amounts of arachidonate metabolites formed in PMNL of asthmatic ($n = 9$) and healthy ($n = 8$) persons. The PMNL (about 7×10^6 cells) were incubated with 8 μM [^{14}C]arachidonic acid in the presence or absence of 10 μM U-60,257 for 2 min, and the metabolites were extracted with ethyl acetate at pH 7.4 and 4.5 and analyzed by TLC. The incubations with PMNL of asthmatic patients without U-60,257 were compared to those with PMNL of healthy persons by Student's t-test. U-60,257 experiments with PMNL of asthmatic subjects were compared to the corresponding incubations without U-60,257 by Student's t-test for paired data: *$2P < 0.05$; **$2P < 0.01$.

0.27 ± 0.03 pmol/ml ($n = 29$), and that of 6-keto-PGF$_{1\alpha}$ was 0.50 ± 0.02 pmol/ml ($n = 29$). In healthy subjects, the corresponding values were 0.20 ± 0.04 pmol/ml ($n = 10$) for TxB$_2$ and 0.55 ± 0.05 pmol/ml ($n = 17$) for 6-keto-PGF$_{1\alpha}$. Also, thromboxane formation during blood clotting was rather similar in asthmatic and healthy subjects, as the serum level of TxB$_2$ was 470 ± 60 pmol/ml ($n = 29$) for asthmatic and 550 ± 50 pmol/ml ($n = 17$) for healthy subjects.

The nonstimulated PMNL of both asthmatic and healthy subjects were incubated with exogenous [^{14}C]arachidonic acid (8 μM) for 2 min. In the ethyl acetate extract at pH 7.4, the following radioactive bands were detected: group I (LTB$_4$), 5-HETE, HHT, mono-HETEs (12- and 15-HETE), group II, arachidonate (Figs. 1, 3). The major metabolites migrated in the vicinity of mono-HETEs (12- and 15-HETE) as a single band and in the vicinity of HHT. The detected amounts of radioactivity were relatively small on the areas of 5-HETE and group I (unlabeled LTB$_4$ migrated in the vicinity of the metabolite group I). Group II, with a R_f value slightly greater than that of 12-HETE, consisted of a small amount of unidentified metabolites. In asthmatic patients, the amount of radioactivity in the areas of LTB$_4$ and 5-HETE was not significantly changed, but in the vicinity of group II, it was slightly decreased ($2P < 0.01$) when compared to control subjects (Fig. 3). In the ethyl acetate extract at pH 4.5, the major metabolites were PGE$_2$ and TxB$_2$. Smaller amounts of PGI$_{2\alpha}$, PGD$_2$, and 6-keto-PGF$_{1\alpha}$ were also formed. In the PMNL of asthmatic patients, there was a slight increase in the formed amounts of PGD$_2$ ($2P < 0.01$) and 6-keto-PGF$_{1\alpha}$ ($2P < 0.05$). The amounts of other arachidonate metabolites in nonstimulated PMNL did not differ significantly between asthmatic and healthy subjects. The amount of cysteine-containing leukotrienes was, in all experiments, under the detection limit (20 pmol/sample) of the bioassay on superfused guinea pig ileum.

U-60,257 (10 μM) had no clear effect on the metabolism of exogenous arachidonic acid in PMNL of asthmatic (Fig. 3) or healthy subjects (data not shown). U-60,257, however, slightly decreased the amount of radioactivity in the unidentified metabolite group II ($2P < 0.05$) (Fig. 3, pH 7.4).

4. DISCUSSION

The present study indicates that the percentages of the eicosanoid precursor fatty acids (dihomogammalinolenic acid, arachidonic acid, and eicosapentaenoic acid) are not changed in plasma phospholipids of asthmatics. Thus, in asthma, the possible changes in the formation of prostaglandins, thromboxanes, and leukotrienes are not caused by changes in the relative amounts of these precursor fatty acids. Obviously the release of arachidonic acid from platelet phospholipids is not changed in asthmatic patients, as thromboxane formation during blood clotting was similar in asthmatic and healthy subjects. The plasma levels of TxB$_2$ and 6-keto-PGF$_{1\alpha}$ were also similar in asthmatic and healthy persons.

The formation of 5-HETE and the metabolites in group I (containing LTB$_4$)

was not significantly changed in asthma (Fig. 3). The amount of cysteine-containing leukotrienes was under the detection limit of the bioassay in all experiments. Thus, the present study indicates that exogenous arachidonic acid is not metabolized in substantial amounts to leukotrienes in nonstimulated PMNL of asthmatic patients. The present results are in accordance with the previous study on the metabolism of arachidonic acid in lung tissue from asthmatics, as only traces of leukotrienes were formed when the lung tissues were incubated with exogenous arachidonic acid alone (Dahlén *et al.*, 1983). However, substantial amounts of leukotrienes were formed in the presence of a specific allergen or the bivalent cation ionophore A23187 (Dahlén *et al.*, 1983).

U-60,257, a prostacyclyin analogue, has been reported to be capable of inhibiting the formation of both LTB_4 (Smith *et al.*, 1982; Dahlén *et al.*, 1983) and the cysteine-containing leukotrienes (e.g., LTC_4, LTD_4) (Dahlén *et al.*, 1983; Bach *et al.*, 1982). As in the present study, the formation of leukotrienes was negligible in nonstimulated PMNL of both asthmatic and healthy persons; consequently, no changes in the formation of these metabolites could be seen in the presence of U-60,257. At 10 μM, U-60,257 has been reported to slightly increase the formation of 5-HETE in human neutrophils (Bach *et al.*, 1982). In the present study, however, the rate of formation of 5-HETE was not changed by U-60,257 in PMNL of asthmatic and healthy subjects.

In the present study, when PMNL were incubated with exogenous $[^{14}C]AA$, the main metabolite band migrated in the vicinity of 12-HETE. Also, 15-HETE is known to migrate in the same area in the TLC system used (König *et al.*, 1982). As the formed proportional amount of 15- and 12-HETE in human PMNL has earlier been reported to be small (Borgeat and Samuelsson, 1979), it is obvious that a part of the metabolites in the mono-HETE band (12- and 15-HETE) represented 12-HETE formed by the small amount of contaminating platelets. Therefore, a part of formed TxB_2 and HHT was obviously also from platelets.

As no significant differences were detected between asthmatic and healthy subjects in the amounts of the main arachidonate metabolites or 5-lipoxygenase products, the present study indicates that the metabolism of exogenous arachidonic acid in non-stimulated PMNL is not significantly changed in asthma. The small increase in the formation of PGD_2 and 6-keto-$PGF_{1\alpha}$ in the PMNL of asthmatics is obviously of minor importance.

5. SUMMARY

The fatty acid composition of plasma phospholipids, and thus also the relative amounts of the eicosanoid precursor fatty acids, arachidonic acid (20 : 4ω6), di-homogammalinolenic acid (20 : 3ω6), and eicosapentaenoic acid (20 : 5ω3), were rather similar in asthmatic and healthy persons. The percentage of arachidonic acid was 8.3 $\pm$ 0.6% of the total fatty acids in plasma phospholipids of healthy subjects; the corresponding value for asthmatics was 9.2 $\pm$ 0.5%. The release of arachidonic acid from platelet phospholipids was obviously unchanged in asthma, as thromboxane formation during blood clotting was similar in asthmatic and healthy subjects. Also, the plasma levels of thromboxane B_2 and 6-keto-$PGF_{1\alpha}$, the stable metabolite of prostacyclin, were similar in asthmatic and healthy persons. When nonstimulated polymorphonuclear leukocytes of asthmatic and healthy persons were incubated with exogenous [^{14}C]arachidonic acid, the metabolite patterns did not significantly differ from each other. The amounts of the 5-lipoxygenase metabolites, including 5-HETE and leukotriene B_4, formed were small.

ACKNOWLEDGMENTS. The excellent technical assistance of Ms. Aili Mäkitalo and Ms. Tarja Laiho is gratefully acknowledged. This study was supported by a grant from Yrjö Jahnsson Foundation, Finland. Synthetic leukotrienes were generous gifts from Dr. J. Rokach (Merck Frosst Canada Inc., Canada), some reference prostaglandins and 5-HETE from Dr. J. Pike (Upjohn, Kalamazoo, MI, USA), and U-60,257 from Dr. M. Bach (Upjohn, Kalamazoo, MI, USA).

REFERENCES

Bach, M. K., Brashler, J. R., Smith, H. W., Fitzpatrick, F. A., Sun, F. F., and McGuire, J. C., 1982, 6,9-Deepoxy-6,9-(phenylimino)-$\Delta^{6,8}$-prostaglandin I_1, (U-60,257), a new inhibitor of leukotriene C and D synthesis. *In vitro* studies, *Prostaglandins* **23**:759–771.

Borgeat, P., and Samuelsson, B., 1979, Arachidonic acid metabolism in polymorphonuclear leukocytes: Effects of ionophore A23187, *Proc. Natl. Acad. Sci. U.S.A.* **76**:2148–2152.

Borgeat, P., Fruteau de Laclos, B., and Maclouf, J., 1983, New concepts in the modulation of leukotriene synthesis, *Br. J. Pharmacol.* **32**:381–387.

Boyum, A., 1976, Isolation of lymphocytes, granulocytes and macrophages, *Scand. J. Immunol.* **5**:9–15.

Dahlén, S.-E., Hedqvist, P., Hammarström, S., and Samuelsson, B., 1980, Leukotrienes are potent constrictors of human bronchi, *Nature* **288**:484–486.

Dahlén, S.-E., Hansson, G., Hedqvist, P., Björck, T., Granström, E., and Dahlén, B., 1983, Allergen challenge of lung tissue from asthmatics elicits bronchial contraction that correlates with the release of leukotrienes C_4, D_4, and E_4, *Proc. Natl. Acad. Sci. U.S.A.* **80**:1712–1716.

Dulbecco, R., and Vogt, M., 1954, Plaque formation and isolation of pure lines with poliomyelitis viruses, *J. Exp. Med.* **99**:167–182.

Ford-Hutchinson, A. W., Bray, M. A., Doig, M. W., Shipley, M. E., and Smith, M. J. H., 1980, Leukotriene B, a potent chemokinetic and aggregating substance released from polymorphonuclear leukocytes, *Nature* **286**:264–265.

Hietanen, E., Ahotupa, M., Heikelä, A., Laitinen, M., and Nienstedt, W., 1982, Dietary cholesterol-induced changes of xenobiotic metabolism in liver. I. Influence of xenobiotic administration on hepatic membrane structure, *Drug Nutrient Interact.* **1**:279–292.

Holroyde, M. C., Altounyan, R. E. C., Cole, M., Dixon, M., and Elliot, E. V., 1981, Bronchoconstriction produced in man by leukotrienes C and D, *Lancet* **11**:17–18.

König, W., Kunau, H. W., and Borgeat, P., 1982, Induction and comparison of the eosinophil chemotactic factor with endogenous hydroxyeicosatetraenoic acids: Its inhibition by arachidonic acid analog, *Adv. Prostaglandin Thromboxane Leukotriene Res.* **9**:301–314.

Norred, W. P., and Wade, A. E., 1972, Dietary fatty acid-induced alterations of hepatic microsomal drug metabolism, *Biochem. Pharmacol.* **21**:2887–2897.

Puustinen, T., and Uotila, P., 1983, The effect of bradykinin, histamine, and leukotrienes B_4, C_4 and D_4 on the formation of 6-keto-prostaglandin $F_{1\alpha}$ and thromboxane B_2 in hamster lungs, *Prostaglandins Leukotrienes Med.* **12**:443–448.

Samuelsson, B., 1983, Leukotrienes: A new class of mediators of immediate hypersensitivity reactions and inflammation, *Adv. Prostaglandin Thromboxane Leukotriene Res.* **11**:1–13.

Smith, R. J., Sun, F. F., Bowman, B. J., Iden, S. S., Smith, H. W., and McGuire, J. C., 1982, Effect of 6,9-deepoxy-6,9-(phenylimino)-$\Delta^{6,8}$-prostaglandin I_1 (U-60,257), an inhibitor of leukotriene synthesis, on human neutrophil function, *Biochem. Biophys. Res. Commun.* **109**:943–949.

Uotila, P., Dahl, M.-L., Matintalo, M., and Puustinen, T., 1983, The effects of aspirin and dipyridamole on the metabolism of arachidonic acid in human platelets, *Prostaglandins Leukotrienes Med.* **11**:73–82.

54

Induction of Suppressor T Cells *in Vitro* by Arachidonic Acid Metabolites Issued from the Lipoxygenase Pathway

NORBERT GUALDE and JAMES S. GOODWIN

1. INTRODUCTION

Oxygenated arachidonic acid derivatives are metabolites that issue from both the cyclooxygenase and the lipoxygenase pathways and act as important local immunoregulators. The cyclooxygenase products, especially the prostaglandins (PGs), are currently being investigated with regard to their role in the immune response. Inhibition of lymphocyte activity is the main function of prostaglandins as immunoregulators, and PGE_2 may be viewed as a macrophage messenger for an inhibitory signal to lymphocytes (Goodwin and Ceuppens, 1983). It is likely that PGE_2 is a biological mediator of a physiological negative feedback mechanism following stimulation by antigens (Goodwin *et al.*, 1983).

The compounds that issue from the lipoxygenase pathway, the eicosanoids, have been described more recently. In this chapter, we review the literature linking hydroperoxyeicosatetraenoic acids (HPETEs), hydroxyeicosatetraenoic acids (HETEs), and leukotrienes (LTs) to regulation of the immune response. First, we briefly describe the lipoxygenase cascade that leads to these eicosanoids.

NORBERT GUALDE ● Immunology and Biochemistry Research Unit, School of Medicine, 87032 Limoges Cédex, France. JAMES S. GOODWIN ● Department of Medicine, Medical College of Wisconsin, Milwaukee, Wisconsin 53226.

2. TRANSFORMATION OF ARACHIDONIC ACID VIA THE LIPOXYGENASE PATHWAY

The term "lipoxygenase" is generally understood to denote the enzymes that catalyze the production of LTs, but the pathway that finally yields numerous LTs also gives rise to many other oxidative derivatives of arachidonic acid (Rabinovitch *et al.*, 1981). Among the cells involved in the immune response, macrophages are more important than lymphocytes as producers of monohydroxylated fatty acids (Rigaud *et al.*, 1979), and one would expect that these compounds, similarly to PGs, serve as physiological macrophage-to-lymphocyte messengers.

Table I lists the monohydroxylated fatty acids found in the supernatant after incubation of mouse peritoneal macrophages with exogenous arachidonic acid. Arachidonic acid leads to six hydroperoxyeicosatetraenoic acid precursors of monohydroxylated eicosatetraenoic acids.

Leukotrienes are produced after oxidation of arachidonic acid at C-5, giving rise to an unstable compound called 5-HPETE. The steps yielding LTs via the 5-lipoxygenase pathway are summarized in Fig. 1. This pathway is the most important mechanism for producing LTs, but novel alternate pathways for leukotriene formation acting with enzymes catalyzing the introduction of oxygen at C-12 and C-15 have also been described (Samuelsson, 1983).

3. BIOLOGICAL ACTIVITIES OF THE LIPOXYGENASE PRODUCTS

The biological activities and physiological roles of the various lipoxygenase products are largely unknown. Leukotriene C_4, LTD_4, and LTE_4 are potent constrictors of bronchial smooth muscle, approximately four orders of magnitude more potent than histamine in this regard (Samuelsson, 1983). Leukotriene C_4 and LTD_4 in doses as low as 10^{-13} M increase microvascular permeability. At slightly higher concentrations (10^{-8} to 10^{-10} M), LTC_4 and LTD_4 constrict terminal arterioles, contributing to vascular leakage (Dahlén *et al.*, 1981). Leukotriene C_4, LTD_4, and

TABLE I. Monohydroxylated Fatty Acid Production by
Mouse Peritoneal Macrophages

Monohydroxylated fatty acids	Percentage
12-OH-5,8,10,14-Eicosatetraenoic acid	52.0
15-OH-5,8,11,13-Eicosatetraenoic acid	15.0
11-OH-5,8,12,14-Eicosatetraenoic acid	10.0
8-OH-5,8,11,14-Eicosatetraenoic acid	8.5
9-OH-5,7,11,14-Eicosatetraenoic acid	8.5
5-OH-6,8,11,14-Eicosatetraenoic acid	5.0

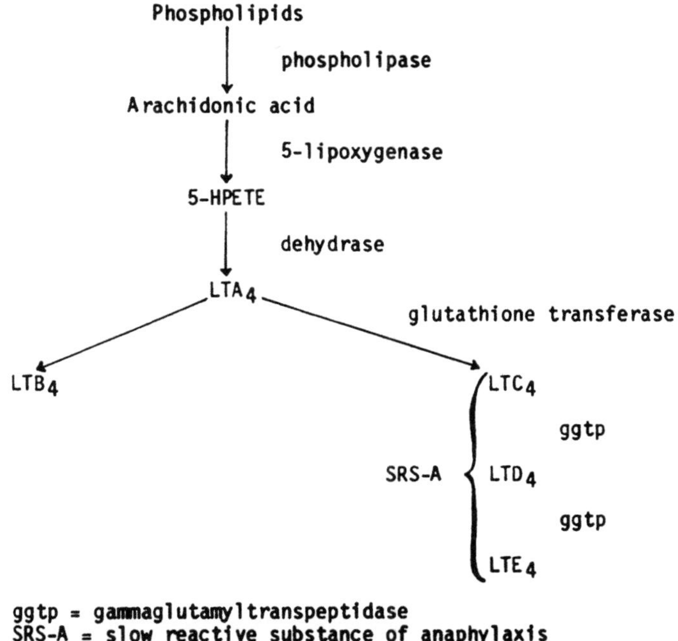

FIGURE 1. Metabolism of arachidonic acid via the lipoxygenase pathway.

LTE$_4$ have also been reported to contract smooth muscle of the stomach and ileum in various species and to have possible cardiac effects (Hammarström, 1983).

Leukotriene B$_4$ is a powerful chemoattractant and chemokinetic agent for polymorphonuclear leukocytes (PMNs) (Ford-Hutchinson *et al.*, 1980) and also stimulates PMN aggregation and adherence to endothelial cells (Goetzl *et al.*, 1983; Lewis and Austen, 1981). In addition, lipoxygenase inhibitors reduce, and LTB$_4$ stimulates, human natural killer (NK) cell activity (Rola-Pleszczynski *et al.*, 1983). Several groups have reported that various arachidonic acid metabolites regulate the synthesis of other arachidonic acid metabolites. For example, LTC$_4$ stimulates leukotriene synthesis in endothelial cells (Benjamin *et al.*, 1983) and 15-HETE in relatively high concentrations ($>10^{-6}$ M) inhibits 5-lipoxygenase activity in some cell types (Bailey *et al.*, 1982; Vanderhoek *et al.*, 1980), although it actually stimulates leukotriene synthesis severalfold in other cell types (Vanderhoek *et al.*, 1982). In addition, inhibition of cyclooxygenase leads to "shunting" of arachidonic acid into the lipoxygenase pathway, resulting in increased leukotriene production in several systems. Indeed, it may well be that some of the physiological effects previously described for cyclooxygenase inhibitors are secondary to the increased production of one lipoxygenase product or another.

Because phospholipase A$_2$ activation, the first step in arachidonic acid metabolism, is probably as ubiquitous a response to stimuli as is adenylate cyclase activation (e.g., Crews *et al.*, 1981), it is safe to assume that the eventual list of biological actions of leukotrienes and other lipoxygenase metabolites of arachidonic acid will be considerably longer than it is today. One area that remains largely uninvestigated is the action of lipoxygenase metabolites in humoral and cellular immune responses. Parker and his co-workers (Kelly and Parker, 1979; Kelly *et al.*, 1979; Parker *et al.*, 1979) found that compounds that inhibit both lipoxygenases and cyclooxygenases inhibit [^{3}H]thymidine incorporation into mitogen-stimulated lymphocytes, whereas pure cyclooxygenase inhibitors stimulate the same response. These investigators concluded that lipoxygenase metabolites are necessary for mitogen-induced lymphocyte proliferation.

Webb and associates (1982) reported that LTB$_4$ and LTE$_4$, in concentrations as low as 10^{-12} M, cause >50% inhibition of phytohemagglutinin (PHA)-induced [^{3}H]thymidine incorporation in mouse splenic T cells, whereas much higher concentrations (10^{-7} M) inhibit the formation of antibody-producing cells against sheep red blood cells (RBC) in Mishell–Dutton cultures. In contrast, in PHA-stimulated cultures of human T cells, Payan and Goetzl (1983) found no inhibition of human T cell proliferation by LTC$_4$ or LTD$_4$, but they did find some inhibition of proliferation and lymphokine production with relatively high concentrations (10^{-7} M) of LTB$_4$. Rola-Pleszczynski *et al.* (1982) also reported a modest inhibition of PHA-stimulated mitogenesis in human lymphocytes by LTB$_4$ over a wide range of concentrations with no obvious dose response. More important, these authors observed that T cells preincubated with concentrations of LTB$_4$ as low as 10^{-12} M suppress the [^{3}H]thymidine incorporation of fresh, autologous T cells in subsequent PHA-stimulated cultures.

In 1980, it was clearly demonstrated by Goodman and Weigle (1980) that 15-HETE, a hydroperoxide derivative of arachidonic acid, inhibits mitogen-induced lymphocyte proliferation *in vitro*. This inhibitory effect was confirmed by Bailey *et al.* (1982) and extended to various HPETEs (Gualde *et al.*, 1981). We examined the effects of HPETEs (Gualde *et al.*, 1981) and HETEs (Gualde *et al.*, 1983a) on lectin-stimulated [^{3}H]thymidine incorporation in mouse splenocytes (Gualde *et al.*, 1983a) and in human peripheral blood mononuclear cells (PBMC) (Gualde *et al.*, 1982). We found that various HPETEs and HETEs diminish [^{3}H]thymidine uptake when added in small concentrations (10^{-6} to 10^{-10} M) to lymphocytes stimulated by concanavalin A (Con A) or PHA. When purified human T cells supplemented by 5% monocytes were stimulated by PHA or Con A in the presence of 15-HPETE, which had been added at the start of the culture, a clear inhibition of proliferation was observed. As shown in Table II, this eicosanoid-induced inhibition depended on the lectin and/or hydroperoxide concentration (Gualde, 1983).

Therefore, it became obvious that many of the molecules produced by the lipoxygenase cascade inhibit the lymphocyte proliferative response to mitogens. Suppressive activity induced *in vitro* by 15-HPETE and by 15-HETE was also observed during the allostimulation procedures. For instance, when C57B1/6 responder splenocytes were cocultured with DBA/2(−)-irradiated stimulator cells, thymidine uptake by proliferating B6 clones was diminished by adding 15-HPETE

TABLE II. 15-Hydroperoxyeicosatetraenoic Acid Inhibition of Human T Cell
Proliferation *in Vitro*

Mitogen	Concentration (μg/ml)	Percentage of inhibition (mean $\pm$ S.D.)		
		4×10^{-8} M 15-HPETE	4×10^{-7} M 15-HPETE	4×10^{-6} M 15-HPETE
PHA	5	20 ± 23	43 ± 26	80 ± 18
PHA	10	6 ± 13	43 ± 34	62 ± 20
Con A	10	44 ± 30	56 ± 25	90 ± 9
Con A	20	41 ± 19	74 ± 18	94 ± 12

or 15-HETE, and the subsequent generation of B6 anti-D2 killer T cells was restricted (Gualde *et al.*, 1981). The immunosuppressive activity of 15-HPETE on human T lymphocytes was studied more carefully when the cells carrying the OKT4 marker of the helper/inducer subset and the lymphocytes positive for the OKT8 marker of the suppressor/cytotoxic subset were sorted and separately stimulated with PHA or Con A in the presence of 15-HPETE (4×10^{-6} to 4×10^{-8} M). It was noticed that, in terms of proliferative reponse to lectins, the OKT4($+$) T cells were less inhibited than the OKT8($+$) T cells (Gualde *et al.*, 1985).

4. INDUCTION OF SUPPRESSOR CELLS BY 15-HPETE AND 15-HETE

We preincubated B6 lymphocytes for 24 hr in culture media with or without 15-HPETE or 15-HETE. When added to cocultures of fresh B6 cells and D2-irradiated stimulator cells, the treated cells partially inhibited the proliferative response of the B6 cells (Aldigier *et al.*, 1984). One possible explanation for this result is that mouse lymphocytes incubated with 15-HPETE or 15-HETE acquire a suppressive activity.

The same phenomenon was observed with human T cells (Gualde *et al.*, 1983b). Purified human T cells preincubated with 15-HPETE no longer supported immunoglobulin production by fresh, autologous, pokeweed mitogen (PWM)-stimulated B cells in the presence of fresh, autologous T lymphocytes.

We extended these observations on the phenotypic nature of eicosanoid-induced mouse and human suppressor cells. When mouse splenocytes were incubated *in vitro* for 48 hr with 15-HETE (10^{-5} to 10^{-8} M final concentration), we noticed an increase in the percentage of cells carrying the Lyt-2($+$) antigen (Mexmain *et al.*, 1984) (Table III).

We also observed that when human T cells were similarly preincubated with 15-HPETE and were treated with an anti-OKT8 monoclonal antibody plus complement, the suppression of IgG production was eliminated. However, when the suppressor OKT8($+$) cells were eliminated before the preincubation, the generation

TABLE III. Modification of Splenocyte Lyt Markers by 15-HETE

Preincubation	Lyt-1(+)2(−) cells	Lyt-1(−)2(+) cells
Medium	30 ± 8	8 ± 3
15-HETE	35 ± 6	14 ± 5

of new suppressor cells by 15-HPETE was not affected (Gualde *et al.*, 1983b).

A possible explanation for the above data is that 15-HPETE and 15-HETE might convert a population of nonsuppressor lymphocytes to a subset of suppressor cells. For instance, 15-HPETE might change OKT8(−) presuppressor T cells into OKT8(+) suppressor cells. This hypothesis is sustained by the following experiment.

T cells and OKT8(−) T cells were incubated separately for 24 hr with or without 15-HPETE. The percentage of cells bearing suppressor surface markers was then determined by flow microfluorograph analysis following incubation with a directly fluoresceinated antibody. In the OKT8(−) T cell populations treated by 15-HPETE, we observed a large increase in suppressor OKT8(+) cells (Gualde *et al.*, 1983b).

The problem was then to determine if the increase in the percentage of cells carrying suppressor surface markers was merely a membrane phenomenon resulting from the expression or relocation of hidden OKT8 or Lyt-2 antigens or if it was a mechanism related to cell proliferation and differentiation. We therefore studied the cell cycle of 15-HETE-treated lymphocytes using flow microfluorograph analysis of nuclear DNA following ethidium bromide staining. We observed that 15-HETE increased the perecentage of cells in the G_2 phase after 41 hr of preincubation and that this increment was preceded from the 17th to the 22nd hour by an increase of cells in the S phase.

5. INDUCTION OF SUPPRESSOR CELLS BY LTB$_4$

We have observed that LTB$_4$ is a potent suppressor of polyclonal Ig production in PWM-stimulated cultures of human peripheral blood lymphocytes, whereas LTC$_4$ and LTD$_4$ have little activity in this system (Atluru and Goodwin, 1984). Preincubation of T cells with LTB$_4$ in nanomolar or picomolar concentrations renders these cells suppressive of Ig production in subsequent PWM-stimulated cultures of fresh, autologous B and T cells. This LTB$_4$-induced suppressor cell is radiosensitive, and its generation can be blocked by cycloheximide. The LTB$_4$-induced suppressor cell is OKT8(+), whereas the precursor for the cell can be OKT8(−). The induction of OKT8(−) T cells with LTB$_4$ for 18 hr results in the appearance of the OKT8(+) on 10–20% of the cells, and this can be blocked by cycloheximide but not mitomycin C.

What is striking about our finding is the very low concentrations of LTB$_4$

required for supressor cell induction. As little as 10^{-12} M caused significant inhibition of IgG production when added to PWM-stimulated cultures or when preincubated with T cells (Table IV). Indeed, LTB_4 is three to six orders of magnitude more potent than either PGE or histamine. It is interesting to contrast the extreme potency of LTB_4 in suppressor cell induction with its relative lack of effectiveness in *in vitro* assays of cellular immunity. Payan and Goetzl (1983) reported a relatively modest inhibition of PHA-induced proliferation and lymphokine generation by 10^{-7} and 10^{-6} M LTB_4, with no effect of lower concentrations. Rola-Pleszczynski and associates (1982) also reported a small degree ($\sim$20%) of inhibition of PHA- or Con-A-induced proliferation of human lymphocytes by LTB_4. In contrast to our findings with LTB_4, we feel that the relatively high concentration of 15-HPETE that was required to induce suppression suggests that perhaps a metabolite such as a novel 14,15-leukotriene is the active agent responsible for suppressor cell generation.

6. MECHANISM(S) OF LIPOXYGENASE PRODUCT-INDUCED IMMUNOSUPPRESSION

There is a good deal of evidence that lipoxygenase arachidonic acid metabolites suppress the immune response, but the mechanism of this immunosuppression is not yet clear.

TABLE IV. Effect of Preincubating T Cells for 18 Hr
with Leukotrienes on the Production of IgG and IgM in
Subsequent PWM Cultures with Fresh Autologous B Cells[a]

Cells	T cells preincubated with	IgG (ng/ml)	Percent inhibition	IgM (ng/ml)	Percent inhibition
B	. . .	93 ± 10	. . .	23 ± 6	. . .
B + T	0	1101 ± 124	. . .	315 ± 39	. . .
B + T	10^{-12} M LTB_4	901 ± 93[b]	18 ± 4	285 ± 31[c]	9 ± 3
B + T	10^{-11} M LTB_4	752 ± 112[b]	32 ± 4	208 ± 27[b]	34 ± 4
B + T	10^{-10} M LTB_4	531 ± 102[b]	52 ± 5	214 ± 14[b]	32 ± 3
B + T	10^{-9} M LTB_4	348 ± 105[b]	68 ± 5	191 ± 19[b]	39 ± 3
B + T	10^{-8} M LTB_4	258 ± 98[b]	76 ± 4	129 ± 17[b]	59 ± 4
B + T	10^{-9} M LTC_4	963 ± 84[b]	12 ± 3	287 ± 46[c]	9 ± 3
B + T	10^{-8} M LTC_4	957 ± 96[b]	13 ± 6	288 ± 33[c]	8 ± 4
B + T	10^{-9} M LTD_4	991 ± 114[b]	10 ± 4	291 ± 34	8 ± 4
B + T	10^{-8} M LTD_4	909 ± 106[b]	17 ± 4	278 ± 41[c]	12 ± 3

[a] T lymphocytes were incubated with leukotrienes LTB_4, LTC_4, or LTD_4 for 18 hr, washed three times with PBS, and cultured with fresh, autologous B cells at 10^5 T cells plus 2.5×10^5 B cells. Viabilities of LTB_4-preincubated cells were >95.0% as determined by trypan blue dye exclusion test. Results are expressed as means ± S.E.M. from experiments on cells from five different subjects.
[b] Significantly different from control by paired *t*-test with $P < 0.001$.
[c] $P < 0.05$.

It is possible that these lipoxigenic metabolites can affect an entire lymphocyte population merely by altering the physical properties of the lymphocytes. We observed that both 15-HPETE and 15-HETE dramatically modify the viscosity of the plasma membranes of splenocytes, as mentioned above (Gualde, 1983). Thus, there is a chance that 15-HPETE and 15-HETE suppress *in vitro* lymphocyte activity by causing rigidity of the lymphocyte membranes, thereby blocking receptor mobility, which is necessary for the initiation of mitogenesis.

It was also observed by Bailey *et al.* (1982) that 15-HETE modifies lymphocyte metabolism somewhat, mostly inhibiting 12-HETE and 5-HETE synthesis. In addition, it is well known that 15-HPETE inhibits the synthesis of PGI_2 (Salmon *et al.*, 1978), and Honn *et al.* (1983) postulated that the increase in tumor metastasis observed after intravenous injection of 15-HPETE is caused by enhanced platelet aggregation resulting from the decrease in PGI_2 synthesis.

Hydroperoxy acids cause a rise in the level of intracellular cyclic guanylate monophosphate (cGMP) (Harbon *et al.*, 1983), a phenomenon usually associated with enhanced cell activities such as lymphocyte proliferation (Strom *et al.*, 1977). According to our preliminary data, 15-HETE increases lymphocyte cGMP (Gualde *et al.*, 1984). If this is true, the effects of 15-HETE on lymphocyte metabolism are quite contradictory, i.e., (1) augmentation of cGMP, which gives rise to proliferation, and (2) inhibition of 5-HETE production, which leads to immunosuppression (Bailey *et al.*, 1982). However, changes in the physical state of the cell membrane or in cell metabolism cannot explain how the sensitivity of human lymphocytes to 15-HPETE is HLA linked (Gualde *et al.*, 1982) or why OKT8(+) cells are less sensitive to 15-HPETE than the OKT4(+) cell subset with regard to mitogen-induced thymidine uptake (Gualde *et al.*, 1985).

Rola-Pleszczynsky *et al.* (1982) obsrved that the immunosuppression induced by LTB_4 is related to the generation of immunosuppressive cells, and it is likewise probable that eicosanoid-induced immunosuppression involves more than cell metabolism or the physical state of the cell membranes. Since we observed contradictory phenomena induced by 15-HPETE and by 15-HETE, namely, inhibition of the proliferative response of lymphocytes versus augmentation of intracellular cGMP and the synthesis of DNA, we speculate that both 15-HPETE and 15-HETE induce differentiation and proliferation of a subset of presuppressor cells, giving rise to suppressor OKT8(+) or Lyt-2(+) lymphocytes. If this hypothesis is correct, the increase in the level of cGMP and in the synthesis of DNA is related to the proliferation of the suppressor subset.

7. CONCLUSION

Although it is clear that LTB_4 and 15-HPETE cause induction of suppressor cells *in vitro,* there is no evidence that endogenous LTB_4 or 15-HPETE play roles in any *in vitro* or *in vivo* model of suppressor cell generation. It is interesting to note, however, that glucocorticosteroids inhibit suppressor cell generation in many *in vitro* and *in vivo* models (e.g., Waldmann *et al.*, 1976; Tosato *et al.*, 1979, 1980). It is currently thought that the action of steroids at the cellular level is mediated by the synthesis of a phospholipase A_2 inhibitory protein, termed lipomodulin (Hirata *et al.*, 1980) or macrocortin (Blackwell *et al.*, 1980), and that this inhibitory protein prevents the release of arachidonic acid from membrane phospholipids. Thus, many or all of the actions of steroids may be mediated by inhibition of arachidonic acid metabolism. If steroids prevent suppressor cell generation by inhibiting arachidonic acid metabolism, then one would expect to find an arachidonic acid metabolite that at physiological concentrations stimulated suppressor cell generation. Because cyclooxygenase inhibitors such as indomethacin do not inhibit suppressor cell generation (Goodwin, 1980; Soppi *et al.*, 1980; Badger *et al*, 1982), it is logical to assume that a lipoxygenase metabolite and not a cyclooxygenase metabolite of arachidonic acid is responsible for the suppressor cell generation. Our findings of suppressor cell generation by physiological concentrations of LTB_4 and higher levels of 15-HPETE are certainly consistent with that concept. Further work should be directed towards the question of whether addition of physiological concentrations of LTB_4 can reverse the inhibition of suppressor cell generation or function caused by corticosteroids.

REFERENCES

Aldigier, J. C., Gualde, N., Mexmain, S., Chable-Rabinovitch, H., Ratinaud, M. H., and Rigaud, M., 1984, Immunosuppression induced *in vivo* by 15 hydroxyeicosatetraenoic acid (15 HETE), *Prostaglandins Med.* **13**:99–107.

Atluru, D., and Goodwin, J. S., 1984, Control of polyclonal immunoglobulin production from human lymphocytes by leukotrienes, *J. Clin. Invest.* **74**:1444–1450.

Badger, A. M., Griswold, D. E., and Walz, D. T., 1982, Augmentation of concanavalin A-induced immunosuppression by indomethacin, *Immunopharmacology* **4**:149–162.

Bailey, J. M., Bryant, R. W., Low, C. E., Pupillo, M. B., and Vanderhoek, J. Y., 1982, Regulation of T lymphocyte mitogenesis by the leukocyte product 15-hydroxyeicosatetraenoic acid (15-HETE), *Cell Immunol.* **67**:112–120.

Benjamin, C. W., Hopkins, N. K., Oglesby, T., and Gorman, R., 1983, Against specific desensitization of leukotriene C_4-stimulated PGI_2 synthesis in human endothelial cells, *Biochem. Biophys. Res. Commun.* **117**:780–787.

Blackwell, G. J., Carnuccio, R., Rosa, M. D., Flower, R. J., Parente, L., and Persico, P., 1980, Macrocortin: A polypeptide causing the anti-phospholipase effect of glucocorticoids, *Nature* **287**:147–149.

Crews, F. T., Morita, Y., McGivney, A., Hirata, F., Siraganian, R. P., and Axelrod, J., 1981, IgE-mediated histamine release in rat basophilic leukemia cells: Receptor activation, phospholipid methylation, Ca^{2+} flux, and release of arachidonic acid, *Arch. Biochem. Biophys.* **212**:561–571.

Dahlén, S. E., Björk, J., Hedqvist, P., Arfors, K. E., Hammarström, S., Lindgren, J. A., and Samuelsson, B., 1981, Leukotrienes promote plasma leakage and leukocyte adhesion in postcapillary venules: *In vivo* effects with relevance to the acute inflammatory response, *Proc. Natl. Acad. Sci. U.S.A.* **78**:3887–3891.

Ford-Hutchinson, A. W., Bray, M. A., Doig, M. V., Shipley, M. E., and Smith, M. J. H., 1980, Leukotriene B, a potent chemokinetic and aggregating substance released from polymorphonuclear leukocytes, *Nature* **286**:264–265.

Goetzl, E. J., Brindley, L. L., and Goldman, D. W., 1983, Enhancement of human neutrophil adherence by synthetic leukotriene constituents of the slow-reacting substance of anaphylaxis, *Immunology* **50**:35–42.

Goodman, M. G., and Weigle, W. O., 1980, Inhibition of lynphocyte mitogenesis by an arachidonic acid hydroperoxide, *J. Supramol. Struct.* **13**:373–383.

Goodwin, J. S., 1980, Modulation of concanavalin-A induced suppressor cell activation by prostaglandin E₂ *Cell. Immunol.* **49**:421–425.

Goodwin, J. S., and Ceuppens, J. L., 1983, Regulation of the immune response by prostaglandins, *J. Clin. Immunol.* **3**:295–315.

Goodwin, J. S., Ceuppens, J. L., and Gualde, N., 1983, Control of the immune response in humans by prostaglandins, in: *Therapeutic Control of Inflammatory Diseases* (I. Otterness, R. J. Capetola, and S. Wong, eds.), Raven Press, New York, pp. 79–92.

Gualde, N., 1983, Les Métabolites Oxygénés de l'Acide Arachidonique et l'Activité des Lymphocytes, Ph.D. Thesis in Immunology, Université Paris VI, Paris.

Gualde, N., Rigaud, M., Rabinovitch, H., Durand, J., Beneytout, J. L., and Breton, J. C., 1981, Inhibition de la réponse lymphocytaire par les métabolites libérés par la lipoxygénase des macrophages chez la souris, *C. R. Seances Acad. Sci.* **293**:359–362.

Gualde, N., Rabinovitch, H., Fredon, M., and Rigaud, M., 1982, Effects of 15-hydroperoxyeicosatetraenoic acid on human lymphocyte sheep erythrocyte rosette formation and response to concanavalin A association with HLA system, *Eur. J. Immunol.* **12**:773–777.

Gualde, N., Chable-Rabinovitch, H., Motta, C., Durand, J., Beneytout, J. L., and Rigaud, M., 1983a, Hydroperoxyeicosatetraenoic acids—potent inhibitors of lymphocyte responses, *Biochim. Biophys. Acta* **750**:429–433.

Gualde, N., Rigaud, M., and Goodwin, J. S., 1983b, Induction of suppressor cells from human peripheral blood T cells by 15 hydroperoxyeicosatetraenoic acid (15 HPETE), *Clin. Res.* **31**:490A.

Gualde, N., Mexmain, S., Aldigier, J. C., Goodwin, J. S., and Rigaud, M., 1984, Eicosanoids and the immune response, in: *Eicosanoids and Cancer* (J. Craste de Paulet, ed.), Raven Press, New York (in press).

Gualde, N., Atluru, D., and Goodwin, J. S., 1985, Effect of lipoxygenase metabolites of arachidonic acid on proliferation of human T cells and T cell subsets, *J. Immunol.* **134**:1125–1129.

Hammarström, S., 1983, Leukotrienes, *Annu. Rev. Biochem.* **52**:355–377.

Harbon, S., Leiber, D., and Vesin, M.-F., 1983, Phospholipase A₂ activation coupled to lipoxygenase and cyclooxygenase pathways in the guinea pig myometrium: Interaction with the cyclic AMP and cyclic GMP systems, *Adv. Prostaglandin Thromboxane Leukotriene Res.* **15**:423–428.

Hirata, F., Schiffmann, E., Venkatasubramanian, K., Salomon, D., and Axelrod, J., 1980, A phospholipase A₂ inhibitory protein in rabbit neutrophils induced by glucocorticoids, *Proc. Natl. Acad. Sci. U.S.A.* **77**:2533–2536.

Honn, K. V., Busse, W. D., and Sloane, B. F., 1983, Prostacyclin and thromboxanes—implications for their role in tumor cell metastasis, *Biochem. Pharmacol.* **32**:1–11.

Kelly, J. P., and Parker, C. W., 1979, Effects of arachidonic acid and other unsaturated fatty acids on mitogenesis in human lymphocytes, *J. Immunol.* **122**:1556–1562.

Kelly, J. P., Johnson, M. C., and Parker, C. W., 1979, Effect of inhibitors of arachidonic acid metabolism on mitogenesis in human lymphocytes: Possible role of thromboxanes and products of the lipoxygenase pathway, *J. Immunol.* **122**:1563–1571.

Lewis, R. A., and Austen, K. F., 1981, Mediation of local homeostasis and inflammation by leukotrienes and other mast cell-dependent compounds, *Nature* **293**:103–108.

Mexmain, S., Gualde, N., Aldigier, J. C., Motta, C., Chable-Rabinovitch, H., and Rigaud, M., 1984, Specific binding of 15 HETE to lymphocytes: Effects on the fluidity of plasmatic membranes, *Prostaglandins Med.* **13**:93–97.

Parker, C. W., Stenson, W. F., Huber, M. G., and Kelly, J. P., 1979, Formation of thromboxane B₂ and hydroxyarachidonic acids in purified human lymphocytes in the presence and absence of PHA, *J. Immunol.* **122**:1572–1577.

Payan, D. G., and Goetzl, E. J., 1983, Specific suppression of human T lymphocyte function by leukotriene B$_4$, *J. Immunol.* **131**:551–553.

Rabinovitch, H., Durand, J., Gualde, N., and Rigaud, M., 1981, Metabolism of polyunsaturated fatty acids by mouse peritoneal macrophages: The lipoxygenase metabolic pathway, *Agents Actions* **11**:580–583.

Rigaud, M., Durand, J., and Breton, J. C., 1979, Transformation of arachidonic acid into 12-hydroxy-5,8,10,14-eicosatetraenoic acid by mouse peritoneal macrophages, *Biochim. Biophys. Acta* **573**:408–412.

Rola-Pleszczynski, M., Borgeat, P., and Sirois, P., 1982, Leukotriene B$_4$ induces human suppressor lymphocytes, *Biochem. Biophys. Res. Commun.* **108**:1531–1537.

Rola-Pleszczynski, M., Gagnon, L., and Sirois, P., 1983, Leukotriene B$_4$ augments human natural cytotoxic cell activity, *Biochem. Biophys. Res. Commun.* **113**:531–537.

Salmon, J. A., Smith, D. R., Flower, R. J., Moncada, S., and Vane, J. R., 1978, Further studies on the enzymatic conversion of prostaglandin endoperoxide into prostacyclin by porcine aorta microsomes, *Biochim. Biophys. Acta* **523**:250–262.

Samuelsson, B., 1983, Leukotrienes: Mediators of immediate hypersensitivity reactions and inflammation, *Science* **220**:568–575.

Soppi, E., Eskola, J., and Ruuskanen, O., 1980, Effects of indomethacin on lymphocyte proliferation, suppressor cell function, and leukocyte migration inhibitory factor (LMIF) production, *Immunopharmacology* **4**:235–242.

Strom, T. B., Lundin, A. P., and Carpenter, C. B., 1977, The role of cyclic nucleotides in lymphocyte activation and function, in: *Progress in Clinical Immunology*, Vol. 3 (R. S. Schwartz, ed.), Grune & Stratton, New York, pp. 115–153.

Tosato, G., Magrath, I. T., Koski, I. R., Dooley, N. J., and Blaese, R. M., 1979, Activation of suppressor T cells during Epstein–Barr virus induced infectious mononucleosis, *N. Engl. J. Med.* **301**:1133–1137.

Tosato, G., Magrath, I. T., Koski, I. R., Dooley, N. J., and Blaese, R. M., 1980, B-cell differentiation and immunoregulatory T cell function in human cord blood lymphocytes, *J. Clin. Invest.* **66**:383–388.

Vanderhoek, J. Y., Bryant, R. W., and Bailey, J. M., 1980, Inhibition of leukotriene biosynthesis by the leukocyte product 15-hydroxyeicosatetraenoic acid, *J. Biol. Chem.* **255**:10064–10065.

Vanderhoek, J. Y., Tare, N. S., Bailey, M., Goldstein, A., and Pluznik, D., 1982, New role for 15-hydroxyeicosatetraenoic acid, *J. Biol. Chem.* **257**:12191–12195.

Waldmann, T. A., Broder, S., Krakauer, R., MacDermott, R. P., Durm, M., Goldman, C., and Meade, B., 1976, The role of suppressor cells in the pathogenesis of common variable hypogammaglobulinemia and the immunodeficiency associated with myeloma, *Fed. Proc.* **35**:2067–2072.

Webb, D. R., Nowowiejski, I., Healy, C., and Rogers, T. J., 1982, Immunosuppressive properties of leukotriene D$_4$ and E$_4$ *in vitro*, *Biochem. Biophys. Res. Commun.* **104**:1617–1622.

Antiinflammatory Drugs and Inhibitors of Cyclooxygenase and Lipoxygenase Systems

55

Novel Aspects of Aspirin Action

G. A. FITZGERALD, I. A. G. REILLY and
A. K. PEDERSEN

1. INTRODUCTION

Aspirin acetylates and irreversibly inhibits the enzyme cyclooxygenase (Roth and
Majerus, 1975). Recent studies (Patrignani *et al.*, 1982; FitzGerald *et al.*, 1983;
Weksler *et al.*, 1983) have clarified the dose–response relationships between aspirin
and the production of prostacyclin (PGI_2) and thromboxane $(Tx)A_2$, the predominant
cyclooxygenase products formed by vascular endothelium and platelets, respectively
and indicate that doses of aspirin less than 100 mg have a more pronounced effect
on Tx formation both *in vivo* (Patrignani *et al.*, 1982; FitzGerald *et al.*, 1983) and
ex vivo (Weksler *et al.*, 1983). Possible explanations for such "biochemical selectivity" have included differential tissue sensitivity to enzyme inhibition by aspirin
and/or accelerated recovery of PGI_2 formation by nucleated endothelial cells, a
phenomenon demonstrable in culture (Jaffe and Weksler, 1979). To address the
possibility that pharmacokinetic variables might also contribute to such dose-related
pharmacodynamic effects of aspirin, we have characterized the dose-related kinetics
of aspirin and identified a selective mode of drug delivery that relies on platelet
acetylation in the presystemic circulation, sparing systemic vascular endothelium
from drug exposure.

2. METHODS

To conduct pharmacokinetic studies of "low"-dose aspirin, we developed a
stable isotope dilution assay of requisite (~10 ng/ml) sensitivity employing gas
chromatography–mass spectrometry (Pedersen and FitzGerald, 1985). The use of

---

G. A. FITZGERALD, I. A. G. REILLY, and A. K. PEDERSEN ● Division of Clinical Pharmacology,
Vanderbilt University, Nashville, Tennessee 37232.

heptadeuterated analogues of aspirin and salicylate as internal standards permitted the performance of classic dose-related bioavailability studies in animals (Pedersen and FitzGerald, 1985) and man (Pederson and FitzGerald, 1984), coadministering native aspirin orally with tetradeuterated aspirin intravenously (Fig. 1). Although the systemic bioavailability of aspirin approximated 50% over the dose range 20 to 1300 mg, we noted a substantial fall in TxB_2 generation in serum *ex vivo* prior to the detection of aspirin in the systemic circulation following dosing with 20 mg. Serum TxB_2 declined further coincident with peak plasma aspirin concentration. These data were consistent with a mixed presystemic and systemic effect of aspirin at this dosage. This interpretation was supported by experiments that demonstrated that inhibition of Tx in serum *ex vivo* was less than that by simulated corresponding peak systemic aspirin concentrations in whole blood *in vitro*.

To test the hypothesis that drug delivery might be designed so that first-pass extraction approached totality, we recently administered 1 mg aspirin every 30 min for 10 hr to volunteers and compared the effects on Tx production to that of a single 20-mg capsule (Reilly and FitzGerald, 1984). Cumulative inhibition of serum TxB_2

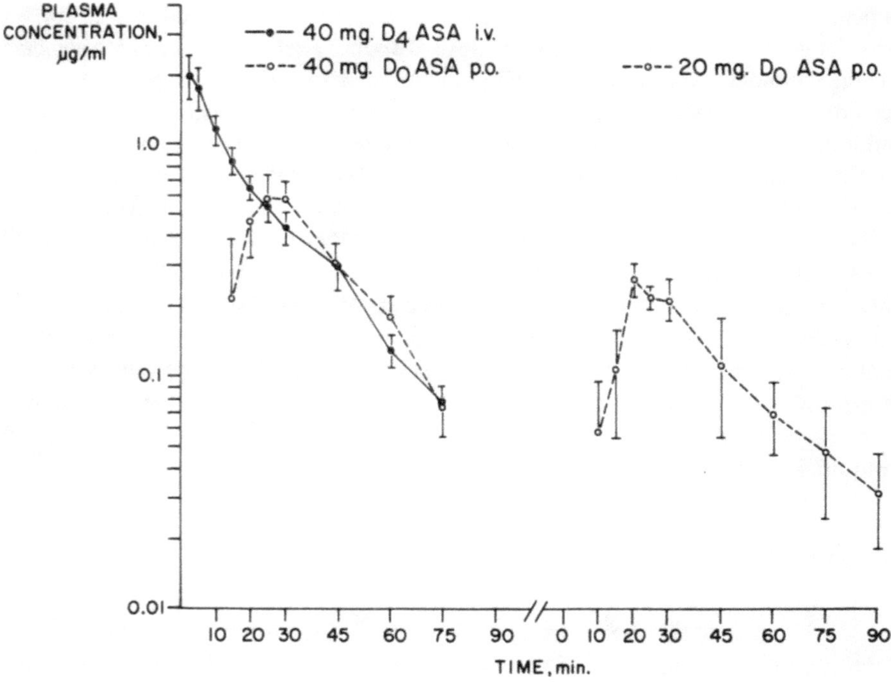

FIGURE 1. Plasma concentrations of native (D_0) and tetra-deuterated (D_4) aspirin followed simultaneous oral and intravenous administration of the native and tetra-deuterated forms respectively. Plasma concentration time curves are also shown following oral administration of 20 mg native aspirin.

formation occurred during the dosing interval with the simulated slow-release preparation, and the extent of Tx depression was ultimately more pronounced than that attained after the 20-mg capsule. Aspirin was only detected in the systemic circulation after the latter formulation. To further support the notion that the slow-release preparation acted predominantly presystemically, recovery of serum TxB_2 after dosing corresponded to platelet life-span but without the 1- to 2-day "lag" corresponding to megakaryocyte exposure to aspirin (Roth and Majerus, 1975).

3. DISCUSSION

The dose-related biochemical selectivity of aspirin may reflect differential tissue sensitivity to enzyme inhibition. However, published data are conflicting on this issue. Similarly, although recovery of PGl_2 after aspirin is accelerated *in vitro*, this may not be the case either *in vivo* (FitzGerald *et al.*, 1983; FitzGerald and Oates, 1984) or in vascular tissues *ex vivo* (Weksler *et al.*, 1983; Hanley *et al.*, 1981). These studies suggest the possibility that pharmacokinetic factors may contribute to pharmacodynamic selectivity *in vivo*.

The importance of preserving PGl_2 biosynthesis during platelet inhibition in man is presently unknown, and factors other than poor biochemical selectivity are likely to have contributed to equivocal results with aspirin in the secondary prevention of myocardial infarction. However, although little PGI_2 is produced under physiological circumstances in man (FitzGerald *et al.*, 1981), production may be increased in syndromes of platelet activation consistant with a local homeostatic role *in vivo* (FitzGerald *et al.*, 1984). The demonstrable efficacy of aspirin (325 mg daily) in unstable angina when therapy is instituted rapidly (Lewis *et al.*, 1983) and the improvement in aortocoronary bypass graft patency with lower doses (100 mg) of aspirin (Lorenz *et al.*, 1984) reflect the utility of this compound as a platelet inhibitor *in vivo*. Should future evidence confirm the importance of coincidentally preserving PGl_2 biosynthesis, it is likely that we can take advantage of aspirin pharmacokinetics to design an ultimately selective mode of drug delivery.

ACKNOWLEDGMENTS. This work was supported by a grant (HL30400) from the National Institutes of Health. Dr. FitzGerald is the recipient of a Faculty Development Award from the Pharmaceutical Manufacturer's Association Foundation. Dr. Pedersen held a fellowship from the Danish Medical Research Council during the course of this work.

REFERENCES

FitzGerald, G. A., and Oates, J. A., 1984, Selective and nonselective inhibition of thromboxane formation, *Clin. Pharmacol. Ther.* **35**:633–640.

FitzGerald, G. A., Brash, A. R., Falardeau, P., and Oates, J. A., 1981, Estimated rate of prostacyclin secretion into the circulation of normal man, *J. Clin. Invest.* **68**:1272–1275.

FitzGerald, G. A., Oates, J. A., Hawiger, J., Maas, R. L., Roberts, L. J., II, Lawson, J., and Brash, A. R., 1983, Endogenous biosynthesis of prostacyclin and thromboxane and platelet function during chronic administration and aspirin in man, *J. Clin. Invest.* **71**:676–688.

FitzGerald, G. A., Smith, B., Pedersen, A. K., and Brash, A. R., 1984, Prostacyclin biosynthesis is increased in patients with severe atherosclerosis and platelet activation, *N. Engl. J. Med.* **310**:1065–1068.

Hanley, S. P., Bevan, J., Cockbill, S. R., and Heptinstall, S., 1981, Differential inhibition by low dose aspirin of human venous prostacyclin synthesis and platelet thromboxane synthesis, *Lancet* **1**:969–971.

Jaffe, E. A., and Weksler, B. B., 1979, Recovery of endothelial cell prostacyclin production after inhibition by low doses of aspirin, *J. Clin. Invest.* **63**:532–535.

Lewis, H. D., Davis, J. W., Archibald, D. G., Steinke, W. E., Smitherman, T. C., Docherty, J. E., Schnaper, H. W., LeWinter, M. M., Linares, E., Pouget, J. M., Sabharwal, S. C., Chester, E., and DeMots, H., 1983, Protective effects of aspirin against acute myocardial infarction and death in men with unstable angina, *N. Engl. J. Med.* **309**:396–403.

Lorenz, R. L., Schacky, C. V., Weber, M., Meister, W., Kotzur, J., Reichart, B., Theisen, K., and Weber, P. C., 1984, Improved aortacoronary bypass patency by low dose aspirin (100 mg daily), *Lancet* **1**:1261–1264.

Patrignani, P., Filabozzi, P., and Patrono, C., 1982, Selective cumulative inhibition of platelet thromboxane production by low-dose aspirin in healthy subjects, *J. Clin. Invest.* **69**:1366–1371.

Pedersen, A. K., and FitzGerald, G. A., 1985, Preparation and analysis of deuterium labelled aspirin: Application to pharmacokinetic studies, *J. Pharm. Sci.* **74**:188–192.

Pedersen, A. K., and FitzGerald, G. A., 1984, Dose related pharmocokinetics of aspirin: Presystemic acetylation of platelet cyclooxygenase in man, *N. Engl. J. Med.* **311**:1206–1211.

Reilly, I. A. G., and FitzGerald, G. A., 1984, Presystemic inhibition of platelet thromboxane formation by simulated slow release, low dose aspirin in man, *Clin. Res.* **32**:320A.

Roth, G. J., and Majerus, P. W., 1975, The mechanism of the effect of aspirin on human platelets. I. Acetylation of a particulate fraction protein, *J. Clin. Invest.* **56**:624–632.

Weksler, B. B., Pett, S. B., Alonso, D., Richter, R. C., Stelzer, P., Subramanian, V., Tack-Goldman, K., and Gay, W. A., 1983, Differential inhibition by aspirin of vascular and platelet prostaglandin synthesis in atherosclerotic patients, *N. Engl. J. Med.* **308**:800–805.

56

Cyclooxygenase Inhibitors
An Overview

RODERICK J. FLOWER

1. THE BACKGROUND

The cyclooxygenase inhibitors, as exemplified by the salicylates, are another example of that group of drugs whose origins lie deep in the folk medicine of many cultures and whose therapeutic efficacy was established empirically long before a plausible reason or explanation for their action emerged.

The exploitation of such drugs tends to have a definite pattern. First, the medicine in the form of a crude plant or animal tissue extract is given to one or more sick people, almost invariably for the wrong reason, and it is found to be effective. Later, often much later, the active principle of the crude extract is isolated, identified, and synthesized. Finally, the mechanism of action of the drug is elucidated, and this generally triggers an intense interest in the drug in question and stimulates efforts to discover other compounds having the same or similar action.

The early history of the salicylates is familiar to most people (see Rodnan and Benedek, 1970, for an excellent review). It is interesting that two of the most frequently employed antiinflammatory drugs, the salicylates and colchicine, were both known to ancient physicians who employed decoctions of willow bark and extracts of roots and bulb of *Colchicum autumnale* to treat gout and other joint pains, as we do today. To understand how and why the salicylates in general, and aspirin in particular, came to be introduced into our present-day pharmacopea we have to return to the 19th century.

The first reliable and quantitative scientific instrument to be put to use in clinical medicine was probably the clinical thermometer. Its introduction in 1861 enabled physicians for the first time to gauge objectively the efficacy or otherwise

RODERICK J. FLOWER ● School of Pharmacy and Pharmacology, The University of Bath, Avon BA2 7AY, United Kingdom.

of their antipyretic treatments, which at this time included such heroic measures as the use of cold baths and alcohol rubs. Actually, it had been realized for some time that quinine, a principle derived from *Cinchona* bark (commonly called *"Peruvian bark"* at the time), exerted a beneficial effect on the febrile patient, but this was generally held to be a simple "tonic" action of the drug. With the advent of the thermometer, however, it became clear that quinine possessed a rather specific antipyretic action. As a consequence, this agent enjoyed wide popularity for many years as an antipyretic agent. Indeed, it would almost be true to say that it was the only really effective agent available. When the blockading tactics adopted by Napoleon made Peruvian bark a scarce and expensive commodity, a search for an alternative drug became imperative.

Fortunately, the bark of the willow, and several other trees, was known to contain antipyretic substances and had in fact been employed as such by the practitioners of folk medicine in many lands since the time of Hippocrates himself. Its use in Europe seems to date from about the middle of the 18th century, and the credit for its rediscovery is usually ascribed to a country parson, the Reverend Edmond Stone of Chipping-Norton in Oxfordshire. He had accidentally tasted willow bark and found its bitter flavor strongly reminiscent of that of the Peruvian bark. In his own words, this similarity

> immediately raised in me a suspicion of its having the properties of the Peruvian bark. As this tree (the willow) delights in a moist or wet soil, where agues chiefly abound, the general maxim that many natural maladies carry their cures along with them or that their remedies lie not far from their causes was so very apposite to this particular case that I could not help applying it; and this might be the intention of Providence I must own had some weight with me (Stone, 1763).

This technique of drug discovery is unfashionable nowadays, but the results obtained by the Reverend Stone seemed unequivocal, and he reported as much in his original letter to the Royal Society. He continued to use the powdered bark for some years "and never failed in the cure, except in a few autumnal and quartan agues. . . ."

The bitter principle of the common white willow (*Salix alba vulgaris*) was a glucoside *salicin*. This was first isolated in a pure form in 1829 by the French pharmacist Leroux, who also demonstrated the antipyretic effect of the pure compound. When hydrolyzed, salicin yields glucose and salicylic alcohol. The latter can be converted to salicylic acid *in vivo* or by chemical manipulation. Salicylic acid was soon being prepared from extracts of other plants too, including oil of wintergreen as well as a relative of the rose, a shrub called *Spiraea ulmaria*. Acetylsalicylic acid itself is said to have been synthesized first by the French chemist Charles Gerhardt in 1853. Gerhardt's drug seems to have been forgotten for almost half a century, but meanwhile progress was made on other fronts: a feasible commercial organic synthesis (from phenol and carbon dioxide) had been discovered by Kolbe and his colleague Lautermann by the mid-1870s, and the sodium salt was discovered to be an effective antipyretic drug. A few years later, in the 1880s, the drug was also in use as a treatment for gout as well as as an antiseptic compound.

The popularity of salicylic acid led to the founding (by one of Kolbe's students) of a company devoted to the commercial synthesis and distribution of salicylic acid. It was probably the success of this venture that prompted Frederick Bayer to actively search for a salicylate derivative of comparable (or, if possible, superior) efficacy to salicylic acid itself. Felix Hoffman was the name of the young chemist at Bayer who was given this task: he also had personal reasons for wanting a more acceptable salicylic acid derivative. His father, who had been taking salicylic acid for many years to treat his arthritis, had recently discovered that he could no longer take the drug without vomiting. Impelled then as much by filial affection as dedication to his job, Hoffman searched through the scientific literature, learned about the work of Gerhardt, and prepared samples of acetylsalicylic acid. After initial laboratory tests, Hoffman's father was given the drug to test and pronounced it effective, and this was later confirmed by a more impartial clinical trial.

The name "aspirin" was given to the drug by Bayer's chief pharmacologist, Hermann Dresser, who was anxious to find a name that could not possibly be confused with salicylic acid itself. At least two accounts are given for Dresser's choice of this name: some authorities maintain that the drug was named after St. Aspirinius, an early Neopolitan bishop, who was the patron saint against headaches. A more prosaic explanation is that the drug was derived from *Spiraea*. This, you will recall, was the botanical name for the genus of plants from which salicylic acid had once been prepared. According to this explanation, the prefix "a" symbolized the acetyl group, and "spirin" indicated the origin of the salicylic acid.

The new synthetic compounds rapidly displaced willow bark and encouraged widespread clinical testing. By the early years of this century, the chief therapeutic actions of aspirin (and sodium salicylate itself) were known to be antipyretic, antiinflammatory, and analgesic effects. With the passing of time, several other drugs were discovered that showed some or one of these effects. Among these were antipyrine, phenacetin, acetaminophen, phenylbutazone, and more recently, the fenamates, indomethacin, and naproxen together with their congeners. On the basis of their pharmacology, all of these drugs were termed "aspirinlike."

In concluding this section, it is worth noting that the active ingredient in willow bark, salicin, would have to undergo several metabolic transformations to become the presumed pharmacophore, salicylic acid. Although it was the first cyclooxygenase inhibitor ever employed, salicylic acid is probably the most complex agent of all, as we shall see.

2. THE NOTION OF CYCLOOXYGENASE INHIBITION

Several ideas were advanced about how the aspirinlike drugs worked. There were attempts to link their action to a biochemical effect, on oxidative phosphorylation, or other aspects of general cellular metabolism. Most of these ideas ultimately foundered, although they have helped us to understand some of the toxic side effects of the drugs (see Smith and Dawkins, 1971, for a comprehensive review).

Collier's contribution to the problem was much more important. He called aspirin an "antidefensive" drug becuse of its ability to prevent the physiological defense mechanisms of pain, fever, and inflammation from functioning normally. Together with his group, Collier demonstrated that aspirin blocked bronchoconstriction in guinea pigs caused by SRS-A and bradykinin injection and also blocked the contraction of the isolated tracheobronchial muscle induced by the same agents.

It was not clear how the bronchoconstrictor response was inhibited by aspirin. Initially, Collier suggested that "A-receptors" (i.e., those that could be blocked by aspirinlike drugs) were involved in the spasmogenic response to these agents, but he later abandoned this concept and wrote instead that the drugs acted ". . . rather by inhibiting some underlying cellular mechanism . . ." (Collier, 1969).

Vane tells us that the idea that the aspirinlike drugs blocked prostaglandin synthesis came to him while writing a review of some experiments in which he and Piper had demonstrated that aspirin prevented the release of "RCS" from guinea pig and dog lung. At the time, "RCS" was thought by Vane to be an intermediate in the generation of prostaglandin biosynthesis, and since agents that released "RCS" presumably did so by causing its synthesis, then "a logical corollary was that aspirin might well be blocking the synthesis of prostaglandins" (Vane, 1972).

What followed is now well known. In 1971, three papers appeared in *Nature* demonstrating that several aspirinlike drugs blocked prostaglandin biosynthesis in a cell-free system (Vane, 1971), in an isolated perfused organ (Ferreira *et al.*, 1971), and in human platelets following oral administration (Smith and Willis, 1971). The importance of this discovery was heightened because of the growing realization that the prostaglandins were involved in the pathogenesis of inflammation and fever and pain.

3. THE DEVELOPMENT OF THE IDEA

What followed was an intense flurry of interest in the biochemical action of the aspirinlike drugs. In the original papers only aspirin, indomethacin, and sodium salicylate had been tested as inhibitors, but it was soon demonstrated that nearly all the commonly used aspirinlike drugs were inhibitors of the cyclooxygenase enzyme (known in those days as prostaglandin synthetase) and, most important, that this inhibition could be achieved by therapeutic concentrations of the drugs (Flower *et al.*, 1972).

There seemed also to be a general correlation between the antiinflammatory activity of the drugs and their anticyclooxygenase activity. Indeed, several quite striking correspondences were observed, particularly between some pairs of enantiomers. The observations of Ham and his colleagues (1972) and of Tomlinson and co-workers (1972) are especially noteworthy. The latter group observed that there was a very impressive correlation between the antienzyme and antiinflammatory actions in the case of naproxen and its enantiomer. Naproxen itself was 150 times more potent than aspirin against bovine seminal vesicle cyclooxygenase and 200 times more potent against adjuvant-induced arthritis in rats. The enantiomer of naproxen was much less potent against the cyclooxygenase (only twice as potent as aspirin) and had almost no activity in the arthritis test. Takeguchi and Sih (1972) also reported similar findings with another series of enantiomeric pairs and indeed took the matter even further and used the anticyclooxygenase test to screen for inhibitors, which were subsequently found to have antiinflammatory properties.

This ability of the cyclooxygenase to distinguish between *dextro-* and *levo-*rotatory isomers is lacking in all other *in vitro* tests of antiinflammatory drugs and reinforced the idea that it could be used to pick out candidate compounds for *in vivo* antiinflammatory compounds. Indeed, in an age where many concerned scientists are trying to find ways of reducing the numbers of animals used for experimental research work, this little enzyme test is of great significance. It is cheap to set up, easy to run, and can not only distinguish between isomers of active drugs but also is unaffected by most compounds that are not aspirinlike antiinflammatory compounds.

Of course, not all enzyme inhibitors discovered this way are antiinflammatory: the test cannot predict whether or not the drug will reach its target when given orally, and neither can it reveal drugs that must first be metabolized to active compounds. A very interesting case in point is the antiinflammatory sulindac, which must be oxidized to its sulfone derivative and then reduced to its sulfide derivative *in vivo* before it becomes active (Van Arman *et al.*, 1976). The biologically active (sulfide) form is more than 500 times more active than sulindac itself as an inhibitor of the cyclooxygenase, again confirming the correlation between the enzyme test and therapeutic activity.

It is extremely difficult to generalize about the nature of the inhibition of the cyclooxygenase by the aspirinlike drugs. Indeed, with one or two exceptions, their mechanism of inhibition is unclear. Ku and Wasvary (1973), Smith and Lands (1971), and Lands *et al.* (1973) made valuable observations on the nature of the inhibition and concluded that the majority of the common aspirinlike drugs exhibited complex "competitive-reversible" kinetics. What can be said is that most of these drugs prevent the initial attack on the substrate by the enzyme.

One drug whose mechanism is reasonably well established is, appropriately, aspirin itself. We now know from the elegant experiments of Roth and Siok (1978) that aspirin acetylates a serine at the active site of the cyclooxygenase.

Given the wide disparity in chemical structure, it would be a little surprising if all these drugs had an identical mode of inhibitory action.

4. THE ANOMALIES

The notion that all aspirinlike antiinflammatories are cyclooxygenase inhibitors and that this is how they exert their clinical effects has been widely accepted, but there are certain anomalies that have not yet been resolved.

Ironically, the chief problem concerns the activity of the archetypal drug, salicylic acid. Even in the early experiments, it was evident that the *in vitro* activity of this drug was extremely low compared with its close relative, aspirin. This was worrying because *in vivo* the two drugs appeared almost equiactive, not only in their therapeutic indications but in their ability to block prostaglandin formation in, for example, inflammatory exudates (Willis *et al.*, 1972) and in man (Hamberg, 1972). The work of the latter author is especially interesting because it clearly demonstrates that oral salicylate reduces the "whole-body" generation of prostaglandins as estimated by the output of metabolites in the urine. In this interesting study, there was a detectable latency in the anticyclooxygenase action of salicylic acid when compared to aspirin or indomethacin.

How could these differences be reconciled? Various authors have suggested that salicylic acid has to be metabolized to an active inhibitor (Willis *et al.*, 1972; Vane, 1972; Blackwell *et al.*, 1975). The metabolism of salicylate has been fairly well investigated, and several metabolites have been tested as putative inhibitors of the cyclooxygenase. Although it is true that certain dihydroxy acid metabolites of salicylic acid were considerably (about 34 times) more potent cyclooxygenase inhibitors than the parent compound (Flower, 1974; Blackwell *et al.*, 1975), because of the low conversion to these compounds *in vivo,* it seems unlikely that the formation of these substances could account for the full observed activity of salicylic acid. It is quite likely that some hitherto undiscovered metabolite, maybe even aspirin itself, is responsible for the anticyclooxygenase action of salicylic acid. It is indeed an irony that the first antiinflammatory used by man, and the simplest drug from the chemical viewpoint, should have such a complex mode of action.

Another problem of a similar nature arose with the drug acetaminophen (paracetamol). This drug is a potent antipyretic and analgesic agent. The drug did not seem active on microsomes from spleen. This seemed to fit with the latter activity but not the former. Flower and Vane (1972) published data suggesting that acetaminophen was more active on brain cyclooxygenase than on that of peripheral tissues and that this explained the apparent differences in biological activity. At the time this provoked a bitter controversy. Among the subsequent literature, about half the papers on the subject have come down in favor of this hypothesis and about half against. The latest contribution to this literature, from Toman *et al.* (1983), clearly shows that acetaminophen when given orally does produce a selective inhibition of brain cyclooxygenase, but the mechanism by which it produces this effect (e.g., selective biodistribution, metabolism within the brain to an active drug, difference in tissue cofactors, greater sensitivity of the enzyme) was not settled.

5. THE LEGACY

Seminal ideas such as the one under discussion have profound effects on the future course of science. In the case of the cyclooxygenase inhibitors, many interesting leads have been uncovered.

Some of these spin-offs have been clinical. The realization that aspirinlike drugs prevent prostaglandin biosynthesis has suggested new clinical uses (Bartter's syndrome, closure of patent ductus) for these drugs as well as explaining some other clinical findings (incidence of prolonged labor, abnormal bleeding time) that commonly follow aspirin ingestion.

Perhaps the most important legacy of this discovery has been what it has told us about the nature of inflammation (and other diseases) and about how we can improve on the aspirinlike drugs. The discovery of the lipoxygenase pathways and more recently the leukotrienes has led to a change in emphasis in our ideas about what we want from an antiinflammatory drug. The realization that although prostaglandins contribute to the signs and symptoms of inflammation, they do not exert any control over the leukocyte traffic that is so crucial to the development of inflammation, and the subsequent discovery that certain of the leukotrienes do have chemotactic properties, have led us to the belief that it would be advantageous to inhibit the leukotriene-forming lipoxygenases as well as the cyclooxygenase.

Such drugs are already available (BW755C and phenidone; see Higgs *et al.*, 1979; Blackwell and Flower, 1978). They were discovered in screening of potential antiinflammatory compounds against a cyclooxygenase and lipoxygenase preparation derived from rat lung, an experimental approach directly suggested by the idea of testing for antiinflammatory activity by using the prostaglandin-forming enzymes *in vitro*. These drugs show a pronounced inhibition of leukocyte migration as well as of prostaglandin biosynthesis and have proved of immense value as experimental tools.

Finally, in this brief survey I shall mention the steroids. When the anticyclooxygenase action of aspirin was first discovered, we were naturally interested to discover what effect, if any, other types of antiinflammatory drugs had on prostaglandin production. The glucocorticoids are the most potent antiinflammatory drugs we possess, and to our surprise, they had no effect on the cell-free cyclooxygenase at all. We concluded in those days that these drugs had a totally different mode of antiinflammatory action. We had to revise our ideas though, because it soon became evident that these drugs did reduce prostaglandin synthesis in intact cells. We now know that this is because of the induction of synthesis and release of an antiphospholipase protein (macrocortin, lipomodulin, renocortin) that, by inhibiting the phospholipase A_2 activity, prevents the release of fatty acid substrates from the cell (Blackwell *et al.*, 1980; Cloix *et al.*, 1983; Hirata *et al.*, 1980). Naturally, not only the prostaglandins are reduced but all eicosanoids (such as the leukotrienes) as well as some other phospholipase-derived products such as lyso-PAF.

The recognition that we may be able to understand the antiinflammatory effects of the steroids in terms of an antienzyme action is only in its infancy but promises to open a new chapter in our understanding of how antiinflammatory drugs work and is the most exciting sequel of this whole chapter of experimental pharmacology.

REFERENCES

Blackwell, G. J., and Flower, R. J., 1978, 1-Phenyl-3-pyrazolidone: An inhibitor of cyclo-oxygenase and lipoxygenase pathways in lung and platelets, *Prostaglandins* **16**:417–425.

Blackwell, G. J., Flower, R. J., and Vane, J. R., 1975, Some characteristics of the prostaglandin synthesizing system in rabbit kidney microsomes, *Biochem. Biophys. Acta* **398**:178–190.

Blackwell, G. J., Carnuccio, R., Di Rosa, M., Flower, R. J., Parente, L., and Persico, P., 1980, Macrocortin: A polypeptide causing the anti-phospholipase effect of glucocorticoids, *Nature* **287**:147–149.

Cloix, J. F., Colard, O., Rothhut, B., and Russo-Marie, F., 1983, Characterization and partial purification of 'renocortins': Two polypeptides formed in renal cells causing the anti-phospholipase-like action of glucocorticoids, *Br. Pharmacol.* **79**:313–321.

Collier, H. O. J., 1969, A pharmacological analysis of aspirin, *Adv. Pharmacol. Chemother.* **7**:333–405.

Ferreira, S. H., Moncada, S., and Vane, J. R., 1971, Indomethacin and aspirin abolish prostaglandin release from the spleen, *Nature* **231**:237–239.

Flower, R. J., 1974, Drugs which inhibit prostaglandin synthesis, *Pharmacol. Rev.* **26**:33–67.

Flower, R. J., and Vane, J. R., 1972, Inhibition of prostaglandin synthetase in brain explains the anti-pyretic activity of paracetamol (4-acetamidophenol), *Nature* **240**:410–411.

Flower, R. J., Gryglewski, R., Herbaczynska-Cedro, K., and Vane, J. R., 1972, The effects of anti-inflammatory drugs on prostaglandin biosynthesis, *Nature* **238**:104–106.

Ham, E. A., Cirillo, K. J., Zanetti, M., Shen, T. Y., and Kuehl, F. A., 1972, Studies on the mode of action of non-steroidal, anti-inflammatory agents, in: *Prostaglandins in Cell Biology* (P. W. Ramwell and B. B. Phariss, eds.), Plenum Press, New York, pp. 345–352.

Hamberg, M., 1972, Inhibition of prostaglandin synthesis in man, *Biochem. Biophys. Res. Commun.* **49**:720–726.

Higgs, G. A., Flower, R. J., and Vane, J. R., 1979, A new approach to antiinflammatory drugs, *Biochem. Pharmacol.* **28**:1959–1961.

Hirata, F., Schiffmann, W., Venkatasubramanian, K., Salomon, D., and Axelrod, J., 1980, A phospholipase A_2 inhibitory protein in rabbit neutrophils induced by glucocorticoids, *Proc. Natl. Acad. Sci. U.S.A.* **77**:2533–2536.

Ku, E. C., and Wasvary, J. M., 1973, Inhibition of prostaglandin synthetase by Su-21524, *Fed. Proc.* **32**:3302.

Lands, W. E. M., LeTellier, P. R., Rome, L. H., and Vanderhoek, J. Y., 1973, Inhibition of prostaglandin biosynthesis, in: *Advances in the Biosciences*, Vol. 9 (S. Bergström and S. Bernhard, eds.), Pergamon Press, Vieweg, Braunschweig, pp. 15–28.

Rodnan, G. P., and Benedek, T. G., 1970, The early history of antirheumatic drugs, *Arthritis Rheum.* **13**:145–165.

Roth, G. R., and Siok, C. J., 1978, Acetylation of the NH_2-terminal serine of the prostaglandin synthetase by aspirin, *J. Biol. Chem.* **253**:3782–3784.

Smith, J. B., and Willis, A. L., 1971, Aspirin selectively inhibits prostaglandin production in human platelets, *Nature* **231**:235–237.

Smith, M. J. H., and Dawkins, P. D., 1971, Salicylate and enzymes, *J. Pharm. Pharmacol.* **23**:729–744.

Smith, W. L., and Lands, W. E. M., 1971, Stimulation and blockade of prostaglandin biosynthesis, *J. Biol. Chem.* **21**:6700–6702.

Stone, E., 1763, An account of the success of the bark of the willow in the cure of agues, *Philosoph. Trans.* **53**:195–200.

Takeguchi, C., and Sih, C. J., 1972, A rapid spectrophotometric assay for prostaglandin synthetase: Application to the study of non-steroidal, antiinflammatory agents, *Prostaglandins* **2**:169–184.

Tolman, E. L., Fuller, B. L., Marinan, B. A., Capetola, R. J., Levinson, S. L., and Rosenthale, M. E., 1983, Tissue selectivity and variability of effects of acetaminophen on arachidonic acid metabolism, *Prostaglandins Leukotrienes Med.* **12**:347–356.

Tomlinson, R. V., Ringold, H. J., Qureshi, M. C., and Forchielli, E., 1972, Relationship between inhibition of prostaglandin synthesis and drug efficacy: Support for the current theory on mode of action of aspirin-like drugs, *Biochem. Biophys. Res. Commun.* **46**:552–559.

Van Arman, C. G., Risley, E. A., Nuss, G. W., Hucker, H. B., and Duggan, D. E., 1976, Pharmacology of sulindac, in: *Clinoril in the Treatment of Rheumatic Disorders: A New Nonsteroidal Anti-*

inflammatory/Analgesic Agent. Proceedings of a Symposium, VIII European Rheumatology Congress (E. C. Huskisson and P. Franchimont, eds.), Raven Press, New York, pp. 9–15.

Vane, J. R., 1971, Inhibition of prostaglandin synthesis as a mechanism of action for aspirin-like drugs, *Nature* **231**:232–235.

Vane, J. R., 1972, Prostaglandins and the aspirin-like drugs, *Hosp. Pract.* **March 1972**:61–71.

Willis, A. L., Davison, P., Ramwell, P. W., Brocklehurst, W. E., and Smith, B., 1972, Release and actions of prostaglandins in inflammation and fever: Inhibition by anti-inflammatory and anti-pyrectic drugs, in: *Prostaglandins in Cellular Biology* (P. W. Ramwell and B. B. Phariss, eds.), Plenum Press, New York, pp. 227–259.

57

Comparative Biochemistry of Lipoxygenase Inhibitors

ROBERT W. EGAN and PAUL H. GALE

1. INTRODUCTION

This chapter describes some studies on the mechanism of 5-lipoxygenase regulation and inhibition and then presents an overview of lipoxygenase inhibitors that are active against intact-cell, broken-cell, and purifed enzymes from a variety of sources. Although a large number of lipoxygenase inhibitors have been reported, many of these are rather nonselective. Some of the lack of selectivity may result from a common mechanism of lipoxygenase inhibition by antioxidants and redox regulation. It is, therefore, not sufficient to define specificity against broken-cell enzymes. A truly diagnostic agent must be specific in cellular systems and *in vivo* to be of utility in defining the role of lipoxygenases in pathophysiology.

Listed in Table I are some of the more common intact cells in which 5-lipoxygenase products have been detected and inhibitors have been studied. Polymorphonuclear leukocytes are isolated from peripheral blood (Palmer and Salmon, 1983) or induced into the peritoneal cavity by intraperitoneal injection of a particulate irritant such as sodium caseinate (Ham *et al.*, 1983). On stimulation, these cells release mono-HETES and di-HETES. It was in this cell that Borgeat and Samuelsson (1979) first identified LTB_4 and established its mechanism of biosynthesis. On the other hand, macrophages synthesize primarily sulfidopeptide leukotrienes (Rouzer *et al.*, 1980). An exception is the human alveolar macrophage, which releases primarily LTB_4 (Fels *et al.*, 1982). RBL-1 cells, used by Jakschik *et al.* (1977) in their studies that identified SRS-A as an arachidonic acid metabolite, can be carried either in ascites culture, in suspension culture, or as an adherent

---

ROBERT W. EGAN and PAUL H. GALE ● Merck Institute for Therapeutic Research, Rahway, New Jersey 07065.

TABLE I. Sources of 5-Lipoxygenase: Intact-Cell Systems

Cell type	Source
Polymorphonuclear leukocytes	Human peripheral blood
	Guinea pig peritoneal
	Rat peritoneal
Macrophage–monocyte	Human alveolar
	Rabbit alveolar
	Mouse pulmonary
	Mouse peritoneal
	Rat peritoneal
Rat basophilic leukemia	Cell culture
Mast cell–mastocytoma	Human lung
	Bone marrow culture
	Dog mastocytoma
	CXBG mouse mastocytoma
Eosinophils	Human peripheral blood
	Horse peripheral blood
Epithelial cells	Human skin
(keratinocytes)	Guinea pig skin
T lymphocyte	Human
	Mouse

layer. The CXBG mouse mastocytoma is carried in ascites culture and was the cell used by Murphy *et al.* (1979) to identify SRS-A as sulfidopeptide leukotrienes.

Leukotriene synthesis in mast cells has subsequently been studied extensively using human lung- (MacGlashan *et al.*, 1982) and bone-marrow-derived mast cells (Razin *et al.*, 1982). Less available, but certainly of great physiological importance, are several other cells that synthesize these leukotrienes such as eosinophils (Weller *et al.*, 1983; Turk *et al.*, 1982), skin cells that are primarily keratinocytes (Voorhees, 1983), and cells of the immune system such as T lymphocytes (Goetzl, 1981). Whereas the former release sulfidopeptide leukotrienes and LTB_4, the latter two produce lipoxygenase products of the di-HETE or mono-HETE type.

Broken-cell preparations are also being used to study lipoxygenases (Table II).

TABLE II. Sources of 5-Lipoxygenase: Broken-Cell Systems

Cell source	Cell fraction
Guinea pig peritoneal PMN	100,000 $\times$ g supernatant fraction, aminobutylsepharose 4B chromatographed
RBL-1 cell	10,000 $\times$ g supernatant fraction
	100,000 $\times$ g supernatant fraction
Mastocytoma P-815	10,000 $\times$ g supernatant fraction
Potato tuber	Purified to homogeneity

Although this table focuses on the 5-lipoxygenase, other cells such as platelets and neutrophils can be used for the 12- and 15-lipoxygenases, respectively. A nonparticulate fraction or an aminobutyl Sepharose 4B chromatographed fraction from sonified guinea pig peritoneal PMN has been used by Ochi *et al.* (1983) to study 5-lipoxygenase. This enzyme is stimulated by Ca^{2+} and a variety of nucleotides, especially ATP. The RBL-1 cell can be lysed by Dounce homogenization (Jakschik *et al.*, 1980) or sonication (Egan *et al.*, 1983). The mastocytoma P-815 is a cell line described by Kawamura *et al.* (1981) in which both cyclooxygenase and 5-lipoxygenase can be induced by *n*-butyrate. The 5-lipoxygenase can be isolated in the 10,000 $\times$ *g* supernatant fraction. Finally, the 5-lipoxygenase from potato tuber purified by Sekiya *et al.* (1977) has been used by Corey *et al.* (1980) to prepare 5-HPETE and most recently was used by Shimizu *et al.* (1984) to demonstrate that this enzyme possesses both lipoxygenase and dehydrase activity.

These are just some of the intact-cell, broken-cell and purified enzyme systems being used to study lipoxygenase inhibitors. For routine evaluation of inhibitors, it is advantageous to use intact cells because of the more immediate correlation with *in vivo* activity. However, to obtain more detailed mechanistic information and to avoid the possibility of metabolism by or lack of penetration into intact cells, broken-cell or purified enzyme preparations provide a more accurate reflection of the action on the enzyme.

2. METHODS

The 5-lipoxygenase is studied in the supernatant fraction obtained from the 100,000 $\times$ *g* centrifugation of sonified RBL-1 cells (Egan *et al.*, 1983). This cell line grows readily in serum and glutamine-enriched Eagle's medium with a doubling time of about 18 hr. Many other enzymes are also present in the soluble fraction. The supernatant, containing 10 mg of protein/ml, is warmed to 30° for 4 min and then incubated with 90 μM [^{14}C]arachidonic acid for an additional 4 min. After acidification, greater than 95% of the products and remaining substrate are extracted into ether in the presence of ammonium chloride and ascorbic acid. Extracts are redissolved in acetone and analyzed either by chromatography on Whatman SG-81 silica-impregnated paper using ether/hexane/acetic acid (50/50/1) as developing solvent or by straight-phase HPLC eluted with hexane/ethanol/acetic acid (95/5/0.1). In a standard reaction mixture, 25% 5-HETE and 30% di-HETEs are formed, both of which reflect 5-lipoxygenase activity. Although intact RBL-1 cells also synthesize PGD_2, no cyclooxygenase products are detected in this preparation.

3. PEROXIDASE DEHYDRASE ACTIVITY

The most prevalent peroxidase in cells is glutathione peroxidase, which reduces H_2O_2 to H_2O and alkyl peroxides to alcohols as it converts reduced glutathione

(GSH) to the disulfide. In intact cells, a specific reductase then utilizes NADPH to reduce the disulfide and create a cycle that removes H_2O_2 at the expense of NADPH. We have examined the role of this peroxidase in the RBL-1 supernatant fraction, where the NADPH required to replenish GSH is no longer present (Table III). In the normal supernatant, there is a minimal effect of added GSH. At 5–10 mM it depresses both 5-HETE, the peroxidase product, and di-HETEs, the dehydrase products, indicating that it inhibits the 5-lipoxygenase very poorly.

The supernatant fraction is then partially depleted of GSH by ammonium sulfate precipitation. At 60% ammonium sulfate, about half the protein is precipitated. Redissolving the precipitate to the original volume reduces the GSH from 450 to 33 μM. When precipitated in the presence of 1 mM dithiothreitol (DTT), this preparation retains virtually all its 5-lipoxygenase activity, representing about a twofold increase in specific activity. At 1 mM, DTT inhibits the 5-lipoxygenase completely, but when the precipitated protein is redissolved in Dulbecco's medium, the DTT is 100 μM, a concentration that does not affect 5-lipoxygenase. In the absence of added GSH, 5-HETE formation by the ammonium-sulfate-precipitated fraction is decreased 50%, but the total arachidonic acid utilization remains the same, and the di-HETE peak increases proportionately. The 5-HETE peak is reinstated to the original percentage by the addition of 5 mM GSH. These data indicate that glutathione peroxidase and the dehydrase compete for the 5-HPETE substrate without altering lipoxygenase activity.

To examine the possibility that this residual peroxidase activity is glutathione peroxidase, the supernatant fraction was precipitated three times with ammonium sulfate using the same procedure. Again, 5-HETE was depressed about 50% by the first precipitation without any loss of 5-lipoxygenase activity and an increase in di-HETE production. However, a second and third precipitation and redissolution

TABLE III. Peroxidase–Dehydrase Competition

Enzyme	Added GSH (mM)	Percent of Radioactivity			Inhibition (%)	
		Di-HETE	5-HETE	AA	5-HETE	Total
RBL-1 sup.	0	34	20	33	—	—
RBL-1 sup.	1	29	20	37	0	9
RBL-1 sup.	5	27	16	43	20	19
RBL-1 sup.	10	26	15	46	25	24
$(NH_4)_2SO_4$ ppt.	0	47	10	33	50	+6[a]
$(NH_4)_2SO_4$ ppt.	1	39	16	34	20	+2
$(NH_4)_2SO_4$ ppt.	5	28	22	38	+10	7
$(NH_4)_2SO_4$ ppt.	10	28	22	39	+10	7
RBL-1 sup.	0	38	21	29	—	—
$(NH_4)_2SO_4$ ppt. I	0	52	10	23	52	+5
$(NH_4)_2SO_4$ ppt. II	0	51	10	26	52	+3
$(NH_4)_2SO_4$ ppt. III	0	45	10	29	52	7

[a] Plus (+) designates stimulation.

caused no change in the ratio of products or 5-lipoxygenase activity despite the ultimate dilution of GSH to below 1 μM. We therefore believe that there are other peroxidase activities in the RBL-1 supernatant that give rise to 5-HETE and a labile oxidant. This oxidant could be the source of peroxidase-dependent irreversible deactivation of the 5-lipoxygenase, a phenomenon that could not be the result of glutathione peroxidase, which cycles oxidants internally and does not generate free oxidizing species (Egan *et al.*, 1976).

4. REDOX REGULATION

The extent of the lipoxygenase reaction can be altered by a variety of redox agents. For example, it was demonstrated with soybean lipoxygenase and cyclooxygenase that depletion of hydroperoxides from the incubation mixture using glutathione peroxidase and reduced glutathione depressed enzyme activity (Lands and Hanel, 1983). This phenomenon also occurs with 5-lipoxygenase, where 0.05 or more units of glutathione peroxidase in the presence of 200 μM glutathione decreases 5-HETE formation in a concentration-dependent fashion (Table IV). In accord with this concept, chemical antioxidants such as nordihydroguaiaretic acid and BW755C also inhibit 5-lipoxygenase. On the other hand, phenol, which modulates other lipoxygenases with well-documented redox regulation, has no effect on 5-lipoxygenase at up to 10 mM, indicating that there is more specificity to chemical inhibitors than would be expected based solely on their ability to depress the hydroperoxide level in the incubation mixture.

Although low concentrations of hydroperoxides are required to initiate lipoxygenases, higher concentrations can cause peroxidase-dependent irreversible deactivation. Indeed, the 5-lipoxygenase is inhibited by preincubation with 200 μM H_2O_2 or 60 μM 15-hydroperoxy PGE_1 (15-HPE_1) even in the presence of endogenous glutathione peroxidase activity. The effect is not observed when hydroperoxide and arachidonic acid are added simultaneously, characteristic of peroxidasedependent inactivation (Egan *et al.*, 1981). The kinetics of the lipoxygenase

TABLE IV. Redox Inhibition of 5-Lipoxygenase

Additive	Concentration	Inhibition (%)
GPX : GSH	0.01 unit : 200 μM	2
	0.05 unit : 200 μM	66
	0.1 unit : 200 μM	97
NDGA	1.7 μM	68
BW755C	4.6 μM	55
Phenol	1 mM	0
H_2O_2	100 μM	33
	200 μM	87
15-HPE_1	60 μM	100

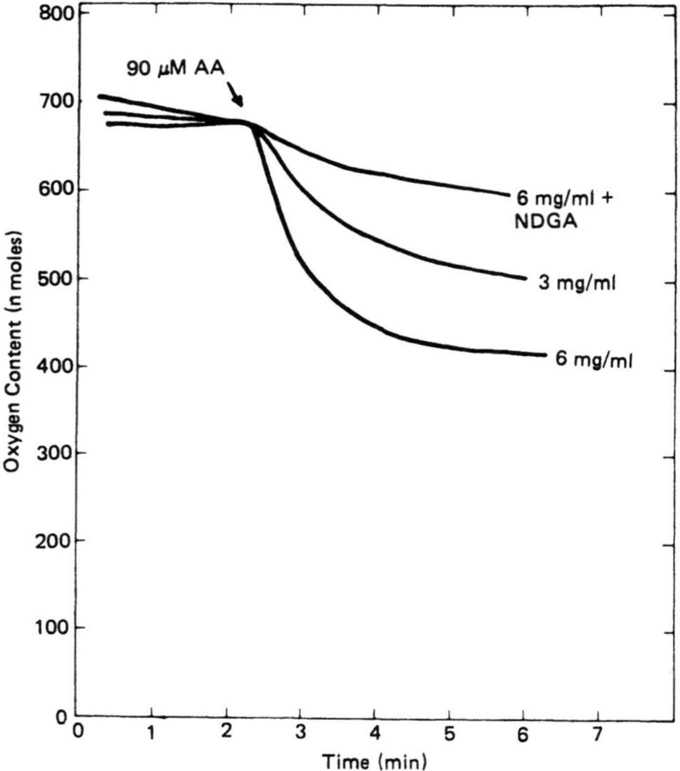

FIGURE 1. Oxygen utilization by RBL-1 lipoxygenase.

reaction are shown in Fig. 1, an oxygen monitor trace. The reaction is initiated by adding 90 µM arachidonic acid to RBL-1 supernatant at 30°C. Substrate oxidation by the 5-lipoxygenase decreases with time, and the reaction ceases irreversibly before all the substrate is consumed. Less enzyme gives less activity, and oxygen uptake is inhibited by NDGA. When the enzyme is normally assayed by a single determination after 4 min, metabolism is maximal, and the reaction product has reached a plateau.

5. SUBSTRATE INDUCED DEACTIVATION

Illustrated in Table V is the apparent inhibition from pretreatment of the 5-lipoxygenase with a substrate. These experiments were performed by incubating enzyme with nonradioactive arachidonic acid prior to the addition of [^{14}C]arachidonic acid to analyze for enzyme activity. With 4-min preincubation, increasing levels of arachidonic acid render the enzyme proportionately less active in the subsequent analytical incubation with [^{14}C]arachidonic acid. Thus, a substrate for the enzyme

TABLE V. Time-Dependent Inhibition
by Substrate

[AA] (μM)	Preincubation (min)	Inhibition (%)
10	4	41
25	4	64
50	4	83
100	4	85
50	0	6
50	0.5	42
50	1	72
50	2	79
50	4	83

will appear as an inhibitor if given sufficient time to react. The lower set of experiments illustrates the time dependence of this phenomenon with 50 μM arachidonic acid. When substrate is added without preincubation, it dilutes the [^{14}C]arachidonic acid and causes a small artifactual decrease in activity. Inhibition increases with preincubation time up to 1–2 min, where the enzyme has finished reacting with the unlabeled arachidonic acid (Fig. 1). This raises a concern about studies using substrate analogues to inhibit 5-lipoxygenase. Because of the kinetics, time-dependent inhibition could be caused by the "inhibitor" being a substrate rather than an actual inhibitor.

6. CLASSES OF INHIBITORS

Although there are specific inhibitors of cyclooxygenase such as aspirin and indomethacin, many 5-lipoxygenase inhibitors are nonselective, possibly as a result of a common mechanism of lipoxygenase inhibition. Shown in Fig. 2 are the potencies of seven symmetrical disulfides against the RBL-1 5-lipoxygenase (Egan *et al.*, 1983). Percentage inhibition was measured relative to a control reaction with no additive. The groups designated by X are listed with the appropriate titration curve. Potency was not altered by incubating the compound with enzyme for 4 min prior to arachidonic acid addition, suggesting no covalent reaction. Particularly potent are those disulfides where X is *ortho*- or *para*-methyl, *ortho*-amino, or hydrogen. Less potent disulfides have *ortho*-acetamido, *ortho*-aldehydo, or a combination of *meta*-carboxyl and *para*-nitro substituents. In fact, the latter is Ellman's reagent, which is commonly used to titrate sulfhydryl groups.

Of most interest are the effects of diphenyl disulfide (DPDS) on other arachidonic-acid-metabolizing enzymes (Table VI). Between 1 and 500 μM, DPDS has no effect on soybean 15-lipoxygenase. Furthermore, it has no effect on vesicular gland cyclooxygenase between 1 and 10 μM, and up to 500 μM, it inhibits only

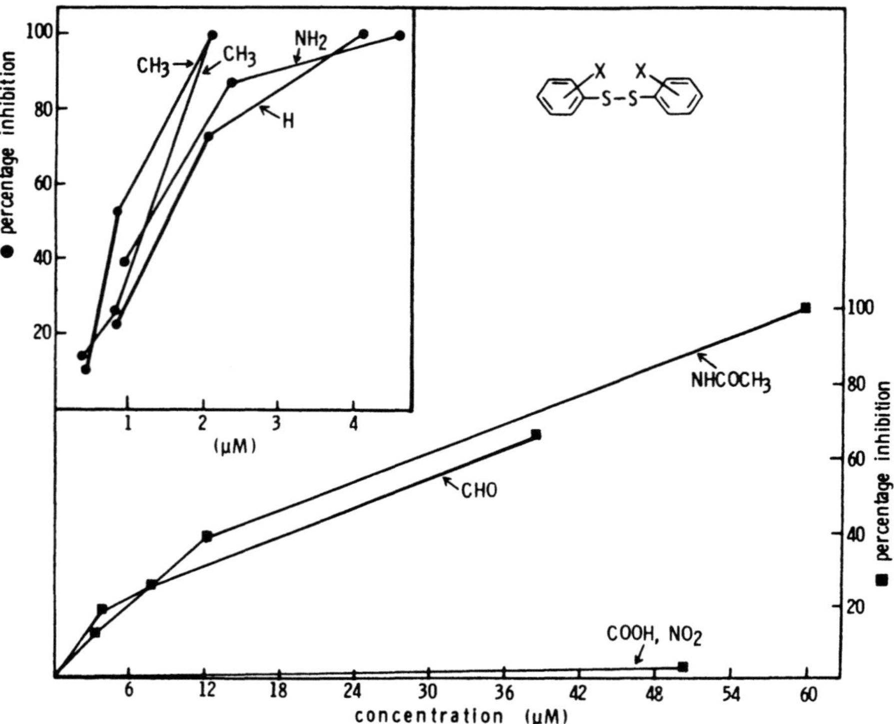

23%, a potency orders of magnitude less than that on the 5-lipoxygenase. On the other hand, diphenyl disulfide is much less effective against zymosan-induced LTC_4 synthesis in intact murine peritoneal macrophages where phospholipids are labeled with [^{3}H]arachidonic acid. In that case, the IC_{50} is 20 µM, and prostaglandin synthesis is inhibited as effectively as LTC_4 synthesis. This nonselectivity and 20-fold decrease in potency suggest that in the macrophage DPDS may not infuse to the enzyme. It may be consumed by indiscriminately reacting with plasma membrane

TABLE VI. Specificity of Diphenyl Disulfide

Enzyme	Activity
5-Lipoxygenase	IC_{50} = 1.5 µM
15-Lipoxygenase (soybean)	<10% at 500 µM
Prostaglandin cyclooxygenase	23% at 500 µM
Murine macrophage	
LTC_4	IC_{50} = 20 µM
PGE_2	IC_{50} = 20 µM

sulfhydryl groups. In considering the specificity of inhibitors, this emphasizes the importance of analyzing their actions on the various pathways within a given cell rather than only on broken-cell enzymes.

Figures 3–5 show several inhibitors of lipoxygenases divided into three categories: antioxidants, substrate analogues, and miscellaneous. It is possible that antioxidants inhibit by depressing peroxides below the level required to support lipoxygenase reaction. However, phenol and some other antioxidants do not inhibit 5-lipoxygenase, suggesting that this may be a class from which specific diagnostic agents can be developed. Nordihydroguaiaretic acid (NDGA) was used by Tappel *et al.* (1953) to inhibit soybean lipoxygenase, where it has an IC_{50} of 5 μM (Sircar *et al.*, 1983). In platelets, NDGA inhibits 12-HETE, HHT, and TxB_2 formation at around 30 μM (Salari *et al.*, 1984). However, as measured by release of products from PMN or RBL-1 cells, it inhibits 5-lipoxygenase at about 500 nM. Gossypol, used in China as a male contraceptive, has an IC_{50} of 0.2 μM against ionophore-stimulated SRS synthesis in RBL-1 cells (Levine, 1983). As with NDGA, it is a symmetrical catechol. Does the symmetry have some unexplored relevance? R-830, another antioxidant, has been reported by Moore and Swingle (1982) to inhibit seminal vesicle cyclooxygenase at 0.5 μM and guinea pig lung lipoxygenase at 20 μM and to possess antiinflammatory properties as measured by carrageenan-induced rat paw edema, adjuvant arthritis, and other classical models.

BW755C and nafazatrom are structurally related to phenidone and are readily

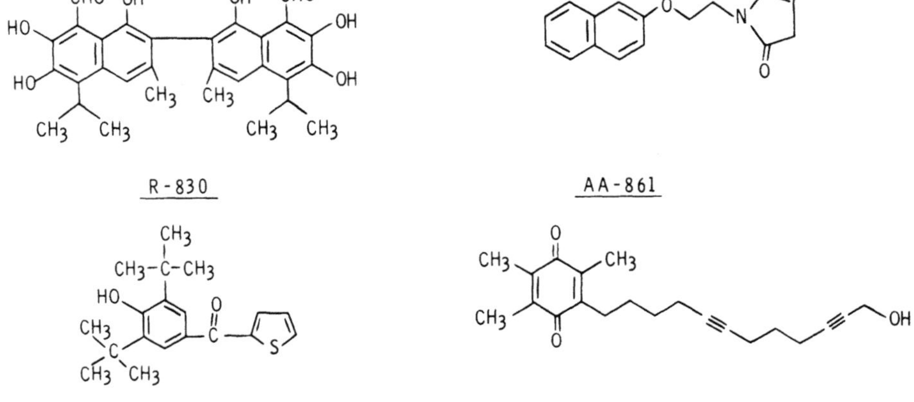

FIGURE 3. Inhibitors of lipoxygenase: antioxidants.

ROBERT W. EGAN and PAUL H. GALE

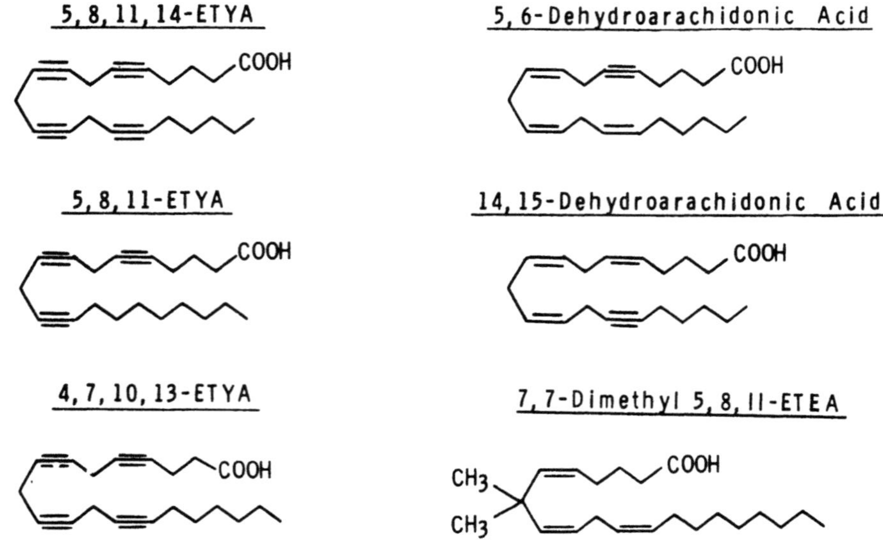

FIGURE 4. Inhibitors of lipoxygenase: substrate analogues.

FIGURE 5. Inhibitors of lipoxygenase: miscellaneous.

oxidized by enzyme systems such as prostaglandin cyclooxygenase (Marnett *et al.*, 1982). BW755C is a dual inhibitor of both cyclooxygenase and lipoxygenase pathways, with IC_{50} values against lipoxygenase in the platelet, PMN, and broken-cell systems between 5 and 30 μM (Salari *et al.*, 1984). A great deal of pharmacology has been done with BW755C in which it has been demonstrated to suppress SRS synthesis (Armour *et al.*, 1981), attenuate antigen-induced bronchospasm (Patterson *et al.*, 1981), decrease neutrophil infiltration into inflammatory exudates (Salmon *et al.*, 1983), decrease the size of myocardial infarctions (Mullane and Moncada, 1982), and prevent PMA induction of ornithine decarboxylase (Nakadate *et al.*, 1982). On the other hand, nafazatrom inhibits 5- and 12-lipoxygenase at 3 μM in B16a tumor cells without altering the cyclooxygenase in the same cell (Honn and Dunn, 1982). This agent possesses antithrombotic, thrombolytic, and antimetastatic activity. AA-861, a substituted quinone, is a competitive inhibitor of the 5-lipoxygenase from sonified guinea pig peritoneal PMN with an IC_{50} of 0.8 μM. In contrast, platelet 12-lipoxygenase and seminal vesicle cyclooxygenase are not significantly inhibited by this agent at 100 μM (Yoshimoto *et al.*, 1982).

Several substrate analogues have also been tested as inhibitors of various lipoxygenases and prostaglandin cyclooxygenase. Listed in Fig. 4 are some of the more common inhibitors of lipoxygenases that are based on the arachidonic acid structure. 5,8,11,14-Eicosatetraynoic acid (ETYA) inhibits 5-lipoxygenase and cyclooxygenase in several intact- and broken-cell systems with an IC_{50} around 5 μM (Salari *et al.*, 1984), although it appears to be somewhat more effective against the 12-lipoxygenase from platelets; 5,8,11-ETYA also inhibits lipoxygenases at 5–10 μM. Several positional isomers of these acetylenic acids have been synthesized, including 4,7,10,13-ETYA, which inhibits platelet lipoxygenase with an IC_{50} of 0.5 μM, although cyclooxygenase is not inhibited comparably until 51 μM (Wilhelm *et al.*, 1981).

Inhibition of individual lipoxygenases and cyclooxygenase by monoacetylenic arachidonic acid molecules has led to the observation that 5,6-dehydroarachidonic acid selectively inhibits broken RBL-1 cell 5-lipoxygenase (Corey and Munroe, 1982). On the basis of time dependence, this inhibition is reported as irreversible. 4,5-Dehydroarachidonic acid also inhibits 5-lipoxygenase (Corey *et al.*, 1983), and 14,15-dehydroarachidonic acid inhibits 15-lipoxygenase (Corey and Park, 1982), apparently acting as suicide substrates with K_i values of 10–20 μM. Substitution of methyl groups (7,7-dimethyl-ETEA) at the sites of hydrogen atom abstraction has led to weak inhibitors with IC_{50}s around 100 μM (Perchonock *et al.*, 1983).

Demonstrated by Fig. 5 is the breadth of compounds that are being studied as inhibitors of leukotriene biosynthesis. U-60,257, discovered by Bach *et al.* (1982), is a PGI_2 analogue with a great deal of biological data supporting its ability to depress leukotriene synthesis. Of most importance in this regard are the reports by Hansson *et al.* (1983) in antigen-challenged human asthmatic lung tissue. U-60,257 has also been described as an inhibitor of liver glutathione transferase, an antagonist of sulfidopeptide leukotrienes, and an inhibitor of lysosomal enzyme release (Bach *et al.*, 1982). Sun and McGuire (1982) have reported that U-60,257 inhibits human peripheral PMN 5-lipoxygenase at 2 μM without affecting 12-lipoxygenase, 15-lipoxygenase, or cyclooxygenase.

Although benoxaprofen inhibits cyclooxygenase more effectively than 5-lipoxygenase, it certainly does inhibit the latter at 50–100 μM (Walker and Dawson, 1979), which is below the levels attained in the circulation after prolonged treatment. Clinical studies with this agent implicate the lipoxygenase pathway in psoriasis. Limited clinical trials by Allen and Littlewood (1983) and by Kragballe and Herlin (1983) indicate a dramatic decline in the symptomology of psoriatic patients treated with benoxaprofen. Furthermore, antipsoriatic drugs such as anthralin and 6-chloro-2,3-dihydroxy-1,4-naphthoquinone inhibit 12-lipoxygenase (Bedord *et al.*, 1983).

5,6-Methano-LTA$_4$ has been reported by Koshishara *et al.* (1982) to inhibit 5-lipoxygenase in mastocytoma P-815 cells with an IC$_{50}$ of 44 μM and to be at least an order of magnitude less potent against the cyclooxygenase in the same cell. The prototype SRS-A antagonist, FPL 55712, has also been described by Casey *et al.* (1983) as an inhibitor of 5-HETE and di-HETE formation in intact RBL-1 cells with an IC$_{50}$ of 20 μM. On the other hand, it did not inhibit PGD$_2$ formation, even at 100 μM. 15-Hydroxyeicosatetraenoic acid (15-HETE) inhibits 5-lipoxygenase in rabbit PMN at 6 μM (Vanderhoek *et al.*, 1980), and verapamil was reported by Levine to suppress SRS synthesis at 1.8 μM in RBL-1 cells (Levine, 1983).

7. SUMMARY

In summary, there are a large number of intact-cell, broken-cell, and purified enzyme systems being used to detect lipoxygenase inhibitors. From these studies, many inhibitors have been reported, several of which are antioxidants. However, since most preparations are not purified, it is important to consider the involvement of other contaminating enzymes when determining the mechanisms of these inhibitors. Lipoxygenase activity is extremely sensitive to the redox state of the system, including both peroxide activation and peroxidase deactivation. Knowledge of the disposition of oxidants within the cell is therefore an essential part of understanding the mechanism of lipoxygenase inhibition and the specificity of many of these compounds.

REFERENCES

Allen, B. R., and Littlewood, S. M., 1983, The aetiology of psoriasis: Clues provided by benoxaprofen, *Br. J. Dermatol.* **109**:126–129.

Armour, C. L., Hughes, J. M., Seale, J. P., and Temple, D. M., 1981, Modification of mediator release from human lung by lipoxygenase inhibitors, *Clin. Exp. Pharmacol. Physiol.* **8**:654–655.

Bach, M. K., Brashler, J. R., Smith, H. W., Fitzpatrick, F. A., Sun, F. F., and McGuire, J. C., 1982, 6,9-Deepoxy-6,9-(phenylimino)-$\delta^{6,8}$-prostaglandin I$_1$ (U-60,257), a new inhibitor of leukotriene C and D synthesis: *In vitro* studies, *Prostaglandins* **23**:759–771.

Bedord, C. J., Young, J. M., and Wagner, B. M., 1983, Anthralin inhibition of mouse epidermal arachidonic acid lipoxygenase *in vitro*, *J. Invest. Dermatol.* **81**:566–571.

Boot, J. R., Sweatmen, W. J. F., Cox, B. A., Stone, K., and Dawson, W., 1982, The antiallergic activity of benoxaprofen [2-(4-chlorophenyl)-α-methyl-5-benoxazole acetic acid]—A lipoxygenase inhibitor, *Int. Arch. Allergy Appl. Immunol.* **67**:340–343.

Borgeat, P., and Samuelsson, B., 1979, Transformation of arachidonic acid by rabbit polymorphonuclear leukocytes, *J. Biol. Chem.* **254**:2643–2646.

Casey, F. B., Appleby, B. J., and Buck, D. C., 1983, Selective inhibition of the lipoxygenase metabolic pathway of arachidonic acid by the SRS-A antagonist, FPL 55712, *Prostaglandins* **25**:1–11.

Corey, E. J., and Munroe, J. E., 1982, Irreversible inhibition of prostaglandin and leukotriene biosynthesis from arachidonic acid by 11,12-dehydro- and 5,6-dehydroarachidonic acids, respectively, *J. Am. Chem. Soc.* **104**:1752–1754.

Corey, E. J., and Park, H., 1982, Irreversible inhibition of the enzymatic oxidation of arachidonic acid to 15-(hydroperoxy)-5,8,11(Z),13(E)-eicosatetraenoic acid (15-HPETE) by 14,15-dehydroarachidonic acid, *J. Am. Chem. Soc.* **104**:1750–1752.

Corey, E. J., Albright, J. O., Barton, A. E., and Hashimoto, S., 1980, Chemical and enzymatic synthesis of 5-HPETE, a key biological precursor of slow-reacting substance of anaphylaxis (SRS), and 5-HETE, *J. Am. Chem. Soc.* **102**:1435–1436.

Corey, E. J., Kantner, S. S., and Lansbury, P. T., Jr., 1983, Irreversible inhibition of the leukotriene pathway by 4,5-dehydroarachidonic acid, *Tetrahedron Lett.* **24**:265–268.

Egan, R. W., Paxton, J., and Kuehl, F. A., Jr., 1976, Mechanism of irreversible self-deactivation of prostaglandin synthetase, *J. Biol. Chem.* **251**:7329–7335.

Egan, R. W., Gale, P. H., Baptista, E. M., Kennicott, K. L., VandenHeuvel, W. J. A., Walker, R. W., Fagerness, P. F., and Kuehl, F. A., Jr., 1981, Oxidation reactions by prostaglandin cyclooxygenase–hydroperoxidase, *J. Biol. Chem.* **256**:7352–7361.

Egan, R. W., Tischler, A. N., Baptista, E. M., Ham, E. A., Soderman, D. D., and Gale, P. H., 1983, Specific inhibition and oxidative regulation of 5-liopoxygenase, *Adv. Prostaglandin Thromboxane Leukotriene Res.* **11**:151–157.

Fels, A. O., Pawlowski, N. A., Cramer, E. B., King, T. K., Cohn, Z. A., and Scott, W. A., 1982, Human alveolar macrophages produce leukotriene B_4, *Proc. Natl. Acad. Sci. U.S.A.* **79**:7866–7870.

Goetzl, E. J., 1981, Selective feed-back inhibition of the 5-lipoxygenation of arachidonic acid in human T lymphocytes, *Biochem. Biophys. Res. Commun.* **101**:344–350.

Ham, E. A., Soderman, D. D., Zanetti, M. E., Dougherty, H. W., McCauley, E., and Kuehl, F. A., Jr., 1983, Inhibition by prostaglandins of leukotriene B_4 release from activated neutrophils, *Proc. Natl. Acad. Sci. USA.* **80**:4349–4353.

Hansson, G., Björch, T., Dahlén, S.-E., Hedqvist, P., Granström, E., and Dahlén, B., 1983, Specific allergen induces contraction of bronchi and formation of leukotriene C_4, D_4 and E_4 in human asthmatic lung, *Adv. Prostaglandin Thromboxane Leukotriene Res.* **12**:153–159.

Honn, K. V., and Dunn, J. R., 1982, Nafazatrom (Bay g 6575) inhibition of tumor cell lipoxygenase activity and cellular proliferation, *FEBS Lett.* **139**:65–68.

Jakschik, B. A., Falkenhein, S., and Parker, C. W., 1977, Precursor role of arachidonic acid in release of slow reacting substance from rat basophilic leukemia cells, *Proc. Natl. Acad. Sci. U.S.A.* **74**:4577–4581.

Jakschik, B. A., Sun, F. F., Lee, L., and Steinhoff, M. M., 1980, Calcium stimulation of a novel lipoxygenase, *Biochem. Biophys. Res. Commun.* **95**:103–110.

Kawamura, M., Koshishara, Y., Senshu, T., and Murota, S.-I., 1981, Enhancement of prostaglandin synthesizing activity by the treatment with sodium *n*-butyrate on cloned mastocytoma cells and various other tissue cell lines, *Prostaglandins* **19**:659–669.

Koshishara, Y., Murota, S., Petasis, N. A., and Nicolaou, K. C., 1982, Selective inhibition of 5-lipoxygenase by 5,6-methanoleukotriene A_4, a stable analog of leukotriene A_4, *FEBS Lett.* **143**:13–16.

Kragballe, K., and Herlin, T., 1983, Benoxaprofen improves psoriasis, *Arch. Dermatol.* **119**:548–552.

Lands, W. E. M., and Hanel, A. M., 1983, Inhibitors and activators of prostaglandin biosynthesis, in: *Prostaglandins and Related Substances* (C. Pace-Asciak and E. Granström, eds.), Elsevier, Amsterdam, pp. 203–223.

Levine, L., 1983, Inhibition of the A-23187-stimulated leukotriene and prostaglandin biosynthesis of rat basophil leukemia (RBL-1) cells by non-steroidal anti-inflammatory drugs, antioxidants, and calcium channel blockers, *Biochem. Pharmacol.* **32**:3023–3026.

MacGlashan, D. W., Jr., Schleimer, R. P., Peters, S. P., Schulman, F. J., Adams, G. K., III, Newball, H. H., and Lichtenstein, L. M., 1982, Generation of leukotrienes by purified human lung mast cells, *J. Clin. Invest.* **70**:747–751.

Marnett, L. J., Siedlik, P. H., and Fung, L. W. M., 1982, Oxidation of Phenidone and BW 755C by prostaglandin endoperoxide synthetase, *J. Biol. Chem.* **257**:6957–6964.

Moore, G. G., and Swingle, K. F., 1982, 2,6-Di-*tert*-butyl-4-(2'-thenoyl)-phenol (R-830): A novel nonsteroidal anti-inflammatory agent with antioxidant properties, *Agents Actions* **12**:674–683.

Mullane, K. M., and Moncada, S., 1982, The salvage of ischaemic myocardium by BW 755C in anaesthetised dogs, *Prostaglandins* **24**:255–266.

Murphy, R. C., Hammarström, S., and Samuelsson, B., 1979, Leukotriene C: A slow reacting substance from murine mastocytoma cells, *Proc. Natl. Acad. Sci. U.S.A.* **76**:4275–4279.

Nakadate, T., Yamamoto, S., Ishii, M., and Kato, R., 1982, Inhibition of 12-O-tetradecanoylphorbol-13-acetate induced epidermal ornithine decarboxylase activity by lipoxygenase inhibitors: Possible role of product(s) of lipoxygenase pathway, *Cacinogenesis* **3**:1411–1414.

Ochi, K., Yoshimoto, T., Yamamoto, S., Taniguchi, K., and Miyamoto, T., 1983, Arachidonate 5-lipoxygenase of guinea pig peritoneal polymorphonuclear leukocytes. Activation by adenosine 5-triphosphate, *J. Biol. Chem.* **258**:5754–5758.

Palmer, R. M. J., and Salmon, J. A., 1983, Release of leukotriene B_4 from human neutrophils and its relationship to degranulation induced by N-formyl-methionyl-leucyl-phenylalanine, serum-treated zymosan and the ionophore A23187, *Immunology* **50**:65–73.

Patterson, R., Pruzanskey, J. J., and Harris, K. E., 1981, An agent that releases basophil and mast cell histamine but blocks cyclooxygenase and lipoxygenase metabolism of arachidonic acid inhibits immunoglobin E mediated asthma in rhesus monkeys, *J. Allergy Clin. Immunol.* **67**:444–449.

Perchonock, C. D., Finkelstein, J. A., Uzinskas, I., Gleason, J. G., Sarau, H. M., and Cleslinski, L. B., 1983, Dimethyleicosatrienoic acids: Inhibitors of the 5-lipoxygenase enzyme, *Tetrahedron Lett.* **24**:2457–2460.

Razin, E., Mencia-Huerta, J. M., Lewis, R. A., Corey, E. J., and Austen, K. F., 1982, Generation of leukotriene C_4 from a subclass of mast cells differentiated *in vitro* from mouse bone marrow, *Proc. Natl. Acad. Sci. U.S.A.* **79**:4665–4667.

Rouzer, C. A., Scott, W. A., Hamill, A. L., and Cohn, Z. A., 1980, Dynamics of leukotriene C production by macrophages, *J. Exp. Med.* **152**:1236–1247.

Salari, H., Braquet, P., and Borgeat, P., 1984, Comparative effects of indomethacin, acetylenic acids, 15-HETE, nordihydroguaiaretic acid and BW 755C on the metabolism of arachidonic acid in human leukocytes and platelets, *Prostaglandins Leukotrienes Med.* **13**:53–60.

Salmon, J. A., Simmons, P. M., and Moncada, S., 1983, The effects of BW 755C and other antiinflammatory drugs on eicosanoid concentrations and leukocyte accumulation in experimentally-induced acute inflammation, *J. Pharm. Pharmacol.* **35**:808–813.

Sekiya, J., Aoshima, H., Kajiwara, T., Togo, T., and Hatanaka, A., 1977, Purification and some properties of potato tuber lipoxygenase and detection of linoleic acid radical in enzyme reaction, *Agric. Biol. Chem.* **41**:827–832.

Shimizu, T., Radmark, O., and Samuelsson, B., 1984, Enzyme with dual lipoxygenase activities catalyzes leukotriene A_4 synthesis from arachidonic acid, *Proc. Natl. Acad. Sci. U.S.A.* **81**:689–693.

Sircar, J. C., Schwender, G. F., and Johnson, E. A., 1983, Soybean lipoxygenase inhibition by nonsteroidal antiinflammatory drugs, *Prostaglandins* **25**:393–396.

Sun, F. F., and McGuire, J. C., 1983, Inhibition of human neutrophil arachidonate 5-lipoxygenase by 6,9-deepoxy-6,9-(phenylimino)-$\delta^{6,8}$-prostaglandin I, (U-60257), *Prostaglandins* **26**:211–221.

Tappel, A. L., Lundberg, W. O., and Boyer, P. D., 1953, Effect of temperature and antioxidants upon the lipoxidase-catalyzed oxidation of sodium linoleate, *Arch. Biochem. Biophys.* **42:**293–304.

Turk, J., Maas, R. L., Brash, A. R., Roberts, J. L., and Oates, J. A., 1982, Arachidonic acid 15-lipoxygenase products from human eosinophils, *J. Biol. Chem.* **257:**7068–7076.

Vanderhoek, J. Y., Bryant, R. W., and Bailey, J. M., 1980, Inhibition of leukotriene biosynthesis by the leukocyte product 15-hydroxy-5,8,11,13-eicosatetraenoic acid, *J. Biol. Chem.* **255:**10064–10066.

Voorhees, J. J., 1983, Leukotrienes and other lipoxygenase products in the pathogenesis and therapy of psoriasis and other dermatoses, *Arch. Dermatol.* **119:**541–547.

Walker, J. R., and Dawson, W., 1979, Inhibition of rabbit PMN lipoxygenase activity by benoxaprofen, *J. Pharm. Pharmacol.* **31:**778–780.

Weller, P. F., Lee, C. W., Foster, D. W., Corey, E. J., Austen, K. F., and Lewis, R. A., 1983, Generation and metabolism of 5-lipoxygenase pathway leukotrienes by human eosinophils: Predominant production of leukotriene C_4, *Proc. Natl. Acad. Sci. U.S.A.* **80:**7626–7630.

Wilhelm, T. E., Sankarappa, S. K., VanRollins, M., and Sprecher, H., 1981, Selective inhibitors of platelet lipoxygenase: 4,7,10,13-Eicosatetraynoic acid and 5,8,11,14-henicosatetraynoic acid, *Prostaglandins* **21:**323–332.

Yoshimoto, T., Yokoyama, C., Ochi, K., Yamamoto, S., Maki, Y., Ashida, Y., Terao, S., and Shiraishi, M., 1982, 2,3,5-Trimethyl-6-(12-hydroxy-5,10-dodecadinyl)-1,4-benzoquinone (AA 861), a selective inhibitor of the 5-lipoxygenase reaction and the biosynthesis of slow-reacting substance of anaphylaxis, *Biochim. Biophys. Acta* **713:**470–473.

58

Inhibitors of Arachidonic Acid Metabolism Inhibit Tumor-Promoter-Stimulated Chemiluminescence in Murine Keratinocytes

SUSAN M. FISCHER and LAUREL M. ADAMS

1. INTRODUCTION

The carcinogenic process can be experimentally divided into stages; this is advantageous in developing an understanding of the molecular changes in each stage that are essential for the development of neoplasias. One of the best-studied multistage models is the mouse skin (Slaga *et al.*, 1980a,b,c). Initiation is the result of a single application of a carcinogen, most commonly 7,12-dimethylbenz[a]anthracene, at a dose that causes no tumors. The second stage, promotion, results from the repetitive treatment with a tumor promoter such as 12-O-tetradecanoylphorbol-13-acetate (TPA). Of all the biochemical and morphological responses of the epidermis to TPA, the most important and relevant appear to be the induction of ornithine decarboxylase, the induction of dark cells, hyperplasia, and the development of an inflammatory state (Slaga *et al.*, 1980b,c,d).

Investigations into the involvement of arachidonic acid metabolism in TPA tumor promotion began with the observation that TPA induces cytotoxicity, inflammation, and vascular permeability changes (Janoff *et al.*, 1970). There is now substantial evidence that TPA induces arachidonic acid metabolism and that PGE_2

SUSAN M. FISCHER and LAUREL M. ADAMS ● The University of Texas System Cancer Center, Science Park, Research Division, Smithville, Texas 78957.

is one of the key prostaglandins induced by TPA (Bresnick *et al.*, 1979; Verma *et al.*, 1980).

Another aspect of tumor promotion affected by the prostaglandins is the TPA induction of ornithine decarboxylase (ODC) activity in mouse skin. Verma *et al.* (1977) demonstrated that not only did prior treatment of the skin with indomethacin depress the extent of induction of ODC by TPA but also the inhibition by indomethacin could be completely overridden by the addition of PGE_1 and PGE_2 but not $PGF_{2\alpha}$ (Verma *et al.*, 1977, 1980). These results strongly implicate a role for particular prostaglandins in the induction of ornithine decarboxylase, although prostaglandins alone did not cause induction.

The above-mentioned studies suggested that tumor promotion could be modified through the application of either exogenous PGs or inhibitors of PG synthesis. Our initial approach to this question was to determine the effect of topical application of the PGs either alone or with TPA on initiated mouse skin. As shown by this laboratory and others (Fischer *et al.*, 1980a,b; Furstenberger and Marks, 1982; Verma *et al.*, 1980), none of the major PGs had tumor-promoting ability when used alone. However, studies in which PGs were used in conjunction with TPA have shown that the various PGs had either inhibitory or enhancing activity depending on the particular PG used. More specifically, $PGF_{2\alpha}$ enhances TPA tumor promotion in the SENCAR mouse by up to 60% with 10-μg applications, whereas PGE_1 reproducibly inhibits tumor promotion at doses as low as 1 μg. Some initial studies on PGE_2 also indicated that the time of application of the prostaglandin with respect to TPA is critical. These findings correlate well with the complete carcinogenesis experiments of Lupulescu (1978), who found that concomitant administration of either PGE_2 or especially $PGF_{2\alpha}$ with 3-methylcholanthrene enhanced the formation of carcinomas in Swiss mice.

Although the results of the above studies suggested that the PGs are involved in promotion, this approach is complicated by the fact that TPA induces PG synthesis on its own, such that the exogenously added PG may affect any normal feedback mechanisms. For this reason, the next series of studies involved the use of inhibitors of various pathways of arachidonic acid metabolism. It was shown many years ago that the steroidal antiinflammatory agents, which have been reported to inhibit phospholipase A_2, are very potent inhibitors of tumor promotion (Schwartz *et al.*, 1977; Slaga *et al.*, 1978), as is the phospholipase inhibitor dibromoacetophenone.

Based on this information, it was hypothesized that the nonsteroidal antiinflammatory agents such as indomethacin should also inhibit promotion, particularly since it has been demonstrated (Verma *et al.*, 1977, 1980) that indomethacin suppresses TPA-induced ODC activity in the CD-1 mouse skin. However, in the SENCAR mouse, indomethacin can enhance promotion at doses of 25 to 100 μg and inhibit at higher doses (Fischer *et al.*, 1980c). In additional tumor experiments, inhibitors were employed that were effective in blocking both the lipoxygenase and cyclooxygenase pathways. These studies show that cyclooxygenase inhibitors (at low doses) such as indomethacin and flurbiprofen enhance promotion, whereas inhibitors of both lipoxygenase and cyclooxygenase such as ETYA and phenidone inhibit promotion. The most potent inhibitors are those that block arachidonic acid release from phospholipids (dibromoacetophenone, dexamethasone). Since these results suggested that the lipoxygenase pathway may be more important than the cyclooxygenase pathway, it was of interest to determine whether indomethacin could cause an elevation of the hydroperoxy fatty acids (HPETEs) in TPA-treated epidermis. This was done by using primary cultures of adult mouse epidermal cells prelabeled with [^{14}C]arachidonic acid and exposing them to TPA with and without indomethacin for appropriate time periods. Thin-layer chromatography, using HPETE references kindly provided by L. Marnett, indicated that indomethacin does cause arachidonic acid to be shunted into the lipoxygenase pathway (Fischer, 1984).

The HPETEs are also interesting in light of the recent work by Slaga *et al.*, (1981) that shows that certain peroxides can act as complete tumor promoters. In particular, he has studied benzoyl peroxide, a widely used antiacne agent, which is also a free-radical-generating compound. Other free-radical-generating peroxides such as lauryl peroxide and chloroperbenzoic acid also have promoting activity (T. J. Slaga, unpublished data). The involvement of free radicals in promotion has also been suggested by somewhat more indirect evidence from studies using various antioxidants as inhibitors of tumor promotion: butylated hydroxyanisole (BHA) (Slaga *et al.*, 1980c), vitamins E and C (Shamberger, 1972), dimethyl sulfoxide (DMSO) (Slaga *et al.*, 1980c), and the superoxide dismutase (SOD) mimetic Cu(II)-(3,4-diisopropylsalicylic acid)$_2$ (CuDIPS) (Kensler *et al.*, 1983) have all been shown to inhibit tumor promotion. In addition, CuDIPS and BHA also inhibit TPA-induced ODC activity (Kensler and Trush, 1983; Kozumbo *et al.*, 1983).

Several *in vitro* studies have been particularly useful in investigating tumor promoter generated free radicals. Goldstein *et al.*, (1981) demonstrated in polymorphonuclear leukocytes (PMNs) that treatment with TPA results in superoxide anion production, as measured by cytochrome *c* reduction. Antipromoters such as dexamethasone, retinoids, and protease inhibitors counteract this effect (Witz *et al.*, 1980). Also using PMNs, Kensler and Trush (1981) measured the TPA stimulation of oxygen free radicals by a chemiluminescence assay and showed that it was inhibitable by SOD and retinoids.

2. CHEMILUMINESCENCE

Using PMNs, Trush *et al.* (1978) have identified three major sources of oxygen radicals in these cells: (1) stimulation of NADPH-dependent oxidative metabolism, (2) degranulation and release of myeloperoxidase, a feature unique to PMNs, and (3) stimulation of lipid peroxidation and prostaglandin synthesis. This latter is of particular interest since work in our laboratory and others (Fischer *et al.*, 1982; Furstenberger and Marks, 1982; Verma *et al.*, 1980) has shown that inhibition of arachidonic acid metabolism can inhibit skin tumor promotion in mice. We have, therefore, developed a chemiluminescence assay for measuring oxygen free radical generation by TPA in the target tissue, i.e., mouse epidermal cells, and have used this assay to determine the relative contribution of arachidonic acid metabolism. The chemiluminescence assay, which is an index of both the generation of and reactions mediated by $O_2^-\cdot$ and $^1O_2$, is a modification of the assay developed by Kensler and Trush (1981). Except where otherwise specified, 10^7 cells isolated from newborn SENCAR mice by the trypsinzation procedure of Yuspa and Harris (1974) were used in each assay. The assays are performed in glass scintillation minivials containing 3 ml Dulbecco's phosphate-buffered saline (PBS) with 0.1% glucose and 1 μg/ml luminol, a chemiluminescence enhancer. An LKB-Rackbeta ambient temperature scintillation counter, set in the chemiluminescence mode, was used to monitor the reaction. After determining the background chemiluminescence (CL), solvent or test reagents were added to start the reaction, and the CL was monitored for up to 30 min at continuous 50-sec intervals. Dimethyl sulfoxide (DMSO) was used as the solvent for TPA and most of the drugs tested. Because of solubility and quenching problems with other solvents (acetone or ethanol), in

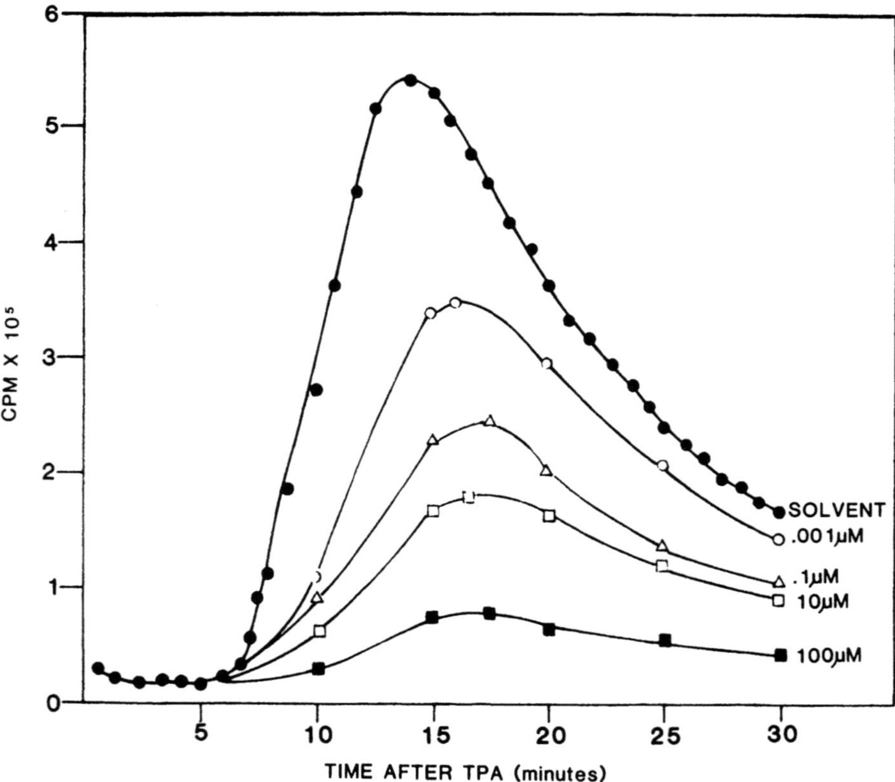

FIGURE 1. Indomethacin inhibition of TPA-induced CL. A representative time course for the stimulation of CL by TPA: TPA (100 ng/ml) was added to 10^7 mouse epidermal cells with and without indomethacin at the doses shown, and the response was monitored.

some experiments it was necessary to preincubate the cells with drugs and then wash them with PBS prior to use in the assay (Fischer and Adams, 1985).

The CL response of the epidermal cells to TPA as compared to a solvent control produces a distinct rise within 5 min and rapidly reaches a peak by 15 min, as shown in Fig. 1. The response then diminishes at a slightly slower rate, although a strong response is still seen at 30 min. The magnitude of the response correlates with the log dose of TPA used (Fig. 2); 1 ng/ml was without significant effect; increasing response was seen from 5 ng/ml up to the highest dose used, 500 ng/ml. The extent of the response is up to 100 times the solvent control; actual values vary somewhat between cell isolations. These TPA concentrations are those commonly used in *in vitro* studies with mouse epidermal cells. Unless otherwise specified, 100 ng/ml TPA was used in subsequent studies. For this dose of TPA, a cell-number dose–response curve was constructed, as shown in Fig. 2. The cell number chosen for the subsequent inhibitor studies described here was 10^7 per 3-ml assay.

A comparison was made of the abilities of a series of tumor promoters of varying tumor-promoting activities to generate CL responses at equimolar doses to

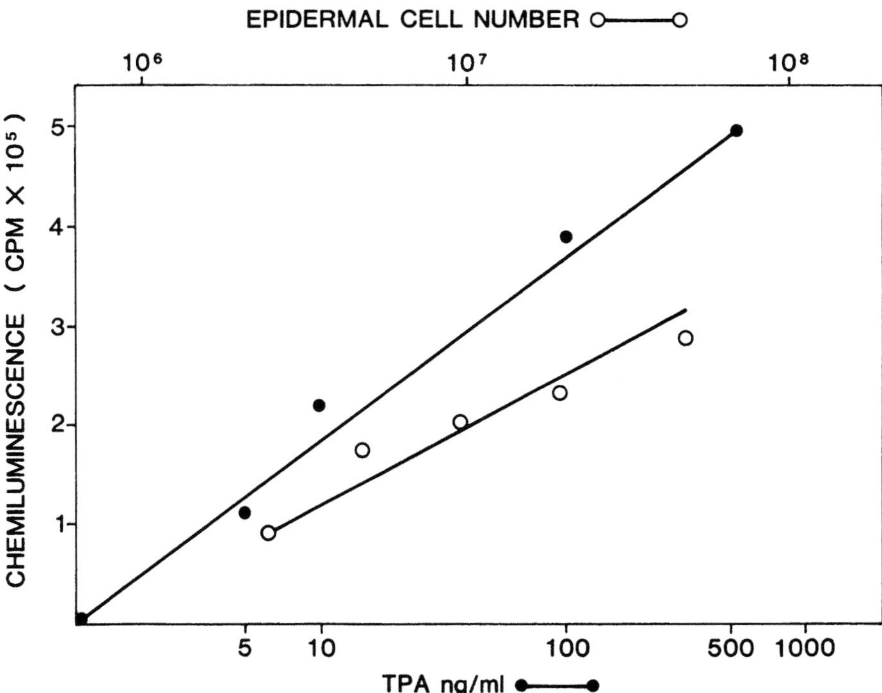

FIGURE 2. Effect of increasing TPA dose and cell number on epidermal cell chemiluminescence: (○-○) 10^7 cells were used per assay with TPA added at the doses shown. The values represent those at the peak of the response. (●-●) The TPA was held constant (100 ng/ml), and the cell number varied.

TPA. As shown in Table I, TPA is the most active of the phorbol esters used. Phorbol and phorbol-13,20-diacetate were both without effect at concentrations equimolar to and ten times greater than TPA. Phorbol-12,13-dibenzoate gave no response at an equimolar dose, a slight rise at ten times, and a more significant response at 100 times. The ranking of the phorbol ester series on the basis of the CL response correlates very well with their tumor-promoting activity. Mezerein, a related diterpene that is a weak complete promoter but strong second-stage promoter (Slaga *et al.*, 1980a), has at least the same capacity as TPA to stimulate CL at equimolar doses.

One of the major oxygen radical species thought to be produced on TPA activation is the superoxide anion (Trush *et al.*, 1978). Its contribution to CL may be direct or indirect, but the significance of its participation can be determined through the use of superoxide dismutase, an enzyme specific for the dismutation of this radical (Bors *et al.*, 1983). As shown in Table II, inclusion of 50 μg/ml SOD in the assay nearly negated the TPA-induced CL response. A SOD mimetic, CuDIPS, also markedly inhibited the TPA response. Neither catalase, which breaks down H_2O_2, nor mannitol, a scavenger for hydroxyl radicals, had an appreciable effect on TPA-induced CL. Retinoic acid, which has been shown to inhibit TPA tumor promotion in mouse skin (Slaga *et al.*, 1980c), was found to inhibit TPA-induced CL as well. Unlike the PMNs, in which Kensler and Trush (1981) showed that retinoic acid alone caused an immediate but transient CL burst; no response to retinoic acid alone was seen in the mouse epidermal cells. The difference between the two cell systems in this regard is not known.

As has been previously described, inhibitors for various parts of the arachidonate pathway have been used in TPA tumor promotion studies. These inhibitors, with the possible exception of indomethacin, are effective in inhibiting promotion. Indomethacin has been reported to cause either enhancement or inhibition depending on the dose used (Fischer *et al.*, 1980c) as well as the strain of mouse used (Furstenberger and Marks, 1982; Verma *et al.*, 1980). In this study, indomethacin

TABLE I. Structure–Function Studies on the Ability of Tumor Promoters and Related Compounds to Induce Chemiluminescence[a]

Compound	Dose (nM)	Peak CL (% TPA, 162 nM)	Relative promoting ability
TPA	162	100	+ + + +
Phorbol-12,13-dibenzoate	16.2	0	
	162	13	+ +
	1620	69	
Phorbol-13,20-diacetate	162	0	±
Phorbol	162	0	−
Mezerein	162	100	±
	1620	303	

[a] Murine epidermal cells (10^7) were used per 3-ml assay; the cpm at the peak of the response were used to determine percentage of the TPA peak.

TABLE II. Free-Radical Modifier Effects on TPA-Induced
Chemiluminescence[a]

Agent	Concentration		Percentage inhibition of TPA response
Superoxide dismutase	155	units/ml	85
CuDIPS[b]	10	μM	65
Catalase	57.6	units/ml	7
	576	units/ml	10
Mannitol	100	μM	11
Ethoxyquin HCl	0.01	μM	25
	0.1	μM	51
	1.0	μM	63
Retinoic acid	10	μM	13
	100	μM	22
Ro-10-9359[c]	10	μM	40
	100	μM	72

[a] Murine epidermal cells (10^7) were used per 3-ml assay, the cpm at the peak of the responses were used to determine percent inhibition. Experiments were done twice on different cell isolations.
[b] Copper(II)(3,5-diisopropylsalicylate)$_2$
[c] Retinoic acid analogue: ethyl-all *trans*-9-(4-methoxy-2,3,6-trimethylphenyl)-3,7-dimethyl-2,4,6,8-nonatetraenoate

gave a dose–response inhibition of TPA-induced CL, as shown in Fig. 1 and Table III, over a dose range of 1 to 100 μM. Phenidone, 5,8,11,14-eicosatetraynoic acid (ETYA) and nordihydroguaiaretic acid (NDGA) inhibited TPA-induced CL in a dose–response manner. Dexamethasone was without effect even when a 2-hr preincubation was used; this may be related to the use of neonatal cells in the assay. Arachidonic acid could not be used since it causes a tremendous CL response by itself in the reaction mixture (no cells) through a direct reaction with luminol. This effect was not observed for any of the other inhibitors either by themselves or with cells in the absence of TPA.

The above finding, that TPA can generate oxygen free radicals in mouse epidermal cells, raises several very interesting questions.

Since leukocyte infiltration is part of the inflammatory state that results from TPA treatment of skin (Slaga *et al.*, 1978), their ability to generate free radicals may contribute to the promotion process. However, the fact that TPA also stimulates oxygen radical production in epidermal cells suggests that it may not be necessary to include leukocytes in a theory on the mechanism(s) of tumor promotion. In epidermal cells, TPA not only stimulates radical production, as shown here, but also, as shown by Solanki *et al.* (1981), diminishes the SOD and catalase levels (detoxification enzymes for $O_2^-\cdot$ and H_2O_2, respectively), making radical damage much more likely.

Since inhibitors of arachidonic acid metabolism, which have been shown to inhibit TPA promotion, also inhibit TPA-induced free radicals that probably arise

TABLE III. Effect of Inhibitors of Arachidonate
Metabolism on TPA-Induced Chemiluminescence[a]

Drug	Concentration (mM)	Percentage inhibition of TPA response
Indomethacin	0.001	36
	0.1	56
	10	67
	100	86
Phenidone	1	58
	10	74
	100	100
NDGA[b]	0.01	0
	0.1	10
	1.0	35
	10	100
ETYA[c]	1	11
	10	16
	100	91

[a] Murine epidermal cells (10^7) were used per 3-ml assay; the cpm at the peak of
the responses were used to determine percent inhibition. Experiments were done
twice on different cell isolations.
[b] Nordihydroguaiaretic acid.
[c] 5,8,11,14-Eicosatetraynoic acid; preincubation required.

from this metabolic pathway, the possibility is raised that the contribution of the
arachidonate cascade to TPA tumor promotion is not necessarily the specific products
that are formed but the oxygenation reactions involved in their biosynthesis.

Perhaps one of the most intriguing of all questions is the identification of the
target(s) of the free radicals. An excess of oxygen radicals has been known for a
long time to have numerous effects, including such events as lipid peroxidation,
polysaccharide depolymerization, enzyme activation or inactivation, and DNA strand
breaks. At least some of these features may be relevant to tumor promotion, which
is thought to involve gene activation. Although little or no work has been done in
this regard on epidermis, several studies using other cells have contributed information. 12-O-Tetradecanoylphorbol-13-acetate (Yotti *et al.*, 1979), the HPETEs
(Fischer, 1984), and the tumor promoter and free radical generator benzoyl peroxide
(Slaga *et al.*, 1981) all inhibit metabolic cooperation as measured in the V-79
cell–cell communication assay (Yotti *et al.*, 1979). Birnboim (1983) recently reported that active oxygen species generated by PMNs in response to TPA can lead
to extensive DNA damage *in situ*. Cerutti *et al.* (1983) have demonstrated the
production in lymphocytes treated with TPA of a clastogenic factor that induces
DNA strand breaks. This is inhibited by indomethacin and ETYA as well as SOD.
Clearly, investigations of this nature are needed in epidermal cells in order to
elucidate the possible mechanisms of free radical involvement in tumor production.

REFERENCES

Birnboim, H. C., 1983, Importance of DNA strand break in tumor promotion, in: *Radioprotectors and Anticarcinogens* (O. F. Nygaard and M. G. Simic, eds.), Academic Press, New York, pp. 539–556.

Bors, W., Saran, M., and Michel, C., 1983, Assays of oxygen radicals: Methods and mechanisms, in: *Superoxide Dismutase,* Vol. II (L. W. Oberley, ed.), CRC Press, Boca Raton, FL, pp. 31–63.

Bresnick, E., Meunier, P., and Lamden, M., 1979, Epidermal prostaglandins after topical application of a tumor promoter, *Cancer Lett.* **7:**121.

Cerutti, P. A., Amstad, P., and Emerit, I., 1983, Tumor promoter phorbol-myristate-acetate induces membrane mediated chromosomal damage, in: *Radioprotectors and Anticarcinogens* (O. F. Nygaard and G. Simic, eds.), Academic Press, New York, pp. 527–538.

Fischer, S. M., 1984, The role of prostaglandins in tumor production, in: *Mechanisms of Tumor Promotion,* Vol. II, *Tumor Promotion and Skin Carcinogenesis* (T. J. Slaga, ed.), CRC Press, Boca Raton, FL, pp. 113–126.

Fischer, S. M., and Adams, L. A., 1985, Tumor promoter-induced chemiluminescence in mouse epidermal cells is inhibited by several inhibitors of arachidonic acid metabolism, *Cancer Res.* **45:**1–6.

Fischer, S. M., Gleason, G. L., Bohrman, J. S., and Slaga, T. J., 1980a, Prostaglandin modulation of phorbol ester skin tumor promotion, *Carcinogenesis* **1:**245–248.

Fischer, S. M., Gleason, G. L., Bohrman, J. S., and Slaga, T. J., 1980b, Prostaglandin enhancement of skin tumor initiation and promotion, *Adv. Prostaglandin Thromboxane Res.* **6:**517–522.

Fischer, S. M., Gleason, G. L., Mills, G. D., and Slaga, T. J., 1980c, Indomethacin enhancement of TPA tumor promotion in mice, *Cancer Lett.* **10:**343–350.

Fischer, S. M., Mills, G. D., and Slaga, T. J., 1982, Inhibition of mouse skin tumor promotion by several inhibitors of arachidonic acid metabolism, *Carcinogenesis* **3:**1243–1245.

Furstenberger, G., and Marks, F., 1982, Studies on the role of prostaglandins in the induction of cell proliferation and hyperplasia and in tumor promotion in mouse skin, in: *Carcinogenesis, Vol. 7: Carcinogenesis and Biological Effects of Tumor Promoters* (E. Hecker, N. E. Fusenig, W. Kunz, F. Marks, and H. W. Thielman, eds.), Raven Press, New York, pp. 325–330.

Goldstein, B. D., Witz, G., Amoruso, M., Stone, D. S., and Troll, W., 1981, Stimulation of human polymorphonuclear leukocyte superoxide anion radical production by tumor promoters, *Cancer Lett.* **11:**257–262.

Janoff, A., Klassen, A., and Troll, W., 1970, Local vascular changes induced by the cocarcinogen phorbol myristate acetate, *Cancer Res.* **30:**2568–2571.

Kensler, T. W., and Trush, M. A., 1981, Inhibition of phorbol-ester stimulated chemiluminescence in human polymorphonuclear leukocytes by retinoic acid and 5,6-epoxy-retinoic acid, *Cancer Res.* **41:**216–222.

Kensler, T. W., and Trush, M. A., 1983, Inhibition of oxygen radical metabolism in phorbol ester-mediated polymorphonuclear leukocytes by an antitumor promoting copper complex with superoxide dismutase mimetic activity, *Biochem. Pharmacol.* **32:**3485–3487.

Kensler, T. W., Bush, D. M., and Kozumbo, W. J., 1983, Inhibition of tumor promotion by a biomimetic superoxide dismutase, *Science* **221:**75–77.

Klein-Szanto, A. J. P., Major, S. M., and Slaga, T. J., 1980, Induction of dark keratinocytes by 12-O-tetradecanoylphorbol-13-acetate and mezerein as an indicator of tumor-promoting efficiency, *Carcinogenesis* **1:**399–406.

Kozumbo, W. J., Seed, J. L., and Kensler, T. W., 1983, Inhibition by 2(3)-*tert*-butyl-4-hydroxyanisole and other antioxidants of epidermal ornithine decarboxylase activity induced by 12-O-tetradecanoylphorbol-13-acetate, *Cancer Res.* **43:**2555–2559.

Lupulescu, A., 1978, Enhancement of carcinogenesis by prostaglandins in male albino Swiss mice, *J. Natl. Cancer Inst.* **61:**97–106.

Schwartz, J. A., Viaje, A., Slaga, T. J., Yuspa, S. H., Hennings, H., and Lichti, U., 1977, Fluocinolone acetonide: A potent inhibitor of mouse skin tumor promotion and epidermal DNA-synthesis, *Chem. Biol. Interact.* **17:**331–347.

Shamberger, R. J., 1972, Increase of peroxidation in carcinogenesis, *J. Natl. Cancer Inst.* **48**:1491–1497.

Slaga, T. J., Fischer, S. M., Viaje, A., Berry, D. L., Bracken, W. M., LeClerc, S., and Miller, D. L., 1978, Inhibition of tumor promotion by anti-inflammatory agents: An approach to the biochemical mechanism of promotion, in: *Carcinogenesis,* Vol. 2, *Mechanisms of Tumor Promotion and Cocarcinogenesis* (T. J. Slaga, A. Sivak, and R. K. Boutwell, eds.), Raven Press, New York, pp. 173–195.

Slaga, T. J., Fischer, S. M., Nelson, K., and Gleason, G. L., 1980a, Studies on the mechanism of skin tumor promotion: Evidence for several stages in promotion, *Proc. Natl. Acad. Sci. U.S.A.* **77**:3659–3663.

Slaga, T. J., Klein-Szanto, A. J. P., Fischer, S. M., Weeks, C. E., Nelson, K., and Major, S., 1980b, Studies on the mechanism of action of anti-tumor promoting agents: Their specificity in two-stage promotion, *Proc. Natl. Acad. Sci. U.S.A.* **77**:2251–2254.

Slaga, T. J., Fischer, S. M., Weeks, C. E., and Klein-Szanto, A. J. P., 1980c, Multistage chemical carcinogenesis, in: *Biochemistry of Normal and Abnormal Epidermal Differentiation* (M. Seije and I. A. Bernstein, eds.), University of Tokyo Press, Tokyo, pp. 193–218.

Slaga, T. J., Klein-Szanto, A. J. P., Triplett, L. L., and Yotti, L. P., 1981, Skin tumor-promoting activity of benzoyl peroxide, a widely used free radical-generating compound, *Science* **213**:1023–1025.

Slaga, T. J., Fischer, S. M., Weeks, C. E., Nelson, K., Mamrack, M., and Klein-Szanto, A. J. P., 1982, Specificity and mechanisms of promoter inhibitors in multistage promotion, in: *Carcinogenesis, Vol. 7: Carcinogenesis and Biological Effects of Tumor Promoters* (E. Hecker, N. E. Fusenig, W. Kunz, F. Marks, and H. W. Thielman, eds.), Raven Press, New York, pp. 19–34.

Solanki, V., Rana, R. S., and Slaga, T. J., 1981, Diminution of mouse epidermal superoxide dismutase and catalase activities by tumor promoters, *Carcinogenesis* **2**:1141–1146.

Trush, M. A., Wilson, M. E., and Van Dyke, K., 1978, The generation of chemiluminescence by phagocytic cells, *Methods Enzymol.* **57**:462–494.

Verma, A. K., Rice, H.M., and Boutwell, R. K., 1977, Prostaglandins and skin tumor promotion inhibition of tumor promoter induced ornithine decarboxylase activity in epidermis by inhibitors of prostaglandin synthesis, *Biochem. Biophys. Res. Commun.* **79**:1160–1166.

Verma, A. K., Ashendel, C. L., and Boutwell, R. K., 1980, Inhibition by prostaglandin synthesis inhibitors of the induction of epidermal ornithine decarboxylase activity, the accumulation of prostaglandins and tumor promotion caused by 12-O-tetradecanoylphorbol-13-acetate, *Cancer Res.* **40**:308–315.

Witz, G., Goldstein, B. D., Amoruso, M., Stone, D. S., and Troll, W., 1980, Retinoid inhibition of superoxide anion radical production by human polymorphonuclear leukocytes stimulated with tumor promoters, *Biochem. Biophys. Res. Commun.* **97**:883–888.

Yotti, L. P., Chang, C. C., and Trosko, J. E., 1979, Elimination of metabolic cooperation in Chinese hamster cells by a tumor promoter, *Science* **207**:1089–1091.

Yuspa, S. H., and Harris, C. C., 1974, Altered differentiation of mouse epidermal cells treated with retinyl acetate *in vitro, Exp. Cell Res.* **86**:95–105.

59

Cyclooxygenase Inhibitors

Beneficial Effects on Tumor Metastasis in Rats

WILL J. KORT, LORETTE O. M. HULSMAN,
PIETER E. ZONDERVAN, and
DICK L. WESTBROEK

1. INTRODUCTION

The treatment of cancer patients can be directed against either primary tumor growth or the occurrence of tumor metastasis. Many ways are available to achieve such a goal. The discovery that cancer cells metabolize prostaglandins gave an important impulse to new research studying the effect of inhibitors of prostaglandin synthesis on tumor growth. Like normal cells, cancer cells synthesize prostaglandins; however, quantitative and qualitative differences have been observed. In tumor-bearing individuals, the balance between prostacyclin (PGI_2) and thromboxane A2 (TxA_2) was sometimes found to be disturbed in favor of TxA_2. Such an imbalance might have important implications, particularly with respect to the capacity of tumor cells to adhere to vessel walls and thus to the formation of metastases. These facts obviously stimulated the study of the effects of nonsteroid antiinflammatory agents such as aspirin and indomethacin on tumor growth and metastasis. The results of such an experiment, carried out in our laboratory, is presented in this chapter.

WILL J. KORT, LORETTE O. M. HULSMAN, and DICK L. WESTBROEK • Laboratory for Experimental Surgery, Erasmus University, 3000 DR Rotterdam, The Netherlands.   PIETER E. ZONDERVAN • Department of Pathological Anatomy I, Erasmus University, 3000 DR Rotterdam, The Netherlands.

2. TUMOR PROSTAGLANDIN SYNTHESIS

2.1. Determination of Prostaglandin Synthesis

As many excellent reviews on tumor prostaglandin synthesis are available (Bennett, 1979; Honn *et al.*, 1983; Karmali, 1983; Smith *et al.*, 1983; Tisdale, 1983), in this chapter we confine this survey to a brief summary of the current models to study aspirin or other treatment protocols with nonsteroid antiinflammatory agents on tumor growth and spread.

Tumor prostaglandin synthesis or prostaglandin synthesis capacity can be determined *in vivo* as well as *in vitro*. *In vitro* experiments have the advantage that tumor growth conditions can be controlled very carefully; however, there are many discrepancies between the results of therapy with cyclooxygenase inhibitors *in vitro* and studies *in vivo* (Honn *et al.*, 1983).

A number of different aspects of prostaglandin synthesis *in vitro* can be established. First, the prostaglandins themselves or their stable metabolites can be determined by means of radioimmunoassays (Caldwell *et al.*, 1972) or bioassays (Bennett *et al.*, 1973). Second, the capacity of some prostaglandins to resorb bone can be determined by culturing tumor cells in a medium in which bone cells are added. (Powles *et al.*, 1976). In such a way, osteolysis can be established as a measure of prostaglandin activity. Another mechanism in which prostaglandins seem to be involved is the adherence capacity of tumor cells *in vitro* (Fantone *et al.*, 1983). The capacity of prostaglandins to inhibit this adherence is used as a measure of prostaglandin-induced antiadherence activity. Finally, tumor cells obtained either from fresh biopsy material or from tissue culture are able to aggregate platelets (Gasic *et al.*, 1976; Evans and Cowie, 1983). This capacity of tumor cells can be determined in an aggregometer in which complete blood or platelet-rich plasma has been added.

To study tumor prostaglandin synthesis *in vivo*, experiments in which a tumor is inoculated and growth or metastasis is determined are used as well. For this purpose a number of experimental tumors are available, some of which are able to metastasize from the inoculated site (Lynch *et al.*, 1978; Blitzer and Huang, 1983); however, prostaglandin synthesis and metastasis have been studied more often in an artificial model by intravenous injection of tumor cells (Honn, 1983).

In addition to the direct approach to establish tumor prostaglandin activity with the methods mentioned above, an indirect approach can be used as well. If drugs that inhibit the cyclooxygenase pathway are given to tumor cells in culture or to tumor-bearing animals, prostaglandin involvement can be shown by the changes that are produced in the growth characteristics of the tumor.

2.2. Cyclooxygenase Inhibitors in the Control of Tumor Growth and Metastasis

Tumor prostaglandin activity has been found to be the cause of clinical symptoms of hypercalcemia and diarrhea, respectively, in mammary tumor patients and in a patient with a thyroid carcinoma (Powles *et al.*, 1976; Barrowman *et al.*, 1975). Therapy with indomethacin resulted in an improvement of the condition of these patients. Therapy with cyclooxygenase inhibitors to prevent tumor spread seems to be indicated as well. Tumor procoagulant activity, almost certainly the result of TxA_2 activity, was associated with metastasis (Gasic *et al.*, 1976; Honn and Meyer, 1981). Tumor cell metastasis was enhanced in tumor-bearing mice when the PGI_2/TxA_2 balance was altered in favor of TxA_2 (Honn *et al.*, 1980). Prostacyclin, however, produced a dose-dependent decrease in pulmonary tumor foci (Honn *et al.*, 1981b).

Also, in our own experiments, beneficial effects of aspirin treatment were seen in rats in which a tumor was inoculated subcutaneously (Kort *et al.*, 1985). In this experiment, the tumor was inoculated in the flank of the animal, and therapy was started immediately after inoculation of the tumor and was continued after removal of the primary graft. Primary tumor growth and number of metastases were found to be decreased compared with controls.

In a recent experiment, the same tumor (BN 175) was inoculated in the foot of the animal in order to facilitate complete removal of the tumor, and therapy was given only after removal of the primary tumor. This experiment was done in 65 inbred Brown Norway (BN) female rats with a body weight of 100–150 g. They were randomly divided into six experimental groups. Group I received aspirin, 20 mg/kg (low dose) i.p. ($n = 10$); group II 200 mg/kg of this drug (high dose) ($n = 10$); group III theophylline, 75 mg/kg s.c. ($n = 10$); group IV low-dose aspirin combined with theophylline, 75 mg/kg ($n = 10$); group V high-dose aspirin together with theophylline, 75 mg/kg ($n = 15$), and group VI saline, 1 ml, pH 5.2, i.p. ($n = 10$), which was the volume of the vehicle given in all experimental groups and had the same pH as the aspirin solutions used.

The tumor was inoculated in pieces of 4 mm^3 subcutaneously into the dorsum of the left foot through a small skin incision, which was closed by tissue adhesive. At day 9, the tumor was removed by means of amputation of the foot. Seventeen days after inoculation of the tumor, the animals were killed by exsanguination, and tumor involvement of paraaortic lymph nodes as judged by macroscopic inspection was scored. The lungs were fixed in Bouin's fixative for 5 days, after which metastases on the surface could be counted. At autopsy, lymph node metastasis was scored positively when there was macroscopic evidence of tumor invasion. Lung metastases were scored by counting the white nodules on the surface of the left lung. An analysis for multiple comparisons according to Newman–Keuls was carried out to determine statistically significant differences. When $P < 0.05$, the difference between the means was considered to be statistically significant.

In group V, the group of animals receiving high doses of aspirin together with theophylline, six animals died shortly after amputation of the foot; all other animals survived the whole treatment period in good health. The six animals dying in group V were lost because of uncontrollable bleeding, probably caused by the strong inhibition of platelet aggregation in this group. When the experiment was terminated, local recurrence of the tumor, i.e., at the site of the amputation, was not seen.

Although negative lymph nodes were found only in groups with aspirin treat-

TABLE I. Effect of Aspirin and Theophylline on Lymph Node
Metastases (Paraaortic Lymph Nodes)

Group	Therapy	Score[a]		
		0	±	+
I	Aspirin, 20 mg/kg	3	2	5
II	Aspirin, 200 mg/kg	2	4	4
III	Theophylline, 75 mg/kg	0	4	6
IV	I + II	0	4	6
V	II + III	2	3	4
VI	Vehicle	0	5	5

[a] Symbols: 0, number of animals with negative paraaortic lymph nodes; ±, number of animals with reactive paraaortic lymph nodes; +, number of animals with positive paraaortic lymph nodes. Treatment was started 9 days after inoculation of the tumor. There were no statistically significant differences.

ment, tumor spread to regional lymph nodes was almost the same in all experimental groups (Table I). No statistically significant differences were observed. It is clear from the results given in Table II that high doses of aspirin showed the strongest inhibition of tumor spread. Treatment with low doses of aspirin, either alone or combined with theophylline, did not significantly decrease the number of metastatic foci in the lungs. Combining theophylline with high doses of aspirin gave a slightly more pronounced effect than did treatment with aspirin, 200 mg/kg, alone.

4. GENERAL DISCUSSION AND PROSPOSAL FOR FURTHER RESEARCH

It is not surprising that many investigators have substantiated the importance of treatment with nonsteroid antiinflammatory agents such as aspirin in reducing primary tumor growth (Bennett, 1980) or metastases (Honn *et al.*, 1981a). However, in this context, only little attention was paid to the amount of the drugs used. This

TABLE II. Effect of Aspirin and Theophylline Treatment on the Number of Lung
Metastases 17 Days after Inoculation of the Tumor[a]

Group	Therapy	Number of metastases
I	Aspirin, 20 mg/kg	43.9 ± 7.7
II	Aspirin, 200 mg/kg	23.1 ± 6.6
III	Theophylline, 75 mg/kg	34.6 ± 5.1
IV	I + III	39.7 ± 9.4
V	II + III	19.6 ± 6.4
VI	Vehicle	57.3 ± 7.6

[a] Treatment was started 9 days after inoculation of the tumor. Statistical significant differences: group V vs. group VI, $P < 0.05$; group II vs. group VI, $P < 0.05$.

is rather surprising, because in studies on the antiaggregatory activity of aspirin, the controversy between high and low doses of aspirin did receive much attention (Moncada and Korbut, 1978). It was pointed out that aspirin in high doses completely blocked the cyclooxygenase pathway, resulting in a blockade of PGI_2 and TxA_2 synthesis. In studies by Ellis *et al.* (1980) this was further substantiated: low doses of aspirin (in humans, 1 mg/kg) resulted in a major decrease of TxA_2 values, whereas PGI_2 synthesis capacity was almost normal.

In studies by Gasic *et al.* (1976), the importance of the ability of certain tumor extracts to aggregate platelets *in vitro* was discussed. Moreover, TxA_2 was found to enhance tumor growth and the survival of metastasizing tumor cells, whereas PGI_2 did the opposite (Honn *et al.*, 1980). In later studies, Honn and Meyer (1981) stressed the importance of the PGI_2/TxA_2 balance for tumor growth and metastasis. These findings suggest that in order to obtain adequate control of tumor growth and metastasis, PGI_2 synthesis should be left unaltered (or, even better, enhanced) and TxA_2 synthesis reduced as much as possible. With regard to these considerations on the dosage of aspirin, treatment with low doses of aspirin alone or possibly in combination with phosphodiesterase inhibitors (to potentiate endogenous prostacyclin) could achieve such a goal. Also, drugs exclusively blocking the enzyme thromboxane synthetase, such as dazoxyben, should be beneficial as well.

However, high levels of PGI_2 in the plasma were not always found to be associated with little tumor spread. For instance, in patients with a carcinoma of the prostate, significantly increased levels of 6-oxo-$PGF_{1\alpha}$ were present in the plasma (6-oxo-$PGF_{1\alpha}$ is the stable endproduct of PGI_2), which implicated high tumor prostacyclin synthesis (Khan *et al.*, 1980). Rising levels of 6-oxo-$PGF_{1\alpha}$ were correlated with disease progression and spread of the tumor to the bones, the latter probably related to the increased bone resorption activity induced by PGI_2 (Bennett *et al.*, 1980). Stringfellow and Fitzpatrick (1979) observed that the metastatic pattern of two melanoma cell lines was dependent on tumor PGD_2 synthesis; i.e., the tumor with low synthesis capacity of PGD_2 was the one with a high number of metastases, whereas the tumor with a low number of metastases could significantly synthesize higher amounts of PGD_2. When indomethacin was given to melanoma-bearing mice, the two tumor types acted identically; the metastatic rate was increased, in particular that of the tumor cell line with the smaller metastatic potential (Stringfellow and Fitzpatrick, 1979; Fitzpatrick and Stringfellow, 1979).

The results from our present study and also those of a previous experiment (Kort *et al.*, 1985) confirmed that aspirin treatment was able to be beneficial in the reduction of tumor metastasis, although it is clear that aspirin alone will not be sufficient to prevent tumor spread completely. Combined treatment with "classical" chemotherapy might give more real improvement in the control of cancer growth and spread. Low doses of aspirin either alone or in combination with theophylline were not effective; this probably implies that in the tumor model used, in order to decrease the number of metastases effectively, a complete blockade is needed of prostaglandin synthesis. This latter is more or less in contrast with the results obtained by Honn (Honn *et al.*, 1981b; Honn, 1983); however, either the tumor model he used or his experimental design may explain this. When treatment is started prior to i.v. inoculation of tumor cells, it is not hard to imagine that better results will be obtained than in a therapeutic regimen and tumor model such as that used in our experiment.

It is clear from our data and data from other groups that prostaglandins are involved in tumor synthesis. In some cases high prostacyclin levels were beneficial (Honn *et al.*, 1983), whereas in other studies they correlated with tumor progress (Khan *et al.*, 1980). Therapy with cycloxygenase inhibitors could decrease the number of metastases (Bennett, 1980) but could also increase this number (String-fellow and Fitzpatrick, 1979). This clearly shows that no single definitive policy of interfering with prostaglandin synthesis can be followed in the control of tumor progression. When a prostaglandin synthesis-inhibiting therapy is considered, tumor prostaglandin synthesis should be monitored meticulously, and therapy should be adjusted to the synthesis values found. Tumor procoagulant activity (probably a TxA_2 effect) seems to be a good parameter of tumor malignancy. Tumor cell aggregation *in vitro* as a method to establish tumor characteristics is presently a subject of research in our laboratory.

REFERENCES

Barrowman, J. A., Bennett, A., Hillenbrand, P., Rollen, K., Pollock, D., and Wright, J. T., 1975, Diarrhoea in thyroid medullary carcinoma: Role of prostaglandins and therapeutic effect of nutmeg, *Br. Med. J.* **3**:11–12.

Bennett, A., 1979, Prostaglandins and cancer, in: *Practical Applications of Prostaglandins and Their Synthesis Inhibitors* (S. M. M. Karim, ed.), MTP Press Limited, Lancaster, pp. 149–188.

Bennett, A., 1980, Prostaglandins and their synthesis inhibitors in cancer, in: *Progress in Cancer Research Therapy*, Vol. 14, *Hormones and Cancer* (S. Iacobelli, H. R. Lindner, R. J. B. King, and M. E. Lippman, eds.), Raven Press, New York, pp. 515–520.

Bennett, A., Stamford, I. F., and Unger, W. G., 1973, Prostaglandin-E_2 and gastric acid secretion in man, *J. Physiol. (Lond.)* **229**:349–360.

Bennett, A., Berstock, D. A., Harris, M., Raja, B., Rowe, D. J. F., Stamford, I. F., and Wright, J. E., 1980, Prostaglandins and their relationship to malignant and benign human breast tumors, *Adv. Prostaglandin Thromboxane Res.* **6**:595–600.

Blitzer, A., and Huang, C. C., 1983, The effect of indomethacin on the growth of epidermoid carcinoma of the palate in rats, *Arch. Otolaryngol.* **109**:719–723.

Caldwell, B. V., Tillson, S. A., Brack, W. A., and Speroff, L., 1972, The effects of exogenous progesterone and estradiol on prostaglandin-F levels in ovariectomized ewes, *Prostaglandins* **1**:217–228.

Ellis, E. F., Jones, P. S., Wright, K. F., Richardson, D. W., and Ellis, C. K., 1980, Effect of oral aspirin dose on platelet aggregation and arterial prostacyclin synthesis: Studies in human and rabbits, *Adv. Prostaglandin Thromboxane Res.* **6**:313–315.

Evans, P. M., and Cowie, F. P., 1983, A species difference in platelet aggregation induced by tumour cells, *Cell. Biol. Int. Rep.* **7**:771–778.

Fantone, J. C., Elgas, L. J., Weinberger, L., and Varani, J., 1983, Modulation of tumor cell adherence by prostaglandins, *Oncology* **40**:421–426.

Fitzpatrick, F. A., and Stringfellow, D. A., 1979, Prostaglandin-D_2 formation by malignant melanoma cells correlates inversely with cellular metastatic potential, *Proc. Natl. Acad. Sci. U.S.A.* **76**:1765–1769.

Gasic, G. J., Koch, P. A. G., Hsu, B., and Niewiarowski, S., 1976, Thrombogenic activity of mouse and human tumors: Effects on platelets, coagulation and fibrinolysis and possible significance for metastases, *Z. Krebsforsch.* **86:**263–277.

Honn, K. V., 1983, Inhibition of tumor cell metastasis by modulation of the vascular prostacyclin/thromboxane-A$_2$ system, *Clin. Exp. Metastasis* **1:**103–114.

Honn, K. V., and Meyer, J., 1981, Thromboxanes and prostacyclin: Positive and negative modulators of tumor growth, *Biochem. Biophys. Res. Commun.* **102:**1122–1129.

Honn, K. V., Cicone, B., Skoff, A., Romine, M., Gossett, O., Thompson, K., and Hodge, D., 1980, Thromboxane synthetase inhibitors and prostacyclin can control tumor cell metastasis, *J. Cell. Biol.* **87:**64a.

Honn, K. V., Bockman, R. S., and Marnett, L. J., 1981a, Prostaglandins and cancer: A review of tumor initiation through tumor metastasis, *Prostaglandins* **21:**833–864.

Honn, K. V., Cicone, B., and Skoff, A., 1981b, Prostacyclin: A potent antimetastatic agent, *Science* **212:**1270–1272.

Honn, K. V., Busse, W. D., and Sloanes, B. F., 1983, Prostacyclin and thromboxanes. Implications for their role in tumor cell metastasis, *Biochem. Pharmacol.* **32:**1–11.

Karmali, R. A., 1983, Prostaglandins and cancer, *Cancer J. Clinicians* **33:**324–332.

Khan, O., Hensby, C. N., and Williams, G., 1980, Prostacyclin levels in patients with prostate cancer: Monitoring the growth and spread of the disease, *Br. J. Surg.* **67:**825.

Kort, W. J., Hulsman, L. O. M., Weijma, I. M., Zondervan, P. E., and Westbroek, D. L., 1985, Beneficial effect of aspirin treatment on primary tumor growth and metastasis of an implanted fibrosarcoma in rats (submitted for publication).

Lynch, N. R., Castes, M., Astoin, M., and Salomon, J. C., 1978, Mechanism of inhibition of tumour growth by aspirin and indomethacin, *Br. J. Cancer* **38:**503–512.

Moncada, S., and Korbut, R., 1978, Dipiridamole and other phosphodiesterase inhibitors act as antithrombotic agents by potentiating endogenous prostacyclin, *Lancet* **1:**1286–1289.

Powles, T. J., Dowsett, M., Easty, G. C., Easty, D. M., and Neville, A. M., 1976, Breast-cancer osteolysis, bone metastases, and anti-osteolytic effect of aspirin, *Lancet* **1:**608–610.

Smith, B. J., Wills, M. R., and Savory, J., 1983, Prostaglandins and cancer, *Ann. Clin. Lab. Sci.* **13:**359–365.

Stringfellow, D. A., and Fitzpatrick, F. A., 1979, Prostaglandin-D$_2$ controls pulmonary metastasis of malignant melanoma cells, *Nature* **282:**76–78.

Tisdale, M. J., 1983, Role of prostaglandins in metastatic dissemination of cancer, *Exp. Cell. Biol.* **51:**250–256.

60

Arylmethyl Phenyl Ethers

A New Class of Specific Inhibitors of 5-Lipoxygenase

STEPHEN M. COUTTS, ATUL KHANDWALA,
RICHARD VAN INWEGEN,
UTPAL CHAKRABORTY, JOHN MUSSER,
JOHN BRUENS, NAVIN JARIWALA,
VERNETTA DALLY-MEADE, RICHARD INGRAM,
THADDEUS PRUSS, HOWARD JONES,
EDWARD NEISS, and IRA WEINRYB

1. INTRODUCTION

Oxidative metabolites of arachidonic acid produced by the 5-lipoxygenase pathway are potent mediators of hypersensitivity and inflammation (see O'Flaherty, 1982; Lewis and Austen, 1984 for reviews). Their proposed roles in the pathophysiology of asthma, chronic inflammation, and psoriasis have prompted an intense search for specific inhibitors of their biosynthesis (Bach, 1984). Three years ago a program was initiated at Revlon Health Care Group to find compounds that would modulate the biosynthesis of leukotrienes or antagonize their spasmogenic actions on smooth muscle. In particular, we sought specific and stable inhibitors of the 5-lipoxygenase (5-LOX) enzyme, utilizing a cellular assay as our primary screening tool.

STEPHEN M. COUTTS, ATUL KHANDWALA, RICHARD VAN INWEGEN, UTPAL CHAKRA-BORTY, JOHN MUSSER, JOHN BRUENS, NAVIN JARIWALA, VERNETTA DALLY-MEADE, RICHARD INGRAM, THADDEUS PRUSS, HOWARD JONES, EDWARD NEISS, and IRA WEINRYB ● Revlon Health Care Group, Research and Development, Tuckahoe, New York 10707.

2. METHODS

The basic elements of the 5-LOX assay are outlined in Fig. 1. Glycogen-elicited rat peritoneal polymorphonuclear leukocytes (PMN) were incubated with 1 μM indomethacin (to inactivate the cyclooxygenase pathway), phosphate-buffered saline plus 1.5 mM Ca^{2+} and Mg^{2+} (pH 7.4), and the compound to be tested, at 30° for 5 min. The cells were then challenged with a mixture of calcium ionophore (1 μM) and [^{14}C]arachidonic acid (4 μM), followed by a 3-min incubation. The

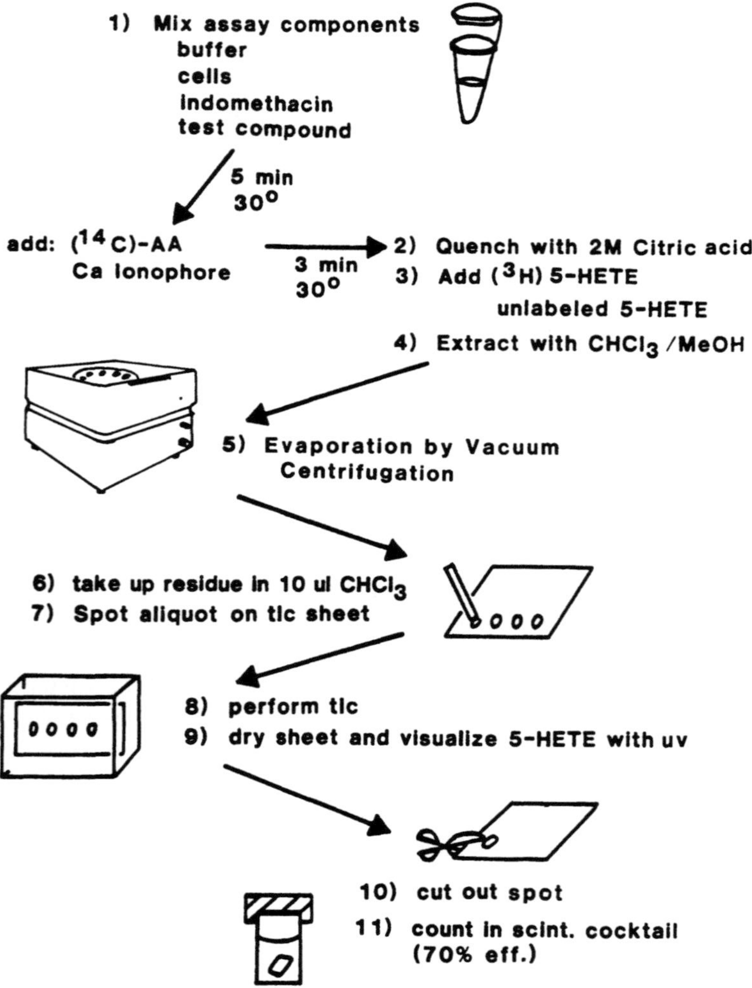

FIGURE 1. Assay for inhibitors of 5-lipoxygenase in rat peritoneal polymorphonuclear leukocytes.

reaction was quenched by the addition of citric acid to reduce the pH to 3.0, followed by the addition of a trace of [^{3}H]5-HETE (*ca.* 10,000 dpm) and excess unlabeled 5-HETE (35 μg) as internal standard and carrier compound, respectively. The assay suspensions were extracted with chloroform/methanol (5/1), and the organic phases were evaporated to dryness *in vacuo.* The residues were reconstituted in chloroform and were spotted on Mylar®-backed silica gel thin-layer sheets, followed by chromatography with the top phase of isooctane/ethyl acetate/acetic acid/water (5/11/1/10). The spots of 5-HETE were visualized by UV light and were cut out and placed in scintillation fluid. The amount of 5-HETE synthesized was quantitated by two-channel scintillation spectroscopy. The same basic assay was employed for quantitating 5-LOX activity in PMN from humans as well as in homogenates of rat basophilic leukemia (RBL) cells. In a similar manner, 12-LOX from rat platelets was also assayed by the inclusion of 15-HETE (which cochromatographs with 12-HETE in this system) in the place of 5-HETE prior to extraction. Glutathione (100 μM) was included in the reaction mixture of platelets or platelet homogenate.

Leukotriene-C$_4$-induced contraction of guinea pig parenchymal strips was carried out according to Drazen *et al.* (1980). Anti-IgE-mediated secretion of histamine from human leukocytes was performed according to Khandwala *et al.* (1982), and the influence of aryleicosanoids on the production of PGE$_2$ by sheep seminal vesicles (SSV) was determined by the method of Flower *et al.* (1973).

3. RESULTS AND DISCUSSION

3.1. Molecular Design of Arylmethyl Phenyl Ethers (Aryleicosanoids)

This chapter will describe the properties of one class of compounds, arylmethyl phenyl ethers, which are specific inhibitors of the 5-LOX enzymes from a number of mammalian species. A strong influence in the chemical design of these molecules, to which we have ascribed the trivial name of aryleicosanoids,* was the report of Vanderhoek *et al.* (1980), which described the ability of 15-HETE to inhibit 5-LOX. The similarity between the aryleicosanoids and 15-HETE may be seen by comparing the structure of one of the first active compounds in the series, REV 5367, with that of 15-HETE (Fig. 2). By redrawing 15-HETE in a form analogous to that of REV 5367, one may visualize that the latter compound has a similar 5,6 and 11,12 *cis* double bond arrangement as well as a benzylic hydroxyl group instead of the hydroxyl group allylic to the *trans* 13,14 double bond in 15-HETE. A unique feature to the design is the inclusion of an arylmethyl–phenyl ether bridge in place of the 8,9 double bond. REV 5367 has an I$_{50}$ of 3 μM in the rat PMN assay compared to a value of 3 μM for 15-HETE. The use of MAACS® software (Mo-

---

* The name "aryleicosanoid" was chosen to reflect the fact that these molecules are aromatic (aryl) analogues of eicosatetraenoic acid.

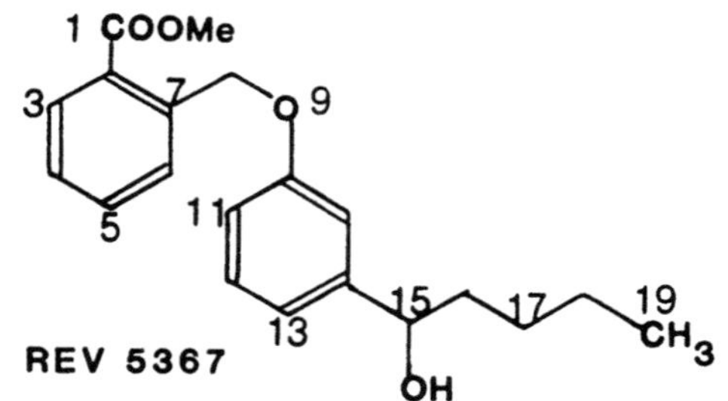

FIGURE 2. Structures of 15-HETE and REV 5367; the structure of 15-HETE has been redrawn to emphasize the potential structural similarity to REV 5367.

lecular Design, Ltd.) allowed us to examine the similarity between 15-HETE and REV 5367 more closely. ORTEP plots of the two molecules, calculating the lowest energy configuration *in vacuo*, are presented in Figs. 3 (ball and stick models) and 4 (space-filling models). Although the configuration chosen by the computer is influenced by the starting conformation chosen by the investigator, the similarity of the two structures suggests that they may indeed present a comparable chemical topography to the 5-LOX enzyme. From a standpoint of rational design, it seemed to us that a competitive inhibitor would stand the greatest chance of being specific for the target enzyme.

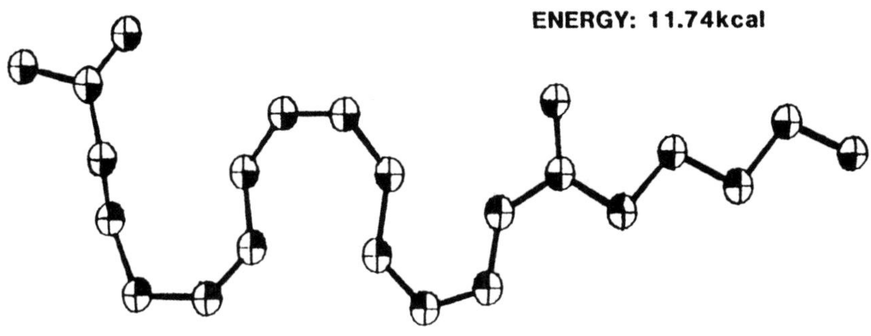

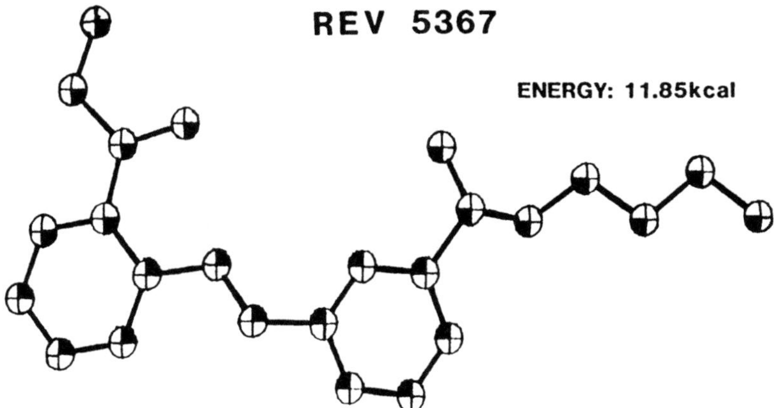

FIGURE 3. ORTEP drawings (ball and stick) of lowest-energy conformations of 15-HETE and REV 5367.

3.2. *In Vitro* Profiles of Aryleicosanoids

The initial success with REV 5367 led to the synthesis and assay of a great many more inhibitors of the aryleicosanoid series. *In vitro* profiles of some of these compounds are presented in Fig. 5. Inhibitory activity against 5-LOX from three sources, rat and human PMN and RBL-1 cells, are compared, as well as inhibition of 12-LOX from intact rat platelets and cyclooxygenase from sheep seminal vesicles (SSV). In addition, the degree to which these compounds antagonized LTC_4-induced contraction of guinea pig parenchymal strips or inhibited IgE-mediated secretion

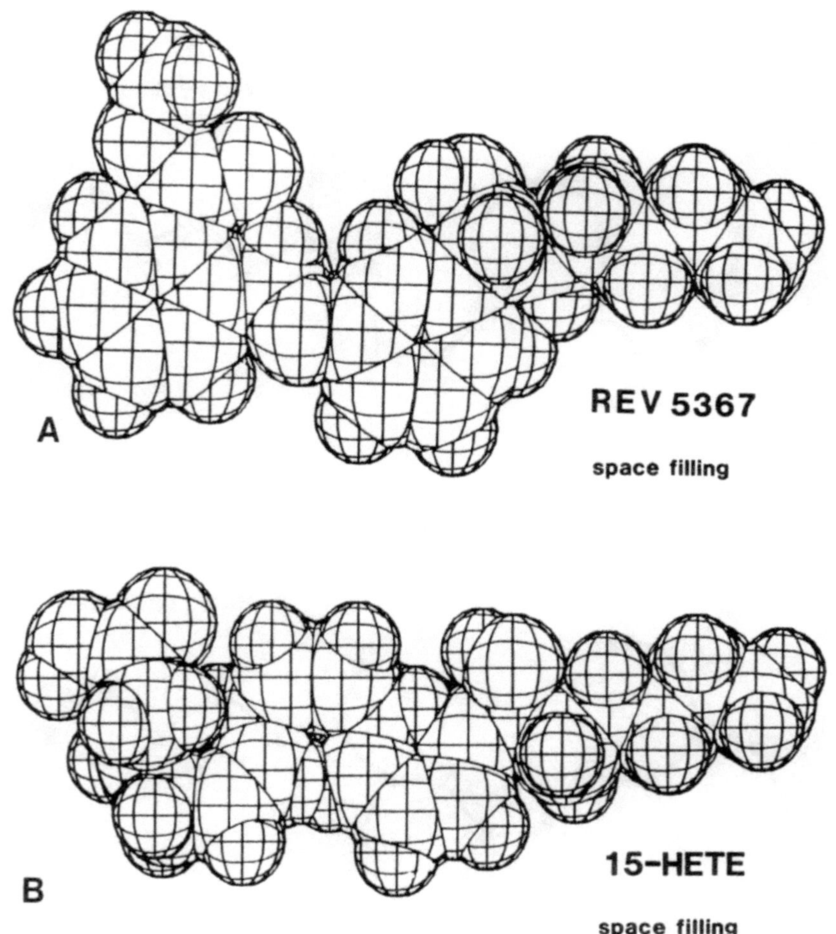

FIGURE 4. ORTEP drawings (space filling models) comparing the lowest energy conformations (*in vacuo*) of 15-HETE (B) and REV 5367 (A); the conformations are the same as pictured in Fig. 3.

of histamine from human leukocytes is also presented. The I_{50} values for inhibiting ionophore-induced production of 5-HETE by rat PMN range 16-fold, from 0.15 μM (REV 6080) to 2.5 μM (REV 5827). BW 755C had an I_{50} of 6 μM by this protocol. It is noteworthy that the molecules with heterocyclic moieties on the left are in general more inhibitory than those with phenyl substituents and that the 15-hydroxy group may be oxidized to a ketone without loss of activity *in vitro*.* Another salient finding from these comparative data is that higher concentrations of aryleicosanoids are required for 50% inhibition of 5-HETE production by human PMN than by rat PMN. The ratio of I_{50} values (human PMN/rat PMN) varies from

	5-LOX			12-LOX	CYC LOX	LTC$_4$ Ant.	HuL
	Rat PMN	Hum. PMN	RBL				
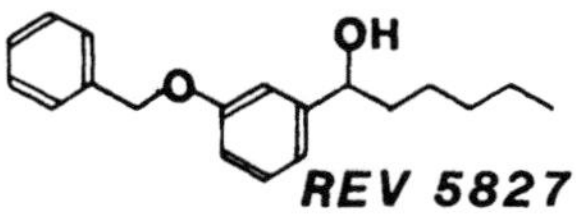 REV 5827	2.5	7	100	43%	2%	10%***	SR
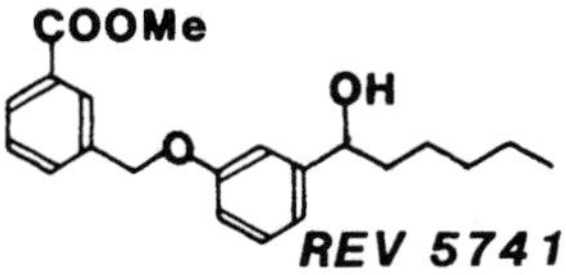 REV 5741	1.4	8	23	80%*	30%	20%**	6
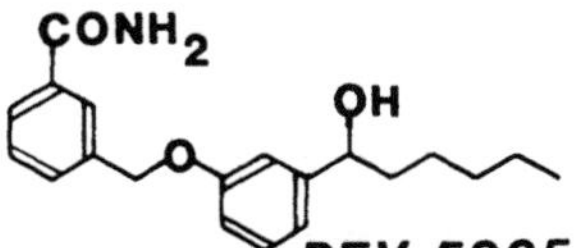 REV 5965	2.0	17	56	8%	-33%	0%*	SR
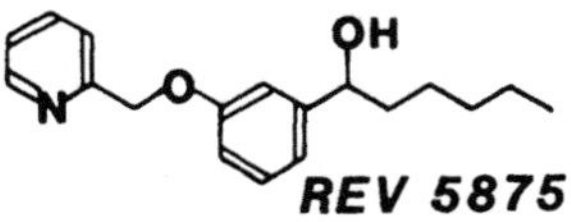 REV 5875	0.5	8	140	0%	0%	37%**	48
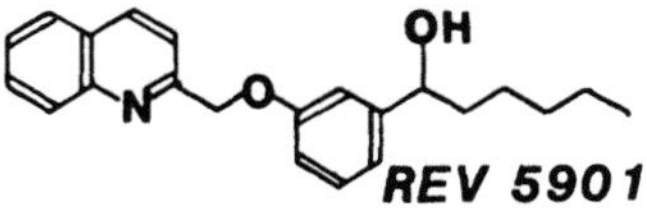 REV 5901	0.16	6	80	-25%	-21%	2	4
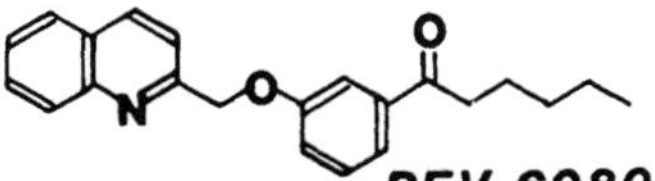 REV 6080	0.15	14	20	-47%	-	15	7

FIGURE 5. The *in vitro* profiles of a selection of six aryleicosanoids. Numerical values in the figure represent I$_{50}$ concentrations or percent inhibition. 5-Lipoxygenase was assayed in rat PMN, human PMN, or in a homogenate of rat basophilic leukemia cells (RBL); compounds were assayed at 100 or 50(*) μM against 12-lipoxygenase (12-LOX) in rat platelets and at 200 μM aginst cyclooxygenase (cyclox) from sheep seminal vesicles. Antagonism of LTC$_4$-induced contraction of guinea pig parenchyma (LTC$_4$ Ant.) was measured in the presence of 10(*), 30 (**), or 100(***) μM test compound. Anti-IgE-mediated release of histamine from human leukocytes (HuL); some compounds caused an increase in spontaneous secretion of histamine (SR); compounds were assayed at 100 μM in this model.

3 to 93 for the six compounds listed. BW 755C, with an I_{50} of 20 μM in the assay using human PMN, has a ratio of 3.3.

A somewhat surprising finding is the fact that the aryleicosanoids are relatively weak inhibitors of cell-free 5-LOX from RBL-1 cells (column 3, Fig. 5). The six aryleicosanoids are 16- to 500-fold less potent against cell-free RBL 5-LOX than against cellular 5-LOX from rat PMN. The difference between the two systems does not arise because the RBL lipoxygenase is a radically different enzyme from that of the rat PMN, since at least one of the compounds at 10 μM completely inhibited the production of 5-HETE in intact RBL cells. Rather, the explanation probably lies in the high uptake of these hydrophobic compounds by cells as well as their low aqueous solubility. These data provide an interesting example that might bear on the strategy one should use in screening for inhibitors: (1) use a cellular and more complex system or (2) a cell-free system that is subject to the limitations of compound solubility and the absence of the native cellular environment for the enzyme.

Of the six compounds, only two, REV 5827 and REV 5741, demonstrated significant inhibition of the 12-LOX enzyme from rat platelets. BW 755C inhibited 50% of 12-HETE formation at 8 μM, showing a lack of specificity between the 5- and 12-LOX enzymes. In addition, only REV 5741 showed significant inhibition of PGE_2 formation by SSV cyclooxygenase when incubated with the microsomal enzyme at 200 μM. By comparison, indomethacin and BW 755C had I_{50} values of 25 and 100 μM, respectively, with this preparation. The heterocyclic members of the series in Fig. 5 thus demonstrate a specificity for 5-LOX, being inactive against both 12-LOX and cyclooxygenase.

An additional and perhaps unexpected activity of the aryleicosanoids was the inhibition demonstrated by certain members of the series of leukotriene-induced contraction of guinea pig parenchymal strips. Parenchymal strips were contracted by either LTC_4 or LTD_4 at 0.2 nM; contractions were standardized by measuring the response to 1 μM histamine. Under the conditions of our assay, FPL 55712 inhibited contraction mediated by either of the leukotrienes with I_{50} values from 0.3 to 0.5 μM. The two quinolinyl derivatives in Fig. 5 (REV 5901 and REV 6080) had I_{50} values of 2 and 15 μM, respectively, against LTC_4-induced contraction. They were equally potent against LTD_4-induced contractions. Although REV 5901 did not inhibit histamine-induced contraction of parenchymal strips at concentrations up to 100 μM, the compound did inhibit methacholine- and $PGF_{2\alpha}$-induced contraction with I_{50} values of 10 and 30 μM, respectively. Thus, the possibility remains that REV 5901 may inhibit leukotriene-induced contraction by an indirect mechanism rather than by direct antagonism of a receptor–leukotriene interaction. However, later studies with highly purified REV 5901 confirmed its potency against leukotrienes, while showing far less inhibitory activity for the drug

---

* An independent confirmation of this finding was published after our discovery; Vanderhoek *et al.* (1982) reported that 15-keto-13-*trans*-5,8,11-*cis*-eicosatetraenoic acid (15-KETE) was as active as 15-HETE in inhibiting 5-LOX from rabbit PMN.

against other spasmogens, implying that inhibition of leukotriene-induced contraction by REV 5901 is due to direct antagonism.

3.3. Mechanism of Inhibition by Aryleicosanoids

Aryleicosanoids represent a class of inhibitors whose eicosanoid-like double bonds are stabilized in aromatic structures, resulting in compounds that possess low potential for oxidation by target lipoxygenases. An important question remained for us to answer, however: are aryleicosanoids competitive inhibitors of lipoxygenases? That is, is the mechanism of inhibition an occupation or blockade of an active site of a lipoxygenase that would otherwise be attaching a hydroperoxy group to a specific position in arachidonic acid?

Our attempts to determine whether or not aryleicosanoids inhibit cell-free RBL 5-LOX in a competitive manner were unsuccessful inasmuch as the self-inactivation of the enzyme did not lend it to a study of steady-state kinetics.

The 12-LOX enzyme from homogenized rat platelets is more tractable, and Fig. 6 shows a Lineweaver–Burk plot for inhibition of 12-HETE formation by REV 5741. The lines are calculated by least-squares regression analysis and yield the kinetic parameters found in Table I. Also included in Table I are data for REV

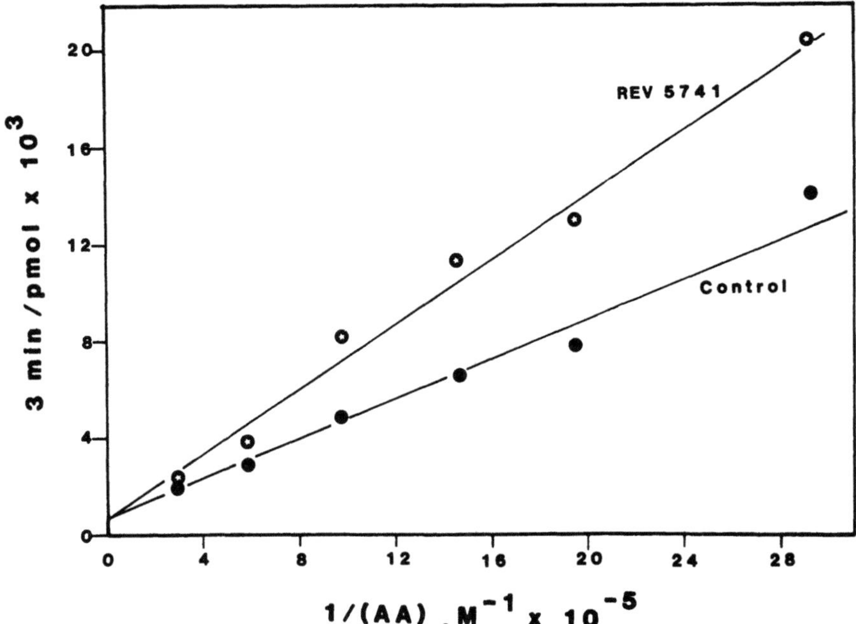

FIGURE 6. Inhibition of rat platelet 12-lipoxygenase by REV 5741; data are graphed as a Lineweaver–Burk plot of reciprocal initial velocity versus reciprocal concentration of arachidonic acid.

TABLE I. Kinetic Parameters Obtained with 12-Lipoxygenase from Rat Platelets

Compound	K_m (μM)	V_{max} (pmol/3 min per mg protein)	K_i (μM)	Mode of inhibition
Control	16 ± 9	2347 ± 760	—	
REV 5741	—	—	68	Competitive
REV 5827	—	—	291	Competitive

5827. Both compounds exhibited competitive inhibition and had apparent inhibition constants of 68 and 291 μM, respectively. The K_m of 16 μM (mean of three determinations) for this enzyme is in close agreement with that reported by Chang *et al.* (1982), 20 μM. We have examined the time course of cell-free 12-LOX with and without 30 or 100 μM REV 5741 and have found that the inhibitor does not prolong a lag phase in the enzyme-catalyzed reaction. Thus, it is quite likely these two aryleicosanoids are binding to the active site of 12-lipoxygenase. By inference, although with less confidence than drawn from the direct measurement above, one might extend these conclusions to the 5-LOX system and propose that aryleicos-anoids are competitive inhibitors of 5-lipoxygenase as well. Although this conclusion is far from proved, the structural similarity of these compounds to substrates for the enzyme and the competitive inhibition displayed by two members of the series against a related enzyme provide strong support for the inference. This inference was later supported by kinetic experiments with cell-free 5-LOX from guinea pig PMN (data not shown) which demonstrated a competitive mode of inhibition for REV 5901 ($K_i = 5$–8 μM) and therefore, presumably, for other members of the aryleicosanoid series, with this enzyme.

3.4. Inhibition of Histamine Release

Peters *et al.* (1981) found that 5-HPETE potentiates the release of histamine from human basophils. Other reports have claimed that an intact lipoxygenase pathway is necessary for IgE-mediated secretion to take place in the same cells (Marone *et al.*, 1980). Thus, the aryleicosanoids presented in Fig. 5 were also examined for their influence on anti-IgE-mediated secretion of histamine from human leukocytes. At a concentration of 100 μM, two of the compounds (REV 5827 and REV 5965), as well as BW 755C, increased the spontaneous release (SR) of histamine. Of the remaining four compounds, three (REV 5741, REV 5901, and REV 6080) had I_{50} values under 10 μM. The concentration–response curves were well behaved, with 80 to 100% inhibition achieved at 30 μM drug concentration. Proxicromil had a mean I_{50} ± S.E. of 52 ± 4 μM in 43 experiments by the same protocol.

These data do not show a correlation between I_{50} values as inhibitors of 5-LOX in human PMN and as inhibitors of IgE-mediated secretion from human basophils and, hence, do not prove or disprove the hypothesis that the 5-lipoxy-genase pathway is a necessary component of stimulus-coupled secretion in baso-phils. However, the data do provide interesting examples of molecules that have multiple pharmacological activities that would make them therapeutically attractive for the treatment of asthma and other disease states associated with immediate hypersensitivity: inhibition of 5-LOX and IgE-mediated secretion as well as antagonism of leukotriene-induced contraction of bronchial smooth muscle.

4. CONCLUSIONS

The aryleicosanoid series of compounds thus meet the *in vitro* objectives we set when the project was first implemented. The class of compounds contains members which are potent inhibitors of 5-lipoxygenase in both human and rodent PMN and lack significant inhibition against rat 12-lipoxygenase or sheep cyclooxygenase. In addition, these compounds possess additional benefits of being inhibitors of mediator release from human basophils and antagonizing the myotropic action of leukotrienes C$_4$ and D$_4$. This attractive *in vitro* profile, together with promising *in vivo* activity against antigen-induced bronchospasm in guinea pigs (Gordon *et al.*, 1984), has prompted us to consider one member of the series, REV 5901, for toxicological evaluation prior to clinical trial.

REFERENCES

Bach, M. K., 1984, Prospects for the inhibition of leukotriene synthesis, *Biochem. Pharmacol.* **33**:515–521.

Chang, W. C., Nakao, J., Orimo, H., and Murota, S., 1982, Effects of reduced glutathione on the 12-lipoxygenase pathways in rat platelets, *Biochem. J.* **202**:771–776.

Drazen, J. M., Austen, K. F., Lewis, R. A., Clark, D. A., Groto, G., Marafat, A., and Corey, E. J., 1980, Comparative airway and vascular activities of leukotrienes C-1 and D *in vivo* and *in vitro*, *Proc. Natl. Acad. Sci. U.S.A.* **77**:4354–4358.

Flower, R. J., Cheung, H. S., and Cushman, D. W., 1973, Quantitative determination of prostaglandins and malondialdehyde formed by the arachidonate oxygenase (prostaglandin synthetase) system of bovine seminal vesicle, *Prostaglandins* **4**:325–341.

Gordon, R. J., Travis, J., Godfrey, H. R., Sweeney, D., Wolf, P. S., Pruss, T. P., Neiss, E., Musser, J., Chakraborty, U., Jones, H., and Leibowitz, M., 1984, *In vivo* activity of a lipoxygenase inhibitor and leukotriene antagonist, REV 5901-A, *Leukotrienes Prostaglandins* **84**:266.

Khandwala, A., Coutts, S., and Weinryb, I., 1982, Antiallergic activity of tiaramide, *Int. Arch. Allergy Appl. Immunol.* **69**:159–168.

Lewis, R. A., and Austen, K. F., 1984, The biologically active leukotrienes: Biosynthesis, metabolism, receptors, function and pharmacology, *J. Clin. Invest.* **73**:889–897.

Marone, G., Hammarström, S., and Lichtenstein, L. M., 1980, An inhibitor of lipoxygenase inhibits histamine release from human basophils, *Clin. Immunol. Immunopathol.* **17**:117–122.

O'Flaherty, J. T., 1982, Biology of disease: Lipid mediators of inflammation and allergy, *Lab. Invest.* **47**:314–329.

Peters, S. P., Siegel, M. I., Kagey-Sabotka, A., and Lichtenstein, L. M., 1981, Lipoxygenase products modulate histamine release in human basophils, *Nature* **292**:455–457.

Vanderhoek, J. Y., Bryant, R. W., and Bailey, J. M., 1980, Inhibition of leukotriene biosynthesis by the leukocyte product 15-hydroxy-5,8,11,13-eicosatetraenoic acid, *J. Biol. Chem.* **255**:10064–10066.

Vanderhoek, J. Y., Bryant, R. W., and Bailey, J. M., 1982, Regulation of leukocyte and platelet lipoxygenases by hydroxyeicosanoids, *Biochem. Pharmacol.* **31**:3463–3467.

61

Relationship between Drug Absorption, Inhibition of Cyclooxygenase and Lipoxygenase Pathways, and Development of Gastric Mucosal Damage by Nonsteroidal Antiinflammatory Drugs in Rats and Pigs

K. D. RAINSFORD, SHEILA A. FOX, and D. J. OSBORNE

1. INTRODUCTION

Many authors have provided evidence for an association between inhibition of prostaglandin (PG) synthesis by the nonsteroidal and antiinflammatory (NSAI) drugs and their propensity to elicit gastric mucosal damage (Boughton-Smith and Whittle, 1983; Peskar *et al.*, 1981, 1984; Rainsford and Willis, 1982; Rainsford, 1984; Rainsford *et al.*, 1984). By comparison, relatively little is known about the relationship between the inhibition by NSAI drugs of the synthesis of various eicosanoids to (1) the time-dependent changes induced in the various mucosal cells by

K. D. RAINSFORD and SHEILA A. FOX ● Pharmacology Department, University of Cambridge, Cambridge CB2 2QD, United Kingdom.   D. J. OSBORNE ● Lilly Research Centre, Erl Wood Manor, Windlesham GU20 6PH, United Kingdom.

these drugs and (2) the pattern of drug uptake into the gastric mucosa. Furthermore, it is not known if the different NSAI drugs with varying potency as PG synthesis inhibitors have the capacity to elicit damage to the mucosal cells in the same way. These aspects are important in order to determine the precise role that inhibition of the synthesis of the different eicosanoids has in the induction of pathological changes in different cells by ulcerogenic NSAI drugs. The studies reported here are an attempt to provide some information in answer to these questions.

2. STUDIES IN RATS

Aspirin (ASA), indomethacin (IND), and benoxaprofen (BEN) have been found to reduce mucosal levels of PGE_2 production from 10 min to 1 hr following oral dosage (Fig. 1). Similar effects have been observed in 6-oxo-$PGF_{1\alpha}$ production, although this was somewhat variable, and the control values amounted to about one-fifth of the PGE_2 content. Even though these drugs were absorbed at somewhat different rates (Fig. 2), the mucosal concentrations of these drugs were clearly adequate to inhibit PG production (based on the *in vitro* potency; see Table I).

In contrast to these drug effects on PG production, the degree of mucosal damage and the time of onset observed both visually (Fig. 3) and electron microscopically (Fig. 4; Table I) differ enormously among these drugs. Thus, aspirin

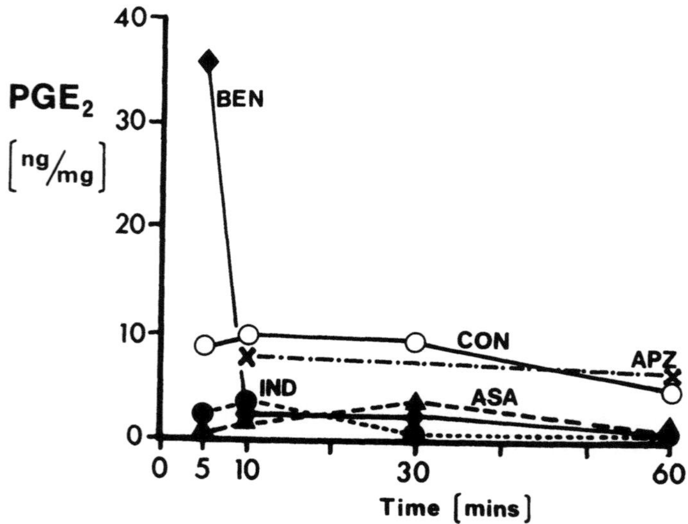

FIGURE 1. Effects of NSAI drugs on gastric mucosal levels of PGE_2, measured by combined gas chromatography/mass spectrometry (Rainsford *et al.*, 1984) or for APZ by radioimmunoassay. Aspirin (ASA, 200 mg/kg), azapropazone (APZ, 100 mg/kg), benoxaprofen (BEN, 100 mg/kg), and indomethacin (IND, 10 mg/kg).

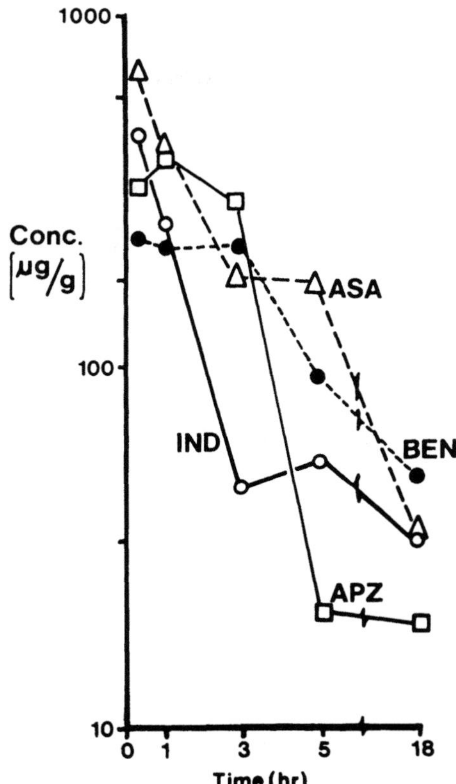

FIGURE 2. Concentrations of drugs in gastric mucosa following oral administration of radiolabeled NSAI drugs in doses as described in Fig. 1 (see also Rainsford *et al.*, 1984).

TABLE I. Comparison of Peak Concentrations of NSAI Drugs in Gastric (Fundic) Mucosa following their Oral Administration to Rats, with the Potencies of these Drugs as Prostaglandin Synthesis Inhibitors *in Vitro*

	Peak concentration[a]		Prostaglandin synthesis
Drug (dose in mg/kg)	$\mu M/g^b$	time	IC_{50} $(\mu M)^c$
Aspirin (200)	4.0	10 min	7.2
Azapropazone (100)	1.7	10–60 min	11.2
Benoxaprofen (110)	2.7	10–180 min	9.8
Indomethacin (10)	1.1	10 min	1.9

[a] Data for peak mucosal concentrations from Rainsford *et al.*, 1984 (see Fig. 2).
[b] Concentrations in g wet weight tissue.
[c] Data from Brune *et al.* (1981), derived from studies of prostaglandin production in mouse peritoneal macrophages stimulated with the phorbol ester TPA.

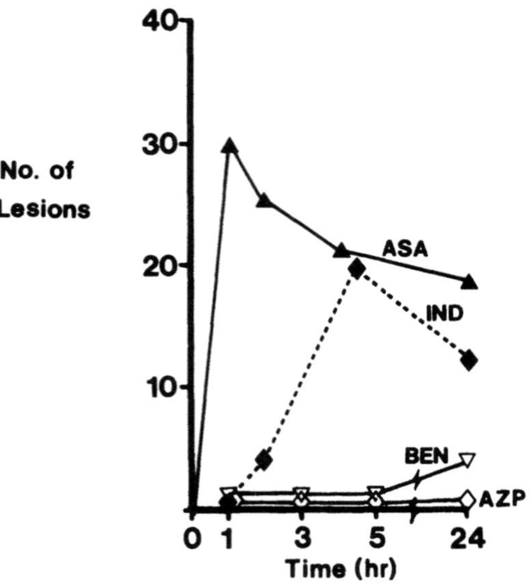

FIGURE 3. Gastric mucosal lesions in stressed (4°C, 3 hr) rats following oral administration of NSAI drugs in doses as detailed in legend to Fig. 1.

typically causes fine structural damage to the mucous, parietal, and microvascular cells by 10 min (see also Harding and Morris, 1976; Rainsford, 1975; Rainsford and Willis, 1982; Rainsford *et al.*, 1984; Robins, 1981), coincident with reduction in PGE_2 content. However, indomethacin only damages mucous cells at 10 min; the changes in parietal and endothelial cells with accompanying extravasation of blood components are only fully evident at 1 hr after oral administration, i.e., well after the reduction in PGE_2 content. Benoxaprofen does not cause appreciable mucosal cell damage (observed both by EM and visually) up to 3 hr after oral dosage, although this drug was well absorbed by this time (Fig. 2) and even reduces PGE_2 content by 10 min. The increase in PGE_2 content at 5 min by benoxaprofen may reflect a prior "cytoprotective" effect of this drug.

Azapropazone does not appear to cause appreciable mucosal damage and, coincidentally, does not affect PGE_2 production (Fig. 1). This is quite a paradoxical situation since the mucosal concentrations of the drug present at up to 1 hr (Fig. 2) are clearly within the range required to inhibit PG production as judged on the effects of azapropazone on PG production in mouse macrophages *in vitro* (Table I).

As shown in Table II, both azapropazone and benoxaprofen markedly reduce the gastric mucosal content of 5-HETE at 1 hr after oral administration of these drugs to rats. Aspirin slightly reduces 5-HETE levels at 1 hr, although not to near the extent noted with the other two drugs. It is tempting to speculate that the lower ulcerogenicity of azapropazone and benoxaprofen might be somehow associated with their capacity to inhibit 5-lipoxygenase activity. The slight reduction in 5-HETE levels induced by aspirin might, however, be considered a factor mitigating against this notion. It is obvious that the effects of azapropazone and benoxaprofen were much more marked than those of aspirin, so this concept is still a possibility. The consequences of inhibiting 5-lipoxygenase activity by drugs that are cyclooxygenase inhibitors might be to balance arachidonate metabolism and prevent excess tissue-destructive oxy-radicals being derived from those hydroperoxyarachidonic acids produced by the diversion of arachidonate through the lipoxygenase pathway. Although gross overproduction of 5-lipoxygenase products does not seem evident

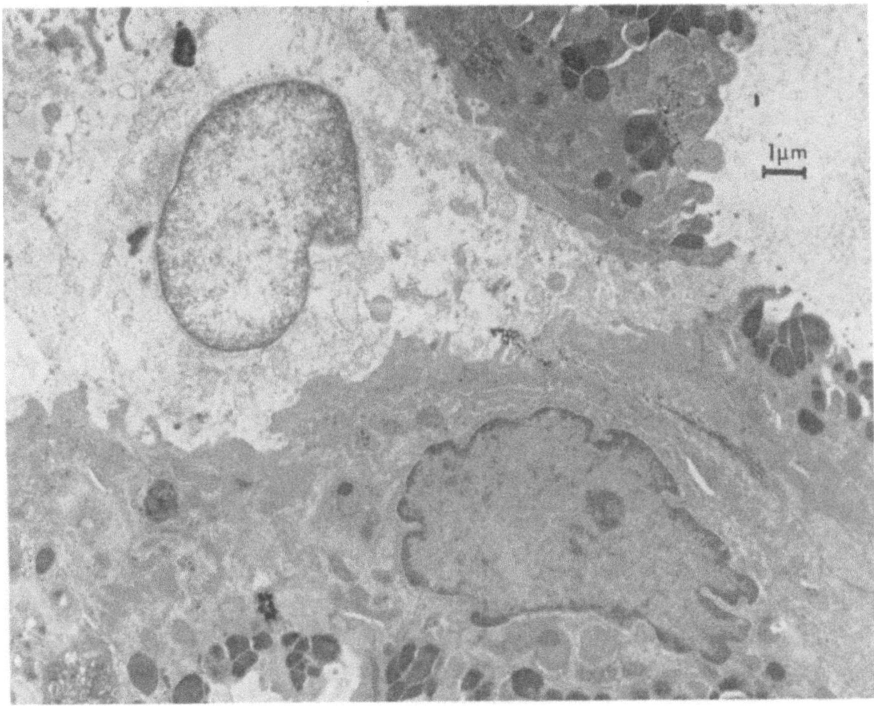

FIGURE 4. Electron microscopic observations of NSAI-induced gastric mucosal damage in rats. Aspirin induced damage to mucous parietal and endothelial cells (with accompanying disruption of interstitial space) as previously described (Rainsford, 1975; Rainsford *et al.*, 1985; Robins, 1981). Indomethacin, shown here, likewise causes damage to these cells, but with a somewhat slower onset of damage. A: Erosion of mucosal and parietal cells. B: Extravasation of blood components and disruption of interstitial space. C: Control for B. No appreciable damage was observed electron microscopically to the deeper regions of mucosa with azapropazone, benoxaprofen, or fenoclofenac. Electron microscopy studies were performed in male Sprague–Dawley rats (180–250 g body weight, fasted and allowed water *ad libitum* for 24 hr) that were dosed orally, in groups of five to eight each with (1) 200 mg/kg aspirin, (2) 100 mg/kg azapropazone, (3) 110 mg/kg benoxaprofen, (4) 10 mg/kg diclofenac, (5) 100 mg/kg fenclofenac, (6) 10 mg/kg indomethacin, (7) 10 mg/kg piroxicam, or (8) 1 ml H_2O (controls). The animals were killed at 10, 60, and, for the slowly absorbed drugs azapropazone, benoxaprofen, and fenclofenac, also at 120 and 180 min. Sections of the fundic mucosa were rapidly excised and fixed in 2% w/v formaldehyde plus 2.5% v/v glutaraldehyde, postfixed in 1% w/v osmium tetroxide, and stained with lead citrate and uranyl acetate.

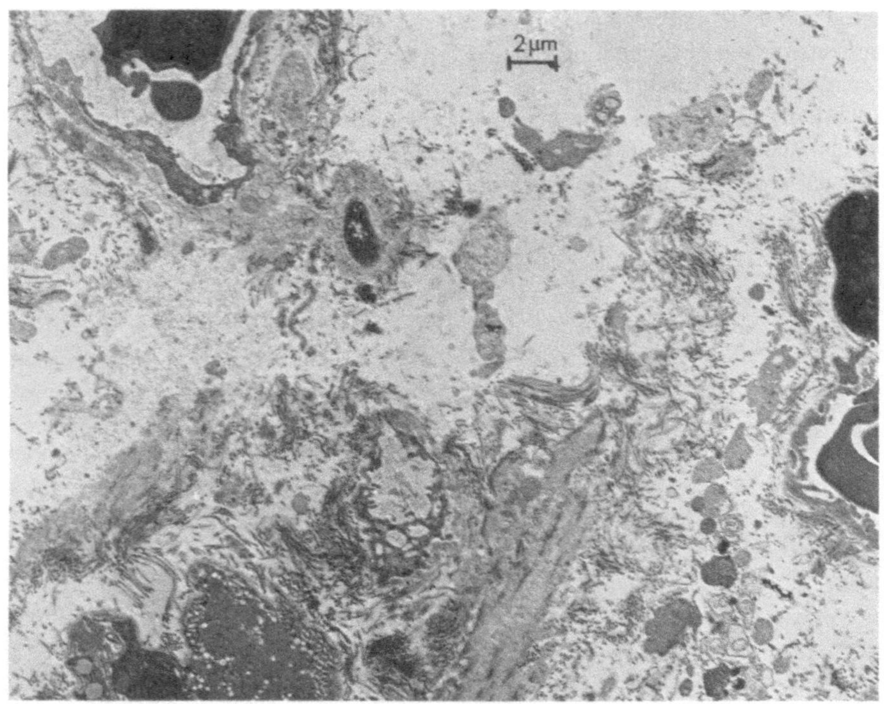

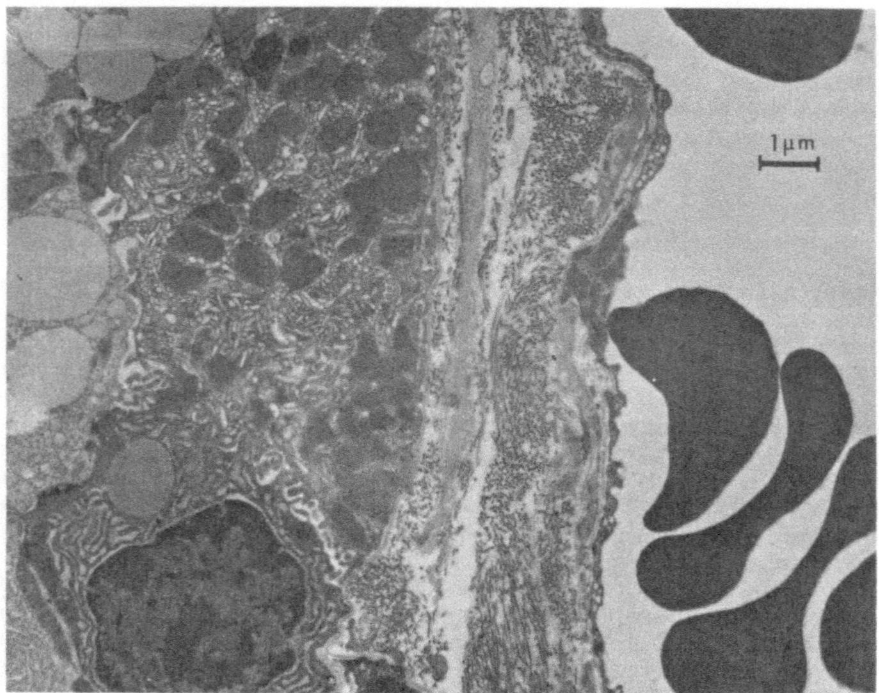

FIGURE 4. (*Cont.*)

TABLE II. Effects of Some Orally Administered NSAI Drugs on the Content of 5-HETE in Rat Gastric (Fundic) Mucosa[a]

Drug (dose in mg/kg)	Time	5-HETE content mean $+$ SD (pg/mg) (N)	Percent change (c.f. control)
Aspirin (200)	1 hr	2.79 ± 1.13 (5)	–57.8%
Azapropazone (100)	1 hr	below det[n] (3)	–100%
	2 hr	7.15 ± 0.95 (3)	N.S.
Benoxaprofen (110)	1 hr	below det[n] (5)	–100%
Control (1 ml H$_2$O/rat)	1 hr	6.60 ± 0.83 (4)	–
	2 hr	3.80 ± 3.28 (3)	–

[a] Content of 5-HETE (5-hydroxyeicosatetraenoic acid) measured by radio-immunoassay of cold acetone extracts prepared and assayed according to manufacturer's (Seragen Inc., Boston) instructions. The drugs were dosed orally to fasted (24 hr) female Sprague–Dawley rats (approx. 300 g body weight) as 1 ml per rat fine aqueous suspensions prepared by homogenizing the drug in distilled water immediately before administration.

with aspirin, it is clear that this drug has much more potent actions on the cyclooxygenase pathway than on 5-HETE production, especially in comparison with other drugs.

3. MICROVASCULAR INJURY

The two potent ulcerogenic agents diclofenac and piroxicam (10 mg/kg) both cause pronounced damage to surface mucous and the nearby microvasculature at 10–60 min following their oral administration (Fig. 5, Table III) in a similar manner to that observed with aspirin and indomethacin. In contrast, fenclofenac, a drug that, like azapropazone and benoxaprofen, has relatively low gastric ulcerogenicity, causes only trace damage to the surface mucous cells at 10–180 min after 100 mg/ kg oral dosing. Thus, these latter low-ulcerogenic drugs could be distinguished

TABLE III. Association of Ultrastructurally-Observed Microvascular Injury by NSAI Drugs in Rats with their Anti-Platelet Effects

Drug	Dose employed (mg/kg)	Assessment of microvascular damage[a]	Inhibition of platelet aggregation[b]
Aspirin	50,200	+ + +	+ + +
Indomethacin	10,20	+	+ +
Diclofenac	10	+	+ +
Azapropazone	100	0	0
Benoxaprofen	110	0	0
Fenclofenac	100	0	0

[a] As observed by electron microscopy in present studies in regions adjacent to surface mucous and gastric pit parietal cells.
[b] From published reports.

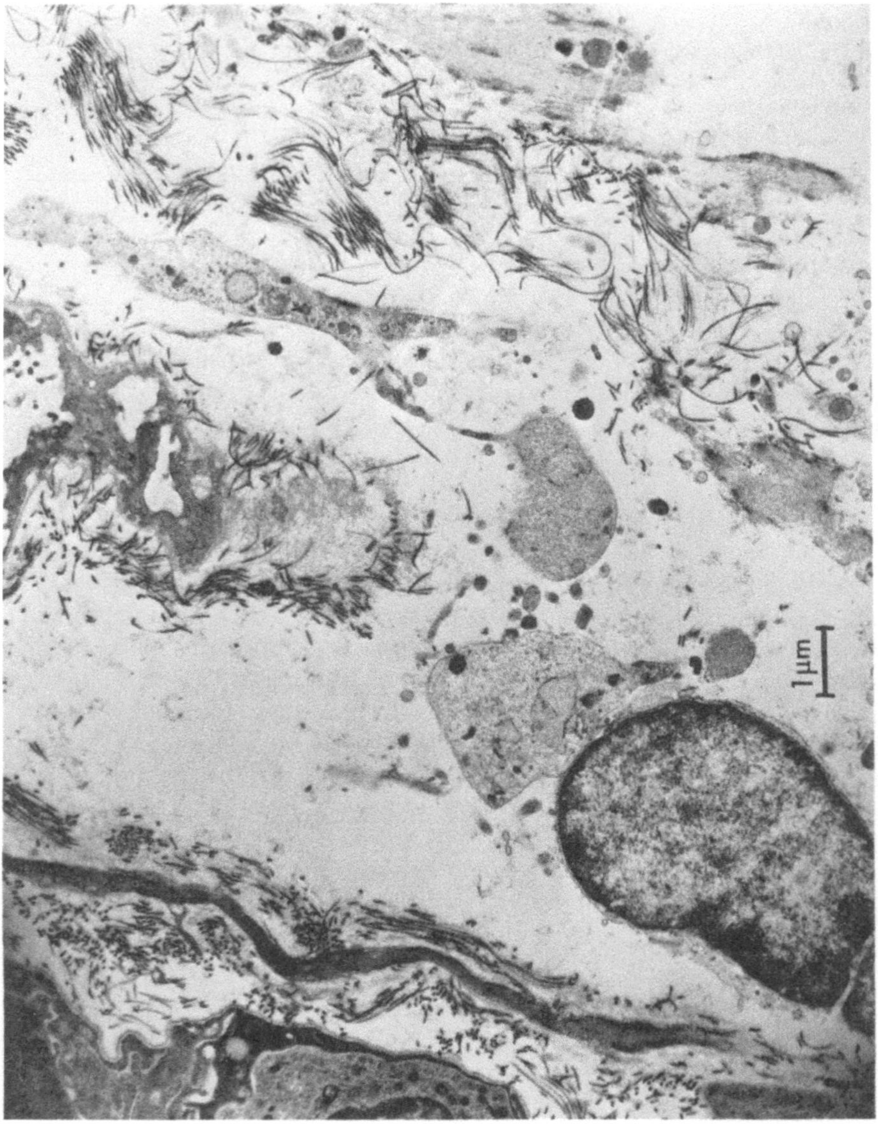

FIGURE 5. Microvascular injury and cellular damage following oral administration of 10 μ/kg piroxicam.

from the more ulcerogenic agents in causing virtually no microscopic damage to the microvasculature. These results parallel those observations with other NSAI drugs reported in the previous section.

An overall summary of the effects of the various NSAI drugs of differing ulcerogenicity on the integrity of the mucosal microvasculature is shown in Table III. It can be seen that there is an apparent relationship between gastric ulcerogenicity of the various NSAI drugs and their propensity to elicit microvascular injury in otherwise undamaged regions of the gastric mucosa. Such microvascular injury, manifest in extravasation of blood compounds into the interstitial space underlying the surface mucous cell layer, might result from a variety of causes. The well-known effect of some NSAI drugs to inhibit platelet aggregation might be considered one factor aiding loss of blood from a previously damaged endothelial cell/basement membrane region. As indicated in Table III, it appears that those drugs inducing microvascular injury also inhibit platelet aggregation. Of course, inhibition of platelet aggregation, which for these NSAI drugs is probably caused by their inhibition of PG cyclooxygenase and thus production of proaggregatory thromboxane A_2 (TxA_2), is balanced by inhibitory effects of these drugs on endothelial cell production of the antiaggregatory PGI_2 (prostacyclin). As noted in the previous section, total mucosal production of this PG is inhibited, albeit variably, by ulcerogenic NSAI drugs at short times after their oral administration. It is conceivable that these two

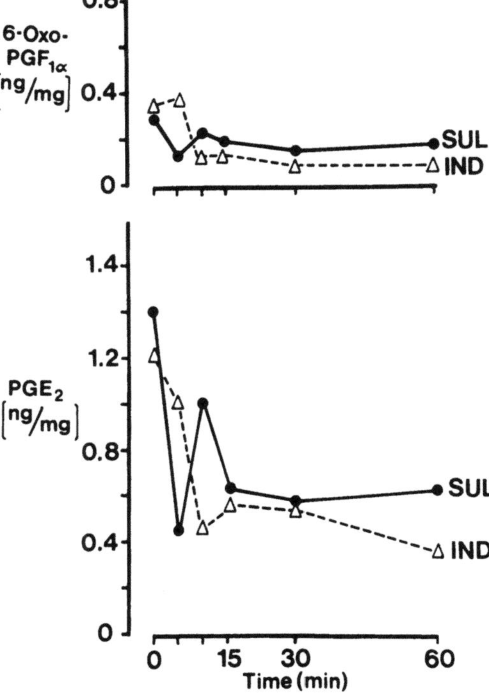

FIGURE 6. Effects of indomethacin (IND, 5 mg/kg p.o.) compared with its prodrug analogue sulindac (SUL, 5 mg/kg p.o.) on the gastric mucosal content of PGE_2 and 6-oxo-$PGF_{1\alpha}$ measured by radioimmunoassay as described by Peskar and co-workers (1978) in biopsies taken from halothane–N_2O-anesthetized pigs (for methods, see Rainsford and Willis, 1982; Peskar *et al.*, 1981). Statistically significant reductions in the content of both prostaglandins were evident at 10 min following administration of both drugs and were maintained for up to 1 hr following drug administration.

effects could virtually cancel out one another assuming equipotent antagonism of platelet TxA$_2$ by PGI$_2$. However, it is possible that the inhibitory effects of NSAI drugs on thromboxane-dependent platelet aggregation could dominate over the antiaggregatory effects of endothelial PGI$_2$ once the endothelium/basement membrane has been disrupted, e.g., from the actions of oxy radicals or, alternatively, histamine released from mast cells such as has been shown with aspirin and salicylic acid (Rainsford, 1984).

4. STUDIES IN PIGS

Intragastric administration of indomethacin (5 mg/kg) to halothane/nitrous oxide-anesthetized pigs reduces the content of both PGE$_2$ and 6-oxo-PGF$_{1\alpha}$ in fundic mucosa initially at 10 min (Rainsford and Willis, 1982), and this is maintained for up to 1 hr following drug administration (Fig. 6). Coincidentally, there is rapid absorption of radiolabeled indomethacin (Fig. 7), although peak levels are not reached until 15 min. No metabolism of indomethacin to its desmethyl or other metabolites appears to occur in the mucosa during these early stages, so that the drug effects are obviously caused by the parent drug. Coincident with the inhibition of PGE$_2$ and 6-keto-PGF$_{1\alpha}$ production at 10 min, there are clear signs of fine structural damage to the fundic mucosal cells. Thus, cell sloughing, karyolysis, and generalized damage are evident in superficial mucous cells (Rainsford and Willis, 1982). Parietal cells and mucosal capillaries are disrupted at the same period,

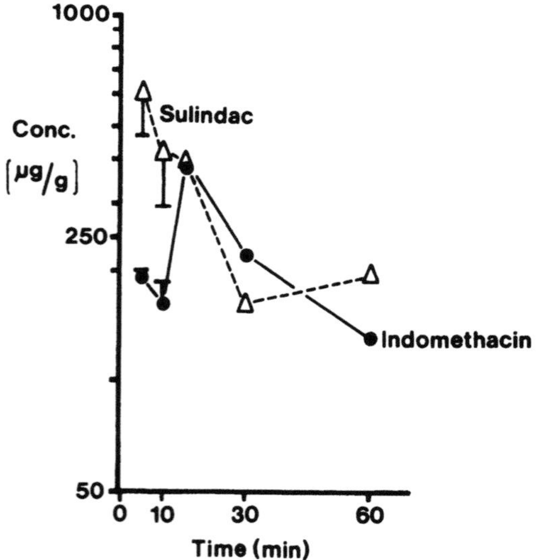

FIGURE 7. Gastric mucosal concentrations of radiolabeled indomethacin compared with sulindac (both given intragastrically at doses of 5 mg/kg). [2-^{14}C]Indomethacin and [^{3}H]sulindac were dosed intragastrically (5 mg/kg with 10 Ci/kg labeled drug) to groups of three halothane/nitrous oxide-anesthetized pigs, as previously described (Rainsford and Willis, 1982). Biopsies (50–100 mg wet weight) were taken at 5, 10, 15, 30, and 60 min, and 0.2 M citric acid (pH 3) homogenates were prepared, from which chloroform extracts were obtained. Aliquots of these concentrated extracts were subjected to thin-layer chromatography on Merck Silica Gel F$_{254}$ plates developed in chloroform/ethyl acetate (95 : 5) together with authentic standards of either indomethacin and its desmethyl metabolite or sulindac and its sulfide and sulfone metabolites. None of these metabolites of indomethacin or sulindac were detectable in the gastric mucosa.

and adjacent to the latter there is appearance of blood cell components in the interstitial tissue (Rainsford and Willis, 1982). These changes progress with time up to 60 min following intragastric administration of indomethacin and are accompanied by efflux of Na^+, K^+, and Cl^- ions into the gastric lumen (Rainsford and Willis, 1982). This suggests that the early drug-induced changes in surface mucosal integrity are not exactly tightly coupled to alterations in mucosal prostaglandin production. Furthermore, it is obvious that, mucosal synthesis of prostaglandins takes some 5–10 min to be affected by the drug despite initial rapid absorption of the drug (Figs. 6, 7).

Identical studies to those above were performed with the indomethacin analogue sulindac, which is a prodrug and undergoes metabolism to its active sulfide metabolite *in vivo*. Sulindac itself is much less potent as an inhibitor of PG synthesis *in vitro* than its sulfide metabolite (Brune *et al.*, 1981). Thus, it seemed possible to employ this drug to discriminate PG-dependent from PG-independent events in the early stages of development of gastric mucosal injury. For this study it was also important to establish the rate of drug absorption and also to ascertain if it is metabolized in the mucosa.

The results of this study (Fig. 7) show that radiolabeled sulindac was rapidly

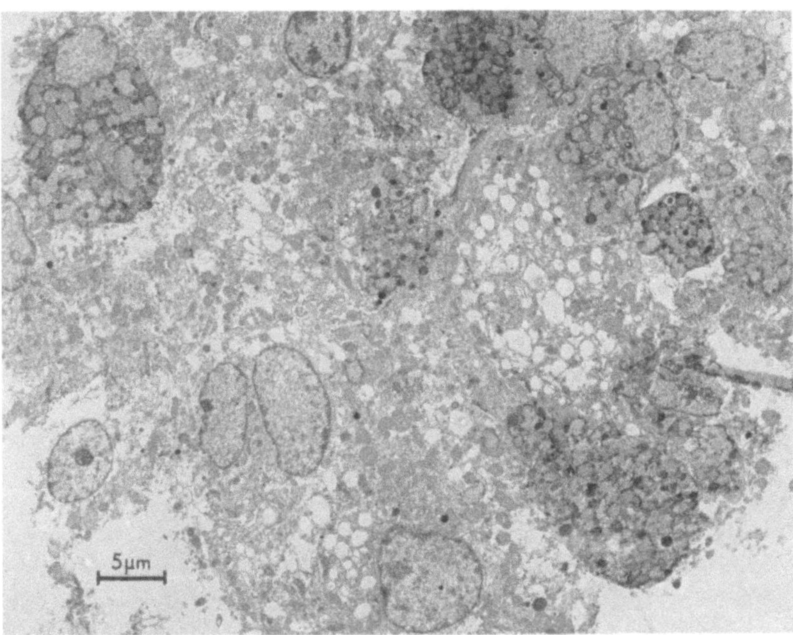

FIGURE 8. Gastric mucosal damage by oral sulindac (5 mg/kg) in biopsies of pigs. A: At 30 min, showing sloughing of mucous and parietal cells together with vacuole formation. B, C: At 5 min, where extensive disruption of interstitial region is evident with accompanying infiltration of blood cells. These changes resemble those seen in pigs dosed with indomethacin (see Rainsford and Willis, 1982).

absorbed, even more so than indomethacin. There was, however, no detectable sulindac sulfide (or, indeed, its other sulfone metabolite) present in the gastric mucosa up to 1 hr after administration of sulindac. It came, therefore, as a surprise that despite the apparent lack of the sulfide metabolite, there was a marked reduction evident by 10–15 min following administration of sulindac in the content of PGE_2 and 6-keto-$PGF_{1\alpha}$ (Peskar *et al.*, 1981) that was comparable to that achieved with the same dose of indomethacin (Fig. 5).

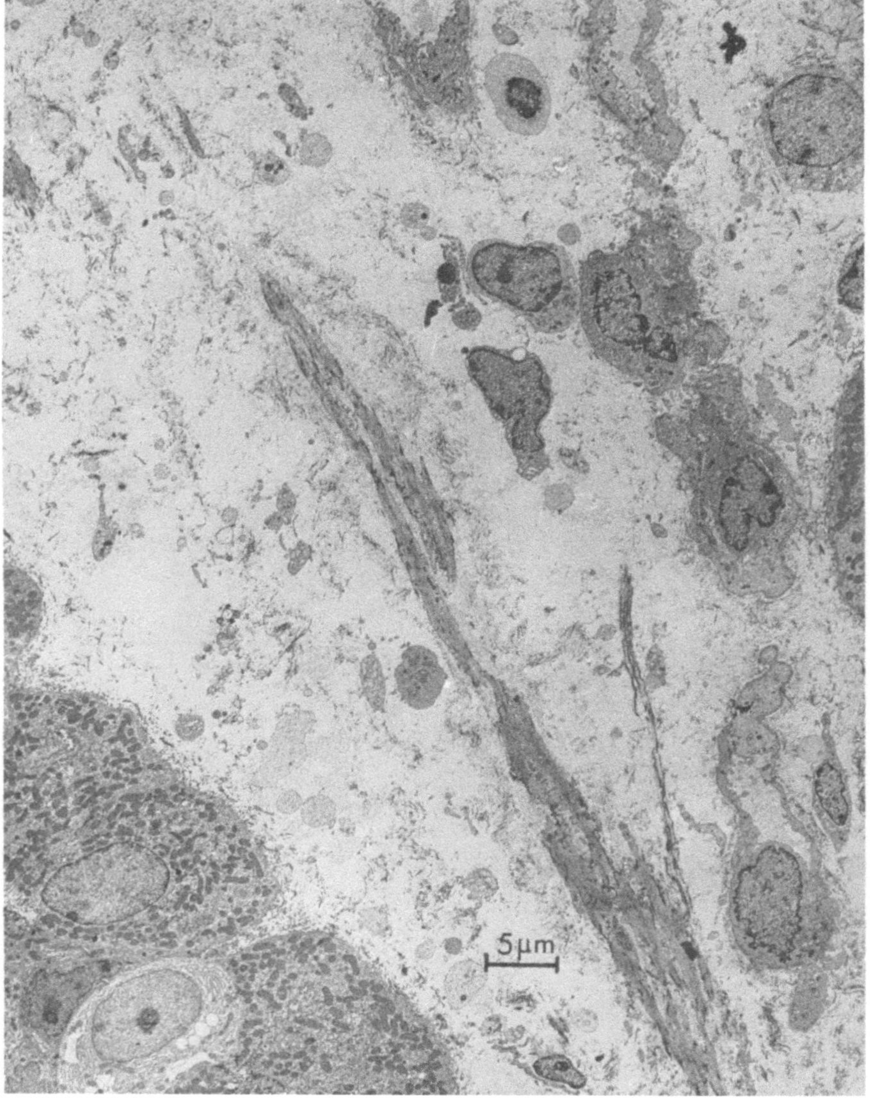

FIGURE 8. (*Cont.*)

Thus, the observed reduction in PGE_2 and 6-oxo-$PGF_{1\alpha}$ levels could only have resulted from the high concentration of the parent drug. Comparison of the effects of these drugs on PGE_2 production by a typical *in vitro* cell system, the mouse macrophage (Brune *et al.*, 1981; see also Table I) shows that sulindac has an IC_{50} of 7.2×10^{-6} M, whereas its active sulfide metabolite has an IC_{50} of 5.8×10^{-9} M, which is comparable with that for indomethacin (IC_{50} 1.9×10^{-9} M). Since mucosal concentrations of about 1 mmol/kg of sulindac are achieved within the

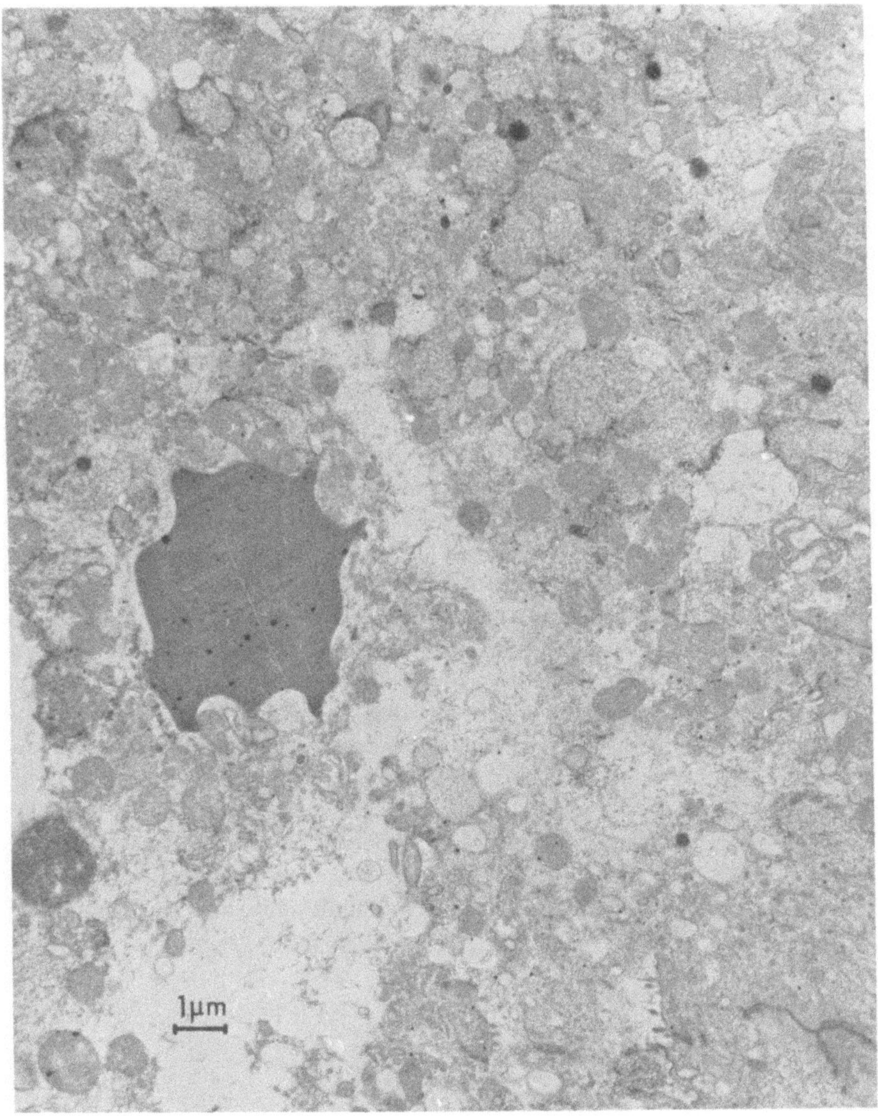

FIGURE 8. *(Cont.)*

first 10 min of oral administration (Fig. 7), it appears that sufficient parent drug could be present to elicit inhibition of gastric mucosal PG production.

Correlated with the reduction in mucosal content of prostaglandins, it was found that sulindac caused damage to mucous cells and disruption of interstitial regions accompanying appearance of blood components at 10–60 min following oral dosage (Fig. 8), although to a somewhat lesser extent than indomethacin. Some parietal cell damage was also evident with sulindac.

These results show that there is a rapid reduction in mucosal content of PGs following intragastric administration of indomethacin and sulindac to pigs and that this parallels rapid drug absorption and the early development of fine structural injury to the mucosal cells. Although drug-induced alterations in mucosal membrane permeability may not be strictly coupled to inhibition of PG production (Rainsford and Willis, 1982), it does appear that the principal structural changes induced by indomethacin and sulindac do coincide with gross changes in PG content. The ultrastructural changes induced by these drugs in pigs (Fig. 8) are the same as those induced in rats by indomethacin, aspirin, and other similar ulcerogenic drugs (Fig. 4).

5. CONCLUSIONS

For the potent ulcerogens such as aspirin and indomethacin, inhibition of prostaglandin production in the gastric mucosa coincides with rapid absorption of these drugs and the early appearance of fine structural damage to mucous, parietal, and microvascular cell components. Less ulcerogenic agents may exhibit variable effects on prostaglandin production even though they are well absorbed.

One of the reasons for the low ulcerogenicity of azapropazone and benoxaprofen might be related to the effects these drugs have on inhibiting 5-lipoxygenase activity in the mucosa, thereby producing balanced inhibition of both cyclooxygenase and lipoxygenase pathways.

There appears to be an association between the propensity of the ulcerogenic NSAI drugs to inhibit platelet aggregation and the appearance of microvascular damage. Those drugs showing little or no tendency to produce fine structural changes in the mucosa do not appear to produce microvascular changes, and neither do they inhibit platelet aggregation. The drug-induced changes in microvascular integrity might, therefore, be related to their effects on platelet aggregation and may contribute to ischemic reactions in the mucosa during the early stages of NSAI-drug-induced mucosal injury.

REFERENCES

Boughton-Smith, N. K., and Whittle, B. J. R., 1983, Stimulation and inhibition of prostacyclin formation in the gastric mucosa and ileum *in vitro* by anti-inflammatory agents, *Br. J. Pharmacol.* **78:**173–180.

Brune, K., Rainsford, K. D., Wagner, K., and Peskar, B. A., 1981, Inhibition by anti-inflammatory drugs of prostaglandin production in cultured macrophages. Factors influencing the apparent drug effects, *Naunyn Schmiedebergs Arch. Pharmacol.* **315:**269–276.

Harding, R. K., and Morris, G. P., 1976, Pathological effects of aspirin and of haemorrhagic shock on the gastric mucosa of the rat, in: *Scanning Electron Microscopy/1976*, Part V, *Proceedings of the Workshop on Advances in Biomedical Applications of the SEM*, ITT Research Institute, Chicago, pp. 253–262.

Peskar, B. A., Anhut, H., Kroner, E. E., and Peskar, B. M., 1978, Development, specificity and some applications of radioimmunoassays for prostaglandins and related compounds, in: *Advances in Pharmacology and Therapeutics, Vol. 7, Biochemical Clinical Pharmacology* (J. P. Tillement, ed.), Pergamon Press, Oxford, pp. 278–286.

Peskar, B. M., Rainsford, K. D., Brune, K., and Gerok, K., 1981, Effekt nichsteroidartiger Antiphlogistika auf Plasma- und Magenmukosakonzentrationen von Prostaglandinen, *Verh. Deut. Ges. Inn. Med.* **87:**833–838.

Peskar, B. M., Weiler, H., and Meyer, C., 1984, Inhibition of prostaglandin production in the gastrointestinal tract by antiinflammatory drugs, in: *Side-Effects of Anti-Inflammatory/Analgesic Drugs* (K. D. Rainsford and G. P. Velo, eds.), Raven Press, New York, pp. 39–50.

Rainsford, K. D., 1975, Electron microscopic observations on the effects of orally-administered aspirin and aspirin–bicarbonate mixtures on the development of gastric mucosal damage in the rat, *Gut* **16:**514–527.

Rainsford, K. D., 1981, Comparison of the gastric ulcerogenic activity of new non-steroidal anti-inflammatory drugs in stressed rats, *Br. J. Pharmacol.* **73:**226P–227P.

Rainsford, K. D., 1984, Mechanisms of gastrointestinal ulceration by nonsteroidal anti-inflammatory/analgesic drugs, in: *Side-effects of Anti-Inflammatory/Analgesic Drugs* (K. D. Rainsford and G. P. Velo, eds.), Raven Press, New York, pp. 51–64.

Rainsford, K. D., and Willis, C., 1982, Relationship of gastric mucosal damage induced in pigs by anti-inflammatory drugs to their effects on prostaglandin production, *Dig. Dis. Sci.* **27:**624–635.

Rainsford, K. D., Fox, S. A., and Osborne, D. J., 1984, Comparative effects of some non-steroidal anti-inflammatory drugs on the ultrastructural integrity and prostaglandin levels in the rat gastric mucosa: Relationship to drug uptake, *Scand. J. Gastroenterol.* **19**(Suppl. 101):55–68.

Robins, P. G., 1981, Ultrastructural observations on the pathogenesis of aspirin-induced gastric erosions, *Br. J. Exp. Pathol.* **61:**497–504.

62

Biosynthesis of Leukotriene B$_4$ in Human Leukocytes

Demonstration of a Calcium-Dependent 5-Lipoxygenase

RODGER M. McMILLAN, DAVID J. MASTERS, WAYNE W. STERLING, and PETER R. BERNSTEIN

1. INTRODUCTION

Arachidonate metabolism via 5-lipoxygenase leads to formation of a family of biologically active molecules, the leukotrienes. The first step in leukotriene biosynthesis is catalyzed by the enzyme 5-lipoxygenase and results in formation of 5-hydroperoxy-6,8,11,14-eicosatetraenoic acid (5-HPETE), which may be converted enzymatically (via the action of a peroxidase) or nonenzymically to the corresponding hydroxy acid, 5-hydroxy-6,8,11,14-eicosatetraenoic acid (5-HETE). Alternatively, 5-HPETE may undergo enzymatic dehydration to form an unstable epoxide, 5,6-epoxy-7,9,11,14-eicosatetraenoic acid, which is termed leukotriene A$_4$ (LTA$_4$). Enzymatic hydrolysis of LTA$_4$ yields 5S,12R-dihydroxy-6,14-*cis*-8,10-*trans*-eicosatetraenoic acid (leukotriene B$_4$, LTB$_4$), which is a potent, stereospecific activator of a variety of human neutrophil responses, including chemokinesis, chemotaxis, and aggregation (Ford-Hutchinson *et al.*, 1980; Goetzl and Pickett, 1981; Malmsten *et al.*, 1980).

A second pathway of LTA$_4$ metabolism leads to formation of leukotrienes that contain a hydroxyl at C-5 and a thiopeptide at C-6 (Murphy *et al.*, 1979). Four

RODGER M. McMILLAN, DAVID J. MASTERS, and WAYNE W. STERLING • ICI Pharmaceuticals Division, Mereside, Alderley Park, Macclesfield, Cheshire SK10 4TG, United Kingdom. PETER R. BERNSTEIN • Medicinal Chemistry Department, Stuart Pharmaceuticals, Division of ICI Americas Inc., Wilmington, Delaware 19897.

peptidyl leukotrienes have been described—LTC_4, LTD_4, LTE_4, and LTF_4—in which the peptides are, respectively, glutathione, cysteine-glycine, glycine, and cysteine-glutamate (Samuelsson, 1983). The peptidyl leukotrienes have powerful spasmogenic activity, particularly in pulmonary, intestinal, and vascular tissue, and collectively LTC_4, LTD_4, LTE_4, and LTF_4 account for the biological activities previously assigned to "slow-reacting substance of anaphylaxis" (SRS-A).

Biosynthesis of LTB_4 is usually accompanied by formation of several stereoisomers. Nonenzymic hydrolysis of LTA_4 forms no LTB_4 but instead produces two other 5,12-di-HETES, which have a *trans* double bond at position C-6 rather than a *cis* double bond as in LTB_4. The two 6-*trans* isomers of LTB_4 differ in their stereochemistry at C-12 and are classified by the order in which they elute on reverse-phase HPLC: thus, 6-*trans*-LTB_4 is termed isomer I, and 12-*epi*-6-*trans*-LTB_4 is termed isomer II (Borgeat and Samuelsson, 1979b). Another isomer of LTB_4, 5*S*,12*S*-dihydroxy-6,10-*trans*-8,14-*cis*-eicosatetraenoic acid, can be formed in leukocytes by the sequential metabolism of arachidonate by 5- and 12-lipoxygenase. Since LTA_4 is not an intermediate in its biosynthesis, the double lipoxygenation product is termed 5*S*,12*S*-di-HETE (Lindgren *et al.*, 1981; Borgeat *et al.*, 1982). The 6-*trans* isomers are 10–100 times less potent than LTB_4 as activators of human neutrophils, and 5*S*,12*S*-di-HETE is virtually devoid of biological activity (Ford-Hutchinson *et al.*, 1981, 1983; Goetzl and Pickett, 1981; Malmsten *et al.*, 1980).

The stereospecificity of neutrophil activation by LTB_4 suggests that its actions may be receptor mediated, and the presence of high-affinity binding sites for LTB_4 on human neutrophils has been reported (Kreisle and Parker, 1983). An inactivation mechanism for LTB_4 also exists. This involves ω-oxidation to yield a trihydroxy acid, 20-hydroxy-LTB_4, which has 10% of the biological activity of LTB_4 in a neutrophil aggregation assay (Ford-Hutchinson, 1983). Further metabolism of 20-hydroxy-LTB_4 yields a dicarboxylic acid, 20-carboxyl-LTB_4, which has less than 1% of the activity of LTB_4 (Ford-Hutchinson *et al.*, 1983).

Although the biological actions of leukotrienes have been extensively studied, comparatively little information is available on the enzymes of the 5-lipoxygenase pathway. In particular, there have been no reports on cell-free preparations of 5-lipoxygenase and related enzymes from human sources. In this chapter, we demonstrate the presence of a calcium-dependent 5-lipoxygenase in human leukocyte cytosol. The characteristics of this enzyme are compared to the previously reported 5-lipoxygenases from animal sources. In addition, we report that the cytosolic fraction of human leukocytes contains all the enzymes necessary for LTB_4 biosynthesis but not its metabolism.

2. MATERIALS AND METHODS

2.1. Materials

Synthetic leukotriene A_4 methyl ester was prepared using the general approach of Rokach *et al.* (1981) as modified by Maduskie and Willard (1983). Synthetic

LTB$_4$ was prepared by modification of the procedure of Corey *et al.* (1981) as reported by Yee (1984). BW755C and 5-HETE were synthesized by Dr. K. H. Gibson (Chemistry Department, ICI Pharmaceuticals Division); 5S,12S-di-HETE was biosynthesized from porcine leukocyte, and 6-*trans* isomers of LTB$_4$ were prepared by hydrolysis of LTA$_4$. [^{14}C]Arachidonic acid (specific activity 60 mCi/ mmol) was purchased from Amersham International. Benoxaprofen was a generous gift from Lilly Research Centre, Surrey, UK; baicalein was obtained from Appin Chemical Company; arachidonic acid, quercetin, phenidone, esculetin, and nor-dihydroguairetic acid (NDGA) were purchased from Sigma Chemical Company (St. Louis, MO).

2.2. Methods

Peripheral human leukocytes (>90% neutrophils) were obtained from anti-coagulated venous blood by a modification of the procedure of Henson (1971). After removal of platelet-rich plasma, red cells were sedimented using 2.5% gelatin in physiological saline at 37°C. After 30 min, the supernatant was aspirated and centrifuged (400 g, 5 min, 25°C). The cell pellet was redissolved in a buffer comprising ammonium chloride (0.15 M), potassium bicarbonate (0.01 M), and EDTA (0.001 M) in order to lyse residual erythrocytes. The resulting leukocytes were washed twice and then resuspended at 6.25 $\times$ 10^7 cells/ml in 0.05 M phosphate buffer, pH 7.0, containing 0.1% gelatin and 1 mM EDTA. Cells were disrupted by Polytron homogenization and sonication; then the homogenate was centrifuged at 105,000 g for 60 min at 4°C. The supernatant (cytosol) was used as the enzyme source. For comparative studies, the high-speed supernatant from RBL-1 cells was prepared using an identical procedure.

For measurement of lipoxygenase activity, glutathione (1 mM) and calcium chloride (2 mM) were routinely added to the cytosolic fraction. Aliquots of cytosol (200 μl) were incubated with drugs (50 μl) for 15 min at 4°C prior to incubation with 15 μM [^{14}C]arachidonate for 20 min at 37°C. Enzyme incubates were terminated by adding four volumes of ethyl acetate. Lipids were separated by thin-layer chromatography using petroleum ether : diethyl ether : acetic acid (50 : 50 : 1). Radioactive metabolites were located by autoradiography and quantified by liquid scintillation counting. In some experiments analysis of di-HETE formation was performed. Ethyl acetate extracts were reconstituted in hexane : ether (95 : 5) and applied to cyanopropyl miniextraction columns. Columns were washed sequentially with 5%, 25%, and 50% ether in hexane, and di-HETEs were finally eluted with ether : methanol (50 : 50). The diHETEs were dissolved in HPLC-grade solvents and separated on reverse-phase HPLC as described in the legend to Fig. 3.

Measurement of LTB$_4$ synthase was carried out using LTA$_4$ as substrate. Optically active LTA$_4$ methyl ester (0.3–3.0 $\times$ 10^{-3} M) was hydrolyzed using 0.25 M lithium hydroxide in dimethoxyethane : water (50 : 50) for 1 hr at 25°C under an atmosphere of argon. The concentration of LTA$_4$ was determined by carrying out a scan of UV absorbance in methanol assuming a molar absorbance coefficient of 51,000 at 279 nm. Human leukocyte cytosol was incubated with LTA$_4$ for 10 min at 37°C, after which lipids were extracted using three volumes of ethyl acetate. Lipids were fractionated on cyanopropyl columns as described above except that the columns were washed with 5% and 25% ether in hexane. The di-HETEs were eluted with ether : methanol (40 : 60) and separated on reverse-phase HPLC as described in the legend to Fig. 4.

3. RESULTS AND DISCUSSION

3.1. 5-Lipoxygenase from Human Leukocytes

Cell-free preparations of 5-lipoxygenase have been reported from animal sources. The first system described was that from rat basophilic leukemia (RBL-1) cells. Jakschik and co-workers (1980a) reported that incubation of radiolabeled arachidonate with the low-speed (10,000 g) supernatant from RBL-1 cells resulted in formation of radiolabeled 5-HETE and 5,12-di-HETES. The characteristics of this 5-lipoxygenase differed in several respects from the lipoxygenases previously described in plants (exemplified by soybean lipoxygenase) and in platelets.

The most obvious difference was the positional specificity of lipoxygenation: arachidonate metabolism by soybean lipoxygenase yielded 15-HETE, whereas platelet lipoxygenases formed 12-HETE. The enzymes also differed in substrate specificity. A range of fatty acids (chain lengths C_{18} to C_{22}) are substrates for the RBL-1 5-lipoxygenase providing they contain *cis* double bonds at C-5, C-8, and C-11 (Jakschik *et al.*, 1980b). In the case of 5-lipoxygenase, the position of oxidation is determined by the distance from the carboxyl terminal, but for other lipoxygenases it is the distance from the methyl terminal that is critical: thus, the soybean and platelet enzymes catalyze lipoxygenation at ω-6 and ω-9, respectively, independent of fatty acid chain length (Galliard and Chan, 1980; Nugteren, 1975). The third and arguably most important difference from other lipoxygenases was that calcium was essential for 5-lipoxygenase activity (Jakschik *et al.*, 1980a). Since the RBL-1 system was first described, cell-free preparations have been obtained from guinea pig neutrophils (Ochi *et al.*, 1983) and from a cloned mouse mastocytoma cell line, P815 (Koshihara *et al.*, 1982). Like the RBL-1 enzyme, both appear to be soluble and calcium dependent, and we now provide evidence for the presence of a calcium-dependent 5-lipoxygenase in human leukocytes.

In initial studies we compared the lipoxygenase products of human leukocytes with those of the well-characterized RBL-1 system. When incubated with the cytosolic fraction (105,000 g supernatant) from RBL-1 cells, [^{14}C]arachidonate was converted to [^{14}C]5-HETE and to more polar radiolabeled metabolites that comigrated with di-HETEs. Metabolism was blocked by BW755C (10^{-5} M), which inhibits both cyclooxygenase and lipoxygenases, but not by indomethacin (10^{-5} M), a selective cyclooxygenase inhibitor. In parallel experiments with the cytosol from human leukocytes, arachidonate was again converted to 5-HETE and to di-HETEs as well as to an additional less polar metabolite that migrated on TLC between 5-

HETE and arachidonate. Formation of all metabolites in the human system was blocked by BW755C but not by indomethacin. The additional product in the human cytosol is not the Δ-lactone of 5-HETE:acidification, which accelerates lactonization, was omitted during lipid extraction, and the additional product was not formed in RBL-1 cytosol with an identical protocol.

Thus, human leukocytes contain a soluble 5-lipoxygenase, but the cytosolic fraction also produces an additional lipoxygenase product that has not been unequivocally identified but, on the basis of TLC mobility, is likely to be either 12-or 15-HETE. Human leukocytes have been shown to release 12-HETE and 15-HETE in addition to 5-HETE (Borgeat and Samuelsson, 1979a), and purified preparations of 15-lipoxygenase and 12-lipoxygenase have been reported from rabbit peritoneal neutrophils and porcine peripheral leukocytes, respectively (Narumiya *et al.*, 1982; Yoshimoto *et al.*, 1982). The cellular source of the additional lipoxygenase has not been determined, and experiments are in progress to determine whether it originates from a different leukocyte type than the 5-lipoxygenase.

In common with the enzymes from RBL-1 cells, P815 cells, and guinea pig neutrophils, the human 5-lipoxygenase has an absolute requirement for calcium (Table I). Since the cytosol was prepared in buffer containing 1 mM EDTA, an excess of calcium (2 mM) was routinely added to study the calcium dependence. Under these conditions, arachidonate was converted to 5-HETE and to the product that comigrated with 12-HETE. When calcium was omitted, synthesis of 5-HETE was reduced to the level in the heat-inactivated enzyme blank, and formation of the additional, unidentified hydroxy acid was also markedly inhibited. The mechanism of activation of 5-lipoxygenase by calcium has been investigated by Parker's group using a purified enzyme from RBL-1 cells (Parker and Aykent, 1982). They reported that in the absence of calcium this enzyme existed in an inactive form with an apparent molecular weight of approximately 90,000. Addition of calcium ions, but not magnesium or manganese, activated the enzyme and increased its molecular weight to 180,000. Addition of the calcium regulatory protein calmodulin

TABLE I. Effect of Calcium on Human 5-Lipoxygenase[a]

	Synthesis of hydroxy acids (ng/ml)	
Treatment	5-HETE	Unidentified HETE
Heat inactivated		
(1 mM EDTA, 2 mM CaCl₂)	15 ± 5	10 ± 5
Control		
(1 mM EDTA, 2 mM CaCl₂)	845 ± 50	935 ± 15
Calcium-free		
(1 mM EDTA)	35 ± 5	165 ± 50

[a] Aliquots of human leukocyte cytosol, with or without added calcium chloride (CaCl₂), were incubated with [¹⁴C]arachidonate (5 μg/ml) for 20 min at 37°C. Lipids were separated by thin-layer chromatography, and HETE formation was quantified by liquid scintillation counting. Results are mean values ± S.E.M. ($n = 3$).

did not increase the sensitivity of the enzyme to calcium, and Parker and Aykent (1982) proposed that calcium ions interact directly with inactive monomers of 5-lipoxygenase to produce an active dimeric form of the enzyme.

Human 5-lipoxygenase was potently inhibited by a range of known lipoxygenase inhibitors. Among the most potent was NDGA, which had an IC_{50} value of approximately 3×10^{-8} M (Fig. 1). The prototype dual inhibitor of lipoxygenase and cyclooxygenase, BW755C, was approximately ten times less potent than NDGA, but benoxaprofen, which has been reported to be a lipoxygenase inhibitor, was virtually inactive against human 5-lipoxygenase at concentrations up to 100 μM (Fig. 1). We have found benoxaprofen to be similarly ineffective as an inhibitor of the soluble 5-lipoxygenase from RBL-1 cells, although it produced weak inhibition of RBL-1 cyclooxygenase (IC_{50} 40 μM). Other potent inhibitors of human 5-lipoxygenase (in descending order of potency) were quercetin, baicalein, phenidone, nafazatrom, and esculetin, all of which had IC_{50} values of less than 10^{-6} M.

5.2. Biosynthesis of LTB$_4$ in Human Leukocytes

Biosynthesis of LTB$_4$ by human leukocytes was first reported by Borgeat and Samuelsson (1979a). In suspensions of human neutrophils, they showed that arachidonate was metabolized to a mixture of monohydroxy acids (5-HETE, 12-HETE, and 15-HETE) and to a number of more polar metabolites including LTB$_4$

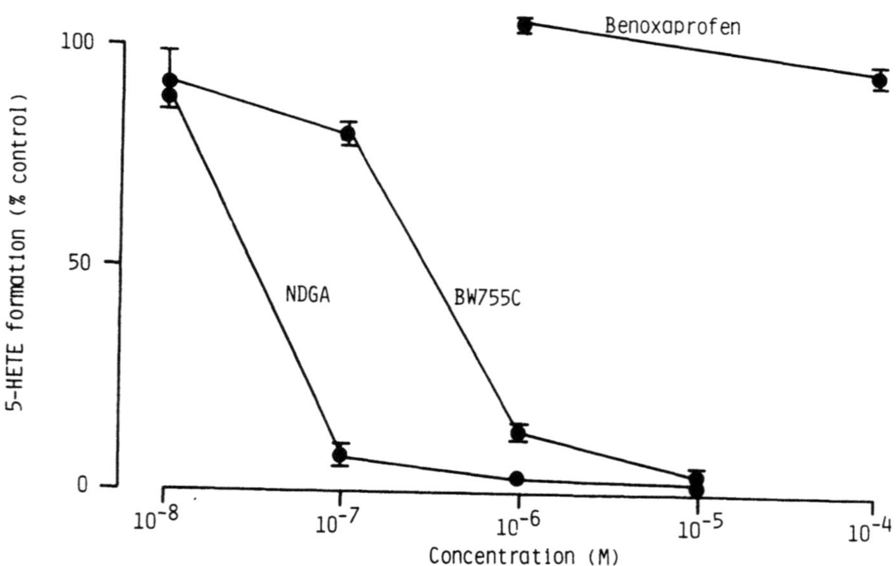

FIGURE 1. Inhibition of human 5-lipoxygenase. Enzyme activity was quantified using 5-HETE synthesis in human leukocyte cytosol following 15 min pre-incubation with drugs. Points represent mean values ±S.E.M. (n = 3).

and its 6-*trans* isomers. In that study, the Karolinska group observed considerable variation in the levels of 5-HETE and 5,12-di-HETEs synthesized from exogenous arachidonate by different preparations of human neutrophils, but they noted that addition of a calcium ionophore, A23187, to the cells produced a marked stimulation in synthesis of 5-lipoxygenase metabolites. In subsequent studies, A23187-stimulated neutrophils have become a classical source of LTB_4 and other lipoxygenase products. The use of such an extreme, and unphysiological, stimulus has been criticized; for example, Smith (1981) described exposure of cells to A23187 as "comparable to chemical homogenization." Evidence for LTB_4 biosynthesis in response to more relevant inflammatory stimuli (phagocytic, chemotactic agents) is still sparse. Release of LTB_4 from neutrophils exposed to formyl peptides and phagocytic stimuli such as urate crystals and serum-treated zymosan has been reported (Palmer and Salmon, 1983; Rae *et al.*, 1982), but in general the levels of LTB_4 released are much lower than those induced by A23187. An interesting report from Clancy *et al.* (1983) suggests that chemotactic agents may stimulate leukotriene biosynthesis by activating 5-lipoxygenase and not by releasing arachidonate via phospholipase activation as is the case with A23187.

Despite the reservations about the relevance of stimulating neutrophils with A23187, the system has provided considerable information about LTB_4 biosynthesis and metabolism. Release of LTB_4 induced by A23187 is accompanied by formation of its 6-*trans* isomers (Borgeat and Samuelsson, 1979b), the double lipoxygenation product 5*S*,12*S*-di-HETE (Lindgren *et al.*, 1981), and of ω-oxidation metabolites of di-HETEs (Hansson *et al.*, 1981). The time course of formation of LTB_4 and its isomers in human leukocytes is shown in Fig. 2. Maximal levels of LTB_4 were achieved within 5 min of exposure to A23187. Thereafter, LTB_4 disappeared rapidly: by 15 min LTB_4 levels were less than 20% of those detected at 6 min. The disappearance of LTB_4 was associated with a rapid increase in the levels of polar metabolites, which were virtually maximal by 15 min. Turnover of 5*S*,12*S*-di-HETE and of 6-*trans* isomers of LTB_4 occurred more slowly than LTB_4: even after 30 min the levels of these isomers had fallen by less than 30% (Fig. 2). These data suggested that there may be a stereospecific inactivation mechanism for LTB_4, and direct evidence for this was recently provided by Powell (1984), who demonstrated that LTB_4 is the preferred substrate for a 20-hydroxylase from human neutrophils. Activation of neutrophils by LTB_4 has previously been shown to be stereospecific (Malmsten *et al.*, 1980; Goetzl and Pickett, 1981; Ford-Hutchinson *et al.*, 1981), and the specificity of metabolism provides further support for a role for LTB_4 as an endogenous neutrophil mediator.

Although a number of groups have studied LTB_4 biosynthesis in intact neutrophils, little information is available on the intracellular localization of LTB_4-synthesizing enzymes. We have found that the cytosolic fraction of human leukocytes contains not only 5-lipoxygenase but also the additional enzymes necessary for LTB_4 biosynthesis: LTA_4 synthase and LTB_4 synthase. Figure 3 shows the profile of di-HETEs produced from incubation of [^{14}C]arachidonate with human leukocyte cytosol. The di-HETE fraction was separated on reverse-phase HPLC

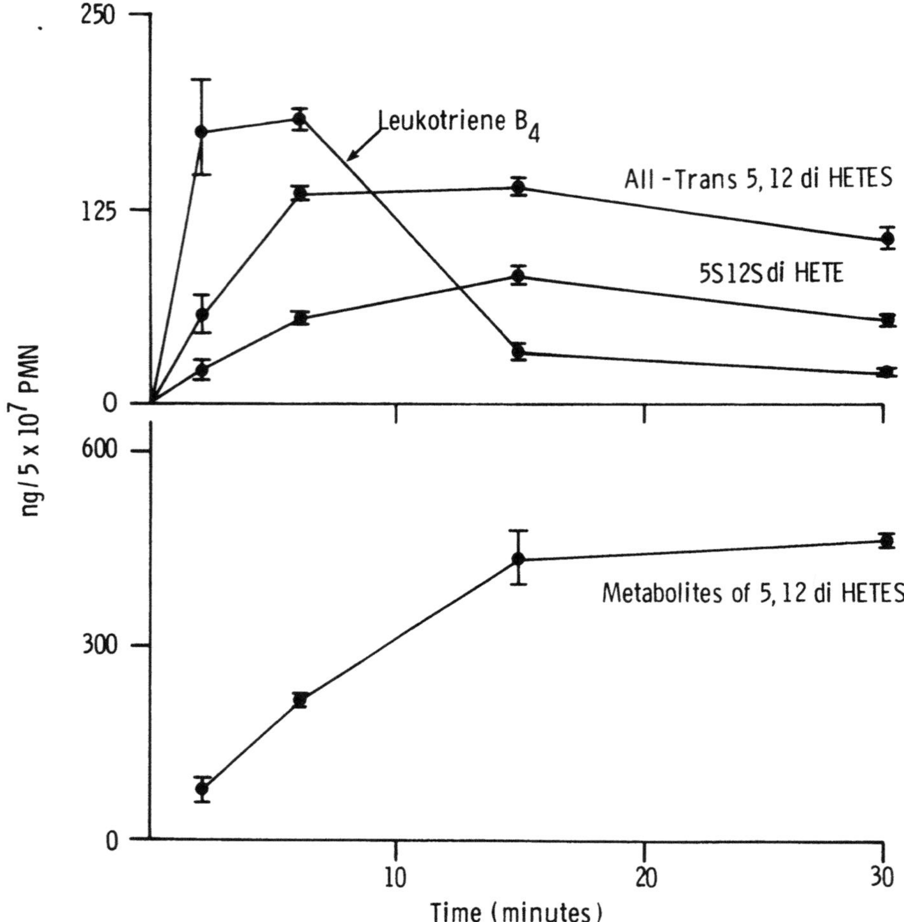

FIGURE 2. Turnover of leukotriene B$_4$ and its isomers in human leukocytes. Suspensions of human leukocytes (5 × 10^7/ml) were incubated for various times with A$_{23187}$ (10 µg/ml). Leukotriene B$_4$, 5S12S diHETE, 6-*trans* isomers (termed "all-*trans* 5,12 diHETEs") and polar metabolites were separated by reverse phase HPLC on Spherisorb ODS1 column using methanol : THF : water : acetic acid (33.5 : 27.5 : 39 : 0.01) at a flow rate of 1.5 ml/min.

using a mobile phase developed in our laboratory to effect resolution of LTB$_4$ from its major biological isomers. The major peak of radioactivity and of UV absorbance (peak II) coeluted with synthetic LTB$_4$ with smaller amounts of 6-*trans* isomers (peak I) and 5S,12S-di-HETE (peak III). Formation of LTB$_4$ and its isomers was rapid, with maximal levels synthesized within 5–10 min at 37°C. No evidence for metabolism of di-HETEs was obtained, which is in accord with the demonstration that 20-hydroxylation of LTB$_4$ is catalyzed by a cytochrome P-450 enzyme, which is presumably of microsomal origin (Shak and Goldstein, this volume, Chapter 10).

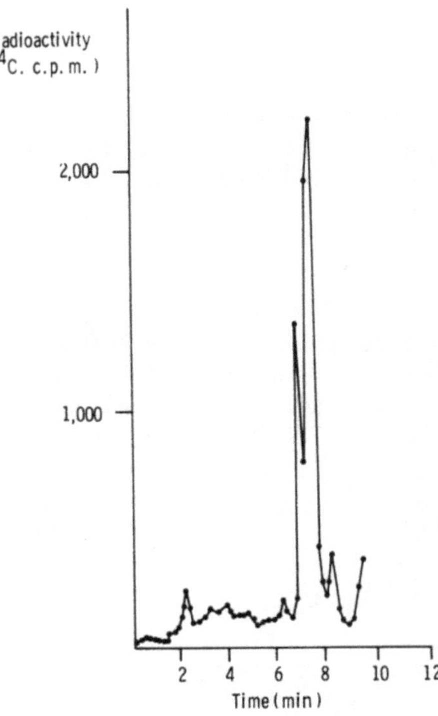

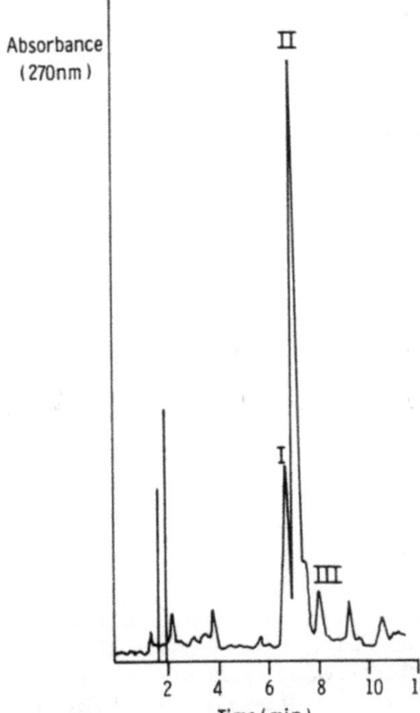

FIGURE 3. Leukotriene B_4 biosynthesis in a cell-free system from human leukocytes. The cytosol from human leukocytes was incubated with [^{14}C]arachidonate (5 μg/ml) for 20 min at 37°C. Formation of leukotriene B_4 and its isomers was monitored by measuring radioactivity and U.V. absorbance (270 nm) following separation on reverse phase HPLC. Chromatography using Spherisorb ODS2 column with a mobile phase of methanol : tetrahydrofuran : water : acetic acid (33.5 : 27.5 : 39 : 0.01) at 1.5 ml/minute. I, retention time of 6-*trans* isomers of Leukotriene B_4; II, retention time of synthetic Leukotriene B_4; III, retention time of 5S12S diHETE.

Direct evidence for the presence of LTB$_4$ synthase was obtained in studies using LTA$_4$. Addition of LTA$_4$ to physiological buffer resulted in formation of two major products that coeluted on reverse-phase HPLC with the 6-*trans* isomers of LTB$_4$ (Fig. 4). Incubation of LTA$_4$ with human leukocyte cytosol at 37°C for 10 min yielded an additional product, which had an identical retention time to synthetic LTB$_4$. At low concentrations of LTA$_4$ (2.8 μM), the major di-HETE formed was LTB$_4$, but at higher LTA$_4$ concentrations (28 μM), the 6-*trans* isomers predominated. These data demonstrate the presence of a soluble LTB$_4$ synthase in human leukocytes as previously reported for RBL-1 cells (Maycock *et al.*, 1982). Our results indicate saturation of LTB$_4$ synthase at high LTA$_4$ concentrations, leading to accumulation of the 6-*trans* isomers of LTB$_4$ by nonenzymic hydrolysis of LTA$_4$. Using the cytosolic fraction from RBL-1 cells, Jakschik and Kuo (1983) reported that LTB$_4$ biosynthesis from arachidonate was maximal at lower substrate concentrations than biosynthesis of LTA$_4$ or 5-HETE. Thus, it appeared that enzymatic hydrolysis of LTA$_4$ was the rate-limiting step in LTB$_4$ biosynthesis. Recently, Fitzpatrick *et al.* (1983) demonstrated the presence in plasma of an enzyme that converts LTA$_4$ to LTB$_4$. This extracellular enzyme is likely to be of secondary

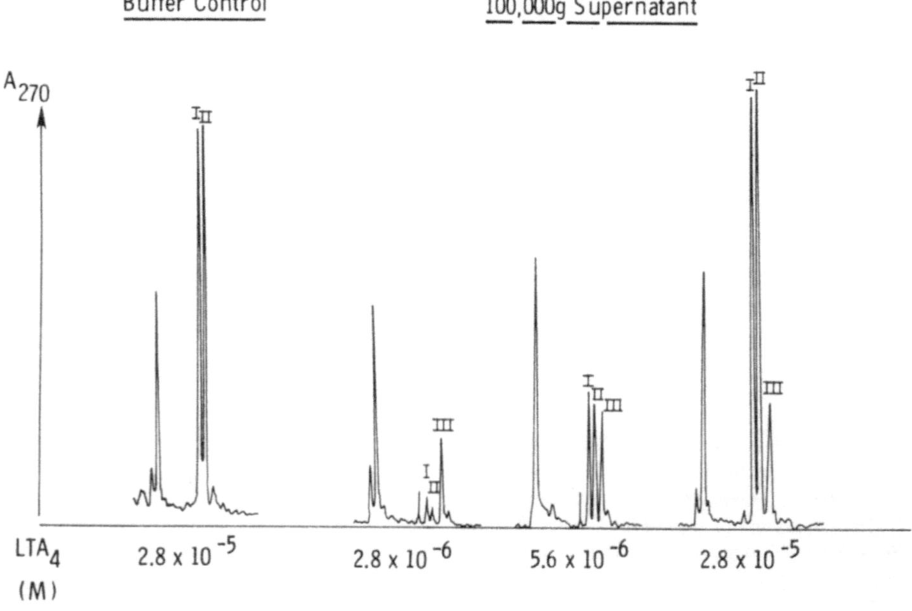

FIGURE 4. Leukotriene B$_4$ biosynthesis in a cell-free system from human leukocytes. The cytosol from human leukocytes was incubated with LTA$_4$ for 20 min at 37°C. Formation of LTB$_4$ and its 6-*trans* isomers was monitored by UV absorbance (270 nm) following separation on reverse-phase HPLC. Chromatography using Spherisorb ODS2 column with a mobile phase of methanol : water : acetic acid (78 : 22 : 0.05) at 1.5 ml/min. I, retention time of 6-*trans* LTB$_4$; II, retention time of 6-*trans*, 12-*epi*-LTB$_4$; III, retention time of synthetic LTB$_4$.

importance to intracellular LTB$_4$ synthase: the close association of 5-lipoxygenase, LTA$_4$ synthase, and LTB$_4$ synthase in the cytosol offers more efficient control of LTB$_4$ biosynthesis.

An important point that remains to be determined is the specificity of the enzyme that catalyzes conversion of LTA$_4$ to LTB$_4$. This enzyme has been referred to as LTB$_4$ synthase, LTA$_4$ hydrolase, or epoxide hydrolase. We prefer to use the term LTB$_4$ synthase until the selectivity of the enzyme is established. Classical epoxide hydrolases generally catalyze conversion of 1,2-epoxides to 1,2-diols. The enzymes are widely distributed and at least two types are known—microsomal and cytosolic forms—which differ in substrate specificity (Hammock and Hasagawa, 1983), pH optima (Ota and Hammock, 1980), and pharmacological sensitivity (Gill and Hammock, 1980). Oliw *et al.* (1982) have shown that microsomal epoxide hydrolase can hydrolyze a series of arachidonic acid epoxides formed by a P-450-linked monooxygenase. An example is shown in Fig. 5: oxygenation of the 5,6 double bond yields 5,6-epoxyeicosatrienoic acid, which in turn is hydrolyzed to the corresponding diol, 5,6-hydroxyeicosatrienoic acid. Epoxides can be formed in this manner across all four double bonds in arachidonate, and each can serve as a substrate for epoxide hydrolase. With the exception of the 8,9-epoxide, they have similar initial rate constants, suggesting a relatively broad substrate specificity for the enzyme. Although LTA$_4$ is also a 5,6-epoxide, its structure is quite distinct since it retains all four double bonds from arachidonate, three of which are conjugated in a highly reactive triene arrangement.

Several lines of evidence suggest that the mechanism of action of LTB$_4$ synthase may be distinct from that of classical epoxide hydrolases. (1) Leukotriene A$_4$ is not converted enzymatically to 5,6-di-HETE but instead forms LTB$_4$, a reaction that involves double bond rearrangement in the triene. (2) Leukotriene B$_4$ possesses a

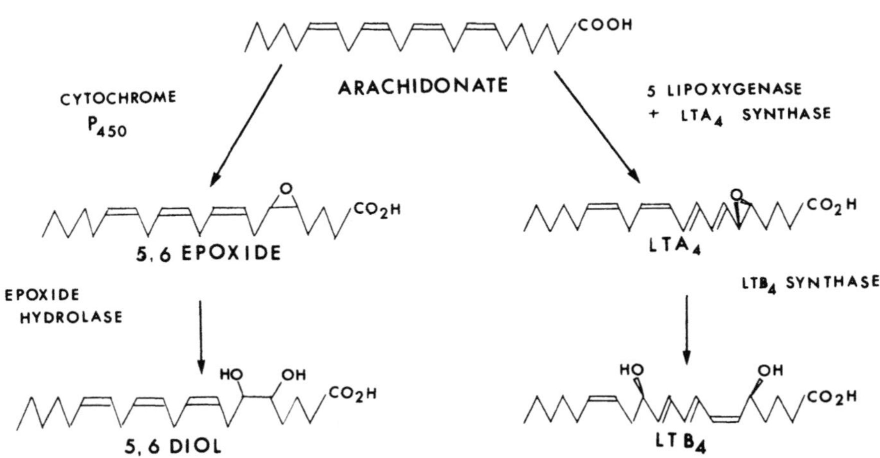

FIGURE 5. Formation of epoxides and diols from arachidonic acid.

cis double bond at C-6, whereas chemical hydrolysis of LTA$_4$ yields diols with 6-*trans* geometry; thus, LTB$_4$ synthase must confer a high degree of structural specificity. (3) 14,15-Leukotriene A$_4$ does not appear to form a product analogous to LTB$_4$, which indicates a degree of substrate specificity that is not apparent with metabolism of arachidonate epoxides by epoxide hydrolase. (4) The apparent molecular weight and terminal amino acid sequence of LTB$_4$ synthase from human leukocytes are reported to differ from those of classical epoxide hydrolases (Samuelsson, this volume, Chapter 1). Further studies are necessary to determine the specificity (or otherwise) of LTB$_4$ synthase, and it will be of particular interest to determine the products of LTA$_4$ metabolism by purified cytosolic epoxide hydrolase.

ACKNOWLEDGMENTS. We are grateful to Dr. Al Willard, to Dr. Ying Yee (Stuart Pharmaceuticals) for supplying synthetic leukotrienes, and to Dr. Keith Gibson for providing synthetic 5-HETE. We are also grateful to Ms. Karen Spruce and Ms. Vanessa Vickers for technical assistance, to Mr. M. Shaw for supplying RBL-1 cells, and to Mavis Brightwell for typing the manuscript.

REFERENCES

Borgeat, P., and Samuelsson, B., 1979a, Arachidonic acid metabolism in polymorphonuclear leukocytes: Effects of ionophore A23187, *Proc. Natl. Acad. Sci. U.S.A.* **76**:2148–2152.

Borgeat, P., and Samuelsson, B., 1979b, Metabolism of arachidonic acid in polymorphonuclear leukocytes. Structural analysis of novel hydroxylated compounds, *J. Biol. Chem.* **254**:7865–7869.

Borgeat, P., de LaClos, B. F., Picard, S., Drapeau, J., Vallerand, P., and Corey, E. J., 1982, Studies on the mechanism of formation of the 5S,12S-dihydroxy-6,8,10,14(*E,Z,E,Z*)-icosatetraenoic acid in leukocytes, *Prostaglandins* **23**:713–724.

Clancy, R. M., Dahinden, C. A., and Hugli, T. E., 1983, Arachidonate metabolism by human polymorphonuclear leukocytes stimulated by N-formyl-Met-Leu-Phe or complement component C5a is independent of phospholipase activation, *Proc. Natl. Acad. Sci. U.S.A.* **80**:7200–7204.

Corey, E. J., Marfat, A., Munroe, J., Kim, K. S., Hopkins, P. B., and Brion, F., 1981, A stereocontrolled and effective synthesis of leukotriene B, *Tetrahedron Lett.* **22**:1077–1080.

Fitzpatrick, F., Haeggstrom, J., Granstrom, E., and Samuelsson, B., 1983, Metabolism of leukotriene A$_4$ by an enzyme in blood plasma: A possible leukotactic mechanism, *Proc. Natl. Acad. Sci. U.S.A.* **80**:5425–5429.

Ford-Hutchinson, A. W., Bray, M. A., Doig, M. V., Shipley, M. E., and Smith, M. J. H., 1980, Leukotriene B, a potent chemokinetic and aggregating substance released from polymorphonuclear leukocytes, *Nature* **286**:264–265.

Ford-Hutchinson, A. W., Bray, M. A., Cunningham, F. M., Davidson, E. M., and Smith, M. J. H., 1981, Isomers of leukotriene B$_4$ possess different biological properties, *Prostaglandins* **21**:143–152.

Ford-Hutchinson, A. W., Rackham, A., Zamboni, R., Rokach, J., and Roy, S., 1983, Comparative biological properties of synthetic leukotriene B$_4$ and its ω-oxidation products, *Prostaglandins* **25**:29–37.

Galliard, T., and Chan, H. W.-S., 1980, Lipoxygenases, in: *The Biochemistry of Plants*, Vol. 4, Academic Press, New York, pp. 131–161.

Gill, S. S., and Hammock, B. D., 1980, Distribution and properties of a mammalian soluble epoxide hydrase, *Biochem. Pharmacol.* **29**:389–395.

Goetzl, E. J., and Pickett, W. C., 1981, Novel structural determinants of the human neutrophil chemotactic activity of leukotriene B, *J. Exp. Med.* **153:**482–487.

Hammock, B. D., and Hasagawa, L. S., 1983, Differential substrate selectivity of murine hapatic cytosolic and microsomal epoxide hydrolases, *Biochem. Pharmacol.* **32:**1155–1164.

Hansson, G., Lindgren, J. A., Dahlen, S.-E., Hedquist, P., and Samuelsson, B., 1981, Identification and biological activity of novel ω-oxidized metabolites of leukotriene B₄ from human leukocytes, *FEBS Lett.* **130:**107–112.

Henson, P. M., 1971, The immunological release of constituents from neutrophil leukocytes. II. Mechanisms of release during phagocytosis, and adherence to non-phagocytosable surfaces, *J. Immunol.* **107:**1547–1557.

Jakschik, B. A., and Kuo, C. G., 1983, Subcellular localisation of leukotriene-forming enzymes, *Adv. Prostaglandin Thromboxane Leukotriene Res.* **11:**141–145.

Jakschik, B. A., Sun, F. F., Lee, L.-H., and Steinhoff, M. M., 1980a, Calcium stimulation of a novel lipoxygenase, *Biochem. Biophys. Res. Commun.* **95:**103–110.

Jakschik, B. A., Sams, A. R., Sprecher, H., and Needleman, P., 1980b, Fatty acid structural requirements for leukotriene biosynthesis, *Prostaglandins* **20:**401–410.

Koshihara, Y., Murota, S.-I., Petasis, N. A., and Nicolaou, K. C., 1982, Selective inhibition of 5-lipoxygenase by 5,6-methanoleukotriene A₄, a stable analogue of leukotriene A₄, *FEBS Lett.* **143:**13–16.

Kreisle, R. A., and Parker, C. W., 1983, Specific binding of leukotriene B₄ to a receptor on human polymorphonuclear leukocytes, *J. Exp. Med.* **157:**628–641.

Lindgren, J.-A., Hansson, G., and Samuelsson, B., 1981, Formation of novel hydroxylated eicosatetraenoic acids in preparations of human polymorphonuclear leukocytes, *FEBS Lett.* **128:**329–335.

Maduskie, T. P., and Willard, A. K., 1983, *17th Middle Atlantic Regional Meeting of the American Chemical Society.* White Haven, PA, p. 162.

Malmsten, C., Palmblad, J., Uden, A.-M., Radmark, O., Engstedt, L., and Samuelsson, B., 1980, Leukotriene B₄: A highly potent and stereospecific factor stimulating migration of polymorphonuclear leukocytes, *Acta Physiol. Scand.* **110:**449–451.

Maycock, A. L., Anderson, M. S., DeSousa, D. M., and Kuehl, F. A., 1982, Leukotriene A₄: Preparation and enzymatic conversion in a cell-free system to leukotriene B₄, *J. Biol. Chem.* **257:**13911–13914.

Murphy, R. C., Hammarström, S., and Samuelsson, B., 1979, Leukotriene C: A slow reacting substance (SRS) from mouse mastocytoma cells, *Proc. Natl. Acad. Sci. U.S.A.* **76:**4275–4279.

Narumiya, S., Salmon, J. A., Flower, R. J., and Vane, J. R., 1982, Purification and properties of arachidonate-15-lipoxygenase from rabbit peritoneal polymorphonuclear leukocytes, in: *Leukotrienes and Other Lipoxygenase Products* (B. Samuelsson and R. Paoletti, eds.), Raven Press, New York, pp. 77–82.

Nugteren, D. H., 1975, Arachidonate lipoxygenase in blood platelets, *Biochim. Biophys. Acta* **380:**299–307.

Ochi, K., Yoshimoto, T., Yamamoto, S., Taniguchi, K., and Miyamoto, T., 1983, Arachidonate 5-lipoxygenase of guinea pig peritoneal polymorphonuclear leukocytes. Activation by adenosine 5′-triphosphate, *J. Biol. Chem.* **258:**5574–5578.

Oliw, E. H., Guengerich, F. P., and Oates, J. A., 1982, Oxygenation of arachidonic acid by hepatic monooxygenases. Isolation and metabolism of four epoxide intermediates, *J. Biol. Chem.* **257:**3771–3781.

Ota, K., and Hammock, B. D., 1980, Cytosolic and microsomal epoxide hydrolases: Differential properties in mammalian liver, *Science* **207:**1479–1480.

Palmer, R. M. J., and Salmon, J. A., 1983, Release of leukotriene B₄ from human neutrophils and its relationship to degranulation induced by N-formyl-methionyl-leucyl-phenylalanine, serum-treated zymosan and the ionophore A23187, *Immunology* **50:**65–73.

Parker, C. W., and Aykent, S., 1982, Calcium stimulation of the 5-lipoxygenase from RBL-1 cells, *Biochem. Biophys. Res. Commun.* **109:**1011–1016.

Powell, W. S., 1984, Properties of leukotriene B₄ 20-hydroxylase from polymorphonuclear leukocytes, *J. Biol. Chem.* **259:**3082–3089.

Rae, S. A., Davidson, E. M., and Smith, M. J. H., 1982, Leukotriene B_4, an inflammatory mediator in gout, *Lancet* **2**:1122–1123.

Rokach, J., Zamboni, R., Lau, C.-K., and Gwriden, Y., 1981, The stereospecific synthesis of leukotriene A_4 (LTA_4), 5-epi-LTA_4, 6-epi-LTA_4 and 5-epi, 6-epi-LTA_4, *Tetrahedron Lett.* **22**:2759–2762.

Samuelsson, B., 1983, Leukotrienes: A new class of mediators of immediate hypersensitivity reactions and inflammation, *Adv. Prostaglandin Thromboxane Leukotriene Res.* **11**:1–13.

Smith, M. J. H., 1981, Leukotriene B_4, *Gen. Pharmacol.* **12**:211–216.

Yee, Y. K., 1984, A short and efficient synthesis of (2S,3R)-oxido-5-*cis*-undecen-1-al. A total synthesis of LTB_4. *14th Northeast Regional Meeting of the American Chemical Society.*

Yoshimoto, T., Miyamoto, Y., Ochi, K., and Yamamoto, S., 1982, Arachidonate 12-lipoxygenase of porcine leukocytes with activity for 5-hydroxyeicosatetraenoic acid, *Biochim. Biophys. Acta* **713**:638–646.

63

Novel Synthesis of Potent Site-Specific Phospholipase A$_2$ Inhibitors

RONALD L. MAGOLDA, WILLIAM C. RIPKA, WILLIAM GALBRAITH, PAUL R. JOHNSON, and MARLA S. RUDNICK

1. INTRODUCTION

Phospholipase A$_2$ (PLA$_2$) is an esterase responsible for the liberation of phospholipid-bound arachidonic acid, a biosynthetic precursor of putative inflammatory mediators. Arachidonic acid is metabolized by cyclooxygenase and lipoxygenase to the corresponding prostaglandins and leukotrienes (Fig. 1). Traditional antiinflammatory therapy has relied on cyclooxygenase and more recently on lipoxygenase blockade (Shen, 1981), but direct control of arachidonic acid pools has remained relatively unexplored. Recent evidence (Hirata *et al.*, 1980; Blackwell *et al.*, 1980; Rothhut *et al.*, 1983) demonstrates that antiinflammatory steroids control polyunsaturated fatty acid release at both cyclooxygense and lipoxygenase pathways by enhancing the production of PLA$_2$ inhibitory proteins (lipomodulin, macrocortin, renocortin). Direct phospholipase A$_2$ site-specific inhibition, therefore, offers new opportunities in antiinflammatory treatment.

2. PHOSPHOLIPASE A$_2$ INHIBITORS

Very few active-site inhibitors of PLA$_2$ are known. Most reported phospholipase A$_2$ inhibitors perturb membranes and thus may indirectly block this membrane

RONALD L. MAGOLDA, and PAUL R. JOHNSON • E. I. DuPont de Nemours and Co., Central Research and Development, Experimental Station, Wilmington, Delaware 19898. WILLIAM C. RIPKA, WILLIAM GALBRAITH, and MARLA S. RUDNICK • E. I. DuPont de Nemours and Co., Biomedical Products Department, Experimental Station, Wilmington, Delaware 19898.

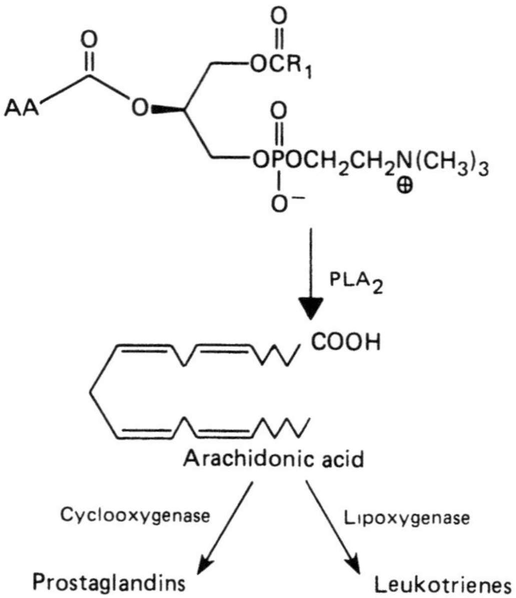

FIGURE 1. Arachidonic acid cascade.

protein's function. Mepacrine, a standard PLA_2 inhibitor, shows a concentration-dependent biphasic profile (Chan *et al.*, 1982) indicative of indirect membrane distortion. Anesthetics, well-known membrane affectors, and indomethacin, a cyclooxygenase inhibitor, have been shown to block phospholipase A_2 by altering membrane fluidity (Vigo *et al.*, 1981; Volwerk *et al.*, 1974; Fan and Shen, 1981; Hwang and Shen, 1981). A good active-site inhibitor had been reported (Wallach and Brown, 1981); however, exact details of binding have not been revealed. More direct inhibition of PLA_2 has been achieved by irreversible alkylation of histidine-48 with *para*-bromophenacyl bromide (BPB), which totally inactivates PLA_2 (Volwerk *et al.*, 1974).

Although the above nonphospholipid inhibitors are effective, phospholipid inhibitors have successfully been shown to occupy and block phospholipase A_2's active site. The unnatural enantomer D-lecithin competitively blocks the active site since the 2-acyl ester is improperly oriented for cleavage (Van Deenan and DeHaas, 1963). Nonhydrolyzable 2-acylamidophosphatidylcholines have also been reported to competitively inhibit phospholipase A_2's action (Chandrakumar and Hajdu, 1982). Simple *n*-alkylphosphorylcholines have been shown to protect PLA_2's active site from histidine-48 alkylation by BPB (Van Dam-Mieras *et al.*, 1975; Teshima *et al.*, 1981); however, no systematic alkyl variations have been investigated. With limited phospholipid and one nonphospholipid active-site PLA_2 inhibitors reported, a systematic study using a variety of *n*-alkylphosphorylcholines to probe PLA_2's lipophilic and electrostatic requirements is warranted.

With the refined (1.7 Å resolution) bovine and (2.6 Å resolution) porcine pancreatic PLA$_2$ X-ray structures reported (Dijkstra *et al.*, 1981, 1983) and a general synthetic method for alkylphospholipids that is flexible enough to accommodate a variety of alkyl substituents, a study could be coupled with computer graphic techniques and enzyme assays to define PLA$_2$'s binding and catalytic requirements.

3. PHOSPHOLIPID SYNTHESIS

With alkylphosphorylcholines selected as initial synthetic targets, a general route that is compatible with a variety of substituents had to be identified. Classical phosphatidylcholine syntheses (Hirt and Berchtold, 1958) require stoichiometric amounts of expensive silver salts and an aqueous step that makes isolation of water-soluble short-chain phospholipids and preservation of moisture-sensitive substituents difficult (Path A, Fig. 2). Another approach (Thuong and Chabrier, 1974) employs 2-chloro-2-oxo-1,2,3-dioxaphospholane (Path B, Fig. 2), a secondary phosphate that reacts sluggishly with hindered alcohols. A more recently reported phosphatidylcholine synthesis combines phosphatidic acid chloride with choline tosylate followed by an aqueous workup (Rosenthal, 1966; Brockerhoff and Ayengar, 1978). Although all of these approaches are ideal for lecithins, they have severe limitations for the preparation of synthetic probes needed for this study.

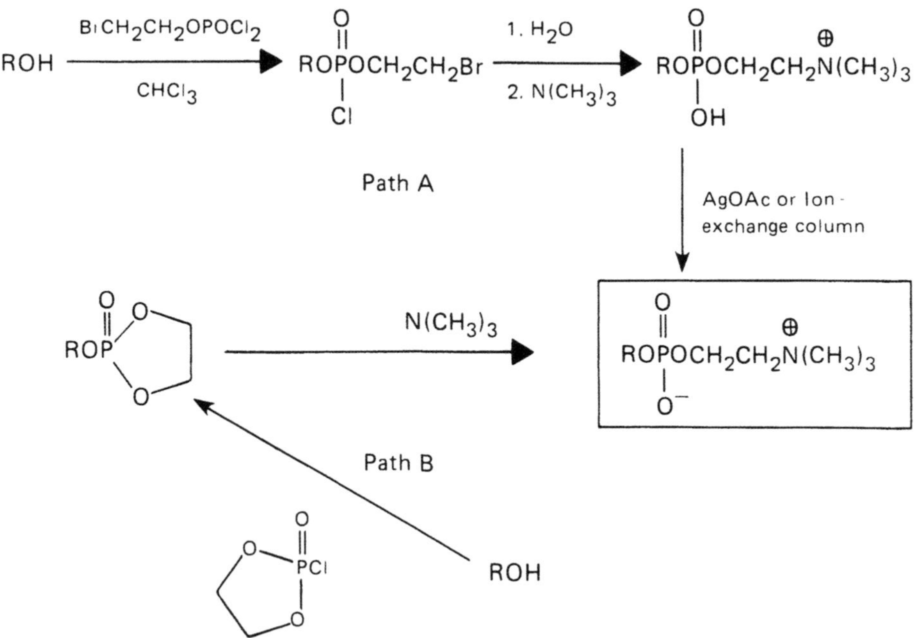

FIGURE 2. Classical phosphorylcholine synthesis.

We have discovered an inexpensive, efficient (two-pot, three-step), anhydrous process (Fig. 3) for making alkylphospholipids that should be suitable for sensitive substrates. Treating the requisite alcohol (**1**) with stoichiometric amounts of phosphorus oxychloride (**2**) and triethylamine under nitrogen in anhydrous ether at 0°C generates in 0.5 hr the dichlorophosphate **3** in quantitative yield. The reaction mixture is filtered to remove the precipitated triethylamine hydrochloride salt and then diluted (0.08–0.1 M), cooled (0°C), and exposed to triethylamine (2 equivalents) and ethylene glycol. After approximately 12 hr, **3** is completely converted into the cyclic phosphate **4**. This step takes advantage of primary-to-secondary phosphate reactivity followed by the facile intramolecular cyclization to the cyclic phosphate **4**, overcoming the typically sluggish secondary-to-tertiary phosphate transformation. Simple filtration to remove triethylamine hydrochloride salt and concentration supplies a stable cyclic phosphate **4** (70–80% yield), which could be used directly in the last step without purification. If purification is desired, **4** could be subjected to silica gel chromatography using nonpolar solvents (dichloromethane, ether). Combining an acetonitrile solution of **4** and trimethylamine (3 equivalents) in a sealed tube for 30 hr at 75°C affords, on cooling, the desired crystalline *n*-alkylphosphorylcholines **5** in good yield (65–70%). Generally, the cool reaction mixture can be recrystallized (acetonile, tetrahydrofuran, acetone) or subjected to column chromatography to give pure phospholipids.

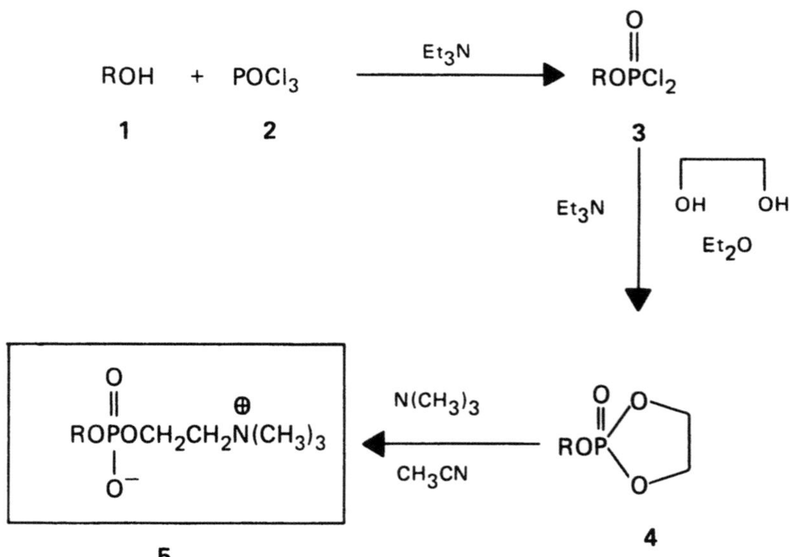

TABLE I. Alkylphosphorylcholine Properties[a]

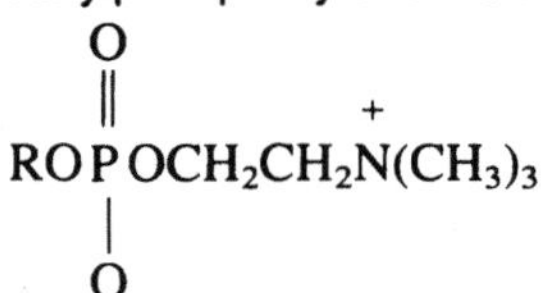

$$ROP(=O)(O^-)OCH_2CH_2N^+(CH_3)_3$$

	R	P.S.	IC$_{50}$ (μM)
Alkyl			
6	C$_6$H$_{13}$	Et$_2$O	75
7	C$_8$H$_{17}$	Et$_2$O	75
8	C$_{12}$H$_{25}$	Et$_2$O	70
9	C$_{18}$H$_{37}$	Et$_2$O	11
10	9Δ-C$_{18}$H$_{35}$	Et$_2$O	14
S-Alkyl			
11	C$_{16}$H$_{33}$S(CH$_2$)$_3$	THF : CH$_3$CN	19
12	C$_{18}$H$_{37}$S(CH$_2$)$_3$	THF : CH$_3$CN	17
13	9Δ-C$_{18}$H$_{35}$S(CH$_2$)$_3$	THF	15
Glycol			
14	C$_{16}$H$_{33}$O(CH$_2$)$_2$	Et$_2$O	10
15	C$_{18}$H$_{37}$O(CH$_2$)$_2$	Et$_2$O	8.8
16	9Δ-C$_{18}$H$_{35}$O(CH$_2$)$_2$	Et$_2$O	6.2
—	Mepacrine	—	17

[a] P.S., phosphorylation solvent; IC$_{50}$, 50% inhibitory concentration of porcine pancreatic PLA$_2$; CH$_3$CN, acetonile; THF, tetrahydrofuran; Et$_2$, O, diethylether.

Several *n*-alkyl- and *n*-alkylglycoletherphosphorylcholines have been efficiently prepared in this manner (Table I). This flexible synthetic procedure can accommodate a variety of aprotic solvents. Although most alcohols are rapidly phosphorylated with phosphorus oxychloride, the S-alkyl alcohols resisted phosphorylation in ether under conditions in which both glycol and alkyl alcohols reacted. The polar nature of the thioether of the S-alkyl substrate may be responsible for micelle formation. The longer alkyl chains favor nonpolar interactions with the solvent (ether), thereby shielding the reactive alcohols from the polar phosphorylating agent. Changing from ether with a low dielectric constant (4.2) to tetrahydrofuran (7.6) or acetonitrile (38.8) should disrupt micelle formation and promote phosphorylation. In fact, as shown in Table I, combinations of tetrahydrofuran and acetonitrile resulted in quantitative phosphorylation. Uneventful cyclic phosphate formation (THF) followed by trimethylamine treatment completed the synthesis of phosphorylcholines.

This approach offers several synthetic advantages. In an efficient two-pot process, moisture-sensitive compounds, substrates prone to form micelles, are easily transformed into phospholipids. Use of reactive and inexpensive reagents (phosphorus oxychloride, ethylene glycol) insures rapid phosphorylation of unreactive alcohols and facile cyclic phosphate formation. Optional cyclic phosphate purification affords synthetic flexibility to prepare more complicated phospholipids.

4. PHOSPHOLIPASE A$_2$ INHIBITION STUDIES

To examine these phospholipids as PLA$_2$ inhibitors, the enzyme assay reported by Hirata (1981) was employed. Determinations were performed below critical micelle concentration (CMC). Porcine pancreatic PLA$_2$ (Sigma, 0.025 μg) and phospholipid inhibitor are preincubated for 2 min at 37°C in buffer (25 mM tris; 25 mM glycylglycine; 25 mM CaCl$_2$; 0.75 mM EDTA) at pH 8.5. Introduction of [^{14}C]arachidonate phosphatidylcholine (0.0096 mM, 0.05 μCi) initiates the enzymatic reaction, which is quenched (dry ice) after 5 min at 37°C. The frozen mixture is thawed and then subjected to silica gel chromatography separating free [^{14}C]arachidonic acid from unreacted [^{14}C]arachidonate phosphatidylcholine. In this manner 50% inhibitor concentrations (IC$_{50}$) were determined for mepacrine and alkylphosphorylcholines shown in Table I.

As previously reported (Van Dam-Mierras, 1975), *n*-alkylphospholipids are good PLA$_2$ inhibitors. Fivefold improved inhibition was observed with the longer-chain (9) versus shorter-chain (6) phosphorylcholines, which is consistent with phospholipase's preference for longer-chain lecithins (DeHaas *et al.*, 1971). Similarly, the S-alkyl series showed no significant improvement in PLA$_2$ inhibition over the simple long-chain phosphorylcholine. However, the glycol series displayed a two- or threefold improvement in inhibition, thus demonstrating PLA$_2$'s selectivity. The S-alkyl series with the misplaced thioether shows virtually no interactions, unlike the glycol phospholipids with the same ether location as the natural lecithins.

5. COMPUTER MODELING

The availability of the bovine pancreatic X-ray structure, permits computer-assisted modeling of potential binding modes for the natural lecithin substrates and our synthetic alkylphospholipids in the active site. Molecular mechanics calculations have been used to refine the initial graphics fits of the ligands bound to this enzyme. The proposed binding conformations of lecithin have then been used to determine possible binding of the related phosphorylcholine inhibitors in the active site. From this preliminary study, short-chain phospholipids do not sufficiently occupy the large active-site lipophilic pocket, consistent with the biochemical results. Similarly, the glycol series provides additional nonbonded interaction with hydrophobic active-site residues. Additional details and implications will be reported later.

6. SUMMARY

In summary, a new efficient and general synthetic route to a variety of unnatural alkylphosphorylcholines has been presented. Significant improvements in PLA$_2$

inhibitors have been discovered by systematically showing longer-chain and iso-electronic phospholipids to be better inhibitors. The bovine pancreatic phospholipase A$_2$'s X-ray structure is used to provide working models of substrate and inhibitor binding that can be probed and tested with synthetic phospholipids. These initial glycol phosphorylcholines are some of the most potent PLA$_2$ inhibitors reported to date and have extended our fundamental understanding of PLA$_2$. This understanding can assist in the future design and synthesis of more potent PLA$_2$ active-site inhibitors.

REFERENCES

Blackwell, G. J., Carnuccio, R., DiRosa, M., FLower, R. J., Parente, L., and Persico, P., 1980, Macrocortin: A polypeptide causing the anti-phospholipase effect of glucocorticoids, *Proc. Natl. Acad. Sci. U.S.A.* **77**:2533–2536.

Brockerhoff, H., and Ayengar, N. K. N., 1978, Improved synthesis of choline phospholipids, *Lipids* **14**:88–89.

Chan, A. C., Pritchard, E. T., Gerrard, J. M., Man, R. Y. K., and Choy, P. C., 1982, Biphasic modulation of platelet phospholipase A$_2$ activity and platelet aggregation by mepacrine (quinacrine), *Biochim. Biophys. Acta* **713**:170–172.

Chandrakumar, N. S., and Hajdu, J., 1982, Synthesis of enzyme-inhibitory phospholipid analogues. Stereospecific synthesis of 2-aminophosphatidylcholines and related derivatives, *J. Org. Chem.* **47**:2144–2147.

DeHaas, G. H., Bonsen, P. P. M., Pieterson, W. A., and Van Deenen, L. L. M., 1971, Studies on phospholipase A and its zymogen from porcine pancreas, *Biochim. Biophys. Acta* **239**:252–266.

Dijkstra, B. W., Kalk, K. H., Hol, W. G. J., and Drenth, J., 1981, Structure of bovine pancreatic phospholipase A$_2$ at 1.7 Å resolution, *J. Mol. Biol.* **147**:97–123.

Dijkstra, B. W., Renetseder, R., Kalk, K. H., Hol, W. G. J., and Drenth, J., 1983, Structure of porcine pancreatic phospholipase A$_2$ at 2.6 Å resolution and comparison with bovine phospholipase A$_2$, *J. Mol. Biol.* **168**:163–179.

Fan, S. S., and Shen, T. Y., 1981, Membrane effects of antiinflammatory agents. I. Interaction of salindac and its metabolites with phospholipid memrane, a magnetic resonance study, *J. Med. Chem.* **24**:1197–1202.

Hirata, F., Carmine, R. D., Nelson, C. A., Axelrod, J., Schiffmann, E., Warabi, A., DeBlas, A. L., Nirenberg, M., Manganiello, V., Vanghan, M., Kumagai, S., Green, I., Decker, J. L., and Steinberg, A. D., 1981, Presence of autoantibody for phospholipase inhibitory protein, lipomodulin, in patients with rheumatic diseases, *Proc. Natl. Acad. Sci. U.S.A.* **78**:3190–3194.

Hirata, F., Schiffmann, E., Venkatasubramamian, K., Salomon, D., and Axelrod, J., 1980, A phospholipase A$_2$ inhibitory protein in rabbit neutrophils induced by glucocorticoids, *Proc. Natl. Acad. Sci. U.S.A.* **77**:2533–2536.

Hirt, R., and Berchtold, R., 1958, Zur Synthese der Phosphatide Eine: Neue Synthese der Lecithine, *Pharmac. Acta Helvetiae* **33**:349–354.

Hwang, S. B., and Shen, T. Y., 1981, Membrane effects of antiinflammatory agents. II. Interaction of nonsteroidal antiinflammatory drugs with liposome and purple membranes, *J. Med. Chem.* **24**:1202–1211.

Rosenthal, A. F., 1966, New, partially hydrolyzable synthetic analogues of lecithin, phosphatidyl ethanolamine and phosphatidic acid, *J. Lip. Res.* **7**:779–781.

Rothhut, B., Russo-Marie, F., Wood, J., DiRosa, M., and Flower, R. J., 1983, Further characterization of the glucocorticoid-induced antiphospholipase protein "renocortin," *Biochim. Biophys. Acta* **117**:878–884.

Shen, T. Y., 1981, Toward more selective antiarthritic therapy, *J. Med. Chem.* **24**:1–17.

Teshima, K., Ikeda, K., Hamaguchi, K., and Hayashi, K., 1981, Binding of monodispersed *n*-alkyl-phosphorylcholines to cobra venom phospholipase A_2, *J. Biochem.* **89**:1163–1174.

Thuong, N. T., and Chabrier, P., 1974, Nouvelle méthode de préparation de la phosphorylcholine, de la phosphorylhomocholine et de leurs dérives, *Bull. Soc. Chem. Fr.* **2(3–4)**:667–676.

Van Dam-Mieras, M. C. E., Slotboom, A. J., Pieterson, W. A., and DeHaas, G. H., 1975, The interaction of phospholipase A_2 with micellar interfaces. The role of the N-terminal region, *Biochemistry* **14**:5387–5394.

Van Deenen, L. L. M., and DeHaas, G. H., 1963, The substrate specificity of phospholipase A, *Biochim. Biophys. Acta* **70**:469–471.

Vigo, C., Lewis, G. P., and Piper, P. J., 1981, Mechanism of inhibition of phospholipase A_2, *Biochim. Pharmacol.* **29**:1–5.

Volwerk, J. J., Pieterson, W. A., and DeHaas, G. H., 1974, Histidine at the active site of phospholipase A_2, *Biochemistry* **13**:1446–1454.

Wallach, D. P., and Brown, V. J., 1981, Studies on the arachidonic acid cascade-I. Inhibition of phospholipase A_2 *in vitro* and *in vivo* by several novel series of inhibitor compounds, *Biochem. Pharmacol.* **30**:1315–1324.

Participants

LAUREL M. ADAMS
The University of Texas System Cancer Center
Science Park, Research Division
Smithville, Texas 78957

M. R. ALIJANI
Department of Surgery
Division of Transplantation
Georgetown University Medical Center
Washington, D.C. 20007

PER ÅLIN
Department of Biochemistry
Arrhenius Laboratory
University of Stockholm
S-106 91 Stockholm
Sweden

MARY ALOSIO
Memorial Sloan-Kettering Cancer Center
New York, New York 10021

K. FRANK AUSTEN
Department of Medicine
Harvard Medical School
    and Department of Rheumatology and
    Immunology
Brigham and Women's Hospital
Boston, Massachusetts 02115

K. S. AUTHI
Department of Biochemistry
Royal College of Surgeons
London, United Kingdom

J. MARTYN BAILEY
Department of Biochemistry
George Washington University School of
    Medicine and Health Sciences
Washington, D.C. 20037

BARBARA J. BALLERMANN
Department of Medicine
Harvard Medical School
    and Department of Rheumatology and
    Immunology
Brigham and Women's Hospital
Boston, Massachusetts 02115

ROBERT M. BARR
Wellcome Laboratories for Skin Pharmacology
Institute of Dermatology
Homerton Grove
London E9 6BX
United Kingdom

SUZANNE K. BECKNER
Laboratory of Cellular and Developmental
    Biology
National Institute of Arthritis, Diabetes,
    Digestive and Kidney Disease
National Institutes of Health
Bethesda, Maryland 20205

E. BENDVOLD
Department of Obstetrics and Gynecology
Karolinska Hospital
S-104 01 Stockholm
Sweden

PETER R. BERNSTEIN
Medicinal Chemistry Department
Stuart Pharmaceuticals
Division of ICI Americas Inc.
Wilmington, Delaware 19897

KERSTIN BERNSTRÖM
Department of Physiological Chemistry
Karolinska Institutet
S-104 01 Stockholm
Sweden

RICHARD S. BOCKMAN
Memorial Sloan-Kettering Cancer Center
New York, New York 10021

PIERRE BORGEAT
Groupe de Recherches sur les Leucotriènes
Laboratoire d'Endocrinologie Moléculaire
Centre Hospitalier de l'Université Laval
Québec GIV 4G2
Canada

L. D. BRACE
Department of Pharmacology
The University of Illinois at Chicago
Health Sciences Center
Chicago, Illinois 60612

MONIQUE BRAQUET
Centre de Recherches du Service de Santé des
    Armées
F-92141 Clamart
France

PIERRE BRAQUET
I. H. B. Research Laboratories
F-92350 Le Plessis-Robinson
France

YVON BROUSSEAU
Département de Pédiatrie
Centre Hospitalier Universitaire
Université de Sherbrooke
Sherbrooke J1H 5N4
Canada

JOHN BRUENS
Revlon Health Care Group
Research and Development
Tuckahoe, New York 10707

ROBERT W. BRYANT
Department of Biochemistry
George Washington University School of
    Medicine and Health Sciences
Washington, D.C. 20037
Present address:
Shering Corporation
Pharmaceutical Research Division
Bloomfield, New Jersey 07003

I. BRYMAN
Department of Obstetrics and Gynecology
University of Göteborg
Göteborg, Sweden

JAMES A. BURKE
Department of Pharmacology
Cornell University Medical College and the
    Rockefeller University
New York, New York 10021

M. BYGDEMAN
Department of Obstetrics and Gynecology
Karolinska Hospital
S-104 01 Stockholm
Sweden

RICHARD D. CAMP
Wellcome Laboratories for Skin Pharmacology
Institute of Dermatology
Homerton Grove
London E9 6BX
United Kingdom

MARIE M. CASSIDY
Department of Physiology
The George Washington University School of
    Medicine and Health Sciences
Washington, D.C. 20037

UTPAL CHAKRABORTY
Revlon Health Care Group
Research and Development
Tuckahoe, New York 10707

MARIE-YVONNE CHAPELAT
I. H. B. Research Laboratories
F-92350 Le Plessis-Robinson
France

ISABEL S. CHEN
Department of Biochemistry
The George Washington University School of
    Medicine and Health Sciences
Washington, D.C. 20037

DAVID CHESTER
Department of Pharmacology
University of Connecticut Health Center
Farmington, Connecticut 06032

RONALD G. COFFEY
Program of Immunopharmacology
University of South Florida Medical College
Tampa, Florida 33612

E. J. COREY
Department of Chemistry
Harvard University
Cambridge, Massachusetts 02138

STEPHEN M. COUTTS
Revlon Health Care Group
Research and Development
Tuckahoe, New York 10707

N. CRAWFORD
Department of Biochemistry
Royal College of Surgeons
London, United Kingdom

M. CROSET
INSERM U 63
Institut Pasteur
Laboratoire d'Hémobiologie
Faculté Alexis Carrel
69372 Lyon Cédex 08
France

CYNTHIA M. CUNARD
Department of Physiology and Biophysics
Georgetown University Medical Center
Washington, D.C. 20007

SVEN-ERIK DAHLÉN
Department of Physiology
Karolinska Institutet
S-104 01 Stockholm
Sweden

VERNETTA DALLY-MEADE
Revlon Health Care Group
Research and Development
Tuckahoe, New York 10707

CAROL DANGELMAIER
Center for Thrombosis Research
Temple Medical School
Philadelphia, Pennsylvania 19140

M. DECHAVANNE
INSERM U 63
Institut Pasteur
Laboratoire d'Hémobiologie
Faculté Alexis Carrel
69372 Lyon Cédex 08
France

RAYMOND A. DEEMS
Department of Chemistry
University of California at San Diego
La Jolla, California 92093

D. DENIS
Department of Pharmacology
Merck Frosst Canada Inc.
Pointe Claire-Dorval
Québec H9R 4P8
Canada

EDWARD A. DENNIS
Department of Chemistry
University of California at San Diego
La Jolla, California 92093

ROGER DUCOUSSO
Centre de Recherches du Service de Santé des
  Armées
F-92141 Clamart
France

ANN M. DVORAK
Department of Pathology
Beth Israel Hospital and Harvard Medical
  School and
The Charles A. Dana Research Institute
Beth Israel Hospital
Boston, Massachusetts 02215

HAROLD F. DVORAK
Department of Pathology
Beth Israel Hospital and Harvard Medical
  School
and The Charles A. Dana Research Institute
Beth Israel Hospital
Boston, Massachusetts 02215

ROBERT W. EGAN
Merck Institute for Therapeutic Research
Rahway, New Jersey 07065

PETER ENEROTH
Department of Obstetrics and Gynecology
Karolinska Hospital
S-104 01 Stockholm
Sweden

DAVID EZRA
Neurobiology Research Unit
Divisions of Cardiology and Clinical
    Pharmacology
Departments of Medicine and Pharmacology
Uniformed Services University of the Health
    Sciences
Bethesda, Maryland 20814

MAURICE B. FEINSTEIN
Department of Pharmacology
University of Connecticut Health Center
Farmington, Connecticut 06032

GIORA FEUERSTEIN
Neurobiology Research Unit
Divisions of Cardiology and Clinical
    Pharmacology
Departments of Medicine and Pharmacology
Uniformed Services University of the Health
    Sciences
Bethesda, Maryland 20814

SUSAN M. FISCHER
The University of Texas System Cancer Center
Science Park, Research Division
Smithville, Texas 78957

GARY FISKUM
Department of Biochemistry
George Washington University School of
    Medicine and Health Sciences
Washington, D.C. 20037

G. A. FITZGERALD
Division of Clinical Pharmacology
Vanderbilt University
Nashville, Tennessee 37232

RODERICK J. FLOWER
School of Pharmacy and Pharmacology
The University of Bath
Avon BA2 7AY
United Kingdom

M. L. FOEGH
Department of Medicine
Division of Nephrology
Georgetown University Medical Center
Washington, D.C. 20007

SHEILA A. FOX
Pharmacology Department
University of Cambridge
Cambridge CB2 2QD
United Kingdom

LYNE GAGNON
Immunology Division
Faculty of Medicine
University of Sherbrooke
Sherbrooke, Québec J1H 5N4
Canada

WILLIAM GALBRAITH
E. I. DuPont de Nemours and Co.
Biomedical Products Department
Experimental Station
Wilmington, Delaware 19898

PAUL H. GALE
Merck Institute for Therapeutic Research
Rahway, New Jersey 07065

STEPHEN J. GALLI
Department of Pathology
Beth Israel Hospital and Harvard Medical
    School
and The Charles A. Dana Research Institute
Beth Israel Hospital
Boston, Massachusetts 02215

ARLYN GARCIA-PEREZ
Department of Biochemistry
Michigan State University
East Lansing, Michigan 48824

MARY E. GERRITSEN
Department of Physiology
New York Medical College
Valhalla, New York 10595

MITCHELL GOLDMAN
Division of Cardiac Surgery
Medical College of Virginia
Richmond, Virginia 23298

IRA M. GOLDSTEIN
Cardiovascular Research Institute and Rosalind
  Russell Arthritis Research Laboratory
Department of Medicine
University of California
San Francisco, California 94102
and Medical Service
San Francisco General Hospital
San Francisco, California 94110

ROBERT E. GOLDSTEIN
Neurobiology Research Unit
Divisions of Cardiology and Clinical
  Pharmacology
Departments of Medicine and Pharmacology
Uniformed Services University of the Health
  Sciences
Bethesda, Maryland 20814

JAMES S. GOODWIN
Department of Medicine
Medical College of Wisconsin
Milwaukee, Wisconsin 53226

GARY G. GORDON
Departments of Ophthalmology and Medicine
New York Medical College
Valhalla, New York 10595

C. GOTTLIEB
Department of Obstetrics and Gynecology
Karolinska Hospital
S-104 01 Stockholm
Sweden

MALCOLM W. GREAVES
Wellcome Laboratories for Skin Pharmacology
Institute of Dermatology
Homerton Grove
London E9 6BX
United Kingdom

NORBERT GUALDE
Immunology and Biochemistry Research Unit
School of Medicine
87032 Limoges Cédex
France

ZHAO-GUI GUO
Hunan Medical College
Changsha, Hunan
People's Republic of China
Present address:
Department of Pharmacology
Cornell University Medical College and the
  Rockefeller University
New York, New York 10021

BENGT GUSTAFSSON
Department of Germfree Research
Karolinska Institutet
S-104 01 Stockholm
Sweden

UGHETTA HACHFELD del BALZO
Department of Pharmacology
Cornell University Medical College and the
  Rockefeller University
New York, New York 10021

ELBA M. HADDEN
Program of Immunopharmacology
University of South Florida Medical College
Tampa, Florida 33612

JOHN W. HADDEN
Program of Immunopharmacology
University of South Florida Medical College
Tampa, Florida 33612

STEPHEN P. HALENDA
Department of Pharmacology
University of Connecticut Health Center
Farmington, Connecticut 06032

MATS HAMBERG
Department of Physiological Chemistry
Karolinska Institutet
S-104 01 Stockholm
Sweden

SVEN HAMMARSTRÖM
Department of Physiological Chemistry
Karolinska Institutet
S-104 01 Stockholm
Sweden

ILAN HAMMEL
Department of Pathology
Sackler School of Medicine
Tel Aviv University
Ramat Aviv, Tel Aviv 69978
Israel

LAURENCE A. HARKER
Roon Research Center for Arteriosclerosis and
    Thrombosis
Scripps Clinic and Research Foundation
La Jolla, California 92037

V. SUSAN HARVEY
Department of Pathology
Beth Israel Hospital and Harvard Medical
    School
and The Charles A. Dana Research Institute
Beth Israel Hospital
Boston, Massachusetts 02215

YUICHI HATTORI
Department of Pharmacology
Hokkaido University School of Medicine
Sapporo, Japan
Present address:
Department of Pharmacology
Cornell University Medical College
    and the Rockefeller University
New York, New York 10021

THEODORE L. HAZLETT
Department of Chemistry
University of California at San Diego
La Jolla, California 92093

G. B. HELFRICH
Department of Surgery
Division of Transplantation
Georgetown University Medical Center
Washington, D.C. 20007

NOREEN J. HICKOK
Memorial Sloan-Kettering Cancer Center
New York, New York 10021

THOMAS H. HINTZE
Department of Physiology
New York Medical College
Valhalla, New York 10595

FUSAO HIRATA
Laboratory of Cell Biology
National Institute of Mental Health
Bethesda, Maryland 20205

TIMOTHY HLA
Department of Biochemistry
George Washington University School of
    Medicine and Health Sciences
Washington, D.C. 20037

CARL E. HOCK
Department of Physiology
Jefferson Medical College
Thomas Jefferson University
Philadelphia, Pennsylvania 19107

TOMAS HÖKFELT
Department of Histology
Karolinska Institutet
S-104 01 Stockholm
Sweden

LORETTE O. M. HULSMAN
Laboratory for Experimental Surgery
Erasmus University
3000 DR Rotterdam
The Netherlands

ANNA-LENA HULTING
Department of Endocrinology
Karolinska Hospital
S-104 01 Stockholm
Sweden

JOHN L. HUSSEY
Department of Surgery
Tulane University School of Medicine
New Orleans, Louisiana 70112

ALBERT L. HYMAN
Departments of Pharmacology, Surgery, and
    Medicine
Tulane University School of Medicine
New Orleans, Louisiana 70112

RICHARD INGRAM
Revlon Health Care Group
Research and Development
Tuckahoe, New York 10707

TERUKO ISHIZAKA
Subdepartment of Immunology
The Johns Hopkins University School of
  Medicine
Good Samaritan Hospital
Baltimore, Maryland 21239

TADASHI ISOMURA
Second Department of Surgery
Kurume University School of Medicine
Kurume 830
Japan

NAVIN JARIWALA
Revlon Health Care Group
Research and Development
Tuckahoe, New York 10707

HELGI JENSSON
Department of Biochemistry
Arrhenius Laboratory
University of Stockholm
S-106 91 Stockholm
Sweden

PAUL R. JOHNSON
I. E. DuPont de Nemours and Co.
Central Research and Development
Experimental Station
Wilmington, Delaware 19898

HOWARD JONES
Revlon Health Care Group
Research and Development
Tuckahoe, New York 10707

T. R. JONES
Department of Pharmacology
Merck Frosst Canada Inc.
Pointe Claire-Dorval
Québec H9R 4P8
Canada

S. M. JOUBERT
Medical Research Council Unit for Preclinical
  Diagnostic Research
University of Natal
Congella, South Africa

PHILIP J. KADOWITZ
Department of Pharmacology
Tulane University School of Medicine
New Orleans, Louisiana 70112

GABOR KALEY
Department of Physiology
New York Medical College
Valhalla, New York 10595

A. KAMBARAN
Medical Research Council Unit for Preclinical
  Diagnostic Research
University of Natal
Congella, South Africa

AKIRA KAWAGUCHI
Division of Cardiac Surgery
Medical College of Virginia
Richmond, Virginia 23298

MORRIS D. KERSTEIN
Department of Surgery
Tulane University School of Medicine
New Orleans, Louisiana 70112

ATUL KHANDWALA
Revlon Health Care Group
Research and Development
Tuckahoe, New York 10707

B. S. KHIRABADI
Department of Physiology and Biophysics
Georgetown University Medical Center
Washington, D.C. 20007

WILL J. KORT
Laboratory for Experimental Surgery
Erasmus University
3000 DR Rotterdam
The Netherlands

RICHARD J. KULMACZ
Department of Biological Chemistry
University of Illinois at Chicago
Chicago, Illinois 60612

M. LAGARDE
INSERM U 63
Institut Pasteur
Laboratoire d'Hémobiologie
Faculté Alexis Carrel
69372 Lyon Cédex 08
France

ALICE Z. LANDRY
Department of Pharmacology
Tulane University School of Medicine
New Orleans, Louisiana 70112

EDUARDO G. LAPETINA
Department of Molecular Biology
The Wellcome Research Laboratories
Burroughs Wellcome
Research Triangle Park, North Carolina 27709

G. C. Le BRETON
Department of Pharmacology
The University of Illinois at Chicago
Health Sciences Center
Chicago, Illinois 60612

TAK H. LEE
Department of Medicine
Harvard Medical School
and Department of Rheumatology and
    Immunology
Brigham and Women's Hospital
Boston, Massachusetts 02115

ALLAN M. LEFER
Department of Physiology
Jefferson Medical College
Thomas Jefferson University
Philadelphia, Pennsylvania 19107

DAVID J. LEFER
Department of Physiology
Jefferson Medical College
Thomas Jefferson University
Philadelphia, Pennsylvania 19017

GORDON LETTS
Merck Frosst, Inc.
Dorval, Québec H9R 4P8
Canada

O. A. LEVANDER
Beltsville Human Nutrition Research Center
Beltsville, Maryland 20705

ROBERTO LEVI
Department of Pharmacology
Cornell University Medical College and the
    Rockefeller University
New York, New York 10021

ROBERT A. LEWIS
Department of Medicine
Harvard Medical School.
and Department of Rheumatology and
    Immunology
Brigham and Women's Hospital
Boston, Massachusetts 02115

LAWRENCE M. LICHTENSTEIN
Department of Medicine
Division of Clinical Immunology
Johns Hopkins University School of Medicine
The Good Samaritan Hospital
Baltimore, Maryland 21239

MICHAEL C. LIN
Laboratory of Cellular and Developmental
    Biology
National Institute of Arthritis, Diabetes,
    Digestive and Kidney Disease
National Institutes of Health
Bethesda, Maryland 20205

B. LINDBLOM
Department of Obstetrics and Gynecology
University of Göteborg
Göteborg, Sweden

JAN ÅKE LINDGREN
Department of Physiological Chemistry
Karolinska Institutet
S-104 01 Stockholm
Sweden

HOWARD L. LIPPTON
Departments of Medicine and Pharmacology
Tulane University School of Medicine
New Orleans, Louisiana 70112

CHOW-ENG LOW
Department of Biochemistry
George Washington University School of
  Medicine and Health Sciences
Washington, D.C. 20037

RICHARD R. LOWER
Division of Cardiac Surgery
Medical College of Virginia
Richmond, Virginia 23298

DONALD W. MacGLASHAN, JR.
Department of Medicine
Division of Clinical Immunology
Johns Hopkins University School of Medicine
The Good Samaritan Hospital
Baltimore, Maryland 21239

YVONNE T. MADDOX
Department of Physiology and Biophysics
Georgetown University Medical Center
Washington, D.C. 20007

RONALD L. MAGOLDA
E.I. DuPont de Nemours and Co.
Central Research and Development
Experimental Station
Wilmington, Delaware 19898

BENGT MANNERVIK
Department of Biochemistry
Arrhenius Laboratory
University of Stockholm
S-106 91 Stockholm
Sweden

DAVID J. MASTERS
ICI Pharmaceuticals Division
Mereside, Alderley Park
Macclesfield, Cheshire SK10 4TG
United Kingdom

GERARD MAUCO
Center for Thrombosis Research
Temple Medical School
Philadelphia, Pennsylvania 19140
Present address:
INSERM U101
Hôpital Purpan
F-31059 Toulouse
France

RODGER M. McMILLAN
ICI Pharmaceuticals Division
Mereside, Alderley Park
Macclesfield, Cheshire SK10 4TG
United Kingdom

DENNIS B. McNAMARA
Department of Pharmacology
Tulane University School of Medicine
New Orleans, Louisiana 70112

STEWART A. METZ
Division of Clinical Pharmacology
University of Colorado Health Sciences Center
Denver, Colorado 80262,
and Denver VA Medical Center
Denver, Colorado 80220

V. C. MORRIS
Beltsville Human Nutrition Research Center
Beltsville, Maryland 20705

JOHN MUSSER
Revlon Health Care Group
Research and Development
Tuckahoe, New York 10707

BARBARA MUZA
Department of Biochemistry
George Washington University School of
  Medicine and Health Sciences
Washington, D.C. 20037

PAUL H. NACCACHE
Department of Pathology
University of Connecticut Health Center
Farmington, Connecticut 06032

EDWARD NEISS
Revlon Health Care Group
Research and Development
Tuckahoe, New York 10707

ELISABETH NORIN
Department of Germfree Research
Karolinska Institutet
S-104 01 Stockholm
Sweden

R. J. NORMAN
Medical Research Council Unit for Preclinical
    Diagnostic Research
University of Natal
Congella, South Africa

A. NORSTRÖM
Department of Obstetrics and Gynecology
University of Göteborg
Göteborg, Sweden

LARS ÖRNING
Department of Physiological Chemistry
Karolinska Institutet
S-104 01 Stockholm
Sweden

D. J. OSBORNE
Lilly Research Centre
Erl Wood Manor
Windlesham GU20 6PH
United Kingdom

NANCY E. OWEN
Department of Biological Chemistry and
    Structure
University of Health Sciences
The Chicago Medical School
North Chicago, Illinois 60064

C. R. PACE-ASCIAK
Research Institute
The Hospital for Sick Children
Toronto, Ontario,
and Department of Pharmacology
University of Toronto
Toronto, Ontario M5G 1X8
Canada

JAMES PASH
Department of Biochemistry
George Washington University School of
    Medicine and Health Sciences
Washington, D.C. 20037

CARLO PATRONO
Department of Pharmacology
Catholic University
I-00168 Rome
Italy

ANDREW PEASE
Department of Biochemistry
George Washington University School of
    Medicine and Health Sciences
Washington, D.C. 20037

A. K. PEDERSEN
Division of Clinical Pharmacology
Vanderbilt University
Nashville, Tennessee 37232

STEPHEN P. PETERS
Department of Medicine
Division of Clinical Immunology
Johns Hopkins University School of Medicine
The Good Samaritan Hospital
Baltimore, Maryland 21239

R. H. PHILPOTT
Medical Research Council Unit for Preclinical
    Diagnostic Research
University of Natal
Congella, South Africa

CLAUDINE POULIOT
Immunology Division
Faculty of Medicine
University of Sherbrooke
Sherbrooke, Québec J1H 5N4
Canada

THADDEUS PRUSS
Revlon Health Care Group
Research and Development
Tuckahoe, New York 10707

KARI PUNNONEN
Department of Physiology
University of Turku
SF-20520 Turku 52, Finland

KATHRYN PYNE
Department of Pathology
Beth Israel Hospital and Harvard Medical
    School,
and The Charles A. Dana Research Institute
Beth Israel Hospital
Boston, Massachusetts 02215

VICTOR RADOUX
Immunology Division
Faculty of Medicine
University of Sherbrooke
Sherbrooke, Québec J1H 5N4
Canada

K. D. RAINSFORD
Pharmacology Department
University of Cambridge
Cambridge CB2 2QD
United Kingdom

PETER W. RAMWELL
Department of Physiology and Biophysics
Georgetown University Medical Center
Washington, D.C. 20007

GUNDU H. R. RAO
Department of Laboratory Medicine and
    Pathology
Medical School
University of Minnesota
Minneapolis, Minnesota 55455

K. REDDI
Medical Research Council Unit for Preclinical
    Diagnostic Research
University of Natal
Congella, South Africa

I. A. G. REILLY
Division of Clinical Pharmacology
Vanderbilt University
Nashville, Tennessee 37232

S. RENAUD
INSERM U 63
Institut Pasteur
Laboratoire d'Hémobiolgie
Faculté Alexis Carrel
69372 Lyon Cédex 08
France

JOHN RHODES
Department of Experimental Immunobiology
Wellcome Research Laboratories
Beckenham, Kent BR3 3BS
United Kingdom

WILLIAM C. RIPKA
E. I. DuPont de Nemours and Co.
Biomedical Products Department
Experimental Station
Wilmington, Delaware 19898

MAREK ROLA-PLESZCZYNSKI
Immunology Division
Faculty of Medicine
University of Sherbrooke
Sherbrooke, Québec J1H 5N4
Canada

ROBERT S. ROSENSON
Department of Medicine
Tulane University School of Medicine
New Orleans, Louisiana 70112

MERRICK I. ROSS
Department of Immunology
Research Institute of Scripps Clinic
La Jolla, California 92037

DAVID M. ROTH
Department of Physiology
Jefferson Medical College
Thomas Jefferson University
Philadelphia, Pennsylvania 19107

CAROL A. ROUZER
Department of Pharmacology
Cornell University Medical College and the
    Rockefeller University
New York, New York 10021
Present address:
Department of Physiological Chemistry II
Karolinska Institutet
S-104 01 Stockholm
Sweden

MARLA S. RUDNICK
E. I. DuPont de Nemours and Co.
Biomedical Products Department
Experimental Station
Wilmington, Delaware 19898

HASSAN SALARI
Groupe de Recherche sur les Leucotriènes
Laboratoire d'Endocrinologie Moléculaire
Centre Hospitalier de l'Université Laval
Québec G1V 4G2
Canada

JOHN SALMON
Prostaglandin Research
Wellcome Research Laboratories
Beckenham, Kent BR3 3BS
United Kingdom

BENGT SAMUELSSON
Department of Physiological Chemistry
Karolinska Institutet
S-104 01 Stockholm
Sweden

S. SATCHITHANANDAM
Department of Biochemistry
The George Washington University School
  of Medicine and Health Sciences
Washington, D.C. 20037

N. W. SCHOENE
Beltsville Human Nutrition Research Center
Beltsville, Maryland 20705

G. E. SCHREINER
Department of Medicine
Division of Nephrology
Georgetown University Medical Center
Washington, D.C. 20007

EDWARD S. SCHULMAN
Department of Medicine
Division of Clinical Immunology
Johns Hopkins University School of Medicine
The Good Samaritan Hospital
Baltimore, Maryland 21239

WILLIAM A. SCOTT
Department of Pharmacology
Cornell University Medical College
  and the Rockefeller University
New York, New York 10021
Present address:
Squibb Institute for Medical Research
Princeton, New Jersey 08540

CHARLES N. SERHAN
Department of Physiological Chemistry
Karolinska Institutet
S-104 01 Stockholm
Sweden
Present address:
Visiting Scientist, Fellow of the Arthritis
  Foundation
Department of Medicine
Division of Rheumatology
New York University
New York, New York 10016

RAMADAN I. SHA'AFI
Department of Physiology
University of Connecticut Health Center
Farmington, Connecticut 06032

STEVEN SHAK
Cardiovascular Research Institute and Rosalind
  Russell Arthritis Research Laboratory
Department of Medicine
University of California
San Francisco, California 94102
and Medical Service
San Francisco General Hospital
San Francisco, California 94110

RON SHAPIRO
Division of Cardiac Surgery
Medical College of Virginia
Richmond, Virginia 23298

ALAN J. SHEPPARD
Division of Nutrition
Food and Drug Administration
Washington, D.C. 20204

WOLFGANG SIESS
Medizinische Klinik Innenstadt
  der Universität München
8000 Munich 2
West Germany

PIERRE SIROIS
Département de Pédiatrie
Centre Hospitalier Universitaire
Université de Sherbrooke
Sherbrooke J1H 5N4
Canada

J. BRYAN SMITH
Center for Thrombosis Research
Temple Medical School
Philadelphia, Pennsylvania 19140

WILLIAM L. SMITH
Department of Biochemistry
Michigan State University
East Lansing, Michigan 48824

A. LOUIS SOUTHREN
Departments of Ophthalmology and Medicine
New York Medical College
Valhalla, New York 10595

WAYNE W. STERLING
ICI Pharmaceuticals Division
Mereside, Alderley Park
Macclesfield, Cheshire SK10 4TG
United Kingdom

K. SVANBORG
Department of Obstetrics and Gynecology
Karolinska Hospital
S-104 01 Stockholm
Sweden

RITVA TAMMIVAARA
Department of Diseases of the Chest
University of Turku
SF-20520 Turku 52, Finland

RICHARD J. ULEVITCH
Department of Immunology
Research Institute of Scripps Clinic
La Jolla, California 92037

PEKKA UOTILA
Department of Physiology
University of Turku
SF-20520 Turku 52, Finland

GEORGE V. VAHOUNY
Department of Biochemistry
The George Washington University School
   of Medicine and Health Sciences
Washington, D.C. 20037

JACK Y. VANDERHOEK
Department of Biochemistry
George Washington University School of
   Medicine and Health Sciences
Washington, D.C. 20037

RICHARD VAN INWEGEN
Revlon Health Care Group
Research and Development
Tuckahoe, New York 10707

D. L. VENTON
Department of Medicinal Chemistry
The University of Illinois at Chicago
Health Sciences Center
Chicago, Illinois 60612

MARIO VOLPI
Department of Physiology
University of Connecticut Health Center
Farmington, Connecticut 06032

HEINRICH KARL WASNER
Diabetes-Forschungsinstitut
Biochemische Abteilung
4000 Düsselforf 1
West Germany

PETER C. WEBER
Medizinische Klinik Innenstadt
   der Universität München
8000 Munich 2
West Germany

IRA WEINRYB
Revlon Health Care Group
Research and Development
Tuckahoe, New York 10707

BERNARD I. WEINSTEIN
Departments of Ophthalmology and Medicine
New York Medical College
Valhalla, New York 10595

BABETTE B. WEKSLER
Division of Hematology–Oncology
Department of Medicine
Cornell University Medical College
New York, New York 10021

SIGBRITT WERNER
Department of Endocrinology
Karolinska Hospital
S-104 01 Stockholm
Sweden

DICK L. WESTBROEK
Laboratory for Experimental Surgery
Erasmus University
3000 DR Rotterdam
The Netherlands

JAMES G. WHITE
Department of Laboratory Medicine and
   Pathology
Medical School
University of Minnesota
Minneapolis, Minnesota 55455

M. WIKLAND
Department of Obstetrics and Gynecology
University of Göteborg
Göteborg, Sweden

N. WIQVIST
Department of Obstetrics and Gynecology
University of Göteborg
Göteborg, Sweden

JOHN WOOD
Department of Experimental Immunobiology
Wellcome Research Laboratories
Beckenham, Kent BR3 3BS
United Kingdom

GEORGE B. ZAVOICO
Department of Pharmacology
University of Connecticut Health Center
Farmington, Connecticut 06032

PIETER E. ZONDERVAN
Department of Pathological Anatomy I
Erasmus University
3000 DR Rotterdam
The Netherlands

Index

MIX
Papier aus verantwortungsvollen Quellen
Paper from responsible sources
FSC® C105338

If you have any concerns about our products,
you can contact us on
ProductSafety@springernature.com

In case Publisher is established outside the EU,
the EU authorized representative is:
Springer Nature Customer Service Center GmbH
Europaplatz 3, 69115 Heidelberg, Germany

Printed by Libri Plureos GmbH
in Hamburg, Germany